Postharvest Handling and Diseases of Horticultural Produce

Postharvest Handling and Diseases of Horticultural Produce

Edited By

Dinesh Singh, Ram Roshan Sharma, V. Devappa, and Deeba Kamil

CRC Press is an imprint of the
Taylor & Francis Group, an **informa** business

First edition published 2021
by CRC Press
6000 Broken Sound Parkway NW, Suite 300, Boca Raton, FL 33487-2742

and by CRC Press
2 Park Square, Milton Park, Abingdon, Oxon, OX14 4RN

ISBN: 9780367492892 (hbk)
ISBN: 9780367774653 (pbk)
ISBN: 9781003045502 (ebk)

Typeset in Times
by KnowledgeWorks Global Ltd.

Contents

Preface

Horticultural produce plays an important role not only in supplying foods protective of human health and well-being but also as a significant source of income. Because fruits and vegetables are perishable in nature, postharvest loss ranges from 15 to 50% of total food produced in developing countries, depending on the produce which starts from harvesting to reach the consumer due to both biotic and abiotic factors. This has led to increased concern over food safety and security in India and abroad. In biotic stress, postharvest diseases are caused mainly by different groups of fungi and bacteria during storage and in the supply chain and are managed by various methods such as refrigeration, controlled atmosphere, chemicals and physical methods. However, for the past two decades, the focus on biological control strategies has emerged as a promising alternative to the use of synthetic chemical pesticides in the pre-harvest period to reduce postharvest disease, especially when harmoniously blended with other already established classic methods, such as refrigeration. Thus, several bio-control agents either as elicitors or as suppressing the pathogen both in storage and transit have been found effective in inhibiting postharvest diseases of horticultural produce. The integrative strategies for management of postharvest diseases include effectively inhibiting pathogen growth, enhancing resistance of hosts and improving environmental conditions, resulting in conditions that are favourable to the host and unfavourable to pathogen growth including biotechnological approaches.

Postharvest Handling and Diseases of Horticultural Produce is a compiled work by many specialists in the areas of horticulture, postharvest technology, plant pathology and microbiology, covering thirteen chapters on postharvest handing and disease management, mycotoxins and health hazards and codex alimentation, thirteen chapters on postharvest diseases of fruits, seven chapters on postharvest diseases of vegetables, one chapter on tuber crops and one chapter on postharvest handling and diseases of cut flowers. This book is first of its kind and has gathered the latest expert knowledge on various dimensions of postharvest handing and management of postharvest produce. We hope this will serve as ready reckoner for postharvest handling and management of diseases of horticultural produce. The book is intended for teachers and students who focus on postharvest pathology and plant pathology in postgraduate programmes. It provides the latest knowledge in the field for researchers, policy makers and all other stakeholders in postharvest handling of horticultural produce, to enable them to face the challenges posed by various fungal and bacterial pathogens of modern agriculture.

We would like to thank all the contributors who expressed the desire to compile the latest information on different areas of postharvest handling and diseases of horticultural produce, for their excellent writing and providing chapters promptly to us for final editing to compile into this comprehensive book. Our thanks also go to CRC Press, New Delhi for publishing the book in the stipulated time.

Dinesh Singh
Ram Roshan Sharma
V. Devappa
Deeba Kamil

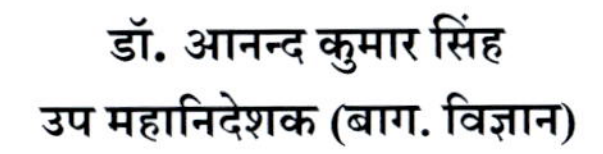

Dr. Anand Kumar Singh
Deputy Director General (Hort. Sci.)

भारतीय कृषि अनुसंधान परिषद
कृषि अनुसंधान भवन-II
पूसा, नई दिल्ली – 110012

INDIAN COUNCIL OF AGRICULTURAL RESEARCH
KRISHI ANUSANDHAN BHAWAN-II
PUSA, NEW DELHI – 110012

दूरभाष सं.:91-11-25842068, फैक्स सं.:91-11-25841976
Ph.: 91-11-25842068(O), Fax: 91-11-25841976
email: ddghort.icar@gov.com
aksingh36@gmail.com

Foreword

There has been a continuous increase in area and production of horticultural crops in India and throughout the world. Research support in the past with respect to pre-harvest management practices and postharvest handling operations has greatly contributed to strengthening this sector. Horticultural crops such as fruits, vegetables, medicinal plants, spices, tuber crops, plantation crops, ornamental plants and flowers etc. have varied adoptability, uses and benefits. These crops give comparatively more economic return per unit area and time scale. In addition, there are several ecological benefits of these crops. Experiences show their potential in nutritional security, livelihood security, employment generation, support to cottage industries, special benefits for women etc. The overall impact of horticultural crops in health and immunity benefits and social sustainability has been emphasized by many scholars in the past. Horticultural crops, in fact, form a core area of Research & Development programmes in many countries. Majority of the horticultural crops are perishable by nature and efficient postharvest handling holds great promise in management and minimization of postharvest losses. However, these losses could be classified into two distinct groups such as quantitative and qualitative. An effective management of postharvest losses in horticultural crops requires an updated scientific approach for pre-harvest and postharvest management operations at field and various stages of handling during transport, storage, processing and distribution. The quantitative and qualitative losses in horticultural crops are both due to biotic and abiotic factors. The biotic factors such as fungi and bacteria cause postharvest diseases in different horticultural crops and lead to great losses to different stakeholders at different stages of value chain. There has always been our quest to refine the technologies and make stakeholders aware of the latest scientific development in the science and technology of handling, transport, storage, processing and distribution of horticultural crops.

I am happy to have read the present book *Postharvest Handling and Diseases of Horticultural Produce* because is a compilation of latest scientific knowledge contributed by many specialists in the area of horticulture, post-harvest technology, plant pathology and microbiology etc. Further, it also covers subjects of topical interest to end users, including postharvest handing & disease management, mycotoxins and health hazard and Codex Alimentarius, postharvest diseases of fruits, vegetables, tuber crops and cut flowers. This book is the first of its kind and very suitably reflects the issues in a comprehensive manner. I hope this will serve as ready reckoner for postharvest handling and management of diseases of horticultural crops.

I appreciate the editors for their hard work in editing the manuscript and the authors of different chapters of the book for incorporating the latest scientific information on the subject. I believe that the scientific literature on postharvest handling and diseases of horticultural produce will be highly useful to all stakeholders and greatly serve its purpose.

(Anand Kumar Singh)

About the Authors

Dr Dinesh Singh (Principal Scientist), ICAR-IARI, New Delhi, earned his MSc (Agriculture) and Ph. D. from the Institute of Agricultural Sciences, BHU, Varanasi. He joined ARS services in January 20, 1997 at IIVR, Varanasi and is presently Principal Scientist in the Division of Plant Pathology, ICAR-IARI, New Delhi. Dr Singh has made significant contributions in the areas of plant pathology, particularly in the areas of postharvest disease management as well as in plant bacteriology. He characterized and deposited 176 isolates of *Ralstonia solanacerarum*, 217 isolates of *Xanthomonas campestris* pv. *campestris*, 6 isolates of *X. oryzae* pv. *oryzae*, 31 isolates of *X. vesicatoria*, 45 isolates of *Bacillus* spp. (*B. subtilis, B. amyloliquifaciens, B. pumilus, B. cereus*) in the ITCC, Division of Plant Pathology, IARI, New Delhi. He developed a multiplex PCR-based sensitive, specific and reliable protocol for simultaneous detection of *R. solanacearum* and *Erwinia carotovora* subsp. *carotovora* from potato tubers. He characterized two races of *Ralstonia solanacerarum* (Race 1 and Race 3) and three races of *X. campestris* pv. *Campestris* (Races 1, 4 and 6) from India. Dr Singh developed a talc-based formulation of bacterial antagonists such as *Pseudomonas fluorescens* DTPF-3, *Bacillus subtilis* DTBS-5 and *B. amyloliquefaciens* DTBA-11 effective against fungal and bacterial diseases of plants. He has guided four Ph. D. and two M. Sc. students in the field of plant bacteriology. He has published 135 research papers in peer-reviewed reputed journals, 44 popular articles, 20 books, 79 book chapters, 5 technical bulletins, 8 training manuals and 2 practical manuals. He has been honoured with awards including NESA Fellowship, NABS Fellow, IPS Fellow, SPPS Fellow, Bioved Fellowship Award, Honorary Fellowship Indian Mycological Society (FIMS) (2020), Dr M.M. Alam Medal, Young Scientist Associate Award, Late Shri P.P. Shinghal Memorial Award, J.P. Varma Memorial Lecture Award, Reviewer Excellence Award, Award for Excellent in Research, Distinguished Scientist in Plant Pathology Award and Best Faculty Award from various professional societies and agencies.

Dr Ram Roshan Sharma was born on 14 March, 1962 in Hamirpur (H.P.). He earned his degree (B. Sc. Agriculture) from HPKV, Palampur and his M. Sc. in Horticulture from IARI, New Delhi. He currently is Principal Scientist in the Division of Food Science and Postharvest Technology, ICAR-Indian Agricultural Research Institute, New Delhi, India. Dr Sharma has developed eight varieties of mango and standardized several production and postharvest management technologies for various fruits including apple, mango, pomegranate, plum, kiwifruit, litchi and Kinnow mandarin. He was previously International Mango Registrar of International Society for Horticultural Science (1999–2004). He is Fellow of four National Academies: the National Academy of Agricultural Science, National Academy of Biological Science, Indian Academy of Horticultural Science and National Environment Science Academy. He is a recipient of the Rajiv Gandhi Award and Education Award of the Government of India, the Dr Rajinder Prasad Award of ICAR twice (2004 and 2010), IARI Best Teacher Award, Dr R.N. Singh award of IARI, New Delhi twice (2004 and 2019). He is a prolific writer and has authored 14 books on different aspects of fruit production and postharvest management and has published more than 180 research papers in journals of national and international repute. He is a member of the editorial board of more than a dozen international journals and is a reviewer for more than 15 international and national journals. His Google Scholar Citation is more than 3500 with h-index of 29 and h-10 index of 70.

Dr V. Devappa was born on 22 September 1970 and earned his bachelor's and master's degrees from the University of Agricultural Sciences, Bangalore and Ph. D. from the University of Agricultural Sciences, Dharwad. He joined the University of Agricultural Sciences, Dharwad in 1997 as Assistant Professor. He is currently Professor and Head, Department of Plant Pathology, College of Horticulture, GKVK, Bengaluru since May 27, 2013. He is well known for his contributions on integrated disease management of vegetable and fruit crops and biological control of soilborne plant pathogens. He played a key role in developing 20 technologies in the field of plant pathology and these technologies are incorporated in the package of practices for horticulture crops in Karnataka. He has 75 publications in addition to 3 books and 10 book chapters. He organized the National Level Symposium on Climate Change and conducted two ICAR short courses on biological control and two workshops pertaining to horticulture crop disease management. He also served as a member of the editorial boards for six online journals. He has guided four Ph.D. and ten M. Sc. students. He also visited various countries including Israel, Spain, Malaysia, Singapore, Thailand, etc. for scientific assignments and earned more than 10 awards for his achievements in the field of plant pathology, including from several scientific societies. He is a life member of more than ten scientific societies, including the Indian Phytopathological Society, Indian Society of Plant Protection Association, *Karnataka Journal of Farm Sciences*, etc.

Dr Deeba Kamil was born on 10 September, 1981 in Lucknow (U.P.) and completed her graduation from Lucknow University, Lucknow and earned her M. Sc. in Agriculture from AAIDU, Naini, Allahabad and Ph. D. from Banaras Hindu University, Varanasi. She currently is Senior Scientist in the Division of Plant Pathology, ICAR-Indian Agricultural Research Institute, New Delhi, India. She has made excellent contributions to plant pathological research. She has been involved in the field of mycology, especially fungal taxonomy, and associated with Indian Type Culture Collection and Herbarium Cryptogamme Indae Orientalis. She also started DNA barcoding of fungal cultures with the genus *Fusarium*. This was the first attempt from *Fusarium* barcoding from India. DNA barcoding of *Trichoderma*, *Colletotrichum*, *Drechslera*, *Phoma*, *Aspergillus* and *Penicillium* is also done. She was successful in developing molecular markers for many *Trichoderma* and *Aspergillus* species. She is currently working on the delineation of different *Fusarium* species complex. She has also tried to produce silver nanoparticles by using different fungi, and worked on their bio-efficacy against plant pathogens. She is a recipient of the MK Patel Young Scientist Award, Woman Scientist Awards and many more societies' awards. She has authored four books on different aspects of plant pathology and has published more than 50 research papers in journals of national and international repute. She is reviewer for many international and national journals.

Contributors

Raghavendra Achari
College of Horticulture
Munirabad (Koppal), Karnataka, India

Mehjabeen Afaque
Division of Plant Pathology
ICAR-Indian Agricultural Research Institute
New Delhi, India

Ajit Kumar Dubedi Anal
Amity Institute of Microbial Technology
Amity University
Noida, Uttar Pradesh, India

Ali Anwar
Division of Plant Pathology
Sher-e-Kashmir University of Agricultural Sciences & Technology of Kashmir
Shalimar Campus
Srinagar, Jammu & Kashmir, India

T. Aravind
Department of Plant Pathology
College of Agriculture
G.B. Pant University of Agriculture
Udham Singh Nagar, Pantnagar
Uttarakhand, India

T. S. Archana
Department of Plant Pathology
University of Horticultural Sciences
Bagalkot, Karnataka, India

S. A. Ashtaputre
Department of Plant Pathology
University of Agricultural Sciences
Dharwad, India

Ram Asrey
Division of Food Science & Postharvest Technology
ICAR-Indian Agricultural Research Institute
New Delhi, India

Pankaj Baiswar
Division of Crop Protection
ICAR Research Complex for NEH Region
Umiam, Meghalaya, India

Saba Banday
Division of Plant Pathology
Sher-e-Kashmir University of Agricultural Sciences & Technology of Kashmir
Shalimar Campus
Srinagar, Jammu & Kashmir, India

T. Bhagyashree
Biochemical Sciences Division
CSIR-National Chemical Laboratory
Pune, Maharashtra, India

Gayatri Biswal
Department of Plant Pathology
College of Agriculture
OUAT
Bhubaneswar, Odisha, India

Tasvina R. Borah
Division of Crop Protection
ICAR Research Complex for NEH Region
Umiam, Meghalaya, India

C. Visalakshi Chandra
ICAR-Central Tuber Crops Research Institute
Sreekariyam
Thiruvananthapuram, Kerala, India

Bharti Choudhary
Division of Plant Pathology
Indian Agricultural Research Institute
New Delhi, India

Shiv Pratap Choudhary
Division of Plant Pathology
ICAR-Indian Agricultural Research Institute
New Delhi, India

Amrita Das
Division of Plant Pathology
ICAR-Indian Agricultural Research Institute
New Delhi, India

Wadzani Palnam Dauda
Division of Plant Pathology
ICAR-Indian Agricultural Research Institute
New Delhi, India

Rubin Debbarma
Division of Plant Pathology
ICAR-Indian Agricultural Research Institute
New Delhi, India

V. Devappa
Department of Plant Pathology
College of Horticulture
UHS Campus, GKVK
Bengaluru, Karnataka, India

Mast Ram Dhiman
ICAR-Indian Agricultural Research Institute
Regional Station, Katrain
Kullu, Himachal Pradesh, India

Vaibhavi C. Dhuri
Biochemical Sciences Division
CSIR-National Chemical Laboratory
Pune, Maharashtra, India

Tahseen Fatima
ICAR-Central Institute for Subtropical Horticulture
Rehmankhera, Kakori
Lucknow, Uttar Pradesh, India

Baradevanal Gundappa
ICAR-Central Institute for Subtropical Horticulture
Rehmankhera, Kakori
Lucknow, Uttar Pradesh, India

H. R. Raghavendra
Division of Food Science and Postharvest Technology
ICAR-Indian Agricultural Research Institute
New Delhi, India

Gurudatt M. Hegde
Department of Plant Pathology
University of Agricultural Sciences
Dharwad, Karnataka, India

Manjunath Hubballi
Department of Plant Pathology
University of Horticultural Sciences
Bagalkot, Karnataka, India

Shaily Javeria
Division of Seed Science and Technology
ICAR-Indian Agricultural Research Institute
New Delhi, India

M. L. Jeeva
ICAR-Central Tuber Crops Research Institute
Sreekariyam
Thiruvananthapuram, Kerala, India

N. Jhansirani
Department of Plant Pathology
College of Horticulture
UHS Campus,GKVK
Bengaluru, Karnataka, India

Alka Joshi
Division of Food Science &Postharvest Technology
ICAR-Indian Agricultural Research Institute
New Delhi, India

Deeba Kamil
Division of Plant Pathology
ICAR-Indian Agricultural Research Institute
New Delhi, India

Swati Kapoor
Punjab Horticultural Postharvest Technology Centre
Punjab Agricultural University
Ludhiana, Punjab, India

Amit Kumar Kesharwani
Division of Plant Pathology
ICAR-Indian Agricultural Research Institute
New Delhi, India

Atul Kumar
Division of Seed Science and Technology
ICAR-Indian Agricultural Research Institute
New Delhi, India

Raj Kumar
ICAR-Indian Agricultural Research Institute
Regional Station, Katrain
Kullu, Himachal Pradesh, India

Sandeep Kumar
ICAR-Indian Agricultural Research Institute
Regional Station, Katrain
Kullu, Himachal Pradesh, India

Vinod Kumar
ICAR-National Research Centre on Litchi
Mushahari
Muzaffarpur, Bihar, India

Nidhi Kumari
ICAR-Central Institute for Subtropical Horticulture
Rehmankhera, Kakori
Lucknow, Uttar Pradesh, India

Sunita J. Magar
Department of Plant Pathology
College of Agriculture
Latur, Maharashtra, India

Bal Vipan Chander Mahajan
Punjab Horticultural Postharvest Technology Centre
Punjab Agricultural University
Ludhiana, Punjab, India

T. Makeshkumar
ICAR-Central Tuber Crops Research Institute
Sreekariyam
Thiruvananthapuram, Kerala, India

S. E. Navyashree
Department of Plant Pathology
College of Horticulture
UHS Campus
Bengaluru, Karnataka, India

Swarajya Laxmi Nayak
Division of Food Science and Postharvest Technology
ICAR-Indian Agricultural Research Institute
New Delhi, India

Joginder Pal
Department of Plant Pathology
Dr YS Parmar University of Horticulture and Forestry Nauni
Solan, Himachal Pradesh, India

K. K. Pandey
ICAR-Indian Institute of Vegetable Research
Varanasi, Uttar Pradesh, India

Govindan Pothiraj
Division of Plant Pathology
ICAR-Indian Agricultural Research Institute
New Delhi, India

Raghavendra Mesta
Department of Plant Pathology
University of Horticultural Sciences
Bagalkot, Karnataka, India

Dinesh Rai
Department of Plant Pathology
RPCAU, Pusa
Samastipur, Bihar, India

Rajesh Kumar Ranjan
Department of Plant Pathology
RPCAU, Pusa
Samastipur, Bihar, India

V. Koteswara Rao
Biochemical Sciences Division
CSIR-National Chemical Laboratory
Pune, Maharashtra, India

Vijay Rakesh Reddy
ICAR-Central Institute for Arid Horticulture
Bikaner, Rajasthan, India

C. G. Sangeetha
Department of Plant Pathology
College of Horticulture
UHS Campus
Bengaluru, Karnataka, India

Shruti Sethi
Division of Food Science and Postharvest Technology
ICAR-Indian Agricultural Research Institute
New Delhi, India

Efath Shahnaz
Division of Plant Pathology
Sher-e-Kashmir University of Agricultural Sciences & Technology of Kashmir
Shalimar Campus,
Srinagar, Jammu & Kashmir, India

Jahagirdar Shamarao
Department of Plant Pathology
University of Agricultural Sciences
Dharwad, Karnataka, India

Veerubommu Shanmugam
Division of Plant Pathology
ICAR-Indian Agricultural Research Institute
New Delhi, India

Amit Sharma
Department of Basic Science
College of Horticulture and Forestry
Dr YS Parmar University of Horticulture and Forestry
Neri, Hamirpur, Himachal Pradesh, India

Anju Sharma
Department of Basic Science
Dr YS Parmar University of Horticulture and Forestry
Nauni, Solan, Himachal Pradesh, India

Manju Sharma
Amity Institute of Biotechnology
Amity University Haryana, Manesar
Gurugram, Haryana, India

Monica Sharma
Department of Plant Pathology
Dr YS Parmar University of Horticulture and Forestry
Neri, Hamirpur, Himachal Pradesh, India

Ram Roshan Sharma
Division of Food Science & Postharvest Technology
ICAR-Indian Agricultural Research Institute
New Delhi, India

Satish K. Sharma
Department of Plant Pathology
Dr YS Parmar University of Horticulture and Forestry
Nauni
Solan, Himachal Pradesh, India

P. K. Shukla
ICAR-Central Institute for Subtropical Horticulture
Rehmankhera, Kakori
Lucknow, Uttar Pradesh, India

Akoijam R. Singh
Plant Pathology, Division of Crop Protection
ICAR Research Complex for NEH Region
Umiam, Meghalaya, India

Dinesh Singh
Division of Plant Pathology
ICAR-Indian Agricultural Research Institute
New Delhi, India

J. Singh
ICAR-Indian Institute of Vegetable Research
Varanasi, Uttar Pradesh, India

Jai Prakash Singh
Fruits and Horticultural Technology
ICAR-Indian Agricultural Research Institute
New Delhi, India

K. P. Singh
Department of Plant Pathology
College of Agriculture
G.B. Pant University of Agriculture
Pantnagar, Udham Singh Nagar, Uttarakhand, India

Pooja Singh
Department of Botany
DDU Gorakhpur University
Gorakhpur, Uttar Pradesh, India

Ravinder Pal Singh
Division of Plant Pathology
ICAR- Indian Agricultural Research Institue
New Delhi, India

Allada Snehalatharani
Horticultural Research Station
Kovvur
Dr YSRHU
Andhra Pradesh, India

S. D. Somwanshi
Krishi Vigyan Kendra
Badnapur, Maharashtra, India

P. Srinivas
Department of Plant Pathology
College of Horticulture
UHS Campus, GKVK
Bengaluru, Karnataka, India

Tirukovalur Sundeep
Department of Plant Pathology
College of Horticulture
UHS Campus, GKVK
Bengaluru, Karnataka, India

A. P. Suryawanshi
Department of Plant Pathology
College of Agriculture
Latur, Vasantrao Naik Marathwada Krishi Vidyapeeth
Parbhani, Maharashtra, India

T. Prameeladevi
Division of Plant Pathology
ICAR-Indian Agricultural Research Institute
New Delhi, India

R. Sudeep Toppo
Division of Plant Pathology
ICAR-Indian Agricultural Research Institute
New Delhi, India

A. N. Tripathi
ICAR-Indian Institute of Vegetable Research
Varanasi, Uttar Pradesh, India

S. S. Veena
ICAR-Central Tuber Crops Research Institute
Sreekariyam
Thiruvananthapuram, Kerala, India

J. U. Vinay
Department of Plant Pathology
University of Agricultural Sciences
Dharwad, Karnataka, India

M. R. Vinay
Department of Plant Pathology
College of Horticulture
UHS Campus, GKVK
Bengaluru, Karnataka, India

N. Vishwambar
Academy of Scientific and Innovative Research (AcSIR)
New Delhi, India

Mahender Singh Yadav
ICAR-National Centre for Integrated Pest Management
Pusa Campus
New Delhi, India

1

Postharvest Losses of Horticultural Produce

Dinesh Singh,[1] Ram Roshan Sharma[2] and Amit Kumar Kesharwani[1]
[1]Division of Plant Pathology, ICAR-Indian Agricultural Research Institute, New Delhi, India
[2]Division of Food Science and Postharvest Technology, ICAR-Indian Agricultural Research Institute, New Delhi, India

CONTENTS

1.1 Introduction

Horticultural crops have improved the financial health of farmers due to higher price per unit of area. These crops are a significant source of income–a means of reducing poverty. Fruits and vegetables are available throughout the year in a wide variety, and they not only good taste but also have favourable attributes of texture, colour, flavour and ease of use. However, in many cases, consumption of fruits and vegetables is still below the dietary guideline goal of 5–10 servings per day. Horticultural produce, particularly fruits and vegetables, are rich sources of critical nutrients for a balanced diet (Yahaya et al. 2015). The constituents provided to the human body by these crops include water, carbohydrates, proteins, fats, fibre, minerals, organic acids, pigments, vitamins and antioxidants, etc. (Table 1.1). They can be used fresh, cooked, hot or cold, canned, pickled, frozen or dried. They are an excellent substitute for meals and are used as snacks, which are relatively low in calories, have no cholesterol, are rich in carbohydrates

TABLE 1.1

Fruits and Vegetables are Recognized as Protective Foods Necessary for the Maintenance of Human Health

Vitamins and Minerals	Function	Sources
Vitamins	Vitamins are essential for the maintenance of health.	Fruits and vegetables supply many vitamins.
i. Vitamin A	Essential for growth and reproduction. Aids in resistance to infections, increases longevity and decreases senility. Deficiency of vitamin A causes night blindness, xerophthalmia, growth retardation, skin roughness and kidney stone formation.	Dates, jackfruit, loquat, mango, orange, papaya, passion fruit, persimmon, walnut, etc.
ii. Vitamin B_1 (thiamine)	Necessary to maintain good appetite and normal digestion. Essential for growth, fertility and lactation as well as proper functioning of nervous system. Deficiency of vitamin B1 causes beriberi disease, paralysis, sensitivity and loss of skin as well as enlargement of the heart and decrease in body temperature.	Almond, apple, apricot, banana, grapefruit, plum, walnut, etc.
iii. Vitamin B_2 (riboflavin)	Important for growth and skin health as well as for respiration in poorly vascularized tissue such as the cornea. Deficiency causes pellagra and alopecia, loss of appetite and weight as well as sore throat, cataract development, swelling of the nose and baldness.	Apple bael, litchi, papaya, pineapple, pomegranate.
iv. Vitamin B_6 (pyridoxin)	Beneficial to the central nervous system. Associated with producing the neurotransmitters serotonin and norepinephrine. It forms myelin. Other functions of pyridoxine are protein and glucose metabolism, and also the manufacture of haemoglobin.	Avocado, banana, carrots, potato, spinach.
vi. Vitamin 3(niacin[nicotinic acid])	Enables anabolic reactions such as synthesis of cholesterol and fatty acids. Plays a critical role in maintaining cellular antioxidant function.	Apple, banana, cashews, cherry, onions, tomatoes, spinach.
vii. Vitamin B_9 (folic acid)	Helps in forming DNA and RNA. Associated with protein metabolism. Plays a key role in breaking down homocysteine, an amino acid that can exert harmful effects in the body. Folate is essential to production of healthy red blood cells which are critical during periods of rapid growth such as pregnancy and fetal development.	Avocado, beets, cabbage, dark green leaves (turnip greens, spinach, romaine lettuce, asparagus, Brussels sprouts, broccoli), lettuce.
viii. Vitamin C	Deficiency of vitamin C causes scurvy disease, pain in joints and swelling of limbs. It also causes tooth decay, unhealthy gums, delay in wound healing and rheumatism.	Aonla, Barbados, ber, cherry, guava, lemon, lime, pineapple, pear, sweet oranges.
ix. Vitamin E (tocopherol)	Plays a role in certain conditions related to aging. The body also needs vitamin E to help keep the immune system strong against viruses and bacteria.	Vegetable oils (palm, olive, nut), seabuckthorn berries, kiwifruit, almonds.
Calcium	Needed for development of bones, heartbeat regulation and controlling of blood clots.	Acid lime, agathi, apricots, beans, cabbage, carrots, dried wood apple, drumstick leaves, fig, green peas, onions, orange, spinach, tomatoes, etc.

TABLE 1.1 ***(Continued)***

Fruits and Vegetables are Recognized as Protective Foods Necessary for the Maintenance of Human Health

Vitamins and Minerals	Function	Sources
Iron	Essential for haemoglobin production. Deficiency of iron causes anaemia, pale lips, eyes and skin, and smooth tongue.	Agati, beans, custard apple, carrot, dried dates, drumstick leaves, grape, guava, pineapple, strawberry, etc.
Phosphorus	Needed to maintain the moisture content of tissues. Essential for development of bones.	Beans, carrots, chilli, cucumber, drumstick leaves, grape, guava, jackfruit, onion, orange, passion fruit.
Proteins	Proteins are bodybuilding foods which are essential for body growth. Its deficiency causes growth retardation and increases susceptibility to many diseases.	Fruits are low in proteins except banana, beans, guava and peas.
Enzymes	Required to control several metabolic activities in the body.	Fig (ficin), papaya (papain), pineapple (bromelin).

Source: FAO (2005); Jaganath and Crozier (2008).

and fibre and are low fat, except for avocados and olives. They are good sources of vitamin C, carotene and vitamin B_6. Fruits and vegetables are relatively low in sodium and high in potassium. A growing body of research has shown that consumption of fruits and vegetables is associated with reduced risk of major diseases. They may delay the onset of age-related disorders, promoting good health. The nutritional value of fruits and vegetables depends on their composition, which shows a wide range of variation depending on the species, cultivar and maturity stage (Table 1.2).

TABLE 1.2

Nutritional Value of Fruits and Vegetables

Product	Carbohydrate	Protein	Fat	Minerals	Calorific value (cal/100 g)	Vit-A (IU/100 g	Vit-B (mg/100 g)	Vit-C (mg/100 g)	Nicotinic Acid (mg/100 g)	Riboflavin (mg/100 g)
	Fruits									
Apple	15.00	0.30	0.40	0.30	56	–	0.03	2	0.2	0.03
Aonla	13.70	0.50	0.10	0.28	59	–	0.03	700	0.2	0.03
Banana	24.0	1.30	0.40	0.80	153	–	0.04	19	0.3	0.03
Guava	8.92	1.40	0.52	0.50	66	–	0.03	300	0.2	0.03
Mango	17.59	0.53	0.28	3.24	50	4800	0.04	24	0.3	0.05
Orange	11.30	0.90	0.20	0.50	49	350	0.05	68	0.3	0.06
Papaya	6.87	0.43	0.10	0.29	40	2020	0.04	46	0.2	0.05
Pear	27.00	1.00	0.25	0.31	47	14	0.02	–	0.2	0.03
Pineapple	9.79	0.42	0.09	4.39	50	60	0.03	63	0.2	0.04
	Vegetables									
Tomato	3.90	0.90	0.20	0.40	21	320	0.04	32	0.4	0.05
Brinjal	5.60	1.00	Trace	0.30	34	5	0.05	23	0.8	0.06
Potato	18.90	2.00	0.10	1.00	99	40	0.10	17	1.2	0.01
Carrot	9.10	1.10	0.20	1.00	47	2000–4300	0.04	3	0.4	0.02
Onion	9.34	1.10	0.10	0.34	51	–	0.08	11	0.4	0.01
Radish	4.20	1.10	0.10	0.90	21	–	0.06	15	0.4	0.02
Sweet potato	27.3	1.30	0.40	1.00	159	–	0.05	–	0.3	0.01
Yam	27.0	1.50	0.20	1.00	79	434	0.06	–	0.7	0.08
Cauliflower	5.30	1.98	0.10	0.42	39	38	0.10	66	0.9	0.08
Cabbage	4.30	1.30	Trace	0.24	33	2000	0.06	134	0.4	0.12
Drumstick	12.50	6.70	1.70	0.81	96	11300	0.06	220	0.8	0.12
Spinach	0.20	2.60	0.60	2.00	32	5500	0.05	48	0.5	0.11
French bean	6.00	1.90	0.30	–	26	221	0.08	14	0.3	0.06
Cucumber	1.20	1.00	0.60	0.40	14	–	0.03	7	0.2	0.02
Lady's finger	7.00	1.90	0.20	0.50	41	58	0.06	116	0.6	0.06
Pea	17.00	6.70	0.40	0.90	109	139	0.25	9	0.8	0.01
Pumpkin	6.50	1.00	0.10	0.42	28	84	0.06	2	0.5	0.04

Horticulture has also played a vital role in human wellbeing. It provides employment opportunities in mushroom cultivation, floriculture, vegetable seed production, etc. This sector has an annual growth rate of more than 6.5%, which constitutes more than 24.5% from out of 8.5% of area to the gross domestic product (GDP) of agriculture. The advantages of horticultural crops are that they supply better food, higher income, year-round occupation, a diversified system of giving an aesthetic touch to life, stimulus to promote intelligence. Horticultural farming promotes the development of natural resources, yields higher returns from land, enhances land values, creates better purchasing power among the people and as a consequence adds to the general prosperity.

1.2 Area and Production of Horticultural Crops

Vegetable production was about 486 million tonnes, where fruit production reached about 392 million tonnes in the world in 2017–2018. India is the second major producer of fruits and vegetables after China in the world. India contributes to 10–14% of world's production of fruits and vegetables, respectively. The Department of Agriculture, Cooperation and Farmers Welfare (Government of India) has released the final estimates of 2018–2019 and also the first advance estimates of 2019–2020 of area and production of various horticultural crops. The total horticulture production of the country was estimated to be 310.74 million tonnes in 2018–2019, which was marginally higher than the horticulture production in 2017–2018. There was an increase in fruits, flowers, spices and honey and a decrease in vegetables, aromatic and medicinal plants and plantation crops. Fruit production was estimated to be around 97.97 million tonnes, compared to 96.45 million tonnes in 2017–2018. Vegetable production was estimated to be around 183.17 million tonnes, which was less than the production of 2017–2018. The country produces a large quantity of fruits and vegetables; however, only about 2% of these products go for processing and >25% is decayed due to lack of proper handling and storage. The remaining products are consumed as fresh. It has been reported that India produces about 41% of the world's mangoes, 23% bananas, 36% green peas and 10% onions.

1.3 Definition of Postharvest

Postharvest can be defined as follows: 'It begins at the movement of separation of the edible commodity from the plant that produced it by a deliberate human act with the intention of starting it on its way to the table. The postharvest period ends when the food comes into the possession of the final consumer'. Postharvest management (PHM) can be defined as 'methods and techniques applied to increase the shelf life and retain quality of fresh or processed horticultural produce'. The postharvest system, as noted by Spurgeon, 'should be thought of as encompassing the delivery of a crop from the time and place of harvest to the time and place of consumption, with minimum loss, maximum efficiency and maximum return for all involved'. The term 'system' designates a dynamic, complex aggregate of logically interconnected functions or operations within a particular sphere of activity, whereas the terms 'chain' and 'pipeline' denote the functional succession of various operations but tend to avoid their complex interactions.

1.4 Postharvest Losses

Postharvest loss can be defined as follows: 'Loss means any change in the availability, edibility, wholesomeness or quality of the produce that prevents it from being consumed by people'. However, Alao (2000) defined postharvest loss of fruits and vegetables as 'that weight of wholesome edible product (exclusive of moisture content) that is normally consumed by human and that has been separated from the medium and sites of its immediate growth and production by deliberate human action with the intention of using it for human feeding but which for any reasons fails to be consumed by human.' Postharvest losses are not only of quantity and quality but also include the appearance of horticultural produce, which affects their market value. It is a general assumption that postharvest losses are greater in fruits and vegetables as compared to cereals and oilseed crops because they have high moisture content, the voluminous and heavy weight of the product and the high respiratory rate of living tissues (Table 1.3). These tissues must be kept alive and healthy throughout the marketing process. These tissues are composed of thousands of living cells, which require care and maintenance (Alao 2000; Mustapha and Yahaya 2006). Relative perishability can be classified into five groups as very high, high, moderate, low and very low shelf life, as given below.

1. **Very highly perishable:** The shelf life of this group of fruits and vegetables is less than 2 weeks. Fruits such as apricot, blackberry, blueberry, cherry, fig, raspberry, strawberry and vegetables including asparagus, bean sprouts, broccoli, cauliflower, green onion, leaf lettuce, mushroom, muskmelon, pea, spinach, sweet corn and tomato (ripe) are highly perishable.
2. **Highly perishable:** The shelf life of this group of fruits and vegetables is 2–4 weeks. Fruits included in this group are avocado, banana, grape (without SO_2 treatment), guava, loquat, mandarin, mango, melons, nectarine, papaya, peach, plum and artichoke. Vegetables in this group include green beans, brussels sprouts, cabbage, celery, eggplant, head lettuce, okra, pepper, summer squash and tomato (partially ripe).
3. **Moderately perishable:** The shelf life of this group of fruits and vegetables is 4–8 weeks. Fruits include apple, pear (some cultivars), grape (SO_2-treated), orange, grapefruit, lime, kiwifruit, persimmon and pomegranate. Vegetables in this group include table beet, carrot, radish and potato (immature).
4. **Low perishable:** The shelf life of this group of fruits and vegetables is 8–16 weeks. Fruits include apple,

TABLE 1.3

Comparison between Non-Perishable and Perishable Crops Related to Their Storage Capacity as well as Postharvest Losses

Sr. No.	Non-Perishable Crops (Cereals, Oilseed and Pulses)	Perishable Crops (Fruits and Vegetables)
1.	Mainly seasonal harvesting and long storage period needed	Short storage period needed; permanent or semipermanent production possibility
2.	Preliminary treatment (except threshing) of the crop before storage	Processing in dried products as an alternative to the shortage of fresh products
3.	Low level of moisture content (10–15% or less) in produce	High level of moisture, about 50–80% in produce
4.	Small size produce, less than 1.0 g	Heavy weight fruits from 5.0 g to 5.0 kg or more, and voluminous
5.	Respiratory activity very low of the stored product and heat limited (Rise the temperature very low)	Respiratory activity high or very high of stored products inducing a heat emission, in particular in tropical climates
6.	Postharvest losses 2–10%	Postharvest losses 15–50%

Source: FAO (1984); Knoth (1993).

pear (some cultivars) and lemon. Vegetables include potato, dry onion, garlic, pumpkin, winter squash, sweet potato, taro, yam and bulbs.

5. **Very low perishable:** Tree nuts, dried fruits and vegetables are very low perishable in nature. The shelf life of this group of fruits and vegetables is >16 weeks.

1.4.1 Types of Postharvest Losses

The different types of postharvest losses based on the different stages of postharvesting are described below.

1.4.1.1 Direct and Indirect Loss

Direct loss of produce is a loss which physically occurs due to decay from packaging materials after harvest and materials therefore do not reach the consumer, while in an indirect loss, produce wastage occurs at the consumer level, such as consumer's refusal to purchase due to poor quality, infrequent visits to the market, etc.

1.4.1.2 Weight Loss

Weight loss is the loss which can be measured by the reduction in the moisture content of the produce during storage or transportation. Sometimes an abnormal increase in the weight of produce through moisture absorption is also a cause of loss, which can occur due to storage of the produce under high relative humidity or rainfall on the produce. Long duration storage, shrinkage of produce, consumption by insects and poor packaging of the produce are some factors that lead to weight loss.

1.4.1.3 Produce Loss

Two types of loss occur in the produce after harvest: qualitative loss and quantitative loss.

1.4.1.3.1 Qualitative Produce Loss

Qualitative loss occurs in the produce due to deterioration of the quality by the degradation of nutrients, texture, taste, shape, etc. For example, the nutritional values of the produce, such as carbohydrates, proteins and vitamins, are reduced by fungal and bacterial infection of fruits and vegetables, as they have a high moisture content, which is favourable for growth of fungi and bacteria. Qualitative losses may be further enhanced by soluble excreta of pests, pesticides and pathogenic organisms.

1.4.1.3.2 Quantitative Food Loss (Physical Loss)

This type of loss occurs by reducing the number (quantity) of produce by weight loss due to involvement of many factors such as an attack of pathogens/insect pests, birds, rodents, etc. Fruits and vegetables may lose weight during storage due to spoilage by fungi and bacteria, and physical damage by insects, birds, rodents etc., biotic stresses and loss of weight due to loss of moisture.

1.4.1.4 Economic Loss

Postharvest loss of fruits and vegetables converted in terms of money is termed as economic loss.

1.4.1.5 Irreducible Loss

This type of loss occurs by the excessive respiration of the produce, shrinkage of the produce due to weight loss and mechanical injuries, etc. The loss can be compensated for by production of extra produce. In other words, the production rate must be higher than that of postharvest losses. For example, if there is loss of 20% of produce, the production rate should be increased to the rate of 25%; for a loss of 40%, the production rate should be increased to the rate of 66%, and so forth.

1.5 Postharvest Losses in Fruits and Vegetables

Many factors such as improper handling during harvesting and transportation, use of poor-quality packaging, postharvest handling of produce at high temperatures and delays in marketing lead to a high level of postharvest losses in fruits and vegetables (Kitinoja and Cantwell 2010; Molla et al. 2010; WFLO 2010; Kitinoja and Al Hassan 2012). Postharvest loss in fruits and vegetables has been reported as ranging from 15% to 50%, which starts from harvesting and ends at final retail marketing (Table 1.4). Such loss leads to a huge waste of seeds and planting materials, land, fertilizers, water, energy, labour and other resources. The magnitude of losses ranges from 0 to 100%, which varies greatly depending on produce and type of handling (Kitinoja and Kader 2002; Ray and Ravi 2005). The losses in major fruits and vegetables are described in Table 1.4.

1.5.1 Mango

Postharvest losses occur in mango commencing at the time of harvesting, packaging, storage, transportation, retailing and consumption. The losses are greater in developing countries because of poor infrastructure and logistics, poor farm handling, lack of knowledge of postharvest handling and a faulty marketing system. Both qualitative and quantitative losses occur in mango fruits. Major postharvest diseases such as anthracnose (*Colletotrichum gloeosporioides*), stem-end rot (*Diplodia natalensis*), bacterial canker/black spot (*Xanthomonas citri* pv. *mangiferaeindicae*) and sooty mould (*Capnodium mangiferae*) affect the quality and quantity of the produce during storage and transportation and marketing. Black rot (*Aspergillus niger*) and Rhizopus rot (*Rhizopus* sp.) are also identified as causes of loss. Moula et al. (2017) reported that the mango fruits were sorted after ripening, which showed that 5.12% of fruits had 10% damaged areas, while 4.81% fruits were damaged 10–50%, and 3.56% of fruits were discarded completely.

In Bangladesh, loss of 25–45% occurs at different stages of postharvest of mango (Hassan 2010). In India, Roy and Pal (1991) reported that 1.3% fruits were discarded in the field and 12–18% of this number was culled fruits which were sold at lower prices. Weight loss during transportation was 3.68%. Total loss was 7.53% during ripening in boxes, and loss during pile ripening was much higher. In Banganapalli mango, Murthy et al. (2002) assessed postharvest losses at different stages of marketing in Andhra Pradesh. They reported that the average loss was 15.6% at the farm level due to immature and small fruits; no loss at wholesale market and during storage; losses were 8.8% and 5.25% in the retail marketing level. The major reason for the loss was pressing injury, which shared about 51% of the fruit damage. Nanda et al. (2010) reported total postharvest loss in mango was 12.74%. They observed that 10.64% loss was found in farm operations such as 4.11% at harvesting, 2.80% during sorting and grading and 2.53% during transportation.

Moula et al. (2017) conducted a survey on mango fruit handling and reported that losses were about 8.44%, 4.93%, 5.46%, 3.19% and 6.82% at farm level, wholesale market including transportation, retail market at storage unit and consumer level, respectively. Overall postharvest losses in mango from harvesting to consumption were quantified as about 34.49%. The aggregate loss occurs due to insects such as fruit fly (0.86%), postharvest diseases (0.43%) and injury (0.51%).

1.5.2 Banana

Banana (*Musa* sp.) is climacteric fruit, which is consumed as both a fruit and a vegetable. There are many problems in banana handling; one major limiting factor in the relatively

TABLE 1.4

Postharvest Loses of Fruits and Vegetables during Different Stages of Handling of Produce in India

Sr. No.	Crops	Harvesting	Collection	Sorting/ Grading	Packaging	Transportation	Total Loss in Storage	Overall Total Loss
1.	Apple	4.56	0.42	4.79	0.10	1.19	1.20	12.26 ± 1.05
2.	Banana	1.33	0.36	0.93	0.44	1.14	2.42	6.60 ± 3.43
3.	Citrus	0.92	0.48	1.79	0.35	1.30	1.54	6.38 ± 1.15
4.	Grapes	0.94	0.24	3.21	0.26	1.93	1.73	8.30 ± 1.00
5.	Guava	4.36	1.20	4.64	0.94	2.77	4.13	18.05 ± 1.12
6.	Mango	4.11	0.68	2.80	0.51	2.53	2.11	12.74 ± 1.57
7.	Papaya	1.45	0.28	1.97	0.23	1.13	2.28	7.36 ± 1.04
8.	Sapota	1.53	0.23	1.43	0.08	1.06	1.46	5.77 ± 0.69
9.	Cabbage	1.08	0.30	1.64	0.27	1.32	2.33	6.94 ± 1.51
10.	Cauliflower	0.84	0.27	1.66	0.18	1.91	2.03	6.88 ± 1.32
11.	Green pea	3.46	1.08	3.30	0.23	0.50	1.70	10.28 ± 1.67
12.	Mushroom	1.37	1.77	4.26	1.64	2.00	1.51	12.54 ± 1.40
13.	Onion	2.70	0.23	1.64	0.14	0.44	2.34	7.51 ± 1.09
14.	Potato	3.18	0.69	2.23	0.10	0.54	2.26	8.99 ± 1.87
15.	Tomato	1.73	1.06	3.24	0.77	3.14	3.04	12.98 ± 1.00
16.	Tapioca	3.61	0.51	1.54	0.53	1.28	1.72	9.19 ± 1.52

Source: Narayana et al. (2014).

shorter shelf life of banana is fungal infection after harvest. The fruit contains high sugar and nutrient elements and low pH values which favour fungal decay (Singh and Sharma 2007). It has been recorded that mechanical damage was main cause for postharvest loss in banana at farm and wholesale levels, while spoilage of fruit was the main reason at the retail level. On average, approximately 20–25% of the harvested fruits are spoiled by fungi during handling postharvest, and nearly 1.6 million bananas are discarded in developing countries everyday (Idris et al. 2015). Zakaria et al. (2009) reported that *Colletotrichum* is causal agent of anthracnose disease of different cultivars of banana. There were 146 fungal isolates of *Colletotrichum musae* and *Fusarium* spp. identified from banana fruit samples collected at farm, wholesale, and retail levels of Jimma town market. Among them, 84% of colonies belonged to *Colletotrichum* spp. and about 16% were *Fusarium* spp. (Kuyu and Tola 2018). Gajanana et al. (2002) observed the losses in banana fruits of 3.9% at the farm level by sorting, 2.19–2.52% during transportation due to distant marketing, 2.52% at wholesale and 75% at retail market storage. Total postharvest loss of 6.60% occurred in banana, of which 1.33% loss was at the harvesting stage, 1.14% during transportation and 2.42% during storage (Nanda et al. 2010).

In Ethiopia, the total postharvest loss of banana was recorded at about 45.78% in which 15.68% of loss was at the farm level, 22.05% at wholesale, including transport from farm gate and ripening, and 8.05% at retailer or consumer levels. The postharvest loss in banana was occurred during transportation from the farm, about 20% due to finger dropwhile the remaining 80% was due to physiological and other mechanical damages (puncturing, compression, bruising and abrasion) (Woldu et al. 2015). Mulualem et al. (2015) from Ethiopia reported total loss of banana was 26.5% at different stages, of which 56.0% of the loss occurred at the retail level, 27.0% at wholesale and 17.0% at farm levels.

1.5.3 Guava

Guava is a highly perishable fruit due to its delicate peel. Guava is harvested and marketed in the raw stage at the wholesale level. It is later ripened at this level in packaging. The fruits have quantitative and qualitative postharvest losses occurring at all stages of handling after harvest (Paltrinieri 2015). About 20–25% of fruits are damaged during handling, which is often not acceptable to the consumer (Kanwal et al. 2016; Omayio et al. 2019). Aggregate loss of 5.17% was reported from Allahabad in 2000–2001. A maximum loss of 3.99% was recorded at the retail level and minimum loss 0.3% at the trader level. However, 10–50% of damaged fruits were shared maximum of 14.48% (Anonymous 2003). The losses in guava fruits are due to faulty harvesting method and pre-harvest factors, i.e., inherent infections prevalent in the orchard. In monsoon season, the loss was mostly due to fruit fly infestation, anthracnose (*Colletotrichum psidii* and *C. gloeosporioides*) and Phytophthora rot (*Phytophthora parasitica*) and diseases, whereas in the winter season, the loss was due to fruit borer infestation, cracking of peel and bird damage.

The overall postharvest loss in guava was 18.05%, of which 13.92% loss was at the time of farm operations, including 4.36% loss during harvesting (Nanda et al. 2010). Physical loss was estimated to be 8.92% of the total harvest, while that in economic terms was 6.04%. Economic loss was less than physical loss because the damaged fruits which were partially degraded fetched some price when sold to poor class people or local processors (Stephen et al. 2016).

1.5.4 Citrus

Mandarin orange is the most important commercial citrus fruit in India, followed by sweet orange, lemon and acid lime. Mandarin is a highly perishable fruit, and postharvest losses in citrus fruits are mainly due to diseases and injuries caused by insects, thorns and rough handling. Fungal pathogens are considered the main cause of citrus diseases, severely affecting postharvest management. Pre-harvest infections include brown rot (*Phytophthora* spp.), Alternaria rot (*Alternaria* spp.), stem-end rot (*Diplodia natalensis* Pole-Evan, *Phomopsis citri* Fawcett), grey mould (*Botrytis cinerea* Pers.) and anthracnose (*Colletotrichum gloeosporioides* Penz.), while postharvest infections include green mould (*Penicillium digitatum* Sacc.), blue mould (*P. italicum* Weh.) and sour rot (*Geotrichum candidum* Link). Green and blue moulds represent the most common and serious diseases and cause significant economic losses during fruit storage and marketing. *P. digitatum* and *P. italicum* particularly attack blood oranges, and infection takes place only through rind wounds, where nutrients are available to stimulate spore germination. The incidence of other pathogens is generally low but becomes a serious problem in warm and wet years (Maria et al. 2017; Agada et al. 2017).

IIHR, Bangalore, conducted a survey to assess the postharvest losses in Coorg Mandarin and they found that of the total fruits that arrived in the market, about 3.5% were discarded and another 37% were sorted out and sold at an 80% reduced price. They reported that the injury during harvesting and insect damage were the main reasons for loss in mandarin at the farm level. The total losses reported in mandarins were 3 and 6.15% in Sikkim and Assam from the farm level to the retail level, respectively (Anonymous 2003). Nanda et al. (2010) recorded 6.38% postharvest loss in citrus fruits, of which 4.84% and 1.54% losses were at farm and storage levels. Ladaniya (2015) reported 25–35% of total losses from the farm level to the retail level in the South Asian regions, and variation was mostly based on season and commodity. Acid lime losses were increased with the onset of monsoon season. Moses (2012) assessed postharvest losses of citrus in the Birim North District, Ghana, and reported that the losses at the farm and wholesale market were 20.2% and 5.6%, respectively, and postharvest losses reduced the income generated from the sale of citrus fruits by farmers (Ayandiji 2010).

1.5.5 Grape

Grape has received major attention due to its value addition to form raisins. Fresh grapes are major export commodities from India to European Union. The losses in grapes are water berries, berry shattering at the farm level, loose berries at retailers' level and spoilage and shrivelling due to fungi such as *Aspergillus, Botrytis, Cladosporium, Fusarium, Penicillium, Mucor* and *Rhizopus* and moisture loss during storage or long-distance transportation, respectively (Sonker et al. 2016). The high sugar content of grapes favours the development of fungal pathogens during transit and storage of fruit, which acts as a substrate and food for these fungi. Spoilage of fruits causes losses of from about 30–40% in developed countries having advanced technologies to >50% in developing countries.

Ladaniya et al. (2005) conducted a study on postharvest handling and losses of grapes in 2000–2001 in Nasik district of Maharashtra. They reported that the aggregate loss was 19–30.9% and in fraction, the loss was 0.00–1.25%, 5.5–8.65% and 12.25–16% at farm, wholesale and retail levels, respectively, when grapes were packed in boxes. However, the fruits packed in bamboo baskets, arriving in Mumbai market, had an average loss of 24.19% due to rupturing, decay and shattering of the fruits. An overall postharvest loss in grapes was recorded at about 8.30% (Nanda et al. 2010). Of these, 3.21%, 1.93% and 5.54% of fruit loss was recorded during sorting and grading at farm level, transportation level and at farm level of storage, respectively. Overall, 8.3% of postharvest loss was found at various stages of postharvest activities. However, maximum loss was recorded at 3.2% during sorting and grading of the produce (Nanda et al. 2012).

Murthy et al. (2014) estimated that the postharvest loss was 3.40% in grapes at the field level for the domestic market, and this was attributed to the practice of cleaning bunches 15 days before harvest. The major components of losses of fruits were water berries and berry drops at the field level. A study in the Bijapur area of Karnataka was conducted by Jadhav et al. (2011) and they reported that the postharvest loss of grapes was 13.25% at the farm level, and maximum losses were recorded due to mummification of berries at the farm level. They reported postharvest losses of 25.41, 20.79 and 77.17 kg at the pre-harvest contractor's level, wholesale level and retail level, respectively, for everyone tonne of produce handled. Fruit loss was observed to be 6.14% during cleaning, grading and sorting and only 0.06% loss was noted at the market level, while 3.75% was seen during the retail process. A total of 16.69% loss was recorded in grapes from harvesting to retailing. Large variations in total losses may be due to varietal impact, climatic conditions of the region, rootstocks, soil and water quality, prevailing disease and insect incidences, handling, packing, transportation and retailing conditions (Sharma et al. 2018).

According to Moghaddasi et al, (2005) about 3 million tonnes of grapes are produced annually and approximately 640,000 tonnes are processed in Iran. Moghaddasi et al. (2005) indicated a total of 30–38% of grapes lost in Iran at various stages such as harvesting, transportation, storage and sorting, wholesale, and retail sale, which transformed into losses and waste of the postharvest chain. In another study, Rajabi et al. (2015) reported that a total of about 53.0% of losses in grapes were recorded in the various stages in Takestan City. However, the major amount (about 46% losses) takes place at 19.0%, 17.6%, 9.0% and 7.0% of losses at the processing stage, agricultural production, postharvest and distribution and consumption stages, respectively.

1.5.6 Papaya

Papaya is one of the major important fruits of tropical and subtropical regions of the world and is highly perishable in nature. Mechanical damage, rapid flesh softening, spoilage, physiological disorders, insect/pest infestation and improper temperature management are the main causes of its postharvest losses (Sivakumar and Wall 2013). In papaya fruits, postharvest handling and market preparation requirements are mainly influenced by susceptibility to certain diseases. The most important postharvest disease of papaya is anthracnose incited by *Colletotrichum gloeosporioides* as small black or light brown spots which gradually enlarge and may coalesce and sink on ripe fruits. Besides this fungus, several other fungi such as *Ascochyta caricae-papayae, Phomopsis caricae-papayae* and *Phytophtora nicotiana* var. *Parasitica* have been recorded during storage and transportation(Nakasone and Paull 1999; Sankat and Maharaj 2001; Gajanana et al. 2010).

CISH, Lucknow, conducted study to quantify the postharvest losses of papaya in Farukhabad and Kanpur districts of Uttar Pradesh. They reported 6.30% loss at the farm level due to cracking and bird damage and 11.47% loss during ripening of the fruit. The total loss of papaya was 24.97% in the papaya marketing system, and 9.00% weight loss was due to physiological disorders. Nanda et al. (2010) reported 7.36% overall postharvest loss in papaya at the field level, of which 5.06% and 2.28% of losses were at the farm level and during storage, respectively. Gajanana et al. (2010) reported that the total postharvest loss in papaya cv. Taiwan 786, which is produced in Ananthpur district of Andhra Pradesh and marketed at Bangalore, was 25.49%, consisting of 1.66%, 4.12%, 8.22% and 11.49% at the field level, transit, market level and retail level, respectively, due to immature and small size of fruits, malformation and harvesting injury at the farm level, and at the market level by bruising and pressing injury.

Azene et al. (2011) reported that postharvest losses in Ethiopia were estimated to be as high as 25–35%. High losses in papaya are mainly due to the lack of packaging and storage facilities as well as poor transportation facilities (Kebede 1991; Wolde 1991). Among environmental factors, temperature is the most important factor in spoilage of papaya fruit and products. Suitable temperature and relative humidity are necessary to reduce high postharvest losses of papaya fruits under African conditions (Workneh et al. 2012).

1.5.7 Strawberry

Strawberry fruits are very delicate and highly perishable in nature due to their extreme tenderness, high vulnerability to mechanical damage, high respiration rate and more prone to fungal decay. Diseases are the greatest cause of postharvest losses in strawberry. They are Botrytis rot or grey mould (*Botrytis cinerea*), Rhizopus rot (*Rhizopus stolonifer*), mucor rot (*Mucor* spp.), anthracnose (*Colletotrichum* spp.) and blue mould (*Penicillium* spp.) (Feliziani and Romanazzi 2016; Kumar and Kaur 2019).

A study was conducted in the United States on postharvest loss of strawberry at the consumer level showed about 20%, and at supermarkets the loss for fresh fruit was estimated to be about 9.5%. The mould growth is often associated with organoleptic concerns by breakdown in fruit texture (Al-Hindi et al. 2011).

Postharvest fruit loss of 20–50% was reported due to spoilage, depending upon fruit maturity, month of harvesting, method, method of packaging and distance of transportation (Mingchi and Kojimo 2005; Panda et al. 2016). Postharvest fruit losses were estimated to be about 10–20% worldwide, but other researchers reported 25–40% losses for the tropics (Ogunleye and Adefemi 2007). The average strawberry postharvest loss was 28.0% in the study area. Thus, the total postharvest strawberry loss was estimated about 6,350.12 tonnes and the total postharvest losses were equal to 7,809,200.00 tonnes (Salami et al. 2010).

1.5.8 Potato

In India, about 90% of potato crop is harvested during January and February from the Indo-Gangetic plains, where the harvest is followed by the rising temperatures of hot and dry summer, and further by the warm and humid rainy season. Since potato tubers contain about 80.0% water and approximately 20.0% solid materials. The potato is a semiperishable commodity which cannot be stored in the absence of refrigeration for more than 3–4 months. At present, postharvest losses of potato are approximately 16% due to pathological and physiological losses; sprouting losses also add to total postharvest losses. Postharvest losses are targeted to reduce to 10% by the year 2050 by introduction of new management strategies and technologies.

Nanda et al. (2010) reported that the postharvest losses in potato were 6.73% and 2.26% at farm level and during storage, respectively, with an aggregate loss of 8.99%. Masum et al. (2013) reported a maximum loss of 5.55% within the 3 months in the Diamant district, which was due to soft rot (2.49%), dry rot (1.05%) and scab (0.97%) diseases of potato. In Mymensingh district, the losses of tuber were about 5.84%, 5.54% and 5.25%, in the months of July, August and September, respectively. Similarly, in Rajshahi district, it was 5.85%, 5.58% and 5.28%, respectively, during the same periods. In Dhaka district, 5.58%, 4.96% and 4.55% losses of potato were recorded in these months, respectively.

Postharvest loss of tubers was 20–25%, which is one of the major problems in the production of potato. The reduction of such huge loss in potato will help small holders in nutrition, food security and income generation (BoFED 2007). Tadesse et al. (2018) assessed postharvest losses of potato along value chain actors in Ethiopia. The study indicated that 21.72%, 0.59%, 0.655% and 1.92% losses were reported at producer, local trader and wholesaler and retailer levels, respectively. Kuyu et al. (2019) conducted a study in two different districts of the Jimma zone (Ethiopia) having different road quality and transportation access. They estimated average harvest- and transportation-related losses of 12.45% and 11.7% in Dedo and Seka districts, respectively.

1.5.9 Onion

There are various kinds of losses in onion, including those that are unmarketable, driage, spoilage, bruising, rotting, microbial decay, transportation, physiological weight loss, sprouting, etc. A study was conducted at IARI, New Delhi and IIHR, Bengaluru under the Indo-USAID project in 1987–1990 to estimate postharvest losses in onion from Rajasthan, Gujarat, Haryana and Punjab. The total loss of onion at the wholesale level was 4.18% and the total physical loss was 18.16% at the market level. Based on survey data conducted in 1989–1990 from Gujarat state on local onion, the economic loss of onion at the market level was 10.62%. The postharvest loss was about 13.0% in small bulb varieties such as Bangalore Rose and Padisu, and in large-sized bulbs, the loss was >30.0% under rainfed areas of Karnataka. Postharvest losses were 22.09% and 55.28% in Arka Nikethan and Arka Kalyan varieties during storage up to 5 months, respectively. It was also reported that postharvest losses were 19.27% in rain-fed onion and 12.76% in irrigated onion, excluding the physiological loss in weight. Kumar et al. (2015) reported diverse species of *Aspergillus*, *Penicillium*, *Alternaria*, *Fusarium*, etc., which attacks onion bulbs during the storage period, and among them, *Aspergillus* spp. was the most virulent fungal pathogen in the field as well as during storage of bulbs.

In difference cultivars of onion, postharvest losses were due to weight loss, sprouting and decay (Baninasab and Rahemi 2006). Nanda et al. (2010) reported that overall postharvest losses in onion were 7.51%, and of which the total loss was 5.17% in farm operations, followed by 2.34% during storage. Gajanana et al. (2011) reported 10.43% postharvest loss in onion (Bellary Red) at field and 2.12% at retailers' level in Karnataka. They reported that 40–50% of the production of onion did not reach to consumers due to postharvest losses caused by sprouting as well as physiological weight loss. The huge loss in onion at producer level was due to faulty storage, lack of adequate transportation facility, drying, poor handling practices of the produce during marketing, decay of bulbs, doubles, bolters, poor packing facilities, injury during harvesting and de-topping. Total losses were 82.65% in the supply chain including at farm level, wholesale and retail level (Sharma 2016). Total losses were significantly less in Bhima Kiran (26.66%) and Bhima Shakti (35.87%) after 4 months of storage (Gorrepati et al. 2018).

In Nigeria, >50% of postharvest losses of onion produce were reported because of poor postharvest handling during the onion value chain, especially during transit (World Vegetable Centre 2018). Emana et al. (2017) reported that postharvest losses affected the quality of produce, which reduced 61% income of producers during the wet season. Postharvest losses are often linked to serious economic impacts including direct financial losses for the growers and also for the marketers. The loss also indicates a waste of productive agricultural resources such as land, water, managerial skills, labour and other inputs that have been proposed as a means of extending the shelf life of onion with additional advantages of ease of transportation, packaging and weight reduction (Kashif et al. 2016; Jolayemi et al. 2018).

1.5.10 Tomato

Postharvest losses in tomatoes occur due to immaturity and over-ripening of fruits, mechanical damage during harvesting, improper packaging and poor transport conditions and decay due to fungi and bacteria (Esguerra et al. 2018; Isaac et al. 2015). Besides these factors, other factors such as physiological deterioration and biological means, i.e., disease and insect pests and rodents and birds, also cause postharvest losses in tomatoes (Asalfew and Nega 2020). Ahmed et al. (2017) reported fungal and bacterial pathogens species associated with postharvest disease in tomato fruits during a survey in Oahu as *Alternaria*, *Botrytis*, *Colletotrichum*, *Fusarium*, *Geotrichum*, *Mucor*, *Stemphyllium*, *Rhizopus* and *Penicillium Acetobacter*, *Gluconobacter*, *Klebsiella*, *Leuconostoc* and *Pectobacterium* were most prevalent. *Alternaria tenuissima*, *Botrytis cinerea*, *Cladosporium fulvum*, *Colletotrichum coccodes*, *Fusarium oxysporum*, *Geotrichum candidatum*, *Rhizopus stolonifer* and *Stemphylium macrosporoideum* were detected and identified on tomato fruits (Cristina et al. 2018).

Nanda et al. (2010) reported a total of 12.98% postharvest loss in tomato, which further broke down to 9.94% at the field level and 3.04% during storage. In another study, the total postharvest loss in tomato was about 19.0% in Karnataka, which comprised at field level 9.43%, market level 4–5% and retail level about 5.0% (Gajanana et al. 2006). Babatola et al. (2008) reported that about 20.0% of losses were incurred by farmers due to transportation delay. In another study, postharvest losses in tomato were recorded with the range of 40.3–55.9% of the total harvestable product; of that 68.6–86.7% of undamaged, edible harvested tomatoes were rejected as out-grades and consequently discarded due to the product not matching the specification, and 71.2–84.1% of produced tomatoes were left in the field and not harvested. However, only 44.1–59.7% of the harvestable crop reached the consumers of the two supply chains, respectively (McKenzie et al. 2017).

1.5.11 Cauliflower

Postharvest diseases in cabbage and cauliflower are most important source of postharvest loss due to the combined effects of rough handling and poor storage facility. Cabbage and cauliflower heads suffer from many fungal and bacterial diseases in fields and storage. Alternaria leaf spot (*Alternaria brassicae*), downy mildew (*Peronospora parasitica*), head rot (*Rhizoctonia solani*), black rot (*Xanthomonas campestris* pv. *campestris*) and bacterial leaf spot (*Pseudomonas maculicola*) are reported, which affect production and marketability of heads of cabbage and cauliflower. Grey mould (*Botrytis cinerea*), Rhizopus soft rot (*Rhizopus stolonifer*), watery soft rot (*Sclerotinia sclerotiorum*) and bacterial soft rot (*Erwinia carotovora* subsp. *carotovora*) are diseases which occasionally cause loss in the field but are important in the later stage of distribution and storage of the heads. In the cruciferous vegetables alone, 18 genera and 32 species have been recorded (Harbola and Khulbe 1994). Harbola and Khulbe (1994) reported the association of many pathogens; namely, *A. tenuissima*, *Aspergillus niger*, *Cordanamusae* and *Fusarium moniliforme* with curd rot of cauliflower. Postharvest bacterial soft rot (*E. carotovora* subsp. *carotovora*) is a major pathogen that causes more loss of production than any other bacterial disease (Agrios 2007). A large list of bacterial and fungal pathogens causes postharvest losses in transit, storage and to the consumer. Bacterial soft-rot, black spot, grey mould and *Cladosporium* rot are common diseases of Cole crops (Cantwell and Suslow 2014).

Total losses in cauliflower during postharvest operations and transportation were recorded 24.9–30.4 and 28.6–35.1%, respectively, and maximum losses occurred during transportation from rural markets to urban markets (Pal et al. 2002). Under ICAR NATP Network project, aggregate postharvest losses were recorded 12.0%, 12.5%, 13.0% and 20.82% in Varanasi, Ranchi, Bhubaneshwar and Guwahati and Cooch behar, respectively. The postharvest losses at different stages such as at field, wholesalers and retailers levels were in the range of 0.5–5.02%, 1.5–8.09% and 5.0–9.5% (Anonymous 2003). Nanda et al. (2010) found that an aggregate postharvest loss was 6.88% in cauliflower, of which 4.85% occurred at farm and 2.03% during storage. The postharvest losses of cauliflower at the growers, 'Bepari', wholesalers and retailers' levels were 4.16%, 9.2%, 10.34% and 10.73%, respectively. The total postharvest loss of cauliflower from harvest to retail sale was estimated to be 34.34% due to the lack of storage facility, rough handling and poor transport system (Ishrat 2010).

1.5.12 Green Peas

Fresh green peas are highly perishable due to high moisture content and sugars under room conditions. However, the shell of the seed protects the peas to a great extent. After shelling, peas undergo rapid deterioration if not refrigerated or frozen. In India, very little information on postharvest losses in green peas is available. An overall postharvest loss in green peas was 10.28%, of which 8.58% loss was recorded at farm levels, and 1.70% at storage (Nanda et al. 2010). As per the information from IIVR, Varanasi, 4.87%, 1.05%, 0.44% and 3.75% losses were recorded at harvesting level, grading and packing, transportation and marketing, respectively, and the total loss was 10.06%. Kader (2011) reported that gradual weight loss of the produce is naturally due to transpiration and respiration.

1.6 Factors Affecting Postharvest Losses

Many factors influence the postharvest losses of fruits and vegetables, including physical, physiological, mechanical and hygienic conditions. A substantial loss occurs between harvesting and consumption of the horticultural products due to these factors (Sani and Alao 2006). In addition, other factors such as insect pest and mite injury, disorder and postharvest diseases also cause postharvest loss of fruits and vegetables. However, among the causes, the most serious postharvest diseases follow mechanical injury (Mustapha and Yahaya 2006; Anonymous 2011). Environmental factors, including temperature, relative humidity and oxygen balance, particularly during storage, are greatly responsible for losses. Temperature and relative humidity are responsible for rendering fruits and vegetables susceptible to pathological attacks. However, the losses in horticultural produce due to physiological and biochemical damage are closely interrelated (Sani and Alao 2006; Williams et al. 1991). The principal causes of postharvest losses in different produce in horticultural crops are described by Kitinoja and Kader (2002), as given below.

1. Mechanical injuries and bruising, e.g. carrots, spinach, artichokes, eggplant, tomato.
2. Improper curing, e.g. sugar beets.
3. Sprouting and rooting, e.g. onions.
4. Water loss (shrivelling), e.g. garlic, lettuce, citrus, squash.
5. Decay, e.g. potato, green onions, okra, snap beans, cauliflower, apples, grapes, stone fruits.
6. Chilling injury (subtropical and tropical root crops), e.g. sweet potato, peppers, bananas.
7. Loss of green colour (yellowing), e.g. chard.
8. Relatively high respiration, e.g. cabbage.
9. Yellowing and other discolouration, e.g. broccoli.
10. Abscission of florets, e.g. cauliflower.
11. Over-maturity at harvest, e.g. cucumbers, melons.
12. Compositional changes, e.g. mango.

1.6.1 Primary Causes of Postharvest Loss

Pathological, mechanical, physiological or environmental factors are the primary factors of losses of horticultural produce, particularly fruits and vegetables (Yahaya 2005).

1.6.1.1 Mechanical Injury

Mechanical damage to horticultural products is caused by careless handling during harvesting, packing, transportation, storage, etc. In addition, insect pests and birds can cause mechanical injury to fruits and vegetables (Figure 1.1). There are many examples in which damage caused by mechanical injuries such as bruising and cracking of fruits and vegetables, which make them more prone to infection by microbes, which increase the rate of water loss and gaseous exchange. The damage caused by mechanical injury to fruits and vegetables is often the result of pressure thrust during transportation. Although this type of damage is often invisible, it causes rupturing of the inner tissues and cells. Such produce degrades faster during the natural ageing process (Alao 2000; Yahaya 2005). The fruits and vegetables may get maximum mechanical damage during harvesting due to improper picking. Tuber and root crops such as potato, sweet potato, etc., can be damaged and become unmarketable due to not following proper methods of digging them out. Fruits such as apples, mango and strawberry, which have soft outer skin, are highly prone to mechanical damage. Damaged fruits are therefore also prone to pathological attacks like rotting (Alao 2000; Yahaya 2005). In general, the fruit quality gradually deteriorates, and the produce fetches lower prices in the market, which is a type of economic loss. Spillage, abrasion, excessive polishing, peeling and trimming of processing operations contribute to the loss of the produce. Puncturing of the containers and defective seals also leads to mechanical injury (Alao 2000; Yahaya 2005). In horticultural products, their high moisture content and soft texture makes them prone to mechanical injury, which can occur at any stage from production to retail marketing due to the following.

FIGURE 1.1 Damaged fruits due to faulty harvesting method by dropping from the tree to the ground.

1.6.1.1.1 Poor Harvesting Practices

- Improper containers and crates used during harvesting and transportation, which may have splintered wood, sharp edges, poor nailing or stapling.
- Over-packing or under-packing of the produce of field or marketing containers.
- Careless handling of produce by dropping or throwing or walking on produce, and during packaging, containers used during the grading, transport or marketing.

Injuries can take many forms:

- Splitting of fruits during harvesting when they are dropped.
- Internal bruising, which is not visible externally due to impact.
- Superficial grazing or scratches affecting the skins and outer layer of cells.
- Crushing of leafy vegetables and other soft produce.

Injuries cutting through or scraping away the outer skin of produce will:

- Allow entry points for the pathogens causing spoilage.
- Enhance water loss from the injured area.
- Cause an increase in the rate of respiration and thus generation of heat.

Bruising injuries which leave the skin intact and may not be visible externally:

- Enhance the rate of respiration of the produce and heat generation.

- Cause internal discolouration due to damaged tissues.
- Have off-flavours due to abnormal physiological reactions in damaged parts.

Damage suffered by packaged produce

a. **From injuries**

  i. Cuts or punctures are caused by sharp objects piercing the package; splinters in bamboo or wooden containers; staples or nails protruding in containers. The effects on the produce are deep punctures or cuts, leading to water loss and rapid decay.

  ii. Impact is caused by throwing or dropping of packages or sudden starting or stopping of transport vehicle, causing load movement, and speeding of vehicle on rough roads. This can cause bursting of packaging and bruising of contents.

  iii. Compression is caused by flimsy or oversized containers, containers overfilled or stacked too high, or both; collapse of stacked containers during transport causes bruising or crushing of contents.

  iv. Vibration is caused by vibration of the vehicle and from rough roads, causing the wooden boxes to come apart, damaging the produce.

b. **From the environment**

  i. Heat damage is caused by exposure of packages to external heat, e.g. direct sunlight, or storage near the heating system; natural build-up of internal heat of produce owing to poor ventilation within packages, in storage or vehicle. Under such conditions, the fruits become overripe or soften, wilts and develops off-flavours; decay develops rapidly; cardboard cartons may become dry and brittle and are easily damaged on impact (Figure 1.2).

  ii. Chilling or freezing damage is caused by low or sub-zero ambient temperatures, exposure of sensitive produce to temperatures below chilling or freezing tolerance level during storage, which causes damage to chilling-sensitive produce, breakdown of frozen produce on thawing, and plastic containers become brittle and may crack.

FIGURE 1.2 Damage of cardboard boxes and fruits during transportation from Abohar, Punjab to Bangalore, Karnataka, India.

  iii. Moisture and free-water damage is caused by exposure to rain or high humidity, condensation on packages and produce moved from cold storage to a damp atmosphere at ambient temperature, and packing wet produce in cardboard containers, with softening and collapse of stacked cardboard containers causing squashing of produce in collapsed containers; decay is promoted in damaged produce.

  iv. Damage from light is caused by plastic sacks and crates not treated with an ultraviolet inhibitor and which eventually break up when exposed to direct sunlight. The plastic sacks disinter grate when they are moved and fracturing of plastic crates can cut or bruise the produce.

c. **From other causes**

  i. Chemical contamination is caused by contamination of containers stored near chemicals, damage to produce by containers treated with preservatives, e.g. boxes made from wood treated with pentachlorphenate (PCP), contamination of produce from boxes affected by mould growth. Containers can cause flavour contamination or surface damage and discolouration of produce in contact with them, decay of produce owing to contaminating moulds; wood-rotting moulds cause collapse of boxes.

  ii. Insect damage is caused by insects present in packed produce; wood-boring insects in wooden boxes affect consumer resistance and cause legal problems due to presence of insects (e.g. spiders, cockroaches) in packed produce, spread of wood-destroying insects in infected boxes.

  iii. Human and animal damage is caused by contamination and eating by rodents and birds as well as pilfering by humans, causing rejection of damaged produce by buyers or inspectors and loss of income through loss of produce.

1.6.1.2 Biological Factors

Rodents, birds, monkeys and other large animals cause direct disappearance of food by consuming it. Sometimes the level of contamination of food by the excreta, hair and feathers of animals and birds is so high that produce is not fit for human consumption. Insects cause both weight losses through consumption of the produce as well as quality losses due to their frass, webbing, excreta, heating and unpleasant odours, which can impart to produce.

1.6.1.3 Chemical Factors

Many of the chemical constituents naturally present in horticultural products react spontaneously to cause loss of colour, flavour, texture and nutritional value. There is one example, Maillard relation, which causes browning and discolouration in dried fruits and other products. There can also be accidental or deliberate contamination of food with harmful chemicals such as pesticides or obnoxious chemicals such as lubricating oil.

1.6.1.4 Biochemical Reactions

A few enzyme-activated reactions can occur in produce during storage that may cause off-flavours, discolouration and softening. Example, unpleasant flavours may develop in frozen vegetables that have not been blanched to inactivate these enzymes before freezing.

There are some instances of changing pectin level during ripening in plum fruits. Insoluble pectin levels decline during ripening and cold storage of plum fruit, with a concomitant increase in soluble pectin levels. There is no significant role of harvest maturity and storage time on the concentration of calcium pectate. This pectic fraction does not appear to influence the development of gel breakdown. However, water-soluble pectin and availability of cell fluids indicate a high gel potential in plums. A significant level of gel breakdown developed only in plums harvested at postoptimum maturity. Higher sugar levels and loss of cell membrane integrity in gel breakdown in fruit probably enhance the formation of pectin sugar gels, as cell fluids bind with pectins in cell walls. The initial response to low temperature is considered to involve physical factors such as membrane alteration and protein/enzyme diffusion. However, physiological changes that lead to losses of structural integrity and overall fruit quality also occur. During softening of tissue, dissolution of the ordered arrangement of cell wall and middle lamella polysaccharides occurs. As the fruit ripens, a substantial portion of its cell wall pectin is converted into a water-soluble form affecting the texture. The major changes involved in softening and chilling injury in peach fruits are the catabolism of cell walls and the development of an intercellular matrix containing pectin. Gel-like structure formation in the cell wall is due to the desertification of pectin without depolymerization which leads to the development of woolliness in peach (Lurie et al. 2003).

1.6.1.5 Microbial Action

Fruits and vegetables are highly prone to spoilage due to microbial infection. The microbial spoilage is mainly caused by species of fungi and bacteria and they contribute significantly to losses of fruits and vegetables during postharvest. The succulent nature of fruits and vegetables makes them easily prone to infection by these microbes (Elias et al. 2010). Postharvest diseases occur on perishable produce rapidly and cause extensive breakdown of the produce, which may sometimes decay the entire package (Alao 2000; Yahaya 2005).

It is generally estimated that about 36% of vegetable spoilage is caused by soft rot bacteria. Similarly, rotting in soft fruits caused by fungi is also very destructive. The most common pathogens causing rot in vegetables and fruits are fungi such as *Alternaria*, *Aspergillus*, *Botrytis*, *Diplodia*, *Monilinia*, *Phomopsis*, *Rhizopus*, *Penicillium*, *Fusarium*, etc. (Singh and Sharma 2007; Sharma et al. 2009). Gram-negative bacteria, such as *Burkholderia*, *Dickeya*, *Erwinia* (*Pectobacterium*) and *Pantoea*, *Pseudomonas*, *Xanthomonas*, and gram-positive bacteria (*Bacillus* and *Clostridium*), are major bacteria which cause spoilage of fruits and vegetables (Table 1.5, Figure 1.3).

TABLE 1.5

Some Important Postharvest Pathogens Causing Diseases in Fruits and Vegetables

Sr. No.	Postharvest Pathogen and Its Species	Diseased Caused	Crops Affected
1.	*Alternaria*, *A. alternata* (Fr.) Keissl., *A. brassicae* (Berk.) Sacc., A. *citri* Ell. and Pierce A. *tenuissima* (Kunze) Wiltshire	Blotch, blight, stem-end rot	*Fruits:* Apple, apricot, avocado, blueberry, Cape gooseberry, carambola, cauliflower, chayote, cherry, citrus, fig, gooseberry, grape, mango, melon (watermelon and muskmelon), nectarine, papaya, passion fruit, pea, pear, persimmon, plum, raspberry, stone fruits, strawberry, watermelon *Vegetables:* Bean, cucumber, eggplant, sweet pepper, potato, squash, tomato
2.	*Aspergillus*, *A. niger*, *A. flavus*	Black mould, Aspergillus rot	*Fruits:* Mango, onion, peach, plum, strawberry
3.	*Botryosphaeria*, *B. ribis* Grossenb. and Duggar Family: Botryosphaeriaceae	Stem-end rot	*Fruits:* Apple, avocado, banana, citrus, grape, kiwifruit, peach, pear
4.	*Botrytis*, *B. cinerea* Pers. Anamorphic Sclerotiniaceae Teleomorph: *Botryotinia*	Soft rot and grey mould	*Fruits:* Apple, apricot, blueberry, cherry, citrus, cucumber, eggplant, fig, garlic, grape, Jerusalem artichoke, kiwifruit, marrow, melon, nectarine, parsnip, peach, pear, pepper, persimmon, plum, pomegranate, raspberry, strawberry *Vegetables:* Asparagus, bean, broccoli, Brussels sprouts, cabbage, carrot, cauliflower, celery, fava bean, fennel, leek, lettuce, onion, pea, potato, pumpkin, squash, sweet potato, taro, tomato, turnip, yam
5.	*Colletotrichum*, *C. aculeatum*, *C. gloeosporioides*, *C. musae*, *C. capsica*, *C. melongenae* Anamorph Family:Glomerellaceae Teleomorph: *Glomerella*	Anthracnose, stem-end rot	*Fruits:* Apple, apricot, avocado, banana, mango, papaya, rambutan, watermelon *Vegetable:* Chilli

(*Continued*)

TABLE 1.5 *(Continued)*

Some Important Postharvest Pathogens Causing Diseases in Fruits and Vegetables

Sr. No.	Postharvest Pathogen and Its Species	Diseased Caused	Crops Affected
6.	*Cladosporium*, *C. herbarum* (Pers.) Link Anamorphic Family: Mycosphaerellaceae Teleomorphs: *Mycosphaerella*, *Venturia*	Scab	*Fruits:* Apple, apricot, banana, Cape gooseberry, carambola, cherry, citrus, date, fig, grape, melon, nectarine, papaya, passion fruit, peach, pear, persimmon, plum, pomegranate, raspberry, watermelon *Vegetables:* Bean, cabbage, cauliflower, cucumber, eggplant, leek, onion, pea, pepper, squash, tomato
7.	*Dothiorella*, *D. gregaria* Sacc, *D. dominicana* Pet. Et. Cif. *D. mangiferae* H. et. P. Syd. et But. Family: Botryosphaeriaceae	Stem-end rot	*Fruits:* Avocado, banana, mango, citrus
8.	*Fusarium*, *F. moniliforme* J. Sheld., *F. oxysporum* E.F. Sm. & Swingle and its special forms, *F. solani* (Mart.) Sacc. Anamorphic Family: Hypocreaceae Teleomorphs: *Gibberella*, *Nectria*	Dry rot and Fusarium rot	*Fruits:* Apple, asparagus, avocado, banana, cape gooseberry, citrus, fig, guava, melon, papaya, parsnip, passion fruit, pear, pineapple, watermelon *Vegetables:* Asparagus, beans, cabbage, carrot, cassava, celery, cucumber, eggplant, garlic, ginger, okra, onion, pea, peppers, potato, squash, sweet potato, taro, tomato, yams
9.	*Geotrichum candidum* Link Family: Dipodascaceae	Sour rot	*Fruits:* Citrus, strawberry *Vegetables:* Carrot, cucumber, potato, pumpkin, tomato
10.	*Gliocephalotrichum micro chlamydosporum* (Mey) Wiley & Simmons	Brown spot	*Fruit:* Rambutan
11.	*Lasiodiplodia*, *L. theobromae (*Pat.) Griffiths & Maubl. (syn. *Botryodiplodia theobromae*) Mitosporic fungi	Dry and soft rot	*Fruits:* Apple, avocado, banana, carambola, citrus, grape, guava, litchi, loquat, mango, mangosteen, melon, papaya, peach, pear, pineapple, rambutan, watermelon *Vegetables:* Cassava, cucumber, eggplant, ginger, onion, pepper, pumpkin, squash, sweet potato, taro, tomato, yam
12.	*Monilinia*, *M. fructicola* (G. Winter) Honey, *M. fructigena* Honey, *M. laxa* (Aderh. & Ruhland) Honey Family: Sclerotiniaceae	Brown and soft rot	*Fruits:* Apple, apricot, blueberry, cherry, nectarine, peach, pear, plum
13.	*Mucor piriformis* Family: Mucaraceae	Mucor rot	*Fruits:* Apple, pear
14.	*Penicillium*, *P. cyclopium* Westling, *P. expansum* Link, *P. digitatum*, *P. italicum* Anamorphic Family: Trichocomaceae Teleomorphs: *Talaromyces Eupenicillium*	Blue and green moulds	*Fruits:* Apple, apricot, avocado, cherry, citrus, fig, grape, Jerusalem artichoke, kiwifruit, mango, melon, nectarine, papaya, passion fruit, peach, pear, persimmon, pineapple, plum, pomegranate, raspberry, strawberry *Vegetables:* Asparagus, carrot, cucumber, garlic, onion, sweet potato, taro, tomato, yam
15.	*Phoma*, *P. exigua* Sacc. and its varieties, *P. lingam* (Tode) Desm. Anamorphic Family: Pleosporaceae Teleomorph: *Pleospora*	Dry rot	*Fruits:* Asparagus, bean, beet, broccoli, Brussels sprouts, cabbage, carrot, celeriac, celery, cauliflower, chayote, eggplant, fennel, guava, kiwifruit, melon, okra, papaya, parsnip, pea, pepper, persimmon, *Vegetables:* Pomegranate, potato, raspberry, sweet potato, tomato, turnip, watermelon
16.	*Pythium*, *P. aphanidermatum* (Edson) Fitzpatrick, *P. ultimum* Trow Family: Pythiaceae	Wet rot	*Fruits:* Banana, melons, parsnip, strawberry *Vegetables:* Beans, beets, broccoli, Brussels sprouts, cabbage, carrot, celery, cucumber, eggplant, endive, garlic, ginger, leek, lettuce, onion, pea, peppers, potato, pumpkin, rhubarb, shallot, squash, sweet potato, taro, tomato, turnip, watermelon, yam
17.	*Phomopsis*, *P. mali* (Schulzer & Sacc.) Died., *P. vexans* Anamorphic Family: *Valsaceae* Teleomorph: *Diaporthe*	Dry rot	*Fruits:* Apple, avocado, blueberry, citrus, grape, guava, kiwifruit, mango, melon, papaya, peach, plum, pomegranate, strawberry, watermelon *Vegetables:* Asparagus, bean, brinjal, cucumber, eggplant, sweet potato, sweet sop

TABLE 1.5 *(Continued)*

Some Important Postharvest Pathogens Causing Diseases in Fruits and Vegetables

Sr. No.	Postharvest Pathogen and Its Species	Diseased Caused	Crops Affected
18.	*Phytophthora, P. cactorum* (Lebert & Cohn) J. Schröter, *P. capsici* Leonian, *P. nicotianae* var. *parasitica*(Dastur) G.M. Waterhouse, *P. palmivora* (E.J. Butler) E.J. Butler, *P. porri* Foister Family: Pythiaceae	Blight and brown rot	*Fruits:* Apple, avocado, banana, breadfruit, citrus, guava, mango, melon, papaya, passion fruit, pear strawberry, watermelon *Vegetables:* Asparagus, beans, cabbage, carrot, cassava, chayote, cucumber, eggplant, fennel, garlic, leek, lettuce, okra, onion, spring onion, pea, peppers, potato, pumpkin, squash, sweet potato, taro, tomato
19.	*Rhizoctonia, R. carotae* Rader, *R. solani* J.G. Kühn *Ceratobasidium* Anamorphic Family: Corticiaceae Teleomorphs: *Thanatephorus, Tricharina*, etc.	Scarf and scab	*Fruits:* Melon, strawberry, watermelon *Vegetables:* Bean, beet, cabbage, carrot, cauliflower, celery, cucumber, eggplant, lettuce, okra, pepper, potato, pumpkin, radish, squash, sweet potato, taro, tomato, turnip, yam
20.	*Rhizopus, R. nigricans*, Ehrenberg *R. oryzae* Went & Prins. Geerl., *R. stolonifer* (Ehrenb.) Vuill. Family: Mucoraceae	Wet rot and Rhizopus rot	*Fruits:* Apple, apricot, avocado, banana, cherry, citrus, fig, grape, guava, litchi, mango, melon, nectarine, papaya, passion fruit, peach, pear, persimmon, pineapple, plum, pomegranate, raspberry, strawberry, watermelon *Vegetables:* Beans, beet, Brussels sprouts, cabbage, carrot, cassava, cauliflower eggplant, garlic, pea, peppers, potato, pumpkin, sweet potato, tomato, yam
21.	*Sclerotinia, S. sclerotiorum* (Lib.) de Bary Family: Sclerotiniaceae	Brown and soft rot	*Fruits:* Apple, banana, citrus, melon, parsley, parsnip, pear, strawberry, watermelon *Vegetables:* Asparagus, cucumber, eggplant, fava bean, fennel, garlic, leek, lettuce, onion, pea, pepper, potato, pumpkin, radish, squash, sweet potato, tomato, turnip
22.	*Sclerotium, S. cepivorum* Berk *S. rolfsii* Sacc. Mitosporic fungi (*S. rolfsii*) Teleomorph: *Corticium*	Soft rot	*Fruits:* Guava, melon, pear, watermelon *Vegetables:* Bean, beet, carrot, cassava, cauliflower, eggplant, endive, garlic, ginger, leek, lettuce, onion, pea, pepper, potato, pumpkin, spring onion, squash, sweet potato, taro, tomato, yam
23.	*Thielaviopsis paradoxa* (De Seynes) V. Hohn. Anamorph Teleomorph: *Ceratocytis paradoxa*	Water blister	*Fruits:* Pineapple, banana, date palm *Vegetables:* Sweet potato
24.	*Burkholderia gladioli* pv. *alliicola, B.* gladioli pv. *agaricicola* Family: Burkholderiaceae (Yabuucchi et al. 1992)	Soft rot, internal browning	*Vegetables:* Onion (Lee et al. 2005), mushroom *Pleurotus* species, sweet potato (Zhang et al. 2020)
25.	*Dickeyadadantii* (previously called *Erwinia chrysanthemi*) Family: Pectobacteriaceae (Samson et al. 2005)	Soft rot	*Fruits:* Pineapple *Vegetables:* Asparagus, broccoli, carrot, celery, eggplant, onion, pepper, potato, radish, sweet potato, tomato
26.	*Erwinia, Pectobacterium carotovorum* ssp. *atrosepticum* (= *E. carotovora* ssp. *atroseptica*), *Pectobacterium carotovorum* ssp. *carotovorum* (= *E. carotovora* ssp. *carotovora*) Family: Enterobacteriaceae	Soft rot	*Fruits:* Avocado, guava, mango, melon, papaya, pineapple, watermelon *Vegetables:* Asparagus, beet, broccoli, Brussels sprouts, cabbage, carrot, cassava, cauliflower, celery, cucumber, fennel, garlic, ginger, leek, lettuce, onion, pea, peppers, potato, pumpkin, radish, rhubarb, squash, sweet potato, taro, tomato, turnip, yam
27.	*Pantoea, Pantoeaananatis* Family: Enterobacteriaceae (Teresa and Stephanus 2009)	Soft rot	*Fruit:* Pineapple *Vegetables:* Onion (Gitaitis and Gay 1997; Yumiko et al. 2005), King oyster mushroom (Kim et al. 2007)
28.	*Pseudomonas, P. marginalis, P. syringae* and their pathovars Family: Pseudomonadaceae	Soft rot	*Fruits:* Apple, avocado, passion fruit, watermelon *Vegetables:* Asparagus, bean, broccoli, Brussels sprout, cabbage, carrot, cauliflower, celery, chicory, cucumber, fennel, ginger, leek, lettuce, melon, okra, onion, parsnip, pea, pineapple, potato, spinach, squash, tomato, turnip
29.	*Ralstonia solanacearum* Family: Ralstoniaceae	Brown rot	*Vegetable:* Potato

(Continued)

TABLE 1.5 *(Continued)*

Some Important Postharvest Pathogens Causing Diseases in Fruits and Vegetables

Sr. No.	Postharvest Pathogen and Its Species	Diseased Caused	Crops Affected
30.	*Xanthomonas*, *X. Campestris* and its pathovars Family: Xanthomonadaceae	Soft rot	*Fruits:* Apricot, cherry, lettuce, mango, melon, nectarine, passion fruit, peach, plum, watermelon *Vegetables:* Bean, broccoli, Brussels sprouts, cabbage, carrot, cauliflower pea, pepper, pumpkin, radish, tomato, turnip
31.	*Bacillus* sp.	Soft rot	*Vegetable:* Potato
32.	*Clostridium* sp.	Soft rot	*Vegetable:* Potato

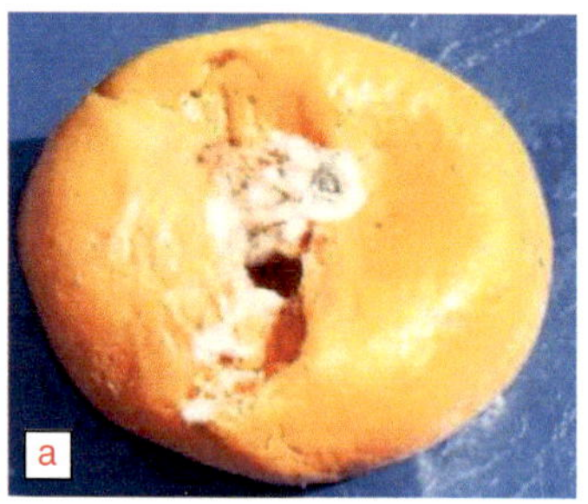

Decay of kinnow fruits caused by *Fusarium* sp.

Rhizopus rot caused by *Rhizopus stolonifer*

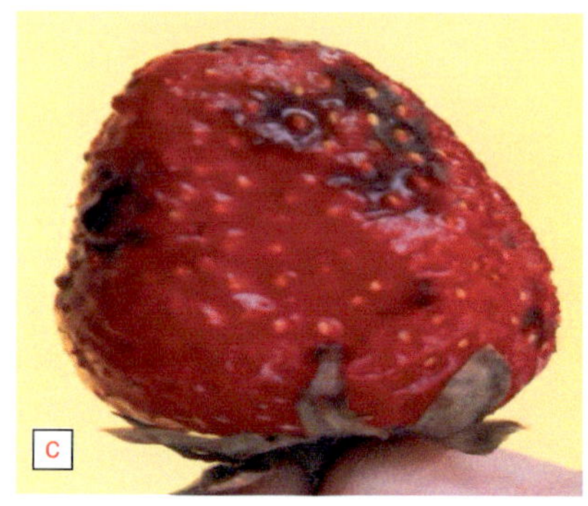

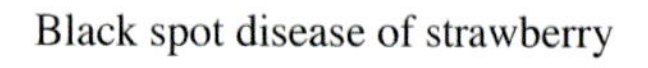
Black spot disease of strawberry

Alternaria rot disease of strawberry

Black rot disease of pineapple fruit

Crown rot disease of banana

FIGURE 1.3 Spoilage of different fruits and vegetables caused by fungi and bacteria.(A)Decay of kinnow fruits caused by *Fusarium* sp.(B)Rhizopus rot caused by *Rhizopus stolonifera*. (C) Black spot disease of strawberry. (D)Alternaria rot disease of strawberry.(E)Black rot disease of pineapple fruit.(F) Crown rot disease of banana.(G) Alternaria rot disease of peach fruits.(H) Rhizopus rot of disease of peach fruit.(I)Soft rot potato caused by *Erwinia* sp.(J)Blue mould rot disease of onion.

Alternaria rot disease of peach fruits

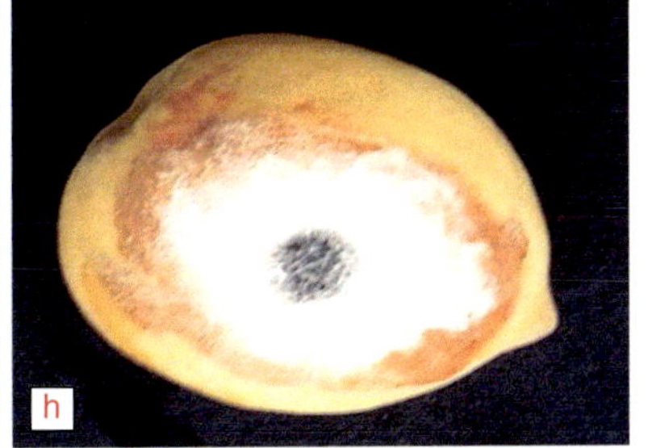

Rhizopus rot of disease of peach fruit

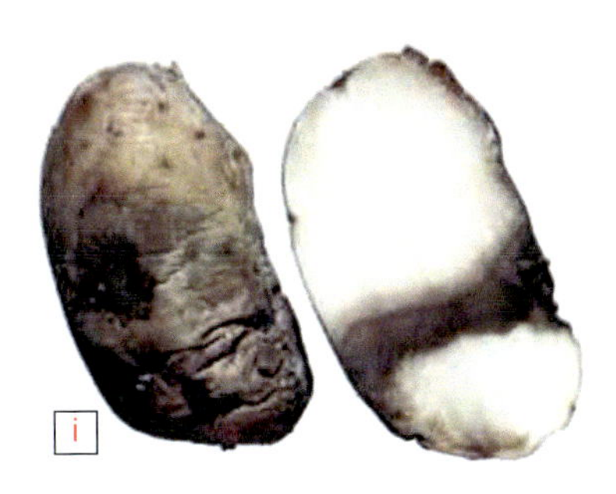

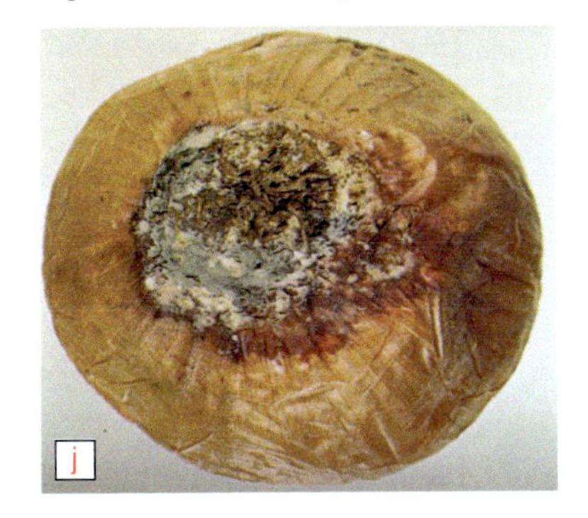

Soft rot potato caused by *Erwinia* sp.

Blue mould rot disease of onion

FIGURE 1.3 *(Continued)*

These spoilage organisms thrive and multiply faster at ambient temperatures and high humidity. Many of these spoilage organisms multiply even at lower temperatures during refrigeration. Refrigeration of vegetables for long periods results in spoilage due to the growth of Psychrotrophic bacteria, which are capable of surviving and multiplying in refrigeration temperatures (Burnett and Beuchat 2001; Vishwanathan and Kaur 2001).

Vegetables are exposed to various types of contaminants. They generally contain heterogeneous microflora; total microbial population depends on the field organisms and other bacteria encountered. Galaxies of pathogens that form surface microflora of vegetables are *Salmonella*, *Aeromonas hydrophila* and *Listeria monocytogenes*. Outbreaks of human listeriosis are often reported, which are associated with vegetables. *L. monocytogenes* has been isolated from many vegetables such as cabbage, cucumber, radish, tomatoes, salad vegetables, bean sprouts, etc. (Francis and O'Beirne 2002; Karen and Harrison 2002).

Postharvest pathogens enter the host tissue through injuries caused by careless handling during harvesting, packaging and transportation, or by insect or other animal damage. They also enter through growth cracks, natural pores in the above- and below-ground parts of plants. Some of the fungal pathogen may directly penetrate the intact skin of the plant and its produce. The time of infection varies with the crop and with different diseases. It can occur in the field before harvest or at any time afterwards. Field infections before harvest may not become visible until after harvest. For example, decay of root crops caused by soil moulds will develop during storage. Similarly, tropical fruits infected at any time during their development may show decay only during ripening. Infection after harvest can occur at any time between the field and the final consumer. It is for the most part the result of invasion of harvesting or handling injuries by moulds or bacteria.

Postharvest diseases may spread in the field before harvest using infected seed or other planting material. Many diseases can survive by using weed plants or other crops as alternate or alternative hosts. They can also spread by means of infected soil carried on farm implements, vehicles, boots, etc. and from crop residue or rejected produce left decaying in the field. Postharvest diseases can also spread by field boxes contaminated by soil or decaying produce or both, contaminated water used to wash produce before packing and decaying rejected produce left lying around packing houses and contaminating healthy produce in packages.

1.6.1.6 Physiological Causes

It is well known that in all living organisms natural respiratory losses occur that account for a significant level of weight loss. Additionally, the process of respiration generates heat. In fruits and vegetables many physiological changes occur in ripening, senescence, wilting and termination of dormancy (e.g. sprouting). These changes may increase the susceptibility of the produce to mechanical damage or infection by the pathogens. A reduction in nutritional level as well as consumer acceptance may also arise with these changes. Physical and chemical changes take place after harvest and continue during storage periods (Salunkhe and Kadam 1998; Porta et al. 2013). Production of ethylene results in premature ripening of certain crops. Postharvest losses, particularly in vegetable crops after picking and before processing, are 'alive'.

The major physiological changes ensuing postharvest period are given below.

Loss of moisture: Loss of moisture leads to fast shrivelling and loss of crispness. Plant tissue becomes mushy and eventually inedible, which leads to loss of soluble carbohydrate and reduction in weight. Moisture loss from surface initiates immediately after harvest and continues during the storage or transportation period (Verma and Joshi 2000).

Loss of stored energy: Vegetables are living tissues and respiration that occurs during the postharvest period requires energy. The stored carbohydrates are degraded for the purpose of energy supply.

Loss of food constituents: Moisture loss during storage is the major factor responsible for nutrient loss. The most labile nutrients that are lost are vitamins.

Fibre development: In vegetables such as beans, knolkhol (kohlrabi) and carrot, pectin degradation takes place as a result of moisture loss. The tissues become hard, and fibre development takes place.

Root and shoot development: Certain vegetables like beans, potatoes and other root vegetables develop roots and shoots during storage.

Loss of nutritional important pigments: Chlorophyll, carotenoids, lycopenes and xanthophylls are degraded during storage. The extents of loss are proportional to period of storage and hostile atmosphere.

Overall quality: Owing to the losses described above, the stored vegetables have poor overall quality.

1.6.1.7 Psychological Factors

In some cases, produce will not be eaten due to religious taboos.

1.6.1.8 Environmental Factors

Environmental factors including temperature, humidity, composition and proportion of gases in controlled atmospheric storage play a vital role in postharvest loss of horticultural produce. The growth of fungi and bacteria is favoured by high temperature and relative humidity, and these may cause various postharvest diseases, while decrease in temperature slows down the rate of microbial attack on different crops, especially when below 5°C during storage and transportation of the produce. High temperature also increases the rate of respiration of horticultural produce, which subsequently results in the breakdown of the inner tissues. In low temperature, chilling injury is observed in some commodities (Table 1.6) as described by Lutz and Hardenburg (1966) but non-freezing temperature above 4°C is mainly observed with tropical and subtropical fruits and vegetables. However, the symptoms caused by chilling injury are described as (i) discolouration: internal, external or both, usually brown or black, (ii) skin piking: sunken spots, especially under dry conditions, (iii) abnormal ripening (fruits): ripening is uneven or fails; off-flavours, and (iv)increase in decay: activity of microorganisms. These injuries may not appear clearly while they are held at chilling

TABLE 1.6

Susceptibility of Fruits and Vegetables to Chilling Injury at Low But Non-Freezing Temperatures

Sr. No.	Horticultural Produce (Minimum Safe Temperature)	Symptoms of Chilling Injury
1.	Aubergines (7°C)	Surface scald, Alternaria rot, decay
2.	Avocados (5–13°C)	Grey discolouration of flesh
3.	Bananas (green/ripe) (12–14°C)	Dull, grey-brown skin colour
4.	Beans (green)(7°C)	Pitting, russeting
5.	Cucumbers (7°C)	Pitting water-soaked spots, decay
6.	Grapefruit (10°C)	Brown scald, piking, watery breakdown
7.	Lemons (13–15°C)	Pitting, membrane stain, red blotch
8.	Limes (7–10°C)	Pitting
9.	Mangoes (10–13°C)	Grey skin scald, uneven ripening
10.	Melons	
	Honeydew (7–10°C)	Pitting failure to ripen, decay
11.	Watermelon (5°C)	Pitting, biker flavour
12.	Okra(7°C)	Discolouration, water-soaked areas, piking
13.	Oranges (7°C)	Pitting brown stain, watery breakdown
14.	Papaya (7°C)	Pitting failure to ripen, off-flavour, decay
15.	Pineapples (7–10°C)	Dull green colour, poor flavour
16.	Potatoes (4°C)	Internal discolouration, sweetening
17.	Pumpkins (10°C)	Decay
18.	Sweet peppers (7°C)	Pitting, Alternaria rot, decay
19.	Sweet potato (13°C)	Internal discolouration, piking, decay
20.	Tomatoes:	
	Mature green (13°C)	Water-soaked softening, decay
21.	Ripe (7–10°C)	Poor colour, abnormal ripening, Alternaria rot, decay

temperature; the symptoms may become visible only when the fruits and vegetables are transferred to room temperature (37°C). In freezing injury, all produce is subjected to freezing at temperatures between 0°C and 20°C. Frozen produce has a water-soaked or glassy appearance, although fewer commodities are tolerant of slight freezing. It is therefore suggested to avoid such temperatures because subsequent storage life is short. It has been observed that produce that has recovered from freezing is highly susceptible to decay.

When fresh horticultural produce is exposed to high temperatures caused by solar radiation, it will deteriorate rapidly. The produce exposed to sunlight after harvest may reach temperatures to about 50°C, which will cause a high rate of respiration. Produce packed and transported without cooling or adequate ventilation will become unusable. Long exposure to tropical sun will cause severe water loss from thin-skinned root crops such as carrots and turnips and also from leafy vegetables. High temperatures during harvesting of the produce can affect the level of decay during postharvest, as reported in grapefruit (Pailly et al. 2004). It is clear that crops such as broccoli, beans, peas and asparagus harvested at an over-mature stage will be prone to excessive accumulation of fibre, which will become more pronounced if weather conditions include high temperatures. Extremely high temperatures during harvesting impact hormonal levels in the marketed, edible portion of horticultural perishables. It is well known that abscisic acid fluctuates during the day. Abscisic acid seems to peak before the maximum light intensity or prior to the onset of the maximum daily temperatures (Fonseca et al. 2005). The symptoms that appear on the fruits and vegetables under extreme high temperature and solar injury are different in various crops, described by Moretti et al. (2010) and listed here.

1. **Apple:** Skin discolouration, pigment breakdown and water-soaked areas.
2. **Avocado:** Skin and flesh browning; increased decay susceptibility.
3. **Snap bean:** Brown and reddish spots on the pod; spots can coalesce to form a water-soaked area.
4. **Bell pepper:** Sunburn–yellowing and in some cases, slight wilting.
5. **Cabbage:** Outer leaves showing a bleached, papery appearance; damaged leaves are more susceptible to decay.
6. **Lettuce:** Damaged leaves assume a papery aspect; affected areas are more susceptible to decay; tip burn is a disorder normally associated with high temperatures in the field and can cause soft rot development during postharvest.
7. **Lime:** Juice vesicle rupture; formation of brown spots on fruit surface.
8. **Muskmelon:** Characteristic sunburn symptoms: dry and sunken areas; green colour and brown spots are also observed on rind.
9. **Pineapple:** Flesh with scattered water-soaked areas; translucent fruit flesh.
10. **Potato:** Black heart–occurs during excessively hot weather in saturated soil; symptoms usually occur in the centre of the tuber as dark-grey to black discolouration.
11. **Tomato:** Sunburn–disruption of lycopene synthesis; appearance of yellow areas in the affected tissues.

Relative humidity (RH) is another environment factor which plays a vital role in postharvest losses of fruits and vegetables. The effects of temperature and RH are comparable and interrelated mainly because the capacity of air to hold moisture varies with the temperature. Aeration in storage containers or in stores has its bearing on RH. Hence, it has crucial role on disease development indirectly. The effects of high RH on decay are also closely related to the effects of temperature. RH near saturation results in lower decay losses for many fruits and vegetables only if the temperature is near 0°C. However, RH <90% does not permit microorganisms to grow on the surface of produce of horticultural crops (Danladi 2000). Therefore, it is clear that all fruits and vegetables have their own specific requirement of heat during storage and transportation. Physical damage to the produce due to tissue breakdown occurs due to improper heat supply during processing and cold storage temperature as well as undesirable gaseous composition of controlled atmosphere of storage.

1.6.2 Secondary Causes of Loss

Inadequate methods of harvesting, transportation, storage and marketing facilities and legislation result in favourable conditions for secondary causes of loss. However, secondary causes of loss are those which load to conditions that encourage a primary cause of loss. They are usually the result of inadequate or non-assistant capital expenditures, technology and quality control. Some examples are given here.

1. Inadequate knowledge about harvesting, packaging and handling skills.
2. Lack of adequate packaging materials/containers for the transport and handling of horticultural produce.
3. Lack of proper storage facilities and cool chain facility to protect the produce.
4. Inadequate drying equipment or poor drying season to reduce postharvest losses.
5. Conventional processing procedures and facility and marketing systems may be responsible for high losses of horticultural produce.
6. Legal standards of the produce can affect the retention or rejection of produce for human use by being too lax or unduly strict.
7. Conscientious, knowledgeable management is essential for maintaining tools in good condition during marketing and storage.
8. Bumper crops are also responsible for overloading the postharvest handling system or exceeding the consumption need and cause excessive wastage.

1.7 Conclusion

Fresh fruits and vegetables are major sources of nutrients, including vitamins, minerals and dietary fibre. These photochemicals play significant roles in human health. Most of the horticultural products are perishable in nature and are highly vulnerable for postharvest damage during handling; mishandling causes injury to the produce which provides entry points for microbial infection. Postharvest losses in horticultural products vary due to poor harvesting methods, rough handling, improper packaging and poor transport conditions. The losses in fruits and vegetables are about 15–50% during the period between harvesting and final retail marketing, which leads to an enormous waste of seeds and planting materials, land, energy, fertilizers, water, labour and other resources. The main causes of postharvest losses include mechanical damage, physiological deterioration and biological causes (i.e. postharvest diseases and insect pests, rodents and birds, and quality deterioration caused by temperature stress). The most common pathogens causing spoilage in vegetables and fruits are fungi such as *Alternaria*, *Aspergillus*, *Botrytis*, *Diplodia*, *Monilinia*, *Phomopsis*, *Rhizopus*, *Penicillium*, *Fusarium*, etc. and bacteria like *Bacillus*, *Burkholderia*, *Clostridium*, *Dickeya*, *Erwinia* (*Pectobacterium*), *Pantoea*, *Pseudomonas* and *Xanthomonas*. These spoilage organisms thrive and multiply more quickly at ambient temperatures and high humidity.

REFERENCES

Agada M, Ojotule U, Mewuese B (2017) Effects of farmers' attitudes towards post-harvest losses of citrus in Ushongo, Benue State, Nigeri A. Int J Agric Environ Res 3(6): 4168–4186.

Ahmed Firas A, Brent SS, Alvarez AM (2017) Postharvest diseases of tomato and natural products for disease management. Afr J Agric Res 12(9): 684–691.

Alao SEL (2000). The importance of post-harvest loss prevention. Paper presented at graduation ceremony of School of Food Storage Technology. Nigerian Stored Products Research Institute, Kano, 1–10.

Al-Hindi RR, Al-Nadaja AR, Mohamed S (2011) Isolation and identification of some fruit spoilage fungi: Screening of plant cell wall degrading enzymes. Afr J Microbiol Res 5: 443–448.

Agrios G N (2007). Plant Pathology, 5th Edition. Amsterdam: Elsevier Academic Press, 2007.

Anonymous (2003) Consolidated Final Report. ICAR Network project on marketing and assessment of postharvest losses in fruits and vegetables in India.

Anonymous (2011) Food and Agricultural Organization FAO Statistical-Database. Rome, Italy.

Asalfew GK, Nega M (2020) Postharvest loss of tomato (*Solanum lycopersicum* L.) in Ethiopia: A review. Acta Sci Agric 4(3): 1–6.

Ayandiji A(2010). Effects of post-harvest losses on income generated in citrus production. African J Food Sci Technol 2: 52–58.

Azene M, Workneh TS, Woldetsadik K (2011) Effects of different packaging materials and storage environment on postharvest quality of papaya fruit. J Food Sci Technol. DOI 10.1007/s13197-011-06076.

Babatola LA, Ojo DO, Lawal OI (2008) Effect of storage condition on tomato (*Lycopersicon esculentum* Mill.) quality and shelf life. J Biol Sci 8(2): 490–493.

Baninasab B, Rahemi M (2006) The effect of high temperature on sprouting and weight loss of two onion cultivars. Am J Plant Physiol 1: 199–204.

BoFED (Bureau of Finance and Economic Development) (2007) Annual statistics for Amhara National Regional State. Bahir Dar, Ethiopia.

Burnett SL, Beuchat LR (2001) Human pathogens associated with raw produce and unpasteurized juices, and difficulties in contamination. J Indust Microbial Biotechnol 27: 104–110.

Cantwell M, Suslow T (2014) Cauliflower: Recommendations for maintaining postharvest quality. http://ucanr.edu/sites/Postharvest_Technology_Center_/Commodity_Resources/Fact_Sheets/Datastores/Vegetables_English/?uid=19&ds=799 (accessed January 18, 2014).

Cristina RA, Pérez JJR, Guillermo HML, Bernardo MA, Omar RPE (2018) Control of phytopathogenic microorganisms of post-harvest in tomato (*Lycopersicon esculentum* Mill.) with the use of citrus extract. J Plant Sci Phytopathol 2:37–43.

Danladi DK (2000) Nigeria agriculture: The efficiency factor announces this article to your friends. 1–3.

Elias SN, Shaw MW, Dewey FM (2010). Persistent symptomless, systemic and seed-borne infection of lettuce by *Botrytis cinerea*. Eur J Plant Pathol 126(1): 61–71.

Emana B, Afari-Sefa V, Nenguwo N, Ayana A, Kebede D, Mohammed H (2017) Characterization of pre- and postharvest losses of tomato supply chain in Ethiopia. Agric Food Sec 6(1): 1–11.

Esguerra E B, Rosa R, Rosendo SR (2018) Post-harvest management of tomato for quality and safety assurance. Guidance for horticultural supply chain stakeholders. FAO, The United Nations, Rome.

FAO (1984). Agricultural Extension: A Reference Manual, 2nd ed. Rome: FAO.

FAO (2005) Production status. Food and Agriculture Organizations of the United Nations, http://www faostat.fao.org (accessed on 08/09/2010).

Feliziani E, Romanazzi G (2016) Postharvest decay of strawberry fruit: Etiology, epidemiology, and disease management. J Berry Res 6(1): 47– 63.

Fonseca JM, Rushing JW, Rajapakse NC, Thomas RL, Riley MB (2005) Parthenolide and abscisic acid synthesis in feverfew are associated but environmental factors affect them dissimilarly. J Plant Physiol 162: 485–494.

Francis GA, O'Beirne D (2002) Effects of vegetable type and antimicrobial dipping on survival and growth of *Listeria innocua* and *E. coli*. Int J Food Sci Technol 37(6): 711.

Gajanana TM, Murthy S., Sudha M (2002) Marketing and postharvest loss assessment of banana var. Poovan in Tamil Nadu. Agric Econ Res Rev 15(1): 56–65.

Gajanana TM, Murthy SD, Sudha M, Dakshinamoorthy V (2006) Marketing and estimation of postharvest losses of tomato crop in Karnataka. Indian J Agric Mark 20(1): 1–10.

Gajanana TM, Sudha M, Saxena AK, Dakshinamoorthy V (2010) Post-harvest handling, marketing and assessment of losses in papaya. Acta Hort 851: 519–526.

Gajanana TM, Murthy S, Sudha M (2011) Postharvest losses in fruits and vegetables in South India – A review of concepts and quantification of losses. Indian Food Packer 65(6): 178–186.

Gitaitis RD, Gay JD (1997) First report of leaf blight, seed stalk rot, and bulb decay of onion by *Pantoeaananas* in Georgia. Plant Dis 81: 1096.

Gorrepati K, Murkute AA, Bhagat Y, Gopal J (2018) Post-harvest losses in different varieties of onion. Indian J Hortic 75(2): 314–318.

Harbola P, Khulbe R D (1994). New records of fungi associated with curd rot of cauliflower in Kumaun Himalaya. Indian J Mycol Plant Pathol 24 (1): 59.

Hassan MK (2010) A guide to postharvest handling of fruits and vegetables. Department of Horticulture,Bangladesh Agricultural University, Mymensingh.

Idris FM, Ibrahim AM, Forsido SF (2015) Essential oils to control *Colletotrichum musae in vitro* and *in vivo* on banana fruits. Amer-Euras J Agric Environ Sci 15(3): 291–302.

Isaac A, Kodzo EK, Etornam KA, Harrison A (2015) An overview of post-harvest losses in tomato production in Africa: Causes and possible prevention strategies. J Biol Agric Healthcare 5 (6): 78–88.

Ishrat Z (2010) Postharvest loss assessment in supply chain and shelf life extension of cauliflower (MS thesis). Bangladesh Food Situation Rep. FPMU, Bangladesh.

Jadhav V, Chinnappa B, Mahadevaiah GS (2011) An economic analysis of post-harvest losses of grapes in Karnataka. Mysore J Agric Sci 45(4): 905–911.

Jaganath I, Crozier A (2008) Overview of health promoting compounds in fruits and vegetables. In: Improving the health-promoting compounds of fruits and vegetables. Tomás Barberán FT, Gil MI (eds.). Woodhead, Cambridge, UK, pp. 3–37.

Jolayemi OS, Nassarawa SS, Lawal OM, MSodipo, OA, Oluwalana IB (2018) Monitoring the changes in chemical properties of red and white onions (*Allium cepa*) during storage. J Stored Products Postharvest Res 9(7): 78–86.

Kader AA (2011) Postharvest biology of tropical and subtropical fruits. In: Postharvest biology and technology of tropical and subtropical fruits: Fundamental issues. Yahia MM (ed.). Woodhead, Oxford, UK, pp. 79–111.

Kanwal N, Randhawa MA, Iqbal Z (2016) A review of production, losses and processing technologies of guava. Asian J Agric Food Sci 4(2): 2321–1571.

Karen M Schuenzel, Harrison Mark A (2002) Microbial antagonists of food borne pathogens on fresh, minimally processed vegetables. J Food Protec 65(12): 1909–1915.

Kashif M, Khan I, Ansar M, Nazir A, Aslam A (2016) Sustainable dehydration of onion slices through novel microwave hydro-diffusion gravity technique. Inno Food Sci Emerging Technol 33:327–332.

Kebede E (1991) Processing of horticultural produce in Ethiopia. Acta Hortic 270:298–301.

Kim MK, Ryu JS, Lee YH, Yun HD (2007) First report of *Pantoea* sp. induced soft rot disease of *Pleurotusergyngii* in Korea. Plant Dis 91: 109.

Kitinoja L, Al Hassan HA (2012) Identification of appropriate postharvest technologies for improving market access and incomes for small horticultural farmers in Sub-Saharan Africa and South Asia. Part 1: Postharvest losses and quality assessments. Acta Hortic (IHC 2010) 934: 31–40.

Kitinoja L, Cantwell M (2010) Identification of appropriate postharvest technologies for improving market access and incomes for small horticultural farmers in Sub-Saharan Africa and South Asia. WFLO Grant Final Report to the Bill & Melinda Gates Foundation. http://ucanr.edu/datastoreFiles/234-1848.

Kitinoja L, Kader AA (2002) Small-Scale Postharvest Handling Practices: A Manual for Horticultural Crops (4th edition). Postharvest horticulture series no. 8E. University of California, Davis Postharvest Technology Research and Information Center, Davis K.

Knoth J (1993) Le stockagetraditionnel de l'ignameet du manioc et son amélioration. Hambourg, GTZ-Post-harvest Project, 95.

Kumar S, Kaur G (2019) Effect of pre and postharvest applications of salicylic acid on quality attributes and storage behaviour of strawberry cv Chandler. J Pharmaco Phytochem 8(4): 516–522.

Kumar V, Sharma N, Sagar A (2015) Post-harvest management of fungal disease of onion. Int J Curr Microbiol App Sci 4(6): 737–752.

Kuyu CG, Tola YB (2018) Assessment of banana fruit handling practices and associated fungal pathogens in Jimma town market, southwest Ethiopia. Food and Nutri 6(3): 609–616.

Kuyu CG, Yetenayet B, Tola G, Abdi G (2019) Study on postharvest quantitative and qualitative losses of potato tubers from two different road access districts of Jimma zone, South West Ethiopia. Heliyon 5(8): https://doi.org/ 10.1016/j.heliyon.2019.e02272.

Ladaniya MS (2015) Postharvest management of citrus fruit in South Asian countries. Acta Hortic 1065: 1669–1676.

Ladaniya MS, Wanjari V, Mahalle BC (2005) Marketing of grapes and raisins and post-harvest losses of fresh grapes in Maharashtra. Ind J Agric Res 39(3):167–176.

Lee CJ, Lee J, Kwor JH, Kim BC (2005) Occurrence of bacterial soft rot of onion plants caused by *Burkholderia gladioli*pv. *alliicola* in Korea. Austra Plant Pathol 34(3):287–292.

Lurie S, Zhou HW, Lers A, Sonego L, Alexandrov S, Shomer I (2003) Study of pectin esterase and changes in pectin methylation during normal and abnormal peach ripening. Physiol Plant 119(2): 287–294.

Lutz JM, Hardenburg RE (1966) The commercial storage of fruits, vegetables and florist and nursery stocks. Agricultural Handbook No. 66. Gross KC, Wang CY, Saltveit M (eds.). USDA, Washington, DC.

Maria CS, Giuseppe A, Naouel A, Francesco G, Giovanni CD (2017) Advance in citrus postharvest management: Diseases, cold storage and quality evaluation. In: Citrus Pathology, Gill H, Garg H (eds.). April 2017. https://doi.org/10.5772/66518

Masum MMI, Islam SMM, Islam MS, Kabir MH (2013) Estimation of loss due to post harvest diseases of potato in markets of different districts in Bangladesh. Afr J Biotechnol 10: 11892–11902.

McKenzie TJ, Singh L, Peterson ID, Steven U Jr (2017) Quantifying postharvest loss and the implication of market-based decisions: A case study of two commercial domestic tomato supply chains in Queensland, Australia. Horticulturae 3: 44. DOI: 10.3390/horticulturae 3030044.

Mingchi L, Kojimo T (2005) Study on fruit injury susceptibility of strawberry grown under different soil moisture to storage and transportation. J Fruit Sci 22: 238–242.

Moghaddasi R, Mehrbanian E, Shariati S (2005) Islamic Republic of Iran (1): Postharvest Management of Fruit and Vegetables in the Asia-Pacific Region. APO seminar on Reduction of Postharvest Losses of Fruit and Vegetables held in India, and Marketing and Food Safety: Challenges in Postharvest Management of Agricultural/Horticultural Products in Islamic Republic of Iran.

Molla MM et al. (2010) Survey on postharvest practices and losses of litchi in selected areas of Bangladesh. Bangladesh J Agric Res 35(3): 439–451.

Moretti CL, Mattos LM, Calbo AG, Sargent S (2010) Climate changes and potential impacts on postharvest quality of fruit and vegetable crops: A review. Food Res Int 43(7):1824–1832.

Moses KA (2012) Assessment of post-harvest losses of citrus in Birim North District. Research project submitted to Kwame Nkrumah University of Science and Technology, Kumasi, 13.

Moula S, Ashok MB, Sudhakara SN (2017). Estimation of post-harvest losses of mangoes at different stages from harvesting to consumption. Int J Curr Microbiol Appl Sci 6 (12): 310–318.

Mulualem AM, Jema H, Kebede W, Amare A (2015) Determinants of postharvest banana loss in the marketing chain of central Ethiopia. Food Sci Quality Manag 37: 52–63.

Murthy SD, Gajanana TM, Sudha M, Subramanyam KV (2002) Postharvest loss estimation in mangoes at different stages of marketing-A methodological perspective. Agric Eco Res Rev 15(2):188–200.

Murthy MRK, Reddy GP, Rao KH (2014) Retail marketing of fruits and vegetables in India: A case study on export of grapes from Andhra Pradesh, India. Euro J Logi Purchasing Supply Chain Manag 2(1): 62–70.

Mustapha Y, Yahaya SM (2006) Isolation and identification of postharvest fungi of tomato (*L. esculentum*) and pepper (*Capsicum annum*) sample from selected irrigated sites in Kano. Biol Environ Sci J Tropics 3: 139–141.

Nakasone HY, Paull RE (1999) Tropical fruits. CABI Publishing, Wallingford, UK, p.45.

Nanda SK, Vishwakarma RK, Bathla HVL, Rai A, Chandra P (2010) Harvest and postharvest losses of major crops and livestock produce in India. All India Coordinated Research Project on Post Harvest Technology (ICAR), Ludhiana.

Nanda SK, Vishwakarma RK, Bathla HVL, Rai A, Chandra P (2012) Estimation of Quantitative Harvest and Postharvest Losses of Major Agricultural Produce in India. AICRP on Postharvest Technology, CIPHET, Ludhiana, India.

Narayana CK, Pandey BK, Malhotra SK, Pandey V (2014) Technical Bulletin: Postharvest Losses in Selected Fruits and Vegetables in India. IIHR, Bengaluru.

Ogunleye RF, Adefemi SO (2007) Evaluation of the dust and methanol extracts of *Garcinia kolae* for the control of *Callosobruchus maculatus* (F.) and *Sitophilus zeamais* (Mots). J Zhejiang UnivSci B8(12): 912–916.

Omayio DG, Abong GO, Okoth MW, Gachuiri CK, Mwangombe AW (2019) Current status of guava (*Psidium Guajava* L) production, utilization, processing and preservation in Kenya: A review. Curr Agric Res 7(3). DOI: http://dx.doi.org/10.12944/CARJ.7.3.07.

Pailly O, Tison G, Amouroux A (2004) Harvest time and storage conditions of "Star Ruby" grapefruit (*Citrus paridisi* Macf.) for short distance summer consumption. Postharvest Biol Technol 34:65–73.

Pal US, Sahoo GR, Khan MK, Sahoo NR (2002) Post-harvest losses on tomato, cabbage and cauliflower. AMA-Agr Mech Asia AF 33(3): 35–40.

Paltrinieri G (2015) Handling of fresh fruits, vegetables and root crops. Food Agric Organ United Nations. 2014: 35–39.

Panda AK, Goyal RK, Godara AK, Sharma VK (2016) Effect of packaging materials on the shelf-life of strawberry cv. Sweet Charlie under room temperature storage. J Appl Natural Sci 8(3): 1290–1294.

Porta R, Rossi-Marquez G, Loredana M, Angela S, Valeria CL, Giosafatto, Marilena E, Prospero DP (2013) Edible coating as packaging strategy to extend the shelf-life of fresh-cut fruits and vegetables. J Biotechnol Biomat 3(4):1–4.

Rajabi S, Farhad L, Maryam O, Seyed JFH (2015) Quantifying the grapes losses and waste in various stages of supply chain. Biol Forum 7(1): 225–229.

Ray RC, Ravi V (2005) Post-harvest spoilage of sweet potato in tropics and control measures. Crit Rev Food Sci and Nutri 45: 623–644.

Roy SK, Pal RK (1991) Multilocational studies to reduce postharvest losses during harvesting, handling, packaging, transportation and marketing of mango in India. Acta Hortic 291: 499–507.

Salami P, Ahmadi H, Alireza K, Mohammad S (2010) Strawberry post-harvest energy losses in Iran. Researcher 2(4): 67– 73.

Salunkhe DK, Kadam SS (1998) Handbook of Vegetable Science and Technology: Production, composition, Storage, and Processing. CRC Press, Boca Raton, FL.

Samson R, Legendre JB, Christen R, Fischer-Le Saux M, Achouak W, Gardan L (2005). Transfer of Pectobacterium chrysanthemi (Burkholder et al., 1953) Brenner et al., 1973 and Brenneria paradisiaca to the genus Dickeya gen. nov. known as *Dickeya chrysanthemi* comb. nov and *Dickeyaparadisiaca* combi. nov. and delineation of four novel species, *Dickeya dianthi* sp nov., *Dickeya dianthicola* sp. nov., *Dickeya dieffenbachiae* sp. nov. and *Dickeya zeae* sp. nov. Int J Syst Evol Microbiol 55:1415–1427.

Sani MY, Alao SEL (2006) Assessment of post-harvest fungi of tomato and pepper in selected irrigation areas of Kano State. Nigeria Int J Res Biosci 53–56.

Sankat CK, Maharaj R (2001) Papaya. In: Postharvest Physiology and Storage of Tropical and Subtropical Fruits, Mitra S (ed.). Faculty of Horticulture, CAB International, West Bengal, India, pp.67–185.

Sharma S(2016) Economic analysis of postharvest losses in onion in Jaipur district of Rajasthan. Asian J Hort 11(1): 124–128.

Sharma AK, Sawant SD, Somkuwar RG, Naik S (2018) Postharvest losses in grapes: Present Indian status. Technical Report, NRC for Grapes, Pune. DOI: 10.13140/RG.2.2.17999.89761/

Sharma RR, Singh D, Singh RB (2009) Biological control of postharvest diseases of fruits and vegetables by microbial antagonists: A review. Biological Cont 50(3): 205–221.

Singh D, Sharma RR (2007) Postharvest Diseases of Fruit and Vegetables and Their Management. Daya Publishing House, New Delhi, India.

Sivakumar D, Wall MM (2013). Papaya fruit quality management during the postharvest supply chain. Food Rev Int 29(1): 24–48.

Sonker N, Pandey AK, Singh P (2016) Strategies to control postharvest diseases of table grape: A review. J Wine Res 27 (2):105–122.

Spurgeon D (1976) Hidden harvest - a systems approach to postharvest technology. Report IDRC 062e. Ottawa: International Development Research Centre, 36pp.

Stephen AJ, Rohin J, Barker N (2016) Study on postharvest losses of guava at different stages of marketing in Allahabad district, U.P. Int J Agric and Environ Res 2(5): 1494–1506.

Tadesse B, Fayera B, Lamirot WM (2018) Assessment of postharvest loss along potato value chain: The case of Sheka Zone, southwest Ethiopia. Agric Food Secur 7:18.

Teresa AC, Stephanus NV (2009) *Pantoea ananatis*: an unconventional plant pathogen. Mol Plant Patho 110(3): 325–335.

Verma LR, Joshi VK (2000) Post harvest technology of fruits and vegetables. Indus Publishing Co., New Delhi.

Vishwanathan P, Kaur R (2001) Prevalence and growth of pathogens on salad vegetables, fruits and sprouts. Int J Hygiene Environ Health 203(3): 205–213.

WFLO (2010) Identification of appropriate postharvest technologies for improving market access and incomes for small horticultural farmers in Sub-Saharan Africa and South Asia. WFLO Grant Final Report to the Bill & Melinda Gates Foundation, March 2010, p. 318.

Williams CN, Uzo JO, Peregrine WT (1991) Vegetables Production in the Tropics. Longman, England, p. 179.

WoldeB (1991). Horticulture marketing systems in Ethiopia. Acta Hortic 270:21–31.

Woldu Z, Mohammed A, Derbew B, Zekarias S, Adam B (2015) Assessment of banana postharvest handling practices and losses in Ethiopia. J Biol Agric Healthcare 5(17): 82–91.

Workneh TS, Azene M, Tesfay SZ (2012) A review on the integrated agro-technology of papaya fruit. Afr J Biotechnol 11(85):15098–15110.

World Vegetable Centre (2018). Increasing production and reducing postharvest losses of onion in Nigeria. https://avrdc.org/increasing-production-and-reducing-postharvest-losses-of-onion-in-nigeria.

Yabuuchi E, Kosako Y, Oyaizu H, Yano I, Hotta H, Hashimoto Y, Ezaki T, Arakawa M (1992) Proposal of *Burkholderia* gen. nov. and transfer of seven species of the genus *Pseudomonas* homology group II to the new genus, with the type species *Burkholderia cepacia* (Palleroni and Holmes 1981) comb. Nov. Microbiol Immunol 36(12):1251–1275.

Yahaya SM (2005) Contribution of harvest to pathogenic and nonpathogenic losses of vegetables grown in Kano State-Nigeria, Bayero University, Kano.

Yahaya SM, Fagwalawa LD, Ali MU, Lawan M, Mahmud S (2015) Isolation and identification of pathogenic fungi causing deterioration of lettuce plant (*Lactucasativa*): A case study of Yankaba and Sharada vegetables markets. J Plant Sci Res 3(1):1–4.

Yumiko T, Toshiyuki M, Yoshiaki C (2005) Bacterial rot of *Allium giganteum* caused by *Erwiniaananas* (= *Pantoeaananatis*), and basal rot of *Belamcanda chinensis* caused by *Aphanomyces* sp. Bull Toyama Agric Res Cent 22:1–6 (Abstract).

Zakaria L, Sahak S, Zakaria M, Salleh B (2009) Characterisation of *Colletotrichum* associated with anthracnose of banana. Trop Life Sci Res 20(2): 119.

Zhang XX, Chen JY, Wang ZY, Zou HD, Luo ZX, Yao ZF, Yang YL, Fang BP, Huang LF (2020) *Burkholderia gladioli* causes bacterial internal browning in sweet potato of China. Austra Plant Pathol 49:191–199.

2

Effect of Pre-harvest Practices on Postharvest Quality and Diseases of Fruits

Ram Asrey[1] and Amrita Das[2]

[1]Food Science & Postharvest Technology Division, ICAR-Indian Agricultural Research Institute, New Delhi, India

[2]Division of Plant Pathology, ICAR-Indian Agricultural Research Institute, New Delhi, India

CONTENTS

2.1 Introduction

A rough estimate of the postharvest losses or spoilage of fruits in developing countries due to mishandling and disease is around 20%; this means that one-fifth of what is produced never reaches the consumer for whom it was grown. Reduction in these losses would be of great significance to growers and would maximize availability to consumers. Injuries during harvesting and handling and disease-causing pathogens are major causes for postharvest losses. These enormous losses may be minimized up to a negligible limit by adopting good pre-harvest and postharvest handling practices by the growers and traders. Infestation of disease in fruits takes place during pre-harvest (on the plant) and postharvest (off the plant) operations including harvesting, sorting, grading and transportation. Pre-harvest infection is generally caused by 'quiescent' or 'latent' infections, where the pathogen initiates infection of the produce at some point in time in the field before harvest and which later enters a period of inactivity or dormancy until the physiological status of the host tissue changes in such a way that infection can proceed.

Only minimal attention has previously been given to the impact of pre-harvest cultural practices on post production quality and shelf life of fruit crops. The quality of horticultural produce at harvest has a major effect on its postharvest life and progression of disease. Host susceptibility to disease, rate of physiological activities (respiration rate, enzymatic reaction) and the formation of health benefitting functional components at maturity often depend on cultivar, environment and cultural factors. The infestation of a particular disease and quality parameter may be affected by several conditions of growth, but it is well known that one factor may predominate and may exert an overriding influence upon the rest. In this chapter we have discussed quality, postharvest diseases and impact of cultural practices on them.

2.2 Pre-harvest Practices

The quality and disease infestation on fruit is closely related to the cultural operations such as soil type, available nutrient, irrigation, rootstock, training, pruning, plant protection, use of bioregulators, etc. (Figure 2.1). These operations may also alter the postharvest physiology of produce by synthesizing pro-disease defence compounds such as carotenoids, xanthophylls, ascorbic acid, tannins and phenolic compounds. These operations minimize disease severity by reducing source of inoculums or by limiting infection (Table 2.1).

TABLE 2.1
Impact of Pre-harvest Practices on Postharvest Diseases

Pre-harvest Factors	Impact on Disease	Reference
Soil	Incidence of canker disease was found to be least in citrus fruits harvested from sandy loam soil-grown trees during storage. Incidence of anthracnose disease was more severe in banana and papaya fruits grown in acidic soil.	Jung et al. (2018) Madani et al. (2014)
Rootstock	Hass avocado grafted on West Indian rootstock (Velvick) significantly reduced the severity of anthracnose and stem-end rot over seedling rootstock. Stem-end rind breakdown (SERB) disease was significantly reduced in Navel orange when these were grown on Cleopatra rootstock.	Willingham et al. (2006) Ritenour et al. (2004)
Training	The vine training system in kiwifruit and grapes can greatly influence the disease incidence and storage life of fruits. It affects the disease cycle and inoculum penetration and establishment inside fruits due to peel performance and accumulation of anthocyanins. Spur-pruned vertical shoot position system suppresses disease infection in grapes. In kiwifruit, modified T-bar training system minimized the incidence of botrytis stem-end rot and other diseases.	Liu et al. (2015) Costa et al. (1992)
Pruning	Tree pruning has a positive impact on storage disease management in fruit crops. Important diseases such as sooty blotch, apple scab, flyspeck, black rot in apple and grey mould rot in grape can be effectively managed by judicious pruning.	Cooley and Autio (2011) Valdés-Gomez et al. (2008)
Irrigation	The choice of irrigation method plays a key role in postharvest disease management. Drip irrigation minimizes the anthracnose, phytophthora and other pathogen dispersals in strawberry and grapes comparing to flood and sprinkler irrigation systems.	Coelho et al. (2008) Garganese et al. (2016)
Macronutrients (N, P, K)	Pronounced effect of reduced nitrogen: calcium ratio was found on the severity and incidence of anthracnose in stored Hass avocado fruits. Pre-harvest application of potassium fertilizers minimizes the risk of brown rot in harvested citrus fruits. The higher level of N and K in fertilizer solution increases disease severity in contract to P and Ca.	Willingham et al. (2006) Ramallo et al. (2019) Nam et al. (2006)
Micronutrients (Ca, Zn, Fe, Bo)	The role of micronutrients in plant defence against pathogens is now a well-proven aspect in several fruit crops. Pre-harvest calcium application (foliar or basal dose) effectively controls the postharvest anthracnose incidence in papaya, mango, banana and strawberry.	Cabot et al. (2019) Nam et al. (2006)

2.2.1 Fertilization

Among production operations, use of fertilizers is the second most important input after planting material. Absorption and assimilation of minerals depends on several subfactors such as plant age, soil type, moisture regime in the root zone and prevailing climatic conditions. For optimum plant performance, flowering, fruiting, growth and fruit maturity, balanced availability of macro and micronutrients are inevitable. Any deviation from the optimum level may cause deficiency or excess of nutrients in the source-sink relationship, which reflects on fruits in the form of various physical/physiological maladies such as bitter pits in apple, spongy tissue in mango, albinism in strawberry and cracking in pomegranate and cherries.

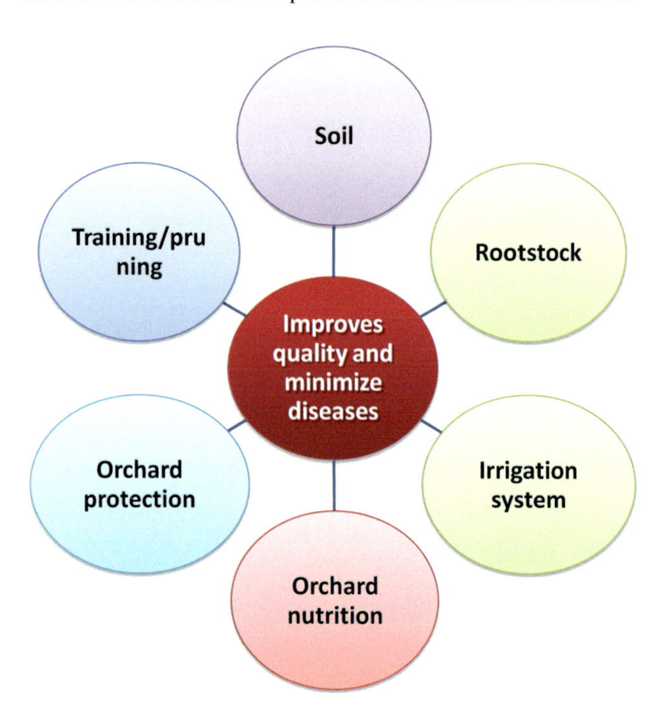

FIGURE 2.1 Pre-harvest practices for fruit disease minimization and quality retention.

Nitrogen (N): Generally, crops that contain high levels of nitrogen have a poor keeping quality than those with lower levels. Excessive nitrogen delays stone fruit maturity, induces poor visual red colour development and inhibits ground colour change from green to yellow. It occurs due to imbalance in essential amino acids (poor-quality protein) generated by the presence of excess soil nitrogen. However, nitrogen deficiency leads to small fruit with poor flavour and unproductive trees (Daane et al. 1995). Tahir et al. (2007) reported that excess nitrogen fertigation to 'Aroma' apple trees adversely affects fruit colouration, flavour quality and causes increased storage decay by *P. malicorticis*. Ca content and Ca/K ratio of fruits are also decreased, leading to greater susceptibility to bruising and fungal rots. Susceptibility to many physiological disorders is also dramatically increased when nitrogen content is too high in fruits. Vitamin C content in juices of oranges, lemons, grapefruit and mandarins is reduced by the application of high levels of nitrogen fertilizer to these crops (Nagy 1980). High nitrogen applications tended to increase ethylene evolution and respiration rate in apple and mango. The presence of excess nitrogen decreases fruit firmness in general. Esters are major contributors to perception of fruit flavour; judicious use of split nitrogen application favours flavour synthesis in fruit.

Phosphorus (P): Phosphorus level in soil and plant does not have much direct impact on internal fruit quality but it emphatically affects the fruit appearance. The phosphorous content in Cox's Orange Pippin apple is positively correlated with soluble solid and acid content. However, a lower level of phosphorous leads to susceptibility of fruits to low-temperature breakdown and senescent breakdown. In 'Jonagold' apple, higher phosphorous content increased fruit firmness and decreased dry matter content (Marcelle 1995). Foliar application of the phosphatic fertilizer FMCP in persimmon fruit enhanced fruit length and diameter, increased hardness and soluble solid content. However, the acidity and infestation caused by disease and insects decreased (Hossain and Ryu 2009). Phosphorous deficiency results in large Valencia oranges, and excess masks the colour development in cranberry.

Potassium (K): Research findings show that there is a relationship between pH regulation, organic acid levels and fruit potassium content. Potassium deficiency causes poor peach fruit colouration and small fruit size in oranges. The application of potassium fertilizer (16 g/plant) in pineapple was shown to decrease internal browning disorder (Soares et al. 2005). The phenolic content and the activities of polyphenoloxidase and phenylalanine ammonia-lyase enzymes were also decreased. Applications of potassium to citrus trees affect the shape of the fruit and increase acidity. Also, in the case of strawberry, acidity increased but it varied between cultivars. Increased potassium fertilization increased vitamin C content in fruit (Nagy 1980).

Calcium (Ca): Calcium affects fruit senescence and quality by changing intracellular and extracellular processes, and the rate of fruit softening depends on fruit Ca content. Ca also plays a regulatory rate in various processes that influence cell function and signal transduction. Cell wall degradation results in softening of fruit. Ca delays softening by delaying the degradation of cell wall polymers. Ca also plays a major role in cell-to-cell adhesion, and this phenomenon is important in the textural quality of the fruit. Fruit Ca concentration is an important factor for determining its quality. Fruit with low Ca content is sensitive to many physiological and pathological disorders and has a short storage life. Ca deficiency symptoms in harvested fruits depend on its transportation distribution within the tissues (Ferguson et al. 1999). A bitter pit in apple fruit is mainly caused by calcium deficiency during fruit growth, and this may be detectable at storage time (Atkinson et al. 1980). In strawberry fruit, pre-harvest spray of Ca was shown to maintain higher firmness, soluble solid content (SSC), titratable acidity, ascorbic acid content and lower incidence of albinism. Further, during storage, they were more resistant to grey mould rot (Singh et al. 2007). Ca deficiency may be overcome by foliar spray or postharvest aqueous dipping treatment (Hewett and Watkins 1991).

Other nutrients: Iron (Fe) and zinc (Zn) deficiency results in reduced fruit size in citrus and both colour and size reduction in peaches. Boron (Bo) deficiency reduces strawberry fruit size and causes external corking of apple fruit. Imbalance in certain nutrients can also have a pronounced impact on fruit shape, e.g. Zn deficiency alters the shape of peach and cherry fruit. Copper (Cu) deficiency affects the shape of citrus fruits (misshape) and kernel filling in walnut. Molybdenum (Mo) deficiency produces fruit misshape disorder in strawberry.

2.2.1.1 Organic Production

It is well accepted that organic and integrated production practices protect soil quality and exert a lower negative impact on environment over chemical farming (Vogeler et al. 2006). Wang et al. (2008) reported that organically grown blueberries were higher in sugars (fructose and glucose), malic acid, total phenolics, total anthocyanins (including delphinidin 3-galactoside, delphinidin 3-glucoside, delphinidin 3-arabinoside, petunidin 3-galactoside, petunidin 3-glucoside and malvidin 3-arabinoside) and antioxidant activity than that from the conventional culture. Organically grown Royal Gala' and 'Fuji' apple trees fruit had a more yellowish colour, higher blush on the fruit peel, and greater flesh firmness (Amarante et al. 2008; Weibel et al. 2000). According to a study, in organic olive oil a higher vitamin E level has been found as compared to conventional production (Gutierrez et al. 1999). Robusta bananas when grown organically ripened faster at 22–25°C than non-organically grown bananas as measured by peel colour change, but in both cases the total soluble solid (TSS) levels were the same.

2.2.2 Irrigation

Water management strategy is an excellent tool for regulating fruit quality, economizing water use, providing a harvestable yield and pest management. Crisosto et al. (1994) found that 'O'Henry' peaches supplied with an optimum amount of irrigation during their growing season produced maximum fruit size; however, higher soluble solid content was obtained by medium water stress during fruit growth before harvest. Deficit irrigation reduces fruit size, but in some cases only slightly. Less frequent irrigation increased concentrations of dietary fibre, protein, vitamin C, Ca, Mg and Mn (Sorensen et al. 1995). In pear-jujube (*Zizyphus jujuba* Mill. cv. Lizao) fruit, moderate and severe deficit irrigation when given at fruit growth stage increased soluble solid content, sugar/acid ratio and organic acid as well as vitamin C content as compared to full irrigation (Cui et al. 2008). 'Salustiano' citrus produced in moderate and severe deficit irrigation maintained higher juice percentage, TSS, titratable acidity and peel thickness than fruit given full irrigation (Tejero et al. 2010). In mango, fruit from early water stress took a significantly longer time to ripen and had much lower chilling injury symptoms after cold storage.

2.2.3 Pruning, Thinning and Girdling

Numerous studies show that improving light penetration into the canopy enhances the fruit composition of grape, raspberry, apple, peach and plum. Judicious pruning generally increases SSC anthocyanins, total soluble phenols, and reduced titratable acidity, malate, pH and potassium content in fruits. Shading results in significant differences in the aroma of fruit. The increased light interception by an individual fruit and its surrounding leaves, the better its quality, including fruit colour, size and flavour. Less pruning in some tropical and subtropical fruits increases light penetration and improves the quality of over-shaded fruit. Fruit thinning increases fruit size but reduces the total production. Therefore, one should maintain balance between yield and fruit size. Fruit thinning increases the size of peaches (Westwood and Balney 1963) and apples (Batjer et al. 1957). Leaving too many fruits on a tree reduces fruit size and soluble solid content in the early ripening 'May Glo' nectarine and late-ripening 'O'Henry' peaches (Crisosto et al. 1995). Ahmad et al. (2006) reported that heavy pruning in plant improved yield as well as quality of fruit. Heavily pruned fruit was found to have better colour development, maximum fruit length, higher fruit weight and a lower amount of seeds/fruit. Further, these fruits had lower peel percentage, lower acidity, higher rag percentage, juice percentage and higher TSS content compared to fruit from moderately pruned and non-pruned trees. The effect on size is based upon the fact that under good production conditions, many fruit crops produce more fruits than desirable. Thinning increases the leaf-to-fruit ratio, giving larger individual fruit (Westwood 1993). The balance between vegetative and fruit growth can be altered by girdling. Girdling has been found to be beneficial in grape and jackfruit. The grapevines that have extra vigour and a record of poor berry and bunch size resulted in improved size and shape and ripening delay in terms of low sugar-to-acid ratio and colour intensity.

2.2.4 Rootstock

The use of rootstock is common in orchards, as they have a major effect on the quality as well as postharvest shelf life of fruit. Remorini et al. (2008) reported that rootstock has a considerable effect on the nutritional quality of peach (cv. Flavorcrest) fruit. Fruits from trees on the rootstock Mr. S 2/5 (natural hybrid of *Prunus cerasifera*) had the highest antioxidant capacity, total phenols, vitamin C and β-carotene content than fruit on Ishtara, GF 677 and Barrierl rootstock. In 'Allen Eureka' lemon, acid content was highest in fruit grafted on Cleopatra mandarin, and TSS was highest on sour orange rootstock (Jaleel et al. 2005). 'Jonagold' apple grown on rootstock PB-4 and M.26 showed lower ethylene production and delayed ripening during storage (Tomala et al. 2008). Grapefruit (cv. Ruby Red) budded on rough lemon when stored at 4°C for 6 weeks showed lower decay loss and chilling injury than budded on *Citrus amblycarpa*.

2.2.5 Tree Age

Limited information is available on the effect of tree age on postharvest quality of fruit. Asrey et al. (2007) reported that guava cv. Allahabad Safeda, harvested from 10-, 15- and 20-year-old trees, fruit from middle-age trees contained higher TSS, total sugar, vitamin C and lowest acidity as compared to other fruit. Fifteen-year-old trees produced Cu- and Mn-rich fruit, while 20-year-old trees yielded Mg- and Zn-rich fruit, but Fe content was highest in the fruit obtained from 10-year-old trees. 'Braeburn' apples harvested from young plants were more susceptible to flesh and core browning over older trees. 'Aroma' apples produced in young trees (less than 4 years) was higher in acid/SSC ratio, and had better colouration and flavour quality as compared to old (more than 20 years) trees. However, fruit produced in old trees was more firm and had delayed ethylene production and better storage potential (Tahir et al. 2007).

2.2.6 Canopy Position

The position of fruit in the canopy also has a considerable effect on quality of fruit. 'Aroma' apples produced outside the canopy had a higher content of dry matter, soluble solids and soluble sugars due to the higher contents of fructose, glucose and sucrose but a somewhat lower amount of titratable acidity than inside fruits. Further, during maturation on the tree, outside fruits developed a red peel colour, while inside fruits remained green, and ethylene production for inside apples was significantly higher compared to outside apples (Nilsson and Gustavsson 2007). 'Jonagold' apples on the top of the canopy were reported to contain higher level of cyanidin 3-galactoside and quercetin 3-glycosides anthocyanins, followed by fruits from the outside and inside of the canopy, but no significant differences in the levels of catechins, phloridzin and chlorogenic acid among fruits from the different canopy positions were found (Awad et al. 2001). This is because the light in the interior of the canopy was poorer in UV-A, blue, green and red light but richer in far-red light than at all other positions, which are very much essential for anthocyanin production. Agabbio et al. (1999) reported that 'Tarocco' oranges harvested from northern, southern and interior parts of the canopy showed significant differences in quality after 10 weeks of storage at 9°C plus 1 week of stipulated marketing period at 21°C. Freshly harvested fruits from southern side showed higher soluble solid concentration, SSC/acid ratio and lower acid levels. Further, after storage, those fruits showed lower respiration rate and decay percentage.

2.2.7 Growth Regulators

Plant growth regulators are powerful horticultural production tools and the effect of several plant growth regulators on postharvest fruit quality has been reported by several researchers. However, with few exceptions, these growth regulators are not used extensively, because the concentration range between obtaining suboptimal and super optimal effects is very narrow.

Pre-harvest gibberellic acid (GA_3) is used commercially to delay peel senescence of citrus fruit, thereby improving on-tree storage and postharvest life. It is also used to increase fruit size in grapes. In orange (*Citrus sinensis* L.), pre-harvest spray of GA_3 increased peel firmness and juice yield, and delayed colour development (Davies et al. 1997). Storage characteristics are also affected by growth regulators. For instance, pre-harvest GA application extended the storage life of the persimmon fruit by delaying both black-spot development and fruit softening when stored at −1°C (Eshel et al. 2000). It also delayed ripening, fruit softening and climacteric rise of respiration (Ben-Arie et al. 1986).

A concentration of 45 ppm naphthalene acetic acid (NAA) sprayed on guava (cv. Red Flesh) increased pulp/seed ratio, TSS, total sugars and ascorbic acid content (Iqbal et al. 2009).

2.2.8 Genotype and Cultivar

In addition to the pre-harvest conditions discussed above, genetics and cultivar selection are major factors involved in postharvest quality and shelf-life outcomes for fruit. Since cultivars vary in genetic makeup, they will also vary in characteristics like size, colour, flavour, texture, nutrition, storage life, processing ability and eating quality. Genetic engineering has made it possible to tailor nutrients, vitamins and pigment-rich fruit crops with longer shelf life. Such transgenic varieties are now commercially grown in mango, papaya, cherry, strawberry and banana.

2.3 Pre-harvest Practices to Manage Postharvest Disease

Postharvest diseases are one of the key reasons for major losses of freshly harvested fruits in storage. Most of the postharvest pathogens primarily exist inside the fruits, showing latent and quiescent infection periods (Table 2.2). The disease gradually develops at various stages of fruit development or after harvesting of crop and ultimately reduces the fruit quality in storage or at transportation, thereby reducing consumer purchase. However, there are some other groups of fungi and bacteria that are also responsible for a postharvest disease in which infection arises through injuries initiated during and after harvest. The best way to ensure the success of a postharvest disease management program is to use integrated disease control measures. Pre-harvest management practice is one of the critical components in the development of integrated disease management strategies to fight postharvest losses. Several pre-harvest management practices may manage many postharvest quality losses by reducing field infections or induce fruit resistance. Different approaches such as bio-control agents, low-risk chemical fungicides and natural antimicrobial substances are utilized as pre-harvest practices against several postharvest diseases.

TABLE 2.2

Major Pathogens Responsible for Pre-harvest Infection in Fruits

Causal Pathogen	Host	Infection Site
Colletotrichum gloeosporioides	Citrus, banana, mango, papaya, avocado, grape, blueberry	Unripe fruit surface
C. musae	Banana	Unripe fruit surface
Lasiodioplodia theobromae	Citrus, mango	Young fruit surface
Phomopsis sp.	Citrus, grape, peach	Flower, young fruit
Alternaria citri	Citrus	Flower, young fruit
Botrytis cinerea	Strawberry, apple, grape, raspberry	Flower, young fruit
Dothiorella dominicana	Mango	Inflorescence
Alternaria alternata	Mango, apricot, persimmon	Fruit cuticle, stomatal opening
Monilinia fructicola	Peach, apricot, nectarine, plum, almond, apple, pear	Flower, peduncle, unripe fruit surface

2.3.1 Synthetic Chemicals

Pre-harvest spraying of fungicide has been found effective to control the postharvest quality of fruits. Systemic or protectant fungicides are commercially applied to reduce the pathogen inoculum as a part of conventional agriculture practices. Efficacious disease management depends on the strategic or repeated application of fungicides as per the nature of disease. Spraying of a protectant fungicide like mancozeb during flowering and fruit development has been found effective to control mango from anthracnose disease (caused by *Colletotrichum gloeosporioides*). Fungicides like ziram, carbendazim iprodione, cyprodinil and fenhexamid reported having curative activity against postharvest diseases when applied just before harvest (Sholberg and Bedford 1999; Singh and Thakur 2003). However, preventive measures are more important than curative ones in pre-harvest treatment, as the development of postharvest decay often arises from an inoculum that survives and accumulates on the fruit in the field or during postharvest storage and shelf life. Early season pre-harvest treatment with fungicides or alternatives has been found to be effective in reducing the pathogen inoculum in fruits such a stable grapes, citrus, strawberry, stone and pome fruit and can thus allow better control of postharvest decay (Sonkar and Ladaniya 1999). Recently, a combination of fungicides with alternate compounds having different modes of action has become available. Such new-generation fungicides belong to the anilinopyrimidine class (cyprodinil, mepanipyrim) and pyridine-carboxamides class, and can control many postharvest diseases such as blue mould and grey mould (Pirgozliev et al. 2006a, 2006b) when sprayed as a pre-harvest treatment. The mode of action of these new combined fungicides consists of disruption of pathogen growth through blocking of essential components required for cell growth, inhibition of fungal respiratory chain complexes and inhibition of enzyme secretion by blocking several biosynthesis pathways. Iprodione belongs to the class of dicarboximides and prevents the germination of spores and the growth of mycelia through inhibition of DNA and RNA synthesis and cell division in fungi. Several fungicides from the fenhexamid group (fenhexamid, fenpyrazamine) and the sterol biosynthesis inhibiting group are also used to control fruit decay-causing pathogens as pre-harvest treatment.

2.3.2 Bio-control Agents

A number of promising bio-control agents (BCAs) can be exploited to control postharvest disease under field conditions before the harvest of crop. To eliminate the negative side effects of chemical use, the use of antagonists is increased to manage postharvest disease. Different mechanisms, such as competition for nutrients and space, antibiosis, parasitism, induction of resistance in the host tissue and production of volatile metabolites, are involved in bio-control management. In the case of pre-harvest application of BCA, the effectivity of antagonists depends on how they colonize the fruit surfaces under field conditions. Natural epiphytic antagonists like yeasts, fungi and bacteria that already exist on fruit surfaces are gaining importance as bio-control agents, particularly against postharvest fungal pathogens. Pre-harvest application of antagonists has been found effective to manage decay in citrus, strawberries and table grapes (Liu et al. 2013; Mari et al. 2014). Several products of these BCAs have been commercially available, including *Pseudomonas syringae* (BioSave), *Cryptococcus albidus* (YieldPlus), *Bacillus subtilis* (Serenade), *Candida sake* (Candifruit), *Pantoea agglomerans* (Pantovital), *Aureobasidium pullulans* (Boni Protect), *Candida oleophila* (Nexy, BioNext), *Bacillus amyloliquefaciens* (Amylo-X, Biogard) and *Metschnikowia fructicola* (Shemer). However, some important factors need to be considered for a successful pre-harvest application of BCA products, such as purity of the antagonists, genetic stability, cell viability and qualities as a good colonizer on fruit surfaces. Integration of BCA with various other substances, such as salts, decontaminating agents and fungicides, also proved promising in pre-harvest application for better control of postharvest decay.

2.3.4 Natural Antimicrobials

Natural antimicrobials, such as salts, chitosan and plant extracts, have been reported as very effective as pre-harvest management practices to control postharvest diseases. Pre-harvest treatments with alternatives to fungicides can reduce the postharvest disease of fruit without negatively affecting the quality parameters of the fruit. Applicable concentration and a proper time for the application are some important factors that influence the effectiveness of the alternative compounds. Organic or inorganic salts such as calcium chloride, sodium carbonate or sodium bicarbonate can successfully reduce postharvest grey mould on table grapes when applied at the pre-harvest stage (Romanazzi et al. 2012). Another natural biopolymer-based compounds, chitosan, controlled storage rot in grapes, strawberry, sweet cherries, citrus fruit and banana effectively when applied in both pre-harvest and postharvest stages. Chitosan induces chitinase activity and elicits phytoalexins and defence barriers in the host tissues. Several chitosan-based commercial compounds, such as Chito Plant and Armour-Zen, are also available the market. Salicylic acid (SA) analogs such as benzothiadiazole have been applied to plant tissues as an alternative or complement to fungicide treatments to control postharvest disease. These plant endogenous hormone-like alternative compounds are shown to be effective in inducing systemically acquired resistance to diseases in fruits such as strawberries, pears, peaches, melons and alternaria rot, blue mould of pears during storage (Cao et al. 2011; Feliziani et al. 2015). Pre-harvest foliar spray of INA (2, 6-dichloronicotinic acid), another analog of SA, significantly reduced postharvest diseases of melons, mango and banana caused by *Colletotrichum* species.

2.4 Conclusions

Pre-harvest practices can have a profound influence on the postharvest quality and shelf life of fruits. In comparison to climatic conditions and cultural practices, genotype plays a key role in determining the overall quality (nutritional profile and cosmetic appeal) of a particular fruit. Fruit growers must use an integrated fruit production system to optimize fruit yield and quality. Future research on those factors that are under human control, including orchard management (cultural practices), cultivar selection and genotype improvement offers great opportunities for postharvest quality. A successful disease control program depends on the usage of multifaceted approaches under integrated disease management practices. Knowledge of methods as well as the right time for application greatly influences the effectiveness of pre-harvest treatments with either fungicides or alternatives. Moreover, knowledge of pathogen interaction with the host, infection process at different levels, the microenvironment, etc., is also equally important to draw the preventive measures. However, evaluation of new-generation fungicides and exploitation of new antagonists or natural antimicrobial compounds are continuous processes for development of better sustainable strategies against postharvest diseases.

REFERENCES

Agabbio M, D'Hallewin G, Mura M, Schirra M, Lovicu G, Pala M (1999) Fruit canopy position effects on quality and storage response of 'Tarocco' oranges. Acta Hortic 485: 19–23.

Ahmad S, Chatha ZA, Nasir MA, Aziz A, Virk NA, Khan AR (2006) Effect of pruning on the yield and quality of kinnow fruit. J Agric Soc Sci 2(1): 51–53.

Amarante CVT, Steffens CA, Mafra AL, Albuquerque JA (2008) Yield and fruit quality of apple from conventional and organic production systems. Pesq Agropec Bras 43(3): 333–340.

Asrey R, Pal RK, Sagar VR (2007) Impact of tree age and canopy height on fruit quality of guava cv. Allahabad Safeda. Acta Hortic 735: 259–262.

Atkinson D, Jackson JE, Sharples RO, Wallery WM (1980) Mineral Nutrition of Fruit Trees. Butterworths, London.

Awad MA, Wagenmakers PS, Jager A (2001) Effects of light on flavonoid and chlorogenic acid levels in the skin of 'Jonagold' apples. Sci Hortic 88: 289–298.

Batjer LP, Billingsley HD, Westwood MN, Rogers BL (1957) Predicting harvest size of apples at different times during the growing season. Proc Am Soc Hortic Sci 70: 46–57.

Ben-Arie R, Bazak H, Blumenfeld A (1986) Gibberellin delays harvest and prolongs storage life of persimmon fruit. Acta Hortic 179: 807–813.

Cao S, Yang Z, Cai Y, Zheng Y (2011) Fatty acid composition and antioxidant system in relation to susceptibility of loquat fruit to chilling injury. Food Chem 127: 1777–1783.

Cabot C, Martos S, Lugany M, Gallego B, Tolra R, Poschenrieder C (2019) A role for zinc in plant defense against pathogens and herbivores. Front Plant Sci. doi.org/10.3389/fpls.2019.01171

Coelho MVS, Palma FR, Café-Filho AC (2008) Management of strawberry anthracnose by choice of irrigation system, mulching material and host resistance. Intl J Pest Manag 54(4): 347–354.

Cooley DR, Autio WR (2011) Summer pruning of apple: Impacts on disease management. Adv Hortic Sci 25(3): 199–204.

Costa G, Biasi R, Giuliani R, Succi F (1992) Comparison of kiwifruit training systems. Acta Hortic 297: 427–434.

Crisosto CH, Johnson RS, Luza JG, Crisosto GM (1994) Irrigation regimes affect fruit soluble solids concentration and rate of water loss of 'O'Henry' peaches. Hortic Sci 29(10): 1169–1171.

Crisosto CH, Mitchell FG, Johnson RS (1995) Factors in fresh market stone fruit quality. Postharvest News Info 6: 217–221.

Cui N, Du T, Kang S, Li F, Zhang J, Wang M, Li Z (2008). Regulated deficit irrigation improved fruit quality and water use efficiency of pear-jujube trees. Agric Water Manag 95: 489–497.

Daane KM, Johnson RS, Michailides TJ, Crisosto CH, Dlott JW, Ramirez HT, Yokota GT, Morgan DP (1995) Nitrogen fertilization affects nectarines fruit yield, storage qualities, and susceptibility to brown rot and insect damage. Calif Agric 49(4): 13–18.

Davies FS, Campbell CA, Zalman GR (1997) Gibberellic acid sprays for improving fruit peel quality and increasing juice yield of processing oranges. Proc Fla State Hortic Soc 110: 16–21.

Eshel D, Ben-Arie R, Dinoor A, Prusky D (2000) Resistance of gibberellin-treated persimmon fruit to *Alternaria alternata* arises from the reduced ability of the fungus to produce endo-1,4-β-glucanase. Phytopathology 90(11): 1256–1262.

Feliziani E, Landi L, Romanazzi G (2015) Pre-harvest treatments with chitosan and other alternatives to conventional fungicides to control postharvest decay of strawberry. Carbohydr Polym 132:111–117.

Ferguson I, Volz R, Woolf A (1999) Pre-harvest factors affecting physiological disorders of fruit. Postharvest Biol Technol 15: 255–262.

Garganese F, Sanzani SM, Ligorio A, Di Gennaro D, Tarricone L, Ippolito A (2016) Effect of irrigation management on field and postharvest quality of organic table grapes. Acta Hortic 1144: 273–278

Gutierrez F, Arnaud T, Albi MA (1999) Influence of ecological cultivation on virgin olive oil quality. JAOCS 76: 617–621.

Hewett EW, Watkins CB (1991) Bitter pit control by sprays and vacuum infiltration of calcium in Cox's Orange Pippin apples. Hortic Sci 26: 284–286.

Hossain MB, Ryu KS (2009) Effect of foliar applied phosphatic fertilizer on absorption pathways, yield and quality of sweet persimmon. Sci Hortic 122:626–632.

Iqbal M, Khan MQ, Jalal-ud-Din Rehman K, Munir M (2009) Effect of foliar application of NAA on fruit drop, yield and physico-chemical characteristics of guava (*Psidiumguajava* L.) red flesh cultivar. J Agric Res 47(3): 259–269.

Jaleel AA, Zekri M, Hammam Y (2005) Yield, fruit quality, and tree health of 'Allen Eureka' lemon on seven rootstocks in Saudi Arabia. Sci Hortic 105: 457–465.

Jung T, Pérez-Sierra A, Durán A, Horta Jung M, Balci Y, Scanu B (2018) Canker and decline diseases caused by soil- and airborne Phytophthora species in forests and woodlands. Persoonia 40: 182–220.

Liu M-Y, Chi M, Tang YH, Song C-Z, Zhu-Mei X, Zhang Z-W (2015) Effect of three training systems on grapes in a wet region of china: yield, incidence of disease and anthocyanin compositions of *Vitisvinifera* cv. Cabernet Sauvignon. Molecules 20: 18967–18987.

Liu J, Sui Y, Wisniewski M, Droby S, Liu Y (2013) Review: Utilization of antagonistic yeasts to manage postharvest fungal diseases of fruit. Int J Food Microbiol 167: 153–160.

Madani B, Muda Mohamed MT, Biggs AR, Kadir J, Awang Y, Tayebimeigooni A, Shojaei TR (2014) Effect of pre-harvest calcium chloride applications on fruit calcium level and post-harvest anthracnose disease of papaya. Crop Protec 55: 55–60.

Marcelle RD (1995) Mineral nutrition and fruit quality. Acta Hortic 383: 219–237.

Mari M, Francesco AD, Bertolini P (2014) Control of fruit postharvest diseases: Old issues and innovative approaches. Stewart Postharvest Rev 1: 1.

Nagy S (1980) Vitamin C contents of citrus fruit and their products: A review. J Agric Food Chem 28:8–18.

Nam et al. (2006) Effects of nitrogen, phosphorus, potassium and calcium nutrition on strawberry anthracnose. Plant Pathol 55: 246–249.

Nilsson T, Gustavsson K (2007) Postharvest physiology of 'Aroma' apples in relation to position on the tree. Postharvest Biol Technol 43: 36–46.

Pirgozliev SP, Errampalli D, Sholberg PL, Stokes SC, De Ell JR, Murr DP (2006a) Effect of pre-harvest pyrimethanil (scalasc, bayercropscience Ltd) application for control of postharvest gray and blue mold in 'Empire' apples, 2004-05. Pest Manag Res Rep 44: 165–170.

Pirgozliev SP, Errampalli D, Sholberg PL, Stokes SC, De Ell JR, Murr DP (2006b) Effect of pre-harvest pyrimethanil (scalasc, Bayer Cropscience Ltd) application for control of postharvest gray and blue mold in 'McIntosh' apples, 2004-05. Pest Manag Res Rep 44: 171–176.

Ramallo AC, Cerioni L, Olmedo GM, Volentini SI, Ramallo J, Rapisarda VA (2019) Control of Phytophthora brown rot of lemons by pre- and postharvest applications of potassium phosphate. Euro J Plant Pathol 154(4): DOI: 10.1007/s10658-019-01717-y.

Remorini D, Tavarini S, Degl'Innocenti E, Loreti F, Massai R, Guidi L (2008) Effect of rootstocks and harvesting time on the nutritional quality of peel and flesh of peach fruit. Food Chem 110:361–367.

Ritenour M, Duo H, Bowman K, Boman B, Stover E, Castle WS (2004) Effect of rootstock on stem-end rind breakdown and decay of fresh citrus. Hort Technology 14(3): 242–248.

Romanazzi G, Lichter A, Mlikota Gabler F, Smilanick JL (2012) Recent advances on the use of natural and safe alternatives to conventional methods to control postharvest gray mold. Postharvest Biol Technol 63: 141–147.

Sholberg PL, Bedford KE (1999) Use of cyprodinil for control of *Botrytis cinerea* on apple. Phytopathology 89: S72.

Singh R, Sharma RR, Tyagi SK (2007) Pre-harvest foliar application of calcium and boron influences physiological disorders, fruit yield and quality of strawberry (*Fragaria x ananassa* Duch.). Sci Hortic 112: 215–220.

Singh D, Thakur AK (2003) Effect of pre harvest sprays of fungicides and calcium nitrate on post-harvest rot of kinnow in low temperature storage. Plant Dis Res 18: 9–11.

Soares AG, Trugob LC, Botrela N, Souzac LF (2005) Reduction of internal browning of pineapple fruit (*Ananascomusus* L.) by pre-harvest soil application of potassium. Postharvest Biol Technol 35: 201–207.

Sonkar RK, Ladaniya MS (1999) Effect of pre-harvest sprays of ethephon, calcium acetate and carbendazim on rind colour, abscission and shelf-life of Nagpur mandarin (*Citrus reticulata*). Indian J Agric Sci 69(2): 130–135.

Sorensen JN, Johansen AS, Kaack K (1995) Marketable and nutritional quality of leeks as affected by water and nitrogen supply and plant age at harvest. J Sci Food Agric 68: 367–373.

Tahir II, Johansson E, Olsson ME (2007) Improvement of quality and storability of apple cv. Aroma by adjustment of some pre-harvest conditions. Sci Hortic 112: 164–171.

Tejero G, Bocanegra JA, Martinez G, Romero R, Duran-Zuazo VH, Muriel-Fernandez JL (2010) Positive impact of regulated deficit irrigation on yield and fruit quality in a commercial citrus orchard [*Citrus sinensis* (L.) Osbeck, cv. *salustiano*]. Agri Water Manag 97(5): 614–622.

Tomala K, Andziak J, Jeziorek K, Dziuban R (2008) Influence of rootstock on the quality of 'Jonagold' apples at harvest and after storage. J Fruit Ornam Plant Res 16: 31–38.

Valdés-Gomez H, Fermaud M, Roudet J, Calonnec A, Gary C. (2008) Grey mould incidence is reduced on grapevines with lower vegetative and reproductive growth. Crop Protec 27(8): 1174–1186.

Vogeler I, Cichota R, Sivakumaran S, Deurer M, McIvor I (2006) Soil assessment of apple orchards under conventional and organic management. Aus J Soil Res 44: 745–752.

Wang SY, Chen C, Sciarappa W, Wang CY, Camp MJ (2008) Fruit quality, antioxidant capacity and flavonoid content of organically and conventionally grown blueberries. J Agric Food Chem 56: 5788–5794.

Weibel FP, Bickel R, Leuthold S, Alfoldi T (2000) Are organically grown apples tastier and healthier? A comparative field study using conventional and alternative methods to measure fruit quality. Acta Hortic 517: 417–426.

Westwood MN (1993) Temperate-Zone Pomology: Physiology and Culture. Timber Press, Portland, OR.

Westwood MN, Balney LT (1963) Non-climatic factors affecting the shape of apple fruit. Nature 200: 802–803.

Willingham SL, Pegg KG, Anderson JM et al (2006) Effects of rootstock and nitrogen fertiliser on postharvest anthracnose development in Hass avocado. Aus Plant Pathol 35: 619–629.

3

Postharvest Handling of Fruits and Vegetables for Disease Management

Bal Vipan Chander Mahajan and Swati Kapoor
Punjab Horticultural Postharvest Technology Centre, Punjab Agricultural University, Ludhiana, India

CONTENTS

3.1 Introduction

Fresh fruits and vegetables are significant sources of nutrition in our daily life. Globally, the horticultural sector has seen a tremendous growth in production and quality. However, despite cultural advancement to improve productivity, tremendous losses occur due to insufficient postharvest facilities and deficient knowledge of technologies used for shelf-life extension of fruits and vegetables. These losses are majorly documented in developing nations or the nations where demand for food is high. Being living entities, fruits and vegetables are prone to quality degradation both physically and biochemically from the time they are harvested because they use up their stored food reserves. It is therefore imperative to strictly monitor and comply with the necessary postharvest applications for quality maintenance from farm to fork.

Some of the principal causes for postharvest losses are physiological deterioration, mechanical damage and spoilage by disease and pests. Among these, the potential reason for losses in horticulture produce is the prevalence of postharvest diseases caused by various bacterial and fungal microorganisms. In a recent study by Nabi et al. (2017), the spoilage index of freshly harvested fruits and vegetables was reported to be around 20–25% due to decay by pathogens during postharvest handling. All living materials are associated with microbial flora that makes them prone to parasite attack, which subsequently leads to pathogenic damage. Microorganisms are widespread in the air, soil and water and infect fresh produce before or after harvest. Damage takes place mainly by penetration of microorganisms at an injured site of the fresh produce; however, some microorganisms are capable of penetrating the unbroken skin of produce to cause infection. Therefore, postharvest pathology deals with practices for protection of crops during harvesting, packing, storing and transportation. All these steps represent integrated management of horticultural produce after harvest that can be applied universally for reducing pathogenic spoilage in fruits and vegetables.

This chapter discuss various postharvest factors, each having its own importance to maintain optimum quality and shelf life of horticultural crops.

3.2 Harvesting

Harvesting is the first and foremost important step to maintain quality of horticulture produce. Type of harvesting, harvesting equipment, harvesting time and handling practices play important roles in the final quality of produce. A cooler climate and maximum moisture content of produce are optimal for harvesting. However, maturity indices of the crop during harvest also determine the quality parameters and end life of the crop. Along with this, pre-harvest factors such as light, temperature, humidity, fertilizer application and market distance also determine the optimum harvesting stage for most crops (El-Ramady 2015). Harvesting time of the produce should be determined based on market distance, and care should be taken to determine maturity stage of the produce, as an immature or over-mature stage may lead to low-quality produce. During harvesting of produce, physical damage or bruising either due to pulling or mechanical handling should be avoided, as the damage, once done, cannot be reversed. Use of clippers or secateurs should be encouraged for harvesting produce instead of pulling action (Figure 3.1A), as the stem end can lead to injury in the fruit due to stem penetration (Figure 3.1B). However, in the case of vegetables, manual harvesting by gentle picking could help reduce produce losses. During harvesting, the produce should be packed in plastic containers such as plastic buckets or baskets because these containers have smooth surfaces and do not damage the produce. The use of gunny bags or wooden baskets should be avoided.

3.2.1 Harvesting Systems

Harvesting of fruits and vegetables is usually divided into two systems: manual harvesting and mechanical harvesting. Harvesting systems are also divided on the basis of produce structure, i.e. root crops and surface crops.

3.2.1.1 Manual Harvesting

This type of harvesting is widely practiced in developing countries and some developed countries. Harvesting is done by hand, where the harvester can judge the maturity stage and quality of the produce prior to harvesting. A wide range of maturity is associated with certain types of crops that need to be harvested multiple times per season. The main prerequisite for this type of harvesting is the use of skilled and properly trained manpower who can harvest the produce with minimum damage.

(a)

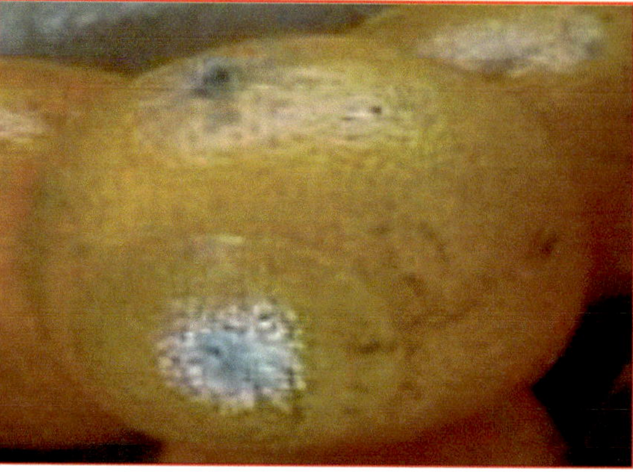

(b)

FIGURE 3.1 (A) Harvesting of kinnow fruit using pulling (L) and secateurs (R) method. (B) Stem penetration into fruit leads to disease initiation.

3.2.1.2 Mechanical Harvesting

Mechanical harvesting systems employ the use of machines to harvest the crops. Many advancements have been made in this field to protect crops from damage during harvesting. However, mechanical harvesting has been successful for only few crops, particularly root crops, as they are harvested once per season, and the soil above the root crop acts as a cushion to protect the produce from machine-based injury. Also, nuts bearing a hard shell/coat are harvested using a mechanical procedure, as the hard shell acts a barrier to withstand mechanical damage, if any, and preserves the integrity of the internal produce. The other type of produce that can be mechanically harvested is fruits and vegetables that are intended to be used for processing, as damage if caused by machines does not affect the final quality of the product because these are processed immediately after harvest and converted into secondary forms such as pulp, slices, powder, etc.

Root crops are usually harvested by two main methods: digging and pulling (Ruiz-Altisent et al. 2004). Digging operations are used for the produce that are uprooted and lifted together with the crop and a large amount of soil, whereas pulling operations are done for root crops like carrot, where leaves above ground are uprooted with the entire plant. Among the two methods, digging offers minimum damage to crops but poses the problem of a large amount of soil, whereas pulling can damage the produce to some extent due to variations in pulling force or pulling speed of the operator.

Various mechanisms such as combing, cutting, vibration, stripping and threshing are used to mechanically harvest surface crops. In the case of leafy vegetables, the most prominent method used is cutting, as these crops need to be harvested many times for maximum yield. Some significant parameters to be considered for cutting action are cutting speed and sharpness of cutting tool, as they indirectly affect the final product quality.

3.3 Precooling

Precooling refers to the process of rapid removal of field heat from horticultural produce to arrest the deterioration and senescence processes (Elansari and Siddiqui 2017). Precooling is an eco-friendly, safe and economical treatment that has been largely researched to compensate the chemical treatments (Sivakumar et al. 2007). Precooling minimizes ethylene production, respiration rate, metabolic processes and microbial activity and thus leads to preservation of quality and prolonged shelf life of fresh produce (Ferreira et al. 2006). During storage, alterations in the phenolic and antioxidant content in the fruits have been observed that has occurred due to various metabolic activities (Amarowicz et al. 2008). The different techniques of precooling, commonly hydrocooling, room cooling, forced-air cooling, ice cooling, etc., have been practiced for a long time and have proven reliable for controlling deterioration in fruits.

The various factors that influence the efficiency of precooling are respiration rate, metabolism and ethylene generation. Precooling helps reduce respiration rate, as witnessed in the case of grapes, where a decrease of 9.5°C in temperature resulted halved respiration rates and doubled the keeping quality (Lider 2004). In the case of strawberries, a lag phase of 2hours in between harvesting and cooling resulted in a 20% fruit loss. Precooling delays the deterioration and rotting process by inhibiting the growth of decay organisms and also decreases wilting, as low temperatures slow down evaporation and transportation rates. The internal metabolism of plants plays an important role in maintaining the quality of product. It has been documented that with every increase of 10°C temperature, the metabolic rate doubles. Ethylene gas is known to accelerate ripening and senescence of produce; therefore, reduction in temperature can help lessen the production and sensitivity of the crop to ethylene gas (Piriyaphansakul and Kanlayanarat 2003). The principal techniques for precooling for highly perishable produce include hydrocooling, room cooling, forced-air cooling, package icing, vacuum cooling and cryogenic cooling, with many variations and alterations (Table 3.1) (Baladhiya and Doshi 2016).

Hydrocooling is an uncomplicated, economical approach to quickly decrease the temperature of products. Hydrocooling has been used since 1923 when it was developed for celery washing. Hydrocooling predominantly exercises the use of chilled or cold water for reduction of the product temperature in bulk or smaller containers before further packaging.

TABLE 3.1
Precooling Methods to Preserve Produce Quality

Precooling Method	Process Description	Specific Crops
Room cooling	Low cost and slow process. Loading of produce in sacks, bins and cartons under cold room conditions. Natural circulation of cold air through the packages.	Potato, onion, garlic, apple, citrus fruits, winter squash, radish, tomato, beans.
Package icing	Produce is packed along with crushed or flaked ice. Must be used with water-tolerant packages such as waxed fibreboard or plastic.	Broccoli, Brussels sprouts, green onion, pea, carrot, radish, turnip, etc.
Hydrocooling	Employs the use of cold water to produce prior to packaging and is of shower and immersion type.	Mango, peach, cherry, sweet corn, asparagus, green onion, artichoke, root crops, watermelon, muskmelon, beans, etc.
Forced-air cooling	Fastest method of cooling. Pushes or pulls air through the vents in storage containers. Cooling time depends on airflow velocity, produce size and temperature difference between air and produce.	Grapes, berries, pears, peach, oranges, strawberries, tomato, other tropical and subtropical fruits, mushrooms, cucumber, brinjal, okra, summer squash, peppers, melons, etc.
Vacuum cooling	Water evaporation from produce takes place under low air pressure, leading to a cooling effect.	Lettuce, cauliflower, celery, cabbage, spinach, mushrooms.

Another benefit of hydrocooling is that it does not dehydrate the products. A comparative study between hydrocooling, air cooling and cooling with liquid nitrogen (LN_2) followed by mechanical refrigeration in mango cv. Amarpali has shown that hydrocooling is faster than air cooling. However, 20% reduction in the time required was observed in LN_2 cooling compared to mechanical refrigeration. Fast cooling systems are significant to achieve rapid temperature reduction in produce (Ravindra and Goswami 2008).

Forced-air or pressure cooling is a modification of room cooling process and is achieved by exposing produce packages to high air pressure on one side than on the other. A variety of forced-aircooling arrangements can be used for cooling produce. The process includes circulation of air at high velocity in refrigerated rooms, movement of forced air through the voids in bulk products during movement through a cooling tunnel on continuous conveyors and encouraging flow to forced air by the pressure differential technique through packed produce. Forced-air precooling is also adapted in a wide range of commodities, and the technique helps to reduce respiration rates and senescence process of the produce. The moisture loss in forced air cooling ranged from very little to significantly high enough to damage the produce. The factors that affect moisture loss include the initial temperature of product, humidity, transpiration coefficient and exposure time to airflow after cooling and waxes or moisture-resistant packaging used (Aswaney 2007). These interventions have shown potential to increase the shelf life of fruits and vegetables.

3.4 Postharvest Treatments

After harvesting, fruits and vegetables continuously respire and therefore quality of the final marketable produce gets degraded over time. Quality maintenance requires the use of certain postharvest treatments to extend the marketable life of fresh horticultural produce. Some important postharvest treatments are discussed below.

3.4.1 Chlorination

Microbial spoilage is one of the major concerns arising during spoilage of fruits and vegetables. In order to minimize these detrimental effects, chlorination has been routinely used to treat surface contamination of freshly harvested produce. Postharvest operations, such as irrigation water, use of harvesting machines/equipment, contact with harvesting bins, human contact and inherent microflora of seeds or cuttings, lead to microbial attachment on surface of produce; therefore, surface decontamination by chlorine can help inactivate pathogenic microflora, such as viruses, moulds, fungi, bacteria, their cysts and other propagules. Of several chlorine salts, sodium or calcium hypochlorite has been used widely due to its easy availability, low cost, ease of handling and high efficiency. Its optimum concentrations are around 100–150 ppm in washing water. Other chlorination techniques include use of chlorine gas that is usually injected at various points in a washing line, which reduces pH of water to 6.5. The gas, usually known as chlorine dioxide, is yellowish-red in colour and is produced by reacting chlorine gas and sodium chlorite or sodium hypochlorite or by using sodium chlorite and hydrochloric acid. The chlorine dioxide has disinfecting power nearly constant within a pH range of 6–10. The concentrations ranging 3–5 ppm of chlorine dioxide were found to be most effective against most microbes in clean water (Suslow 2000). However, owing to its hazardous and explosive nature, proper installation and safety programs are necessary, which make it costly.

The following points should be considered while exploiting the maximum potential of chlorine (Ivey et al. 2015).

- Rinsing the produce with potable water before treatment with chlorine is advisable, as organic debris and soil lessen the effectiveness of chlorine.
- Water temperature between 55 and 120°F increases the effectiveness of chlorine.
- Similarly, a pH between 6 and 7 increases effectiveness of the chlorine. Sodium bisulphate or vinegar can be used to lower the pH and soda ash (sodium carbonate and/or washing soda) can be used to increase pH.
- Different pH values alter the effect of chlorine solutions and make them corrosive; pH value less than 5 leads to the formation of toxic chlorine gas, and for pH values greater than 8, chlorine solution loses its effectiveness for microbial inhibition.
- Effectiveness of the sanitizer to inhibit or kill pathogens can be measured from its oxidative reduction potential (ORP). An ORP value of 650–750 m kV of the sanitizing solutions is known of kill the surface spoilage bacteria, *Escherichia coli* and *Salmonella* within a few seconds, whereas the spoilage yeasts, *Mucor*, *Botrytis*, *Penicillium* and *Phytophthora* are killed within 1–2 minutes.
- After chlorine treatment, the produce should be rinsed with clean, potable water.

3.4.2 Wax Coatings

According to rule 2.3.6 FSSAI Regulation, fresh fruits and vegetables having coatings of waxes, mineral oils and colours cannot be sold, except the coating of bee wax (white and yellow), shellac and carnauba wax, but the name of wax must be noted on the package. The modified atmosphere created by the application of edible coatings helps in regulation of the metabolic process and reduces the moisture loss and oxidation of fruits. Edible coating application on fresh produce minimizes moisture loss by creating a partial barrier to the movement of moisture from the surface of the fruit during postharvest storage. Establishment of the modified atmosphere around the produce acts as a gas barrier, which slows down respiration, enzymatic oxidation and senescence and preserves colour and texture and lessens physical and mechanical impacts, and to some extent microbial growth. Gago et al. (2003) studied the effect of edible coatings on postharvest quality of mandarin cv. 'Clemenules' and reported delayed dehydration in coated fruits as compared to uncoated fruits, but no effect on sensory score was observed in both coated and uncoated fruits.

FIGURE 3.2 Waxing operation of kinnow fruit to maintain postharvest shelf life.

Salvador et al. (2003) reported that the shellac wax slowed down the Physiological loss in weight (PLW) as well as softening, and enhanced green colour retention in apple cultivar 'Reinette'. Kumpoun et al. (2013) examined the response of coating materials on shelf life and quality changes of tangerine cv. 'Sai Nam Peung' during storage and reported that non-coated fruits showed wilting signs after 8 days at an ambient temperature, whereas the coated fruits maintained their freshness. Jorge et al. (2013) reported increased shelf life of the strawberries with edible coating, indicating that shelf life of uncoated strawberries was 10 days, whereas coated strawberries had a shelf life of 15 days. The study concluded that edible coating had a considerable effect in checking the senescence of the fruits. Palou et al. (2014) studied the antifungal properties of some edible coatings such as beeswax and shellac on oranges and mandarin. Their study revealed significant reduction in fruit weight loss and caused sustained firmness in the coated fruits during cold storage (Figure 3.2). A study also reported that the changes in the internal gas composition ascribed to the coatings have no effect on the sensory scores of coated oranges and mandarins.

Rising issues on deleterious health effects of chemical-based coatings on food products have created a demand for natural-based edible coatings in fresh fruits and vegetables. In the past few years, there has been a great deal of research on the development of edible coatings to prevent disease incidence in fresh fruits and vegetables. Anthracnose was controlled with application of lemongrass oil in chitosan-based films in green bell peppers (Ali et al. 2015). Usage of *Thymus vulgaris* essential oil was found to control fruit rot caused by *Rhizopus stolonifer*, *Phytophthora citrophthora* and *Botrytis cinerea* (Camele et al. 2010). Oranges treated with essential oils of *Ocimum canum*, *Mentha arvensis* and *Zingiber officinale* were found to have promising effects on blue mould rot (Tripathi and Dubey 2004). Aloe vera gel is one of the widely accepted bio preservatives owing to its antimicrobial properties. Papaya fruits coated with aloe vera-based coating showed only 27% disease incidence as compared to uncoated fruits, where 100% spoilage was observed (Brishti et al. 2013). With respect to natural volatiles as inhibitors of diseases, methyl jasmonate, a major derivative of the plant hormone jasmonic acid, was reported to efficaciously suppress postharvest disease incidence of loquat and some other fruits (Cao et al. 2008).

3.4.3 Acid Treatments

Microbial growth, deterioration of texture, enzymatic and non-enzymatic browning in harvested fruits and vegetables were slowed down by the application of calcium and ascorbic acid-based solutions, due to their chelating nature, and the acids inhibit the activity of polyphenol oxidase. Ascorbyl, the ascorbic acid, directly scavenges the damaging radicals (Yamaguchi et al. 1999). Moreover, acid treatments lead to destabilization of the browning process by lowering the pH of the pericarp (Lichter et al. 2000) and reducing production of *o*-quinones caused by the action of polyphenol oxidase (PPO) on phenolic compounds (Robert et al. 2003). Ascorbic acid (AA) and citric acid (CA) were reported for their anti-browning activity in minimally processed vegetables and fruits. Furthermore, some of the organic acids act predominantly as fungistatics, while some others act potently as bacteriostatics (Nehal et al. 2008). Organic acids have been used widely to retard the initiation of browning both alone and in combination with other chemicals. The most commonly used acids are AA, oxalic acid and CA. These are weak organic acids and generally recognized as safe (GRAS) (Suttirak and Manurakchinakorn 2010).

3.4.4 Ethylene Inhibitors/Growth Regulators/Fungicide Treatments

Amenoethoxyvinyl glycine (AVG), 1-methyl cyclopropane (1-MCP), silver thiosulphate, silver nitrate, benzothiadiazole, cycloheximide, etc. are a few chemicals which are known for the inhibition of ethylene production and/or its activity during ripening and storage of fruits. 1-MCP is a new chemical used widely for extension of shelf stability and quality of fruits in several countries (Watkins 2006). The working principle involves the blockage of ethylene receptors and prevention of the effects of ethylene on plant tissues for longer periods (Sisler and Serek 1997). El-Ramady et al. (2015) reported that the application of fungicides or growth regulators like GA3 could increase the storage life of fruits. The application of 1-MCP helps in the inhibition of the softening and ripening process and delays disease outburst on fruits and vegetables. 'Qingnai' plum treated with 1-MCP resulted in decreased formation of ethylene along with lower respiration rates and delayed softening (Luo et al. 2009). Similarly, a dose of 1 µL/L of 1-MCP for 10 hours at 20°C deferred softening in kiwifruit (Mao et al. 2007). In loquat fruits, using 625 ppb of 1-MCP in combination with MAP at 5°C achieved the best results to control browning and significantly inhibited the occurrence of total psychrotrophic bacterial count, total aerobic mesophilic count and total yeast and mould count (Oz and Ulukanli 2011). However, the efficiency of 1-MCP application varies with respect to horticultural crops and depends on cultural practices, harvesting maturity and postharvest practices. Pre-harvest treatment of benzo(1,2,3)thiadiazole-7-carbothioic acid S-methyl ester (BTH) in strawberry was found useful to prevent powdery mildew in the fruit, while its use as postharvest treatment enhances the anthocyanin and phenolic content of strawberries (Anttonen et al. 2003; Cao et al. 2011). The presence of polyphenols in fruits and vegetables prevents them from biotic and abiotic stress conditions. BTH as a postharvest treatment is known to increase polyphenols in mango, peach, strawberry, banana and loquat (Ruiz-Garcia and Gomez-Plaza 2013).

3.4.5 Thermal Treatments

Thermal treatments include the following.

3.4.5.1 Hot Water Treatment

Fruits can be immersed in hot water before storage or marketing to prevent numerous postharvest diseases and enhance the colour of fruit peel (El-Ramady et al. 2015). The restriction in the development of decay by hot water treatment is primarily due to removal of inoculum by physical means and direct inhibition of pathogen growth at the fruit surface. Hot water brushing leads to increased physical barriers by uniform redistribution of the epicuticular wax layer (part of the constitutive defence system) and notable decrease in cuticular cracks (Fallik 2004; Ben-Yehoshua 2003)

3.4.5.2 Vapour Heat Treatment

Vapour heat treatment (VHT) was found to be effective in controlling infection caused by fruit flies in fruits after harvesting. The stacking of the boxes in humid and heated rooms was accomplished by injecting steam. Vapour heat governs the principle of quarantine treatment that involves heating of commodities at air temperatures ranging from 40 to 50°C surrounded by saturated water vapour to kill insect eggs and larvae before shipment. The water vapour condensation on the cooler fruit surface leads to a heat transfer mechanism (Lu et al. 2007).

3.5 Packaging

Appropriate packing of produce will guarantee secure transportation from the field to the storage and end users while reducing damage during transportation. The produce should be packed in such a way as to restrict collision with each other during transportation. A wide array of containers such as wooden boxes, corrugated fibreboard boxes (CFB) and plastic crates are used in the transportation and dissemination of fruits and vegetables. All packages must ensure some amounts of ventilation to prevent physiological injuries to the produce. Traditionally, wooden crates and boxes were used as a primary packaging material. However, due to disadvantages such as bruising of fruits and vegetables, CFB boxes were introduced, with the first patents being granted to England in 1856.

Comparatively CFB cartons made a significant new entry into the field of packaging. Various advantages such as light weight, less damaging to the fruits, easy handling and printing and improvement of product image have been associated with CFB. The packing of produce in gunny bags or wooden baskets tied with jute cloth should be avoided, as these methods lead to bruising and result in huge postharvest losses. For retail marketing, fruits and vegetables must be adequately packed into consumer packs using paper moulded trays, wrapped with shrink film or cling film. The principal plastic materials for modified atmospheric packaging include polybutylene, high density polyethylene, low density polyethylene, polypropylene, polyvinylchloride, ethylene vinyl acetate ionomer, polyvinylidene chloride, rubber hydrochloride (pliofilm) and polystyrene. Cautious handling and proper packaging of fruits play vital roles in maintaining freshness and quality of fruits as well as avoiding damage or decay during transit (Semeerbabu et al. 2007). Consistent physiological and biochemical changes occur after the harvest of fruits, until the fruit becomes unfit for consumption. Packaging helps in maintaining high relative humidity in the headspace of packaging, thereby reducing water loss from fresh produce (Petrisor et al. 2010).

3.5.1 Principal Factors Determining the Success of Packaging of Fruits and Vegetables

3.5.1.1 Moisture Loss

All fruits and vegetables respire through a natural diffusion phenomenon, leading to moisture removal from the surface. This is dependent on numerous factors, including relative humidity and temperature, air velocity of the environment, surface area and moisture content of the fruit. The difference in the texture of produce is visible when moisture loss occurs beyond 5%, leading to shrivelling of the crop. According to Siddiqui (2016), marketability of the produce is reduced as a result of wilting, shrinking and wrinkling. Packages must therefore be designed in such a way, or suitable materials be used, that prevent moisture migration through produce but at the same time avoid condensation inside the packages as a result of produce respiration.

3.5.1.2 Gas Composition

Modified atmosphere packaging affects physiological as well as biochemical modifications in fruits and vegetables during ripening and storage through reducing the concentration of O_2 and increasing the levels of CO_2 in the headspace. In general, lower O_2 concentration achieved by modified atmospheric packaging results in lowering the rate of physiological as well as biochemical changes. These changes result in reducing respiratory metabolism and other biochemical processes, which helps in retaining the chemical quality characteristics of the produce (Workneh and Osthoff 2010). However, a balance between O_2 and CO_2 is necessary to maintain the quality of produce, as high CO_2 levels may cause tissue damage, leading to increase in respiration, whereas low O_2 levels beyond tolerance limits result in anaerobic respiration, leading to off-flavours and off-odours (Pongener and Mahajan 2017).

3.5.1.3 Temperature and Relative Humidity

Each fruit and vegetable has its optimum temperature for quality maintenance and postharvest shelf life. Increase in the temperature leads to proportional increase in the respiration rate, which leads to reduced shelf life of the produce. Environmental conditions such as temperature and relative humidity have a direct effect on maintaining quality of fruits and vegetables during storage. Ideally, the fresh produce should be stored at a low temperature that does not cause chilling injury. Any deviation from ideal conditions could be harmful for the produce. A minimal difference in the temperature of stored products and cooling coil is required. However, water can be sprayed into controlled chambers to attain high relative humidity values during storage (Ramjan and Ansari 2018). Nowadays, sensors have been developed that are applied to packages of fresh fruits and vegetables to show any temperature abuse during the supply chain.

3.5.2 Types of Packaging Material

For quality maintenance throughout the transportation chain from farm to fork, packaging of fresh produce is one of the crucial steps. Bags, baskets, crates, hampers, bulk bins, cartons and palletized containers are frequently used for transporting, handling and marketing of fresh fruits and vegetables. Packaging acts as a tool for extension of shelf life by reducing or preventing moisture loss in fresh produce. The primary functions of a package are containment and protection of the produce. Packaging also symbolizes value addition and assurance of quantity and quality.

Different types of packaging materials are widely available, and the specific application depends upon the shape and perishability of the fruit to be packed. Based on this, fruits can be classified as hard fruit, soft fruit, root vegetables, stem products and green vegetables (Matche 2005), and their respective packaging requirements are elucidated in Table 3.2. Packaging requirements for retaining high-quality produce are summarized as follows.

- Rough packages such as wooden crates and baskets should be lined with cardboard inserts.

TABLE 3.2

Packaging Requirements of Some Fruits and Vegetables

Commodity	Characteristics	Packaging Requirement
Soft fruits: Cherries, grapes, blueberries, strawberries, raspberries, plums	Highly perishable, easily subject to anaerobic spoilage. Susceptible to bruising and crushing.	Semi-rigid containers with a cover of cellophane, cellulose acetate, polystyrene. Polythene bags with ventilation holes.
Hard fruits: Apples, bananas, citrus fruits, peaches, pears, tomatoes	Less perishable, low respiratory rates.	Open traywithplastic film overwrap or sleeve. Bagged in perforated polyethylene film or nets.
Stem products: Celery, rhubarb, asparagus	Highly perishable, easily loses moisture.	Packed in moisture-proof cellophane or polyethylene with ventilation. Banded or sleeved with shrink film.
Root vegetables: Carrots, turnips, radishes, onions, beets, yams, potatoes	Semi-perishable, long-period storage.	Packed in polyethylene bags with perforations.

- Package reinforcing with folder dividers and corner supports are used for heavy produce.
- Large packages should be avoided, as the produce suffers more damage during handling.
- Single- or double-layered shallow packages should be used for delicate produce such as grapes, berries, ripe stone fruits and summer squash.
- Under filling or over-filling of the packages should be avoided.
- About 5% of the surface area per side should be vented for adequate ventilation of packages.
- Packaging materials such as trays, cups, pads, wraps and liners should be used for immobilization of the produce.
- Perforated plastic film liners should be used to decrease the rate of water loss from produce like cherries.
- Use of netted bags for garlic and onions will create the lower humidity environment essential for proper handling.
- Care must be taken while using fillers and liners in packages so as to prevent blockage of ventilation holes.

3.6 Storage

The main aim of storage is to control the rate of transpiration, respiration, ripening, disease infection and also undesirable biochemical changes. Proper temperature management can be a very effective tool in ensuring that produce remains in good

TABLE 3.3

Recommended Storage Conditions for Fruits and Vegetables

Name of Commodity	Temp (°C)	RH (%)	Approximate Shelf Life
Apple	0–2	90–95	1–6 months
Asian pear	0–2	90–95	2 months
Grape	0–2	90–95	2 months
Guava	6–10	90–95	2–3 weeks
Lemon	10–13	85–90	1–6 months
Mandarin (Kinnow)	5–6	90–95	2 months
Mango	13–15	85–90	2–4 weeks
Papaya	7–13	90–95	1–3 weeks
Peach	0–1	90–95	2–4 weeks
Plum and prune	0–1	90–95	2–4 weeks
Cabbage	0–1	90–95	2–3 months
Carrots (topped)	0–1	90–95	3–6 months
Cauliflower	0–1	90–95	3–4 weeks
Cucumber	10–11	85–90	10–14 days
Eggplant	10–12	90–95	1–3 weeks
Okra	7–10	90–95	7–10 days
Peppers	7–10	90–95	2–3 weeks
Tomato	10–13	90–95	1–3 weeks
Potato (seed)	0–1	90–95	5–6 months
Onion	0–1	65–70	4–5 months
Garlic	0–1	65–70	5–6 months

condition throughout storage and transportation. Cold chain technology plays a vital role in establishing controlled low temperature conditions to maintain quality of produce during transportation. There have been numerous attempts to make this technology accessible throughout India for reduction in losses occurring due to high temperatures. Following harvest, certain storage guidelines need to be followed to maximize quality and minimize spoilage. Recommended storage conditions pertaining to temperature and relative humidity for some fruits and vegetables are outlined in Table 3.3.

General prerequisites for storage of high quality of horticultural produce (Mahajan and Kapoor 2019) include the following.

- Storage of high-quality produce, free of decay, damage and of proper maturity (not over-ripe or under-mature) is advisable.
- Knowledge of proper relative humidity, temperature and ventilation of each commodity must be known prior to storage.
- Tuber crops must be cured before storage.
- Over-stacking of containers or loading too close together in storage rooms should be avoided.
- Adequate ventilation must be provided in the storage room.
- Storage rooms must be sanitized prior to storing fresh commodities.
- Well-ventilated and adequately strong containers are necessary to withstand stacking. Stacking beyond the stacking strength of packages should be avoided.
- Crops should be stored in a dark room. This is especially important for potatoes, to avoid stimulation of solanine, a toxic compound not destroyed by cooking, under light conditions.
- Storage of ethylene-sensitive commodities with those that produce ethylene must be avoided.
- Collective storage of strong odour-emitting (apples, garlic, turnips, cabbages, onions and potatoes) with odour-absorbing commodities should be avoided.
- Regular inspection for signs of injury, damage, water loss and disease is required. Removal of damaged or diseased produce is required to prevent spread of these problems.

3.7 Conclusion

The horticultural sector has emerged as the best option for diversification of agriculture to meet the need for food, nutrition and healthcare. Enormous losses have been witnessed during postharvest operations and are a serious matter to be addressed wisely. During the glut season a substantial quantity of produce goes waste and growers bear the burden of distress sales. The prime sites where postharvest losses occur are in the farmers' fields, storage, packaging areas, transportation and wholesale or retail markets. The preventive techniques discussed in this chapter to maintain quality of produce could be interconnected to form an integrated approach to managing horticultural crops after harvest. This integrated approach, including harvesting, grading, waxing, packaging and storage, will not only help in reducing postharvest losses but farmers and traders also can get better prices for their produce in the market, and consumers will get better quality produce.

REFERENCES

Ali A, Noh NM, Mustafa MA (2015) Antimicrobial activity of chitosan enriched with lemongrass oil against anthracnose of bell pepper. Food Packag Shelf Life 3: 56–61.

Amarowicz R, Estrella I, Hernández T, Troszyńska A (2008) Antioxidant activity of extract of adzuki bean and its fractions. J Food Lipids 15:119–136.

Anttonen M, Hukkanen A, Tiilikkala K, Karjalainen R (2003) Benzothiadiazole induces defence responses in berry crops. Acta Hortic 567:177–182.

Aswaney M (2007) Forced air pre-cooling of fruits and vegetables. Air Conditioning Refrig J January-March: 57–62.

Baladhiya C, Doshi J (2016) Precooling techniques and applications for fruits and vegetables. Internat J Proc Post Harv Technol 7 (1): 141–150.

Ben-Yehoshua S (2003) Effects of postharvest heat and UV applications on decay, chilling injury and resistance against pathogens of citrus and other fruits and vegetables. Acta Hortic 599:159–173.

Brishti FH, Misir J, Sarker A (2013) Effect of biopreservatives on storage life of papaya fruit (*Carica papaya* L.). Int J Food Stud 2: 126–136.

Camele I, De Feo V, Altieri L, Mancini E, De Martino L, Luiqi Rana G (2010) An attempt of postharvest orange fruit rot control using essential oils from Mediterranean plants. J Medicinal Food 13: 1515–1523.

Cao SF, Hu ZC, Zheng YH, Yang ZF, Lu BH (2011) Effect of BTH on antioxidant enzymes radical-scavenging activity and decay in strawberry fruit. Food Chem 125: 145–149.

Cao SF, Zheng YH, Yang ZF, Tang SS, Jin P, Wang KT, Wang XM (2008) Effect of methyl jasmonate on the inhibition of *Colletotrichum acutatum* infection in loquat fruit and the possible mechanisms. Postharvest Biol Technol 49: 301–307.

Elansari AM, Siddiqui MW (2017) Recent advances in postharvest cooling of horticultural produce. In: Siddiqui MW, Ali A (Eds.) Postharvest Management of Horticultural Crops. Apple Academic Press, Waretown, NJ, pp. 1–68.

El-Ramady HR, Domokos-Szabolcsy É, Neama AA, Hussein ST, Miklós F (2015) Postharvest management of fruits and vegetables storage. In: Lichtfouse E (Ed.). Sustainable Agriculture Reviews. Springer International, Switzerland.

Fallik E (2004) Prestorage hot water treatments (immersion, rinsing and brushing). Postharvest Biol Technol 32(2):125–134.

Ferreira MD, Brecht JK, Sargent SS, Chandler CK (2006) Hydrocooling as an alternative to forced-air cooling for maintain fresh market strawberry quality. Hortic Technol 16: 659–666.

Gago PMB, Rojas C, Del RMA (2003) Edible coating effect on postharvest quality of mandarins cv. ‘Clemenules’. Acta Hortic 989: 91–94.

Ivey MLL, Malekian F, Fontenot K (2015) Wash water chlorine disinfection: Best Practices to ensure on-farm food safety. Available at https://www.lsu.edu/agriculture/plant/extension/hcpl-publications/8_Pub.3448-WashWaterChlorineDisinfection.pdf.

Jorge TDS, Soares AG, Fonseca MJDO, Barboza HTG, Jumior MF, Oliveira LC, Motta GF, Brito CD, Silva MHGE, Barbosa WJ (2013) Evaluation of packaging and edible coating on postharvest strawberry. Acta Hortic 71: 10–17.

Kumpoun W, Chuttong B, Uthaibutra J (2013) Development of bee wax coating materials for ‘Sai Nam Pueng’ tangerine fruit. Acta Hortic 989: 117–120.

Lichter A, Dvir O, Rot I, Akerman M, Regev R, Wiesblum A, Fallik E, Zauberman G, Fuchs Y. (2000) Hot water brushing: An alternative method to SO_2 fumigation for color retention of litchi fruit. Postharvest Biol Technol 18: 235–244.

Lider M. (2004) General Viticulture. University of California Press, Berkeley, CA.

Lu J, Vigneault C, Charles MT, Raghavan GSV (2007) Heat treatment application to increase fruit and vegetable quality. Stewart Postharvest Rev 3: 1–7.

Luo Z, Xie J, Xu T, Zhang L (2009) Delay ripening of ‘Quingnai’ plum (Prunus salicina L) with 1-methylcyclopropene. Plant Sci 177(6): 705–709.

Mahajan BVC, Kapoor S (2019) Storage systems for fruits and vegetables: A practical approach. In: Joshi VK (Ed.). Technology of Handling, Packaging, Processing, Preservation of Fruits and Vegetables: Theory and Practicals. New India Publishing Agency, New Delhi, India. pp. 151–160.

Matche RS (2005) Packaging aspects of fruits and vegetables. Plant Food Package 20: 115–132.

Mao L, Wang G, Que F (2007) Application of 1-methylcylopropene prior to cutting reduces wound responses and maintains quality in cut kiwifruit. J Food Eng 78(1): 361–365.

Nabi SU, Raja WH, Kumawat KL, Mir JI, Sharma OC, Singh DB, Sheikh MA (2017) Post harvest diseases of temperate fruits and their management strategies-A review. Int J Pure App Biosci 5(3): 885–898.

Nehal EL, Mougy S, Nadia E, Gamal GF, Kareem AE (2008) Use of organic acids and salts to control postharvest diseases of lemon fruits in Egypt. Arch Phytopathol Plant Prot 41: 467–476.

Oz AT, Ulukanli Z (2011) Effects of 1-methylcyclopropene (1-MCP) and modified atmosphere packing (MAP) on postharvest browning and microbial growth of loquat fruit. J Appl Bot Food Qual 84: 125–133.

Palou L, Gago MB, Perez SA (2014) Edible composite coatings formulated with antifungal GRAS compounds: A novel approach for postharvest preservation of fresh citrus fruit. Acta Hortic 1053: 143–149.

Petrisor C, Ilie A, Barbulescu A, Petcu A, Dumitru M (2010) Effects of refrigeration and modified atmosphere packaging on quality of strawberries. Scientific Papers of the R.I.F.G. Pitesti, 15: 109–112.

Piriyaphansakul S, Kanlayanarat S (2003) Effects of hydrocooling on fruit quality and storage life of ‘Namdokmai’ mango (*Mangifera indica* L.). In ACIAR Proc No. 50, pp. 430–433.

Pongener A, Mahajan BVC (2017). Advances in packaging of fresh fruits and vegetables. In: Siddiqui MW, Ali A (Eds.). Postharvest Management of Horticultural Crops–Practices for Quality Preservation. Apple Academic Press, Canada, pp. 231–264.

Ramjan MD, Ansari MT (2018) Factors affecting of fruits, vegetables and its quality. J Med Plants Stud 6(6): 16–18.

Ravindra MK, Goswami TK (2008) Comparative performance of precooling methods for the storage of mangoes (*Mangifera Indica* L. cv. Amrapali). J Food Process Eng 31(3): 354–371.

Robert C, Soliva F, Martın OB (2003) New advances in extending the shelf life of fresh-cut fruits: A review. Trends Food Sci Technol 14: 341–353.

Ruiz-Altisent M, Ortiz-Cañavate J, Valero C (2004) Fruit and Vegetables Harvesting Systems. In: Dris R and Jain SM (Eds.). Pre-harvest Practice. Kluwer Academic, Dordrecht, pp. 261–285.

Ruiz-Garcia Y, Gomez-Plaza E (2013) Elicitors: A tool for improving fruit phenolic content. Agriculture 3: 33–52.

Salvador ML, Jaime P, Oria R (2003) Use of edible coatings to reduce water loss and maintain quality of Reinette apple. Acta Hortic 600: 701–705.

Semeerbabu MT, Kudachikar VB, Revathy B, Ushadevi A, Matche RS, Ramana KVR (2007) Effect of postharvest treatments on shelf life and quality characteristics of litchi fruit (*Litchi chinensis* Sonn.) stored under modified atmosphere packaging at low temperature. J Food Sci Technol 13: 1–19.

Siddiqui MW (2016) Eco-friendly Technology for Postharvest Produce Quality. Academic Press, Elsevier Science, New York, NY, p. 324.

Sisler EC, Serek M. (1997) Inhibitor of ethylene responses in plants at the receptor level: Recent developments. Physiol Plant 100: 577–582.

Sivakumar D, Korsten L, Zeeman, K (2007) Postharvest management on quality retention of litchi during storage. Fresh Produce 1: 66–75.

Suslow T (2000) Fruit and vegetable processing. In: McLaren D. (Ed.). Chlorination in the Production and Post-Harvest Handling of Fresh Fruits and Vegetables. Food Processing Center, University of Nebraska, Lincoln, pp. 2–15.

Suttirak W, Manurakchinakorn S (2010) Potential application of ascorbic acid, citric acid and oxalic acid for browning inhibition in fresh-cut fruits and vegetables. Walailak J Sci Technol 7: 5–14.

Tripathi P, Dubey N. (2004) Exploitation of natural products as an alternative strategy to control postharvest fungal rotting of fruit and vegetables. Postharvest Biol Technol 32: 235–245.

Watkins CB (2006) The use of 1-MCP on fruits and vegetables. Biotechnol Adv 24: 389–409.

Workneh TS, Osthoff G (2010) A review on integrated agro-technology of vegetables. Afr J Biotechnol 9: 9307–9327.

Yamaguchi F, Yoshimura Y, Nakazawa H, Ariga A (1999) Free radical scavenging activity of grape seed extract and antioxidants by electron spin resonance spectrometry in an H_2O_2/NaOH/DMSO system. J Agri Food Chem 47: 2544–2548.

4

Postharvest Treatments for Horticultural Produce

Ram Roshan Sharma, H. R. Raghavendra and Shruti Sethi
Division of Food Science and Postharvest Technology, ICAR-Indian Agricultural Research Institute, New Delhi, India

CONTENTS

4.1 Introduction

Fruits and vegetables are essential for the well-being of mankind as they are rich and cheap sources of several vitamins and minerals. Hence, these commodities have been designated as ‘protective foods’ by nutritionists. These commodities are metabolically active and perishable in nature, and therefore liable for severe postharvest losses, from the time of harvesting until the produce reaches consumers. Of the postharvest losses, most of the spoilage of fruits and vegetables is due to microbes, which infest these commodities during production, harvesting and subsequent handling. It has been estimated that nearly 20–25% of the harvested horticultural produce is spoiled by microbes, even in developed countries such as the United States, United Kingdom, Japan and France (Singh and Sharma 2007). However, postharvest losses are much more severe in developing countries primarily because of inadequate infrastructure for transportation and storage of produce. Fungicides are primarily used for controlling or reducing incidence of diseases in produce. However, now there has been a dramatic shift from usage of chemical fungicides to safer, eco- and farmer-friendly approaches for reducing the loss in the harvested horticultural produce caused by microbes.

Considering the economics involved in postharvest decay loss in horticultural produce, and development of resistance of pathogens to synthetic fungicides, an urgent need was felt globally to develop safer approaches for controlling postharvest diseases of horticultural produce. As a result, several physical, gaseous and chemical treatments have been standardized for perishable commodities (fruits and vegetables) to maintain freshness, retain quality and reduce postharvest decay loss (Mahajan et al. 2014). The recommended postharvest treatments can also be combined with appropriate storage temperatures for achieving better results in the frontier areas of postharvest management. This chapter deals with the postharvest treatments, which can be used to maintain freshness, enhance quality and reduce losses in fresh produce by controlling diseases (Table 4.1).

TABLE 4.1

Major Postharvest Treatments Used in Harvested Fruits and Vegetables

Treatment Type	Name of Treatment	Benefits of Treatment	Benefited Fruits	Benefited Vegetables
Physical treatments	Heat treatment	Delay in ripening, reduction in chilling injury, control of insect attack, reduction in decay	Grape, kiwifruit, peach, plum, mango, etc.	Beans, celery, lettuce, spinach, potato, tomato, carrot, asparagus, broccoli, etc.
	Edible coating	Minimizes moisture loss by providing a partial barrier, creating modified atmosphere, preserving colour, maintaining texture, retaining aroma	Pear, strawberry, apple, etc.	Carrot, celery, etc.
	Irradiation	Meets quarantine requirements, a safe approach for inhibition of sprouting of tubers, bulbs and roots	Strawberry, mango, etc.	Potato, onion, etc.
Gaseous treatments	Ozone (O_3)	Easy to use efficiently and can be installed into existing cold storage, washing system, and has good efficacy than chlorine	Onion, kiwifruit, plum, peach, apple, cherry, table grapes, etc.	Carrot, garlic, potato, etc.
	1-MCP(1-methylcyclopropene)	Reduces ethylene production, maintains peel colour, fruit cell wall integrity and develops aroma and flavour	Avocado, kiwifruit, mango, papaya, apple, pear, peach, nectarine, plum, persimmon, pineapple, etc.	Cucumber, broccoli, tomato, bell pepper, etc.
	Ethylene	Commercial ethylene enhances production, enhances ripening process, maintains and improves fruit colour and quality	Mango, citrus, avocado, persimmon, kiwifruit, banana, etc.	Tomato
	Controlled atmosphere (CA) storage	Slows down the physiological and metabolic activities by retarding senescence; reduces decay	Persimmon, pomegranate, banana, cranberry, apple, pear, strawberry, cherry, kiwifruit, avocados, mango, nectarine, peach, plum, etc.	Broccoli, asparagus, cabbage, etc.
	Modified atmosphere packaging (MAP)	Delay senescence, slows down rate of respiration and deterioration of produce	Cherry, fresh cut fruits, apple, banana, strawberry, etc.	Carrots, salad, leafy green vegetables
Chemical treatments	Nitric oxide (NO)	Reduces respiration rate, counters ethylene biosynthesis, water loss, browning; reduces incidence of postharvest diseases	Mango, apple, kiwifruit, pear, peach, plum, papaya, banana, strawberry, loquat, etc.	Tomato
	Anti-microbial and anti-browning agents	Retard deterioration and microbial growth of produce and maintain texture; prevent browning of cut surface of fresh produce	Orange, apple, strawberry, grape, fresh cut produce.	Tomato, celery, lettuce, etc.
	Sulphur dioxide	Retards deterioration, prevents postharvest decay	Banana, grape, fig, litchi, lemon, apple, blueberry	–

4.2 Major Postharvest Treatments

Several treatments are currently in use to improve storage or shelf life and reduce the incidence of postharvest diseases (Mahajan et al. 2014) as shown in Table 4.1.

4.2.1 Physical Treatments

4.2.1.1 Heat Treatment

In the present era of increased consumer awareness that most chemicals used for controlling pests and diseases in harvested produce can cause serious issues to health (Singh and Sharma 2007),an urgent need has been felt by scientists to develop effective but consumer-friendly physical, gaseous or chemical treatments for quality enhancement of fresh horticultural produce. Accordingly, several technologies, such as irradiation, use of GRAS (generally regarded as safe) additives, hypobaric storage, controlled atmosphere (CA) storage or modified atmosphere packaging (MAP), etc., have been developed and standardized to reduce disease incidence and increase storage life. Since no chemical is involved in these technologies there should be no consumer concern about pesticidal residue (Mahajan

TABLE 4.2

Beneficial Effects of Heat Treatment on Different Fruits and Vegetables

Fruit/Vegetable	Aim	Treatment Type	Temperature (°C)	Time	Reference
Apple	Control of diseases, inhibition of ripening	HWRB	55	15 sec	Fallik et al. (2001)
Avocado	Control of diseases, improvement in fruit quality	HWT	40–42	20–30 sec	Hofman et al. (2002)
Lemon	Control of diseases, decay resistance	HWT	52–53	2 min	Nafussi et al. (2001)
Litchi	Control of decay	HWRB	55	20 sec	Lichter et al. (2000)
Orange	Control of diseases, better fruit quality	HWRB	56	20 sec	Porat et al. (2000)
Plum	Control of diseases, resistance to chilling injury	HWT	45–50	30–35 sec	Abu-Kpawoh et al. (2002)
Asparagus	Reduces geotropism	HWT	47.5	2–5 sec	Paull and Chen (2000)
Green onion	Growth inhibition	HWT	55	2 min	Cantwell et al. (2001)
Melon	Decay control, ripening inhibition	HWRB	59	15 sec	Fallik et al. (2001)
Potato	Sprouting inhibition, better quality	HWT	57.5	20–30 sec	Ranganna et al. (1998)
Sweet pepper	Induction of resistance to chilling, enhanced production of polyamines	HWT	45	15 sec	Gonzalez-Aquilar et al. (1999)
Tomato	Induction of resistance to chilling, reduction in decay	HWT	39	60 sec	McDonald et al. (1999)

et al. 2014). Of these technologies, heat treatment appears to be a quite promising method for postharvest decay control (Droby 2006; El-Ghaouth et al. 2004; Zhu 2006) in horticultural produce (Table 4.2).

Hot water treatment (HWT) was first reported in 1922 for controlling citrus fruit decay (Fawcett 1922). During prestorage heat treatment, the produce is exposed for a short time with the objective to get rid of those pathogens which remain present on fruit peel or in outer layers under the peel of fruit. In HWT, water is used as a preferable medium as it is a better heat transferring medium than air. In this treatment, the harvested produce is treated with heat in several ways including hot water dips, dry hot air or vapour heat (Lurie 1998) and/or by rinsing and brushing in hot water (Fallik et al. 1996). Vapour heat treatment (VHT) has been developed primarily for controlling insects, whereas hot dry air technique is used for both microbial and insect control (Lurie 1998). Heat treatment is also applied to inhibit the ripening process in fruits and also for inducing resistance to CI and reducing peel damage during storage. Hence, HWT extends the storability and market value of the harvested produce (Paull and Chen 2000; Woolf 1997) (Table 4.2).

4.2.1.1.1 Methods of Hot Water Treatment

Hot water dips: Hot water dip or immersion is one of the methods of HWT, which is usually practiced for quarantine purposes. In this method, fruits are dipped in hot water maintained at 43–49° C for 30 minutes to 2 hours (Fallik 2003) in a tank having a water circulation system, a heat exchanger unit and a thermostat for controlling temperature (Lurie 1998). The exposure time of hot water depends on fruit size. For larger fruit, longer time is usually required, and vice-versa. This treatment has been reported to be most effective against Mexican fruit fly without any adverse effects on fruit quality. A HWT treatment of the produce at 49°C for 20 minutes has been approved by several European countries for tropical fruits such as mango and papaya (Lurie 1998). Incidence of blue mould in grapefruit can be reduced by dipping fruits in hot water (50°C) for 2 minutes (Table 4.2).

Hot water rinsing and brushing (HWRB): This method was first employed commercially in 1996. It is primarily used to clean and disinfect harvested fruits and vegetables. In this method, fruits are rinsed with tap water and then brushed to remove dust, dirt, microbial and/or pesticidal load (Mahajan et al. 2014). The process is continued in hot water at temperatures of 48°C and 63°C for 15–25 seconds, depending on the fruit or its cultivar (Ilic et al. 2001; Lichter et al. 2000; Smilanick et al. 2003) (Table 4.2).

Vapour heat treatment (VHT): This treatment has been standardized to control fruit flies in harvested produce. It consists of stacking the produce in a hot and humidified chamber by steam injection. The exposure time and the temperature in the VHT chamber are maintained to kill all the stages of insect growth without any damage to the fruit (Pantastico 1975). In India, VHT has been standardized for the control of stone weevil and fruit fly in mango.

4.2.1.1.2 Advantages of Heat Treatments

Heat treatment is preferable for harvested produce because of its (i) relative ease of use, shorter exposure time, easy monitoring of temperatures, (ii) killing of decay-causing microbes on the peel, (iii) reduction and delay in CI, (iv) delay in the ripening process due to inactivation of enzymes by heat, (v) killing of insects and (v) controlling the onset of microbial decay.

4.2.1.2 Edible Coatings

The external appearance of fruits or vegetables plays an important role in consumer acceptability. Fruits with attractive colour, glossiness and which are russet-free usually fetch good price in the market. Physical appearance of the commodity is the first attribute that consumers observe. Although no correlation exists between visual appearance and edible quality, external appearance is a powerful agent that affects the marketing value of produce (Dhall 2013). Many buyers prefer produce with a shiny or glossy appearance. A few fruits and vegetables have natural waxy covering, which provides an attractive appearance and also protects them from microbial infestation and/or insect attack. However, this waxy covering is usually disturbed during harvesting and postharvest handling (Chauhan et al. 2014). Hence, various types of artificial and natural waxes are applied to harvested horticultural produce to increase its cosmetic appeal, such as shining and colour of the product (Dhall 2013). This was first attempted in 1922, and since then, a lot of work has been done on this aspect. Waxing is preferable only to good quality products, as it cannot improve quality in inferior products. This treatment only acts as a supplement to produce natural waxes on the surface of products, which were lost during harvesting and subsequent handling stages.

Edible coating is defined as the applying of a thin layer of wax or edible material over the fruit or vegetable, which can be consumed without any harm. This coating acts as a barrier to oxygen, moisture and microbes. As a result, there is minimum loss of moisture from the produce during storage. It also creates a modified atmosphere (MA) around the product, which helps in slowing down several processes of the produce such as respiration, senescence and oxidation (Chauhan et al. 2014). It also improves appearance, reduces postharvest decay and helps to retain flavour and functional compounds (Skurtys et al. 2005; Chauhan et al. 2014). In the last few years, there has been a significant improvement in the research work on the development of coating materials from natural sources as well (Donhowe and Fennema 1993; Warriner et al. 2009). Several edible coatings have been developed for different fruits and vegetables (Tables 4.3 and 4.4). However, stringent efforts and focussed research are required for recommending specific edible coatings for specific fruits or vegetables or their specific varieties to understand the variation in shelf life of different commodities (Skurtys et al. 2005; Dhall 2013).

4.2.1.2.1 Methods of Application

Edible coatings are applied by dipping, brushing, extrusion, spraying or solvent casting. However, dipping is the simplest and most common method of applying edible coatings on harvested produce. It consists of dipping fruits or vegetables in coating solution for a short time of 1–2 minutes (Valverde et al. 2005). Brushing is commonly used for beans and highly perishable fruits such as strawberry and berries. The extrusion method is the best technique for industrial purposes (Valverde et al. 2005). Other methods are also used in the food industry.

4.2.1.2.2 Benefits of Edible Coatings

Improved appearance: Coated fruits or vegetables generally retain colour and freshness for a longer period. Such produce exhibits greater glossiness than noncoated produce (Dhall 2013).

Reduction in moisture loss: In the process of waxing/edible coating, the surface of a fruit or a vegetable is coated with a thin layer, which blocks the pores of the cuticle and reduces loss of water (Dhall 2013; Kumar and Sethi 2018). Edible coating can usually reduce loss of moisture by nearly 30–40% or more. Reduction in moisture loss helps to keep the produce fresh and nutritive for longer time.

Reduction in economic loss: Water is the major component of all horticultural produce, which is usually between 80%and 90% of total weight. After harvesting, fresh produce starts losing moisture through the processes of respiration and transpiration. Due to moisture loss, the produce shrinks and becomes unattractive to the consumer. Since the loss of water is lower after edible coating, economic loss is necessarily less (Dhall 2013).

Reduction in postharvest decay: Edible coating establishes a physical barrier for entry of microbes into the product. In addition, edible coatings also create a hydrophobic surface, which is not conducive to growth and development of microbes. Hence, decay loss is significantly reduced in the coated horticultural produce.

Extension on shelf or storage life: Horticultural produce is a living commodity, which continues its metabolic activities (e.g. respiration, transpiration, etc.) even after harvesting.

TABLE 4.3

Different Types of Coatings for Horticultural Produce

S. No.	Fruit/Vegetable	Edible Coating Used	Reference
1	Apple	Neem oil, marigold flower extract, guar gum, *Aloe vera*	Chauhan et al. (2014)
2	Grape	*Aloe vera* gel	Chauhan et al. (2014)
3	Fresh cut pineapple	Pectin and alginate	Mantilla (2012)
4	Strawberry	Sodium alginate and calcium alginate gel	Moayednia et al. (2010)
5	Banana	Polyvinyl acetate (PVA) carboxy methyl cellulose (CMC), tannins	Senna et al. (2014)
6	Plum	CMC, pectin	Panahirad et al. (2015)
7	Potato	chitosan, coconut oil	Saha et al. (2014)
8	Cantaloupe	*Aloe vera* gel, pectin, chitosan	Yulianingsih et al. (2013)
9	Tomato	Vegetable oil, cellulose gum, emulsifier	Athmaselvi et al. (2013)

TABLE 4.4

Beneficial Effects of Some Recommended Coatings on Fruits and Vegetables

Edible Coating	Constituents of Coating	Fruit/Vegetable	Effects of Coating	Reference
Casein	Calcium caseinate	Bell pepper	Physical barrier to O_2 and CO_2 gas	Lerdthanangkul and Krochta (1996)
Cellulose	MC, glycerol	Carrot	Improved carotene, extended storage life	Li and Barth (1998)
Chitosan	Chitosan, Tween 80	Strawberry, bell pepper, cucumber	Acts as an anti-microbial	El-Ghaouth et al. (1991)
		Apple, pear, peach, plum	Acts as a gas barrier	Elson and Hayes (1985)
		Litchi	Reduced browning and weight loss	Zhang and Quantick (1997)
		Carrot	Reduced decay and improved appearance	Cheah et al. (1997)
		Tomato	Reduced ethylene rate, respiration rate and increased titratable acidity	El-Ghaouth et al. (1992)
		Pear	Reduced ethylene production and delayed ripening	Li and Yu (2000)
Nature-Seal™	Cellulose	Mango, tomato	Delayed ripening	Nisperos-Carriedo et al. (1992)
		Carrot	Retarded discolouration and carotene loss	Chen et al. (1996)
Semperfresh™	Sucrose esters, unsaturated fatty acids, CMC and glycerides	Zucchini	Reduction in moisture loss and CO_2	Avena-Bustillos et al. (1994)
		Apple	Reduced colour changes, retained acidity and maintained quality	Drake et al. (1987)
		Banana	Decreased ethylene production and delayed chlorophyll loss	Nisperos-Carriedo et al. (1992)
Prolong	Sucrose, fatty acids, CMC, glycerides	Mango	Reduction in ripening and weight loss	Motlagh and Quantick (1998)
		Pear	Retention of firmness, green peel colour and titratable acidity	Farber et al. (2003)
Tal Prolong	Sucrose, fatty acids, CMC and glycerides	Mango	Delayed ripening with extended shelf-life	Nisperos-Carriedo et al. (1992)
Zein	Corn zein protein	Tomato	Delayed colour changes, loss of firmness and weight	Park et al. (1994)

When edible coating is applied, it creates a modified atmosphere condition, which decreases oxygen and increases carbon dioxide (Dhall 2013; Kumar and Sethi 2018). As a result, there is a reduction in respiration rate, which helps to increase the storage or shelf life of the produce and increases availability of the produce for longer time in the market.

Decreasing susceptibility to chilling injury: Several fruits and/or vegetables of subtropical and tropical regions are susceptible to CI if stored at a low temperature. By inactivating enzymes, waxing/coating reduces the severity of CI. However, edible coating does not completely eliminate CI in commodities, which are susceptible to CI (Dhall 2013; Kumar and Sethi 2018).

4.2.1.2.3 Limitations of Edible Coatings

- Edible coatings create a gaseous barrier around the produce, which may cause anaerobic respiration. This may disturb ripening process in some fruits and/or vegetables.
- Few edible coatings are hygroscopic in nature and favour the growth of microbes.
- Thicker edible coating prohibits oxygen exchange resulting in development of off-flavour in the produce.
- Improperly applied coatings can result in burning of the fruit or vegetable surface. Similarly, tissue damage occurs if the temperature is kept too high.
- At high temperatures, wax whitens the surface (chalking) of the produce, giving an unappealing look.

4.2.1.3 Irradiation

Several food-borne diseases are caused by bacteria, which cause havoc in several countries. Good practices adopted after harvesting the produce (e.g. washing with ozonated or chlorinated water) can reduce microbial load to some extent, but it is not possible to completely eliminate it by primary processing. Hence, several decontamination methods have been standardized and developed; the most versatile method is irradiation of food with ionizing radiation (Mahajan et al. 2014). It is a safe, environment-friendly and energy-efficient method for preservation of food. Irradiation can effectively eliminate the potential

TABLE 4.5

Irradiation Doses and Their Uses or Effects

Low Dose Application	Medium Dose Application	High Dose Application
Up to 1 kGy	1–10 kGy	Above 10 kGy
Low doses of irradiation disrupt cells, which inhibit sprouting and rooting of tubers (potato), bulbs (onions) and roots (carrot) and delay senescence	Extends shelf life	Complete sterilization, used in canning of food Kill several broad spectrum fungi and bacteria and also pests
	Reduces microorganism level	Decontaminate certain food additives such as spices

non-sporulating bacteria like Salmonella, *Staphylococcu saureus*, Campylobacter, *Escherichia coli* or *Listeria monocytogenes* without affecting sensory and nutritional quality of produce (Reddy et al. 2018a). The unique feature of this technology is that it can also be applied to a frozen food. Similarly, irradiation can be effectively utilized in combination with other techniques of preservation for obtaining a safe and high quality food with enhanced shelf life (Mahajan et al. 2014).

4.2.1.3.1 Irradiation of Food

Irradiation may be defined as the process of exposure of biological materials to radiation, particularly mutagenic radiations. In this process, several foods are exposed to irradiation to eliminate microbes or insects (Mahajan et al. 2014, Reddy et al. 2018a). It is also called 'cold pasteurization', as the microorganisms are deactivated in low temperature rather than heat, which is used in other methods such as pasteurization, sterilization or blanching. This method of food preservation helps to increase shelf life, improves food safety and reduces the use of preservatives, fumigants and additives (Mahajan et al. 2014; Reddy et al. 2018a).

4.2.1.3.2 Mode of Action

Irradiation is a method of food preservation in which prepackaged or packed or frozen food is exposed to electromagnetic rays. During this process, an electron is emitted from an atom or molecule which causes ionization, hence it is usually called ionizing radiation (Mahajan et al. 2014; Reddy et al. 2018a). Produce is directly exposed to radiant energy generated either from γ rays or e-beam that enters the produce and breaks its molecular bonds, including the DNA (Farkas and Mohacsi-Farkas 2011; Lado and Yousef 2002). For ionizing radiation, cobalt-60 or caesium-137 is usually used. Sometimes, electron beams are also generated through machines. Ionizing radiation inhibits cellular reproduction and kills microbes and pests, and as a result, food is not only safe but also has a longer shelf life (Farkas 2006; Reddy et al. 2018a).

4.2.1.3.3 Irradiation Doses

Determining a dose of irradiation is called 'dosimetry'. Irradiation doses are usually measured in kilograys (1 kGy = 1000 Gy) and its effects are dose-dependent (Mahajan et al. 2014). For irradiation of fresh horticultural produce, medium- and high-irradiation doses are not usually recommended because they can adversely affect the sensory attributes of produce (colour, texture and/or flavour) (Table 4.5). Furthermore, such doses may also accelerate senescence of the produce by denaturing the proteins and DNA (Reddy et al. 2018a). Thus, 10-kGy irradiation dose is safe and promising as per international food-related organizations (FAO/IAEA/WHO 1981, 1999).

4.2.1.3.4 Effect of Irradiation Dosage on Microbes

Irradiation is a potent treatment for eliminating microbes such as bacteria, yeasts and moulds which cause food spoilage. Irradiation also controls insects and parasites in food, and hence reduces storage losses, extends shelf life and improves food safety (Mahajan et al. 2014). Accordingly, irradiation doses have been standardized and recommended for specific purposes in some fruits and vegetables (Tables 4.6 and 4.7).

The major objective of using this novel technique on fresh horticultural produce is to increase shelf life and reduce decay. In addition, this technology is used for killing, sterilizing or preventing further spread of insect pests of quarantine

TABLE 4.6

Effective Irradiation Doses Approved by FSSAI, India

Commodity	Objective	Minimum Recommended Dose (kGy)	Maximum Allowable Dose (kGy)
Potato	Sprout inhibition	0.06	0.15
Onion	Sprout inhibition	0.03	0.09
Garlic	Sprout inhibition	0.03	0.15
Ginger	Sprout inhibition	0.03	0.15
Mango	Disinfection (quarantine)	0.25	0.75
Spices	Microbial decontamination	6	14
Raisin, dried dates or figs	Extension of shelf life, decay control	2.5	4

TABLE 4.7

Irradiation Doses for Achieving Desirable Benefits in Specific Fruits and Vegetables

Purpose	Dose (kGy)	Produce
Sprouting inhibition	0.06–0.20	Onion, garlic, potato
Disinfestation of insects (including quarantine)	0.15–1.0	Cereals, legumes, fresh and dried fruits
Parasitic disinfestation	0.3–1.0	Fresh pork, fish, fruits
Delay of ripening	0.5–1.0	Fresh fruits
Extension of shelf life	1.0–3.0	Fruits and vegetables
Inactivation of spoilage organisms	1.0–7.0	Raw and frozen foods, spices
Increasing properties	3.0–7.0	Increase juice yield (grapes)
Reduction in sprouting	0.3–0.5	Potato
Controlling grey mould	2	Strawberry
Controlling grey mould	1–1.25	Tomato
Controlling brown scald	1–1.5	Apple
Controlling fruit fly	0.75	Papaya

importance (Reddy et al. 2018a). However, in spite of several benefits of this technology, it did not achieve a great success in India, primarily because of political issues and fear of irradiation among consumers. Hence, extensive research is required for testing and demonstrating this technology to remove the fear of consuming irradiated food.

4.2.2 Gaseous Treatments

4.2.2.1 1-MCP (1-Methylcyclopropene)

As the ripening process in fruits increases, ethylene production also increases, which leads to reduction in postharvest quality of fruits by increasing the respiration rate, production of volatiles, loss of firmness and softness in texture, ultimately leading to spoilage of fruits by attack of saprophytic microbes. Hence, there was an urgent need for a proper technology to control the ethylene action (Mahajan et al. 2014). There are more strategies to regulate the ethylene action, such as1-MCP (1-methylcyclopropene), AVG (1-aminoethoxyvinylglicine), NO (nitric oxide), MAP, CA storage, STS (silver thiosulphate), NBD (2,5-norbornadiene), salicylic acid, hexanal and many more. Among these, discovery of 1-MCP has been regarded as the most effective molecular tool for controlling the ripening process in fruits.

1-MCP was synthesized in 1996. It is a cyclopropene derivative, and is now artificially synthesized and commercially used as a plant growth regulator. It inhibits ethylene by blocking the receptor sites of ethylene in a plant cell. Structurally, it is similar to the natural plant hormone ethylene (Blankenship and Dole 2003). 1-MCP is a non-toxic gas, with 54 molecular weight and C_4H_6 chemical formula, which activates at very low concentrations with negligible amount of residue. 1-MCP has nearly 10 times higher affinity for ethylene receptor than ethylene (Blankenship and Dole 2003).

4.2.2.1.1 Commercial Applications of 1-MCP

1-MCP has been standardized for commercial use on different fruits (e.g. mango, papaya, banana, plantain, avocado, apple, pear, plum, peach, nectarine, date, kiwifruit, persimmon, pineapple), vegetables (e.g. broccoli, cucumber, melon, pepper, squash, tomato) and flowers (e.g. rose, carnation, gerbera, chrysanthemum, gladiolus) (Blankenship and Dole 2003). The concentration and time of exposure of the produce to 1-MCP are given in Table 4.8. It has been recommended for the following benefits.

***Delay in ripening*:** The major role of 1-MCP is to delay fruit ripening (Claire et al. 2003) by lowering ethylene production (Zisheng 2007), chlorophyll degradation (Fan and Mattheis 1999), declining of nutrients, and finally, it maintains good quality of fruits.

***Alleviation of chilling injury*:** Chilling injury is a major problem in tropical fruits when stored at a low temperature. In some fruits such as mango, banana, guava and pineapple, chilling injury is a major problem when fruits are transferred from cold storage to normal conditions. 1-MCP cannot directly control chilling injury but it reduces its symptoms such as flesh mealiness, flesh discolouration, internal browning, etc., by retarding the ripening process (Singh and Pal 2007).

***Control of postharvest fruit/vegetable rot*:** 1-MCP also retards postharvest decay in some fruits such as leak rot in strawberry (Jiang et al. 2001), anthracnose in guava (Singh and Pal 2007) and diseases caused by *Rhizopus* sp.

4.2.2.2 Ozone

Recent developments in the area of postharvest management of fresh horticultural produce indicate that ozone can be used as a sanitizer for fruits and vegetables. Ozone is a naturally occurring gas, which is very pungent and has very high oxidizing property. It has nearly a 1.5 time higher oxidizing capacity than chlorine and about 30,000 times than hypochlorous acid.

Ozone has three atoms of oxygen, in which central O_2 is attached to the other two at equal distance (Mahajan et al. 2014; Reddy et al. 2018b). It is a pale blue gas at very high concentrations. Its distinct odour is noticeable at very low concentrations, which may be associated with a 'clean' smell.

TABLE 4.8

Time and Concentration Conditions of 1-MCPfor Different Fruits and Vegetables

Crop	1-MCP Concentration	Temperature (°C)	Time (hours)	Effects on Produce	Reference
Pineapple	0.1 μL/1	20	18	Reduced browning, slowed ascorbic acid and ethylene production	Selvarajah et al. (2001)
Papaya	25 μL/1	20	14	Slight increase in blemishes, delayed ripening	Hofmanet al. (2001)
Orange	100 nL/1	25	6	Reduced ethylene-induced degreening, inhibited abscission	Porat et al. (1999)
Strawberry	5–15 μL/1	20	2–18	Improved shelf life, maintained colour, lowered phenol content	Tian et al. (2000)
Mango	25 μL/1	20	14	Improved shelf life	Hofman et al. (2001)
Banana	0.1 μL/1	20–24	6–24	Delayed fruit ripening process and colour-related changes in peel	Harris et al. (2000)
Pear	2.4 μL/1	2	16	Reduced chilling and ethylene production; slowed down softening	Baritelle et al. (2001)
Avocado	25 μL/1	22	6–48	Inhibited fruit ripening, maintained higher firmness, reduced activity of degrading enzymes	Jeong et al. (2002)

Its applications include water treatment, washing of produce, cleaning of surfaces and equipment and use in CA storage. It has been approved as a GRAS food additive by the Food and Drug Administration (FDA) and the USDA (1997), which allows its direct contact with food as an antimicrobial agent.

4.2.2.2.1 Mode of Action

The mode of action of ozone is similar to chlorine (another oxidant). It disinfects the food items by directly oxidizing and destroying the cell wall of microorganisms, which causes cellular components to leak out of the cell. The result is protoplasmic destruction, which damages different constituents of cell such as nucleic acids and also breaks down the carbon–nitrogen bonds, thereby leading to de-polymerization (Reddy et al. 2018b).

4.2.2.2.2 Benefits of Ozonation

Ozone can be used as a broad-spectrum anti-microbial agent, as a supplementary to reduce and degrade pesticide residues, for wastewater treatment and as an environment-friendly treatment without any harm to humans. It has the following applications in the food industry (Reddy et al. 2018b).

- It brings significant reduction in microbial load of *E.coli* and *Salmonella*.
- It is quite effective against several spoilage microbes.
- It is a potent sanitizer and is a greener alternative to chlorine.
- It brings oxidation of ethylene, thus delays ripening of fruits and vegetables.
- It is helpful in extending the storage or shelf life of horticultural produce.
- It can be easily applied to fruit juices.

Different effects of ozone treatments on fresh horticultural produce are presented in Table 4.9.

4.2.2.2.3 Ozone Toxicity

Ozone toxicity is a valuable criterion for its approval in the food industry. It is always important to check people who come in contact with ozone. Exposure of human beings to ozone may cause dizziness, headache and a burning sensation in eyes or throat, and sometimes coughing. In chronic cases, it causes weakness, decreased memory, bronchitis and excitability in muscles (Hoof 1982; Reddy et al. 2018b). At practical and safe concentrations, it rapidly attacks bacterial cell walls more effectively than chlorine, especially against thick-walled microbial spores. Its treatment has been recommended for carrots, asparagus, onions, garlic, apples, kiwifruit, cherries, plums, peaches, grapes and potatoes (Reddy et al. 2018b). However, extensive research is still needed to determine its potential concentrations and limits so that it can be effectively used in fresh horticultural produce.

4.2.2.3 Ethylene

Endogenous production of ethylene and its application on fresh horticultural produce exhibit beneficial as well as deleterious

TABLE 4.9

Effects of Ozone Treatment on Different Fruits and Vegetables

Fruit/Vegetable	Effects	Reference
Fruits and vegetables	Increased shelf life	Rice et al. (1982)
Blackberry	Decreased fungal deterioration	Beuchat (1992)
Grape	Decreased fungal deterioration	Beuchat (1992)
Onion	Mould and bacterial counts were significantly reduced without changes in chemical or sensory attributes	Song et al. (2000)
Shredded lettuce	Decrease in bacterial count	Kim et al. (1999)

effects. If applied exogenously, it enhances ripening, improves colour and/or quality in mango, banana, persimmon, avocado, kiwifruit and citrus fruits (Mahajan et al. 2014). Its deleterious effects on harvested fruits and vegetables are shorter storage/shelf life, softening of fruits and enhancement in senescence, discoloration and yellowing, and increased susceptibility to diseases. Therefore, ethylene management plays an important and crucial role in significantly enhancing the shelf life of harvested produce. Most commercial and useful ethylene management strategies include storing the produce under cold conditions, changing ethylene biosynthesis and action, minimizing exposure of produce to ethylene during and after harvesting or postharvest handling and/or by controlling temperature and atmospheric gas composition (Mahajan et al. 2014). For this, biosynthesis inhibitors of ethylene such as AVG or ethylene binders such as 1-MCP may use with other management practices such as CA storage or MAP.

4.2.2.4 Controlled Atmosphere Storage

Air contains 20.9% O_2, 78.1% N_2 and 0.03% of CO_2 along with other traces of gases. For normal respiration of fruits, oxygen is absorbed from the surrounding atmosphere and CO_2 is released. If the concentration of oxygen reduces, it leads to reduction in respiration which increases the shelf life of commodity. Therefore, in CA storage, the concentration of O_2 is reduced and that of CO_2 is increased to slow down the respiration rate of the produce (Bodbodak and Moshfeghifar 2016). CA storage means adjusting and monitoring the specified concentration of CO_2 and O_2 in a gas-tight store at optimum temperature. In most cases, the CO_2 concentration is higher and O_2 is lower, which are standardized for specific fruits or vegetables and the purposes of the user and/or conditions of CA storage (Bodbodak and Moshfeghifar 2016).

Every fruit or vegetable has a specific requirement for O_2 and CO_2 concentrations for maintaining appropriate quality and shelf life. In CA storage, oxygen level is usually kept at 0–10% and CO_2 at 0–20% (Table 4.10).

CA storage technology is commercially used in only a few fruits and vegetables, the major fruit being apple. However, it is also used to a limited extent in other fruits such as avocados, banana, pomegranate, orange, kiwifruit, olive, persimmon, strawberry, nuts and dried fruits, and in some vegetables such as cabbage, okra, tomato and onion (Table 4.10). Major limitations of CA storage are: it is a very costly technology, problems may arise during non-availability of gases, off-flavours may develop at very low oxygen levels and irregular ripening may take place in some fruits such as banana, pear, tomato, etc.

4.2.2.5 Modified Atmosphere Storage

In modified atmosphere (MA) storage, the concentration of oxygen is also reduced and that of carbon dioxide is increased, but the degree of control over the gas concentration is lower than CA storage. In MA, normally the commodity is packed in different types of polymeric films and containers, and regulation of the gases is achieved by following methods.

Active MA: In this method, the commodity is packed and sealed, and gas is removed slightly, creating a vacuum. The package atmosphere is replaced with the desired mixture of gases (Mahajan et al. 2014). This ensures faster creation of the desired atmosphere inside the package. Ethylene absorber and CO_2 scrubbers are also used to avoid build-up of these gases (Mahajan et al. 2014). In active MAP, the desired atmosphere is created in the packed produce before sealing; however, the final concentrations of oxygen and carbon dioxide are maintained by the produce.

Passive MA: In this method, the commodity is packed in sealed films. As a continuous respiration of produce inside the package and nonpermeability of packaging materials, the concentration of oxygen reduces and concentration of carbon dioxide increases automatically (Mahajan et al. 2014). In passive MAP, the O_2 and CO_2 concentrations are mediated by the weight of the produce and its rate of respiration. This is influenced by storage temperature, level of perforations in the package, surface area of the produce and permeability and thickness of polyfilms to gases used as a packaging material.

Oxygen absorbers: Ferrous oxide is used.

CO_2 absorbers: Hydrated lime, activated charcoal and MgO are used to absorb the produced carbon dioxide.

TABLE 4.10

Recommended CA Storage Conditions for Fruits and Vegetables

Fruit/Vegetable	Temperature (°C)	O_2 (%)	CO_2 (%)
Apple, apricot, peach, plum	0–5	1–3	0–5
Strawberry	0–5	10	15–20
Persimmon	0–5	3–5	5–8
Nuts and dried fruits	0–25	0–1	0–100
Lemon, mango, papaya, pineapple	10–15	5	5–10
Banana	12–15	2–5	2–5
Avocado	5–13	2–5	3–10
Orange	5–10	10	5
Olive	8–12	2–5	5–10
Cucumber, okra, tomato	8–12	3–5	0
Green onion	0–5	1–4	10–20
Broccoli, cabbage, cauliflower, leek, lettuce, spinach	0–5	1–5	2–20

Ethylene absorbers: $KMnO_4$ on celite, vermiculite, silica gel, etc. are used to absorb ethylene inside the package.

4.2.3 Chemical Treatments

4.2.3.1 Anti-Microbial and Anti-Browning Agents

During the last few decades, we have witnessed several outbreaks of food-borne diseases causing several deaths. This has highlighted the importance of food safety and has given regulatory agencies, producers and/or consumers reason for serious consideration of this issue. Many fresh-cut fruits and vegetables or their processed products become microbially contaminated, which causes such outbreaks (Mahajan et al. 2014). Anti-microbial agents and anti-browning agents are able to eliminate microbial contamination, which causes health concerns for human beings (Mahajan et al. 2014). Chemical agents such as chlorine (Cl) or Cl-based solutions, organic acids, hydrogen peroxide (H_2O_2), peroxyacetic acid (PAA) and electrolysed water are used commercially for this purpose.

4.2.3.1.1 Chlorine-Based Solutions

Sodium hypochloride (NaOCl), a chlorine-based solution, is a commonly used disinfectant in fruits and vegetables, primarily because it is considered a potent oxidizing agent and is quite inexpensive (Mahajan et al. 2014). Its efficacy largely depends on chlorine concentrations. At higher concentrations, it may cause adverse effects on taste and odour in the treated produce. In addition, chlorine-based compounds are least effective in the reducing microbial load on fresh horticultural produce. However, detergents, surfactants and solvents, alone or in combination with physical treatment like brushing, can reduce microbial load (Mahajan et al. 2014).

4.2.3.1.2 Peroxyacetic Acid, Hydrogen Peroxide and Organic Acids

Peroxyacetic acid ($C_2H_4O_3$), also called peracetic acid, is a strong oxidizing agent and produces no harmful effects. It is effective against *E. coli* on apples, strawberries, lettuce, etc. Hydrogen peroxide (H_2O_2) exhibits both bactericidal and sporicidal ability. It also generates hydroxyl radicals (Mahajan et al. 2014). Treatment of apple, orange, grape, prune and tomato with H_2O_2 extends their shelf life and reduces fruit decay significantly. H_2O_2 treatment is generally given for a long period of time, which may cause injury to the produce (Mahajan et al. 2014). Some organic acids, vitamin C, and some Ca-based solutions are also used to slow down the browning reactions, softening and microbial load in the horticultural produce.

4.2.3.2 Nitric Oxide

Nitric oxide (NO) is regarded as a highly reactive free radical gas. It acts on several physiological processes of fruits and vegetables such ripening and senescence, etc. Endogenous concentrations of NO usually decrease with progress in maturation and senescence processes of fruits and vegetables, which can be modulated by exogenous NO application. NO delays the climacteric upsurge in many tropical fruits by impeding the ripening and senescence processes and suppressing ethylene, and delaying fruit ripening (Mahajan et al. 2014). NO gas is produced by sodium nitroprusside (SNP) and S-nitrosothiol compounds. In this process, produce is dipped in their solutions and NO gas is produced, which brings beneficial benefits as a fumigant. NO binds with 1-aminocyclopropane-1-carboxylic acid (**ACC**) and ACC oxidase, which forms a stable complex, thereby reducing ethylene production. It also inhibits ethylene biosynthesis, regulates gene expression and ameliorates oxidative stress in several fruits and vegetables. NO has been used successfully in apple, pear, peach, plum, mango, banana, papaya, kiwifruit, loquat, strawberry and tomato. However, commercial application of NO largely depends upon the development of an effective and controlled system for NO release in the produce (Mahajan et al. 2014).

4.2.3.3 Sulphur Dioxide

Sulphur dioxide (SO_2) is used commercially in grapes to prevent postharvest decay loss during storage, by either fumigation before fruit packaging or using in-package pads containing sodium metabisulphite. It has also been used to reduce postharvest fruit decay in apple, banana, litchi, blueberry and lemon. Although this is a potential technology, the recommended concentration of SO_2 may induce injury in the treated produce, and the sulphite residue on the fruits may create some health issues to the consumer. However, it is quite popular because of the multifaceted advantages of antiseptic action and cost effectiveness (Mahajan et al. 2014).

4.2.4 Emerging Technology

4.2.4.1 Cold Plasma Technology

Heat processing of food has been used for several decades and it is still a major food preservation technique used in the food industry globally. However, the use of harsh heat treatments leads to undesirable and harmful or adverse effects on nutrition and sensory parameters of processed food. In recent years, cold plasma technology has emerged as a decontaminating technique for fruits and vegetables (Sharma et al. 2018). It is a non-thermal and novel technology which appears to be an effective and alternative technique for decontamination of food and enhancement of its storage life. Plasma technology makes use of ionized gas molecules that are dissociated by an energy input. The particles are generated either at a low or a high temperature. Accordingly, it is referred to as cold or thermal plasma technology, respectively (Kudra and Mujumdar 2009). Cold plasma is generated at atmospheric pressure in which argon gas is transformed into plasma at a radio frequency of 27 MHz or by electric discharge between two electrodes, which are separated by the dielectric barriers. Microbial load in cold plasma is reduced by direct destruction of DNA via UV radiation and/or volatilization of compounds present in surface of spores by UV photons or erosion of surface of spores by adsorption of free radicals (Dey et al. 2016). This technology is quite effective against microbes such as *E.coli*, *Salmonella typhimurium*, *S.aureus* and *L.monocytogenes*. Precise understanding of cold plasma mechanism by advance research can lead to universal acceptance of cold plasma technology at commercial scale as an effective and potential alternative to other food processing technologies (Wan et al. 2009).

4.3 Conclusion

A wide range of physical, gaseous and chemical treatments are used to maintain freshness, quality and reduction of postharvest decay in harvested horticultural produce. Some of the treatments may only be applicable to a specific commodity and/or to control specific microbe. Postharvest management of produce using CA and MA storage along with appropriate management of temperature is the basis for maintaining functional, sensory and nutritional attributes of the produce and thereby reduce decay during postharvest handling or storage. These techniques can be supplemented by sanitizers such as chlorine and/or ozone or other treatments such as HWT, SO_2, irradiation, anti-microbial agents, and/or edible coatings depending upon the need of the specific horticultural produce. Some of the latest technologies include management of ethylene by oxidation and/or inhibition, and modulation of ripening using NO. However, stringent efforts are needed to focus on research for development of delivery systems, which will not only improve the efficacy of postharvest management system but also address the safety issues.

REFERENCES

Abu-Kpawoh JC, Xi YF, Zhang YZ, Jin YF (2002) Polyamine accumulation following hot-water dips influences chilling injury and decay in 'Friar' plum fruit. J Food Sci 67: 2649–2653.

Athmaselvi KA, Sumitha P, Revathy B (2013) Development of *Aloe vera* based edible coating for tomato. Int Agrophy 27(4):369–375.

Avena-Bustillos RJ, Cisneros Zevallos LA, Krochta JM, Saltveit ME (1994) Application of casein-lipid edible film emulsions to reduce white blush on minimally processed carrots. Postharvest Biol Technol 4:319–329.

Baritelle AL, Hyde GM, Fellman JK, Varith J (2001) Using 1-MCP to inhibit the influence of ripening on impact properties of pear and apple tissue. Postharvest Biol Technol 23:153–160.

Beuchat LR (1992) Surface disinfection of raw produce. Dairy Food Environ Sanitation 12(1):6–9.

Blankenship SM, Dole JM (2003) 1-Methylcyclopropene: A review. Postharvest Biol Tecnol 28(1):1–25.

Bodbodak S, Moshfeghifar M (2016) Advances in controlled atmosphere storage of fruits and vegetables. In: Eco-friendly technology for postharvest produce quality. CABI, New York, pp. 39–76.

Cantwell MI, Hong G, Suslow TV (2001) Heat treatments control extension growth and enhance microbial disinfection of minimally processed green onions. Hortic Sci 36:732–737.

Chauhan S, Gupta KC, Agrawal M (2014) Application of biodegradable *Aloe vera* gel to control post-harvest decay and longer the shelf life of grapes. Int J Curr Microbiol Appl Sci 3(3):632–642.

Cheah LH, Page BBC, Shepherd R (1997) Chitosan coating for inhibition of Sclerotinia rot of carrots. New Zeal J Crop Hortic 25:89–92.

Chen XH, Campbell CA, Grant LA, Li P, Barth M (1996) Effect of NatureSeal^{IR} on maintaining carotene in fresh-cut carrots. Proc Fla State Hortic Soc 109:258–259.

Claire PA, Weksler A, Lurie S (2003) Responses of 'Anna', a rapidly ripening summer apple, to 1-methylcyclopropene. Postharvest Biol Technol 27(2):163–170.

Dey A, Rasane P, Choudhury A, Singh J, Maisnam D, Rasane P (2016) Cold plasma processing: A review. J Chem Pharm Sci 9:2980–2984.

Dhall RK (2013) Advances in edible coatings for fresh fruits and vegetables: A review. Crit Rev Food Sci Nutr 53:435–450.

Donhowe IG, Fennema O (1993) The effects of plasticizers on crystallinity, permeability, and mechanical properties of methylcellulose films. J Food Process Pres 17(4):247–257.

Drake SR, Fellman JK, Nelson JW (1987) Postharvest use of sucrose polyesters for extending the shelf-life of stored 'Golden Delicious' apples. J Food Sci 52:1283–1285.

Droby S (2006) Improving quality and safety of fresh fruit and vegetables after harvest by the use of bio-control agents and natural materials. Acta Hortic 709:45–51.

El-Ghaouth A, Arul J, Ponnampalam R, Boulet M (1991) Chitosan coating effect on storability and quality of strawberries. J Food Sci 56:1618–1620.

El-Ghaouth A, Ponnampalam R, Castaigne F, Arul J (1992). Chitosan coating to extend the storage life of tomatoes. Hortscience 27:1016–1018.

El-Ghaouth A, Wilson CL, Wisniewski ME (2004) Biologically based alternatives to synthetic fungicides for the postharvest diseases of fruit and vegetables. In: Diseases of fruit and vegetables, Vol. 2. Kluwer Academic Publishers, the Netherlands, pp. 11–535.

Elson CM, Hayes ER (1985) Development of the differentially permeable fruit coating Nutri-Save^{R} for the modified atmosphere storage of fruit. In: Proceedings of the 4th National Controlled Atmosphere Research Conference: Controlled Atmosphere for Storage and Transport of Perishable Agricultural Commodities, Raleigh, North Carolina, pp. 248–262.

Fallik E (2003) Prestorage hot water treatments (immersion, rinsing and brushing). Postharvest Biol Technol 32(2):125–134.

Fallik E, Tuvia Alkalai S, Copel A, Wiseblum A, Regev R (2001) A short water rinse with brushing reduces postharvest losses-4 years of research on a new technology. Acta Hortic 553:413–416.

Fallik E, Aharoni Y, Yekutieli O, Wiseblum A, Regev R, Beres H, Barlev E (1996) A method for simultaneously cleaning and disinfecting agricultural produce. Israel Patent Application No. 116965.

Fan X, Mattheis JP (1999) Impact of 1-methylcyclopropene and methyl jasmonate on apple volatile production. J Agric Food Chem 47(7):2847–2853.

FAO/IAEA/WHO (1981) Wholesomeness of irradiated food. Technical Report Series 659. Geneva, Switzerland: Joint FAO/IAEA/WHO Expert Committee.

FAO/IAEA/WHO (1999) High-dose irradiation: Wholesomeness of food irradiated with doses above 10 kGy. Report of a Joint FAO/IAEA/WHO study group. World Health Organization Technical Report Series 890:1–197.

Farber JN, Harris LJ, Parish ME, Beuchat LR, Suslow TV, Gorney JR, Garrett EH, Busta FF (2003) Microbiological safety of controlled and modified atmosphere packaging of fresh and fresh-cut produce. Comp Rev Food Sci Food Saf 2:142–160.

Farkas J (2006) Irradiation for better foods. Trends Food Sci Technol 17:148–52.

Farkas J, Mohacsi-Farkas C (2011) History and future of food irradiation. Trends Food Sci Technol 22(2-3):121–126.

Fawcett HS (1922) Packing house control of brown rot. Citrograpth 7:232–234.

Gonzalez-Aguilar GA, Cruz R, Baez R, Wang CY (1999) Storage quality of bell peppers pretreated with hot water and polyethylene packaging. J Food Qual 22:287–299.

Harris DR, Seberry JA, Wills RBH, Spohr LJ (2000) Effect of fruit maturity on efficiency of 1-methylcyclopropene to delay the ripening of bananas. Postharvest Biol Technol 20:303–308.

Hofman PJ, Jobin-Decor M, Meiburg GF, Macnish AJ, Joyce DC (2001) Ripening and quality responses of avocado, custard apple, mango and papaya fruit to 1-methylcyclopropene. Aus J Exp Agric 41:567–572.

Hofman PJ, Stubbings BA, Adkins MF, Meiburg GF, Woolf AB (2002) Hot water treatments improve 'Hass' avocado fruit quality after cold disinfestation. Postharvest Biol Technol 24:183–192.

Hoof FV (1982). Professional risks associated with ozone. In: Masschelein WJ (Ed.). Ozonation manual for water and wastewater treatment. CABI, pp. 200–201.

Ilic Z, Polevaya Y, Tuvia-Alkalai S, Copel A, Fallik E (2001) A short pre-storage hot water rinse and brushing reduces decay development in tomato, while maintaining its quality. Trop Agric Res Ext 4:1–6.

Jeong J, Huber DJ, Sargent SA (2002) Influence of 1-methylcyclopropene (1-MCP) on ripening and cell-wall matrix polysaccharides of avocado (*Persea americana*) fruit. Postharvest Biol Technol 25:241–256.

Jiang Y, Joyce DC, Terry LA (2001) 1-Methylcyclopropene treatment affects strawberry fruit decay. Postharvest Biol Technol 23(3):227–232.

Kim JG, Yousef AE, Dave S (1999) Application of ozone for enhancing the microbiological safety and quality of foods: A review. J Food Protect 62(9):1071–1087.

Kudra T, Mujumdar AS (2009) Advanced drying technologies. CRC Press, Boca Raton, FL, pp.212–218.

Kumar P, Sethi S (2018) Edible coating for fresh fruit: A review. Int J Curr Microbiol App Sci 7(5): 2619–2626.

Lado BH, Yousef AE (2002) Alternative food-preservation technologies: efficacy and mechanisms. Microbiol Infec 4(4):433–440.

Lerdthanangkul S, Krochta JM (1996) Edible coating effects on post harvest quality of green bell peppers. J Food Sci 61:176–179.

Li P, Barth MM (1998) Impact of edible coatings on nutritional and physiological changes in lightly processed carrots. Postharvest Biol Technol 14:51–60.

Li H, Yu T (2000) Effect of chitosan on incidence of brown rot, quality and physiological attributes of postharvest peach fruit. J Sci Food Agric 81:269–274.

Lichter A, Dvir O, Rot I, Akerman M, Regev R, Wiseblum A, Fallik E, Zauberman G, Fuchs Y (2000) Hot water brushing: An alternative method to SO_2 fumigation for colour retention of litchi fruit. Postharvest Biol Technol 8:235–244.

Lurie S (1998) Postharvest heat treatments of horticultural crops. Hortic Rev 22:91–121.

Mahajan PV, Caleb OJ, Singh Z, Watkins CB, Geyer M (2014) Postharvest treatments of fresh produce. Philos Trans Math Phys Eng Sci 13:372–380.

Mantilla N (2012) Development of an alginate-based antimicrobial edible coating to extend the shelf-life of fresh-cut pineapple. Doctoral dissertation, Texas A&M University, US.

McDonald RE, McCollum TG, Baldwin EA (1999) Temperature of hot water treatments influences tomato fruit quality following low-temperature storage. Postharvest Biol Technol 16:147–155.

Moayednia N, Ehsani MR, Emamdjomeh Z, Asadi MM, Mizani M, Mazaheri AF (2010) A note on the effect of calcium alginate coating on quality of refrigerated strawberries. Irish J Agric Food Res 165–170.

Motlagh HF, Quantick PC (1998) Effect of permeable coatings on the storage life of fruits. Int J Food Sci Technol 23:99–105.

Nafussi B, Ben-Yehoshua S, Rodov V, Peretz J, Ozer BK, Dhallewin G (2001) Mode of action of hot-water dip in reducing decay of lemon fruit. J Agric Food Chem 49(1):107–113.

Nisperos-Carriedo MO, Baldwin EA, Shaw PE (1992) Development of an edible coating for extending postharvest life of selected fruits and vegetables. Proc Ann Meet Fla State Hortic Soc 104:122–125.

Panahirad SM, Nasser RN, Hassani B, Ghanbarzadeh N (2015) Plum shelf life enhancement by edible coating based on pectin and carboxymethyl cellulose. J Biodiver Environ Sci 423–430.

Pantastico EB (1975) Postharvest physiology, handling and utilization of tropical and subtropical fruits and vegetables. Avi Publishing Company, Inc., Westport, CA.

Park HJ, Chinnan MS, Shewfelt RL (1994) Edible corn-zein film coatings to extend storage life of tomatoes. J Food Process Preserv 18:317–331.

Paull RE, Chen NJ (2000) Heat treatment and fruit ripening. Postharvest Biol Technol 21:21–38.

Porat R, Daus A, Weiss, B, Cohen, L, Fallik E, Droby S (2000) Reduction of postharvest decay in organic citrus fruit by a short hot water brushing treatment. Postharvest Biol Technol 8:151–157.

Porat R, Weiss B, Cohen L, Daus A, Goren R, Droby S (1999) Effects of ethylene and 1-methylcyclopropene on the postharvest qualities of 'Shamouti' oranges. Postharvest Biol Technol 15:155–163.

Ranganna B, Raghavan GSV, Kushalappa, AC (1998) Hot water dipping to enhance storability of potatoes. Postharvest Biol Technol 3:215–223.

Reddy VRS, Sharma RR, Gajanan G (2018a) Use of irradiation for postharvest disinfection of fruits and vegetables. In: Siddique MW (Ed.). Postharvest disinfection of fruits and vegetables. Academic Press, London, pp. 121–136.

Reddy VRS, Sudhakar Rao DV, Sharma RR (2018b) Ozone treatments. In: Pareek S (Ed.). Novel postharvest treatments of fresh produce. CRC Press, Taylor and Francis Group, Boca Raton, FL, pp. 217–241.

Rice RG, Farguhar JW, Bollyky LJ (1982) Review of the applications of ozone for increasing storage times of perishable foods. Ozone Sci Eng 4:147–163.

Saha A, Gupta RK, Tyagi YK (2014) Effects of edible coatings on the shelf life and quality of potato (*Solanum tuberosum* L.) tubers during storage. J Chem Pharmacol Res 6:802–809.

Selvarajah S, Bauchot AD, John, P (2001) Internal browning in cold-stored pineapples is suppressed by a postharvest application of 1-methylcyclopropene. Postharvest Biol Technol 23(2):167–170.

Senna MM, Al Shamrani KM, Al Arifi AS (2014) Edible coating for shelf-life extension of fresh banana fruit based on gamma irradiated plasticized poly (vinyl alcohol)/carboxymethyl cellulose/tannin composites. Mater Sci Applic 5:395–399.

Sharma RR, Reddy VRS, Sethi S (2018) Cold plasma technology for surface disinfection. In: Siddique MW (Ed.). Postharvest disinfection of fruits and vegetables Academic Press, London, pp. 197–210.

Singh SP, Pal RK (2007) Response of climacteric-type guava to postharvest treatment with 1-MCP. Postharvest Biol Technol 47(3):307–314.

Singh D, Sharma RR (2007) Postharvest diseases of fruit and vegetables and their management. In: Prasad D (Ed.). Sustainable pest management. Daya Publishing House, New Delhi, India.

Skurtys OP, Velasquez O, Henriquez S, Matiacevich EJ, Osorio P (2005) Wetting behaviour of edible coating (*Opuntia ficusindica*) and its application to extend strawberry (*Fragaria ananassa*) shelf life. Food Chem 91(4):751–756.

Smilanick JL, Sorenson D, Mansour M, Aieyabei J, Plaza P (2003) Impact of a brief postharvest hot water drench treatment on decay, fruit appearance, and microbe populations of California lemons and oranges. Hortic Technol 13:33–338.

Song J, Fan L, Hildebrand PD, Forney CF (2000) Biological effects of corona discharge on onions in a commercial storage facility. Hortic Technol 10(3):608–612.

Tian MS, Prakash S, Elgar HJ, Young H, Burmeister DM, Ross GS (2000). Responses of strawberry fruit to 1-methylcyclopropene (1-MCP) and ethylene. Plant Growth Regul 32:83–90.

USDA. Code of Federal Regulations, Title 9, Part 381.66—poultry products; temperatures and chilling and freezing procedures. (1997). Office of the Federal Register National Archives and Records Administration, Washington, DC.

Valverde JM, Valero D, Martinez Romero D, Guillen F, Castillo S, Serrano M (2005) Novel edible coating based on *Aloe vera* gel to maintain table grape quality and safety. J Agric Food Chem 53(20):7807–7813

Wan J, Conventry J, Swiergon P, Sanguansir P, Versteeg C (2009) Advances in innovative processing technologies for microbial inactivation and enhancement of food safety pulsed electric field and low-temperature plasma. Trends Food Sci Technol 20:414–424.

Warriner K, Huber A, Namvar A, Fan W, Dunfield K (2009) Recent advances in the microbial safety of fresh fruits and vegetables. Adv Food Nutr Res 57:155–208.

Woolf AB (1997) Reduction of chilling injury in stored 'Hass' avocado fruit by 38°C water treatments. Hortic Sci 32:1247–1251.

Yulianingsih R, Maharani DM, Hawa LC, Sholikhah L, Ukpabi UJ, Chijioke U, Helmiyati S (2013) Physical quality observation of edible coating made from *Aloe vera* on cantaloupe (*Cucumis melo* L.) minimally processed. Pak J Nutr 12(9):800–805.

Zhang DL, Quantick PC (1997) Effects of chitosan coating on enzymatic browning and decay during post-harvest storage of litchi fruit. Postharvest Biol Technol 12:195–202.

Zhu SJ (2006) Non-chemical approaches to decay control in postharvest fruit. Advances in postharvest technologies for horticultural crops. Research Signpost, Trivandrum, India, 297–313.

Zisheng L (2007) Effect of 1-methylcyclopropene on ripening of postharvest persimmon fruit. Food Sci Tech 40(2):285–291.

5

Management of Postharvest Diseases of Fruits and Vegetables through Chemicals

Dinesh Singh and Ravinder Pal Singh
Division of Plant Pathology, ICAR-Indian Agricultural Research Institute, New Delhi, India

CONTENTS

5.1 Introduction

Fruits and vegetables are important because of their potential use for nutrition and as antioxidants and dietary fibre supplements. They provide major dietary sources of vitamins, sugar, organic acid and minerals which have beneficial health effects. An additional characteristic of fruits and vegetables is that they are available in various shapes, colours, aromas, flavours and qualities that provide good sensory experience in human diet. There is increasing awareness among consumers of the need for fresh produce consumption rather than processed products because of the health benefits, as demonstrated by medical research, resulting in positive support by the media and health agencies. The perishable nature of fruits and vegetables makes them susceptible to loss in postharvest management. Freshly produced horticultural products are imperative at the international level due to free trade agreements and globalization of trade. Prolonged transportation and supply phases generate heavy losses. Therefore, it is important to care for fresh produce by using advanced techniques after harvesting (Wu 2010).

Most fruits have resistance to fungal attack when unripe, and the process of infection is stopped almost as soon as it begins, but the spores of fungal organisms remain alive and enter a quiescent phase. Quiescent infections, however, are macroscopically visible, although mycelia development is arrested after infection and resumes only as the host reaches maturity or senescence. Some postharvest spoils result from pre-harvest inactive diseases, particularly in tropical and subtropical locales where natural conditions in the field are especially conducive to organic product contamination. The restraining of rot developed from pre-harvest hidden infections is difficult with postharvest treatments. However, a few reports have shown successful control of latent infections by postharvest applications (Spadaro 2011).

Postharvest management in fruits and vegetables provides viable and plausible methods of dealing with advancements that delay the pace of senescence in the collected plant materials. Postharvest management also plays an important role in the time period between the farmer and consumer. The use of advanced methods of postharvest management helps to reduce the losses between harvest and utilization of produce. It also maintains the quality and texture of food.

The perishable nature of fresh fruits and vegetables is imposed by their high moisture content, active metabolism and ample nutrients. These attributes make them susceptible to dehydration, mechanical injury, pathological breakdown and environmental stresses. Postharvest losses can occur at any time during the production and marketing of fruits and vegetables.

Postharvest losses of perishables in developing nations have been evaluated to be in the range of 5–50% or greater. Losses of 20–30% in fruits and vegetables due to postharvest diseases were estimated by various researchers. In India, almost 20–50% of transient perishables are lost due to postharvest infections. It is evaluated that the major causes of these losses are due to improper postharvest handling. During transportation and storage stage, 20–50% higher losses occur in developing nations, while 5–25% losses occur in developed nations. Understanding the cause of fast deterioration in fruits and vegetables helps to determine effective strategic technologies to facilitate delay senescence and conserve product quality (Wu 2010).

5.2 Pathogens Causing Postharvest Diseases

A few microorganisms, bacteria and fungi, are capable of causing diseases in fruits and vegetables. However, it is notable that the major postharvest losses are caused by fungi, such as *Alternaria*, *Aspergillus*, *Botrytis*, *Colletotrichum*, *Diplodia*, *Monilinia*, *Penicillium*, *Phomopsis*, *Rhizopus*, *Mucor*, *Sclerotinia*, and bacteria, such as *Erwinia* and *Pseudomonas* (Table 5.1) (Sharma et al. 2009). Most of these organisms are weak microbes in that they can only attack already damaged produce. Physical damage provides a significant role in postharvest deterioration and is the main cause of losses. Different types of injuries can be reported before and after the harvest of fruits and vegetables. Climate, creepy crawlies, flying creatures, rodents and farm implements can cause injury. Wounds to fruits usually occur when produce is dropped onto a solid surface before or after packing, but injury is not always immediately visible. Wounding may also occur later, but is just seen remotely (e.g. apple) or might be clearly seen just on stripping (e.g. potato). Pressure wounding may result from the overloading of mass produce in storage houses or from stuffing of the bundling/packaging (e.g. grape). Vibration damage can occur in under-filled packs, particularly during significant distance road transportation. Different microorganisms attack the damaged produce, bringing about a dynamic rot which may influence the whole batch.

5.3 The Infection Process

Microorganisms either contaminate the produce while still juvenile on the plant (precollect disease) or during the harvesting and ensuing activities (postharvest disease). The postharvest disease process is enormously aided by mechanical wounds to the peel of the produce; for example, fingernail

TABLE 5.1

Chemicals Used for Control of Postharvest Diseases of Fruits and Vegetables

Fungicide	Group of Chemicals	Crops	Postharvest Disease (Pathogen)	References
Biphenyl	Fungicides (aromatic hydrocarbons)	Citrus	Green (*Penicillium digitatum*) and blue (*P. italicum*) mould rot	Nagy et al. (1982)
Dicloran	Chlorophenyls, nitroanilines	Stone fruits	Soft-watery rot (*Rhizopus stolonifer*)	Ravetto and Ogawa (1972)
		Kiwifruit	*Alternaria* and *Botrytis*	Heggen et al. (1980); Sommer et al. (1983)
		Tomato	*Botrytis* rot (*B. cinerea*)	Chastagner and Ogawa (1979)
Sodium orthophenylphenate	–	Citrus	Green (*P. digitatum*) and blue mould rot (*P. italicum*)	Eckert (1978)
		Litchi	Fruit rot	Prasad and Bilgrami, (1973)
		Kiwifruit	*Alternaria alternata* and grey mould (*B. cinerea*)	Heggen et al. (1980); Sommer et al. (1983)
		Tomato	*Geotrichum* and *Rhizopus*	Krochta et al. (1977)
		Carrot	*Botrytis* and other storage rots	Wells and Merwarth (1973)
Sec-butylamine	–	Citrus	Green mould rot (*P. digitatum*)	Eckert (1977)
		Apple	*P. expansum* and *Gloeosporium*	Eckert and Kolbezen (1970); Little et al. (1980)
		Blueberry	*Botrytis*, *Alternaria* and *Colletotrichum*	Ceponis and Cappellini (1978), (1985)
Sulphur dioxide	–	Grape	Grey mould	Ahmed et al. (2018)
Sodium lauryl sulphate	–	Tomato	*Geotrichum* and *Rhizopus*	Krochta et al. (1977)
Chlorothalonil	Chloronitriles (phthalonitriles)	Tomato	*Alternaria alternata* and grey mould (*B. cinerea*)	Eckert and Ogawa (1988)
Thiabendazole and carbendazim (>500 mg/L)	Benzimidazoles	Stone fruits	Brown rot (*M. fructicola*)	Eckert (1990)
		Apple, pear	Blue mould (*Penicillium expansum*), grey mould (*Botrytis cinerea*) and lenticel rot (*Gloeosporium* spp.)	Hardenburg and Spalding (1972); Sitton and Pierson (1983)
		Citrus	Green and blue mould rot and stem-end rot	Eckert et al. (1979)
		Citrus	Stem-end rot	Brown (1983)
		Mango, banana, papaya	Anthracnose	Spalding and Reeder (1972)
		Banana	Crown rot	Shillingford (1977)
		Pineapple	Black rot (*Ceratocystis paradoxa*)	Frossard (1968), (1970)
		Mango	Anthracnose	Spalding and Reeder (1975)
		Papaya	*Colletotrichum*, *Ascochyta*, and *Gloeosporium* rots	Balkan et al. (1976)
		Avocado	Stem-end rot (*Botryodiploida*, *Alternaria* and *Dothiorella*)	Zauberman et al. (1975)
	Thiabendazole	Peach, nectarines, sweet cherry	Brown rot (*M. fructicola*)	Daines (1970); Koffman and Kable (1975)
		Papaya	*Colletotrichum*, *Ascochyta* and *Gloeosporium*	Balkan et al. (1976)
Thiophanate-methyl	Thiophanates	Banana	Crown rot and anthracnose	Griffee and Pinegar (1974); Long (1971)
		Citrus	*Penicillium* decay and *Diplodia* stem-end rots	Peiser (1973)
Guazatine	Guanidines	Citrus	*Penicillium* spp., sour rot (*Geotrichum candidatum*)	Brown (1979); Eckert et al. (1981)
			Diplodia stem-end rot	Brown (1983)
		Melon	Sour rot and Alternaria rots	Morris and Wade (1983)
		Tomato	*Mucor mucedo* and *M. piriformis*	Moline and Kuti (1984)
			Fusarium semitectum	Roy (1981)

(*Continued*)

TABLE 5.1 *(Continued)*

Chemicals Used for Control of Postharvest Diseases of Fruits and Vegetables

Fungicide	Group of Chemicals	Crops	Postharvest Disease (Pathogen)	References
Iprodione	Dicarboximides	Apple	Penicillium rot (*P. expansum*)	Heaton (1980)
		Apple, pear	*Gloeosporium*, *Botrytis* and *Penicillium*	Little et al. (1980)
		Stone fruits	Brown rot (*M. fructicola)* and *Rhizopus* rots (*R. stolonifer*)	Heaton (1980)
			Rhizopus, *Alternaria*, *Botrytis* and *Monilinia*	Jones et al. (1973); Jones (1975)
		Mango	*Alternaria* black spot	Prusky et al. (1983)
		Strawberry, blueberry	*Botrytis* rot (*B. cinerea*)	Dennis (1983); Bartz and Eckert (1987)
Imazalil	Imidazoles	Apple, pear, persimmon	*Alternaria* rot	Eckert and Ogawa (1985)
		Apple, pear	*Gloeosporium*, *Penicillium* and *Botrytis*	Little et al. (1980)
		Pome fruits	Blur mould rot (*P. expansum*)	Prusky and Ben-Arie (1985)
		Stone fruit	Brown rot (*M. fructicola*)	Bompeix et al. (1979); Dijkhuizen et al. (1983)
		Persimmons	Black spot (*A. alternata*)	Prusky et al. (1981)
		Citrus	Green (*P. digitatum*) and blue (*P. italicum*) mould rot	Gutter et al. (1981)
		Tomato	Grey mould (*B. cinerea*)	Maas and MacSwan, (1970)
		Banana	Crown rot and *Colletotrichum* rot	Frossard et al. (1977)
		Mango	Anthracnose and stem-end rots	Spalding (1982)
		Tomato, pepper	*A. alternate*	Miller et al. (1984); Spalding (1980)
Vinclozolin	Dicarboximides	Strawberry, blueberry	*Botrytis* rot (*B. cinerea*)	Dennis (1983); Bartz and Eckert (1987)
		Cabbage	*Botrytis* and *Alternaria*	Jennrich (1985)
Metalaxyl	Acylalanines	Citrus	Brown rot, Phytophthora rot	Eckert et al. (1981); Cohen (1981)
		Apple	Phytophthora rot *(Phytophthora syringae*)	Edney and Chambers (1981)
Fosetyl aluminium	Ethyl phosphonates	Citrus	Brown rot (*Phytophthora parasitica*, *P. citrophthora*)	Eckert et al. (1981)
			Green mould (*P. digitatum*)	Gaulliard and Pelossier (1983); Gutter (1983)
		Apple	Phytophthora rot (*P. syringae*)	Edney and Chambers (1981)
Prochloraz	Imidazoles	Mango, papaya	Anthracnose and stem-end rot	Muirhead et al. (1982)
		Citrus	*Penicillium digitatum* and *P. italicum*	Brown (1981), (1983)
			Diplodia and Phomopsis stem-end rot	Brown (1981), (1983)
		Papaya	Anthracnose and stem-end rot	Muirhead et al. (1982)
		Avocado	Anthracnose and stem-end rot (*Botryodiploida*, *Alternaria*, and *Dothiorella*)	Muirhead et al. (1982); Schaffer et al. (2013)
		Stone fruit	Brown rot (*M. fructicola)*	Bompeix et al. (1979); Dijkhuizen et al. (1983)
		Persimmon	Black spot (*A. alternata*)	Prusky et al. (1981)
Fludioxonil	Phenylpyrroles	Tarocco orange	*Penicillium* rot (*P. digitatum*)	Schirra et al. (2005)
		Apple	Blue mould rot (*P. expansum*)	Adaskaveg and Forster (2004); Errampalli (2004)
Fenpropimorph and flutriafol		Lemon	Sour rot, blue mould and green mould	Cohen (1989)
Pyrimethanil	Anilino-pyrimidine	Apple	Blue mould (*Penicillium* spp.) and grey mould (*B. cinerea*)	Sholberg et al. (2005)
		Lemon	Green mould (*P. digitatum*)	Kanetis et al. (2007)

TABLE 5.1 *(Continued)*

Chemicals Used for Control of Postharvest Diseases of Fruits and Vegetables

Fungicide	Group of Chemicals	Crops	Postharvest Disease (Pathogen)	References
Etaconazole	Triazoles	Citrus	Sour rot and Penicillium decay	Brown (1979), (1983); Eckert et al. (1981)
			Diplodia and Phomopsis stem-end rots	Brown (1981), (1983)
		Mango	Anthracnose and stem-end rots	Spalding (1982)
		Tomato, pepper	*A. alternata*	Miller et al. (1984); Spalding (1980)
Propiconazole	Triazole	Peach, nectarine	Sour rot (*Geotrichum candidum*)	Yaghmour et al. (2012)
		Citrus	Sour rot and Penicillium decay	Eckert et al. (1981); Worthing et al. (1983); McKay et al. (2012)
Azoxystrobin + fludioxonil	–	Lemon	Green mould (*P. digitatum*)	Kanetis et al. (2007); McKay et al. (2012)
Imazalil + pyrimethanil		Lemon	Green mould (*P. digitatum*)	Kanetis et al. (2007)

scratches and scraped areas, improper handling, bug punctures and cut stems. The disease may occur by direct infiltration of the cuticle or passage through stomata, lenticels and injury or abscission scratch tissue. In addition to the physiological state of the produce, the temperature and the arrangement of the periderm influence these two types of infection, that is pre-harvest and postharvest disease.

5.3.1 Pre-harvest Infection

Pre-harvest contamination of fruits and vegetables may take place through a few avenues; for example, direct infiltration of the peel, disease through normal openings on the produce and infection through damaged parts. A few types of pathogenic fungi can start the infection process on the outside of floral parts, and on well, developing fruits. The contamination is then captured, which stays quiet until after harvest, when the opposition of the host diminishes and conditions become great for the development of the microbe, that is when the fruit begins to ripen or its tissues become senescent. Such 'inert contaminations' are significant in the postharvest damage of numerous tropical and subtropical fruits; for example, anthracnose of papaya and mango, stem-end decay of citrus, and crown decay of banana. For instance, spores of *Colletotrichum* grows in dampness on the outer surface of fruits and the end of the germ tube swells within a few hours of germination and forms a structure known as appressorium, which could conceivably infiltrate the fruit peel before the disease is stopped.

5.3.2 Postharvest Infection

Numerous fungi that cause significant wastage of produce cannot infiltrate the unblemished peel of produce, yet they promptly attack by means of any break point in the peel. The damage is infinitesimal but it is adequate for microbes to develop on the crop. In addition, the cut stem is a frequent point of entry for microorganisms, and stem-end decay is a significant type of postharvest waste of numerous fruits and vegetables. For instance, postharvest disease by *Sclerotina* and *Colletotrichum* is common in numerous fruits through direct infiltration of the peel. For disease of postharvest produce, various parts of the plants are tainted, i.e. flower disease, stem-end infection and quiescent contamination, as given below.

5.3.2.1 Floral Infection

Disease by microorganisms can occur through floral parts in numerous fruits. For instance, *Botrytis cinerea* on black current and raspberry, *Monilinia laxa* on plums and *Lasiodiplodia theobromae* on citrus fruits contaminate the produce at floral parts. In anthracnose of mango, an extra-fruit contamination may emerge from quiescent diseases at the base of the ovary.

5.3.2.2 Stem-End Infection

Endophytic colonization of the inflorescence is a significant method of infection for mango by *Dothiorella dominicana*. Colonization by stem-end decay fungi, *Lasiodiplodia theobromae* and *Phomopsis citri*, of the peduncle and pedicel of citrus is confined by wound cuticle and periderm. This pathogen does not enter the fruit until abscission occurs. Postharvest management with the development of the hormone 2,4-dichlorophenoxyacetic (2,4-D) corrosive was consequently acquainted to prevent abscission of the buttons (Eckert and Eaks 1989).

5.3.2.3 Quiescent Infection

The period between introductory contamination and presence of disease indication is known as the inactive or quiescent stage. The term alludes to a 'quiescent' or lethargic parasitic relationship, which after a period of time changes to a functioning one. A fungus may quiescent at commencement of germination, germ tube stretching, aspersorium development, infiltration or ensuing colonization. According to Swinburne (1973), the inability to grow or to develop beyond any appearing stage is due to unfavourable physiological conditions incidentally forced by the host, either straight forwardly on the microorganism or in a roundabout way by adjustment in its pathogenic capacity.

5.4 Factors Involved in Postharvest Diseases of Fruits and Vegetables

A significant factor for crumbling of produce is uninterrupted metabolism, development, water misfortune, physiological issues, mechanical harm and pathological crash. The appropriate conditions of produce provide a significant role being developed of contamination by the microbes and reduce postharvest wastage of the produce. The various types of factors that support the postharvest infections are presented here.

5.4.1 Mode of Infection

The advancement of postharvest disease can occur before harvest, during harvest or after harvest. The diseases that exist during handling and remain until aging of produce are common among tropical fruit crops. For instance, a wide range of tropical and subtropical fruits, including mango, avocado, papaya and banana, are contaminated by anthracnose that emerges from taciturn diseases (latent infection), which occurred before harvest. Anthracnose is also an important postharvest disease of a number of vegetables (e.g. bean) and temperate fruits (e.g. strawberry). Various species of *Colletotrichum* can cause anthracnose. Some species (e.g. *C. musae*) are host-specific, while others can assault a wide range of vegetables and fruits (e.g. *C. gloeosporioides*). The development of conidia on the cuticle of the host tissue is the beginning of infection, which initially produces a germ tube and appressorium to develop the symptoms of the disease, and fungus shows a quiescent life cycle during infection process. However, some studies represent the conflicts related to the quiescent stage of the fungus. Furthermore, the behaviour of pathogenic fungi in terms of the host is different, which restricts the detection of appressorial germination through ancient techniques. For instance, earlier studies in avocado reported the ungerminated appressoria as the quiescent phase of *C. gloeosporioides*. The extensive research in the earlier two decades reveals that the infection will begin after the germination of appressoria, and after the development of appressoria the pathogenic fungus stops its growth and enters into a quiescent phase until the ripening of fruit is not begins. However, colonization of fruit tissue begins during ripening, which leads to the formation of anthracnose symptoms.

5.4.2 Continuous Metabolism and Growth of Vegetables and Fruits

The perishable nature of vegetables and fruits is due to their living biological systems and promotes metabolic and developmental changes even after harvesting. Improved postharvest handling techniques and operations are useful to retain the quality of vegetables and fruits until their utilization. The reduction of metabolic activities of vegetables and fruits also improves their value. In regard to the metabolic action of green products, the respiration rate is related to commodity decay.

5.4.2.1 Respiration

Respiration is involved in enzymatic oxidation of natural substrates with energy production, resulting in O_2 utilization and production of CO_2 and water, present in all of the metabolic activity of the tissue. The respiration pace of harvest products is the inversion extent of the capacity of life, which exhibits that a higher respiration rate will decrease storability of the product. This is because of the utilization of reserved food stores by produce. Moreover, outside variables such as temperature and other physiological boundaries also affect the respiration rate of produce after harvest.

5.4.2.2 Temperature Quotient (Q_{10})

The portion of the rate of an effect at one temperature (T_1) versus the rate at that temperature plus 10°C [(rate at T_1 + 10°C)/ rate at T_1] is the determination usually used for respiration to give a general evaluation of the impact of temperature on the general metabolic rate of produce. Within the 5–25°C range, the speed of respiration expands 2- to 2.5-fold for each 10°C ascent in temperature for nearly all harvested produce; i.e. Q_{10} = 2.0–2.5. The gas composition, for example, ethylene, O_2 or CO_2, encompassing the agricultural produce, likewise applies an incredible effect on both its respiratory and general metabolic rate. Postharvest techniques are primarily concerned about hindering the rate of respiration for maintaining quality and boosting storage life.

5.4.2.3 Ethylene Production

- Ethylene is a simple olefin and also a gaseous plant hormone that assumes a significant role in the elicitation of the respiration and senescence processes in harvested vegetables and fruits (under 0.1 μL/L). Ethylene production in higher plants begins from methionine via the following reactions.
- Methionine conversion by S-AdoMet synthetase into S-adenosyl-L-methionine (S-AdoMet).
- S-AdoMet converts into 1-aminocyclopropane-1-carboxylic acid (ACC) which is a quick precursor of ethylene and a rate-limiting step catalysed by ACC synthase (ACS) in this pathway.
- Degradation of ACC by ACC oxidase (ACO) to deliver ethylene. The physiological reactions (fruit ripening, senescence, abscission of plant organ and chlorophyll degradation) of harvested products are regulated through ethylene.

In addition, the presence of ethylene can be unsafe to fresh vegetables and fruits.

5.4.2.4 Ripening

The breakdown of the cellular probity of tissues in many fruit products demonstrates the ripening which is a process of senescence. The ripening of fruits is a part of the developmental phase representing maturation after harvest. During the transformation of unripe fruit into a ripening product, several biochemical and physiological parameters play an important role to change the quality of products, including change of colour, firmness, flavour and aroma. The ripening of fruits is categorized into non-climacteric or climacteric, which is based on environmental conditions.

Climacteric fruits such as banana, apple and tomato are defined by the dramatic expansion of respiration rate and ethylene discharge during maturity. In **non-climacteric fruits** (citrus, grape and strawberry), respiration declines steadily without the significant role of ethylene production. The important role of the physiological nature of produce is important in postharvest technology. For instance, the longer the storage time during marketing, climacteric fruits are harvested at the ripening stage, which promotes normal ripening after harvesting.

Inversely, non-climacteric fruits are stored after fully ripening. These fruits produce a lower amount of ethylene after harvesting. The control of ethylene is very helpful to increase the shelf life life of both non-climacteric and climacteric fruits.

5.4.2.5 Water Loss

The percentage of water in vegetables and fruits is ~80–90% as per fresh weight. Most of the parts of the plant are very sensitive to water loss. The minimum loss of water content, i.e. up to 5%, has had unfavourable consequences for appearance, attractive weight and surface nature of numerous perishable goods. In this way, the main cause of deterioration is desiccation that occurs during the postharvest period. The major reason behind loss of moisture of fresh horticultural products is stomatal respiration. However, some other types of water misfortune incorporate stem alarm, lenticels and cracks due to mechanical damage. The transpiration rate is influenced by many morphological and anatomical conditions such as ratio of surface area, ripening stage and surface damage. For instance, the larger the surface of produce, these vegetables will easily lose a greater percentage of their water content.

5.4.2.6 Physiological Disorders

Physiological disorders of vegetables and fruits originated from inexpedient environmental conditions during the post- and pre-harvesting periods. Lower temperatures can also create respiratory or nutritional disorders in harvested produce. Inappropriate temperatures may aggravate the typical metabolism of the preserved products. Chilling injury (CI) is caused by low temperatures, i.e. <10–13°C, yet non-frigid temperature is watched regularly with tropical- and some subtropical-origin vegetables and fruits. CI to plants produce potential symptoms such as lesions on the surface, discoloration of internal and external tissue, accelerated decay, water-soaking of tissues and abnormal ripening. The manifestations of CI may not be obvious while the produce is held at chilling temperature; however, it is observable subsequent to being moved to room temperature. Freezing injury occurs because of keeping the products below their frigid temperatures. Ice crystals are formed in very low temperature which breakdown the tissues and thus cause the total loss of the produce.

Nutritional misfortune that began from pre-harvest, mineral lopsidedness sometimes show up after harvest. Calcium is related to more postharvest-related insufficiency issues than some other minerals. Harsh pit of apples and bloom end decay of tomato are notable calcium lack symptoms in horticultural crops. Respiratory issues are related to exceptionally low O_2 (<1%) and additionally high CO_2 (>20%) concentrations in or around harvested produce in packaging/storage condition.

5.4.2.7 Mechanical Damage

Mechanical injury of vegetables and fruits as a consequence of inappropriate harvesting and handling is one of the most widely recognized and serious deformities of horticultural products. It not only influences appearance properties (skin and flesh lesions and browning) but also provides sites for water loss and microbe infection. Moreover, physical injury animates respiration in plant tissues and ethylene production, which can prompt senescence.

5.4.2.8 Pathological Decay

As noted above, vegetables and fruits contain a wide range of natural substrates and high water activity and thus are susceptible substrates to microbial decay. Bacteria and fungi colonization and disease development cause a significant loss of fresh produce. The acidic tissue of fruits prompts their decay predominately by fungi, though both fungi and bacteria generally attack vegetables with a pH above 4.5. The most widely recognized pathogens causing spoilage in vegetables and fruits are of fungal species, including *Botrytis*, *Alternaria*, *Collectotrichum*, *Diplodia*, *Monilinia*, *Botryosphaeria*, *Penicillium*, *Sclerotinia Rhizophus*, and *Phomopsis*, and the bacterial species such as *Erwinia* and *Pseudomonas*. It has been observed in fruits and vegetables during postharvest life that there is considerable resistance to potential pathogens. During senescence, stresses or maturity stage, due to CI and mechanical damage the produces are easily contaminated by microbes. Please check and suggest. Albeit most microorganisms absolutely depend on physical injury or physiological breakdown of the commodity to attack the host tissues, some – for example *Colletotrichum* – are able to effectively infiltrate the skin of healthy product. Microbial disease can occur before and after harvest. Latent/quiescent infection is the period in which a product is infected before harvest with no noticeable symptoms developing until the pathogens are reactivated by the onset of favourable conditions, such as fruit maturity or appropriate temperatures. Diseases with latent infection, for example anthracnose diseases of tropical natural products brought about by *Colletotrichum gloeosporioides*, frequently cause fast and sever postharvest rot since the contaminated produce cannot be sorted through effectively before storage.

5.4.3 Environmental Conditions

The environmental conditions are suitable factors for pathogen replication and disease development. The pathogen easily causes disease during wet, warm and humid climatic conditions. In addition, the postharvest environmental factors including temperature, moisture and air also influence the initiation and development of symptoms of infectious fruits.

5.4.3.1 Temperature

Temperature is a significant factor that plays the main role in controlling postharvest diseases. Generally, a lower temperature is used to store vegetables and fruits, which also reduces the chances of growth of microorganisms. High temperatures and humidity favour the decay of postharvest produce. However, the storage temperature of postharvest produce is an important environmental factor that helps produce to develop resistance against plant pathogens.

5.4.3.2 Humidity

An important and effective environmental factor is relative humidity, which is required to keep and maintain balance in harvested fruits and vegetables during storage. The saturated atmosphere or moisture on the fruit surfaces provides favourable conditions for the germination of some fungal organisms that also facilitates direct penetration into fruit. High humidity also supports the development of fungal diseases with favourable temperatures. However, harvested fresh fruits and vegetables require optimum humidity levels at nearly 95% relative humidity to hold moisture and avoid loss of tissue turgidity and shrinkage.

5.4.3.3 Maturity

Fruits and vegetables are generally preserved prior to their maturity in order to secure adequate time for marketing and long-distance transportation. Ripening of fruit makes it susceptible to colonization by a range of fungi. During storage conditions, fruits and vegetables do not have their own mechanisms to protect themselves. Therefore, the spores of fungal organisms are deposited on the surfaces of fruits during the growing season and are a reason for decay of the fruits during storage. However, specific plant pathogens easily make ripe fruits and vegetables susceptible to invasion including bacteria and fungi, due to sufficient moisture and rich sources of nutrients. In this context, an additional cause of infection is full maturation. Full maturation helps pathogens such as fungi (*Botrytis cinerea*, *Monilinia* spp., *Penicillium* spp. or *Rhizopus* spp.) to injure fruits and vegetables easily.

5.4.4 Wounds and Bruises

The harvesting of fresh fruit and vegetables creates wounds, including cuts, punctures, bruises and abrasions, which are primary sites of infection by fungal spores. Other sources of injury include climate, creepy crawlies, feathered creatures, rodents and farm implements. However, damage that can occur due to handling of produce before, after and during packaging is also a factor in initiation of infection. Afterward, wounding may occur but it is seen just externally (e.g. apple) or it might be apparent just on stripping (e.g. potato). Pressure wounding may result from the overstocking of bulk produce in storage houses or from stuffing of the bundling (e.g. grape). Another important cause of damage to the produce is vibration via mechanical means or by road transportation. Vibration can create damage to postharvest produce and can open sites of infection by microorganisms, resulting in progressive decay. Fresh injuries or wounds also promote germination of fungal spores by supplementation of nutrients and required humidity. For instance, substantial moisture helps to germinate the conidia of *Monilinia fructicola*.

5.5 Management of Postharvest Diseases Through Chemicals

5.5.1 Synthetic Fungicides

Synthetic fungicides are utilized to control the postharvest infection of vegetables and fruits as another alternative, and natural or synthetic chemicals should be used. There are a few fungicides used to guarantee the quality of fruits over extended periods. According to Sholberg and Gaunce (1996), acidic acid controls *Penicillium* and *Botrytis* rot and in addition it also lessens the break of berry in two separate years similarly. An extra impact of using 8 mg/mL acidic acid in changed packaging for 74 days at 0°C was accounted for by Moyls et al. (1996) which indicated the decrease in percentage rot of grape from 94% in the control to 2%.

The choice of an antimicrobial compound for a particular postharvest application relies on (a) the effectiveness of the microorganism against the chemical agent; (b) the capacity of the agent to infiltrate through surface hindrances of the host to the contamination site; and (c) the tolerant limit of the crop to the chemical agent, from the standpoint of both phytotoxicity and any adverse effect upon the quality of the product. Fungicides might be applied to harvested products either in a vaporous state or blended in a liquid formulation, depending on the physical and chemical properties of the fungicide, and similar treatment must be taken with care for the harvest (Eckert and Ogawa 1985).

In recent years, a few acceptable fungicides (azoxystrobin, fludioxonil, trifloxystrobin, pyrimethanil and cyprodinil) that created an insignificant hazard to human and natural well-being have been enlisted for the control of citrus postharvest decay. The adequacy of fludioxonil tested on Tarocco orange fruit applied 24 hours before with *P. digitatum* showed high therapeutic activity in the fruit of many cultivars, harvested at various degrees of development (Schirra et al. 2005). Discovering options that are generally acknowledged and financially suitable has been a challenge (Wisniewski et al. 2016). Elective techniques of fungicide treatments incorporate the use of physical treatments, generally recognized as safe (GRAS) compounds, natural antimicrobial compounds and biological control agents. Their inhibitory impact on decay control is for the most part a result of direct restraint of microorganisms. Moreover, a few of these methodologies have indicated the capacity to upgrade barrier instruments on citrus fruit tissues, inducing infection resistance and making an important contribution to *Penicillium* rot control. All things considered, because of low persistence and lack of preventive impact on previous diseases, the substitute strategies should be used in permutation, as a feature of a coordinated administration program, able to further reduce postharvest rot and to expand the fruit's shelf life.

5.5.1.1 Pre-harvest Chemical Treatments

In most cases, control of postharvest diseases should start before harvest in the field or orchard. Control of postharvest disease by postharvest disinfection is not very effective because most fungicides do not penetrate deeply into tissues due to the reduced concentration of fungicides in deep-seated infections. Therefore, the use of broad-spectrum fungicides is an effective strategy to reduce postharvest infections in fruits developed in the field. In addition, germination of fungal spores and formation of appressoria on fruits are controlled by spraying with fungicides (Singh and Thakur 2003). For control of brown rot caused by *M. fructicola*, propiconazole in a different formulation has been and continues to be used as a pre-harvest treatment. A pre-harvest mist/spray application of 8 mM SA (salicylic acid) or 3 mM MeJ (methyl jasmonate) to 'Lane Late' orange proved to be effective in reducing colony/lesion diameter, wound rotting and spore mass density of *P. digitatum* in contrast to postharvest treatment. We accept this is the main report showing the adequacy of two natural elicitors in controlling the green form in sweet orange (Iqbal et al. 2012).

Standard treatments for control of postharvest infection of papayas are field sprays of chlorothalonil, mancozeb or benomyl applied to the fruit on a 10- to 14-day plan throughout the growing season, and hot-water dipping after harvest. Biweekly orchard sprays control inert infections of *Colletotrichum* and disease by other organisms, for example, *Phytophthora* and *Alternaria*. The field sprays also diminish the degree of inoculum of *Botryodiplodia*, *Mycosphaerella* and other wound pathogens (Alvarez et al. 1977; Balkan et al. 1976). These fungicide sprays are essential to diminish disease pressure on the fruit during harvest; the hot-water treatment alone cannot provide satisfactory control of fruit decay during postharvest (Alvarez et al. 1977; Aragaki et al. 1981). In areas that have a significant issue with lenticel spoils during storage in pears and apples, the orchards are sprayed with fungicides in the late summer to reduce the production of inoculum of *Nectria galligena* and *Gloeosporium* spp. and to guard the fruit lenticels from inert infection (Corke and Sneh 1979; Dennis 1983).

Pristine (a premixed formulation of boscalid and pyraclostrobin) as a pre-harvest management was assessed for control of postharvest blue and grey mould in cultivars Fuji and Red Delicious apples during 2004–2006. Perfect (0.36 g/L of water) was applied 1, 7 or 14 days before harvest. Perfect applied inside about 14 days before gather might be a powerful option to postharvest fungicides for control of postharvest grey and blue mould in Fuji and Red Delicious apples (Xiao and Boal 2009).

In many cases, control of postharvest infection should begin prior to harvest in the field or orchard. The chance of controlling well-settled microbes by postharvest disinfection is exceptionally less since most fungicides cannot enter deeply into the tissues, and powerful convergences of the fungicide would not arrive at deeply situated diseases. Consequently, the viable method to decrease diseases started in the field, including quiescent contaminations, is the application of broad-range defensive fungicides to the fruits on the plant itself.

Pre-harvesting spray of citrus fruits in the groove with fixed copper compounds to reduce early infections of brown rot (*Phytophthora citrophthora*) in the fruit strip is quite common. For this situation, the preventive spray controls the infiltration of the fungal zoospores into the fruits, which are on the tree. Results of the preventive sprays rely upon the duration of disease; since germination of the zoospores relies upon water, the sprays should be applied preceding the rains. Defensive sprays in the plantation have been broadly used to control anthracnose (*C. gloeosporioides*) in different subtropical and tropical fruits, since this fungus enters into the mature fruits on the tree, and builds up a quiescent disease. Mancozeb spray on papaya in fields has been accounted to diminish postharvest Rhizopus delicate decay, presumably by lessening field-initiated fruit infection caused by *Phomopsis* and *Colletotrichum* species; injury brought about by these fungi may serve as courts of infection for *R. stolonifer*, which expects wounds to enter the host.

5.5.1.2 Sanitation

Vegetables and fruits that are damaged at the time of harvesting, packaging, transportation and have prevailing with regard to maintaining a strategic distance from disease caused by microbes are still at risk of coming into contact with microorganisms during storage and packaging. Since ailment improvement requires the presence of a given microorganism alongside an accessible injury for infiltration, a decrease in both of these variables will prompt the concealment of ailment advancement (Barkai-Golan 2005). Watchful harvesting, arranging, packaging or transportation, including preventing the fruit from falling at all stages, can limit injury to the produce. With respect to evasion of wounds, physiological wounds are likewise brought about by oxygen inadequacy, heat, cold and other ecological burdens, which inclines the commodity to assault by wound microorganisms. An overall decrease in wounds additionally lessens the odds of contamination of vegetables and fruits by microbes during travel and storage; such factors should likewise be contemplated while packaging or storing away the vegetables or fruits (Barkai-Golan 2005).

5.5.1.3 Postharvest Chemical Treatments

Wounds to produce occurred during harvesting, packaging and handling are significant sites of intrusion by postharvest wound microbes; the sealing of wounds by synthetics will extensively diminish rot. Types of wounds include wounds made in cutting off the harvest from the plant or cuts made intentionally during taking care of methods – for example, stem cuts in banana hands or petiole cuts in celery planned for sending out. Other likely sites of contamination are the natural openings in the host surface – for example, lenticels and stomata – whose vulnerability to disease is expanded by injury or following washing the commodity in water. A productive sterilization cycle should arrive at the pathogenic microorganisms accumulated in each of those sites (Eckert and Ogawa 1985).

The choice of a chemical compound for the administration of a particular postharvest sickness fundamentally relies on (i) the effectiveness of the microbe against the chemical; (ii) the capacity of the chemical to enter through surface hindrances of the host to the disease arena; and (iii) the tolerant limit of the crop to the compound, because of the the phytotoxicity just as unfavourable impact upon the nature of the produce. Presently, most fungicides or chemicals are generally applied in liquid formulations, which are sprayed onto the fruits and conveyed precisely in packing houses. In certain fruits, synthetic fungicides are mixed in liquid wax formulations and in hydrocarbon-solvent-base waxes and then are additionally used. The fungicides applied in liquid formulations to control injury infection are classified into two categories according to their dissemination on the outside of the treated products: (i) thiabendazole (non-ionic insoluble in water) that is applied as uniform cover over the whole surface of the treated product and (ii) water-soluble salts (e.g. SOPP, sodium carbonate and sec-butylamine) that are applied as moderately concentrated solutions (0.5–3.0%) in water (Eckert and Ogawa 1985). Fruit injury uptakes liquid solutions of these salts, but the intact cuticle is not penetrable to such polar mixture. After treatment, the crop is washed gently with fresh water to expel a large portion of the fungicide from the outside of the fruit; however, a critical build up stays at the harmed portion to hinder development of microbes there. A favourable result of uniform non-discriminating exposure of the crop with a fungicide is that the entire surface is secured somewhat against diseases and wounds produced after the treatment is applied. In postharvest management, various synthetic chemicals, for example, thiabendazole, SOPP, prochloraz, biphenyl (diphenyl), carbendazim, dicloran, ridomil, iprodione, etc., have been applied to effectively reduce postharvest decay in vegetables and fruits (Table 5.1). Aluminium-containing salts gave solid restraint of all the tested microbes (*B. cinerea*, *A. solani*, *F. sambucinum*, *R. stolonifer* and *P. sulcatum*) with insignificant inhibitory concentration of 1–10 mM. Aluminium sulphate and aluminium chloride are commonly the most effective, hindering mycelial development of microorganisms by as much as 47% and 100%, individually, at a salt concentration of 1 mM. Applied at 5 mM, aluminium sulphate additionally gave 28% and 100% hindrance of dry decay and depression spots, respectively. Aluminium chloride (5 mM) controls dry decay by 25%, though aluminium lactate (5 mM) diminished depression spot injuries by 86%. These outcomes demonstrate that different aluminium-containing salts may give an option in contrast to the utilization of fungicides to control these microorganisms (Kolaei et al. 2013). Youssef et al. (2014) confirmed that both bicarbonate and sodium carbonate exert a straight antifungal impact on *P. digitatum* and induce citrus fruit guard systems to postharvest rot.

5.5.2 Plant Growth Regulators

Plant development controllers are known to hinder senescence and onset of fruit decay. It has been shown that maleic hydrazide and indole acidic acid (IAA) were best against *Aspergillus* and *Rhizopus* decays of papaya fruits, whereas planofix (NAA; use at 0.01%) stop all decay except *Fusarium* decay in post inoculation treatment. Tak et al. (1985) concluded that maleic hydrazide (100 ppm), hydrogenated ground nut oil and rovral (500 ppm) showed the best results both as pre-and post-inoculation treatment to control fruit decay of apple brought about by *A. alternata*.

5.5.3 Generally Regarded as Safe Chemicals

GRAS (generally regarded as safe) synthetic chemicals are delegated as food preservatives and substances, as recorded by the United States Food and Drug Administration. These include natural compounds isolated from plants, creatures or microorganisms including a few volatiles and phenolic compounds, essential oils, peptides, plant extricates, alkaloids, antibiotics, lectins, latex or chitosan, propolis (Barkai-Golan 2001; Troncoso-Rojas and Tiznado-Hernández 2007) and various synthetic chemicals like ammonium molybdate or calcium polysulfide. These synthetic chemicals are an option in contrast to customary fungicides for postharvest infection control, having insignificant toxicological consequences for mammals and the atmosphere. In California, for more than 75 years the standard strategy for cleaning oranges or lemons was to dip fruits for 2–4 minutes in a warmed (43°C) solution of 4% borax (sodium tetraborate decahydrate) and 3% sodium carbonate within a day or two after harvest. Detergent was typically included, and the fruits were washed with a water spray to expel salt deposits from the surface. The borax shower treatment was surrendered in view of residue issues and removal of rinse water containing boron. Sodium carbonate and sodium bicarbonate treatment stay in like manner use to today, since they are successful and modest food preservatives permitted without any limitations for some, applications including organic agriculture (Smilanick et al. 1995, 1997, 1999). Moreover, they can be applied viably through high-pressure washer sprayers; low capacity spray applications over rotating brushes are avoided because their adequacy is lower and calcium carbonate scale aggregates on the brushes. Despite the fact that their adequacy is lower in mandarins than other citrus fruits, great control of penicillium moulds and reasonable control of harsh decay, brought about by *Geotrichum citri-aurantii*, is acquired with these treatments, particularly when heated solution and delayed inundation times are utilized (Palou et al. 2001, 2002; Smilanick et al. 1997, 1999; Smilanick and Sorenson 2001). The action of carbonate salts adjacent to penicillium rot is indistinct. It was shown to be due in part to the attendance of an alkaline residue in injury, though equimolar solutions of the same pH organized from SC or SBC were more efficient than those organized from potassium or ammonium salts, which recommended that the sodium cation and any factors may be significant (Smilanick et al. 1999). Interestingly, it was found in other work (Zhang and Swingle, 2003) that the viability of potassium bicarbonate in opposition to green moulds was equal to that of SBC at a similar concentration. Besides carbonates, other regular food additives have been assessed for the control of citrus fruits or blue moulds. Some short-chain organic acids – for example, acidic, formic or propionic acid – have been tested as fumigants (Sholberg and Gaunce 1995; Sholberg, 1998) and a few organic acid salts – for example, potassium sorbate, sodium propionate or sodium benzoate – have been applied as liquid solutions (Palou et al. 2002). Among over 40 food preservatives and less-toxic synthetic compounds tested, PS and SB were the best on citrus fruits (Palou et al. 2002). They were about equivalent in movement to one another and to sodium carbonate. PS ($C_6H_7O_2K$) was first evaluated against fungicide-resistant strains of *Penicillium* spp. (Nelson et al. 1983) and it has been applied to citrus fruit in commercial packinghouses to inhibit rot, even though its use for this application is uncommon and some regulatory supports may not be current (Smilanick et al. 2008). Dipping of fruit in warmed solution is the best technique for application (Wild 1987; Smilanick et al. 2008). Preferences of PS are that *P. italicum* and *P. digitatum* grew practically zero resilience after delayed and repeated introduction to it (Schroeder and Bullerman 1985), and removal of utilized solutions would have less administrative issues than the sodium salts SC or SBC (Smilanick et al. 2008). All in all, shortcomings related to the utilization of GRAS salt solutions include absence of preventive movement, restricted constancy (Palou et al. 2001, 2002; Smilanick et al. 2008), risk of fruit damage or weight and immovability losses during long-haul stockpiling when treated fruits have not been washed, decrease of treatment viability by high-pressure water washing or flushing, and removal issues related to high pH and sodium or potassium content (Smilanick et al. 1997; 1999, Larrigaudiere et al. 2002; Palou et al. 2007). In addition, chlorine (200 µg/mL) should be added and kept to kill conidia of *Penicillium* spp. in the solution and on fruit surface (Smilanick et al. 1999; 2002). Some of these issues could be dealt with by the advancement of new innovations – for example, the consolidation of antifungal GRAS compounds as ingredients of new edible coatings or synthetic waxes.

Postharvest disease imposed by *B. cinerea* is constrained by the use of a sub-lethal degree of ethanol in a mix with potassium sorbates. The use of 0.5–1.0% potassium sorbate with 10–20% ethanol, respectively, essentially hastened the restraint of spore germination of *B. cinerea* (Sharma et al. 2009). Be that as it may, dips and fumes of ethanol have been extremely viable to control the postharvest infection of fruits such as table grapes, citrus fruits and peaches (Lichter et al. 2002; Karabulut et al. 2004). The impact of ethylene on postharvest rot has been centred on the infection of non-climacteric fruits (Palou et al. 2002). Archbold et al. (1999) revealed that fumigation of grape, dark berry, and strawberry with (E)-2-hexenal, a compound with antifungal action that is normally delivered in plant tissue, at the time of postharvest essentially diminishes the deterioration of tested fruits. Fallik et al. (1998) reasoned that the inhibitory impact of (E)-2-hexenal against *B. cinerea* at higher fumes phase cause serious harm to the fungal cell films and cell wall, which brought about breakdown and crumbled hyphae.

Jasmonate, a plant hormone, plays a significant function as signal molecules in plant safeguard reactions against pathogenic assault. A compound olepine in Jasmonate isolated from oxygenase-subordinate oxidation of unsaturated fats safeguards against plant pathogens (Kondo et al. 2000). The use of jasmonate in grape during postharvest stifles green mould rot (Dorby et al. 1999). Another intense antimicrobial specialist is ozone which used to reduce the disease of *R. stolonifer*. The ozone in table grapes initiated the creation of resveratrol and pterostilbene phytoalexins, which is useful to oppose the contamination of *R. stolonifer* in berries (Sarig et al. 1996). β-glycosidase chemical has been diminished in grapes in the presence of ethanol (Sarig et al. 1996). The control can be credited to ozone extraction of pterostilbene and resveratrol, which are associated with the overall resistance of grapes to fungal rot (Jeandet et al. 1991; Sarig et al. 1996). A list of GRAS chemicals used for postharvest treatment of fruits and vegetables is given in Table 5.2.

5.5.4 Natural Chemical Compounds

Natural compounds occurring in plant products are important sources of antifungal activity with low toxicity to human beings and safe to the environment due to being biodegradable. They can be used as substitutes for synthetic fungicides to manage the postharvest diseases of fruits and vegetables. According to their characteristics, they are grouped as flavour compounds, glucosinolates, acetic acid essential oils, plant extracts, jasmonates, chitosan, latex, etc. (Knight et al. 1997). The quality of high volatility of natural compounds makes them preferable to use as biofumigation of fruits with their off-odours or off-flavours (Mari et al. 2016). Some of natural compounds used in postharvest disease managements are briefly described below.

5.5.4.1 Flavour Compounds

Flavour compounds are volatile secondary metabolites resembling fat and low-water solubility. The advantages in the use of such flavour compounds include easy adsorption, less chance of off-odours, a harmless nature in mammalian systems, and a high efficacy even when applied in low concentrations.

5.5.4.1.1 Acetaldehyde

This is a volatile compound produced by plant organs that accumulates during ripening of fruits. Postharvest pathogens such as *Erwinia carotovora*, *Pseudomonas flourescens*, *Monilinia fructicola*, *Penicillium* sp. and yeast are inhibited by acetaldehyde on several fruits and vegetables (Sharma and Alemwati 2010). Wilson et al. (1987) detailed that 9 volatile compounds out of 16 occurring in plum and peach fruits inhibited spore germination of *B. cinerea* and *M. fructicola*. The volatiles most effective in hindered spore germination were benzyl alcohol, benzaldehyde, gamma-valerolactone and gram-caprolactone. Strawberry mature fruits exposed to 5000 µL/L acetaldehyde for 1 hour or with 1500 µL/L for 4 hours slowdown fungal rot during storage for 4 days at 5°C following

TABLE 5.2

GRAS Chemicals Used for Postharvest Treatment of Fruits and Vegetables

GRAS Chemicals	Crops	Postharvest Disease (Pathogen)	Dose	References
Hydrogen peroxide Santosil-25	Eggplant, red pepper	*Alternaria alternata*, *Botrytis cinerea*	5000 µL/L	Fallik et al. (1994)
Hydrogen peroxide	Red Globe grape	*B. cinerea*	30–35% solution of liquid	Forneyl et al. (1991)
	Bell pepper fruits	Fruit rot	15 mM for 3 min	Bayoumi (2008); Thakur et al. (2017)
	Jiashi muskmelon	Decay	3%–5%	Chen et al. (2015)
	Carrot	Decay	0.5% for 3 min	Isaac and Maalekuu (2013)
	Tomato	Decay	5–15 mM	Al-Saikhan and Shalaby (2019)
	Cherry tomato			Islam et al. (2018)
Acetic acid	Apple, pear	*P. expansum*, *B. cinerea*	2–4 mg/L	Sholberg and Gaunce (1995)
	Tomato, grape, kiwifruit	*B. cinerea*	2–4 mg/L	Sholberg and Gaunce (1995)
	Stone fruits	Brown rot (*M. fructicola*), Rhizopus rot (*R. stolonifer*)	1.4 mg/L	Sholberg and Gaunce (1996)
	Sweet cherry	*M. fructicola*, *P. expansum*		Chu et al. (2001)
	Apple	Blue mould rot (*P. expansum*)		Radi et al. (2010)
	Grape	Fruit rot		Venditti et al. (2017)
Peracetic acid	Plum	Brown rot (*M. fructicola*)	1000 µg/mL	Mari et al. (1999)
	Nectarine, plum	Brown rot (*M. laxa*)	2500 µg/mL for 5 min	Mari et al. (1999)
	Strawberry	Grey mould (*B. cinerea*)	0.5–2.0 mL/L	Abd-Alla et al. (2011)
Citrocide® PLUS, a PAA (15% w/w peracetic acid + 23% H_2O_2 w/w, Productos Citrosol S.A.)	Tomato	Decay	0.20%, dipping the tomato for 30 s	Mottura et al. (2019)
Propionic acid	Pear	Black spot (*A. alternata*)	9 g/L	Liu et al. (2013)
Chlorine dioxide (ClO_2)	Nectarine,plum	Brown rot (*M. laxa*)	100 µg/mL for 5 min	Mari et al. (1999)
	Grape	*Botrytis* rot (*B. cinerea*)	–	Zoffoli et al. 1999
Sodium bicarbonate ($NaHCO_3$)	Citrus	Green mould (*P. digitatum*)	2–3% (w/v) for 60–150 s	Smilanick et al. (1997)
	Tomato	Grey mould rot (*B. cinereal*)	300 mM	Alaoui et al. 2017
Sodium carbonate (Na_2CO_3, soda ash)	Citrus	Green mould (*P. digitatum*)	2–3% (w/v) for 60–150 sec	Smilanick et al. 1997
	Tomato	Grey mould rot (*B. cinereal*)	300 mM	Alaoui et al. (2017)
	Yellow pitahaya	Black rot (*A. alternata*)		Vilaplana et al. (2018)
Calcium chloride	Peach	Rhizopus rot (*R. solonifer*)	2.0%	Singh (2005); Salem et al. (2016)
	Strawberry	Decay	–	Rosen and Kader (1989)
Sodium benzoate	Orange	*G. candidatum*, Green mould (*P. digitatum*)	4.0%	El-Mougy et al. (2008); Palou et al. (2018)
Potassium sorbate	Citrus	Phomopsis stem-end rot (*P. citri*) and *Penicillium* rot	2.0%	Smilanick et al. (2008)
		G. candidatum, green mould (*P. digitatum*)	2.0%	El-Mougy et al. (2008); Youssef et al. (2017)
	Stone fruit	Brown rot (*M. laxa*)	15.0 g/L for 120 s	Gregori et al. (2008)
Potassium carbonate	Tomato	Grey mould rot (*B. cinerea*)	300 mM	Alaoui et al. (2017)
Sodium metasulphite	Tomato	Grey mould rot (*B. cinerea*)	300 mM	Alaoui et al. (2017)
Sodium salicylate	Tomato	Grey mould rot (*B. cinerea*)	300 mM	Alaoui et al. (2017)

1 day at 20°C. These treatments additionally provoked a modest amount of various fruit volatiles – for example, ethyl butyrate, ethyl acetate and ethanol – which expanded fruit fragrance and improved its taste (Pesis and Avissar 1990). The efficacy of acetaldehyde vapour and of a number of other aliphatic aldehydes produced naturally by sweet cherry (cv. Bing) was evaluated in *P. expansum* inoculated fruits (Mattheis and Roberts 1993). It has also been reported that high concentrations of acetaldehyde, propanal and butanal suppressed conidial germination but resulted in extensive stem browning and

fruit phytotoxicity, which increased with the aldehyde concentration. The degree of inhibition decreased with increasing aldehyde molecular weight.

5.5.4.1.2 Hexenal

Hexenal is another flavour compound that has strong antifungal activity (Hamilton-Kemp et al. 1992; Fallik et al. 1998). Use of hexenal is reported to control various postharvest pathogens. For example, fumigation of 'Crimson Seedless' table grapes with (E)-2-hexenal resulted in efficient control of mould (Archbold et al. 1999) and inhibited hyphal growth of *P. expansum* and *B. cinerea* in vitro and on apple slices (Song et al. 1996). Six-carbon (C6) aldehydes have also been found to inhibit hyphal growth of *A. alternata* and *B. cinerea*, suggesting that hexenal and similar aldehydes have the potential to be used against postharvest decay pathogens of fresh horticultural produce (Hamilton-Kemp et al. 1992). Almenar et al. (2007) studied the efficacy of inclusion complex β-cyclodextrin hexanal against postharvest pathogens like *Colletotrichum acutatum*, *A. alternata* and *B. cinerea* and found that the sustained release of hexanal from the complex could reduce or prevent postharvest diseases in berries. Furthermore, the best control of blue mould of apple and pear caused by *P. expansum* was achieved by treating the fruits with 12.5 μL/L of trans-2-hexenal (Neri et al. 2006). Similarly, continuous controlled release of hexanal effectively suppressed grey mould of tomato caused by *B. cinerea* (Utto et al. 2008).

5.5.4.1.3 Acetic Acid

Acetic acid is a metabolic intermediate that occurs naturally in many fruits. It is considered an effective fumigant for surface-sterilising a wide range of horticultural produce. Acetic acid vapour in pure form, or as vinegar, is a very effective treatment for reducing postharvest decay in several horticultural products. Its effectiveness in preventing postharvest fruit decay caused by postharvest pathogens like *P. digitatum*, *B. cinerea*, *P. expansum*, *M. fructicola* and *R. stolonifer* is well documented. For example, Sholberg and Gaunce (1995) reported acetic acid as a very effective postharvest fumigant for controlling the decay caused by *M. fructicola* and *R. stolonifer* on peaches, Alternaria rot of sweet cherries and brown rot caused by *M. fructicola* on apricots. Fumigation with acetic acid prevented postharvest decay in table grapes (Sholberg and Gaunce 1996; Moyls et al. 1996). These results suggest the use of acetic acid vapour as a potential replacement for SO_2 fumigation, which is currently employed on a commercial scale for decay control in stored grapes. The best postharvest control of decay caused by *P. digitatum* in 'Fremont' and 'Fairchild' mandarins was achieved by combining curing and acetic acid vapour fumigation (Venditti et al. 2009). Acetic acid vapour also proved to be very effective in disinfecting d'Anjou pear stems and fruit surfaces (Sholberg et al. 2004). In addition, the application of acetic acid vapour to treat larger volumes of perishable fruits requires a higher level of monitoring on its concentration for use at a safe level.

5.5.4.1.4 Jasmonates

Jasmonates are naturally occurring plant growth regulators, belonging to a class of olypines which have been implicated in the regulation of various processes of the development of plants and their responses against environmental stresses. They play an important role as signal molecules in plant defence responses against pathogen attack. They participate in the development of defence response by the activation of genes encoding antifungal proteins (thionin and osmotin) and are involved in phytoalexin biosynthesis. Methyl jasmonate (MeJA) is a derivative of jasmonic acid that is responsible for the induction of resistance in plants during the attack of fungal pathogens. It has been used in postharvesting of several fruits due to its effects against pathogens. For instance, MeJA suppressed *B. cinerea* in strawberry (Moline et al. 1997) and *P. digitatum* in 'Marsh Seedless' grapefruit (Droby et al. 1999). The treatment of loquat fruit with MeJA reduced the incidence of anthracnose rot caused by *C. acutatum* during postharvest (Cao et al. 2009). Furthermore, MeJA along with the antagonistic yeast *Pichia membranifaciens*, significantly inhibited the growth of *C. acutatum* which causes postharvest anthracnose rot in loquat fruit (Cao et al. 2009). Yao and Tian (2005) reported that pre- and postharvest use of MeJA reduced the occurrence of brown rot of sweet cherries caused by *M. fructicola* during storage. Furthermore, the exposure of papaya fruit to MeJA vapours at 10^{-5} or 10^{-4} M, 20°C for 16 hours inhibits the fungal decay caused by *C. gloeosporioides* (Gonzalez-Aguilar et al. 2003). However, the use of 1 μmol/L MeJA to treat strawberries significantly restricted the decay of fruit (by *B. cinerea*) during storage conditions by the activation of defence-responsive genes (Zhang et al. 2006).

5.5.4.1.5 Glucosinolates

Glucosinolates are a natural substance produced by crucifers. They have potential antimicrobial activity and inhibit the growth and spread of postharvest pathogens (Fenwick et al. 1983). Postharvest antipathogenic activity of glucosinolates, particularly of allylisothiocyanate (AITC), has been studied by various workers (Ishiki et al. 1992; Mari et al. 1996). Mari et al (2002, 2003) reported that the AITC has vapour-phase antifungal activity against blue mould rot in apple caused by *P. expansum*. The use of isothiocyanates (0.03 mg/mL) is very effective in inhibiting the growth of *A. alternata*in vitro, while the combination of 0.56 mg/mL isothiocyanates with low-density polyethylene bags is much more effective in contrast with commercial fungicides to control fungal rot on bell pepper with no deleterious effect on fruit quality (Troncoso et al. 2005). An additional effect of AITC is the retardation of blueberry decay during storage at 10°C (Wang et al. 2010).

5.5.4.1.6 Propolis

Propolis is a natural resin substance produced from the buds and bark of poplar and coniferous plants. The importance of this resin is its action against plant pathogens as an antibacterial and antifungal agent (Tosi et al. 1996). Lima et al. (1998) elucidated the use of propolis to inhibit the growth of postharvest pathogens such as *B. cinerea* and *P. expansum* and control postharvest decay in fruits.

5.5.4.1.7 Fusapyrone and Deoxyfusapyrone

Fusapyrone is a metabolite produced by *Fusarium semitectum*. It is widely used to control decay due to its lower toxicity level to animals and its lack of phytotoxic effects

(Altomare et al. 2000). Applied at 100 µg/mL, it inhibited the growth of *B. cinerea* on grape (Altomare et al. 1998).

5.5.4.2 Chitosan

Chitosan, a deacylated form of chitin, is a natural biodegradable compound derived from the animal such as crustaceous shells of crabs and shrimps. It is used as a coating for fruits or vegetables and regulates gas moisture and exchange around the product. Chitosan is used as antifungal agent and is known for the induction of a defensive response in plants during pathogenic attacks. Chitosan stimulates the formation of structural defence barriers such as callose synthesis, thickening of host cell wall formation of papillae and plugging of some intercellular spaces with fibrillar material, which are probably impregnated with antifungal phenolics compounds in bell peppers and tomatoes (Wilson et al. 1994). Several studies reveal that it holds potential for commercial postharvest decay control. For instance, it induced resistance in harvested apples rather than by direct inhibition of the pathogen (Capdeville et al. 2002). It is very effective to prevent the maceration of host tissues by *B. cinerea* (El Ghaouth et al. 1997) and also reducing the level of polygalacturonases in *B. cinerea* by cytological damage in fungal hyphae.

5.5.4.3 Essential Oils`

Essential oils derived from various plants have antipathogenic activity and control phytopathogenic microorganisms (Mihaliak et al. 1991; Meepagala et al. 2002). The antifungal activity of essential oils of *Monarda citriodora* var. *citriodora* and *Melaleuca alternifolia* on postharvest pathogens has been evaluated by Bishop and Thornton (1997), while Tzortzakis and Economakis (2007) reported the antifungal activity of lemon-grass essential oil against key postharvest pathogens like *C. coccodes*, *B. cinerea*, *Cladosporium herbarum*, *R. stolonifer* and *A. niger*. Similarly, the essential oil of *Pimentadioica* has been reported to inhibit pathogens like *F. verticillioides*, *P. expansum*, *P. brevicompactum*, *A. flavus* and *A. fumigatus* (Zabka et al. 2009). The essential oil produced by *Thymus vulgaris* has antifungal properties. These antifungal properties make it valuable to restrict plant pathogens such as *B. cinerea* and *R. Stolonifer* that is grown during the storage of strawberry (Reddy et al. 1998). Cassia oil at 500 µL/L significantly inhibited *A. alternata* of cherry tomatoes stored at 20°C for 3 days (Feng et al. 2008). The essential oils of *Caesulia axillaris* and *Mentha arvensis* controlled the blue mould rot of oranges caused by *P. italicum* (Varma and Dubey 2001). *M. spicata* and *Lippias caberrima* essential oils, as well as pure (d)limonene and R-(–)-carvone, amended coatings applied postharvest to 'Tomango' oranges resulted in excellent control of decay caused by *P. digitatum* (Plooy et al. 2009). Similarly, *Cassia* and thyme essential oils exhibited strong inhibitory and antifungal activity against postharvest pathogen (*A. alternata*) of cherry tomato (Feng and Zheng 2007).

Essential oil of *Lippia scaberrima* caused inhibition of mycelia growth of postharvest pathogens of mango such as *Botryosphaeria parva* and *C. gloeosporioides*, and use of wax coating, enriched with essential oil, reduced the fungal infection during storage (Regnier et al. 2008). The postharvest decay in peach, kiwifruit, orange and lemon caused by *B. cinerea*, *Monilinia laxa* and *P. digitatum*, was controlled effectively with using laurel oil, as evidenced by postharvest decay inhibition in peach, kiwifruit, orange and lemon. The combination of eugenol (2 mg/mL) and soy lecithin (50 mg/mL) restricts the development of disease in apple caused by *P. expansum*, *P. vagabunda*, *B. cinerea* and *M. fructigena* to less than 7%, 6%, 4% and 2%, respectively, after 6 months of storage at 2°C (Amiri et al. 2008). Cinnamon oil enrichment significantly reduced the development of colonies and formation of spores in postharvest pathogens like *C. coccodes*, *B. cinerea*, *C. herbarum*, *R. stolonifer* and *A. niger* (Tzortzakis 2009). Carvacrol vapour treatment of table grapes resulted in significant reduction in postharvest decay and spoilage of stored fruits (Romero et al. 2007).

5.5.4.4 Gel

Aloe vera gel has antifungal activity against postharvest pathogens like *P. digitatum*, *P. expansum*, *B. cinereal* and *A. alternata*. The natural gel inhibited spore germination and mycelial growth of *P. digitatum* and *A. alternate*, which are most sensitive species. Sake and Barkai-Golan (1995) gel was applied on grapefruit to delay lesion development; it suppressed growth of *P. digitatum* and also significantly decreased decay during storage. Dried latex and *A. vera* extractives obtained by hexane, ethyl acetate and methanol showed higher antifungal activity against *Colletotrichum* species than *F. solani*. In addition, aloin and aloe-emodin are important compounds used against *C. gloeosporioides* and *Cladosporium cucumerinum* (Sebastian et al. 2011).

5.5.4.5 Latex

Latex is another natural fungicide used to prevent several postharvest diseases in banana, papaya and other fruits due to its harmless properties (Adikaram et al. 1996). Hevein is a chitin-binding protein produced by latex of the rubber tree (*Hevea brasiliensis*). It is an antifungal compound that interferes with fungal growth by binding or cross-linking newly synthesized chitin chain. It has antifungal activity against *B. cinerea*, and species of *Fusarium* and *Trichoderma* in in-vitro conditions (van Parijs et al. 1991). The water soluble faction of papaya latex can completely digest the condia of many fungi, including many postharvest pathogens (Indrakeerthi and Adikaram 1996). Sibi et al. (2013) reported that latex is potential source of antifungal activity against postharvest pathogens. Although, latex of *Thevetia peruviana* and *Artocarpus heterophyllus* was most potential against fungi *A. fumigates*, *A. niger*, *A. terreus*, *F. solani*, *P. digitatum* and *R. arrizus* isolated from fruits and vegetables.

5.5.4.6 Plant Extracts

Plant extract use as a bioactive compound has a great potential to control plant pathogenic fungi. The preservative nature of some plant extracts has been known for centuries, and the antimicrobial properties of extracts of aromatic plants have received renewed attention for only a decade or so. Recent research supports the use of plant extracts in the management and control of postharvest diseases in horticultural produce. The aqueous extract of leaves of garlic creeper (*Adenocalymna alliaceum* Miers) is reported to have antifungal activity (Rana et al. 1999). Inhibitory activity of 7-geranoxy coumarin, a compound isolated from tissue of 'Star Ruby' grapefruit, has been used against postharvest pathogens like *P. italicum* and *P. digitatum* (Agnioni et al. 1998). Furthermore, an antifungal compound kaempferol (extracted from *Acacia nilotica*) has an effective response against *P. italicum* (Tripathi et al. 2002). Presently, a citrus seed extract (Lonlife™, Citrex) (250 ppm) is widely used to treat postharvest diseases. This organic fungicide provides satisfactory control of crown and peel rot in several commodities in organic niche markets (Jansen et al. 1995).

5.5.4.7 Lectins

Lectins are a class of sugar-binding proteins which are widely found in nature. Their occurrence in plants has been well known since the end of the 20th century. There is a gap in the understanding of their function. However, two theories support their use:(i) they play a role recognizing the symbiotic relationship between leguminous plants and nitrogen-fixing bacteria, and (ii) they also defend plants against phytopathogenic fungi and animals (Sharon, 1997). The interaction of fungal hyphae with lectins was first reported by Mirelmann et al. (1975). Another compound, wheat germ agglutinin (WGA) is a lectin that protects wheat from plant pathogens. During pathogenesis, it inhibits hyphal growth and germination of spores by binding to hyphal tips and hyphal septa of *Trichoderma* spp. The strategy of wheat protection by WGA suggested that this compound is effective against chitin-containing fungi. An additional report suggests the non-functional nature of WGA during pathogenesis by chitinless *Phytopthora citrophthora* (Barkai-Golan et al. 1978).

It has been observed in some research that soybean agglutinin (SBA) (binds D-galactose and N-acetyl-D-galactose) and peanut agglutinin (PNA) (binds D-galactose) compounds specifically bind galactose containing chitinous cell walls of fungi. Other fungi, such as *B. cinerea*, *F. moniliforme* and *G. candidum*, showed bonding with SBA itself. The capability of SBA and PNA to bind to fungal surfaces of *Aspergillus* and *Penicillium* species is an acceptable concurrence with the occurrence of galactose seen in few species of these genera, for example *P. digitatum* and *P. italicum* (Grisaro et al. 1968) or *A. niger* (Bardalaye and Hordin 1976). The authoritative of fungi to concanavalin A (Con A) is more difficult to represent since this lectin inadequately responds with β-linked glucans and with chitin (Sharon and Lis 1989). It is conceivable, however, that a positive response with Con A shows the nearness of little amounts of α-connected D-glucose (or D-mannose) residue on the fungal surface (Barkai-Golan et al. 1978). A growth interruption at the time of germination of spores was accounted for three fungi species, including the postharvest microorganism *Botryodiplodia theobromae* (Brambl and Gade 1985). ensuing study reported that, notwithstanding, the antifungal activity of WGA was undoubtedly due to polluting chitinases in the lectin arrangements used, which are known as intense inhibitors of fungal growth (Schlumbaum et al. 1986).

5.6 Integrated Postharvest Disease Management

The control of stem-end rot of mango has been achieved by the use of HWT and fungicide dips (azoxystrobin 150–175 ppm, carbendazim 312.5 ppm and tebuconazole 125–156 ppm) that significantly reduced disease and extended marketable life for 8 days (Akem et al. 2013). Arrebola et al. (2009) reported that *Bacillus amyloliquefaciens* in combination with thyme and lemon grass essential oils improved the beneficial effect of modified atmosphere packaging in retaining the overall fruit quality of peaches during storage. The combination of MAP with eugenol and thymol on table grapes reduced microbial spoilage up to 56 days of storage (Valero et al. 2006). The extract of seeds of grapefruit, alone or in combination with chitosan, significantly reduced postharvest grey mould rot of 'Red Globe' table grapes and maintained their keeping quality (Xu et al. 2007).

5.7 Methods of Application of Chemicals

Various methods are available to apply fungicides on fruits to control postharvest diseases, as described here.

5.7.1 Dipping

The dipping method is a simple and preventive procedure used to treat fruits and vegetables with a suitable concentration of chemical in the water that creates toxicity to disease-causing fungi. The shower of diluted chemicals is used to treat the produce, and this technique is called 'cascade application'. The influence of dips is improved by the addition of additives in the formulation such as Teepol or Triton-x-100 (as wetting agents), which help to reduce the surface tension and improve the coating of chemicals on the produce. Furthermore, to increase the effectiveness of fungicides, some acids, such as citric acid, are used to lower their pH (Narayanasamy 2006).

5.7.2 Electrostatic Sprays

The spreading technique of pesticides including breaking of pesticide solution into fine droplets and passing an electrical charge before spraying gives an additional advantage to increase the effectiveness of pesticides. There is no loss of biological activity of the materials with such sprays. The principle on which they work is that all particles have the same electrical charge and thus repel each other. They are attracted to the crop, and form a thin, even layer/coat. This method is being used in several fruits in many advanced countries (Narayanasamy 2006).

5.7.3 Fumigation

Fumigation of chemicals is an important and promising method to effectively control the postharvest diseases of some fruits. The harvested fruits are immediately treated with fumigation to restrict infection of injuries. This method also helps in the transport of the produce on long distances (Avissar et al. 1989).

5.7.4 Chemical Pads

Chemical pads are an effective approach towards the restriction of postharvest disease on banana by the use of paper pads. The paper pads are impregnated with fungicidal chemical agents that play a major role in restricting postharvest diseases during the transportation of fruits and vegetables. These paper pads are known as crown pads and are used for the prevention of fungal infections on the cut crowns of fruits (Burden and Wills 1989; Ahmed et al. 2018).

5.8 Conclusion

The quality and extended shelf life of postharvest fruits and vegetables depend on the use of better handling practices, precautions and effective treatment strategies. The use of these strategies has not increased the quality of the harvested produce but is very effective to retain its quality and enhance shelf life. In this context, the use of synthetic chemical fungicides is a different approach to successfully maintain the postharvest diseases caused by fungi and bacteria during storage and transportation of fruits and vegetables. Besides synthetic fungicides, natural plant products such as essential oils lectins and GRAS chemicals are also useful to control postharvest diseases of these crops. Postharvest application of synthetic fungicides may be minimized by using high-tech postharvest handling practices such as harvesting, precooling, cleaning or disinfecting, sorting and grading, packaging, storage and transporting, which might play a significant role in maintaining the quality and extending the shelf life of fruits and vegetables after harvest. Also, the use of appropriate postharvest treatment methods like refrigeration, postharvest physical treatment, proper storage facilities including modified atmosphere packaging and GRAS chemical application will also be vital to reduce postharvest losses particularly caused by diseases.

5.9 Future Research Thrust Area

1. Pre-harvest practices affect quality of produce; hence more emphasis may be given to this aspect, including plant protection practices.
2. New-generation fungicides which have fewer pesticide residue problems will be tried for postharvest treatment to reduce the incidence of disease.
3. Residue on treated produce can be accurately measured, and the safety of the treatment can be assessed by accepted toxicological criteria to try to remain below Maximum residue limit (MRL).
4. Alternative to synthetic chemical fungicides like GRAS chemical natural plant products may be explored for postharvest disease management.
5. Development of induced resistance in fruits and vegetables due to bio-elicitors or chemicals to reduce microbial infection should be investigated.
6. Short-lived antimicrobial agents could protect fresh wounds against invasion by a fast-growing pathogen while permanent resistance develops.
7. Critical enzymes involved in the pathogenicity of postharvest pathogens should provide another focus for postharvest disease control.

REFERENCES

Abd-Alla MA, Abd-El-Kader MM, Abd-El-Kareem F, El-Mohamedy RSR (2011) Evaluation of lemongrass, thyme and peracetic acid against gray mold of strawberry fruits. J Agric Technol 7(6): 1775–1787.

Adikaram NKB, Indrakeerthi SRP, Charmalie A, Menike RR, Ajani K (1996) Antifungal activity in fruit and postharvest disease In: Proc Australian Postharvest Hortic Conf Sci Technol Fresh Food Revo, Melbourne, Australia, pp. 381–385.

Agnioni A, Cabras P, Dhallewin G, Pirisi FM, Reniero F, Schirra M (1998) Synthesis and inhibitory activity of 7-geranoxy coumarin against *Penicillium* species in citrus fruits. Phytochemistry 47: 1521–1525.

Ahmed S, Sergio RR, Allan RD, Muhammad S, Osmar JCJ, Ciro Hideki S, Reginaldo TS (2018) Effects of different sulfur dioxide pads on botrytis mold in 'Italia' table grapes under cold storage. Horticulturae 4: 29.

Akem C, Opina O, Dalisay T, Esguerra E, Ugay V, Palacio M, Juruena M, Fueconcillo G, Sagolili J (2013) Integrated disease management of stem end rot of mango in the Southern Philippines. ACIAR Proc Series 139: 104–110.

Almenar E, Auras R, Rubino M, Harte B (2007) A new technique to prevent the main post harvest diseases in berries during storage: Inclusion complexes β-cyclodextrin-hexanal. Int J Food Microbiol 118: 164–172.

AL-Saikhan MS, Shalaby TA (2019) Effect of hydrogen peroxide (H_2O_2) treatment on physicochemical characteristics of tomato fruits during post-harvest storage. Aust J Crop Sci 13(05): 798–802.

Altomare C, Perrone G, Stornelli C, Bottalico A (1998) Quaderni della Scuola di specializzazione in Viticoltura ed Enologia. Univ Torino Ital 22: 59–66.

Altomare C, Perrone G, Zonno MC, Evidente A, Pengue R, Fanti F, Polonelli L (2000) Biological characterization of fusapyrone and deoxyfusapyrone, two bioactive secondary metabolites of *Fusarium semitectum*. J Nat Prod 63: 1131–1135.

Alvarez AM, Hylin JW, Ogata JN (1977) Postharvest diseases of papayas reduced by biweekly orchard sprays. Plant Dis Rep 61:731–735.

Amiri A, Dugas R, Pichot AL, Bompeix G (2008) *In vitro* and *in vivo* activity of eugenol oil (*Eugenia caryophylata*) against four important postharvest apple pathogens. Int J Food Microbiol 126: 13–19.

Aragaki M, Kimoto WS, Uchida JY (1981) Limitations of hot water treatment in the control of Phytophthora fruit rot of papaya. Plant Dis 65:744–745.

Archbold DD, Hamilton-Kemp TR, Clements AM, Collins RW (1999) Fumigating 'Crimson Seedless' table grapes with (E)-2-hexenal reduces mold during long-term postharvest storage. HortSci 34: 705–707.

Arrebola E, Sivakumar D, Bacigalupo R, Korsten L (2009) Combined application of antagonist *Bacillus amyloliquefaciens* and essential oils for the control of peach postharvest diseases. Crop Protec 30: 1–9.

Avissar I, Marinansky R, Pesis E (1989) Postharvest decay control of grape by acetaldehyde vapour. Acta Hort 258: 655–660.

Balkan HA, Cupertino FP, Dianese JC, Takatsu A (1976) Fungi associated with pre- and postharvest fruit rots of papaya and their control in central Brazil. Plant Dis Rep 60: 605–609.

Bardalaye PC, Hordin JH (1976) Galactosaminogalactan from cell walls of Aspergillus niger. J Bacteriol 125: 655–669.

Barkai-Golan R (2001) Postharvest diseases of fruits and vegetables. Development and control. Amsterdam, The Netherlands: Elsevier Science BV; 2001.

Barkai-Golan R, Mirelmann D, Sharon N (1978) Studies on growth inhibition by lectins of *Penicillia* and *Aspergilli*. Arch Microbiol 116: 119–124.

Barkai-Golan R (2005) Postharvest diseases of fruits and vegetables. A Division of Reed Elsevier.India, Pvt. Limited, New Delhi.

Bartz JA, Eckert JW (1987) Bacterial diseases of vegetable crops after harvest. In: Weichmann J (Ed.). Postharvest physiology of vegetables. New York, Dekker, pp. 351–376.

Bayoumi YA (2008) Improvement of postharvest keeping quality of white pepper fruits (*Capsicum annuum* L.) by hydrogen peroxide treatment under storage conditions. Acta Biol Szegediensis 52(1): 7–15.

Bishop CD, Thornton IB (1997) Evaluation of the antifungal activity of the essential oils of *Monarda citriodora* var. *citriodora* and *Melaleuca alternifolia* on the post harvest pathogens. J Essential Oil Res 9: 77–82.

Bompeix G, Coeffic M, Greffier P (1979). Lutte contre les pourritures des peches a Monilia sp., Botrytis sp. Rhizopus sp. Fruits 34: 423–430.

Brambl R, Gade W (1985) Plant seed lectins disrupt growth of germinating fungal spores. Physiol Plant 64: 402–408.

Brown GE (1979) Biology and control of *Geotrichum candidum*, the cause of citrus sour rot. Proc Fla State Hortic Soc 92:186–189.

Brown GE (1981) Investigations with experimental citrus postharvest fungicides in Florida. Proc Int Soc Citric 2:815–818.

Brown GE (1983) Control of Florida citrus decays with guazatine. Proc Fla State Hortic Soc 96: 335–337.

Burden OJ, Wills RBH (1989) Prevention of postharvest food losses: Fruits, vegetables and root crops. Food and Agriculture Organization of the United Nations, Rome, Training Series 17(2): 157.

Cao S, Zheng Y, Wang K, Tang S, Rui H (2009) Effect of yeast antagonist in combination with methyl jasmonate treatment on postharvest anthracnose rot of loquat fruit. Biol Control 50: 73–77.

Capdeville G, De Wilsoon CL, Beer SV, Aist JR (2002) Alternative disease control agents induce resistance to blue mould in harvested Red Delicious apple fruit. Phytopathology 92: 900–908.

Ceponis M, Cappellini RA (1978) Relationship of chemical application time to postharvest disease control of blueberries. Plant Dis Rep 62: 1005–1007.

Ceponis M, Cappellini RA (1985) Reducing decay in fresh blueberries with controlled atmospheres. HortSci 20: 228–229.

Chastagner GA, Ogawa JM (1979) A fungicide-wax treatment to suppress *Botrytis cinerea* and protect fresh market tomatoes. Phytopathology 69: 59–63.

Chen G, Chen J, Feng Z, Mao X, Guo D (2015) Physiological responses and quality attributes of Jiashi muskmelon (Cucurbitaceae, *Cucumis melo* L.) following postharvest hydrogen peroxide treatment during storage. Eur J Hortic Sci 80(6): 288–295.

Chu CL, Liu WT, Zhou T (2001) Fumigation of sweet cherries with thymol and acetic acid to reduce post harvest brown rot and blue mold rot. Fruits 56: 123–130.

Cohen E (1981) Metalaxyl for postharvest control of brown rot of citrus fruit. Plant Dis 65: 672–675.

Cohen E (1989) Evaluation of fenpropimorph and flutriafol for control of sour rot, blue mould and green mould in lemon fruit. Plant Dis 73:807–809.

Corke ATK, Sneh B (1979) Antisporulant activity of chemicals towards fungi causing cankers on apple branches. Ann Appl Bio 91: 325–330.

Daines RH (1970) Effects of fungicide dip treatments and dip temperatures on postharvest decay of peaches. Plant Dis Rep 54: 764–767.

Dennis C (1983) Post-harvest pathology of fruits and vegetables. Academic Press, New York, p. 264.

Dijkhuizen JP, Ogawa JM, Manji BT (1983) Activity of captan and prochloraz on benomyl-sensitive and benomyl-resistant isolates of *Molilinia fructicola*. Plant Dis 67: 407–409.

Droby S, Porat R, Cohen L, Weiss B, Shapira B, Philosoph-Hadas S, Meir S (1999) Suppressing green mould decay in grapefruit with postharvest jasmonate application. J Am Soc Hortic Sci 124: 184–188.

Eckert JW (1977) Control of postharvest diseases. In: Siegel MR, Sisler HD (Eds). Antifungal compounds. Marcel Dekker, New York, pp. 269–352.

Eckert JW (1978) Pathological diseases of fresh fruits and vegetables. In: Hultin HO, Milner N (Eds). Postharvest biology and biotechnology. Food and Nutrition Press, Westport, CT, pp. 161–209.

Eckert JW (1990) Recent development in the chemical control of postharvest diseases. Acta Hortic. 269: 477–494.

Eckert JW, Bretschneider BF, Ratnayake M (1981) Investigations on new postharvest fungicides for citrus fruits in California. Proc Int Soc Citric 2: 804–810.

Eckert JW, Eaks IL (1989) Postharvest disorders and diseases of citrus fruit. In: Reuther W, Calavan EC, Carman G (Eds). The citrus industry. University of California Div Agric Sci, Oakland, 5: 179–260.

Eckert JW, Kolbezen MJ (1970) Fumigation of fruits with 2-aminobutane to control certain postharvest diseases. Phytopathology 60: 545–550.

Eckert JW, Kolbezen MJ, Rahm ML, Eckard KJ (1979). Influence of benomyl and methyl-2-benzimidazole carbamate on the development of *Penicillium digitatum* in the pericarp of orange fruit. Phytopathology 69: 934–939.

Eckert JW, Ogawa JM (1985) The chemical control of postharvest harvest diseases: Subtropical and tropical fruits. Annu Rev Phytopathol 23: 421–454.

Eckert JW, Ogawa JM (1988) The chemical control of postharvest diseases: Deciduous fruits, berries, vegetables and root/ tuber crops. Annu Rev Phytopathol 26: 433–469.

Edney KL, Chambers DA (1981) Postharvest treatments for the control of *Phytophthora syringae* storage rot of apples. Ann Appl Bio 97: 237–341.

Edney KL, Chambers DA (1981) The use of metalaxyl to control *Phytophthora syringae* of apple fruits. Plant Pathol 30: 167–170.

El Ghaouth A, Arul J, Wilson C, Benhamou N (1997) Biochemical and cytological aspects of the interactions of chitosan and Botrytis cinerea in bell pepper fruit. Postharvest Biol Tec 12:183–194.

El-Mougy NS, El-Gamal NG, Abd-El-Kareem F (2008) Use of organic acids and salts to control postharvest diseases of lemon fruits in Egypt. Arch Phytopathol Plant Prot 41: 467–476.

Errampalli D (2004) Effect of fludioxonil on germination and growth of *Penicillium expansum* and decay in apple cvs. Empire and Gala. Crop Prot 23: 811–817.

Fallik E, Aharoni Y, Grinberg S, Copel A, Klein JD (1994) Postharvest hydrogen peroxide treatment inhibits decay in eggplant and sweet red pepper. Crop Prot 13: 451– 454.

Fallik E, Archbold DD, Hamilton-Kemp TR, Clements AM, Collins RW, Barth ME (1998) (E)-2-Hexenal can stimulate *Botrytis cinerea* growth in vitro and on strawberry fruit *in vivo* during storage. J Amer Soc Horti Sci 123: 875–881.

Alaoui FT, Askarne L, Boubaker H, El Hassane B, Ben A (2017) Control of gray mold disease of tomato by postharvest application of organic acids and salts. Plant Pathol J 16: 62–72.

Feng W, Zheng X (2007) Essential oils to control *Alternaria alternatain vitro* and *in vivo*. Food Control 18: 1126–1130.

Feng W, Zheng X, Chen J, Yang Y (2008) Combination of cassia oil with magnesium sulphate for control of postharvest storage rots of cherry tomatoes. Crop Protec 27: 112–117.

Fenwick GR, Heaney RK, Mullin WJ (1983) Glucosinolates and their breakdown products in food and food plants. CRC Crit Rev Food Sci Nutr18: 123–201.

Forneyl CF, Rij RE, Denis-Arrue R, Smilanick JL (1991) Vapor phase hydrogen peroxide inhibits postharvest decay of table grapes. Hort Sci 26(12):1512–1514.

Frossard P (1968) Essais de desinfection des pedoncules d'ananas contre Ie *Thielaviopsis paradoxa*. Fruits 23: 207–215.

Frossard P (1970) Desinfection des ananas contre *Thielaviopsis paradoxa*. Fruits 25: 785–791.

Frossard P, Laville E, Plaud G (1977) Etude des traitements fongicides appliques aux bananes apres recolte. III. Action de l'imazalil. Fruits 32: 673–676.

Gaulliard JM, Pelossier R (1983) Efficacite de phosethyl Al entrempagedesagrumes (fruits) contre *Phytophthora parasitica* agent de la pourriturebruneetcontre *Penicillium digitatum*. Fruits 38: 693–697.

Gregori R, Borsetti F, Neri F, Mari M, Bertolini P (2008) Effects of potassium sorbate on postharvest brown rot of stone fruit. J Food Protec 71 (8): 1626.

Gonzalez-Aguilar GA, Buta JG, Wang CY (2003) Methyl jasmonate and modified atmosphere packaging (MAP) reduce decay and maintain postharvest quality of papaya 'Sunrise'. Postharvest Biol Tec 28: 361–370.

Griffee PJ, Pinegar JA (1974) Fungicides for control of the banana crown rot complex: *in vivo* and *in vitro* studies. Trap Sci 16:107–120.

Gutter Y (1983) Supplementary antimold activity of phosethyl AI, a new brown rot fungicide for citrus fruits. Phytopathol Z 107: 301–308.

Gutter Y, Shachnai A, Schiffmann NM, Dinoor A (1981) Chemical control in citrus of green and blue molds resistant to benzimidazoles. Phytopathology 102: 127–138.

Hamilton-Kemp TR, McCracken Jr CT, Loughrin JH, Anderson RA, Hildebrand DF (1992) Effect of some natural volatile compounds on the pathogenic fungi *Alternaria alternata* and *Botrytis cinerea*. J Chem Ecol 18: 1083–1091.

Hardenburg RE, Spalding DH (1972) Postharvest benomyl and thiabendazole treatments, alone and with scald inhibitors, to control blue and gray mold in wounded apples. J Am Soc Hortic Sci. 97: 154–158.

Heaton JB (1980) Control of postharvest decay of clingstone peaches with postharvest fungicidal dips. Queensland J Agri Animal Sci 37: 155–159.

Heggen B, Dave B, Kaplan HJ (1980) Effective fungicidal control of postharvest disease on kiwifruit (Abs.). HortSci 15: 92.

Indrakeerthi SRP, Adikaram NKB (1996) Papaya latex, a potential postharvest fungicide. In: Proc. Australian Postharvest Hortic Conf 'Science and Technology for the Fresh Food Revolution', Melbourne, Australia, pp. 423–427.

Iqbal Z, Singh Z, Khangura R, Ahmad S (2012) Management of citrus blue and green moulds through application of organic elicitors. Austra Plant Pathol 41:69–77.

Isaac O, Maalekuu BK (2013) Effect of some postharvest treatments on the quality and shelf life of three cultivars of carrot (*Daucus carota* L.) during storage at room temperature. Am J Food Nutr 3(2): 64–72.

Ishiki K, Tokuora K, Mori R, Chiba S (1992) Preliminary examination of allyl isothiocyanate vapour for food preservation. Biosci Biotechnol Biochem 56: 1476–1477.

Islam MZ, Mele MA, Hussein KA, Kang H-M (2018) Acidic electrolyzed water, hydrogen peroxide, ozone water and sodium hypochlorite influence quality, shelf life and antimicrobial efficacy of cherry tomatoes. Res J Bio Technol 13:4.

Jansen LG, Florentino CO, Cruz JN, Gomez JJ, Van den Berg Y (1995) Algunos Estudiossobre Citrex (Lonlife) en la Republica Dominicana, Santo Domingo. Republica Dominicana, Citrex Dominicana, SA.

Jeandet P, Bessis R, Gautheron B (1991) The production of resveratrol (3,5,4′-trihydroxystilbene) by grape berries in different developmental stages. Am J Enol Vitic 42: 41–44.

Jennrich H (1985). Ronilan-smoke, a new vincozolin formulation for the control *of Botrytis cinerea* and other diseases in greenhouses and storage rooms. Med Fac Landbouww Rijsuniv Gent 50: 1227–1233.

Jones AL (1975) Control of brown rot of cherry with a new hydantoin fungicide and with selected fungicide mixtures. Plant Dis Rep 59: 127–129.

Jones AL, Burton CL, Tennes BR (1973) Postharvest fungicide and heat treatments for brown rot control on stone fruits. Mich State Univ Agric Exp Sta Res Rep. 209.

Kanetis L, Forster H, Adaskaveg JE (2007) Comparative efficacy of the new postharvest fungicides azoxystrobin, fludioxonil, and pyrimethanil for managing citrus green mold. Plant Dis 91:1502–1511.

Karabulut OA,Gabler FM, Mansour M, Smilanick JL (2004) Postharvest ethanol and hot water treatments of table grapes to control gray mold. Postharvest Biol Technol 34(2): 169–177.

Knight SC, Anthony VM, Brady AM, Greenland AJ, Heaney SP, Murray DC, Powell KA, Schulz MA, Spinks CA, Worthington PA, Youle D (1997) Rationale and perspectives on the development of fungicides. Annu Rev Phytopathol 35: 349–372.

Koffman W, Kable PF (1975) Improved control of brown rot in harvested sweet cherries by triforine dip treatments. Plant Dis Rep 59: 586–590.

Kolaei EA, Cenatus C, Tweddell RJ, Avis TJ (2013) Antifungal activity of aluminium-containing salts against the development of carrot cavity spot and potato dry rot. Ann Appl Biol 163(2): 311–317.

Kondo A, Liu Y, Furuta M, Fujita Y, Matsumoto T, Fukuda H (2000) Preparation of high activity whole cell biocatalyst by permeabilization of recombinant flocculent yeast with alcohol. Enzyme Microbial Technol 27(10): 806–811.

Krochta JM, Carlson RA, Ogawa JM, Manji BT (1977) Harvesting into foam reduces tomato losses. Food Technol 31:42–46.

Larrigaudiere C, Pons J, Torres R, Usall J (2000) Storage performance of clementines treated with hot water, sodium carbonate and sodium bicarbonate dips. J Horti Sci Biotechnol 77(3): 314–319.

Lichter A, Zutkhy Y, Sonego L, Dvir O, Kaplunov T, Sarig P, Ben-Arie R (2002) Ethanol controls postharvest decay of table grapes. Postharvest Biol Technol 24: 301–308.

Lima G, De Curtis F, Castoria R, Pacifica S, De Cicco V (1998) Additives and natural products against post harvest pathogens compatibility with antagonistic yeasts. In: Plant Pathology and Sustainable Agriculture. Proceedings of the Sixth SIPaV Annual Meeting, Campobasso, 1998: 17–18 September.

Little CR, Taylor HJ, Peggie ID (1980) Multiformulation dips for controlling storage disorders of apples and pears. I. Assessing fungicides. Sci Hort 13: 213–219.

Liu J, Sui Y, Wisniewski M, Droby S, Liu Y (2013) Utilization of antagonistic yeasts to manage postharvest fungal diseases of fruit. Int J Food Microbiol 167: 153–160.

Long PG (1971) Evaluation of benomyl, thiabendazole, benzene thiophanate and methyl thiophanate for control of banana stem end rot disease (*Gloeosporium musarum*). J Exp Agric Anim Husb 11:559–561.

Maas JL, Mac Swan IC (1970). Postharvest fungicide treatments for reduction in Penicillium decay of Anjou pears. Plant Dis Rep 54: 887–890.

Mari M, Bertoii P, Prateiia GC (2003) Non-conventional methods for the control of post harvest pear diseases. J Appl Microbiol 94: 761–766.

Mari M, Cembali T, Baraldi E, Casalini L (1999) Peracetic acid and chlorine dioxide for postharvest control of *Monilinia laxa* in stone fruits. Plant Dis 83(8): 773–776.

Mari M, Leoni O, Lori R, Cembali T (2002) Antifungal vapour-phase activity of allyl isothiocyanate against *Penicillium expansum* on pears. Plant Pathol 51: 231–236.

Mari M, Leoni O, Lori R, Marchi A (1996) Bioassay of glucosinolate derived isothiocyanates against post harvest pear pathogens. Plant Pathol 45: 753–760.

Mari M, Neri F, Spadoni A (2016) Natural compounds: An alternative in postharvest disease control. Acta Horti 1144: 385–390.

Mattheis J, Roberts RG (1993) Fumigation of sweet cherry (*Prunus avium* 'Bing') fruit with low molecular weight aldehydes for postharvest decay control. Plant Dis 77: 810–814.

McKay AH, Förster H, Adaskaveg JE (2012) Efficacy and application strategies for propiconazole as a new postharvest fungicide for managing sour rot and green mold of citrus fruit. Plant Dis 96: 235–242.

Meepagala KM, Sturtz G, Wedge DE (2002) Antifungal constituents of the essential oil fraction of Artemisia dracunculus L. var. dracunculus. J Agri Food Chem 50: 6989–6992.

Mihaliak CA, Gershenzo J, Croteau R (1991) Lack of rapid monoterpene turnover in rooted plants, implications for theories of plant chemical defense. Oecologia 87: 373–376.

Miller WR, Spalding DH, Risse LA, Chew V (1984) The effects of an imazalil-impregnated film with chlorine and imazalil to control decay of bell peppers. Proc Fla State Hortic Soc 97:108–111.

Mirelmann D, Galun E, Sharon N, Lotan R (1975) Inhibition of fungal growth by wheat germ agglutinin. Nature 256: 414–416.

Moline HE, Buta JG, Saftner RA, Maas JL (1997) Comparison of three volatile natural products for the reduction of post harvest diseases in strawberries. Adva Strawberry Res 16: 43–48.

Moline HE, Kuti JO (1984) Comparative studies of two *Mucor* species causing postharvest decay of tomato and their control. Plant Dis 68(6): 524–526.

Morris SC, Wade NL (1983) Control of postharvest disease in cantaloupes by treatment with guazatine and benomyl. Plant Dis 67: 792–794.

Mottura MC, Perelló R, Orihuel-Iranzoa B (2019) Effects of postharvest application of Citrocide® PLUS, a peracetic acid-based formulation, on tomato decay control. Acta Hortic 1256: 407–412.

Moyls AL, Sholberg PL, Gaunce AP (1996) Modified atmosphere packaging of grapes and strawberries fumigated with acetic acid. HortSci 31: 414–416.

Muirhead IF, Pitzell RD, Davis RD, Peterson RA (1982) Postharvest control of anthracnose and stem-end rots of Puerte avocados with prochloraz and other fungicides. Aust J Exp Agric AnimHusb 22:441–446.

Nagy S, Wardowski WF, Heam CJ (1982) Diphenyl absorption and decay in 'Dancy' and 'Sunburst' tangerine fruit. J Am Soc Hortic Sci 107:154–157.

Narayanasamy P (2006) Postharvest pathogens and disease management. John Wiley, Somerset, NJ, p. 578. Hardcover. ISBN: 978-0-471-74303-3.

Nelson PM, Wheeler RW, McDonald PD (1983) Potassium sorbate in combination with benzimidazoles reduces resistant *Penicillium digitatum* decay in citrus. Proc Int Soc Citric 2: 820–823.

Neri F, Mari M, Menniti AM, Brigati S, Bertolini P (2006) Control of *Penicillium expansum* in pears and apples by trans-2-hexenal. Postharvest Biol Technol 41: 101–108.

Palou L, Marcilla A, Rojas-Argudo C, Alonso M, Jacas J, del Río MA (2007) Effects of X-ray irradiation and sodium carbonate treatments on postharvest *Penicillium* decay and quality attributes of clementine mandarins. Postharvest Biol Technol 46: 252–261.

Palou L, Moscoso-Ramírez PA, Montesinos-Herrero C (2018) Assessment of optimal postharvest treatment conditions to control green mold of oranges with sodium benzoate. Acta Horti 1194: 221–226.

Palou L, Smilanick JL, Usall J, Vinas I (2001) Control of postharvest blue and green molds of oranges by hot water, sodium carbonate, and sodium bicarbonate. Plant Dis 85: 371–376.

Palou L, Usall J, Muñoz JA, Smilanick JL, Vinas I (2002) Hot water, sodium carbonate, and sodium bicarbonate for the control of postharvest green and blue molds of clementine mandarins. Postharvest Biol Tec 24: 93–96.

Palou L, Usall J, Smilanick JL, Aguilar MJ, Vinas I (2002) Evaluation of food additives and low-toxicity compounds as alternative chemicals for the control of *Penicillium digitatum* and *Penicillium italicum* on citrus fruit. Pest Manag Sci 58:459–466.

Peiser P duT (1973) Post-harvest decay in citrus fruits: New fungicides show promise of control. Citrus Subtrop Fruit J 474:7–8.

Pesis E, Avissar I (1990) Effect of postharvest application of acetaldehyde vapour on strawberry decay, taste and certain volatiles. J Sci Food Agr 52: 377–385.

Plooy WD, Regnier T, Combrinck S (2009) Essential oil amended coatings as alternatives to synthetic fungicides in citrus postharvest management. Postharvest Biol Technol 53: 117–122.

Prasad SS, Bilgrami RS (1973) Investigations on diseases of litchi. III. Fruit rots and their control by postharvest treatments. Indian Phytopathol 26: 523–527.

Prusky D, Ben-Arie R (1985) Effect of imazalil on pathogenicity of *Penicillium* spp. causing storage rot of pome fruits. Plant Dis 69: 416–418.

Prusky D, Ben-Arie R, Guelfat-Reich S (1981) Etiology and histology of *Alternaria alternata* of persimmon fruits. Phytopathology 71(11): 24–28.

Prusky D, Fuchs Y, Yanko U (1983) Assessment of latent infections as a basis for control of postharvest disease of mango. Plant Dis 67: 816–618.

Radi M, Jouybari HA, Mesbahi G, Farahnaky A, Amiri S (2010) Effect of hot acetic acid solutions on postharvest decay caused by *Penicillium expansum* on Red Delicious apples. Sci Hortic 126(4): 421–425.

Rana BK, Taneja V, Singh UP (1999) Antifungal activity of an aqueous extract of leaves of garlic creeper (*Adenocalymna alliaceum* Miers.). Pharmaceut Biol 37: 13–16.

Ravetto DJ, Ogawa JM (1972) Penetration of peach fruit by benomyl and 2,6-dichloro-4-riitroaniline fungicides (Abst.). Phytopathology 62: 754.

Reddy MVB, Angers P, Gosselin A, Arul J (1998) Characterization and use of essential oils from *Thymus vulgaris* against *Botrytis cinerea* and *Rhizopus stolonifer* in strawberry fruits. Phytochemistry 47 (8): 1515–1520.

Regnier T, Plooy WD, Combrink S, Botha B (2008) Fungitoxicity of *Lippiascaberrima* essential oil and selected terpenoid components on two mango postharvest spoilage pathogens. Postharvest Biol Technol 48: 254–258.

Romero DM, Guillen F, Valverde JM, Bailen G, Zapata P, Serrano M, Castillo S, Valero D (2007) Influence of carvacrol on survival of *Botrytis cinerea* inoculated in table grapes. Int J Food Microbiol 115:144–148.

Rosen J C, Kader A A (1989). Postharvest physiology and quality maintenance of sliced pear and strawberry fruits. *J Food Sci* 54: 656–659.

Roy MK (1981) Guazatine in the control of *Fusarium semitectum* in tomato fruits. Indian Phytopathol 34: 241–243.

Salem EA, Youssef K, Sanzani SM (2016) Evaluation of alternative means to control postharvest Rhizopus rot of peaches. Sci Hortic 198: 86–90.

Sarig P, Zahavi T, Zutkhi Y, Yannai S, Lisker N, Ben AR (1996) Ozone for control of post-harvest decay of table grapes caused by *Rhizopus stolonifer*. Physiol Mol Plant Pathol 48(6): 403–415.

Schaffer B, Wolstenholme BN, Whiley AW (2013) The avocado: Botany, production and uses. 2nd ed. CAB Intl. Press, Wallingford, 560 p.

Schirra M, D'Aquino S, Palma A, Marceddu S, Angioni A, Cabras P, Scherm B, Migheli Q (2005) Residue level, persistence and storage performance of citrus fruit treated with fludioxonil. J Agric Food Chem 53: 6718–6724.

Schlumbaum A, Mauch F, Vogeli U, Boller T (1986) Plant chitinases are potent inhibitors of fungal growth. Nature 324: 365–367.

Schroeder LL, Bullerman LB (1985) Potential for development of tolerance by *Penicillium digitatum* and *Penicillium italicum* after repeated exposure to potassium sorbate. Appl Envir Microbiol 51: 919–923.

Sebastian E, Nidiry J, Girija G, Lokesha AN (2011) Antifungal activity of some extractives and constituents of *Aloe vera*. Res J Medicinal Plants 5: 196–200.

Sharma RR, Alemwati P (2010) Natural products for postharvest decay control in horticultural produce: A review. Stewart Postharvest Rev 6 (4):1–9.

Sharma RR, Singh D, Singh R (2009) Biological control of postharvest diseases of fruits and vegetables by microbial antagonists: A review. Biol Control 50: 205–221.

Sharon N (1997) Functional aspects of microbial and plant lectins. Nova Acta Leopoldina NF75 301: 13–26.

Sharon N, Lis H (1989) Lectins. Chapman & Hall, New York. https://doi.org/10.1016/0307-4412(90)90245-J.

Shillingford CA (1977) Control of banana fruit rots and of fungi that contaminate washing water. Trop Sci 19:197–203.

Sholberg PL (1998) Fumigation of fruit with short-chain organic acids to reduce the potential of postharvest decay. Plant Dis 82:689–693.

Sholberg PL, Bedford K, Stokes S (2005) Sensitivity of *Penicillium* spp. and *Botrytis cinerea* to pyrimethanil and its control of blue and gray mold of stored apples. Crop Prot 24:127–134.

Sholberg PL, Gaunce AP (1996) Fumigation of stonefruit with acetic acid to control postharvest decay. Crop Protec 15(8): 681–686.

Sholberg PL, Gaunce AP (1995) Fumigation of fruit with acetic acid to prevent postharvest decay. Hort Sci 30: 1271–1275.

Sholberg PL, Shephard T, Randall P, Moyls L (2004) Use of measured concentrations of acetic acid vapour to control postharvest decay in d'Anjou pears. Postharvest Biol Technol 32: 89–98.

Sibi G, Rashmi Wadhavan, Singh S, Shukla A, Dhananjaya K, Ravikumar KR, Mallesha H (2013) Plant latex: A promising antifungal agent for post harvest disease control. PJBS 16: 1737–1743. Antifungal activity of an aqueous extract of leaves of garlic creeper (Adenocalymna alliaceum Miers.).

Singh D (2005) Interactive effect of *Debaryomyces hansenii* and calcium chloride to reduce Rhizopus rot of peaches. J Mycol Plant Pathol 35(1): 118–121.

Singh D, Thakur AK (2003) Effect of pre harvest spraying of fungicides and calcium nitrate on fruit rot and occurrence of mycoflora on Kinnow during low temperature storage. Plant Dis Res 18(1): 9–11.

Sitton JW, Pierson CF (1983) Interaction and control of Alternaria stem decay and blue mold of d'Anjou pears. Plant Dis 67: 904–907.

Smilanick JL, Aiyabei J, Mlikota Gabler F, Doctor J, Sorenson D, Mackey B (2002) Quantification of the toxicity of aqueous chlorine to spores of *Penicillium digitatum* and *Geotrichumcitri-aurantii*. Plant Dis 86: 509–514.

Smilanick JL, Mackey BE, Reese R, Usall J, Margosan DA (1997) Influence of concentration of soda ash, temperature, and immersion period on the control of postharvest green mold of oranges. Plant Dis 81: 379–382.

Smilanick JL, Mansour MF, Mlikota Gabler F, Sorenson D (2008) Control of citrus postharvest green mold and sour rot by potassium sorbate combined with heat and fungicides. Postharvest Biol Technol 47: 226–238.

Smilanick JL, Margosan DA, Henson DJ (1995) Evaluation of heated solutions of sulfur dioxide, ethanol and hydrogen peroxide to control postharvest green mold of lemons. Plant Dis 79: 742–747. Smilanick JL, Margosan DA, Mlikota Gabler F, Usall J, Michael IF (1999) Control of citrus green mold by carbonate and bicarbonate salts and the influence of commercial postharvest practices on their efficacy. Plant Dis 83: 139–145.

Smilanick JL, Sorenson D (2001) Control of postharvest decay of citrus fruit with calcium polysulfide. Postharvest Biol Technol 21:157–168.

Sommer JF, Fortlage RJ, Edwards DC (1983) Minimizing postharvest diseases of kiwifruit. Cal Agric 37–38: 16–18.

Song J, Leepipattanawit R, Deng W, Beaudry RM (1996) Hexenal vapor is a natural, metabolizable fungicide: inhibition of fungal activity and enhancement of aroma biosynthesis in apple slices. J Amer Soc Hortic Sci 121: 937–942.

Spadaro D (2011) Biological control of postharvest diseases of fruits and vegetable. In: UNESCO-EOLSS Joint Committee (Ed.). Agricultural Sciences, Encyclopedia of Life Support Systems (EOLSS), developed under the auspices of the UNESCO. EOLSS Publishers, Oxford. http://www.eolss.net.

Spalding DH (1980) Control of Alternaria rot of tomatoes by postharvest application of imazalil. Plant Dis 64: 169–171.

Spalding DH (1982) Resistance of mango pathogens to fungicides used to control postharvest diseases. Plant Dis 66:1185–1186.

Spalding DH, Reeder WF (1972) Postharvest disorders of mangos as affected by fungicides and heat treatments. Plant Dis Rep 56: 751–753.

Spalding DH, Reeder WF (1975) Low-oxygen high-carbon dioxide-controlled atmosphere storage for control of anthracnose and chilling injury of avocados. Phytopathology 65: 458–460.

Swinburne TR (1973) The resistance of immature Bramley's seedling apples to rotting by *Nectrixgalligena* Bres. In: Byrde RJW, Cutting CV (Eds.). Fungal pathogenicity and the plant's response. Academic Press, London, New York, pp. 365–382.

Tak SK, Verma OP, Pathak VN (1985) Control of Alternaria rot of apple fruits by postharvest application of chemicals. Indian Phytopathol 38(3): 471–474.

Thakur KS, Jyoti K, Kumar S, Gautum S (2017) Improvement of postharvest keeping quality of bell pepper (*Capsicum annum* L.) fruits treated with different chemicals following cold storage. Int J Curr Microbiol App Sci 6(7): 2462–2475.

Tosi B, Donini A, Romagnoli C, Bruni A (1996) Antimicrobial activity of some commercial extracts of propolis prepared with different solvents. Phytopthora Res 10: 335–336.

Tripathi P, Dubey NK, Pandey VB (2002) Kaempferol: The antifungal principle of *Acacia nilotica* Linn. Del J Indian Bot Soc 81: 51–54.

Troncoso R, Espinoza C, Sanchez-Estrada A, Tiznado ME, Garcia HS (2005) Analysis of isothiocyanates present in cabbage leaves extract and their potential application to control Alternaria rot in bell peppers. Food Res Int 38:701–708.

Troncoso-Rojas R, Tiznado-Hernández ME (2007) Natural compounds to control fungal postharvest diseases. In: Troncoso-Rojas R, Tiznado-Hernández ME, González-León A (Eds.). Recent advances in alternative postharvest technologies to control fungal diseases in fruits and vegetables. Transworld Research Network, Trivandrum, Kerala, India, pp. 127–156.

Tzortzakis NG (2009) Impact of cinnamon oil-enriched on microbial spoilage of fresh product. Innov Food Sci Emerg 10: 97–102.

Tzortzakis NG, Economakis CD (2007) Antifungal activity of lemongrass (*Cympopogon citratus* L.) essential oil against key postharvest pathogens. Innov Food Sci Emerg 8: 253–258.

Utto W, Mawson AJ, Bronlund JE (2008) Hexanal reduces infection of tomatoes by *Botrytis cinerea* whilst maintaining quality. Postharvest Biol Technol 47: 434–437.

Valero D, Valverde JM, Romero DM, Guillen F, Castillo S, Serrano M (2006) The combination of modified atmosphere packaging with eugenol or thymol to maintain quality, safety and functional properties of table grapes. Postharvest Biol Technol 41: 317–327.

Van Parijs J, Broekaert WF, Goldstein IJ, Peumans WJ (1991) Hevein: An antifungal protein from rubber tree (*Hevea brasiliensis*) latex. Planta 183: 258–262.

Varma J, Dubey NK (2001) Efficacy of essential oils of *Caesulia axillaris* and *Mentha arvensis* against some storage pests causing biodeterioration of food commodities. Int J Food Microbiol 68: 207–210.

Venditti T, Dore A, Molinu MG, Aggabio M, D'hallewin G (2009) Combined effect of curing followed by acetic acid vapour treatments improves postharvest control of *Penicillium digitatum* on mandarins. Postharvest Biol Technol 54: 11–114.

Venditti T, Ladu G, Cubaiu L, Myronycheva O, D'hallewin G (2017) Repeated treatments with acetic acid vapors during storage preserve table grapes fruit quality. Postharvest Biol Technol 125:91–98.

Vilaplana R, Alba P, Valencia-Chamorro S (2018) Sodium bicarbonate salts for the control of postharvest black rot disease in yellow pitahaya (*Selenicereus megalanthus*). Crop Prot 114: 90–96.

Wang SY, Chen CT, Yin JJ (2010) Effect of allyl isothiocyanate on antioxidants and fruit decay of blueberries. Food Chem 120: 199–204.

Wang Y, Ren X, Song X, Yu T, Lu H, Wang P, Wang J, Zheng XD (2010) Control of postharvest decay on cherry tomatoes by marine yeast *Rhodosporidium paludigenum* and calcium chloride. J Appl Microbiol 109(2): 651–656.

Wells JM, Merwarth FL (1973) Fungicide dips for controlling decay of carrots in storage for processing. Plant Dis Rep 57: 697–700.

Wild BL (1987) Fungicidal activity of potassium sorbate against *Penicillium digitatum* as affected by thiabendazole and dip temperature. Sci Hortic 32:41–47.

Wilson CL, EI-Ghaouth A, Chalutz E, Droby S, Stevens C, Lu JY, Kahn VA, Arul J (1994) Potential of induced resistance to control postharvest diseases of fruit and vegetable. Plant Dis 78: 837–844.

Wilson CL, Franklin JD, Otto BE (1987) Fruit volatiles inhibitory to *Monilinia fructicola* and *Botrytis cinerea*. Plant Dis 71: 316–319.

Wisniewski M, Droby S, Norelli J, Liu J, Schena L (2016) Alternative management technologies for postharvest disease control: The journey from simplicity to complexity. Postharvest Biol Technol 122: 3–10.

Worthing CR, Walker BS (1983) British Crop Protection Council The Pesticide Manual. 7th ed. British Crop Protection Council, Croydon, UK; 695 pp.

Wu CT (2010) An overview of postharvest biology and technology of fruits and vegetables technology on reducing postharvest losses and maintaining quality of fruits and vegetables. 2010 AARDO Workshop on Technology on Reducing Post-harvest Losses and Maintaining Quality of Fruits and Vegetables 2–11.

Xiao CL, Boal RJ (2009) Pre-harvest application of a Boscalid and Pyraclostrobin mixture to control postharvest gray mold and blue mold in apples. Plant Dis 93:185–189.

Xu W, Huang K, Gua F, Qu W, Yang J, Liang Z, Luo Y (2007) Postharvest grapefruit seed extract and chitosan treatments of table grapes to control *Botrytis cinerea*. Postharvest Biol Technol 46: 86–94.

Yaghmour MA, Bostock RM, Adaskaveg JE, Michailides TJ (2012) Propiconazole sensitivity in populations of *Geotrichum candidum*, the cause of sour rot of peach and nectarine, in California. Plant Dis 96:752–758.

Yao H, Tian S (2005) Effects of pre- and post-harvest application of salicylic acid or methyl jasmonate on inducing disease resistance of sweet cherry fruit in storage. Postharvest Biol Technol 35: 253–262.

Youssef K, Hashim AF, Margarita R, Alghuthaymi MA, Abd-Elsalam KA (2017) Fungicidal efficacy of chemically-produced copper nanoparticles against *Penicillium digitatum* and *Fusarium solani* on citrus fruit. Philipp Agric Sci 100: 69–78.

Youssef K, Sanzani SM, Ligorio A, Ippolito A, Terry LA (2014) Sodium carbonate and bicarbonate treatments induce resistance to postharvest green mould on citrus fruit. Postharvest Biol Technol 87: 61–69.

Zabka M, Pavela R, Slezakova L (2009) Antifungal effect of *Pimenta dioica* essential oil against dangerous pathogenic and toxinogenic fungi. Indus Crops and Products 30:250–253.

Zauberman G, Schiffmann-Nadel M, Fuchs Y, Yanko V (1975) Laluttecontre les pourritures de I' avocat et son effet sur Iechangement de la flore des champignons pathogenes des fruits. Fruits 30: 503–504.

Zhang J, Swingle PP (2003) Control of green mold on Florida citrus fruit using bicarbonate salts. Proc Florida State Hort Soc 116:375–378.

Zhang FS, Wang XQ, Ma SJ, Cao SF, Li N, Wang XX, Zheng YH (2006) Effect of methyl jasmonate on postharvest decay in strawberry fruit and the possible mechanisms involved. Acta Hortic 712(2): 693–698.

Zoffoli JP, Latorre BA, Rodrıguez EJ, Aldunce P (1999) Modified atmosphere packaging using chlorine gas generators to prevent *Botrytis cinerea* on table grapes. Postharvest Biol Technol 15: 135–142.

6

Bio-control of Postharvest Pathogens Affecting Perishable Horticultural Commodities

Ram Roshan Sharma[1], **Dinesh Singh**[1], **Vijay Rakesh Reddy**, **Shruti Sethi**[2] **and H. R. Raghavendra**[1]
[1]*ICAR-Indian Agricultural Research Institute, New Delhi, India*
[2]*ICAR-Central Institute for Arid Horticulture, Bikaner, Rajasthan, India*

CONTENTS

6.1 Introduction

Several postharvest diseases and disorders infect horticultural produce during handling until it reaches the consumption. A major loss of this valuable produce due to postharvest diseases may occur at any stage of handling from harvest to consumption. Fresh fruits and vegetables are vulnerable to postharvest losses, as they cause 20–30% losses despite using modern infrastructure and storage facilities (Singh and Sharma 2007, Sharma et al. 2009). Due to improper management of postharvest storage facilities, a wide range of horticultural crops are affected by postharvest diseases. The disease inoculums could be acquired from the field during the growth period, at harvest time, during postharvest handling, storage, transportation or during marketing, and sometimes even after purchase by the consumer (Singh et al. 2017). Postharvest losses lead to greater economic losses as compared to field losses (Eckert 1978). Some reports across the globe have revealed that there is approximately 10–30% and 40–50% losses of harvested fresh horticultural produce due to postharvest spoilage in developed and developing countries, respectively, due to minimal refrigeration facilities, transportation infrastructure, and other problems (Salunkhe et al. 1991). Availability and application of reliable techniques in the proper place and time could prevent and manage postharvest spoilage. Furthermore, the production of mycotoxins by *Fusarium*, *Penicillium*, etc. can create an unwanted impact on consumer health.

6.2 Major Pathogens Responsible for Postharvest Diseases

Many pathogens such as bacteria and fungi are accountable for instigative diseases in perishables like fruits and vegetables. Some of the prime bacterial and fungal genera responsible for major postharvest losses include *Erwinia* sp., *Pseudomonas* sp., *Aspergillus* sp., *Alternaria* sp., *Bortytis* sp., *Colletotrichum* sp., *Diplodia* sp., *Monilinia* sp., *Penicillium* sp., *Phomopsis* sp., *Rhizopus* sp., *Mucor* sp., *Sclerotinia* sp., etc. (Table 6.1) (Sharma et al. 2009, Singh and Sharma 2018). Physical damage can play a major role in deterioration of the produce after harvest and it is the prime cause for postharvest losses, as most pathogens are weak but can easily invade injured produce. Fresh produce sustains various injuries during harvest or in later stages triggered by factors such as weather, birds, rodents, insects and/or farm implements. Most of the damage is caused when produce items are dropped on a hard surface during postharvest handling without showing any visible damage marks. A few crops, like apple, exhibit external bruising a little later, and in crops like potato it might be noticed only after peeling. Certain produce like grapes develop compression bruising mainly due to over-stacking in the storehouses. The injured produce is amenable to attack by numerous microbes, thus resulting in progressive decay of the fresh produce, affecting the quality of total produce (Snowdon 1990).

TABLE 6.1

Major Pathogens Causing Diseases in Fruits and Vegetables after Harvest

Fruit/Vegetable	Name of the Disease	Causal Agent	Reference (s)
Apple	Bulls eye rot	*Neofabraea* species	Michalecka et al. (2017)
Apple, pear, peach, plum, cherry, raspberry, strawberry, kiwifruit, grape, citrus, persimmon, pumpkin, squash, cucumber, cabbage, cauliflower, tomato, pepper, eggplant, lettuce, broccoli, pea, beans, carrot, onion, potato, sweet potato, etc.	Gray mould	*Botrytis cinerea*	Masih and Paul (2002),Thomas et al. (2012)
Avocado, banana, mango, citrus, etc.	Crown rot, finger rot, stalk rot, stem-end rot	*Botryodiplodia theobromae*	Barkai-Golan (2005), Renganathan and Muthukumar (2012)
Avocado, banana, stone fruits, pome fruits, tomato, melon, etc.	Pink mould	*Trichothecium roseum*	Wang et al. (2008), Bello (2008)
Avocado, mango, banana, citrus fruits, etc.	Stem-end rot, crown rot, stalk rot, finger rot	*Diplodia natalensis*	Sriram and Poornachandra (2013)
Avocado, papaya, grapes, cherry, raspberry, strawberry, apple, pear, pepper, tomato, eggplant, pepper, pumpkin, melon, pea, bean, sweet potato, etc.	Watery white rot	*Rhizopus stolonifer*	Frances et al. (2006), Fabrício et al. (2010)
Banana	Anthracnose crown rot	*Colletotrichum musae*	Sakinah et al. (2013)
Banana	Crown rot, cigar-end rot	*Verticillium theobromae*	Igeleke and Ayanru (2006)
Banana	Crown rot	*Fusarium pallidoroseum*, *Acremonium* spp.	Barkai-Golan (2005), Umaña-Rojas and Garcia (2011), Renganathan and Muthukumar (2012)
Citrus	Black pit	*Pseudomonas syringae*	Mirik et al. (2005)
Citrus	Brown rot	*Phytophthora citrophthora*	Pane et al. (2001), Vicent et al. (2012)
Citrus fruits	Blue mould	*Penicillium italicum*	Ramírez et al. (2013b)
Citrus fruits	Stem-end rot	*Alternaria citri*	Hiroshi et al. (2007)
Citrus, banana, avocado, mango, etc.	Anthracnose	*Colletotrichum gloeosporioides*	Singh and Thakur (2002); Robert et al. (2012), Lima et al. (2013), Sellamuthu et al. (2013)
Citrus, garlic, onion, pumpkin, squash, cucumber, melon, tomato, brinjal, pea, cabbage, cauliflower, lettuce, celery, broccoli, etc.	Watery white rot, watery soft rot	*Sclerotinia sclerotiorum*	Mei et al. (2012), Margaret et al. (2013)
Exclusively citrus fruits	Green mould	*Penicillium digitatum*	Porat et al. (2000a, 2000b), Ncumisa et al. (2013)

TABLE 6.1 ***(Continued)***

Major Pathogens Causing Diseases in Fruits and Vegetables after Harvest

Fruit/Vegetable	Name of the Disease	Causal Agent	Reference (s)
Grape, papaya, cherry, apple, peach, pear, onion, garlic, potato, sweet potato, beans, carrot, cabbage, cucurbitaceous vegetables, cabbage, cauliflower, broccoli, etc.	Dark spots, fruit rot, sooty mould	*Alternaria alternata*	Barkai-Golan (2005), Yin et al. (2012), Kadam (2012)
Lettuce, tomato, grape, pome fruits, papaya	Black lesion, dark spots	*Stemphylium botryosum*	Barkai-Golan (2005), Llorente et al. (2010), Toselli et al. (2012)
Onion	Bulb rot	*Pantoea agglomerans, P. ananatis, P. allii*	Vahling-Armstrong et al. (2016)
Onion, garlic, pepper, tomato, eggplant, watermelon, squash, pumpkin, artichoke, celery, cabbage, corn, carrot, sweet potato, potato, etc.	Dry or soft rot	*Fusarium* spp.	Barkai-Golan (2005), Sriram et al. (2010), Hou et al. (2012), Thuy et al. (2013)
Onion, garlic, tomato, melon, grape, date, etc.	Black rot	*Aspergillus niger*	Barkai-Golan (2005), Irkin and Korukluoglu (2007), Storari et al. (2012), Ramírez et al. (2013b)
Papaya	Stem-end rot, dry black rot	*Phoma caricae-papayae*	Martins et al. (2010)
Papaya, persimmon, avocado, mango	Stem-end rot; black spot	*Alternaria alternata*	Prusky et al. (2006)
Pea, bean, onion, cucumber, melon, squash, asparagus, cabbage, cauliflower, lettuce, celery, spinach, etc.	Spots or bacterial soft rot	*Pseudomonas syringae*	Leonard et al. (2010), María et al. (2011)
Pear	Bitter rot	*C. fructicola*	Li et al. (2013)
Pineapple, banana	Black heart, brown rot	*Fusarium moniliformae*	Barkai-Golan (2005)
Pineapple, banana	Black rot, stalk rot, crown rot, soft rot	*Ceratocystis paradoxa*	Yadahalli et al. (2007)
Pome and stone fruits	Brown rot	*Monilinia laxa*	Barkai-Golan (2005), Zhu and Guo (2010), Marietta et al. (2012)
Potato	Dry rot	*Fusarium sambucinum*	Bojanowski et al. (2013)
Stone and pome fruits	Blue mould	*Penicillium expansum*	Barkai-Golan (2005), Palou et al. (2013), Masoud et al. (2013b)
Stone fruits	Brown rot	*Monilinia fructicola*	Barkai-Golan (2005), Yin et al. (2013), Sisquella et al. (2013)
Stone fruits and pome fruits		*Colletotrichum gloeosporioides*	Janisiewicz et al. (2003), Masoud et al. (2013a)
Strawberry, raspberry and tomato	Watery soft rot	*Mucor pyriformis*	Borve and Vangdal (2007), Borve et al. (2008)
Tomato, citrus fruits	Sour rot	*Geotrichum candidum*	Talibi et al. (2012a, 2012b), Thornton et al. (2010)
Tomato, pepper, melon, apple, pear, peach, plum, stone fruits, grape, date, papaya	Olive-green mould, sooty mould	*Cladosporium herbarum*	Latorre et al. (2011)

6.3 Management of Postharvest Spoilage

Management of postharvest spoilage consists of prevention and control measures for avoidance of pathogenic microbes until the product reaches consumption. It also includes exclusion of infected produce from healthy ones and methods of developing or incorporating resistance to host and good sanitation measures to eliminate spoilage. Methods for controlling postharvest spoilage and effective eradication of inoculums include physical, chemical or biological approaches. Initially, target organisms have a low resistance capacity to fungicides that are simple element, organic compounds and systematic in nature, while the fungicides which are advanced, complex compounds, non-systematic in plant tissue have a high resistance potential to target organisms (Zhu 2006). Nowadays, there is a global trend of shifting from the use of synthetic fungicides to safer and eco-friendly approaches for reducing postharvest spoilage of fruits and vegetables (Mari et al. 2007). Among the various non-chemical, eco-friendly techniques in controlling of diseases, use of bio-control agents or antagonists like bacteria, yeast and fungi has gained popularity. In this chapter, we have elaborated biological methods for managing of postharvest diseases in perishable commodities.

6.4 Bio-control of Postharvest Pathogens

Various effective and potential approaches for controlling postharvest spoilage of horticultural produce exist through biological control such as antagonistic microorganisms, constitutive or induced resistance, and through use of plant derived natural extracts. A brief description of these strategies is given hereunder.

6.4.1 Developing Resistant Varieties to Pathogens

Using advanced breeding methods to develop varieties, which are resistant to disease-causing microbes, is assumed to be one of the most dependable methods for disease control. However, special consideration is required in developing resistant crosses or varieties to control postharvest spoilage organisms of horticultural commodities. Before releasing of resistant varieties, certain attributes or characteristics need to be introduced into the susceptible variety from the varieties, which were proved to possess resistance to the specified postharvest spoilage organisms (Singh and Sharma 2018). In general, the variety having thin skin, high sugar content, less tannin and other factors make a crop susceptible to postharvest pathogens. There is a great need for plant breeders to recognize varieties that are resistant to diseases occurring after harvest, which are totally different from those exhibiting resistance under field conditions. The major postharvest disease of potato, soft rot (*Erwinia* sp.), has been screened using somatic hybridization through protoplasmic fusion between a wild, non-tuber bearing, diploid species, *Solanum brevidens*, and a tetraploid potato, *S. tuberosum*. Resistance ability was found in *S. tuberosum* that could be easily transferred using a sexual hybridization process.

6.4.2 Induction of Resistance After Harvest

Aspects that are accountable for incorporating resistance in perishable commodities after harvesting operation are discussed below.

6.4.2.1 Wound Healing

Wound healing, also known as wound recovering, is a cell-mediated response resulting in the deposition of lignin substances. For instance, in lemon, the resistance to *Geotrichum candidum,* after wounding has direct correlation with the quantities of lignin deposited in their outer peel (Baudoin and Eckert 1985). Furthermore, Skene (1981) observed that the fruits of apple heal faster immediately after harvesting, which is far better than fruits receiving wounds during storage (Skene 1981).

6.4.2.2 Enzyme Inhibitors

Many research findings have reported that alteration and inhibition of certain pectinolytic enzymes in pear, peach and plum after harvest lead to effective control of postharvest diseases (Fielding 1981; Abu-Goukh and Labavitch 1983).

6.4.2.3 Resistance Induction

It has been widely acknowledged that ethylene triggers enzyme activities, such as phenyl ammonia lyase (PAL) in fruits, that induces the biosynthesis of lignins, phenols and phytoalexins through a shikimic acid pathway. All these secondary metabolites help in incorporation of resistance to disease (Kuc 1982). Host plant defensive mechanisms can be provoked by diverse approaches. For example, use of certain physical as well as bio-agents could induce disease resistance in certain commodities after harvest. Similarly, small dosages of UV-C (<280 nm) could persuade resistance against certain rots caused during storage and also extend storage life of a few harvested commodities such as apple, citrus and peach (Chalutz et al. 1992; Lu et al. 1991). In onion and sweet potato also, postharvest rot can be controlled by incorporating resistance using low doses of UV (Stevens et al. 1990). Swinburne (1973, 1978) has revealed that postharvest rots of apple fruits can be controlled by using an elicitor, benzoic acid. Now, it is time for discovering the diversity and suitability of various physical/chemical triggers for inducing resistance in harvested fruits and vegetables. Efforts are needed for popularization of the identified elicitors for ecological control of postharvest pathogens in perishable commodities.

Natural plant products, derived fungicides and low dosage of UV-C light also render fruits resistant to disease. For example, some fruit waxes having the ingredient chitosan, reduce the growth of pathogens under *in vitro* conditions such as *Alternaria alternata*, *Botrytis cinerea*, *Colletotrichum gloeosporioides* and *Rhizopus stolonifer* (El-Ghaouth and Arul 1992; El-Ghaouth et al. 1992b). Kendra et al. (1989) and Wilson et al. (1994) identified the triggering effects of chitosan in a few hosts through production of phytoalexins. When used as an edible coating, it reduces disease incidence, delays ripening and stimulates defence mechanisms.

Various microbes such as *Pichia guilliermondi* (US-7) have been identified to be antagonistic in nature and control different kinds of rots in fruits such as apples, citrus and peach by inducing phytoalexins (e.g. scoparone) and through triggering of PAL activity in citrus peel (Wilson and Wisniewski 1992). Wilson et al. (1994) reported the development of a chitinase layer in apple and deposition of papillae along the cell walls of the host, which was induced by *Candida* sp. Ultimately, all these affect disease incidence and provide resistance to host cells. When disease resistance is induced, it results in an increased level of phenolic compounds and also has antioxidant properties in the plant tissue which are highly beneficial to humans. Moreover, induction of resistance preserves the natural microflora which is rich in potential bio-control agents and thus provides a combined approach in the eradication of postharvest decay (Romanazzi et al. 2016).

6.4.2.4 Polyamine Biosynthesis Inhibitors

Polyamines play a major role in biological systems and help in biosynthesis and functions of nucleic acids. Polyamines possess aliphatic nitrogenous bases with low molecular weight (Galston and Kaur-Sawhney 1982). Putrescine, spermine, spermidine and cadaverine are the most common polyamines occurring extensively among prokaryotes as well as eukaryotes. Mycological polyamines can be controlled easily through inhibition of ornithine decarboxylase enzyme without disturbing the growth and development of plant tissue. The difloromethyl ornithine is the effective inhibitor of ornithine decarboxylase with no effect on arginine decarboxylase' (Metcalf et al. 1978; Kallio et al. 1981). Thus, incidence of postharvest diseases can be avoided by using these inhibitors.

6.4.3 Bio-control Through Plant-Based Products

Horticultural commodities possess various potential bioactive compounds with antimicrobial properties, which need to be explored for bio-control of postharvest pathogens. However, such compounds extracted could be used on the commodities obtained from same plants or other plant species (Sharma and Pongener 2010). Dubey and Kishore (1988) revealed that *Citrus medica* and *Ocimum* leaves contain essential oil (active at concentration between 500 and 2000 ppm) having the ability to protect various stored commodities from biodegradation of *A. versicolor* and *A. flavus*. The growth of *B. cinerea* and *M. fruticola* was completely inhibited by compounds such as ethyl benzoate, benzaldehyde and methyl salicylate. The fruit rot caused by *B. cinerea* in strawberry and by *R. stolonifer* in raspberry could be easily managed by acetaldehyde vapours (Prasad and Stadelbacher 1973). Smilanick et al. (1995) reported that immersion of lemon fruits in ethanol solution (≥10% w/v) at 44° C reduces the occurrence of green mould by 80% without affecting the fruit quality. Some of the plant products used for controlling postharvest decay are discussed hereunder.

6.4.3.1 Acetic Acid

Acetic acid in varying concentrations was proved to constrain the development of various spoilage microbes and thus has the scope to be used for minimization of postharvest spoilage caused to fruits and vegetables. For instance, the conidia of *B. cinerea* on apple fruits were controlled by lower concentrations of acetic acid without any phytotoxic effects (Sholberg and Gaunce 1995).

6.4.3.2 Jasmonates

Jasmonates belong to the class of olypines and are derived naturally through oxidation of fatty acids mediated by oxygenase enzyme. In nature they occur in the form of jasmonic acid and methyl jasmonate. The incidence of green mould caused by *Penicillium digitatum* in grapefruit *cv.* Marsh Seedless was significantly reduced by the postharvest application of jasmonates (Droby et al. 1999). Jasmonates are soluble in water soluble and hence preferable for use in solution form such as a drip or drench.

6.4.3.3 Glucosinolates

Glucosinolates exhibiting antimicrobial activity are derived from crucifereae plants (Fenwick et al. 1983). When glucosinolates undergo hydrolysis, they produce sulphate ions and D-glucose and many other component gases such as nitrile, isothiocyanate (ITC) and thiocyanate with the potential applications to be used with gaseous treatment or modified atmosphere packaging(MAP) before storage. Pear fruits exposed to allyl-isothiocyanate (AITC), a naturally derived component from mustard, exhibited effective inhibition against blue mould as well as thiabendazole-resistant strains (Mari et al. 2002).

6.4.3.4 Propolis

Propolis is a thick resinous compound procured from the bark and leaf buds of conifer plants (Tosi et al. 1996). It displays very good antifungal and antibacterial properties, thus preventing pathogens such as *Botrytis cineriai* and *Penicillium expansum* (Lima et al. 1998).

6.4.3.5 Plant Extracts

The extract of *Datura stramonium*, *Lawsonia inermis*, *Eucalyptus globulus* and *Punica granatum* plants were more effective in lemon fruit rot (Babu and Reddy 1986). The leaf extract of *tulsi* (Ocimum) effectively restricted the germination of spores and their growth, total proteins, activities of cellulolytic and pectinolytic enzymes in several postharvest pathogens. The concentration of phenols can affect the growth of fungi at a lower concentration (3–5 μg/mL).

6.4.3.6 Essential Oils

Some essential oils exhibit antimicrobial properties, and when applied over perishable commodities improve their shelf life through control of postharvest pathogens. Several reports have been published indicating the role of essential oils in effective control of decay in perishable commodities either by spraying or dipping (Smid et al. 1994; Dixit et al. 1995)For instance, an essential oil extracted from thyme (*Thymus capitatus*), i.e. thymol, has the ability to decrease the occurrence of diseases such as grey mould and brown mould in apricot when used at a lower concentration of 2–4 mg/L (Liu et al. 2002). Tripathi (2001) reported that volatile oils extracted from *Zingiber officinale*, *Ocimum canum* and *Mentha arvensis* were perceived to enhance storage life of citrus by controlling blue mould incidence.

6.4.4 Microbial Contenders for Controlling Postharvest Diseases

Spoilage of perishable commodities after harvest has also been successfully controlled by several microbial antagonists (Van Lenteren et al. 2018). Microbial contenders are generally used through two different methods. The first method is through management and promotion of contending microbes already present over the perishable commodities, and the second method is through artificial induction of desired antagonistic microbes counter to the postharvest pathogens.

6.4.4.1 Natural Microbial Contenders

These are the microbes already present on the outer peel of perishable commodities and have the ability to restrict infection by postharvest pathogens. Several antagonistic microbes have been identified and isolated from surfaces of fruits such as citrus, apples, etc., and they were again applied over the affected fruits to act as bio-control agents. Chalutz and Wilson (1990) reported that when concentrated washings of citrus fruit surface were plated on a suitable media, only yeast and bacteria appeared. After dilution of washings, there was an appearance of rot fungi on agar medium indicating the suppression of that pathogen by the yeast and bacteria.

6.4.4.2 Artificially Introduced Microbial Contenders

Compared to other methods of biological control of postharvest spoilage, use of artificial contenders is a highly effective and efficient approach (Sharma et al. 2009). A number of investigators have tried various bio-control agents for control of postharvest pathogens (Table 6.2).

TABLE 6.2

Microbial Contenders Used for Controlling Various Postharvest Pathogens

Microbial Contender/Antagonist	Disease and Its Causal Agent	Fruit/Vegetable	Reference(s)
Aureobasidium pullulans	Soft rot (*Monilinia laxa*)	Grape	Barkai-Golan (2001)
	Botrytis rot (*Botrytis cinerea*)	Grape	Schena et al. (2003)
	Penicillium rots (*Penicillium* spp.)	Citrus	Wilson and Chalutz (1989)
	Green mould (*Penicillium digitatum*)	Kinnow	Adikaram et al. (2018)
Aureobasidium pullulans PL5	Monilinia rot (*M. laxa*)	Peach	Zhang et al. (2010)
Bacillus subtilis	Anthracnose (*Colletotrichum gloeosporioides*)	Mango	Govender et al. (2005)
	Alternaria rot (*Alternaria alternata*)	Litchi	Jiang et al. (2001)
	Alternaria rot (*A. alternata*)	Muskmelon	Yang et al. (2006)
	Green mould (*P. digitatum*)	Citrus	Singh and Deverall (1984)
	Brown rot (*Lasiodiplodia theobromae*)	Peach, plum and Nectarine	Pusey and Wilson (1984)
Bacillus pumilus	Anthracnose (*C. gloeosporioides*)	Mango	Kefialew and Ayalew (2008)
	Gray mould (*B. cinerea*)	Pear	Mari et al. (1996)
Lactobacillus plantarum	Green mould (*P. digitatum*)	Citrus	Matei et al. (2015)
Pantoea agglomerans	Green (*P. digitatum*) and blue mould (*P. italicum*)	Citrus	Torres et al. (2007)
	Penicillium rot (*P. expansum*)	Apple	Morales et al. (2008)
Pseudomonas fluorescens	Decay	Mango	Barman et al. (2017)
	Green mould rot (*P. digitatum*)	Citrus	Wang et al. (2018)
P. syringae	Blue mould (*P. expansum*)	Apple	Zhou et al. (2002)
	Gray mould (*B. cinerea*)	Apple	Zhou et al. (2001)
Streptomyces globisporus JK-1	Gray mould rot (*B. cinerea*)	Tomato	Li et al. (2012)
S. violascens	Sour rot (*Geotrichum citriaurantii*)	Citrus	Choudhary et al. (2015)
Aureobasidium pullulans	Soft rot (*M. laxa*)	Grape	Barkai-Golan (2001)
	Botrytis rot (*B. cinerea*)	Grape	Schena et al. (2003)
	Penicillium rots (*Penicillium* spp.)	Citrus	Wilson and Chalutz (1989)
	Green mould (*P. digitatum*)	Kinnow	Adikaram et al. (2018)
A. pullulans PL5	Monilinia rot (*M. laxa*)	Peach	Zhang et al. (2010)
Candida ernobii	Stem end rot (*Diplodia natalensis*)	Citrus	Liu et al. (2010)
C. guilliermondii	Gray mould (*B. cinerea*)	Tomato	Saligkarias et al. (2002)
	Gray mould (*B. cinerea*)	Peach, nectarine	Tian et al. (2002)
C. oleophila	Anthracnose (*C. gloeosporioides*)	Papaya	Gamagae et al. (2003)
	Penicillium rots (*P. digitatum, P. italicum*)	Citrus	Lahlali et al. (2005)
	Penicillium rot (*P. expansum*)	Apple	El-Neshawy and Wilson (1997)
	Crown rot (*C. musae*)	Banana	Lassois et al. (2008)
C. sake	Gray mould, Rhizopus rot, Penicillium rot (*P. expansum*)	Pear, apple	Torres et al. (2006), Morales et al. (2008)
C. sake CPA-1	Botrytis bunch rot (*B. cinerea*)	Grape	Calvo-Garrido et al. (2017)
C. stellimalicola	Blue mould rot (*P. italicum*)	Citrus	da Cunha et al. (2018)
C. tropicalis YZ1, YZ27	Anthracnose (*C. musae*)	Banana	Zhimo et al. (2016)

TABLE 6.2 *(Continued)*
Microbial Contenders Used for Controlling Various Postharvest Pathogens

Microbial Contender/Antagonist	Disease and Its Causal Agent	Fruit/Vegetable	Reference(s)
Cryptococcus laurentii	Bitter rot (*Glomerella cingulata*)	Apple	Blum et al. (2004)
	Gray mould (*B. cinerea*)	Tomato	Xi and Tian (2005)
	Gray mould (*B. cinerea*)	Pear	Zhang et al. (2005)
	Brown rot (*M. fructicola*)	Cherry	Qin et al. (2006)
	Rhizopus rot (*R. stolonifer*) and grey mould (*B. cinerea*)	Peach	Zhang et al. (2007)
Debaryomyces hansenii	Green and blue mould (*P. digitatum, P. italicum*)	Citrus	Singh (2002)
D. nepalensis	Anthracnose	Mango	Luo et al. (2015)
Hanseniaspora uvarum (*Kloeckera apiculata*)	Grey mould (*B. cinerea*)	Grapes	Liu et al. (2010)
Kloeckera apiculata	Blue mould decay (*P. italicum*)	Citrus	Liu et al. (2013a, 2013b)
Kluyveromyces marxianus	Green mould (*P. digitatum*)	Citrus	Geng et al. (2011)
Metschnikowia fructicola AP47	Brown rot (*M. laxa*)	Peach	Zhang et al. (2010)
M. pulcherrima Disva 267	Brown rot	Sweet cherries	Lucia et al. (2014)
M. pulcherrima	Anthracnose	Mango	Tian et al. (2018)
Meyerozyma guilliermondii strain 443	Anthracnose	Papaya	Lima et al. (2014)
Pseudozyma fusiformata AP6	Brown rot (*M. laxa*)	Peach	Zhang et al. (2010)
Papiliotrema aspenensis DMKU-SP67	Anthracnose (*C. gloeosporioides*)	Mango	Konsue et al. 2020
Pichia fermentans	Green and blue moulds (*P. digitatum, P. italicum*)	Citrus	Comitini et al. (2009)
P. membranifaciens	Green mould (*P. digitatum*)	Citrus	Spadaro and Droby (2016)
Rhodotorula glutinis	Alternaria rot (*A. alternata*), Penicillium rot (*P. expansum*)	Jujube	Tian et al. (2005)
	Blue rot (*P. expansum*),grey mould (*B. cinerea*)	Pear	Zhang et al.(2008)
	Blue mould (*P. expansum*),grey mould (*B. cinerea*)	Apple	Zhang et al. (2009)
Rhodosporidium paludigenum	Black rot (*A. alternata*)	Cherry tomato	Wang et al. (2010)
Saccharomyces cerevisiae Disva 599	Brown rot	Sweet cherry	Lucia et al.(2014)
S. cerevisiae	Anthracnose (*C. musae*)	Banana	Zhimo et al. (2016)
Torulaspora indica DMKU-RP35	Fruit rot (*L. theobromae*)	Mango	Konsue et al. (2020)
Wickerhamomyces anomalus strain 422	Anthracnose	Papaya	Lima et al. (2014)
W. anomalus Disva 2	Brown rot	Sweet cherry	Lucia et al. (2014)
Talaromyces tratensis KUFA 0091	Fruit rot (*L. theobromae*)	Mango	Suasard et al. (2019)
Trichoderma harzianum	Anthracnose (*C. musae*)	Banana	Devi and Arumugam (2005)
	Grey mould (*B. cinerea*)	Grape, kiwifruit, pear, strawberry	Batta (2007)

6.5 Modes of Action of Microbial Contenders

Microbial contenders act by competing for nutrients, induction of resistance or through direct parasitism to suppress the initiation and spread of postharvest pathogens during handling (Table 6.3).

6.6 Enhancing the Bioefficacy of Microbial Contenders

Postharvest diseases should be efficiently controlled during handling and storage. Commercial control of postharvest pathogens during handling of produce must be efficient, contrasting with the regulation of field/soilborne diseases. As we cannot achieve this level of efficient control with bio-control agents alone, we need to supplement with other modes of control in an integrated manner to achieve effective results. Some of the following approaches have been suggested.

6.6.1 Complementation with Low Dose of Fungicides

Bioagent products such as ASPIRE and BIOSAVE-110 were found to work effectively against the targeted postharvest pathogens only when complemented with low doses of fungicides. For example, application of *Candida oleophila* along with low amounts of thiabendazol at 200 mg/mL could

TABLE 6.3

Modes of Action of Bio-control Agents Used Against Different Postharvest Pathogens

Antagonist	Crop	Disease	Reference(s)
Antibiotic Production			
Pseudomonas cepacia	Apple	Blue mould	Janisiewicz et al. (1991)
	Apple	Mucor rot	Janisiewicz and Roitman (1988)
	Pear	Grey mould	Janisiewicz and Roitman (1988)
	Pear	Blue mould	Janisiewicz and Roitman (1988)
Bacillus subtilis	Apricot	Brown rot	Pusey et al. (1988)
	Cherry	Brown rot	Utkhede and Sholberg (1986)
	Citrus	Sour rot	Singh and Deverall (1984)
	Citrus	Green mould	Singh and Deverall (1984)
	Peach	Brown rot	Pusey et al. (1988)
	Nectarine	Brown rot	Pusey et al. (1988)
	Plum	Brown mould	Pusey et al. (1988)
Enterobacter aerogenes	Cherry	Alternaria rot	Utkhede and Sholberg (1986)
Bacillus amyloliquefaciens	Citrus	Postharvest diseases	Arrebola et al. (2010)
Trichoderma spp.	Citrus	Sour rot	De-Matos (1983)
Nutritional Competition (N)			
Debaromyes hansenii	Grapes	Grey mould	Chalutz et al. (1988)
	Grapes	Rhizopus rot	Chalutz et al. (1988)
	Tomato	Grey mould	Chalutz et al. (1988)
	Tomato	Rhizopus rot	Chalutz et al. (1988)
E. cloacaei	Peach	Rhizopus rot	Wisniewski et al. (1988)
Cryptococcus laurentii	Strawberry	Grey mould	Castoria et al. (1997)
Host Resistance (HR) Induction			
Pseudomonas cepacia	Apple	Blue mould	Janisiewicz (1987)
Aceromonium breve	Apple	Grey mould	Janisiewicz (1987)
Nutritional Competition plus Host Resistance Induction (N+HR)			
Debaromyes hansenii	Apple	Grey mould	Wisniewski et al. (1988)
	Citrus	Green mould	Chalutz and Wilson (1990)
	Citrus	Blue mould	Chalutz and Wilson (1990)
	Citrus	Sour rot	Chalutz and Wilson (1990)

effectively control decay in citrus fruits, which was comparable to commercial fungicide treatment (Brown and Chambers 1996; Droby et al. 1999).

6.6.2 Complementation with Physical Treatments

The blue mould of citrus and grey mould of apple could be eliminated with hot air treatment at 38°C for 4 days (Fallik et al. 1995). Huang et al. (1995) noticed that complementation of bio-control agent, *Pseudomonas glathei,* with heat treatment retards germination of conidia of *P. digitatum.* Irradiation with UV-C treatment ($\lambda < 280$ nm) of root crops (carrot or sweet potato) can be used to induce resistance and reduce postharvest decay (Stevens et al. 1990).

6.6.3 Addition of Generally Regarded as Safe Chemicals

The generally regarded as safe (GRAS) chemicals cannot cause harm to human health. The major GRAS chemicals are calcium carbonate, calcium bicarbonate, calcium propionate, EDTA, etc., which have been used for antagonist activity for controlling of postharvest pathogens. For instance, $CaCl_2$ at 2% has shown a defensive role for antagonistic yeast of pome fruits and controlled Rhizopus rot in peach fruits. Blue and grey moulds of apple and pear can be reduced by application of $CaCl_2$ at 2% with yeast contending *Candida* spp. (Singh 2005; Zhang et al. 2005)

6.6.4 Application of Mixed Cultures

A single strain microbial contender does not possess broad-spectrum activity for controlling pathogens. Hence, to improve the activity, it has been suggested to use antagonistic mixtures in perishable horticultural commodities (Schisler et al. 1997; Fukui et al. 1999). For example, the quantity of total biomass of the antagonists required for controlling decay in apple fruits was significantly lower when the contending bioagents were applied in mixtures rather than individually (Janisiewicz and Bors 1995; Leibinger et al. 1997).

6.6.5 Addition of Plant Extracts and Nutrients

Nitrogen-containing substances (L-proline and L-asparagine) or sugar analogues such as 2-deoxy-D-glucose enhance the contending ability of the bioagents (Janisiewicz 1996; El-Ghaouth et al. 2000). The combination of *Candida saitoana* with a low dose of 0.2 (w/v) of 2-deoxy-D-glucose can control decay caused by *B. cinerea*, *P. expansum* and *P. digitatum* of apple, orange and lemon, respectively (El-Ghaouth et al. 2000).

6.7 Conclusion

Postharvest pathogens cause huge spoilage, resulting in significant losses to harvested and stored perishable horticultural commodities. These losses are avoided by using synthetic fungicides, but due to health and environment concerns, scientists have explored the possibilities of using biological approaches. The approaches such as use of resistant varieties, induction of resistance, use of plant extracts and essential oils and microbial antagonists are slowly becoming popular. Bioagents have now been identified for several postharvest diseases, and several bio-products have also been developed. There is great future for this technology, yet its application under field conditions needs further validation.

REFERENCES

Abu-Goukh AA, Labavitch HM (1983) The *in vivo* role of 'Bartlett' pear fruit polygalacturonase inhibitors. Physiol Plant Pathol 23: 123–35.

Adikaram N, Singh D, Jayasinghe L (2018) *Aureobasidium pullulans* suppression of green mould (*Penicillium digitatum*) development in mandarin var. 'Kinnow' through multiple modes of action. In: ICPP, Boston Plant Health in Global Economy, July 29-August 03, 2018.

Arrebola E, Jacobs R, Korsten L (2010) Iturin A is the principal inhibitor in the bio-control activity of *Bacillus amyloliquefaciens* PPCB004 against postharvest fungal pathogens. J Appl Microbiol 108(2): 386–395.

Babu KJ, Reddy SM (1986) Efficacy of some indigenous plant extracts in the control of lemon rot by two pathogenic fungi. Natl Acad Sci Lett 9:133–134.

Barkai-Golan R (2001) Postharvest diseases of fruit and vegetables; development and control. Elsevier Sciences: Amsterdam, Netherlands.

Barkai-Golan R (2005) Postharvest diseases of fruits and vegetables. A Division of Reed Elsevier India, Pvt. Limited, New Delhi, India.

Barman K, Asrey R, Singh D, Patel VB, Sharma S (2017) Effect of *Pseudomonas fluorescens* formulations on decay and quality of mango (*Mangifera indica*) fruits during storage. Indian J Agric Sci 87(9): 1214–1218.

Batta YA (2007) Control of postharvest diseases of fruit with an invert emulsion formulation of *Trichoderma harzianum* Rifai. Postharvest Biol Technol 43(1):143–150.

Baudoin ABAM, Eckert JW (1985) Development of resistance against *Geotrichum candidum* in lemon peel injuries. Phytopathol 74:174–179.

Bello GD (2008) First report of *Trichothecium roseum* causing postharvest fruit rot of tomato in Argentina. Aust Plant Dis Notes 3(1):103–104.

Blum LEB, Amarante CVT, Valdebenito-Sanhueza RM, Guimaraes LS, Dezanet A, Hack-Neto P (2004) Postharvest application of *Cryptococcus laurentii* reduces apple fruit rots. Fitopatol Brasi 29(4):433–436.

Bojanowski A, Avis TJ, Pelletier S, Tweddell RJ (2013) Management of potato dry rot. Postharvest Biol Technol 84: 99–109.

Borve J, Meland M, Sekse L, Stensvand A (2008) Plastic covering to reduce sweet cherry fruit cracking affects fungal fruit decay. Acta Hortic 795:485–488.

Borve J, Vangdal E (2007) Fungal pathogens causing fruit decay on plum (*Prunus domestica* L.) in Norway. Acta Hortic 734:367–370.

Brown GE, Chambers M (1996) Evaluation of biological products for the control of postharvest diseases of Florida citrus. Proc Florida State Hortic Soc 109:278–282.

Calvo-Garrido C, Josep U, Rosario T, Neus T (2017) Effective control of Botrytis bunch rot in commercial vineyards by large-scale application of *Candida sake* CPA-1. Bio Control. DOI: 10.1007/s10526-017-9789-9.

Castoria R, Curtis F, Lima G, Cicco V (1997) β-1,3-Glucanase activity of two saprophytic yeasts and possible mode of action as bio-control agents against postharvest diseases. Postharvest Biol Technol 12(3):293–300.

Chalutz E, Ben-Arie R, Droby S, Cohen L, Weiss B, Wilson CL (1988) Yeasts as bio-control agents of postharvest diseases of fruit. Phytoparasitica 16:69–75.

Chalutz E, Droby S, Wilson CL, Wisniewski ME (1992) UV-induced resistance to postharvest diseases of citrus fruit. J Photochem Photobiol 15:367–374.

Chalutz E, Wilson CL (1990) Postharvest bio-control of green and blue mold and sour rot of citrus fruits by *Debaryomyces hansenii*. Plant Dis 74:134–137.

Choudhary B, Nagpure A, Gupta RK (2015) Biological control of toxigenic citrus and papaya-rotting fungi by *Streptomyces violascens* MT7 and its extracellular metabolites. J Basic Microbiol 55:1343–1356.

da Cunha T, Ferraz LP, Wehr PP, Kupper KC (2018) Antifungal activity and action mechanisms of yeasts isolates from citrus against *Penicillium italicum*. Int J Food Microbiol 276: 20–27.

De-Matos AP (1983) Chemical and microbiological factors influencing the infection of lemons by *Geotrichum candidum* and *Penicillium digitatum*. PhD thesis, University of California, Riverside, p. 106.

Devi AN, Arumugam T (2005) Studies on the shelf life and quality of Rasthali banana as affected by postharvest treatments. Orissa J Hortic 33(2): 3–6.

Dixit SN, Chandra H, Tewari R, Dixit V (1995) Development of botanical fungicide against blue mold of mandarins. J Stored Prod Res 31:165–172.

Droby S, Porat R, Cohen L, Weiss B, Shapira B, Philosoph-Hadas S, Meir S (1999) Suppressing green mold decay in grapefruit with postharvest jasmonate application. J Amer Soc Hortic Sci 124:184–188.

Dubey NK, Kishore N (1988) Book of Abstracts: Exploitation of higher plant products as natural fumigants. In: Proc Fifth International Cong on Plant Pathol, Kyoto, Japan, p. 423.

Eckert JW (1978) Pathological diseases of fresh fruits and vegetables. In: Postharvest Biology and Biotechnology. Food and Nutrition Press, Westport, CT, pp. 161–209.
El-Ghaouth A, Arul J (1992) Potential use of chitosan in postharvest preservation of fresh fruits and vegetables. Proc Int Symp Physiol Basis Postharvest Technol. University of California, Davis, p. 50.
El-Ghaouth A, Arul J, Asselin A, Benhamou N (1992) Antifungal activity of chitosan on postharvest pathogens: induction of morphological and cytological alterations in *Rhizopus stolonifer*. Mycol Res 96:769–779.
El-Ghaouth AJ, Smilanick M, Wisniewski, Wilson CL (2000) Improved control of apple and citrus fruit decay with a combination of *Candida saitoana* with 2-deoxy-D-glucose. Plant Dis 84:249–253.
El-Neshawy SM, Wilson CL (1997) Nisin enhancement of bio-control of postharvest diseases of apple with *Candida oleophila*. Postharvest Biol Technol 10 (1):9–14.
Fabrício PG, Marise CM, Geraldo JSJ, Silvia AL, Lilian A (2010) Postharvest control of brown rot and *Rhizopus* rot in plums and nectarines using carnauba wax. Postharvest Biol Technol 58(3): 211–217.
Fallik E, Grinberg S, Gambourg M, Lure S (1995) Prestorage heat treatment reduces pathogenicity of *Penicillium expansum* in apple fruit. Plant Pathol 45:92–97.
Fenwick GR, Heaney RK, Mullin WJ (1983) Glucosinolates and their breakdown product in food and food plants. CRC Crit Rev Food Sci Nutr 18:123–201.
Fielding A (1981) Natural inhibitors of fungal polygalacturonases in infected fruit tissues. J Gen Microbiol 123:377–381.
Frances J, Bonaterra A, Moreno MC, Cabrefiga J, Badosa E, Montesinos E (2006) Pathogen aggressiveness and postharvest bio-control efficiency in *Pantoea agglomerans*. Postharvest Biol Technol 39(3): 299–307.
Fukui R, Fukui H, Alverez AM (1999) Comparisons of single versus multiple bacterial species on biological control of anthurium blight. Phytopathol 89:366–373.
Galston AW, Kaur Sawhney R (1982) Polyamines: Are they a new class of plant growth regulators? In: Wareing PF, ed. Plant growth substances. Academic Press, New York, pp. 451–564.
Gamagae SU, Sivakumar D, Wilson-Wijeratnam RS, Wijesundra RLC (2003) Use of sodium bicarbonate and *Candida oleophila* to control anthracnose in papaya during storage. Crop Protec 22(5):775–779.
Geng P, Chen S, Hu M, Rizwan-ul-Haq M, Lai K, Qu F, Zhang Y (2011) Combination of *Kluyveromyces marxianus* and sodium bicarbonate for controlling green mold of citrus fruit. Int J Food Microbiol 151:190–194.
Govender V, Korsten L, Sivakumar D (2005) Semi-commercial evaluation of *Bacillus licheniformis* to control mango postharvest diseases in South Africa. Postharvest Biol Technol 38(1):57–65.
Hiroshi K, Sarunya N, Hiroyuki Y, Kazuya A (2007) Overexpression of citrus polygalacturonase-inhibiting protein in citrus black rot pathogen *Alternaria citri*. J Plant Physiol 164(5): 527–535.
Hou Y, Hu X, Zhou B (2012) Hot pepper growth promotion and inhibition of Fusarium wilt (*Fusarium oxysporum*) with different crop stalks. Afr J Agric Res 7(35): 5005–5011.
Huang Y, Deverall BJ, Morris SC (1995) Postharvest control of green mold on oranges by a strain of *Pseudomonas glathei* and enhancement of its bio-control by heat treatment. Postharvest Biol Technol 13:129–137.
Igeleke CL, Ayanru DKG (2006) Sugar and amino acid contents of fruit and foliar tissues from two cultivars of plantain (*Musa paradisiaca*) susceptible and resistant to cigar-end rot disease caused by *Verticillium theobromae*. J Biol Sci 6(5): 916–920.
Irkin R, Korukluoglu M (2007) Control of *Aspergillus niger* with garlic, onion and leek extracts. Afr J Biotechnol 6(4): 384–387.
Janisiewicz WJ (1987) Postharvest biological control of blue mold on apple. Phytopathol 77:481–485.
Janisiewicz WJ (1996) Ecological diversity, niche overlap, and coexistence of antagonists used in developing mixture for bio-control of postharvest diseases of apples. Phytopathol 86: 473–479.
Janisiewicz WJ, Bors B (1995) Development of a microbial community of bacterial and yeast antagonists to control wound invading postharvest pathogens of fruits. Appl Environ Microbiol 61: 3261–3267.
Janisiewicz WJ, Leverentz B, Saftner RA, Reed AN, Camp MJ (2003) Control of bitter rot and blue mold of apples by integrating heat and antagonist treatments on 1-MCP treated fruit store under CA conditions. Postharvest Biol Technol 29: 129–143.
Janisiewicz WJ, Roitman J (1988) Biological control of blue mold and gray mold on apple and pear with *Pseudomonas cepacia*. Phytopathol 78: 1697–1700.
Janisiewicz WJ, Yourman L, Roitman J, Mahoney N (1991) Postharvest control of blue mold and gray mold of apples and pears by dip treatment with pyrrolnitrin, a metabolite of *Pseudomonas cepacia*. Plant Dis 75:490–494.
Jiang YM, Zhu XR, Li YB (2001) Postharvest control of litchi fruit rot by *Bacillus subtilis*. LWT-Food Sci Technol 34(7): 430–436.
Kadam KS (2012) Studies on sensitivity of *Alternaria alternata* isolates against Aureofungin. Bioinfolet 9(2):162–165.
Kallio A, McCann LP, Bey P (1981) DL- alpha-difluoromethyl arginine: A potent enzyme activated irreversible inhibitor of bacterial arginine decarboxylases. Biochem 20:3163–3166.
Kefialew Y, Ayalew A (2008) Postharvest biological control of anthracnose (*Colletotrichum gloeosporioides*) on mango (*Mangifera indica*). Postharvest Biol Technol 50(1): 8–11.
Kendra DF, Christian D, Hadwiger LA (1989) Chitosan oligomers from *Fusarium solani* / Pea interactions, chitinase/β-glucanase digestion of sporelings and from fungal wall chitin actively inhibit fungal growth and enhance disease resistance. Physiol Mol Plant Pathol 3(5):215–230.
Konsue W, Tida D, Savitree L (2020) Biological control of fruit rot and anthracnose of postharvest mango by antagonistic yeasts from economic crops leaves. Microorganisms 8(3): 317.
Kuc J (1982) Induced immunity to plant disease. Bioscience 32:854–860.
Lahlali R, Serrhini MN, Jijakli MH (2005) Development of a biological control method against postharvest diseases of citrus fruit. Commun Agric Appl Biol Sci 70(3): 47–58.

Lassois L, de Bellaire L, Jijakli MH (2008) Biological control of crown rot of bananas with *Pichia anomala* strain K and *Candida oleophila* strain O. Biol Control 4(3)410–418.

Latorre BA, Briceño EX, Torres R (2011) Increase in *Cladosporium* spp. populations and rot of wine grapes associated with leaf removal. Crop Prot 30(1):52–56.

Leibinger W, Breuker B, Hahn M, Mendgen K (1997) Control of postharvest pathogens and colonization of the apple surface by antagonistic microorganisms in the field. Phytopathol 87:1103–1110.

Leonard SO, Els HMN, Harrie K, Johnny V, Gijs K (2010) The role of crop waste and soil in *Pseudomonas syringae* pathovar *porri* infection of leek (*Allium porrum*). Appl Soil Ecol 46(3): 457–463.

Li HN, Jiang JJ, Hong N, Wang GP, Xu WX (2013) First report of *Colletotrichum fructicola* causing bitter rot of pear (*Pyrusb retschneideri*) in China. Plant Dis 97(7): 1000.

Lima NB, de A, Batista MV, De Morais Jr MA, Barbosa MAG, Michereff SJ, Hyde KD, Câmara MPS (2013) Five *Colletotrichum* species are responsible for mango anthracnose in north-eastern Brazil. Fungal Diversity 61:75–88.

Lima G, De Curtis F, Castoria R, Pacifica S, De Cicco V (1998) Additives and natural products against postharvest pathogens compatibility with antagonistic yeast. In: Plant pathology and sustainable agriculture. Proceedings of the Sixth SIPa V Annual meeting, Campobasso.

Lima JRDE, Marto F, Viana P, Lima FA, Pieniz V, Rocha L (2014) Efficiency of a yeast-based formulation for the bio-control of postharvest anthracnose of papayas. Summa Phytopathol 40:203–211.

Liu WT, Chu CL, Zhou T (2002) Thymol and acetic acid vapours reduce postharvest brown rot of apricot and plums. Hortic Sci 37:151–156.

Liu HM, Guo JH, Liu P, Cheng YJ, Wang BQ, Long CA, Deng BX (2010) Inhibitory activity of tea polyphenol and *Candida ernobii* against *Diplodia natalensis* infections. J Appl Microbiol 108(3):1066–1072.

Liu J, Sui Y, Wisniewski M, Droby S, Liu Y (2013). Utilization of antagonistic yeasts to manage postharvest fungal diseases of fruit. Int J Food Microbiol 167(2):153–160.

Llorente L, Vilardell A, Vilardell P, Pattori E, Bugiani R (2010) Control of brown spot of pear by reducing the overwintering inoculum through sanitation. Eur J Plant Pathol 128(1):127–141.

Lu JY, Stevens C, Khan VA, Kabwe MK, Wilson CL (1991) The effect of ultraviolet irradiation on shelf life and ripening of peaches and apples. J Food Qual 14:299–305.

Lucia O, Erica F, Maurizio C, Francesca C (2014) Bio-control of postharvest brown rot of sweet cherries by *Saccharomyces cerevisiae* Disva 599, *Metschnikowia pulcherrima* Disva 267 and *Wickerhamomy cesanomalus* Disva 2 strains. Postharvest Biol Technol 96: 64–68.

Luo S, Wan B, Feng S, Shao Y (2015) Bio-control of postharvest anthracnose of mango fruit with *Debaryomyces nepalensis* and effects on storage quality and postharvest physiology. J Food Sci 80(11): 2555–2563.

Margaret BU, Ming PY, Patrick MF, Surinder SB, Shashi KB, Prabhjot SS, Huang Yi, Phillip AS, Martin JB (2013) New sources of resistance to *Sclerotinia sclerotiorum* for crucifer crops. Field Crops Res 154:40–52.

Mari M, Guizzardi M, Pratella GC (1996) Biological control of gray mold in pears by antagonistic bacteria. Biol Contl 7(1):30–37.

Mari M, Lori R, Leoni O, Lori R, Cembli, T (2002). Antifungal vapour–phase activity of allyl isothiocyanates against *Penicillium expansum*. Plant Pathol 51:231–236.

Mari M, Neri F, Bertolini P (2007). Novel approaches to prevent and control postharvest diseases of fruits. Stewart Postharvest Rev 3(6):1–7.

Marietta P, Andras S, Laszló P (2012) *Monilinia* Species in Hungary: Morphology, culture characteristics, and molecular analysis. Trees-Struct Funct 26(1):153–164.

María PL, Cristina T, Laura C, Purificación L, Ismael R, José MB, Vicente C (2011) Identification of defence metabolites in tomato plants infected by the bacterial pathogen *Pseudomonas syringae*. Environ Exp Bot 74:216–228.

Martins DMS, Blum LEB, Sena MC, Dutra JB, Freitas LF, Lopes LF, Yamanishi OK, Dianese AC (2010). Effect of hot water treatment on the control of papaya (*Carica papaya* L.) postharvest diseases. Acta Hortic 864: 181–186.

Masih EI, Paul B (2002) Secretion of beta-1,3-glucanases by the yeast *Pichia membranifaciens* and its possible role in the bio-control of *Botrytis cinerea* causing grey mold disease of the grapevine. Curr Microbiol 44(6): 391–395.

Masoud A, Ibrahim T, Hilde N (2013) Impact of harvesting time and fruit firmness on the tolerance to fungal storage diseases in an apple germplasm collection. Postharvest Biol Technol 82:51–58.

Matei A, Cornea CP, Matei S, Matei GM, Rodino S (2015) Comparative antifungal effect of lactic acid bacteria strains on *Penicillium digitatum*. Bull UASVM Food Sci Technol 72: 226–230.

Mei J, Wei D, Disi JO, Ding Y, Liu, Qian W (2012). Screening resistance against *Sclerotinia sclerotiorum* in *Brassica* crops with use of detached stem assay under controlled environment. Eur J Plant Pathol 134(3): 599–604.

Metcalf BW, Bey P, Danzin C, Jung MJ, Casara MJ, Verert JP (1978) Catalytic irreversible inhibition of mammalian ornithine decarboxylase E.C.4.1.1.117 by substrate and product analogs. J Am Chem Soc 100:2551–2553.

Michalecka M, Bryk H, Poniatowska A, Pulawska J (2017) Identification of *Neofabraea* species causing bull's eye rot of apple in Poland and their direct detection in apple fruit using multiplex PCR. Plant Pathol 65(4):643–654.

Mirik M, Baloglu S, Aysan Y, Cetinkaya-Yildiz R, Kusek M, Sahin F (2005) First outbreak and occurrence of citrus blast disease, caused by *Pseudomonas syringae* pv. *syringae*, on orange and mandarin trees in Turkey. Plant Pathol 54(2):238.

Morales H, Sanchis V, Usall J, Ramos AJ, Marín S (2008) Effect of bio-control agents *Candida sake* and *Pantoea agglomerans* on *Penicillium expansum* growth and patulin accumulation in apples. Int J Food Microbiol 122(1–2): 61–67.

Ncumisa SN, Arno E, Paul HF (2013) Evaluation of curative and protective control of *Penicillium digitatum* following imazalil application in wax coating. Postharvest Biol Technol 77: 102–110.

Palou L, Montesinos-Herrero C, Taberner V, Vilella-Esplá J (2013) First report of *Penicillium expansum* causing postharvest blue mold of fresh date palm fruit (*Phoenix dactylifera*) in Spain. Plant Dis 97(6):846.

Pane A, Nicosia MGLD, Cacciola SO (2001)First report of *Phytophthora citrophthora* causing fruit brown rot of Feijoa in Italy. Plant Dis 85(1): 97.

Porat R, Daus A, Weiss B, Cohen L, Fallik E, Droby S (2000a) Reduction of postharvest decay in organic citrus fruit by a short hot water brushing treatment. Postharvest Biol Technol 18:151–157.

Porat R, Pavoncello D, Peretz Y, Weiss B, Cohen L, Ben-Yehoshua S, Fallik E, Droby S, Luri S (2000b) Induction of resistance against *Penicillium digitatum* and chilling injury in Star Ruby grapefruit by a short hot water brushing treatment. J Hortic Sci Biotechnol 75: 428–432.

Prasad K, Stadelbacher GJ (1973) Control of postharvest decay of fresh raspberries by acetaldehyde vapour. Plant Dis Report 57: 795–797.

Prusky D, Kobiler I, Akerman M, Miyara I (2006) Effect of acidic solutions and acidic prochloraz on the control of postharvest decay caused by *Alternaria alternata* in mango and persimmon fruit. Postharvest Biol Technol 42(2):134–141.

Pusey PL, Hotchkiss MW, Dulmage HT, Banumgardner RA, Zehr EI (1988) Pilot tests for commercial production and application of *Bacillus subtilis* (b-3) for postharvest control of peach brown rot. Plant Dis 72:622–626.

Pusey PL, Wilson CL (1984) Postharvest biological control of stone fruit brown rot by *Bacillus subtilis*. Plant Dis 68:753–756.

Qin GZ, Tian SP, Xu Y, Chan ZL, Li BQ (2006) Combination of antagonistic yeasts with two food additives for control of brown rot caused by *Monilinia fructicola* on sweet cherry fruit. J Appl Microbiol 100(3): 508–515.

Ramírez PAM, Montesinos-Herrero C, Palou L (2013) Control of citrus postharvest *Penicillium* molds with sodium ethylparaben. Crop Prot 46:44–51.

Renganathan P, Muthukumar A (2012) Evaluation of fungicides against *Colletotrichum musae* and *Botryodiplodia theobromae*, the cause of banana crown rot. Indian J Plant Prot 40(3): 240.

Robert JM, Katherine H, Pamela DR, Teresa ES (2012) New report of *Colletotrichum gloeosporioides* causing postbloom fruit drop on citrus in Bermuda. Can J Plant Pathol 34(2):187–194.

Romanazzi G, Sanzani SM, Bi Y, Tian S, Martínez PG, Alkan N (2016) Induced resistance to control postharvest decay of fruit and vegetables. Postharvest Biol Technol 122:82–94.

Sakinah MAI, Suzianti IV, Latiffah Z (2013) First report of *Colletotrichum gloeosporioides* causing anthracnose of banana (*Musa* spp.) in Malaysia. Plant Dis 97(7):991.

Saligkarias ID, Gravanis FT, Epton HAS (2002) Biological control of *Botrytis cinerea* on tomato plants by the use of epiphytic yeasts *Candida guilliermondii* strains 101 and US 7 *and Candida oleophila* strain I-182: In vivo studies. Biol Control 25(2): 143–150.

Salunkhe DK, Bolin HR, Reddy NR (1991) Storage, processing, and nutritional quality of fruits and vegetables. Volume I. Fresh fruits and vegetables (No. Ed. 2). CRC Press, Boca Raton, FL.

Schena L, Nigro F, Pentimone IA, Ippolito A (2003) Control of postharvest rots of sweet cherries and table grapes with endophytic isolates of *Aureobasidium pullulans*. Postharvest Biol Technol 30(3): 209–220.

Schisler DA, Slinigger PJ, Bothast RJ (1997) Effect of antagonist cell concentration and two strain mixtures on biological control of Fusarium dry rot of potatoes. Phytopathol 87:177–183.

Sellamuthu PS, Mafune M, Sivakumar D, Soundy P (2013) Thyme oil vapour and modified atmosphere packaging reduce anthracnose incidence and maintain fruit quality in avocado. J Sci Food Agric 9(12):3024–3031.

Sharma RR, Pongener A (2010) Natural products for postharvest disease control of horticultural produce: A review. Stewart Postharvest Rev 4:1–9.

Sharma RR, Singh D, Singh R (2009) Biological control of postharvest diseases of fruits and vegetables by microbial antagonists: A review. Biol Control 50: 205–221.

Sholberg PL, Gaunce AP (1995) Fumigation of fruit with acetic acid to prevent postharvest decay. HortSci 30:1271–1275.

Singh D (2002) Bioefficacy of *Debarymyces hansenii* on the incidence and growth of *Penicillium italicum* on Kinnow fruits in combination with oil and wax emulsions. Ann Plant Protec Sci 10(2): 272–276.

Singh D (2005) Interactive effect of *Debaryomyces hansenii* and calcium chloride to reduce Rhizopus rot of peaches. Indian J Mycol Plant Pathol 35(1):118–121.

Singh V, Deverall BJ (1984) *Bacillus subtilis* as a control agent against fungal pathogens of citrus fruit. Trans Brit Mycol Soc 83:487–490.

Singh D, Sharma RR (2007) Postharvest diseases of fruit and vegetables and their management. Daya Publishing House New Delhi, India.

Singh D, Thakur AK (2002) Suppression of green mould mold rot caused by *Penicillium digitatum* in Kinnow fruits by hot water immersion treatment. Indian Phytopathol 55 (3):282–285.

Singh BK, Yadav KS, Verma A (2017) Impact of postharvest diseases and their management in fruit crops: an overview. J Bio Innov 6(5):749–760.

Singh, D, Sharma RR (2018). Postharvest diseases of fruits and vegetables and their management. In: Postharvest disinfection of fruits and vegetables. Academic Press, pp. 1–52.

Sisquella M, Casals C, Viñas I, Teixidó N, Usall J (2013) Combination of peracetic acid and hot water treatment to control postharvest brown rot on peaches and nectarines. Postharvest Biol Technol. 83:1–8.

Skene DS (1981) Wound healing in apple fruits: The anatomical response of Cox's orange Pippin at different stages of development. J Hortic Sci 56:145–153.

Smid EJ, Witte Y, de Vrees O, Gorris LMG (1994) Use of secondary plant metabolites for the control of postharvest fungal diseases on flower bulbs. Acta Hortic 368:523–530.

Smilanick JL, Margosan DA, Henson DJ (1995). Evaluation of heated solution of sulfur dioxide, ethanol, and hydrogen peroxide to control postharvest green mold lemons. Plant Dis 79:742–747.

Spadaro D, Droby S (2016) Unraveling the mechanisms used by antagonistic yeast to control postharvest pathogens on fruit. Acta Hortic 1144:63–70.

Sriram S, Poornachandra SR (2013) Biological control of postharvest mango fruit rot caused by *Colletotrichum gloeosporioides* and *Diplodia natalensis* with *Candida tropicalis* and *Alcaligenes feacalis*. Indian Phytopathol 66(4):375–380.

Sriram S, Savitha MJ, Ramanujam B (2010) *Trichoderma*-enriched coco-peat for the management of *Phytophthora* and *Fusarium* diseases of chilli and tomato in nurseries. J Biol Control 24(4):311–316.

Stevens C, Khan VA, Tang AY, Lu, J (1990) The effect of ultraviolet radiation on mold rots and nutrients of stored sweet potatoes. J Food Prot 53:223–326.

Storari MD, Broggini GAL, Bigler L, Cordano E, Eccel E, Filippi R, Gessler C, Pertot I (2012) Risk assessment of the occurrence of black Aspergilli on grapes grown in an Alpine region under a climate change scenario. Eur J Plant Pathol 134(3): 631–645.

Suasard S, Eakjamnong W, Dethoup T (2019) A novel biological control agent against postharvest mango disease caused by *Lasiodioplodia theobromae*. Eur J Plant Pathol 155:583–592.

Swinburne TR (1973) The resistance of immature Bramley's seedling apples to rotting by *Nectrix galligena* Bres. Academic Press, London, New York, pp. 365–382.

Swinburne TR (1978) Post-infection antifungal compounds in quiescent or latent infections. Ann Appl Biol 89:322–324.

Talibi I, Askarne L, Boubaker H, Boudyach EH, Msanda F, Saadi B, Ait Ben Aoumar A (2012a) Antifungal activity of some Moroccan plants against *Geotrichum candidum*, the causal agent of postharvest citrus sour rot. Crop Prot 35:41–46.

Talibi I, Askarne L, Boubaker H, Boudyach EH, Msanda F, Saadi B, Ait Ben Aoumar A (2012b) Antifungal activity of Moroccan medicinal plants against citrus sour rot agent *Geotrichum candidum*. Lett Appl Microbiol 55:155–161.

Thomas V, Johan K, Annemie G, Bart MN, Maarten LATMH (2012) Evaluation of fast volatile analysis for detection of *Botrytis cinerea* infections in strawberry. Food Microbiol 32(2): 406–414.

Thornton CR, Slaughter DC, Davis RM (2010). Detection of the sour-rot pathogen *Geotrichum candidum* in tomato fruit and juice by using a highly specific monoclonal antibody-based ELISA. Int J Food Microbiol 143(3):166–172.

Thuy TTT, Chi NTM, Yen NT, Anh LTN, Te LL, Waele DD (2013) Fungi associated with black pepper plants in Quang Tri Province (Vietnam), and interaction between *Meloidogyne incognita* and *Fusarium solani*. Arch Phytopathol Plant Prot 46(4): 470–482.

Tian SP, Fan Q, Xu Y, Qin GZ, Liu HB (2002) Effect of bio-control antagonists applied in combination with calcium on the control of postharvest diseases in different fruit. Bulletin-OILB/SROP 25(10):193–196.

Tian YQ, Li W, Jiang ZT, Jing MM, Shao YZ (2018). The preservation effect of *Metschnikowia pulcherrima* yeast on anthracnose of postharvest mango fruits and the possible mechanism. Food Sci Biotechnol 27(1):95–105.

Tian SP, Qin GZ, Xu Y (2005) Synergistic effects of combining bio-control agents with silicon against postharvest diseases of jujube fruit. J Food Prot 68(3):544–550.

Torres R, Nunes C, Garcia JM, Abadias M, Vinas I, Manso T, Olmo M, Usall J (2007) Application of *Pantoea agglomerans* CPA-2 in combination with heated sodium bicarbonate solutions to control the major postharvest diseases affecting citrus fruit at several Mediterranean locations. Eur J Plant Pathol 118(1):73–83.

Torres R, Teixido N, Vinas I, Mari M, Casalini L, Giraud M, Usall J (2006) Efficacy of *Candida sake* CPA-1 formulation for controlling *Penicillium expansum* decay on pome fruit from different Mediterranean regions. J Food Prot 69(11): 2703–2711.

Toselli M, Sorrenti G, Quartieri M, Baldi E, Marcolini G, Solieri D, Marangoni B, Collina M (2012) Use of soil-and foliar-applied calcium chloride to reduce pear susceptibility to brown spot (*Stemphylium vesicarium*). J Plant Nutr 35(12):1819–1829.

Tosi B, Donini A, Romagnoli C, Bruni A (1996) Antimicrobial activity of some commercial extracts of propolis prepared with different solvents. Phytother Res 10:335–336.

Tripathi P (2001) Evaluation of some plant products against fungi causing postharvest diseases of some fruits. PhD thesis, Department of Botany, Banaras Hindu University, Banaras, India.

Umaña-Rojas G, García J (2011) Frequency of organisms associated with crown rot of bananas in integrated and organic production systems. Acta Hortic 906: 211–217.

Utkhede RS, Sholberg PL (1986) *In vitro* inhibition of plant pathogens: *Bacillus subtilis* and *Enterobacter aerogenes in vivo* control of two postharvest cherry diseases. Can J Microbiol 32:963–967.

Vahling-Armstrong C, Dung JKS, Humann JL, Schroeder BK (2016) Effects of postharvest onion curing parameters on bulb rot caused by *Pantoea agglomerans, Pantoea ananatis* and *Pantoea allii* in storage. Plant Pathol 65(4):536–544.

Van Lenteren JC, Bolckmans K, Kohl J, Ravensberg WJ, Urbaneja A (2018) Biological control using invertebrates and microorganisms: Plenty of new opportunities. Biol Control 63(1):39–59.

Vicent A, Botella-Rocamora P, López-Quilez A, Roca E, Bascón J, Garcia-Jimenez J (2012) Relationships between agronomic factors and epidemics of *Phytophthora* branch canker of citrus in Southwestern Spain. Eur J Plant Pathol 133(3): 577–584.

Wang Z, Jiang M, Chen K, Wang K, Muying D, Zala´n ZT, Hegy F, Kan J (2018). Bio-control of *Penicillium digitatum* on postharvest citrus fruits by *Pseudomonas fluorescens*. J Food Qual https://doi.org/10.1155/2018/2910481.

Wang Y, Li X, Bi Y, Ge Y, Li Y, Xie F (2008) Postharvest ASM or Harpin treatment induce resistance of muskmelons against *Trichothecium roseum*. Agric Sci China 7(2):217–223.

Wang Y, Ren X, Song X, Yu T, Lu H, Wang P, Wang J, Zheng XD (2010) Control of postharvest decay on cherry tomatoes by marine yeast, *Rhodosporidium paludigenum* and calcium chloride. J Appl Microbiol 109(2): 651–656.

Wilson CL, Chalutz E (1989) Postharvest bio-control of *Penicillium* rots of citrus with antagonistic yeasts and bacteria. Sci Hortic 40:105–112.

Wilson CL, EI- Ghaouth A, Chalutz E, Droby S, Stevens C, Lu JY, Kahn VA, Arul J (1994) Potential of induced resistance to control postharvest diseases of fruit and vegetable. Plant Dis 78: 837–844.

Wilson CL, Wisniewski ME (1989) Biological control of postharvest diseases of fruits and vegetables: an emerging technology. Annu Rev Phytopathol 27:425–441.

Wisniewski M, Wilson CL, Chalutz E, Hershberger W (1988) Biological control of postharvest diseases of fruit: inhibition of *Botrytis* rot on apples by an antagonistic yeast. Proc EMSA Mtg 46:290–291.

Xi L, Tian SP (2005) Control of postharvest diseases of tomato fruit by combining antagonistic yeast with sodium bicarbonate. Sci Agric Sinica 38(5):950–955.

Yadahalli KB, Adiver SS, Kulkarni S (2007) Effect of pH, temperature and relative humidity on growth and development of *Ceratocystis paradoxa* – A causal organism of pineapple disease of sugarcane. Karnataka J Agric Sci 20(1):159–161.

Yang DM, Bi Y, Chen XR, Ge YH, Zhao J (2006) Biological control of postharvest diseases with *Bacillus subtilis* (B1 strain) on muskmelons (*Cucumis melo* L. cv. Yindi). Acta Hortic 712(2):735–739.

Yin Y, Bi Y, Li Y, Wang Y, Wang D (2012) Use of thiamine for controlling *Alternaria alternata* postharvest rot in Asian pear (*Pyrus bretschneideri* Rehd. cv. *Zaosu*). Int J Food Sci Technol 47(10):2190–2197.

Yin LF, Chen SN, Yuan NN, Zhai LX, Li GQ, Luo CX (2013) First report of peach brown rot caused by *Monilinia fructicola* in Central and Western China. Plant Dis 97(9):1255.

Zhang D, Lopez-Reyes JG, Spadaro D, Garibaldi A, Gullino ML (2010) Efficacy of yeast antagonists used individually or in combination with hot water dipping for control of postharvest brown rot of peaches. J Plant Dis and Protec 117(5): 226–232.

Zhang H, Wang L, Dong Y, Jiang S, Zhang H, Zheng X (2008) Control of postharvest pear diseases using *Rhodotorula glutinis* and its effects on postharvest quality parameters. Int J Food Microbiol 126 (1–2):167–171.

Zhang H, Wang L, Ma L, Dong, Y, Jiang S, Xu B, Zheng X (2009) Bio-control of major postharvest pathogens on apple using *Rhodotorula glutinis* and its effects on postharvest quality parameters. Biol Control 48(1):79–83.

Zhang HY, Zheng XD, Chengxin F, Yufang X (2005) Postharvest biological control of gray mold rot of pear with *Cryptococcus laurentii*. Postharvest Biol Technol 35:79–86.

Zhang H, Zheng XD, Yu T (2007) Biological control of postharvest diseases of peach with *Cryptococcus laurentii*. Food Control 18(4):287–291.

Zhimo VY, Bhutia DD, Saha J (2016) Biological control of postharvest fruit diseases using antagonistic yeasts in India. J Plant Pathol 98(2): 275–283.

Zhou T, Chu, CL, Liu WT, Schneider KE (2001) Postharvest control of blue mold and gray mold on apples using isolates of *Pseudomonas syringae*. Can J Plant Pathol 23: 246–252.

Zhou T, Northover J, Schneider KE, Lu XW (2002) Interactions between *Pseudomonas syringae* MA-4 and cyprodinil in the control of blue mold and gray mold of apples. Can J Plant Pathol 24(2): 154–161.

Zhu SJ (2006) Non-chemical approaches to decay control in postharvest fruit. Advances in postharvest technologies for horticultural crops. Research Signpost, Trivandrum, India, pp. 297–313.

Zhu XQ, Guo LY (2010) First report of brown rot on plum caused by *Monilina polystroma* in China. Plant Dis 94(4): 478.

7

Endophytes for Postharvest Disease Management in Vegetables and Fruits

Veerubommu Shanmugam, Govindan Pothiraj and Wadzani Palnam Dauda
Division of Plant Pathology, ICAR-Indian Agricultural Research Institute, New Delhi, India

CONTENTS

7.1 Introduction

Vegetables and fruits are a major source of minerals and vitamins and thus serve as part of a balanced diet. In addition to direct consumption, vegetables and fruits are also used for medicinal and religious purposes (Parthasarathy et al. 2017). Though the production of vegetables and fruits has registered a growth of 4.14% in production, 3.2% in area and a meagre 0.91% in productivity, it has not improved the nutritional status of the people. The availability of vegetables and fruits per capita is only about 207 g/day for vegetables and 104 g/day for fruits, which is significantly less than 300 g/day and 120 g/day, respectively, the recommended levels for vegetables and fruits (Gajanana et al. 2011). This is because, in comparison to durable field crop products, vegetables and fruits are highly perishable due to their high moisture content (up to 95%) and soft texture. This makes them prone to damage during harvesting and postharvest operations like packaging, transportation and storage. In addition, vegetables and fruits are metabolically active and are prone to short shelf life. All these factors reduce the marketable value of vegetables and fruits, causing an extensive loss in quantity and quality. The losses incurred by the produce after harvest till consumption through handling, storage, processing and marketing are termed postharvest losses (Grolleaud 2002), which have been estimated to be about 30–50% or more in developing countries (Sharma et al. 2009). During storage, various mechanical and physiological factors apart from environmental conditions such as relative humidity, temperature and oxygen balance together with pathogen and pest infestation account for the major losses. Among them, postharvest disease development is one of the most serious factors for being aided by other primary factors and in spoiling the produce and reducing its shelf life. Though bacteria and fungi are major causes of postharvest losses, viruses also cause a phenomenal loss in yield (Narayanasamy 2005). Among the pathogens, fungi account for a major role, mainly due to their adaptive lifestyles, which enable them to grow and develop under storage conditions (Pétriacq et al. 2018).

Significant loss of postharvest produce due to fungi is attributed to *Mucor, Rhizopus, Phomopsis, Penicillium, Monilinia, Diplodia, Colletotrichum, Botrytis, Aspergillus, Alternaria*, and *Sclerotinia*, that of bacteria to *Pseudomonas* and *Erwinia* (Sharma et al. 2009). Certain genera such as *Fusarium, Alternaria* and *Penicillium*, by virtue of their abilities to produce mycotoxins, reduce the quantity and quality of the produce in addition to imparting health hazards in humans and animals (Singh et al. 2017) but mostly in cereal grains. In general, most of the postharvest pathogens are weak and can only invade the damaged produce caused by mechanical injuries to the produce peel. However, certain pathogens such as *Phomopsis, Monilinia, Alternaria, Lasiodiplodia, Colletotrichum, Botryosphaeria* and *Botrytis* cause postharvest diseases due to pre-harvest infection of the immature produce on the plant. Such infection occurs during flower or fruit development either by direct penetration of the cuticle or entry through natural openings such as lenticels, stomata, or damage such as abscission of scar tissue or wounds. Upon penetration, these pathogens remain quiescent until after harvest and cause infection only upon the ripening of fruits or when the fruit tissues become senescent due to a decrease in the resistance (Barkai-Golan 2005). These infections, called 'latent infections', are common in subtropical and tropical fruits, such as anthracnose of papayas and mangoes, stem-end rot of citrus and crown rot of bananas. The extent of postharvest infection of produce depends on the type of commodity, the susceptibility of cultivar to postharvest disease, conditions in the storage environment such as relative humidity, temperature, atmosphere composition, etc., maturity and ripeness of produce, disease control treatments, methods of handling the produce and storage hygiene (Singh et al. 2017).

To complement the process of increasing production and to keep the produce alive and healthy in the marketing process as well as disease-free seeds and seed material production, several prophylactic measures have been advocated. The produce is harvested at the appropriated stage considering the distance to the market. The produce is packaged in bamboo baskets, gunny bags and wooden crates. In advanced countries, vegetables are trimmed after cleaning, pretreated and packed in transparent bags after precooling. Since the rate of infection of the produce is largely determined by the temperature, humidity and ventilation during transit and storage, and the fact that the produce continues to respire even after harvest producing heat, the produce is stored at near-freezing temperatures under refrigeration. This increases the storage life of the products by minimizing dehydration, rooting and respiration, and prevents bacterial and fungal rots (Aked 2002). In addition to the preventive measures, postharvest diseases of the produce are commonly managed by physical, chemical and biological methods with varying degrees of success to reduce the incidence and spread during storage. Physical methods include the usage of gamma radiation, low temperature and heat to treat the produce during storage (Mahajan et al. 2014). Chemical control of postharvest losses is usually resorted by using synthetic pesticides applied in the orchard or after harvesting and/or food preservatives in storage (Spadaro and Droby 2016). The success of any method depends on a load of initial inoculum, the level of infection of the host tissue, the rate of infection, the humidity and temperature, and the extent of penetration of the host tissue by the chemical (Sharma and Kulshrestha 2015). The development of synthetic pesticides is costly, and their prolonged use leads to the development of pesticide-resistant pathogen races. Also, there are increasing concerns about the effect of synthetic pesticides on both humans and the environment, and increasing interest among consumers for organic produce. Indiscriminate use of pesticides has also aggravated the incidence of several minor fruit diseases mainly due to their effect on limiting competition from saprophytes (Narayanasamy 2005). Among biological methods, though the development of disease-resistant cultivars could be an option, the wide host range of pathogens together with high genetic variability and its potential to survive longer (Sidharthan et al. 2018) cause more difficulties in effective control of the disease by resistance breeding. These necessitate the adoption of alternative disease control methods, which are non- or less phytotoxic, leaving no toxic residue, are environmentally-friendly, safer and easier to use, economical, and that lie within the scope of local laws on food additives. Considering the needs, biological control of postharvest diseases presents a potential option (Dukare et al. 2018).

7.2 Bio-control of Postharvest Diseases

Bio-control of postharvest diseases involves the application of antagonistic microbes and naturally derived compounds to enhance the inherent resistance of the produce, applied on-field or postharvest produce to minimize and/or stabilize the levels of phytopathogen population below economic injury levels (Narayanasamy 2005). Contrary to bio-control of field diseases, the confined storage structure provides stable temperature and humidity conditions, favouring the microbial antagonists to control of postharvest diseases. In addition, the feasibility of the application of the bioproduct is greater for being limited to the required site in the harvested produce. Use of a relatively high-cost bioproduct for harvested vegetables and fruits is also justified due to the high market value of the produce. Application of an antagonistic microbe for control of postharvest diseases was first reported against Botrytis rot in strawberry using the fungal endophyte, *Trichoderma* spp. However, Pusey and Wilson (1984) reported the classical bio-control of brown rot of stone fruits using *Bacillus subtilis*. Thereafter, a range of antagonistic microbes, fungi, bacteria and actinomycetes, have been used to contain postharvest infection. The major antagonistic microbes to control pathogens are *Rhizobium* spp., *Bradyrhizobium japanicum, Streptomyces* spp., *S. griseoviridis, Bacillus* spp., *Pseudomonas* spp., *Enterobacter* spp., *Lysobacter* spp., *Pantoea* spp., *Cladorrhinum foecundissimum, Coniothyrium minitans, F. oxysporum* f. sp. *niveum, Gliocladium virens*,

Glomus spp., *Penicillium* spp., *Sporidesmium sclerotivorum*, *Talaromyces flavus*, *Vertilicillium biguttatum*, *Trichoderma* spp., *Aspergillus* spp., *Laetisaria arvalis*, *Pythium* spp., *Rhizoctonia* spp. and *Chaetomium globosum* (Whipps 1997; Dukare et al. 2018). Bio-control of postharvest diseases is usually accomplished by promoting and managing native microbes that exist on the surfaces of vegetables and fruits, termed epiphytes, and upon introducing potent microbial antagonists (Sharma et al. 2009) of rhizosphere and phyllosphere. Despite the advantages, bio-control agents (BCAs) often exhibit significant variability in their performances due to their narrow range of activity and sensitiveness to variation in temperature and relative humidity. The performance of the antagonistic microbes therefore needs to be improved to a level comparable to the pesticides (Schena et al. 2003). This can be achieved by bioprospecting new BCAs from different ecological niches inhibiting different pathogens for operating at environments in which they are used. In the search for new antagonists, use of native microbes that reside inside the tissue, termed endophytes, are highly promising. The use of endophytic microbes comprising bacteria, fungi and yeast to control postharvest diseases of vegetables and fruits is discussed here.

7.3 Endophytes: Trustworthy Tenants of Plants

Endophytes are bacterial or fungal endosymbionts residing in plant tissues for some part of their life cycles without causing diseases, and are isolated from the plant tissues upon surface disinfection (Coombs and Franco 2003). Different symbiotic lifestyles occur among endophytes. Azevedo and Araújo (2007) gave a more comprehensive definition, describing endophytes as the microbes that colonize host plant tissues internally without causing visible damage and/or visible external structures and can or cannot be successfully cultured. This definition was later modified by Mendes and Azevedo (2008) to classify endophytes as type I for not producing external structures, and type II for producing external structures as in the case of nitrogen-fixing bacteria forming root nodules and fungi-plant mycorrhizal associations. Later, Hardoim et al. (2015) proposed that the term 'endophyte' should be used as a habitat only and not as a function by including all microbes that are able to colonize the inner plant tissues. The ability to avoid the defensive response of the host by inducing the expression of different plant features might be mediated by a fine-tuned equilibrium at the molecular level of interactions (Ludwig-Müller 2015). Currently, with advances in molecular biology, Gaiero et al. (2013) proposed that this definition should also include the sequences of unculturable endophytes but not isolated. Endophytes – existing over 400 million years, as determined by fossil records – indicate their ancient association with host plant habitat transitions (Schardl et al. 2008).

7.3.1 Entry of Endophytes into Host Plants

Endophytes are universally present in all species of plants and are transmitted either vertically, from parent to offspring, or horizontally from individual to unrelated individual. In the former, the endophytes colonize the host through seeds/vegetative propagules or pollen, where they have already established (Truyens et al. 2014). Such endophytes are often mutualistic since their reproductive fitness is closely linked to that of their host plants. In horizontal transmission, the endophytes are acquired due to colonization of plants from the atmosphere, soil and insects (Frank et al. 2017). This type of transmission is more common due to the acquisition of endophytes from a diverse environment, which enables the plants to better respond to a changing environment (Carroll 1988). In the acquisition of endophytes from the soil, the best-studied mode of acquisition, the microbes need to be rhizosphere competent for increased rhizoplane colonization and infection of the host plant. In comparison to specialized rhizosphere microbes, the endophytes interact with roots in an efficient way (Ali et al. 2014a, 2014b). Root cracks, wounds, primary and lateral root hair cells, and root cell hydrolysis, in addition to stomata, lenticels and germinating radicals, are the common modes of entry (Santoyo et al. 2016). Upon entry, the endophytic microbes colonize the intercellular spaces in the root tissues and can be microscopically visualized by tagging with fluorescently labelled tags. True endophytes are identified by their ability to re-infect seedlings which are disinfected (Sansanwal et al. 2017). Unlike the rhizospheric populations, though endophytes are also conditioned by abiotic and biotic factors of hosts, they are better protected than the rhizosphere microbes (Rosenblueth and Martinez-Romero 2006).

7.3.2 Interactions of Endophytes with Host Plants

An endophyte is competent only if it can successfully colonize the host plant and modulate the host physiology to selectively favour its establishment and maintenance of the association. Colonization is an interactive process, which occurs either due to communication between the host plant and the endophyte or between the endophyte and the host plant (Rosenblueth and Martinez-Ramirez 2006). In either case, it is exerted by the host plant for its own evolutionary and ecological benefit (Sørensen and Sessitsch 2006; Hardoim et al. 2008). Such communication by the host plants not only involves symbiotic endophytes but also with commensalistic, mutualistic and pathogenic endophytes (Bais et al. 2006). However, successful entry and colonization of the host tissues are favoured by traits deployed by the microbes to sense compounds such as amino acids and carbohydrates (Bacilio-Jimenez et al., 2003), organic acids (De Weert 2002), etc. secreted by plant roots (De Weert 2002). These compounds also play significant roles in interactions of the belowground community (Bais et al. 2004). A successful endophyte on moving towards the active exosphere zone in the rhizosphere will enhance its physiological metabolism to enable competition, growth and better nutrient acquisition. Endophytic microbes also increase their numbers by repeated cell divisions to establish a microcolony and invade the root tissues by elaborating enzymes for effective degradation of plant cell wall. In comparison to plant pathogens, endophytes secrete limited levels of enzymes to degrade the plant cell wall (Elbeltagy et al. 2000). Some endophytes also enter host tissues, without producing enzymes, through root cracks and wounds caused by soil herbivores or phytopathogens. Within the roots, the competent endophytes are able to spread to other parts of the plant after passing through the Casparian strips in the endoderm. Invasion within the plant tissues, i.e. endosphere, occurs due to the production of endoglucanases (Reinhold-Hurek et al. 2006) and endopolygalacturonases (Elbeltagy et al. 2000). Endophytes are proposed to have genomes different from those of rhizosphere microbes, although no distinct group of genes for endophytic lifestyle have been identified. Though some genes have been identified for their putative roles in endophytic lifestyle (Ali et al. 2014b), only a few have been validated for endophytic colonization. Upon colonization, endophytes exhibit different symbiotic lifestyles ranging from antagonism to mutualism and hence are referred to as a continuum. Latent pathogens inclusive of latent saprophytes and mutualistic species seem to represent a relatively small proportion of endophytic assemblages (Zabalgogeazcoa 2008). Mutualistic interaction of endophytes with plants influences plant health and growth and hence the plants select their microbiome colonizers, including those living with the plant tissues, for their benefit (Hardoim et al. 2008; Marasco et al. 2012; Rashid et al. 2012).

7.3.3 Role of Endophytes in Plant Health and Growth

In colonization of plant hosts by endophytes, the reasons for the lack of defence mechanisms in plants against endophytes, allowing them to coexist, are not yet understood. Nevertheless, the association between plants and endophytes is beneficial to both parties. While endophytes derive nutrients from the host plants, they benefit the plants by promoting growth either directly or indirectly. Hence, plants colonized by endophytes are healthier than those not colonized by them, as the colonized plants are able to tolerate the biotic and abiotic stresses inflicted by the environment (Zhang et al. 2006). The endophytes facilitate direct plant growth promotion either through the acquisition of essential nutrients such as nitrogen, phosphorus and iron or through modulating the levels of auxin, cytokinin and gibberellin hormones in the plant. Some endophytes synthesise 1-aminocyclopropane-1-carboxylate (ACC) deaminase, the enzyme that reduces the level of the phytohormone ethylene in all higher plants by cleaving its precursor, ACC. Indirect plant growth promotion occurs due to the inhibition of pests by the endophytes and decreasing the damage caused by them to plants (Gaiero et al. 2013). In addition to promoting the growth and development of plants and protecting them from pathogens, endophytes are also used as adjuncts in phytoremediation of a variety of environments (Ryan et al. 2008). The biomolecules elaborated by the endophytes in association with plants exhibit industrial and medical applications for possessing antimicrobial, antifungal, anticarcinogenic, immunosuppressant or antioxidant activities (Zhang et al. 2006).

7.3.4 Endophytes as BCAs

Microbes with an endophytic lifestyle are protected from the influence and fluctuations of the environment, which may threaten their ability to survive besides reducing their bio-control efficacy (Card et al. 2016). Hence, the use of endophytic microbes has an advantage over their synthetic analogues due to their safety to the environment and their ability to remain stable over a range of temperatures and pH, making them ideal for application in agriculture and food production (Rodrigues et al. 2006). These advantages make endophytes a suitable tool for bio-control to deliver a stable control effect. Endophytes that are seed transmitted do not demand development of formulations and delivery for application, thereby providing an added advantage in commercialization. They create a 'barrier effect' during plant tissue colonization, which gives them the advantage to 'outsmart' phytopathogens from colonizing the plant subsequently, and have also been reported to secrete antagonistic chemicals against their competitors, such as pathogenic organisms (Schardl et al. 2013; Sansanwal et al. 2017). The emergence of endophytic microbes as an eco-friendly substitute to synthetic chemicals has remarkably improved fungal disease management (Bacilio-Jimenez et al. 2001) and the incidence of nematodes and insects. Endophytic bacteria constitutes *Acetobacter diazotrophicus*, *Herbaspirillum seropedicae*, *H. rubrisubalbicans*, *Serratia* spp., *Bacillus* spp., *Enterobacter* spp., *Agrobacterium radiobacter*, *Burkholderia gladioli*, *B. solanacearum*, *Pseudomonas putida*, *Pseudomonas fluorescens*, *Achromobacter xylosoxidans*, *Pseudomonas aeruginosa*, *Micrococcus* spp. and *Flavobacterium* spp. (Varma et al. 2017). The most common genera of fungal endophytes are *Trichoderma* spp., *Epicoccum nigrum*, *Penicillium* spp., *Xylaria*, *Alternaria*, *Cladosporium*, *Fusarium* spp., *Scolecobasidium humicola*, *Fusarium oxysporum*, *Chaetomium globosum*, *Cladosporium cladosporioides*, *Aspergillus*, *Aspergillus niger*, *Curvularia*, *Gilmaniella*, *Arthrobotrys foliicola*, *Colletotrichum graminicola*, *Fusarium verticillioides*, *Saccharomyces cerevisiae*, *Alternaria alternata*, *Trichoderma koningii*, *Aspergillus flavus*, *Acremonium zeae* and others (Varma et al., 2017). Besides, certain entomopathogenic fungi – *Paecilomyces* spp. and *Beauveria bassiana* – were also reported as endophytic microbes in tobacco and cotton (Sword et al. 2012; Ek-Ramos et al. 2013) and are used to control storage diseases. Since most of the endophytic genera also inhabit the rhizosphere, the endophyte microbiome is postulated to a subpopulation of the rhizosphere inhabiting bacteria (Germida et al. 1998; Marquez-Santacruz et al. 2010).

7.3.5 Bio-control Attributes of Potent Endophytic Microbes for Controlling Postharvest Diseases

In general, the desirable characteristics of endophytes vary by the bio-control system. However, certain characteristics are essential for any endophyte to be deployed for managing postharvest diseases. They can be successful bioagents only if they can act on the pathogens under different environmental conditions and exert their functions together with other factors, such as storage conditions, compatibility with usual postharvest practices, treatment, additives and cultivar (Narayanasamy 2005). Besides, antibiotics or any toxic metabolites produced by the BCAs should not inflict health hazards. The BCAs should be able to colonize and multiply rapidly at injury sites caused by improper handling of the produce to reach desirable concentrations; protection against postharvest pathogens entering through injuries is accomplished by field application of the BCA at the harvesting or pre-harvest stage for early colonization and protection of the products. Hence, potent BCAs are expected to tolerate high temperatures, UV radiation, low nutrient availability and dry conditions that frequently occur during growing seasons. The BCA should also be able to colonize senescent flower parts either before or during pathogen infection, as the pathogens may infect the host tissues any time after flowering and by remaining quiescent, proceed through different stages of fruit development in the field and after harvest till ripening (Irtwange 2006).

7.3.6 Isolation and Selection of Antagonistic Endophytic Microbes for Bio-control of Postharvest Pathogens/Diseases

7.3.6.1 Isolation of Endophytic Microbes

Isolation of endophytes is essential to utilize them as microbial inoculants and to extract secondary metabolites to control postharvest diseases (Schulz et al. 2002; Strobel 2003; Schulz and Boyle 2005). The method of isolation should be sensitive enough to recover endophytes but to eliminate epiphytes (Hallmann et al. 2006). Endophytic microbes are isolated from plant tissues after surface sterilization (Schena et al. 2003). The tissues are ground, and the ground tissue is plated on a suitable media and incubated at 24°C. Surface sterilization can be substituted with vacuum or pressure extraction. Non-culturable endophytes are identified by cloning the target gene fragments amplified by polymerase chain reaction (PCR) from the sample and comparing the sequences with those in data banks (Shanmugam et al. 2011). The target sequences are usually 16S rRNA for bacteria, 28S rRNA or the internal transcribed region (ITS) for fungi. The microbial community structure and their temporal and spatial variations due to storage or field environmental factors are then analyzed by fingerprinting techniques such as denaturing or temperature gradient gel electrophoresis (DGGE/TGGE), terminal restriction fragment length polymorphism (T-RFLP) and PCR-single-strand-conformation polymorphism (Hallmann et al. 2006). The extent of microbial colonization can be quantified by real-time PCR (Schena et al. 2004), which is a useful technique for non-culturable endophytes.

7.3.6.2 Selection of Endophytic Microbes

The isolated microbes are tested against each of the pathogens by standard confrontation assays under in vitro conditions. However, since many potent bio-control microbes have been identified by screening them on fruit directly, despite the bioagents not inhibiting pathogen development in vitro, in situ tests on vegetables and fruits are imperative to identify antagonistic microbes. Potent microbes can be identified by using different criteria such as inhibition of rot expansion by greater than 75% and reduction in the number of wounds on infection to less than 50%. Further screening is needed to determine the minimum effective concentration for each of the potent antagonistic microbes. About 20% of the potent antagonists are further tested on additional fruits (Narayanasamy 2005). The promising ones are further evaluated for their ability to survive on vegetables and fruits at the wound site under different conditions of temperature, relative humidity and composition of gases during storage and for compatibility with other postharvest treatments, additives and antagonists (Janisiewicz 1998). The bioactive substances can then be isolated and characterized from the culture filtrates of potential endophytic BCAs using bioassay-guided fractionation and spectroscopic methods.

7.3.7 Endophytes to Control Postharvest Diseases of Vegetables and Fruits

Endophytes are applied either before or during harvesting and storage to control postharvest diseases of vegetables and fruits.

7.3.7.1 Pre-harvest Application of Endophytes

Postharvest diseases often occur due to pre-harvest infection in the field, while the produce is still immature on the plant. The biomanagement strategy employing endophytes should therefore ensure the reduction of pathogen growth and mycotoxin production in the produce to promote the general condition of crops. However, the biomanagement practice needs to be optimized under field conditions, which is often difficult due to challenging environmental conditions. Despite this, pre-harvest application of endophytes has been promising in controlling several pathogens such as *Penicillium digitatum* and *Penicillium italicum* on orange (Luo et al., 2019; Camañas et al., 2008), *Colletotrichum acutatum* on citrus fruits (Lopes et al. 2015) and *Colletotrichum musae* on banana (Silva and De Costa, 2014). Pre-harvest application of endophytic microbes isolated from the flesh of sweet cherries (Loi et al. 2000; Schena et al. 2000) minimized rot of sweet cherries and table grapes by 47% and 38%, respectively, during storage (Schena et al. 2003).

7.3.7.2 Postharvest Application of Endophytes

The controlled temperature and humidity in environments where harvested vegetables and fruits are preserved offer a better milieu than field conditions for the effect of BCAs on harvested vegetables and fruits (Nunes and Nunes 2012). In postharvest applications, the antagonistic endophytes are applied directly to the produce by pulverization or immersion in solution (Dukare et al. 2018). Endophytic bacteria have increasingly been used as bioagents to control a plethora of plant pathogens. Among the bacterial endophytes, strains of the genera *Pantoea*, *Bacillus* and *Pseudomonas* are commonly employed to control diseases after harvest. Since the efficiency of endophytic bacteria to control wound-infesting pathogens was recognized, *Rhizopus stolonifer* and *Monilinia laxa* in stone fruits was first reported by Pratella et al. (1993), a vast majority of endophytic bacteria identified as potential BCAs against postharvest diseases of fresh-cut fruit/vegetables belongs to *Bacillis* spp. (Sharma et al. 2009). Control of postharvest diseases of vegetables and fruits by other bacteria and fungi, though efficient, has been scantly reported. Examples of endophytic microbes effective against postharvest diseases are given in Table 7.1.

TABLE 7.1

Endophytes Used as Bio-control Agents against Postharvest Diseases

Endophytic Microbes	Target Pathogen	Target Disease	Fruit	Reference
Bacillus subtilis	*Botrytis cinerea*	Grey mould	Apple	Sholberg et al. (1995)
Bacillus pumilus and *Bacillus amyloliquefaciens*	*Botrytis cinerea*	Grey mould	Pear	Bacon et al. (2001); Mari et al. (1996)
Bacillus megaterium ENB-86	*Colletotrichum capsici*	Fruit rot	Chilli	Ramanujam et al. (2012)
Pseudomonas aeruginosa, *Burkholderia cepacian*, *Bacillus tequilensis*	*Phytophthora capsica*, *Botrytis cinerea*, *Fusarium oxysporum*, *Colletotrichum acutatum*	Anthracnose, leaf blight, grey mould	Chilli	Paul et al. (2012)
Bacillus lentimorbus	*Botrytis cinerea*	Grey mould	Fruits	Cheng et al. (2015)
Bacillus velezensis	*Colletotrichum musae*	Anthracnose	Banana	Damasceno et al. (2019)
Lactobacillus spp.	*Penicillium expansum*, *Xanthomonas campestris*, *Monilinia laxa*, *Botrytis cinerea*, *Erwinia carotovora*	Grey mould, soft rot	Apple	Trias et al. (2008)
Pseudomonas cepacian, *Pseudomonas corrugata*	*Monilinia fructicola*	Brown rot	Nectarine, peach	Smilanick et al. (1993)
Pseudomonas fluorescens isolate 1100-6	*Penicillium expansum* or *Penicillium solitum*	Blue mould	Apple	Etebarian et al. (2005)
Pseudomonas fluorescens	*Botrytis mali*	Grey mould	Apple	Mikani et al. (2008)
	Monilinia laxa or *Monilinia fructicola*	Brown rot	Stone fruit	Harvey et al. (1972)
Cladosporium, *Fusarium*, *Alternaria*	*Colletotrichum acutatum*, *Phytophthora capsica*	Anthracnose, blight	Chilli pepper	Paul et al. (2012)
Penicillium spp.	*Botrytis cinerea*	Botrytis rot	Grape	Noumeur et al. (2016)
Phaeosphaeria nodorum	*Monilinia fructicola*	Twig blight of stone fruits, brown rot, blossom blight	Plum	Pimenta et al. (2012)
Phomopsis columnaris	*Colletotrichum*, *Botrytis*	Anthracnose, Fruit rot	–	Singh et al. (2011)
Trichoderma harzianum	*Colletotrichum musae*	Anthracnose	Banana	Devi and Arumugam (2005)
Trichoderma harzianum	*Botrytis cinerea*	Rot	Tomato	Malmierca et al. (2015)
Trichoderma spp.	*Lasiobasidium theobromae*, *Rhizopus* spp.	Fruit rot	Mango	Pathak (1997)
Aureobasidium pullulans	*Monilinia laxa*, *Botrytis cinerea*	Botrytis rot, soft rot	Sweet cherry, table grape	Schena et al. (2003); Barkai-Golan (2001)

7.3.8 Managing Postharvest Products' Microbiome for Bio-control

The limited studies on the survival and role of endophytic microbes within host tissues in harvested commodities indicate the presence of natural bio-control mechanisms wherein the pathogens remain as silent endophytes and can cause diseases only when the composition of microbiota alters due to physiological changes in the host and postharvest treatments (Droby and Wisniewski 2018). This was indicated by a study on stem-end rot in mango during storage and ripening (Diskin et al. 2017). It was observed that the abundance of stem-end pathogens and endophytic microbes was influenced by fruit ripening and postharvest treatment. The microbial communities were more diverse in the stem ends that did not develop disease relative to stem ends that were diseased. Several other studies also reported the effect of genotype on the microbiome (Shanmugam et al. 2011; Balint et al. 2013; Liu et al. 2018), although such effect is not directly correlated with postharvest produce and bio-control but added support to the co-evolution concept (Droby and Wisniewski 2018). Highlighting the findings of Colla et al. (2017) and Mazzola and Freilich (2017) on the effect of soil rhizosphere manipulation in suppressing soilborne diseases, Droby and Wisniewski (2018) indicated that manipulating the microbiome of postharvest produce may enable efficient bio-control of postharvest diseases.

7.3.9 Mechanism of Action of Endophytic BCAs

In general, the rhizosphere and endophytic microbes exhibit similar mechanisms of plant growth promotion. In either case, the antagonistic microbes operate multiple mechanisms to exhibit wide-spectrum antagonistic activity against several pathogens. Since the functioning and proliferation of rhizosphere microbes are often subjected to changing soil conditions and the presence of competing microbes, the performance of endophytes is relatively more stable for being not influenced by those factors (Glick 2012). Likewise, many of the mechanisms operated for plant disease bio-control also apply to postharvest bio-control. It is important to understand the mechanism of action of antagonists in order to develop more practical methods for efficient application and to establish a standard for expected antagonistic properties, thereby making a financial investment towards its eventual commercialization worthwhile. The mechanisms operated by the endophytes in bio-control include competition for space and nutrients, direct inhibition through mycoparasitism and antibiosis, and induction of host resistance by activating its own defence mechanisms.

7.3.9.1 Competition for Nutrients and Space with Phytopathogens

The composition of a plant microbiome is determined mainly by competitive exclusion. In addition, it is also identified to be a possible mechanism exerted by the endophytes against the pathogens and prevent the colonization of host tissues (Latz Meike et al. 2018). This is because the endophytes rapidly colonize inter- or intracellular host tissues either locally or systemically and establish an ecological niche. The ecological niche deprives nutrients and space to the pathogen (Gao et al. 2010). Mango leaves treated with fungicides were deprived of fungal endophytes, which allowed other fungi to establish in the niche and cause infection (Mohandoss and Suryanarayanan 2009). This principle of competitive exclusion and ability to adapt to the host environment allows preemptive colonization of host tissues by endophytes on the postharvest produce. Competition is not an exclusive mechanism operated by endophytes because the mechanism is often localized with limited effect on pathogens infecting other host parts. Also, competitive efficiency is greatly limited by high disease potential. Competitive exclusion can be investigated by observing the level of host plant colonization by the endophyte through microscopy and by quantifying the endophyte biomass in the plant tissue and correlating to disease severity (Latz Meike et al. 2018).

While competing for certain nutrients like iron, the microbes produce siderophores. Siderophores are extracellular, low-molecular-weight (500–1000 Da) iron(III) transporting agents, which selectively complex iron(III) ions in the environment with very high affinity. The microbes deploy specific mechanisms to uptake iron from the ferric siderophore complexes. Bacteria produce siderophores belonging to functional groups, carboxylates (i.e. rhizobactin), catecholates (i.e. enterobactin) and hydroxamates (i.e. ferrioxamine B) or a combination of the main functional groups (i.e. pyoverdine). Most fungi produce hydroxamate-type siderophores of the ferrichrome family (i.e. ferrichrome), except for the polycarboxylate rhizoferrin produced by Zygomycetes. The contribution of siderophores in bio-control of antagonistic bacteria has been demonstrated (Kloepper et al. 1980). However, the role of siderophores in bio-control of postharvest diseases in stored vegetables and fruits is seldom reported. The bio-control yeast *Rhodotorula glutinis* effective against *P. expansum* causing blue rot in harvested apples produces rhodotorulic acid, a hydroxamate-type siderophore. Likewise, the bacterium *Rahnella aquatilis* effective against *B. cinerea* causing grey rot in storage apples produces enterochelin, a catecholate siderophore. In bio-control of blue and grey moulds, the apple wounds were significantly better controlled by a combination of antagonistic yeast/bacteria with siderophore than by the antagonistic agent alone (Calvente et al. 1999; Ferramola et al. 2013).

Endophytic bacteria form complex multicellular or multispecies assemblies, including smaller aggregates and biofilms, by physically interacting with plant host surfaces or endophytic fungi through adhesins such as surface proteins and polysaccharides. The duration, extent and result of interactions are greatly favoured by the adherent populations conformation. The specificity of the interaction is the recognition between lectins and their cognate carbohydrates. The formation of biofilm requires cell-cell communication (quorum sensing) between the bacteria colonizing the host tissues (Danhorn and Fuqua 2007). Quorum sensing signals are 'intercepted' by signal degrading molecules by a process called quorum quenching.

7.3.9.2 Mycoparasitism

Fungal endophytes inhibit or kill pathogenic fungi by attacking them. During parasitism, the fungal endophytes coil and penetrate the hyphae of pathogenic fungi by producing hook-like and appressorium-like structures and undergo branching and sporulation, produce resting bodies and cause lysis. Mycoparasitism is commonly exhibited by *Trichoderma*. Endophytic *Trichoderma* parasitizing hyphae of *Rhizoctonia solani* and fungal pathogens of cocoa has been reported earlier (Grosch et al. 2006; Evans et al. 2003; Bailey et al. 2008; Mejía et al. 2008). *Acremonium strictum*, a novel endophytic fungus isolated from *Dactylis glomerata* and some grasses, displays mycoparasitism against *Helminthosporium solani* (Rivera Varas et al. 2007; Sánchez Márquez et al. 2007).

7.3.9.3 Antibiosis

Production of antimicrobial compounds is another important mechanism of bio-control, wherein the antagonistic microbes secrete antibiotic substance(s), lytic enzymes, volatiles or toxins that target and kill the pathogen. The coexistence and evolution of endophytes together with their host plants significantly influence the physiology of plants by activating silent gene clusters to produce novel secondary metabolites in plants (Jia et al. 2016) from similar precursors (Abdalla and Matasyoh 2014; Deepika et al. 2016). Many secondary metabolites are antimicrobial, which includes antibiotics, lytic enzymes to degrade cell wall, volatiles or toxins (Nagorska et al. 2007; Maksimov et al. 2015). These antimicrobial compounds disrupt protein synthesis and pathogen cell structure and inhibit the function of respiratory enzymes (Waewthongrak et al. 2015).

7.3.9.3.1 Antibiotics

Antibiotics are low-molecular-weight heterogeneous organic compounds elaborated by antagonistic microbes to disrupt the structure and function of pathogen cell walls, disrupt protein synthesis and pathogens cell structure, and inhibit the function of respiratory enzymes (Lastochkina et al. 2019).

TABLE 7.2

Antibiotics Produced by Endophyte Microbes

Endophytes	Host	Antibiotics	Target Pathogens	Reference
Ampelomyces spp.	*Urospermum picroides*	A 3-O-methylalaternina and Altersolanol	*Enterococcus faecalis aureus* and *Staphylococcus epidermidis*	Aly et al. (2008)
Muscodor albus	Tropical trees	2-butanone, Aciphyllene, Tetrohydofuran and 2-methyl furan	*Stachybotrys chartarum*	Atmosukarto et al. (2005)
Phomopis cassiae	*Cassia spectabilis*	Sesquiterpenes and Cadinane	*Cladosporium cladosporioides* and *Cladosporium sphaerospermum*	Silva et al. (2006)
Verticillium spp.	*Rehmannia glutinosa*	Ergosterol peroxide and Massariphenone	*Pyricularia oryzae* P-2b	You et al. (2009)

Antibiotics secreted by endophytes varies according to the species and strain (Ongena and Jacques 2008; Maksimov and Khairullen 2016). Significant antibiotics produced by antagonistic bacteria are pyrrolnitrin, secreted by *Pseudomonas cepacia*; antifungal peptide iturin, secreted by *B. subtilis*; and trichothecene, produced by *Myrothecium roridum* (Torres et al. 2014). Antifungal activity of metabolites produced by a bacterial endophyte of mountain-cultivated ginseng plants, *Burkholderia stabilis* EB159 against *Cylindrocarpon destructans* causing root rot of ginseng, has been recently reported (Kim et al. 2020). Antibiotics produced by endophytic microbes are listed in Table 7.2.

7.3.9.3.2 Volatiles

Though endophytes produce a range of secondary metabolites to survive in the plant environment, most studies have focussed on fungal metabolites. Endophytic fungi produce an array of secondary metabolites, including peptides, terpenoids, alkaloids, phenols, quinones and xanthones (Korpi et al. 2009; Li et al. 2018). Several species of *Muscodor*, *Noduliosporium* and *Trichoderma* regulate plant defences (Garnica-Vergara et al. 2016) by producing volatile organic compounds (VOCs) (Chen et al. 2016). These VOCs largely include the chemical groups, terpenoids, phenylpropanoids, benzenoid compounds, fatty acid and amino acid derivatives (Dudareva et al. 2013; Delory et al. 2016). VOCs are hydrophobic, organic molecules of <300 Da size with a high vapour pressure of ≥0.01 kPa at 20°C (Pagans et al. 2006) and serve as defence molecules against herbivores and pathogens. They also serve as a signal molecule for communication between plant parts or plants, and attraction of pollinators, seed dispersers and other beneficial agents (Farré-Armengol et al. 2016; Kaddes et al. 2016; Delory et al. 2016). VOCs, by a virtue of their abilities to diffuse in the air, reach inaccessible habitats in a closed environment (Morath et al. 2012) like storage and thus complement bio-control of postharvest diseases. The use of VOCs from fungi for bio-control of postharvest diseases in vegetables and fruits is termed mycofumigation (Strobel et al. 2001), whose efficiency depends on the type of postharvest disease, the fungal species and the amount of inoculum used (Kaddes et al. 2019).

The concept of mycofumigation emerged from the discovery of potential antimicrobial VOCs from the endophytic fungus, *Muscodor albus* isolated from *Cinnamomum zeylanicum* (Gomes et al. 2015). The compounds exhibited antifungal activity against many fungal species, including *Penicillium verrucosum*, *P. digitatum*, *Aspergillus fumigatus*, *A. carbonarius*, *A. flavus*, *A. niger*, *A. ochraceus*, *Geotrichum candidum*, *Monilinia fructicola*, *F. graminearum*, *Fusarium culmorum*, *Colletotrichum acutatum*, *B. cineréal* and *Rhizopus* sp., which produce mycotoxin and decay postharvest produce (Strobel et al. 2001; Mercier and Jiménez 2004; Strobel 2011). Later, VOCs were also discovered from several other species of *Muscodor* (Strobel 2006). In addition to *Muscodor* species, antimicrobial volatiles produced by several other fungi such as *Myrothecium inundatum*, *Phomopsis* sp., *Hypoxylon* sp., *Nodulisporium* sp., *Bionectria ochroleuca*, *Schizophyllum commune*, *Gloeosporium* sp. and *Gliocladium* sp. were also reported. These volatiles inhibited the growth of several fungi associated with postharvest diseases, such as *Aspergillus ochraceus*, *A. flavus*, *A. fumigatus*, *B. cinerea*, *C. capsici*, *C. gloeosporioide*, *C. lagenarium*, *C. musae*, *G. candidum*, *Penicillium digitatum*, *Penicillium expansum*, *Pythium ultimum*, *Sclerotinia sclerotiorum*, and *Phytophthora palmivora*, albeit with relatively less in vitro antimicrobial effect (Gomes et al. 2015). Most of these endophytes were obtained from tropical plants used in alternative medicines, such as *Aegle marmelos*, *Ananas ananassoides*, *Cinnamomum* spp. and *Myroxylon balsamum*. Though most of them belong to ascomycetes, production of VOCs by the basidiomycete *Oxyporus latemarginatus* has also been reported (Gomes et al. 2015). The volatile compound identified as 5-pentyl-2-furaldehyde inhibited a range of pathogenic fungi including the postharvest pathogens *R. solani* and *B. cinerea*, which causes root rot of moth orchids and postharvest apple decay, respectively (Lee et al. 2009).

In general, though the volatiles produced by *Muscodor* species contain mostly low-molecular-weight esters, alcohols and acids, they vary for different species of the genus. Nevertheless, the mixture of VOCs produced by a majority of *Muscodor* species has antimicrobial activity (Ezra and Stobel 2003; Suwannarach et al. 2013). The volatile compounds released by the endophytes enhance the plant defence mechanisms and enable them to compete with microbes that compete for the same ecological niche (Strobel 2011). The compound is also speculated to impair the replication and/or transcription process in *Escherichia coli* in addition to increasing fluidity of the cell membrane by inciting morphological changes (Daisy et al. 2002). In another study, *Trichoderma arundinaceum* exhibiting antifungal activity against *B. cinerea* produced trichodiene, which induces the expression of defence genes for salicylic and jasmonic acid production and by interacting with hydrolytic enzymes (Malmierca et al. 2012). VOCs from strains of endophytic fungi, *Hypoxylon anthochroum*, act synergistically and exhibit antifungal effect individually in a concentration-dependent manner against *Fusarium oxysporum* in cherry tomato fruits. The compounds alter the permeability of the cell membrane in addition to damaging the morphology of hypha, and inhibit respiration (Medina-Romero et al. 2017).

In addition to filamentous fungi, the production of antifungal VOCs by bacteria and some yeasts is also documented. Volatile substances released by an endophytic bacterium *Bacillus thuringiensis* were antifungal to *Fusarium sambucinum* infecting potato tubers (Zouaoui et al. 2001). Bio-control of postharvest diseases is also achieved by VOCs of *B. amyloliquefaciens* CPA-8 against *M. fructicola* infecting cherry fruit (Gotor-Vila et al. 2017); *B. thuringiensis* and *B. pumilus* against mango anthracnose (Zheng et al. 2013); *B. amyloliquefaciens* PPCB004 against *Penicillium crustosum* causing postharvest decay in citrus (Eva et al. 2010); *Bacillus amyloliquefaciens* LI24 and PP19, *B. licheniformis* HS10, *Exiguobacterium acetylicum* SI17 and *B. pumilus* PI26 against *Peronophythora litchi* causing downy blight in litchi (Zheng et al. 2019). VOCs produced by *Arthrobacter agilis* UMCV2 and identified to be dimethylhexadecylamine were antifungal to *B. cinerea* and *Phytophthora cinnamomi*. The compound was 12 times higher inhibitory than that caused by the fungicide Captan (Velázquez-Becerra et al. 2013). Volatile compound mixtures emitted by the yeasts *Metschnikowia pulcherrima*, *Aureobasidium pullulans*, *Candida intermedia*, *Wickerhamomyces anomalus* and *Saccharomyces cerevisiae* inhibit fungal growth causing postharvest decay in vegetables and fruits (Gomes et al. 2015).

7.3.9.3.3 Lytic Enzymes

Lysis is the process of partial or complete cell destruction incited by enzymes (Shanmugam, 2005). Chitin and glucan are polysaccharides constituting the cell wall of fungi. Antagonistic microbes secrete extracellular hydrolases such as β-1,3-glucanase and chitinase to act on glucan and chitin and lyse the pathogenic fungal hyphae (Mardanova et al. 2017). *B. subtilis* antagonistic to *F. oxysporum*, a postharvest pathogen of yam (*Dioscorea* spp.) tubers, induced fungal cell wall lysis by producing extracellular chitinases (Swain et al. 2008; Pieterse et al. 2014). Kilani-Feki et al. (2013) observed that harvested tomatoes treated with *B. subtilis* V26 displaying improved resistance to grey mould incited by *B. cinerea* up to 79% during storage was due to the production of chitosanases and proteases. Postharvest citrus mould caused by *Penicillium digitatum* was effectively controlled by the endophytes *B. subtilis* and *Agrobacterium radiobacter* through lytic enzymes, glucanases and chitinases (Mohammadi et al. 2017). Development of grey mould of fruits caused by *B. cinerea* was significantly inhibited by endophytic *B. lentimorbus*. Alpha- and beta-glucosidase produced by the bacterium is speculated for inhibition (Cheng et al. 2015). In addition to chitinases and glucanases, production of proteases, lipases, cellulases, xylanases and mannanases were also correlated with antifungal activity for several endophytes (Aktuganov et al. 2007; Damasceno et al. 2019). Likewise, an acidic peptide produced by a *B. subtilis* strain endophytic in apple significantly controlled the disease during storage (Sholberg et al. 1995).

7.3.9.4 Induction of Systemic Resistance in Plants

Bio-control microbes trigger induced systemic resistance (ISR) in plants due to the recognition of their surface and effector molecules by the host plant. ISR varies from systemic acquired resistance (SAR) induced by pathogens in not inducing a hypersensitive reaction in the host plants (Ramanathan et al. 2002; Shanmugam and Kanoujia 2011). ISR is salicylic acid-independent and induces a potentiated defence capacity, called 'priming', through JA/ET pathway. Priming does not improve phytohormone synthesis or induce hormone-responsive genes in systemic tissues. Rather, it improves hormone sensitivity (Conrath et al. 2006; Pastor et al. 2013). SAR depends on the salicylic acid pathway, which commonly induces pathogenesis-related proteins. Though SAR and ISR follow distinct signalling pathways, both activate their corresponding defence responses through the regulatory protein NPR-1. ISR triggered defence responses include cell-wall reinforcement, oxidative burst, accumulation of defence-related enzymes and secondary metabolite production (Nie et al. 2017).

An overview of literature indicates that the defence mechanisms elicited by *Bacillus* spp. in harvested vegetables and fruits control postharvest decay by accumulating phytoalexins (Maksimov et al. 2011). In postharvest bio-control of *Alternaria alternata* infecting melon using *B. subtilis* EXWB1, the bacterium suppresses ethylene production and respiration in the produce and maintains freshness (Wang et al. 2010). In addition to endophytes, the antifungal compounds produced by them in host cells also induce the defence mechanisms and control diseases in harvested vegetable and fruit disease (Lastochkina et al. 2019). The antibiotics iturin and fungicin produced by *B. subtilis* trigger ISR in plants by inducing the genes for phenylpropanoid metabolism (Falardeau et al. 2013). The bio-control effect of *B. subtilis* strains 10-4 and 26D either alone or with SA on *Fusarium oxysporum* and *Phytophthora infestans* infecting potato during storage is dose dependent. The strains protect the tubers against reactive oxygen by increasing ascorbic acid content and decreasing lipid peroxidation and proline accumulation induced in tubers by the pathogen (Lastochkina et al. 2020). Though activation of induced resistance is expected to alter fruit growth and quality negatively, many studies of inducing resistance are correlated with the induction of antioxidant mechanisms of the fruit, enhancing the quality for human consumption.

Among endophytic fungi, species of *Neotypodium* are the most studied for inducing host resistance. However, the effect of induced resistance in controlling postharvest diseases is seldom reported.

7.3.10 Bioproducts Based on Endophytes for Control of Postharvest Diseases

Great progress has been made to register and commercialize bioproducts to control postharvest diseases caused by *Fusarium sambucinum*, *Rhizopus stolonifera*, *B. cinerea*, *P. italicum*, *P. digitatum* and *P. expansum* (Spadaro and Droby 2016). Interestingly, a majority of bioproducts registered or commercialized to manage diseases in postharvest produce are mainly bacteria and yeast-based due to their ability to tolerate adverse climatic conditions in the field and during storage. In addition, they also adapt to micro-environmental conditions such as low pH, high sugar concentration and high osmotic pressure present in the wounded fruit tissues. Yeasts, unlike filamentous fungi, are not known to produce mycotoxins or allergenic spores, and due to simple nutritional requirements can colonize dry surfaces for longer periods (Droby et al. 2016). Additionally, their ability to grow fast on cheaper substrates in bioreactors favours their largescale production (Spadaro et al. 2010).

The first generation of bioproducts commercially available was based on the yeasts *Candida oleophila* (Aspire™, Ecogen, Langhorne, PA, USA), *Cryptococcus albidus* (YieldPlus™, Lallemand, Montreal, Canada) and *Candida sake* (Candifruit™, IRTA, Lleida, Spain), and on the bacteria *Pseudomonas syringae* (BioSave™, JET Harvest, Longwood, FL, USA) (Droby et al. 2016). The bioproducts, Aspire, YieldPlus and Candifruit, though, were discontinued due to commercial deficiencies. Newer yeast-based bioproducts, *C. oleophila* (Nexy™, Leasafre, Lille, France) and *Aureobasidium pullulans* (BoniProtect™, Bio-Ferm, Tulln, Austria), have also been registered. Among them, Nexy is recommended to control wound pathogens infecting postharvest produce of pome fruits, citrus and banana (Lahlali et al. 2011) and *Aureobasidium pullulans* (BoniProtect, Bio-Ferm, Tulln, Austria) to control wound pathogens infecting pome fruit during pre-harvest and storage (Lima et al. 2015). *Metschnikowia fructicola* (Shemer™, Bayer, Leverkusen, Germany) originally registered in Israel for control of both pre- and postharvest diseases on various vegetables and fruits, including sweet potatoes, peppers, citrus fruit, peaches, grapes, strawberries and apricots, was acquired by Bayer CropScience (Germany) to use as a postharvest bio-control product (Spadaro and Droby 2016).

Among the bacteria-based postharvest products, BioSave, though originally registered for controlling diseases of pome and citrus fruits during storage, was later used for potatoes, sweet potatoes and cherries (Janisiewicz and Peterson 2004). The other commercial bio-control product, Avogreen™ (University of Pretoria, Pretoria, South Africa), based on *B. subtilis*, was used in South Africa to control *Cercospora* spot of avocado during storage, with no commercial success (Demoz and Korsten 2006). Amylo-X™ (Biogard CBC, Grassobbio, Italy) is a *B. amyloliquefaciens*-based product and was used to manage bacterial and fungal diseases of many vegetables (Carmona-Hernandez et al. 2019). An endophytic *B. cereus* CE3 formulated as a wettable powder but not commercialized effectively controlled *Endothia parasitica* and *F. solani* causing postharvest fruit rot (Cheng et al. 2015). Apart from *Pseudomonas* and *Bacillus* based formulations, a *Pantoea agglomerans* CPA-2™ (Pantovital, Domca, Granda, Spain)

TABLE 7.3

Commercial Bioproducts of Endophytes for Postharvest Disease Control

Microorganism	Commercial Product	Target Diseases and Pathogens	Fruit	Country
Bacillus subtilis	Serenade	Fire blight, late blight, brown rot, powdery mildew	Apple, vegetables, pear, grape	USA
Bacillus subtilis FZB 24	Rhio-plus	Root rot, powdery mildew	Vegetables	Germany
Cryptococcus albidus	Yield plus	*Mucor*, *Botrytis*, *Penicillium*	Pome fruit	South Africa
Pantoea agglomerans	Pantovital	*Monilinia Penicillium*, *Botrytis*	Citrus fruit, pome fruit	Spain
Pseudomonas fluorescencs-A 506	Blight Ban A 506	Soft rots, fire blight	Apple, pear, strawberry, potato	USA
Pseudomonas syringae	BioSave	*Mucor*, *Penicillium*, *Botrytis*	Potato, sweet potato, Pome, cherry, citrus fruit	USA
Pseudomonas syringae (strain 10 LP, 110)	Biosave 10 LP, 110	Sour rot, blue and grey mould, mucor	Apple, pear, citrus, cherry, potato	USA
Aureobasidium pullulans	Boniprotect	*Monilinia Penicillium*, *Botrytis*	Pome fruit	EU
Candida oleophila	Nexy	*Penicillium*, *Botrytis*	Pome fruit	Belgium
Candida oleophila (strain 1–182)	Aspire	Blue, grey, and green moulds	Apple, pear, citrus	USA
Candida sake	Candifruit	*Botrytis*, *Rhizopus*, *Penicillium*	Pome fruit	Spain
Metschnikowia fructicola	Shemer	*Rhizopus*, *Aspergillus*, *Botrytis*, *Penicillium*	Sweet potato, grape, strawberry	Netherlands
Ampelomyces quisqualis	AQ-10	Powdery mildew	Strawberry, apple, grape, cucurbit, tomato	USA
Coniothyrium minitans	Contans WG, Intercept WG	Neck and basal rots	Onion	Germany

based formulation despite tested effectively against pome and citrus postharvest diseases was not commercialized (Torres et al. 2014). Commercial bioproducts of endophytes registered for control of postharvest diseases are listed in Table 7.3.

Inconsistency in performance under commercial conditions, inefficient formulations that reduce the efficiency of antagonistic cells, cost and regulatory barriers in the registration of the bio-control agents that discourages their dissemination is often cited as reasons for the limited use of bioproducts.

7.4 Conclusion

Development of green technologies for controlling postharvest diseases, although gaining momentum with the deployment of antagonists for bio-control of diseases infecting postharvest produce, very few antagonists have been commercialized. Hence, certain factors need to be considered from designing to development of an endophyte-based bio-control product.

In postharvest control, the environment in which the produce is to be stored and the nature and economic importance of target pathogens need to be clearly determined to decide the parameters for selection of endophytic BCAs. These include the adaptability of the bioagents to the environment during storage, their efficiency in terms of efficacy and cost over chemical control, and plant protection from pathogens without influencing the plant's microbiome. In established or latent pathogenic infections, often the BCAs are unable to exert their fullest potential in postharvest disease control. In such cases, early application of BCAs before harvest in the field before pathogen infection would be the next best option. In either case of pre- or postharvest application, the bio-control agent should exert broad-spectrum activity to control a broad range of pathogens infecting the same postharvest produce, because if the BCA is effective against a particular pathogen, it may prevent a minor disease from developing into a major one. Understanding the microbial symbionts commonly associated with a crop, ideal growth conditions for establishment of bioagents in the plant system, the role of bioagents in promoting plant growth and controlling diseases, and the ability of bioagents to produce bioactive metabolites would enable identification of potential endophytes. Among the microbes, filamentous fungi and endophytic bacteria being extensively utilized, endophytic yeasts by virtue of the convenience by which they are cultured in comparison to filamentous fungi, they may serve as potential alternatives for exploitation of the metabolites.

One of the reasons for the poor performance and inconsistency of bio-control agents under commercial conditions is lack of understanding of mechanisms of actions of BCAs. From a commercial viewpoint, the performance and efficacy of antagonists with complex modes of action make antagonists depend more on producing the formulation, packing, application and storage. Detailed investigations of the mechanisms of action on antagonism and on physiology of postharvest produce and preservation under pathogenic infection are important to develop suitable formulation and application methods, and to obtain registration.

The interactions of endophytes with the hosts may at times result in a pathogenic infection, if host susceptibility and/or endophyte virulence increase. The pathogenicity risk can be avoided by exploiting the secondary metabolites responsible for antagonistic action. Adaptation of standard protocols for isolation and structural elucidation to define modes of action enable preparation of the appropriate formulations for commercial use and minimize the additional expenditure realized in synthesis of chemical compounds. Since some of the metabolites like VOCs are emitted in limited quantities, identification and overexpression of key genes involved in their biosynthesis will enable large-scale production of these compounds.

REFERENCES

Abdalla MA, Matasyoh JC (2014) Endophytes as producers of peptides: An overview about the recently discovered peptides from endophytic microbes. Nat. Prod. Bioprospect 4(5): 257–270.

Aked J (2002) Maintaining the post-harvest quality of fruits and vegetables. In: Fruit and vegetable processing. Woodhead Publishing, Cambridge, pp. 119–149.

Aktuganov GE, Galimzyanova NF, Melentev AI, Yu Kuzmina L (2007) Extracellular hydrolases of strain *Bacillus* sp. 739 and their involvement in the lysis of micromycete cell walls. Microbiology 76(4): 413–420.

Ali S, Charles TC, Glick BR (2014a) Amelioration of high salinity stress damage by plant growth-promoting bacterial endophytes that contain ACC deaminase. Plant Physiol Biochem 80: 160–167.

Ali S, Duan J, Charles TC, Glick BR (2014b) A bioinformatics approach to the determination of genes involved in endophytic behavior in *Burkholderia* spp. J Theor Biol 343: 193–198.

Aly AH, Edrada ER, Wray V, Muller EG, Kozytska S, Hentschel U, Proksch P, Ebel R (2008) Bioactive metabolites from the endophytic fungus *Ampelomyces* sp. isolated from the medicinal plant *Urospermum picroides*. Phytochemistry 69: 1716–1725.

Atmosukarto I, Castillo U, Hess WM, Sears J, Strobel G (2005) Isolation and characterization of *Muscodor albus* I-41.3s, a volatile antibiotic producing fungus. Plant Sci 169: 854-861.

Azevedo JL, Araujo WL (2007) Diversity and applications of endophytic fungi isolated from tropical plants. Fungi: Multifaceted Microbes. CRC Press, Boca Raton, FL, pp. 189–207.

Bacilio-Jimenez M, Aguilar-Flores S, Del Valle M V, Perez A, Zepeda A, Zenteno E (2001) Endophytic bacteria in rice seeds inhibit early colonization of roots by *Azospirillum brasilense*. Soil Biol Biochem 33: 167–172.

Bacilio-Jiménez M, Aguilar-Flores S, Ventura-Zapata E, Pérez-Campos E, Bouquelet S, Zenteno E (2003) Chemical characterization of root exudates from rice (*Oryza sativa*) and their effects on the chemotactic response of endophytic bacteria. Plant Soil 249(2): 271–277.

Bacon CW, Yates IE, Hinton DM, Meredith F (2001) Biological control of Fusarium moniliforme in maize. Environ Health Perspect 109(Suppl 2): 325–332.

Bailey BA, Bae H, Strem MD, Crozier J, Thomas SE, Samuels GJ, Holmes KA (2008) Antibiosis, mycoparasitism, and colonization success for endophytic *Trichoderma* isolates with biological control potential in *Theobroma cacao*. Biol Control 46(1): 24–35.

Bais HP, Park SW, Weir TL, Callaway RM, Vivanco JM (2004) How plants communicate using the underground information superhighway. Trends Plant Sci 9(1): 26–32.

Bais HP, Weir TL, Perry LG, Gilroy S, Vivanco JM (2006) The role of root exudates in rhizosphere interactions with plants and other organisms. Annu Rev Plant Biol 57: 233– 266.

Balint M, Tiffin P, Hallström B, O'Hara RB, Olson MS, Fankhauser JD, Piepenbring M, Schmitt I (2013) Host genotype shapes the foliar fungal microbiome of balsam poplar (*Populus balsamifera*). PLOS ONE 8(1): e53987.

Barkai-Golan R (2005) Postharvest diseases of fruits and vegetables. A Division of Reed Elsevier India, Pvt. Limited, New Delhi.

Barkai-Golan, R (2001) Postharvest diseases of fruit and vegetables: Development and control. Elsevier Sciences, Amsterdam, the Netherlands.

Calvente V, Benuzzi D, De Tosetti MIS (1999) Antagonistic action of siderophores from *Rhodotorula glutinis* upon the postharvest pathogen *Penicillium expansum*. Int Biodeterior Biodegrad 43: 167–172.

Camañas TP, Viñas I, Usall J, Torres R, Anguera M, Teixidó N (2008) Control of postharvest diseases on citrus fruit by pre-harvest applications of bio-control agent *Pantoea agglomerans* CPA-2. Part II. Effectiveness of different cell formulations. Postharvest Biol Technol 49: 96–106.

Card S, Johnson L, Teasdale S, Caradus J (2016) Deciphering endophyte behaviour: The link between endophyte biology and efficacious biological control agents. FEMS Microbiol Ecol 92(8): 1–19.

Carmona-Hernandez S, Reyes-Pérez JJ, Chiquito-Contreras RG, Rincon-Enriquez G, Cerdan-Cabrera CR, Hernandez-Montiel LG (2019) Bio-control of postharvest fruit fungal diseases by bacterial antagonists: A review. Agronomy 9(3): 121.

Carroll G (1988) Fungal endophytes in stems and leaves: From latent pathogen to mutualistic symbiont. Ecology 69: 2–9.

Chen JL, Sun SZ, Miao CP, Wu K, Chen YW, Xu LH, Zhao LX (2016) Endophytic *Trichoderma gamsii* YIM PH30019: A promising bio-control agent with hyperosmolar, mycoparasitism, and antagonistic activities of induced volatile organic compounds on root-rot pathogenic fungi of *Panax notoginseng*. J Gin Res 40(4): 315–324.

Cheng H, Li L, Hua J, Yuan H, Cheng S (2015) A preliminary preparation of endophytic bacteria CE3 wettable powder for biological control of postharvest diseases. Notulae Botanicae Horti Agrobotanici Cluj-Napoca 43(1): 159–164.

Colla G, Cardarelli M, Bonini P, Rouphael Y (2017) Foliar applications of protein hydrolysate, plant and seaweed extracts increase yield but differentially modulate fruit quality of greenhouse tomato. Hortic Sci 52: 1214–1220.

Conrath U, Beckers GJM, Flors V, García-Agustín P, Jakab G, Mauch F, Mauch-Mani B (2006) Priming: Getting ready for battle. Mol Plant Microbe Interact 19(10): 1062–1071.

Coombs JT, Franco CMM (2003) Isolation and identification of actinobacteria from surface-sterilized wheat roots. Appl Environ Microbiol 69(9): 5603–5608.

Daisy B, Strobel G, Ezra D, Castillo U, Baird G, Hess WM (2002) *Muscodor vitigenus anam.* sp. nov., an endophyte from *Paullinia paullinioides*. Mycotaxon 84: 39–50.

Damasceno CL, Duarte EAA, Dos Santos LBPR, de Oliveira TAS, de Jesus FN, de Oliveira LM, Soares ACF (2019). Postharvest bio-control of anthracnose in bananas by endophytic and soil rhizosphere bacteria associated with sisal (*Agave sisalana*) in Brazil. Bio-control 137: 104016.

Danhorn T, Fuqua C (2007) Biofilm formation by plant-associated bacteria Annu Rev Microbiol 61: 401–422.

De Weert S (2002) Flagella-driven chemotaxis towards exudate components is an important trait for tomato root colonization by *Pseudomonas fluorescens*. Mol Plant Microbe Interact 15: 1173–1180.

Deepika VB, Murali TS, Satyamoorthy K (2016) Modulation of genetic clusters for synthesis of bioactive molecules in fungal endophytes: A review. Microbiol Res 182: 125–140.

Delory BM, Delaplace P, Du Jardin P, Fauconnier ML (2016) Barley (*Hordeum distichon* L.) roots synthesise volatile aldehydes with a strong age-dependent pattern and release (e)-non-2-enal and (e, z)-nona-2, 6-dienal after mechanical injury. Plant Physiol Biochem 104: 134–145.

Demoz BT, Korsten L (2006) *Bacillus subtilis* attachment, colonization, and survival on avocado flowers and its mode of action on stem-end rot pathogens. Biol Control 37: 68–74.

Devi AN, Arumugam T (2005) Studies on the shelf life and quality of rasthali banana as affected by postharvest treatments. Orissa J Hortic 33(2): 3–6.

Diskin S, Feygenberg O, Maurer D, Droby S, Prusky D, Alkan N. (2017) Microbiome alterations are correlated with occurrence of postharvest stem-end rot in mango fruit. Phytobiomes 1(3): 117–127.

Droby S, Wisniewski M (2018) The fruit microbiome: A new frontier for postharvest bio-control and postharvest biology. Postharvest Biol Tec 140: 107–112.

Droby S, Wisniewski M, Teixidó N, Spadaro D, Jijakli MH (2016) *The science, development, and commercialization of postharvest bio-control products*. Postharvest Biol Technol 122: 22–29.

Dudareva N, Klempien A, Muhlemann JK, Kaplan I (2013) Biosynthesis, function and metabolic engineering of plant volatile organic compounds. New Phytol 198(1): 16–32.

Dukare AS, Paul S, Nambi VE, Gupta RK, Singh R, Sharma K, Vishwakarma RK (2018) Exploitation of microbial antagonists for the control of postharvest diseases of fruits: A review. Crit Rev Food Sci Nutr 59(9): 1498–1513.

Ek-Ramos MJ, Zhou W, Valencia CU, Antwi JB, Kalns LL, Morgan GD, Kerns DL, Sword GA (2013) Spatial and temporal variation in fungal endophyte communities isolated from cultivated cotton (*Gossypium hirsutum*). Plos One 8(6): e66049.

Elbeltagy A, Nishioka K, Suzuki H, Sato T, Sato YI, Morisaki H, Mitsui H, Minamisawa K (2000) Isolation and characterization of endophytic bacteria from wild and traditionally cultivated rice varieties. Soil Sci Plant Nutr 46: 617–629.

Etebarian HR, Sholberg P L, Eastwell KC, Sayler RJ (2005). Biological control of apple blue mold with *Pseudomonas fluorescens*. Can J Microbiol 51: 591–598.

Eva A, Dharini S, Lise K (2010) Effect of volatile compounds produced by *Bacillus* strains on postharvest decay in citrus. Biol Control 53: 122–128.

Evans HC, Holmes KA, Thomas SE (2003) Endophytes and mycoparasites associated with an indigenous forest tree, Theobroma gileri, in Ecuador and a preliminary assessment of their potential as bio-control agents of cocoa diseases. Mycol Prog 2: 149–160.

Ezra D, Strobel GA (2003) Effect of substrate on the bioactivity of volatile antimicrobials produced by *Muscodor albus*. Plant Sci 165: 1229–1238.

Falardeau J, Wise C, Novitsky L, Avis TJ (2013) Ecological and mechanistic insights into the direct and indirect antimicrobial properties of *Bacillus subtilis* lipopeptides on plant pathogens. J Chem Eco 139(7): 869–878.

Farré-Armengol G, Filella I, Llusia J, Peñuelas J (2016) Bidirectional interaction between phyllospheric microbiotas and plant volatile emissions. Trends Plant Sci 21: 854–860.

Ferramola MIS, Benuzzi D, Calvente V, Calvo J, Sansone G, Cerutti S, Raba J (2013) The use of siderophores for improving the control of postharvest diseases in stored fruits and vegetables. In: Microbial pathogens and strategies for combating them: Science, technology and education. Formatex, Badajoz, Spain, pp. 1385–1394.

Frank A, Saldierna Guzmán J, Shay J (2017) Transmission of bacterial endophytes. Microorganisms 5: e70.

Gaiero JR, Mc Call CA, Thompson KA, Day NJ, Best AS, Dunfield KE (2013) Inside the root microbiome: Bacterial root endophytes and plant growth promotion. Am J Bot 100(9): 1738–1750.

Gajanana TM, Sreenivasa Murthy D, Sudha M (2011) Post-harvest losses in fruits and vegetables in South India-A review of concepts and quantification of losses. Indian Food Packer 65(6): 178–187.

Gao FG, Dai CC, Liu XZ (2010) Mechanisms of fungal endophytes in plant protection against pathogens. Afr J Microbiol Res 4(13): 1346–1351.

Garnica-Vergara A, Barrera-Ortiz S, Muñoz-Parra E, Raya-González J, Méndez-Bravo A, Macías-Rodríguez L, López-Bucio J (2016) The volatile 6-pentyl-2H-pyran-2-one from *Trichoderma atroviride* regulates *Arabidopsis thaliana* root morphogenesis via auxin signaling and ethylene insensitive 2 functioning. New Phytol 209(4)1496–1512

Germida JJ, Siciliano SD, Freitas JR, Seib AM (1998) Diversity of root-associated bacteria associated with field grown canola (*Brassica napus* L.) and wheat (*Triticum aestivum* L.). FEMS Microbiol Ecol 26: 43–50.

Glick BR (2012) Plant growth-promoting bacteria: Mechanisms and applications. Scientifica (Cairo) 2012: 963401

Gomes AAM, Queiroz MV, Pereira OL (2015) Mycofumigation for the biological control of post-harvest diseases in fruits and vegetables: A review. Austin J Biotechnol Bioeng 2(4): 1051.

Gotor-Vila A, Teixidó N, Di Francesco A, Usall J, Ugolini L, Torres R, Mari M (2017) Antifungal effect of volatile organic compounds produced by *Bacillus amyloliquefaciens* CPA-8 against fruit pathogen decays of cherry. Food Microbiol 64: 219–225.

Grolleaud M (2002) Post-harvest losses: Discovering the full story. Overview of the phenomenon of losses during the post-harvest System. FAO, Compendium on post-Harvest Operations, Rome, Italy.

Grosch R, Scherwinski K, Lottmann J, Berg G (2006) Fungal antagonists of the plant pathogen *Rhizoctonia solani*: Selection, control efficacy and influence on the indigenous microbial community. Mycol Res 110(12): 1464–1474.

Hallmann J, Berg G, Schulz B (2006) Isolation procedures for endophytic microorganisms. In: Schulz BJE, Boyle CJC, Sieber TN (Eds.). Microbial root endophytes (Soil biology) vol. 9. Springer, Berlin, Heidelberg.

Hardoim PR, Overbeek LS, Berg G, Pirttilä AM, Compant S, Campisano A, Sessitsch A (2015) The hidden world within plants: Ecological and evolutionary considerations for defining functioning of microbial endophytes. Microbiol Mol Biol Rev 79(3): 293–320.

Hardoim P, Van Overbeek LS, van Elsas JD (2008) Properties of bacterial endophytes and their proposed role in plant growth. Trends Microbiol 16: 463–471.

Harvey JM, Smith WL, Kaufman J (1972) Market diseases of stone fruits: Cherries, peaches, nectarines, apricots, and plums. Agricultural Handbook No. 414, USDA-ARS,Washington, DC.

Irtwange SV (2006) Application of biological control agents in pre- and postharvest operations. Agricultural Engineering International: The CIGR Ejournal. Invited Overview No. 3. Vol. VIII. February, 2006.

Janisiewicz WJ (1998) Bio-control of postharvest diseases of temperate fruits challenges and opportunities. In: Plant–microbe interactions and biological control. Boland GJ, Kuykendall LD (Eds.). Marcel Dekker, New York, pp. 171–197.

Janisiewicz WJ, Peterson DL (2004) Susceptibility of the stem pull area of mechanically harvested apples to blue mold decay and its control with a bio-control agent. Plant Dis 88: 662–664.

Jia M, Chen L, Xin HL, Zheng CJ, Rahman K, Han T, Qin LP (2016) A friendly relationship between endophytic fungi and medicinal plants: A systematic review. Front Microbiol 7: 906

Kaddes A, Fauconnier ML, Sassi K, Nasraoui B, Jijakli MH (2019) Endophytic fungal volatile compounds as solution for sustainable agriculture. Molecules 24(6): 1065.

Kaddes A, Parisi O, Berhal C, Ben Kaab S, Fauconnier ML, Nasraoui B, Jijakli MH, Massart S, De Clerck C (2016) Evaluation of the effect of two volatile organic compounds on barley pathogens. Molecules 21: 1124.

Kilani-Feki O, Frikha F, Zouari I, Jaoua S (2013) Heterologous expression and secretion of an antifungal *Bacillus subtilis* chitosanase (CSNV26) in *Escherichia coli*. Bioprocess Biosyst Eng 36: 985–992.

Kim H, Mohanta TK, Park YH, Park SC, Shanmugama G, Park JS, Jeon J, Bae H (2020) Complete genome sequence of the mountain-cultivated ginseng endophyte *Burkholderia stabilis* and its antimicrobial compounds against ginseng root rot disease. Biol Control 140: 104126.

Kloepper J, Leong J, Teintze M (1980) Enhanced plant growth by siderophores produced by plant growth-promoting rhizobacteria. Nature 286: 885–886.

Korpi A, Järnberg J Pasanen AL (2009) Microbial volatile organic compounds. Crit Rev Toxicol 39: 139–193.

Lahlali R, Raffaele B, Jijakli MH (2011) UV protectants for *Candida oleophila* (strain O), a bio-control agent of postharvest fruit diseases. Plant Pathol 60: 288–295.

Lastochkina O, Baymiev A, Shayahmetova A, Garshina D, Koryakov I, Shpirnaya I, Pusenkova L, Mardanshin I, Kasnak C, Palamutoglu R (2020) Effects of endophytic *Bacillus subtilis* and salicylic acid on postharvest diseases (*Phytophthora infestans*, *Fusarium oxysporum*) development in stored potato tubers. Plants 9(1): 76.

Lastochkina O, Seifikalhor M, Aliniaeifard S, Baymiev A, Pusenkova L, Garipova S, Kulabuhova D, Maksimov I (2019) *Bacillus* spp.: Efficient biotic strategy to control postharvest diseases of fruits and vegetables. Plants 8(4): 97.

Latz Meike AC, Jensen B, Collinge DB, Jørgensen HJL (2018) Endophytic fungi as bio-control agents: Elucidating mechanisms in disease suppression. Plant Ecol Divers. DOI: 10.1080/17550874.2018.1534146.

Lee SO, Kim HY, Choi GJ, Lee HB, Jang KS, Choi YH, Kim JC (2009) Mycofumigation with *Oxyporus latemarginatus* EF069 for control of postharvest apple decay and rhizoctonia root rot on moth orchid. J Appl Microbiol 106: 1213–1219.

Li R, Li J, Zhou Z, Guo Y, Zhang T, Tao F, Hu X, Liu W (2018) Antibacterial and antitumor activity of secondary metabolites of endophytic fungi Ty5 from *Dendrobium officinale*. J Biobased Mater Bioenergy 12: 184–193.

Lima G, Sanzani SM, De Curtis F, Ippolito A (2015) Biological control of postharvest diseases. In: Wills RBH, Golding J (Eds.). Advances in postharvest fruit and vegetable technology. CRC Press, Boca Raton, FL, pp. 65–81.

Liu J, Abdelfattah A, Norelli J, Burchard E, Schena L, Droby S, Wisniewski M (2018) Apple endophytic microbiota of different rootstock/scion combinations suggests a genotype-specific influence. Microbiome 6(1): 18.

Loi MC, Ballero M, Loddo M, Sponga F, Ariu A (2000) Variazioni della presenza endofitica in specie vegetali della Sardegna. II. Contributo. Mic Ital 3: 11–16.

Lopes MR, Nadjara KM, Pompeo FL, Silva AC, Kupper KC (2015) *Saccharomyces cerevisiae*: A novel and efficient biological control agent for *Colletotrichum acutatum* during pre-harvest. Microbiol Res 175: 93–99.

Ludwig-Müller J (2015) Plants and endophytes: Equal partners in secondary metabolite production? Biotechnol Lett 37: 1325–1334.

Luo J, Xia W, Cao P, Xiao ZA, Zhang Y, Liu M, Zhan C, Wang N (2019) Integrated transcriptome analysis reveals plant hormones jasmonic acid and salicylic acid coordinate growth and defense responses upon fungal infection in poplar. Biomolecules 12. DOI:10.3390/biom9010012.

Mahajan PV, Caleb OJ, Singh Z, Watkins CB, Geyer M (2014) Postharvest treatments of fresh produce. Philos Trans A Maths Phys Eng Sci 372(2017): 20130309.

Maksimov IV, Abizgil'dina RR, Pusenkova LI (2011) Plant growth promoting rhizobacteria as alternative to chemical crop protectors from pathogens. Appl Biochem Microbiol 47: 333–345.

Maksimov IV, Khairullin RM (2016) The role of *Bacillus* bacterium in formation of plant defense: Mechanism and reaction. In: Gupta VK, Sharma GD, Tuohy MG, Gaur R (Eds.). The handbook of microbial bioresources. CAB International, Galway, Irish Republic, pp. 56–80.

Maksimov IV, Veselova SV, Nuzhnaya TV, Sarvarova ER, Khairullin RM (2015) Plant growth-promoting bacteria in the regulation of plant resistance to stress factors. Rus J Plant Physiol 62: 715–726.

Malmierca MG, Barua J, McCormick SP, Izquierdo-Bueno I, Cardoza RE, Alexander NJ, Hermosa R, Collado IG, Monte E, Gutiérrez S (2015) Novel aspinolide production by *Trichoderma arundinaceum* with a potential role in *Botrytis cinerea* antagonistic activity and plant defence priming. Environ Microbiol 17(4): 1103–1118.

Malmierca MG, Cardoza RE, Alexander NJ, McCormick SP, Hermosa R, Monte E, Gutiérrez S (2012) Involvement of Trichoderma trichothecenes in the bio-control activity and in the induction of plant defense related genes. Appl Environ Microbiol 78: 4856–4868.

Malmierca MG, McCormick SP, Cardoza RE, Alexander NJ, Monte E, Gutiérrez S (2015) Production of trichodiene by *Trichoderma harzianum* alters the perception of this bio-control strain by plants and antagonized fungi. Environ Microbiol 17: 2628–2646.

Marasco R, Rolli E, Ettoumi B, Vigani G, Mapelli F, Borin S, Abou-Hadid AF, El-Behairy UA, Sorlini C, Cherif A, Zocchi G, Daffonchio D (2012) A drought resistance-promoting microbiome is selected by root system under desert farming. Plos One 7: e48479.

Mardanova AM, Hadieva GF, Lutfullin MT, Khilyas IV, Minnullina LF, Gilyazeva AG, Martinez-Klimova E, Rodríguez-Peña K, Sánchez S (2017) Endophytes as sources of antibiotics. Biochem Pharmacol 134: 1–17.

Mari M, Guizzardi M, Brunelli M, Folchi A (1996) Postharvest biological control of grey mould (*Botrytis cinerea* Pers.:Fr.) on fresh-market tomatoes with *Bacillus amyloliquefaciens*. Crop Prot 15(8): 699–705.

Marquez-Santacruz HA, Hernandez-Leon R, Orozco-Mosqueda MC, Velazquez-Sepulveda I, Santoyo G (2010) Diversity of bacterial endophytes in roots of Mexican husk tomato plants (*Physalis ixocarpa*) and their detection in the rhizosphere. Gen Mol Res 9: 2372–2380.

Mazzola M, Freilich S (2017) Prospects for biological soil-borne disease control: Application of indigenous versus synthetic microbiomes. Phytopathology 107: 256–263.

Medina-Romero YM, Roque-Flores G, Macías-Rubalcava ML (2017) Volatile organic compounds from endophytic fungi as innovative postharvest control of *Fusarium oxysporum* in cherry tomato fruits. Appl Microbiol Biotechnol 101(22): 8209–8222.

Mejía LC, Rojas EI, Maynard Z, Bael SV, Arnold AE, Hebbar P, Herre EA (2008) Endophytic fungi as bio-control agents of *Theobroma cacao* pathogens. Biol Control 46(1): 4–14.

Mendes R, Azevedo JL (2008) Valor biotecnológico de fungos endofíticos isolados de plantas de interesse econômico. In: Maia LC, Malosso E, Yano-Melo AM (Eds.). Micologia: avanços no conhecimento. Editora Universitária da UFPE, Recife, BR, pp. 129–140.

Mercier J, Jiménez JI (2004) Control of fungal decay of apples and peaches by the biofumigant fungus *Muscodor albus*. Postharvest Biol Technol 31: 1–8.

Mikani A, Etebarian HR, Sholberg PL, Gormanb DT, Stokes S, Alizadeh A (2008) Biological control of apple gray mold caused by *Botrytis mali* with *Pseudomonas fluorescens* strains. Postharvest Biol Technol 48: 107–112.

Mohammadi P, Tozlu E, Kotan R, Senol Kotan M (2017) Potential of some bacteria for biological control of postharvest citrus green mould caused by *Penicillium digitatum*. Plant Protect Sci 53(3): 134–143.

Mohandoss J, Suryanarayanan TS (2009) Effect of fungicide treatment on foliar fungal endophyte diversity in mango. Sydowia 61(1): 11–24.

Morath SU, Hung R, Bennett JW (2012) Fungal volatile organic compounds: A review with emphasis on their biotechnological potential. Fungal Biol Rev 26(2–3): 73–83.

Nagorska K, Bikowski M, Obuchowski M (2007) Multicellular behavior and production of a wide variety of toxic substances support usage of *Bacillus subtilis* as a powerful biocontrol agent. Acta Biochim Pol 54: 495–508.

Narayanasamy P (2005) Postharvest pathogens and disease management. Wiley, Hoboken, NJ.

Nie P, Li X, Wang S, Guo J, Zhao H, Niu D (2017) Induced systemic resistance against *Botrytis cinerea* by *Bacillus cereus* AR156 through a JA/ET- and *NPR1*-dependent signaling pathway and activates PAMP-triggered immunity in arabidopsis. Front Plant Sci 8: 238.

Noumeur SR, Mancini V, Romanazzi G (2016) Activity of endophytic fungi from Artemisia absinthium on Botrytis cinerea. Acta Hortic 1144: 101–104.

Nunes N, Nunes CA (2012) Biological control of postharvest diseases of fruit. Eur J Plant Pathol 133: 181–196.

Ongena M, Jacques P (2008) *Bacillus* lipopeptides: Versatile weapons for plant disease bio-control. Trends Microbiol 16(3): 115–125.

Pagans E, Font X, Sánchez A (2006) Emission of volatile organic compounds from composting of different solid wastes: Abatement by biofiltration. J Hazard Mater 131: 179–186.

Parthasarathy S, Rajalakshmi J, Narayanan P, Arunkumar K, Prabakar K (2017) Bio-control potential of microbial antagonists against post-harvest diseases of fruit crops: A review. Res Rev Bot Sci 6(1): 17–23.

Pastor V, Luna E, Ton J, Cerezo M, Garcia-Agustin P, Flors V (2013) Fine tuning of reactive oxygen species homeostasis regulates primed immune responses in Arabidopsis. Mol Plant Microbe Interact 26: 1334–1344.

Pathak VN (1997) Postharvest fruit pathology: present status and future possibilities. Indian Phytopath 50: 161–185.

Paul NC, Deng JX, Sang HK, Choi YP, Yu SH (2012) Distribution and antifungal activity of endophytic fungi in different growth stages of chili pepper (*Capsicum annuum* L.) in Korea. Plant Pathol J 28(1): 10–19.

Pétriacq P, López A, Luna E (2018) Fruit decay to diseases: Can induced resistance and priming help? Plants 7:77.

Pieterse CM, Zamioudis C, Berendsen RL, Weller DM, Van Wees SC, Bakker PA (2014) Induced systemic resistance by beneficial microbes. Annu Rev Phytopathol 52: 347–375.

Pimenta RS, Moreira Da Silva JF, Buyer JS, Janisiewicz WJ (2012) Endophytic fungi from plums (*Prunus domestica*) and their antifungal activity against *Monilinia fructicola*. J Food Protect 75(10): 1883–1889.

Pratella G, Mari M, Guizzardi F, Folchi A (1993) Preliminary studies on the efficiency of endophytes in the biological control of the postharvest pathogens *Monilinia laxa* and *Rhizopus stolonifer* in stone fruit. Postharvest Biol Technol 3: 361–368.

Pusey PL, Wilson CL (1984) Postharvest biological control of stone fruit brown rot by *Bacillus subtilis*. Plant Dis 68:753.

Ramanathan A, Shanmugam V, Raguchander T, Samiyappan R (2002) Induction of systemic resistance in ragi against blast disease by *Pseudomonas fluorescens*. Ann Plant Protect Sci 10(2): 313–318.

Ramanujam B, Hemannavar V, Basha H, Rangeshwaran R (2012) Post harvest fruit bioassay of phylloplane, pomoplane and endophytic microbes against chilli anthracnose pathogen, *Colletotrichum capsici* (Syd.) E.J. Butler & Bisby. J Biol Control 26(1): 62–69.

Rashid S, Charles TC, Glick BR (2012) Isolation and characterization of new plant growth-promoting bacterial endophytes. Appl Soil Ecol 61: 217–224.

Reinhold-Hurek B, Maes T, Gemmer S, Van Montagu M, Hurek T (2006) An endoglucanase is involved in infection of rice roots by the not-cellulose-metabolizing *Endophyteazoarcus* sp. strain BH72. Mol Plant Microbe Interact 19(2): 181–188.

Rivera-Varas V, Thomas AF, Neil CG, Gary AS (2007) Mycoparasitism of *Helminthosporium solani* by *Acremonium strictum*. Biol Control 97(10): 1331–1337.

Rodrigues GS, Campanhola C, Rodrigues I, Frighetto RTS, Valarini PJ, Ramos-Filho LO (2006) Gestão ambiental de atividades rurais: estudo de caso em agroturismo e agricultura orgânica. Agric Sao Paulo 53(1): 17–31.

Rosenblueth M, Martínez-Romero E (2006) Bacterial endophytes and their interactions with hosts. Mol Plant Microbe Interact 19(8): 827–837.

Ryan RP, Germaine K, Franks A, Ryan DJ, Dowling DN (2008) Bacterial endophytes: Recent developments and applications. FEMS Microbiol Lett 278 (1): 1–9.

Sánchez Márquez S, Bills GF, Zabalgogeazcoa I (2007) The endophytic mycobiota of the grass Dactylis glomerata. *Fungal Divers* 27: 171–195.

Sansanwal R, Ahlawat U, Priyanka, Wati L (2017) Role of endophytes in agriculture. Chem Sci Rev Lett 6(24): 2397–2407.

Santoyo G, Moreno-Hagelsieb GM, Orozco-Mosqueda C,Glick BR (2016) Plant growth-promoting bacterial endophytes. Microbiol Res 183: 92–99.

Schardl CL, Craven KD, Speakman S, Stromberg A, Lindstrom A, Yoshida R (2008) A novel test for host-symbiont codivergence indicates the ancient origin of fungal endophytes in grasses. Syst Biol 57(3): 483–498.

Schardl CL, Young CA, Pan J, Florea S, Takach JE, Panaccione DG (2013) Currencies of mutualisms: Sources of alkaloid genes in vertically transmitted epichloae. Toxins 5: 1064–1088.

Schena L, Nigro F, Pentimone I, Ligorio A, Ippolito A (2003) Control of postharvest rots of sweet cherries and table grapes with endophytic isolates of *Aureobasidium pullulans*. Postharvest Biol Technol 30: 209–220.

Schena L, Ippolito A, Nigro F, Pentimone I, Salerno M (2000) Efficacy of endophytic isolates of *Aureobasidium pullulans* in controlling storage rots of sweet cherries. In: Proc Fifth Cong Euro Foundation for Plant Pathol, Taormina-Giardini Naxos (Catania), Italy, Sept 18(22): 527–530.

Schena L, Nigro F, Ippolito A, Balotelli D (2004) Real-time quantitative PCR: A new technology to detect and study phytopathogenic and antagonistic fungi. Eur J Plant Pathol 110: 893–908.

Schulz B, Boyle C (2005) The endophytic continuum. Mycol Res 109: 661–687.

Schulz B, Boyle C, Draeger S, Römmert AK, Krohn K (2002) Endophytic fungi: A source of biologically active secondary metabolites. Mycol Res 106: 996–1004.

Shanmugam V (2005) Chitinases in defence against phytopathogenic fungi. Prasad D (Ed.). Crop protection-management strategies, Daya Publishing House, New Delhi, p. 403.

Shanmugam V, Kanoujia N (2011) Biological management of vascular wilt of tomato caused by *Fusarium oxysporum* f. sp. *lycopersici* by plant growth-promoting mixture. Biological Control 57(2): 85–93.

Shanmugam V, Verma R, Rajkumar S, Naruka DS (2011) Bacterial diversity and soil enzyme activity in diseased and disease-free apple rhizosphere soils. Ann Microbiol 61: 765–772.

Sharma M, Kulshrestha S (2015). *Colletotrichum gloeosporioides*: An anthracnose causing pathogen of fruits and vegetables. Biosci Biotech Res Asia 12(2): 1233–1246.

Sharma RR, Singh D, Singh R (2009) Biological control of postharvest diseases of fruits and vegetables by microbial antagonists: A review. Biol Control 50: 205–221.

Sholberg PL, Marchi A, Bechard (1995) Bio-control of postharvest diseases of apple using *Bacillus* spp. isolated from stored apples. Can J Microbiol 41: 247–252.

Sidharthan KV, Aggarwal R, Shanmugam V (2018) Selection and characterization of the virulent *Fusarium oxysporum* f. sp. *lycopersici* isolate inciting vascular wilt of tomato. Int J Curr Microbiol App Sci 7(2): 1749–1756.

Silva YY, De Costa DM (2014) Potential of pre-harvest application of *Burkholderia spinosa* for biological control of epiphytic and pathogenic microorganisms on the phyllosphere of banana (*Musa* spp.). Trop Agric Res 25: 443–454.

Silva GH, Teles HL, Zanardi LM, Marx Young MC, Eberlin MN, Hadad R, Araújo ÂR (2006) Cadinane sesquiterpenoids of *Phomopsis cassiae*, an endophytic fungus associated with *Cassia spectabilis* (Leguminosae). Phytochemistry 67(17): 1964–1969.

Silva GH, Teles HL, Zanardi LM, Marx Young MC, Eberlin MN, Hadad R, Pfenning LH, Costa-Neto CM, Castro-Gamboa I, Bolzani YS, Araújo AR (2006) Cadinane sesquiterpenoids of *Phomopsis cassiae*, an endophytic fungus associated with *Cassia spectabilis* (Leguminosae). Phytochemistry 67: 1964–1969.

Singh SK, Strobel GA, Knighton B, Geary B, Sears J, Ezra D (2011) An endophytic Phomopsis sp. possessing bioactivity and fuel potential with its volatile organic compounds. Microb Ecol 61(4): 729–739.

Singh BK, Yadav KS, Verma A (2017) Impact of postharvest diseases and their management in fruit crops: an overview. J Bio Innov 6(5): 749–760.

Smilanick JL, Denis-Arrue R, Bosch JR, Gonzalez AR, Henson D, Janisiewicz WJ (1993) Control of postharvest brown rot of nectarines and peaches by *Pseudomonas* species. Crop Prot 12: 513–520.

Sørensen J, Sessitsch A (2007) Plant-associated bacteria lifestyle and molecular interactions. Modern Soil Microbiol 211–236.

Spadaro D, Ciavorella A, Dianpeng Z, Garibaldi A, Gullino ML (2010) Effect of culture media and pH on the biomass production and bio-control efficacy of a *Metschnikowia pulcherrima* strain to be used as a biofungicide for postharvest disease control. Can J Microbiol 56: 128–137.

Spadaro D, Droby S (2016) Development of bio-control products for postharvest diseases of fruit: The importance of elucidating the mechanisms of action of yeast antagonists. Trends Food SciTech 47: 39e49.

Strobel GA (2003) Endophytes as sources of bioactive products. Microbes Infect 5: 535–544.

Strobel GA (2006) Harnessing endophytes for industrial microbiology. Curr Opin Microbiol 9: 240–244.

Strobel GA, Dirkse E, Sears J, Markworth C (2001) Volatile antimicrobials from *Muscodor albus*, a novel endophytic fungus. Microbiology 147: 2943– 2950.

Strobel GA (2011) *Muscodor* species – endophytes with biological promise. Phytochem Rev10: 165–172.

Suwannarach, N, Kumla J, Bussaban B, Hyde KD, Matsui K, Lumyong L (2013) Molecular and morphological evidence support four new species in the genus Muscodor from northern Thailand. Ann Microbiol l63: 1341–1351.

Swain MR, Ray RC, Nautiyal CS (2008) Bio-control efficacy of *Bacillus subtilis* strains isolated from cow dung against postharvest yam (*Dioscorea rotundata* L.) pathogens. Curr Microbiol 57: 407.

Sword G, Ek-Ramos MJ, Lopez DC, Kalns L, Zhou W, Valencia C (2012) Fungal endophytes and their potential for bio-control in cotton. In: Entomological Society of America Annual Meeting, 2012.

Torres R, Solsana C, Viñas I, Usall J, Plaza P, Teixidó N (2014) Optimization of packaging and storage conditions of a freeze dried *Pantoea agglomerans* formulation for controlling postharvest diseases in fruit. J Appl Microbiol 117: 173–184.

Trias R, Baneras L, Montesinos E, Badosa E (2008) Lactic acid bacteria from fresh fruit and vegetables as bio-control agents of phytopathogenic bacteria and fungi. Int Microbiol 11(4): 231–236.

Truyens S, Weyens N, Cuypers A, Vangronsveld J (2014) Bacterial seed endophytes: Genera, vertical transmission and interaction with plants. Environ Microbiol 7: 40–50.

Varma PK, Uppala S, Pavuluri K, Jaya Chandra K, Chapala MM, Kumar VKK (2017) Endophytes: Role and functions in crop health. In Singh DP et al. (Eds.). Plant-microbe interactions in agro-ecological perspectives. Springer. DOI 10.1007/978-981-10-5813-4-15.

Velázquez-Becerra C, Macías-Rodríguez LI, López-Bucio J, Flores-Cortez I, Santoyo G, Hernández-Soberano C, Valencia-Cantero E (2013) The rhizobacterium *Arthrobacter agilis* produces dimethylhexadecylamine, a compound that inhibits growth of phytopathogenic fungi *in vitro*. Protoplasma 250(6): 1251–1262.

Waewthongrak W, Pisuchpen S, Leelasuphakul W (2015) Effect of *Bacillus subtilis* and chitosan applications on green mold (*Penicillium digitatum* Sacc.) decay in citrus fruit. Postharvest Biol Technol 99: 44–49.

Wang Y, Xu Z, Zhu P, Liu Y, Zhang Z, Mastuda Y, Xu L (2010) Postharvest biological control of melon pathogens using *Bacillus subtilis* EXWB1. J Plant Pathol 92(3): 645–652.

Whipps JM (1997) Developments in the biological control of soil-borne plant pathogens. Adv Bot Res 26: 1–133.

You F, Han T, Wu J, Huang B, Qin L (2009) Antifungal secondary metabolites from endophytic *Verticillium* sp. Biochem Syst Ecol 37(3): 162–165.

Zabalgogeazcoa I (2008) Fungal endophytes and their interaction with plant pathogens. Spanish J Agric Res 6(Special issue): 138–146.

Zhang HW, Song YC, Tan RX (2006) Biology and chemistry of endophytes. Nat Prod Rep 23(5): 753.

Zheng L, Situ J, Zhu Q, Xi P, Zheng Y, Liu H, Zhou X, Jiang Z (2019) Identification of volatile organic compounds for the bio-control of postharvest litchi fruit pathogen *Peronophythora litchii*. Postharvest Biol Technol 155: 37–46.

Zheng M, Shi J, Shi J, Wang Q, Li Y (2013) Antimicrobial effects of volatiles produced by two antagonistic *Bacillus* strains on the anthracnose pathogen in postharvest mangos. Biol Control 65(2): 200–206.

Zouaoui NS, Chérif M, Fliss I, Boudabbous A (2001) Evaluation of bacterial isolates from salty soils and *Bacillus thuringiensis* strains for the bio-control of fusarium dry rot of potato tubers. J Plant Pathol 83(2): 101–117.

8

Management of Postharvest Diseases of Fruits and Vegetables through Yeasts

Bharti Choudhary and Dinesh Singh
Division of Plant Pathology, ICAR-Indian Agricultural Research Institute, New Delhi, India

CONTENTS

8.1 Introduction

Fruits and vegetables are considered important commodities for all developing and developed nations. They are the source of various nutrients such as minerals and vitamins and provide health benefits to humans as the potential source of antioxidants and anticancerous properties. There has been consumer shift in enhanced preference to fruits and vegetables due to these properties.

Postharvest decay of fruits and vegetables causes significant economic losses. The losses are estimated to reach up to 25% of the total production levels in developed countries and more than 50% in developing countries (Nunes 2012). Infection by pathogens may occur during the various stages, including during growing season, harvesting, handling and storage, transportation and after purchase by the end consumer. Pathogenic disease is initiated by infection which is followed by the manifestation of symptoms in the host. The major reason for the susceptibility of fruits and vegetables to various bacterial and fungal pathogens is their very high moisture and nutritional content along with low pH (Choudhary et al. 2015). Another major reason is that after harvest, fruits and vegetables loose most of their intrinsic resistance that protects them from various pathogens. Considering the situation in the field before the harvesting process, most pathogens are incapable of causing damage in the host, as they require a wound or cut for penetration. Only after the outer cuticle is damaged do fruits and vegetables become susceptible to various pathogens. Alao (2000) defined postharvest losses of fruits and vegetables as 'that weight of wholesome edible product (exclusive of moisture content) that is normally consumed by human and that has been separated from the medium and sites of its immediate growth and production by deliberate human action with the intention of using it for human feeding but which for any reasons fails to be consumed by human.'

The majority of food losses occur in the following food supply chains. The first round of losses takes place during agricultural production. Mechanical damage, spillage during harvesting operations (threshing of crops or fruit picking) are the major reasons. The second round of losses takes place during postharvest handling and storage operations. This includes damage caused due to spillage and deterioration during handling, storage, and distribution between the fields and the distribution network. The third round of losses involves processing. This includes losses during the industrial or domestic processing of vegetables and fruits. The fourth stage of losses occurs in the distribution network, such as losses in the marketplace which includes wholesale markets, retailers, etc. The final round of losses takes place at the end consumer level. These losses include waste during the household consumption level. The pathogens that cause the major postharvest diseases in fruits and vegetables belong to various species, such as *Alternaria*, *Aspergillus*, *Botrytis*, *Cladosporium*, *Fusarium*, *Geotrichum*, *Gloeosporium*, *Mucor*, *Monilinia*, *Phytopthora*, *Penicillium*, *Rhizopus* and *Trichothecium*, amongst others (Barkai-Golan 2001).

These postharvest losses of fruits and vegetables can be reduced to a great extent by pre- and postharvest treatments. For years, several conventional and nonconventional methods have been employed to control postharvest losses by fungi, including synthetic fungicides. Synthetic fungicides have been widely used because of their effectiveness against most pathogenic fungi, easy applicability and low cost (Tripathi and Dubey 2004; Tripathi et al. 2013). However, the use of postharvest fungicides is in a downward trend because of their environmental and toxicological risks. In 2010, it was reported that postharvest fungicide treatment has been banned in some countries due to its adverse effects. Also, the excessive use of fungicides has resulted in the development of fungal resistant strains which are becoming more and more intolerant to the fungicides. Recent trends in the organic farming and sustainable agriculture approach have further led to the development of other methods to control postharvest pathogens.

Currently, postharvest disease management procedures are moving towards biological control of postharvest decay of fruits and vegetables. Bio-control agents (BCAs) have several advantages over synthetic fungicides, including less toxicity to non-target organisms and faster degradation (Thakore 2006). Extensive research is being carried out on application of antagonistic microorganisms to control spoilage of fruits and vegetables across the globe. In the last few decades, several studies have been done to highlight the potential of these biological control agents as a viable alternative to synthetic fungicides against various postharvest diseases of fruits and vegetables.

In the past three decades, many BCAs have been explored and investigated against different postharvest fungal pathogens. In particular, several fungi and bacteria, isolated from fruit surfaces as epiphytic microbial population, have been employed for control of postharvest pathogens (Chan and Tian 2005; Cirvilleri et al. 2005; Cirvilleri 2008; Filonow et al. 1996) either individually or in combinations (Panebianco et al. 2015). Ideal properties of the screened antagonist should involve non-pathogenicity for fruits, plants and vegetables along with non-production of secondary metabolites hazardous to human health.

Biological control using antagonistic yeasts has shown to be a promising approach to manage postharvest fruit and vegetable decay. Their simple nutritional requirements coupled with a variety of mechanisms of action against pathogens make them an ideal choice for postharvest disease management of fruits and vegetables.

8.2 Pathogens Causing Postharvest Diseases

Various pathogens such as bacteria and fungi are responsible for causing diseases in fruits and vegetables. Some of the fungal species that are known to cause major postharvest losses include *Alternaria*, *Aspergillus*, *Botrytis*, *Monilia*, *Penicillium*, *Rhizopus*, *Mucor* and others.

8.3 Development of Postharvest Bio-control Agents

Development of a postharvest BCA is a complex process, which involves various costs and long duration spans. Nunes (2012) has provided a framework for the development of postharvest biological control, which is shown in Figure 8.1. The development process involves two major sub-processes of discovery followed by commercial development. The first sub-process involves isolation, screening, efficacy testing and laboratory to pilot trails. Possible modes of action of the microorganisms, their growth requirements and factors for enhancement of bio-control activity are also studied. The second sub-process includes commercial development of the biological control agent, involving commercial trials, formulations and registrations.

8.4 Mechanism of Action for Bio-control Agents

Information related to an antagonist's mode of action is not completely clear because of the complexity involved in the interaction between host, pathogen, antagonist and other microorganisms. Various bio-control mechanisms have been proposed as discussed briefly hereunder.

8.4.1 Antibiosis

Bacterial species producing antibiotics are known to be potent BCAs due to their effective ability to control pathogens. *Bacillus subtilis* is known to produce iturin, which is a strong antifungal peptide (Gueldner et al. 1988) along with gramicidin S (Edwards and Seddon 2001). *Pseudomonas cepacia* produces antibiotic pyrrolnitrin, which is a highly potential BCA against *Botrytis cinerea* and *Penicillium expansum*, which are major pathogens for pome fruits (Janisiewicz et al. 1991). The only concern for using these BCAs in food products is because of the chances of development of human pathogens resistant to these compounds and the possibility of resistance generation in fruit and vegetable pathogens.

8.4.2 Competition for Nutrition and Space

Some microorganisms, particularly yeast, acts by competing for the space or utilization of the nutrients with the pathogens (Spadaro et al. 2002). Several studies have shown yeast to successfully compete with pathogens in terms of nutrient utilization and render them inactive. Formation of an extracellular polysaccharide capsule helps them to adhere to the surface of fruits and compete for space.

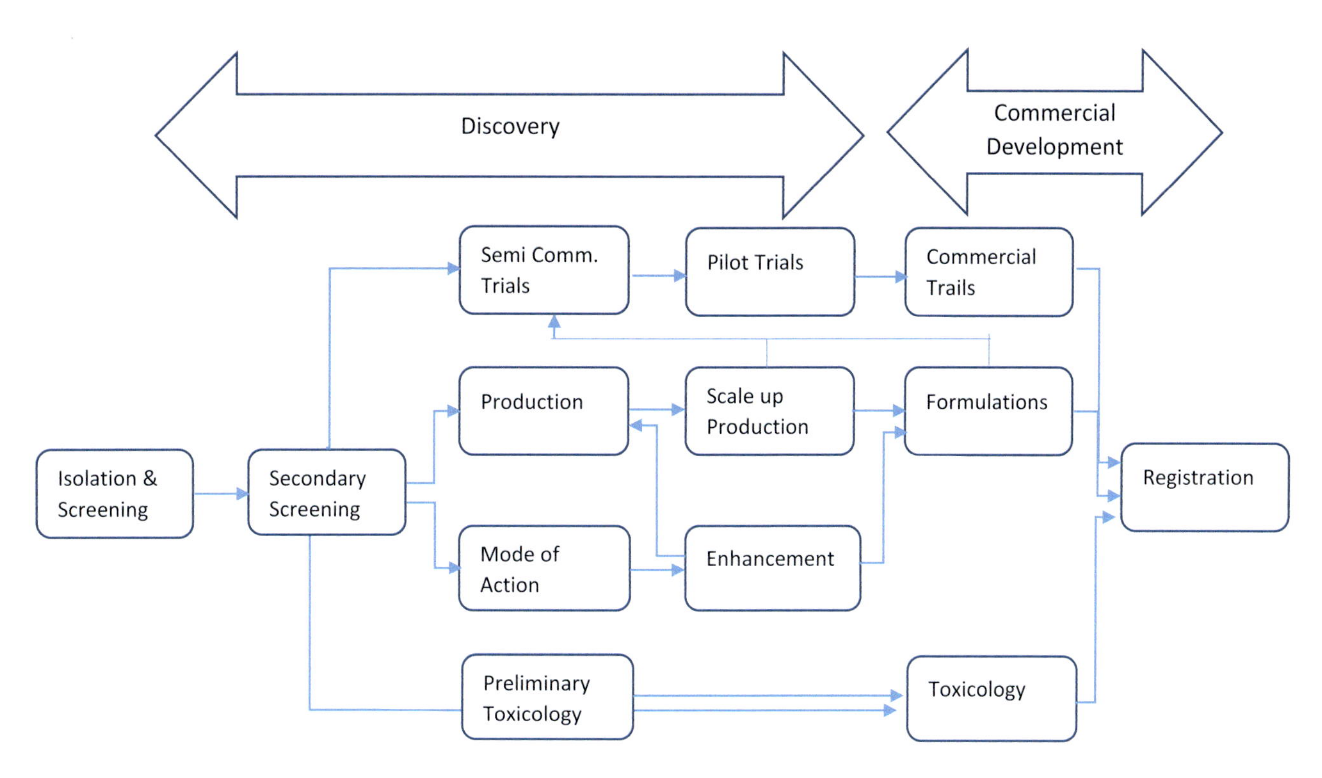

FIGURE 8.1 Flow diagram for development and commercialization of bio-control agent(s).

8.4.3 Direct Parasitism

Parasitism is another mode where the antagonist and pathogen interact directly. *Pichia guilliermondii* shows high enzymatic activity of β-1,3 glucanase, which results in degradation of the fungal cell wall. Some yeast cells such as *Cryptococcus laurentii* and *S. roseus* are able to reduce the postharvest decay on apples caused by *B. cinerea* through parasitism.

8.4.4 Induction of Resistance in Host Tissues

In some cases, BCAs can interact with host tissue even before the pathogen inoculation. Several antagonistic yeasts are effective in enhancing the resistance in fruit and vegetable skins. A few *Candida* strains are also known to provide resistance in apple through chemical and osmotic changes in apple's tissue, thereby enhancing the fruit's life.

8.5 Yeasts as Potential Bio-control Agents

Yeasts have been considered the model organism for various biotechnological applications. However, the use of antagonist yeasts as BCAs still has a long way to go considering the potential. As of now, only a few yeast-based postharvest products are available in the market. However, the strong antifungal activities of yeast combined with advantageous properties (antagonistic, culturability, formulatability, applicability and stress resistance) for the application makes them the ideal choice for postharvest management (Freimoser et al. 2019).

8.5.1 Yeast Properties – Advantageous for Potent Bio-control Applications

One of the most important properties that makes yeast as potent BCAs is its morphology, which is the single-most reason for its wide culturability in fermenters, superior formulation characteristics and numerous applications usages. Like bacteria, the single-cell dynamics of yeast help in biofilm formation, along with adhesion, which results in enhanced bio-control activity (Pandin et al. 2017; Fanning and Mitchell 2012). Another important aspect similar to bacteria involves their growth characteristics and bio-control activities. The bio-control potential of yeast does not carry the risk of taking or passing on the plasmid-based antibiotic resistance or other toxicogenic genes which are more prevalent in the case of bacteria. Another advantage of yeasts is the low frequency of horizontal gene transfer as compared to the other prokaryotes due to their more complex genome organization (Richards et al. 2011; Fitzpatrick 2012). Yeasts have been used for food and beverage industry for thousands of years. Yeasts are widely used in the food industry mainly due to their safe nature. In most cases these identified 'food industry yeasts' belong to the same genus and species as those intended for bio-control and hence are considered safe for the bio-control process. Several antagonistic yeast strains are effective against various pathogens of fruits and vegetables (Table 8.1). Currently, some

TABLE 8.1

Major Yeasts Exploited for Control of Postharvest Disease of Fruits and Vegetables

Yeast Strain	Pathogen	Host	Reference (s)
Candida oleophila	*Colletotrichum*	Papaya	Gamagae et al. (2003)
	Gloeosporioides	Peach	Karabulut and Baykal (2004)
	Botrytis cinerea	Citrus	Lahlali et al. (2004, 2005)
	Penicillium digitatum, P. italicum		
Candida oleophila (I-182)	*P. digitatum*	Grape	Droby et al. (2002)
Pichia membranaefaciens	*Monilinia fructicola*	Apple	Chan and Tian (2005)
	P. expansum		
	Rhizopus stolonifer		
Cryptococcus laurentii (LS-28)	*B. cinerea*	Apple	Lima et al. (2011)
	P. expansum		
Cryptococcus laurentii	*Botrytis cinerea*	Peach	Zhang et al. (2007a, 2007b)
	R. stolonifer	Strawberry	Qin and Tian (2004)
	Alternata alternata	Jujube	Tian et al. (2005)
	P. expansum	Tomato	Xi and Tian (2005)
	B. cinerea	Apple	Blum et al. (2004)
	Glomerella cingulata	Cherry	Qin et al. (2006)
	M. fructicola	Pear	Zhang et al. (2003)
	P. expansum	Strawberry	
Candida sake (CPA-1)	*P. expansum*	Pear	Torres et al. (2006)
	P. expansum	Apple	Morales et al. (2008)
Aureobasidium pullulans (PL5, PI1)	*B. cinerea, P. expansum*	Apple	Zhang et al. (2010)
	M. laxa	Stone fruits	Zhang et al. (2012)
	B. cinerea	Table grapes	Parafati et al. (2015)
Cryptococcus albidus KKUY0017 *Wickerhamomyces anomalus* KKUY0051	*P. expansum*	Apple	Hashem et al. (2014)
Aureobasidium pullulans (L1, L8)	*B. cinerea, C. acutatum, P. expansum*	Apple	Di Francesco et al. (2015)
	P. digitatum, P. italicum	Orange	
Cryptococcus laurentii (LS-28)	*B. cinerea, P. expansum*	Apple	Castoria et al. (2003)
Saccharomyces cerevisiae (M25)	*P. expansum*	Apple	Scherm et al. (2003)
Debaryomyces hansenii	*R. stolonifera*	Peach	Mandal et al. (2007)
	P. italicum	Lemon	Hernández-Montiel et al. (2010, 2018)
	C. gloeosporioides	Papaya	
	P. digitatum, P. italicum	Citrus	Singh (2002)
Lachancea thermotolerans (751)	*A. carbonarius*	Grape	Fiori et al. (2014)
Metschnikowia pulcherrima	*B. cinerea, A. alternata, P. expansum*	Apple	Saravanakumar et al. (2009)
Pichia guilliermondii	*C. capsici*	Chillies	Chanchaichaovivat et al. (2007)
	R. nigricans	Tomato	Zhao et al. (2008)
Pichia guilliermondi (M8)	*B. cinerea*	Apple	Zhang et al. (2011)
Cryptococcus albidus	*M. fructicola, P. expansum*	Apple	Chan and Tian (2005)
Pichia membraneafaciens	*P. expansum*	Peach	Chan et al. (2007)
Candida oleophila	*P. digitatum, P. italicum*	Citrus	Lahlali et al. (2004, 2005)
	C. gloeosporioides	Papaya	Gamagae et al. (2003)
	B. cinerea	Peach	Karabulut and Baykal (2004)
Candida guilliermondii	*Botrytis cinerea*	Nectarine	Tian et al. (2002)
	Botrytis cinerea	Peach	Tian et al. (2002)
	Botrytis cinerea	Tomato	Saligkarias et al. (2002)
Candida oleophila	*C. musae*	Banana	Lassois et al. (2008)
	B. cinerea	Tomato	Saligkarias et al. (2002)
Rhodotorula glutinis	*P. expansum*	Apple	Zhang et al. (2009)
	B. cinerea	Apple	Zhang et al. (2009)
	P. expansum	Jujube	Tian et al. (2005)
Metschnidowia pulcherrima *Curibasidium pallidicorallinum* *Candida saitoana*	*Phytophthora infestans*	Potato	Hadwiger et al. (2015)

TABLE 8.2

Commercially Available Yeast Biological Control Products to Manage Fruit and Vegetable Diseases

Bio-control Agent	Target Organism or Disease	Host	Product	Manufacturer or Distributor
Candida oleophila strain I-182	Postharvest diseases	Various fruits, vegetables, flowers, ornamentals, other plants	Aspire®	Ecogen, USA
C. oleophila strain O	*Botrytis cinerea* (grey mould), *Penicillium expansum* (blue mould	Apples and pears	*Nexy*®	European Food Safety Authority (EFSA)
Aureobasidium pullulans strain DSM 14940 (CF 10) & DSM 14941 (CF 40)	*Erwinia amylovora* (fire blight disease)	Pome fruit	Blossom-Protect®	EuropeanFood Safety Authority (EFSA)
Aureobasidium pullulans strain DSM 14940 (CF 10) and DSM 14941 (CF 40)	Postharvest diseases	Apple	Boni-Protect®	
Cryptococcus albidus	*B. cinerea* *P. expansum*	Pome citrus fruits	Yield plus® (withdrawn from the market)	Anchor Bio-Technologies (South Africa)
Saccharomyces cerevisiae	Powdery and downy mildew	Grapes, fruits and vegetables	Romeo® with Cerevisane® as the active ingredient	European Food Safety Authority (EFSA)

yeast-based commercial formulations have been developed, which are being used worldwide (Table 8.2).

8.6 Mechanisms for Bio-control Activity of Yeasts

Understanding of the mechanisms involved in the bio-control activity of yeast is the most important step for the development of successful BCA. Multiple mechanisms for yeast bio-control have been studied in detail, which involves competition for nutrients and space with the pathogen, through enzyme secretion, toxin production, use of volatile organic compounds (VOCs), mycoparasitism and resistance induction in plants are amongst the major reasons (Droby et al. 2009; Punja and Utkhede 2003; Wisniewski and Droby 2012).

8.6.1 Competition for Nutrients and Space

All organisms compete against one another and with the host for the nutrient availability and space, which leads to their primary mode of action as BCAs. Most of the organisms synthesize iron-binding molecules to deprive the competing organisms, pathogens and other parasites for iron, which is an essential requirement for most of the organisms (Johnson 2008; Barber and Elde 2015). Also, for yeast species, iron is considered the most important nutrient for their growth, and competition for iron is one of the most important reasons for the mode of action. Fungal pathogen deprivation of iron was suggested as one of several mechanisms by which yeast antagonizes plant pathogenic fungi (Gore-Lloyd et al. 2019; Sipiczki 2006).

Biofilm formation is another important aspect which is a specific strategy for competing for space with other organisms. Biofilms are microbial communities that live and grow on surfaces. They either comprise of a single species or multispecies, depending upon the morphology. The formation of yeast biofilm starts with the adhesion of individual cells to a surface. This involves secretion of extracellular matrix and formation of hyphae by yeast cells (Cavalheiro and Teixeira 2018). This mechanism is considered one of the most important processes for bio-control.

8.6.2 Secretion of Enzymes

The activity of enzyme secretion degrading cellular components in the host pathogen scenario has been very well studied as a bio-control mechanism of yeasts. Secreted enzymes such as chitinases, glucanases or proteases, which are mentioned here, are important antagonistic yeasts for bio-control activity. Chitinases are known for degradation of the fungal cell wall. The results for these have been studied in the bio-control yeasts of genera *Candida*, *Debaryomyces*, *Metschnikowia*, *Meyerozyma*, *Pichia*, *Saccharomyces* and *Tilletiopsis*. Chitinase activity has been monitored in the presence of pathogenic cells for their bio-control activity (Bar-Shimon et al. 2004; Junker et al. 2019; Lopes et al. 2015; Pretscher et al. 2018). Chitinases have shown bio-control activity against plant pathogenic fungi and are widely studied as potential biopesticides in various applications (Nagpure et al. 2014).

Glucanase is another enzyme involved in the cell adhesion, cell wall modification and killer toxin resistance (Adams 2004). In *Wickerhamomyces anomalus*, the deletion of the two exo-β-glucanase genes (PaEXG1 and PaEXG2) reduced the bio-control activity on fruits against *Botrytis cinerea*, showing the importance of enzyme activity in bio-control mechanism. Lipolytic activity of lipases is frequently found during screening of extracellular enzymatic activity in yeast and yeast-like strains. Lipase activity has also been detected in pathogenicity of yeast such as *Candida*, *Cryptococcus* or *Malassezia* species (Mayer et al. 2013; Park et al. 2013). Proteases have been scarcely studied so far in terms of their bio-control potential in yeast.

TABLE 8.3

Toxins Produced by Various Yeast Species

Name of Toxin	Yeast	Reference
Flocculosin	*Pseudozyma flocculosa*	Mimee et al. (2009)
Aureobasidins	*Aureobasidium pullulans*	Takesako et al. (1991)
WaF17.12	*Wickerhamomyces anomalus*	Cappelli et al. (2014)
2-propylacrylic acid, 2-methylene succinic acid	*Aureobasidium pullulans*	Zain et al. (2009)
Liamocin	*Aureobasidium pullulans*	Price et al. (2017)
Mycocins	*S. cerevisiae*	Rima et al. (2012)
Unidentified toxins	*Tilletiopsis* sp.	Urquhart and Punja (2002)

The activity of proteases has been detected in the later growth stages (after 6–8 days of growth), and a small function of bio-control was hypothesized (Bar-Shimon et al. 2004).

8.6.3 Toxin Production

Compared to bacteria and fungi, yeast is known for lower production of secondary metabolites, which is why there is less concern for biosafety compared to other organisms (Freimoser et al. 2019). Some of the key toxins produced by various yeast species that result in the bio-control of various bacteria and fungi are provided in Table 8.3.

8.6.4 Production of Volatile Organic Compounds

Volatile organic compounds (VOCs) are molecules with low water solubility and high vapour pressure. VOC includes various molecules which belong to the category of hydrocarbons, alcohols, aldehydes, ketones, esters, phenols, heterocyclic compounds and other benzene derivatives (Morath et al. 2012). Various studies have depicted the roles of these volatile compounds produced by yeast species against pathogens, including postharvest pathogens and mycotoxin producing fungi. Details of the volatile organic compounds produced by yeast species against pathogenic agents for bio-control mechanisms are shown in Table 8.4.

8.6.5 Direct Parasitism

Parasitism has been studied quite a bit less in yeasts than other organisms. In one study, *P. guilliermondii* was shown to strongly attach to hyphae of the plant pathogen *B. cinerea* and to cause hyphal collapse, which prevented disease induction (Wisniewski et al. 1991). Other examples studied include *Pseudozyma aphidis* parasiting the pathogen *Podosphaera xanthii* and *B. cinerea* (Calderon et al. 2019). The genus *Saccharomyces* has also been studied for bio-control of different *Penicillium* species for prevention of diseases.

8.6.6 Indirect Mechanisms

Various indirect mechanisms including the induction of resistance by the yeast have also been studied as a potential source of bio-control against pathogens. Plants possess an inbuilt immune system that recognizes and responds to the presence of various microorganisms. Bio-control yeasts can invoke systemic resistance in plants against various pathogens which is suggested to contribute to their bio-control activity. Various yeast species including *S. cerevisiae*, *R. paludigenum*, *C. saitoana*, *C. oleophila* and *Metschnikowia* have shown the potential through the indirect mechanism as potential BCAs. In some studies, bio-control yeasts such as *C. laurentii* and *R. glutinis* have been used along with inducers such as salicylic acid or rhamnolipids to increase the effectiveness of bio-control process (Yan et al. 2014; Zhang et al. 2007a).

8.7 Future Thrusts

As the reduced consumption of pesticides on fruits and vegetables is becoming a trend, there is a huge potential for alternative methods, including bio-control through yeasts. In the environment, these bio-control yeasts interact intra-specifically as well as with other species of organisms including the plant. Various complex interactions between the organisms determine whether the pathogen will infect the host, or the bio-control yeast will suppress the plant pathogen. Understanding the complex interaction between the organisms therefore is the most important priority for the development of effective BCAs. Recent advances in computational biology, molecular biology, biochemistry and others have opened new avenues for

TABLE 8.4

Volatile Organic Compounds Produced by Yeast Species and Target Pathogens

Yeast Species	Volatile Organic Compound	Target Pathogen	Reference
Pichia kluyveri, *Hanseniaspora uvarum*, *Pichia anomala*	2-phenylethyl acetate	*Aspergillus ochraceus*	Masoud et al. (2005)
Pichia anomala	2-phenylethanol	*Aspergillus flavus*	Hua et al. (2014)
Aureobasidium pullulans	2-phenylethanol	*B. cinerea*, *C. acutatum*, *P. expansum*, *P. digitatum*, *P. italicum*	Di Francesco et al. (2015)
Sporidiobolus pararoseus	2-ethyl-1-hexanol	*B. cinerea*	Huang et al. (2012)
Candida intermedia	1,3,5,7- cyclooctatetraene, 3-methyl-1-butanol and others	*B. cinerea*	Huang et al. (2011)
Aureobasidium pullulans	Unidentified volatiles	*B. cinerea*, *P. digitatum*	Parafati et al. (2015)

understanding the structural and functional dynamics between plant pathogens and BCAs. Better understanding of these will lead to increased consortia of microbial antagonists. Also, usage of taxonomically divergent but functionally complementary strains of BCAs will show the way for an effective and multi-targeted bio-control process for a wide range of host pathogens.

Currently very few studies have been done related to the gene level (deletion, overexpression or transgene) effectiveness of yeast as a BCA. These genetic tools have the potential to understand the organisms and mechanisms involved in bio-control to the next level. Another scope for yeast as a BCA lies in the application in soil against soilborne plant pathogens. Currently very few bio-control options are available at the commercial level. Strengthening the science, technology, regulation and legislature fields will play a crucial role in the commercialization aspect of yeast as a BCA in the near future.

REFERENCES

Adams DJ (2004) Fungal cell wall chitinases and glucanases. Microbiology 150: 2029–2035.

Alao SEL (2000) The importance of postharvest prevention. Paper presented at graduation ceremony of School of Food Storage Technology, Nigerian Stored Products Research Institute, Kano, pp. 1–10.

Barber MF, Elde NC (2015) Buried treasure: Evolutionary perspectives on microbial iron piracy. Trends Genet 31: 627–636.

Barkai-Golan R (2001) Postharvest diseases of fruit and vegetables: Development and control. Elsevier, Amsterdam.

Bar-Shimon M et al (2004) Characterization of extracellular lytic enzymes produced by the yeast bio-control agent *Candida oleophila*. Curr Genet 45: 140–148.

Blum LEB, Amarante CVT, Valdebenito-Sanhueza RM, Guimaraes LS, Dezanet A, Hack-Neto P (2004) Postharvest application of *Cryptococcus laurentii* reduces apple fruit rots. Fitopatol Brasileira 29: 433–436.

Calderon CE, Rotem N, Harris R, Vela-Corcia D, Levy M (2019) *Pseudozyma aphidis* activates reactive oxygen species production, programmed cell death and morphological alterations in the necrotrophic fungus *Botrytis cinerea*. Mol Plant Pathol 20: 562–574.

Cappelli A, Ulissi U, Valzano M, Damiani C, Epis S, Gabrielli MG, Ricci I (2014) A *Wickerhamomyces anomalus* killer strain in the malaria vector *Anopheles stephensi*. Plos One 9(5): 95988.

Castoria R, Caputo L, De Curtis F, De Cicco V (2003) Resistance of postharvest bio-control yeasts to oxidative stress: A possible new mechanism of action. Phytopathology 93: 564–572.

Cavalheiro M, Teixeira MC (2018) Candida biofilms: threats, challenges, and promising strategies. Front Med (Lausanne) 5: 28.

Chan Z, Qin G, Xu X, Li B, Tian S (2007) Proteome approach to characterize proteins induced by antagonist yeast and salicylic acid in peach fruit. J Proteome Res 6: 1677–1688.

Chan Z, Tian S (2005) Interaction of antagonistic yeasts against postharvest pathogens of apple fruit and possible mode of action. Postharvest Biol Technol 36: 215–223.

Chanchaichaovivat A, Ruenwongsa P, Panijpan B (2007) Screening and identification of yeast strains from fruit and vegetables: Potential for biological control of postharvest chilli anthracnose (*Colletotrichum capscii*). Biol Control 42: 326–335.

Choudhary B, Nagpure A, Gupta RK (2015)Biological control of toxigenic citrus and papaya-rotting fungi by *Streptomyces violascens* MT7 and its extracellular metabolites. J Basic Microbiol 55: 1343–1356.

Cirvilleri G (2008). Bacteria for biological control of postharvest diseases of fruits. In: Ait Barka E, Clement C (Eds.). Plantmicrobe interactions, Research Signpost, 37/66, pp. 1–29.

Cirvilleri G, Bonaccorsi A, Scuderi G, Scortichini M (2005) Potential biological control activity and genetic diversity of *Pseudomonas syringae* pv. syringae strains. J Phytopathol 153: 654–666.

Di Francesco A, Ugolini L, Lazzeri L, Mari M (2015) Production of volatile organic compounds by *Aureobasidium pullulans* as a potential mechanism of action against postharvest fruit pathogens. Biol Control 81: 8–14.

Droby S, Vinokur V, Weiss B, Cohen L, Daus A, Goldschmidt EE, Porat R (2002) Induction of resistance to *Penicillium digitatum* in grapefruit by the yeast bio-control agent *Candida oleophila*. Phytopathology 92: 393–399.

Droby S, Wisniewski M, Macarisin D, Wilson C (2009) Twenty years of postharvest bio-control research: Is it time for a new paradigm? Postharvest Biol Technol 52: 137–145.

Edwards SG, Seddon B (2001) Mode of antagonism of *Brevibacillus brevis* against *Botrytis cinerea* in vitro. J Appl Microbiol 91: 652–659.

Fanning S, Mitchell AP (2012) Fungal biofilms. PLOS Pathog 8: e1002585.

Filonow AB, Vishniac HS, Anderson JA, Janisiewicz WJ (1996) Biological control of *Botrytis cinerea* in apple by yeasts from various habitats and their putative mechanisms of antagonism. Biol Control 7: 212–220.

Fiori S, Urgeghe PP, Hammami W, Razzu S, Jaoua S, Migheli Q (2014) Bio-control activity of four non- and low-fermenting yeast strains against *Aspergillus carbonarius* and their ability to remove ochratoxin A from grape juice. Int J Food Microbiol 189: 45–50.

Fitzpatrick DA (2012) Horizontal gene transfer in fungi. FEMS Microbiol Lett 329: 1–8.

Freimoser FM, Rueda-Mejia MP, Tilocca B, Migheli Q (2019) Bio-control yeasts: Mechanisms and applications. World J Microbiol Biotechnol 35: 154.

Gamagae SU, Sivakumar D, Wijeratnam RW, Wijesundera RLC (2003) Use of sodium bicarbonate and *Candida oleophila* to control anthracnose in papaya during storage. Crop Prot 22: 775–779.

Gore-Lloyd D, Sumann I, Brachmann AO et al. (2019) Snf2 controls pulcherriminic acid biosynthesis and antifungal activity of the bio-control yeast *Metschnikowia pulcherrima*. Mol Microbiol 112: 317–332.

Gueldner RC, Reilly CC, Pusey RL, Costello CE, Arrendale RF, Cox RH, Himmelsbach DS, Crumley FG, Culter HG (1988) Isolation and identification of iturins as antifungal peptides in biological control of peach brown rot with *Bacillus subtilis*. J Agric Food Chem 36: 366–370.

Hadwiger LA, McDonel H, Glawe D (2015) Wild yeast strains as prospective candidates to induce resistance against potato late blight (*Phytophthora infestans*). Am J Potato Res 92: 379–386.

Hashem M, Saad AA, Hesham AEL, Al-Qahtani FMH, El-Kelani M (2014) Bio-control of apple blue mold by new yeast strains: *Cryptococcus albidus* KKUY0017 and *Wickerhamomyces anomalus* KKUY0051 and their mode of action. Bio-control Sci Technol 24: 1137–1152.

Hernández-Montiel LG, Gutierrez-Perez ED, Murillo-Amador B, Vero S, Chiquito-Contreras RG, Rincon-Enriquez G (2018) Mechanisms employed by *Debaryomyce shansenii* in biological control of anthracnose disease on papaya fruit. Postharvest Biol Technol 139: 31–37.

Hernández-Montiel LG, Ochoa JL, Troyo-Diéguez E, Larralde-Corona CP (2010) Bio-control of postharvest blue mold (*Penicillium italicum* Wehmer) on Mexican lime by marine and citrus *Debaryomyce shansenii* isolates. Postharvest Biol Technol 56: 181–187.

Hua SS, Beck JJ, Sarreal SB, Gee W (2014) The major volatile compound 2-phenylethanol from the bio-control yeast, *Pichia anomala*, inhibits growth and expression of aflatoxin biosynthetic genes of *Aspergillus flavus*. Mycotoxin Res 30: 71–78.

Huang R, Che HJ, Zhang J, Yang L, Jiang DH, Li GQ (2012) Evaluation of *Sporidiobolus pararoseus* strain YCXT3 as bio-control agent of *Botrytis cinerea* on post-harvest strawberry fruits. Biol Control 62: 53–63.

Huang R, Li GQ, Zhang J, Yang L, Che HJ, Jiang DH, Huang HC (2011) Control of postharvest Botrytis fruit rot of strawberry by volatile organic compounds of *Candida intermedia*. Phytopathology 101: 859–869.

Janisiewicz WJ, Yourman L, Roitman J, Mahoney N (1991) Postharvest control of blue mold and gray mold of apples and pears by dip treatments with pyrrolnitrin, a metabolite of *Pseudomonas cepacia*. Plant Dis 75: 490–494.

Johnson L (2008) Iron and siderophores in fungal-host interactions. Mycol Res 112: 170–183.

Junker K, Chailyan A, Hesselbart A, Forster J, Wendland J (2019) Multi-omics characterization of the necrotrophic mycoparasite *Saccharomycopsis schoenii*. Plos Pathog 15: e1007692.

Karabulut OA, Baykal N (2004) Integrated control of postharvest diseases of peaches with a yeast antagonist, hot water and modified atmosphere packaging. Crop Prot 23(5): 431–435.

Lahlali R, Serrhini MN, Jijakli H (2004) Efficacy assessment of *Candida oleophila* (strain O) and *Pichia anomala* (strain K) against major postharvest diseases of citrus fruits in Morocco. Comm Appl Biol Sci Ghent Univ 69: 601–609.

Lahlali R, Serrhini MN, Jijakli MH (2005) Studying and modelling the combined effect of temperature and water activity on the growth rate of *P. expansum*. Int J Food Microbiol 103: 315–322.

Lassois L, de Bellaire L, Jijakli MH (2008) Biological control of crown rot of bananas with *Pichia anomala* strain K and *Candida oleophila* strain O. Biol Control 45: 410–418.

Lima G, Castoria R, De Curtis F, Raiola A, Ritieni A, De Cicco V (2011) Integrated control of blue mold using new fungicides and bio-control yeasts lowers levels of fungicide residues and patulin contamination in apples. Postharvest Biol Technol 60: 164–172.

Lopes MR, Klein MN, Ferraz LP, da Silva AC, Kupper KC (2015) *Saccharomyces cerevisiae*: A novel and efficient biological control agent for *Colletotrichum acutatum* during pre-harvest. Microbiol Res 175: 93–99.

Mandal G, Singh D, Sharma RR (2007) Effect of hot water treatment and bio-control agent (*Debaryomyce shansenii*) on shelf life of peach. Indian J Hort 64: 25–28.

Masoud W, Poll L, Jakobsen M (2005) Influence of volatile compounds produced by yeasts predominant during processing of *Cofea arabica* in East Africa on growth and ochratoxin A (OTA) production by *Aspergillus ochraceus*. Yeast 22: 1133–1142.

Mayer FL, Wilson D, Hube B (2013) *Candida albicans* pathogenicity mechanisms. Virulence 4: 119–128.

Mimee B, Labbe C, Belanger RR (2009) Catabolism of focculosin, an antimicrobial metabolite produced by *Pseudozyma focculosa*. Glycobiology 19: 995–1001.

Morales H, Sanchis V, Usall J, Ramos AJ, Marín S (2008) Effect of bio-control agents *Candida sake* and *Pantoea agglomerans* on *Penicillium expansum* growth and patulin accumulation in apples. Int J Food Microbiol 122 (1–2): 61–67.

Morath SU, Hung R, Bennett JW (2012) Fungal volatile organic compounds: A review with emphasis on their biotechnological potential. Fungal Biol Rev 26: 73–83.

Nagpure A, Choudhary B, Gupta RK (2014) Chitinases: In agriculture and human healthcare. Crit Rev Biotechnol 34: 215–232.

Nunes CA (2012) Biological control of postharvest diseases of fruit. Eur J Plant Pathol 133: 181–196.

Pandin C, LeCoq D, Canette A, Aymerich S, Briandet R (2017) Should the biofilm mode of life be taken into consideration for microbial bio-control agents? Microbiol Biotechnol 10: 719–734.

Panebianco S, Vitale A, Polizzi G, Scala F, Cirvilleri G (2015) Enhanced control of postharvest citrus fruit decay by means of the combined use of compatible bio-control agents. Biol Control 84: 19–27.

Parafati L, Vitale A, Restuccia C, Cirvilleri G (2015) Bio-control ability and action mechanism of food-isolated yeast strains against *Botrytis cinerea* causing post-harvest bunch rot of table grape. Food Microbiol 47: 85–92.

Park M, Do E, Jung WH (2013) Lipolytic enzymes involved in the virulence of human pathogenic fungi. Mycobiology 41: 67–72.

Pretscher J et al. (2018) Yeasts from different habitats and their potential as bio-control agents. Fermentation 4: 31.

Price NP, Bischof KM, Leathers TD, Cosse AA, Manitchotpisit P (2017) Polyols, not sugars, determine the structural diversity of anti-streptococcal liamocins produced by *Aureobasidium pullulans* strain NRRL 50380. J Antibiot 70: 136–141.

Punja ZK, Utkhede RS (2003) Using fungi and yeasts to manage vegetable crop diseases. Trends Biotechnol 21: 400–407.

Qin GZ, Tian SP (2004) Bio-control of postharvest diseases of jujube fruit by *Cryptococcus laurentii* combined with a low dose of fungicides under different storage conditions. Plant Disease 88(5): 497–501.

Qin GZ, Tian SP, Xu Y, Chan ZL, Li BQ (2006) Combination of antagonistic yeasts with two food additives for control of brown rot caused by *Monilinia fructicola* on sweet cherry fruit. J Appl Microbiol 100: 508–515.

Richards TA, Leonard G, Soanes D, Talbot N (2011) Gene transfer into the fungi. Fungal Biol Rev 25: 98–110.

Rima H, Steve L, Ismail F (2012) Antimicrobial and probiotic properties of yeasts: From fundamental to novel applications. Front Microbiol 3: 421.

Saligkarias ID, Gravanis FT, Epton HAS (2002) Biological control of *Botrytis cinerea* on tomato plants by the use of epiphytic yeasts *Candida guilliermondii* strains 101 and US 7 and *Candida oleophila* strain I-182: in vivo studies. Biol Control 25: 143–150.

Saravanakumar D, Spadaro D, Garibaldi A, Gullino ML (2009) Detection of enzymatic activity and partial sequence of a chitinase gene in *Metschnikowia pulcherrima* strain MACH1 used as post-harvest bio-control agent. Eur J Plant Pathol 123: 183–193.

Scherm B, Ortu G, Muzzu A, Budroni M, Arras G, Migheli Q (2003) Bio-control activity of antagonistic yeasts against *Penicillium expansum* on apple. J Plant Pathol 85: 205–213.

Singh D (2002) Bioefficacy of *Debaryomyce shansenii* on the incidence and growth of *Penicillium italicum on* Kinnow fruit in combination with oil and wax emulsions. Ann Plant Protect Sci 10: 72–276.

Sipiczki M (2006) Metschnikowia strains isolated from botrytized grapes antagonize fungal and bacterial growth by iron depletion. Appl Environ Microbiol 72: 6716–6724.

Spadaro D, Vola R, Piano S, Gullino ML (2002) Mechanisms of actions and efficacy of four isolates of the yeast *Metschnikowia pulcherrima* active against postharvest pathogens on apples. Postharvest Biol Technol 24: 123–134.

Takesako K et al. (1991) Aureobasidins, new antifungal antibiotics. Taxonomy, fermentation, isolation, and properties. J Antibiot (Tokyo) 44: 919–924.

Thakore Y (2006) The biopesticide market for global agricultural use. Ind Biotechnol 3: 194–208.

Tian SP, Fan Q, Xu Y, Qin GZ, Liu HB (2002) Effect of bio-control antagonists applied in combination with calcium on the control of postharvest diseases in different fruit. BullOILB/SROP 25: 193–196.

Tian SP, Qin GZ, Xu Y (2005) Synergistic effects of combining bio-control agents with silicon against postharvest diseases of jujube fruit. J Food Protec 68: 544–550.

Torres R, Teixido N, Vinas I, Mari M, Casalini L, Giraud M, Usall J (2006) Efficacy of *Candida sake* CPA-1 formulation for controlling *Penicillium expansum* decay on pome fruit from different Mediterranean regions. J Food Protec 69: 2703–2711.

Tripathi P, Dubey NK (2004) Exploitation of natural products as an alternative strategy to control postharvest fungal rotting of fruit and vegetables. Postharvest Biol Technol 32: 235–245.

Tripathi A, Sharma N, Sharma V, Alam A (2013) A review on conventional and non-conventional methods to manage post-harvest diseases of perishables. Researcher 5: 6–19.

Urquhart EJ, Punja ZK (2002) Hydrolytic enzymes and antifungal compounds produced by *Tilletiopsis* species, phyllosphere yeasts that are antagonists of powdery mildew fungi. Can J Microbiol 48: 219–229.

Wisniewski M, Biles C, Droby S, McLaughlin R, Wilson C, Chalutz E (1991) Mode of action of the postharvest bio-control yeast, *Pichia guilliermondii*.1. Characterization of attachment to *Botrytis cinerea*. Physiol Mol Plant 39: 245–258.

Wisniewski M, Droby S (2012) Biopreservation of food and feed by postharvest bio-control with microorganisms. In: Sundh I, Wilcks A, Goettel MS (Eds.). Beneficial microorganisms in agriculture, food and the environment. CABI International, Oxfordshire, pp. 57–66.

Xi L, Tian SP (2005) Control of postharvest diseases of tomato fruit by combining antagonistic yeast with sodium bicarbonate. Scientia Agric Sinica 38: 950–955.

Yan F, Xu S, Chen Y, Zheng X (2014) Effect of rhamnolipids on *Rhodotorula glutinis* bio-control of *Alternaria alternata* infection in cherry tomato fruit. Postharvest Biol Technol 97: 32–35.

Zain M, Awaad A, Razzak A, Maitland D, El-Sayed N, Sakhawy M (2009) Secondary metabolites of *Aureobasidium pullulans* isolated from Egyptian soil and their biological activity. J Appl Sci Res 5: 1582–1591.

Zhang D, Spadaro D, Garibaldi A, Gullino ML (2010) Efficacy of the antagonist *Aureobasidium pullulans* PL5 against postharvest pathogens of peach, apple and plum and its modes of action. Biol Control 54: 172–180.

Zhang D, Spadaro D, Garibaldi A, Gullino ML (2011) Potential bio-control activity of a strain of *Pichia guilliermondii* against grey mold of apples and its possible modes of action. Biol Control 57: 193–201.

Zhang H, Wang L, Ma L, Dong Y, Jiang S, Xu B, Zheng X (2009) Bio-control of major postharvest pathogens on apple using *Rhodotorula glutinis* and its effects on postharvest quality parameters. Biol Control 48: 79–83.

Zhang H, Zheng X, Fu C, Xi Y (2003) Biological control of blue mold rot of pear by *Cryptococcus laurentii*. J Hortic Sci Biotechnol 78: 888–893.

Zhang H, Zheng X, Wang L, Li S, Liu R (2007a) Effect of antagonist in combination with hot water dips on postharvest Rhizopus rot of strawberries. J Food Eng 78: 281–287.

Zhang H, Zheng XD, Yu T (2007b) Biological control of postharvest diseases of peach with *Cryptococcus laurentii*. Food Control 18: 287–291.

Zhang D, Spadaro D, Valente S, Garibaldi A, Gullino ML (2012) Cloning, characterization, expression and antifungal activity of an alkaline serine protease of *Aureobasidium pullulans* PL5 involved in the biological control of postharvest pathogens. Int J Food Microbiol 153: 453–464.

Zhao Y, Tu K, Shao X, Jing W, Su Z (2008) Effects of the yeast *Pichia guilliermondii* against *Rhizopus nigricans* on tomato fruit. Postharvest Biol Technol 49: 113–120.

9

Management of Postharvest Diseases in Fruits and Vegetables Using Botanicals

Gayatri Biswal[1] and Dinesh Singh[2]
[1]Department of Plant Pathology, College of Agriculture, OUAT, Bhubaneswar, Odisha, India
[2]Division of Plant Pathology, ICAR-Indian Agricultural Research Institute, New Delhi, India

CONTENTS

9.1 Introduction

Horticultural crops are an important source of minerals and vitamins as well as dietary fibres, thus they called 'protective' foods. They play a vital role in maintaining the nutritional security of the world population, just as food crops (cereal crops) play an important role in food security. Despite the increasing trend in production of horticultural produce, per capita availability of fruits and vegetables is lower in under-developing and developing countries than in developed countries. This may be due to mishandling of the produce, insufficient storage facilities and lack of proper transportation facilities, and hence postharvest losses of horticultural produce are from 15% to 40% of total produce. The producers do not get the maximum price during surplus production and fragmented supply chain. Spoilage of fruits and vegetables by biochemical changes and microbial infections is common.

There have been many reports about the use of lot of chemicals in the form of waxing and coating of horticultural products to minimize postharvest losses due to microbial infections and ultimately helps prolong shelf life. The products are harmful to human consumption, soil and plant health as well as the environment. Eating of wax or chemical coated fruits directly without peeling are detrimental to health. Some fruits like like guava, apple, plum, pear, grape and ber are usually consumed with peels. If we treat these fruits with chemicals or coated with wax to increase their shelf life and these are while directly consumed the chemicals present on the fruit surface are taken to our stomach. Thus, it has now become important to find some other option of chemical control to manage postharvest diseases, which should be safe, nontoxic, eco-friendly and not disturb the ecosystem. Various eco-friendly approaches are applied like physical treatments (hot water treatment, ultraviolet and Gamma rays irradiation, low temperature storage),

TABLE 9.1

Botanicals Used for Managing Postharvest Diseases of Vegetables and Fruits

Name of Botanicals	Botanical Name	Compound	Class	Activity of Compounds
Neem	*Azadirachta indica* Juss	Azadirachtin	Terpenoid	Bacteria, fungi
Onion	*Allium cepa* L.	Allicin	Sulfoxide	Bacteria, fungi
Clove	*Syzygium aromaticum* L.	–	–	–
Garlic	*Allium sativum*	Allicin	Sulfoxide	Bacteria, fungi
Ginger	*Zingiber officinale* Rosc.	Gingerol, paradol, zingerone	–	Microbes
Eucalyptus	*Eucalyptus globules*	Tanin	Polyphenol	Fungi, bacteria, viruses
Turmeric	*Curcumalonga*	Curcumin	Terpenoids	Bacteria, fungi, protozoa
Thyme oil	*Thymus vulgaris*	Caffeic acid	Terpenoids	Fungi, bacteria, viruses
Black pepper	*Piper nigrum* L.	Piperine	Alkaloid	Fungi
Dhatura	*Datura stramonium* L.	Hyoscyamine, scopolamine	Alkaloids	Fungi
Paradise apple	*Malus pumila* Mill.	Phloretin	Flavonoids	General
Castor bean	*Ricinus communis* L.	Ricinine, ricininoleic	Alkaloids	Fungi
Ashwagandha	*Withania somnifera* Dunal.	Withafarin A	Lactone	Bacteria and fungi
Bael	*Aegle marmelos* L.	Essential oil	Terpenoids	Fungi

use of resistant varieties and biological control (microbes, and botanicals). Among them, botanical-based compounds which are specific in their activity against different groups of microbes responsible for postharvest decaying are listed in Table 9.1, and these have no adverse effects on soil, plant and human health. Overall, they are environmentally safe. This chapter deals with botanicals that have been utilized for management of postharvest diseases of fruits and vegetables.

Advantages of botanicals

- Botanicals used in plant disease management are more sustainable.
- They reduce incidence of postharvest diseases of fruits and vegetables by nonchemical methods.
- They are easily biodegradable in nature.
- They are relatively inexpensive and are easily available locally.
- This method of plant disease management is a component of organic farming.
- They are nontoxic and of the low cost.
- There is less chance for development of resistance against pathogens.
- They also act as an insect repellent and antifeedant.
- They can be used as an alternative to the chemical control method of postharvest diseases.
- They are safer to non-targeted organisms.

Disadvantages of botanicals

- They have a limited range of effectiveness.
- Their efficacy may be decreased when the botanicals are used independently under commercial application.
- Extraction methods are not standardized for all the antimicrobials, so their efficacy against pathogens is affected.
- They are rapidly degraded; hence they have less residual effect.
- Most studies are in vitro efficacy. There is a need to standardize protocol for mass production and application in large scale.
- Development of formulations is needed.
- They are less effective or specific to restricted plants as compared to chemicals.
- Fewer formulations of botanicals are available in the market.

9.2 Compounds of Plant Origin in Postharvest Disease Management

Some locally grown plants contain bioactive compounds which are antifungal and antibacterial. Bioactive compounds that are taken out from plants and can be applied on infested or diseased plants and their parts are called botanicals or botanical pesticides (Table 9.1). Some of the different forms of botanicals used are given below.

9.2.1 Plant Extracts

Plant extracts like neem, garlic, eucalyptus, turmeric, tobacco and ginger are used to treat the various horticultural products after harvest to manage diseases caused mostly by fungi during storage and transportation. For extraction, suitable solvents are used to obtain concentrated active compounds. These plants have antimicrobial properties, or they may induce disease resistance by activating a signal transduction system in plants. There are reports which show that more than four lakhs compounds based on origin of plant-bioactive organic chemicals have been reported possess actvites against pests and diseases (Koparde et al. 2019)." These compounds have antifungal (Mishra et al. 2020), antibacterial (EL-Kamali and EL-Amir 2010) and antiviral propeties (Vlietinck and Mvukiyumwami 1995). Greater than 10,000 compounds have been reported that possess activity against diseases and insect pests.

Among them, 342 compounds have fungicidal properties, 92 have bactericidal properties and 90 have antiviral properties. Plant extracts possess nature specificity, biodegradability and minimal/no phytotoxicity. They have minimum residual toxicity in the ecosystem. Plant extracts can be applied on fruits and vegetables by various methods such as spraying, fumigation and dipping at the concentration of 5–15%, based on plant genotypes and age of plants, to decrease/minimize the activities of the microbes that cause decay in vegetables and fruits in the postharvest stage and enhance shelf life.

9.2.2 Essential Oils

Different types of volatile compounds are present in the essential oils extracted from different plant parts. The essential oils are obtained from various plants like nettle, thyme, eucalyptus, rue, lemongrass and tea tree. The composition of the essential oil often varies between species and may be collected from different sources in the same plant. These oils have antimicrobial properties and also are able to induce defence mechanisms in plants against diseases caused by various pathogens. They can be used as an alternative to fungicides and antibiotics. The essential oils used most widely against postharvest pathogens are terpenoids and aromatic compounds.

9.2.3 Volatile Oils

Volatile compounds are small molecular weight, organic compounds and have enough vapour pressure in ambient temperature. Volatile oils of black pepper, clove, oregano (*Origanum vulgare* L.), nutmeg (*Myristica fragrans* Houtt.) and thyme are used for management of postharvest diseases of fruits and vegetables (Table 9.3). They have antimicrobial properties to suppress the growth of postharvest pathogens. They are grouped as generally recommended as safe (GRAS) chemicals. These volatile oils are isolated by hydro distillation or steam. They contain terpenoids-monoterpenes (C10), sesquiterpenes (C15) and diterpenes (C20). These volatile oils are used against postharvest pathogens of fruits and vegetables.

9.2.4 Gel

The gel derived from *Aloe vera* (Tourn. Ex Linn.) is found to be effective against postharvest pathogens of grape as well as other pathogens like *Penicillium digitatum*, *P. expansum*, *Botrytis cinerea* and *Alternaria alternata* that cause diseases in fruits and vegetables.

9.2.5 Latex

Latex is a natural fungicide. Hevein compound is isolated from the latex of the rubber plant (*Hevea brasiliensis*). It can be used for control of *B. cinerea* fungal pathogen. It is safe and the latex is found to be effective against postharvest diseases of banana, papaya and other fruits. The papain obtained from papaya is a water-soluble fraction and has been used against postharvest pathogens (Parthasarathy et al. 2016).

9.3 Extraction of Bioactive Compounds from Plants

Bioactive fractions are separated from inactive/inert components of plant tissue using suitable solvents; extraction methods as given in Table 9.1. The solvents permeate into the solid plant tissues and solubilize bioactive compounds of similar polarity. The plant extract quality is based mainly on factors such as planting material solvent type and method of extraction.

9.3.1 Selection of Plant

Healthy plants are selected randomly from the local area where they are grown naturally/cultivated. Generally, fresh or dried plant materials are used for the extraction of secondary metabolites. The samples are prepared as reported by Tiwari et al. (2005) using fresh plant tissues for plant extract preparation. Plant parts are usually utilized/processed in the dry form (or as an aqueous extract) by traditional healers, because water content within different plant tissues is varied, which affects the concentration of active compound. Hence, before extraction the plants/plant parts must be air dried for an invariant weight, and the plant samples may be kept in the oven at 40°C for 72 hours to dry (Salie et al. 1996). Underground parts of plants like tuber, root, rhizome, bulb, etc. are widely used as compared to the above-ground parts like leaves, buds, flowers, inflorescence, fruits and barks while collecting bioactive compounds having antifungal and antibacterial activities.

9.3.2 Selection of Solvents

Different types of solvents are used in the extraction procedure, which is the main factor for the extraction of bioactive compounds from plants. A good solvent should possess low toxicity, evaporate easily at low heat and promote rapid physiologic absorption of the extract. It should also have preservative action and inability to cause the extract to complex or dissociate (Gurjar et al. 2012). There is the possibility of the presence of solvent in traces along with bioactive compounds at the end product of the extraction. The solvent should not be toxic, and it should also not intervene with the bioassay procedure (Ncube et al. 2008). Solvents are to be selected based on the targeted compounds to be extracted from plants. Crude or alcoholic extractions of these bioactive compounds followed by other solvents should be done in the initial plant screening for their antimicrobial activities. Water is a universal solvent that is used to extract plant active compounds having antimicrobial activity. However, water is primarily used as traditionally healers. Water-soluble phenolics are important antioxidant compounds. However, organic solvents are found more consistent antimicrobial activity as compared to water extracts (Parekh et al. 2005). Tannins and other phenolics are extracted better in aqueous acetone than in aqueous methanol. In another study, Harmala et al. (1992) evaluated 20 different solvents and found that only chloroform resulted in the best solvent to extract nonpolar biological active compounds. Almost all identified antimicrobial compounds extracted from plants are aromatic or saturated organic compounds, and they are generally obtained through initial ethanol or methanol extraction methods. Thus, methanol, ethanol and water are the most commonly used solvents to preliminary screening of plants for antimicrobial activities; other solvents include dichloromethane, acetone and hexane. Eloff (1998a) used different types of extracts based on their ability to solubilize bioactive compounds from plants having antimicrobial activity. Acetone received the highest overall rating based on rate of extraction, ease of removal, and toxicity in bioassay. Focus should be on standardization of the extraction procedure and solvent system to minimize variability in antimicrobial efficacy. Different types of solvents are used for extraction bioactive compounds from plant parts as described by Cowan (1999).

- Acetone: Flavanols
- Chloroform: Flavonoids, terpenoids
- Dichloromethanol: Terpenoides
- Ethanol: Alkaloids, flavanol, tannins, terpenoids
- Ether: Alkaloids, coumarins, terpenoids
- Methanol: Flavanes, saponins, tannins, terpenoids
- Water: Saponins, tannins, terpenoids

9.3.3 Methodology for Extraction

Different factors govern the extraction methods, which are mainly plant tissues particle size, ratio of plant sample and solvent, type and quantity of solvent to be used, pH of the solvent, duration of extraction period and temperature. Grinding of plant materials (dry or wet) to make them into finer particles is important; the principle behind this is that increasing the surface area of extraction materials ultimately increases the bioactive compounds to be released. The size of the particles is more important than duration of extraction. Eloff (1998b) recorded 5-minute extractions of very fine particles, i.e., 10 μm diameter size, which provide higher quantity than less finely ground material by using a shaking machine for 24 hours. Generally, solvent-to-sample ratio of 10:1 (v/w) solvent to dry weight ratio is used (Green 2004). The homogenization of plant tissue in the solvent is the most widely extraction method used by researchers. Dried or wet, fresh plant parts are ground in a blender to render fine particles and a certain quantity of solvent is added. The samples are shaken vigorously for 5–10 minutes or left for 24 hours, and then the extract is filtered. The filtrate then may be dried under reduced pressure. The dried samples are re-dissolved in the solvent to determine the concentration. For clarification of the extract, some workers, however, centrifuged the filtrate at approximately 20,000 × g for 30 minutes.

Another common method of extraction used is serial exhaustive extraction. In this method, successive extraction with solvents was done to increase the polarity from a nonpolar (hexane) to a more polar solvent (methanol) to ensure that a wide polarity range of bioactive compound could be extracted (Green 2004). Sharma et al. (2009) worked on bioefficacy of phytoextracts like *Withania somnifera*, *Allium sativum*, *Aloe barbadensis*, *Lawsonia inermis*, *Eucalyptus amygdalina* and *Curcuma longa* in controlling blue mould fruit rot (*P. italicum)* by pre- and post-inoculation methods. Each plant extract was prepared in sterile distilled water and 15% concentration was prepared. The 100% concentration of plant extracts was prepared by taking fresh leaves of *W. somnifera*, *A. sativum*, *A. barbadensis*, *L. inermis*, *E. amygdalina* and rhizome of *C. longa* and thoroughly washing in running tap water, washing with sterile distilled water and then air drying. The plant materials were then macerated with an equal volume of sterile distilled water (1:1) in a Waring blender. The samples were homogenized for 5 minutes and filtered through double-layered muslin cloth. Then extract was centrifuged for 20 minutes at 5000 rpm. The filtrate was used as plant extract solution of 100% concentration. The fruits were then dipped in the 15% aqueous extract of plants, followed by air drying. The treated fruits were packed in pre-sterilized perforated polythene bags, partially sealed with paper pins and incubated at 25 ± 1°C for 6 days. It was reported that *E. amygdalina* (15%) showed the most effective treatment to reduce rot incidence caused by *P. italicum* (73–76%) followed by *A. sativum* (68.72%).

9.4 Natural Bioactive Compounds Extracted from Plants for Postharvest Disease Management

For a long time, researchers have credited the importance of secondary metabolites in plant resistance to manage diseases of fruits and vegetables. However, not much attention has been given to utilize these metabolites having antimicrobial properties to manage postharvest diseases in fruits and vegetables. The higher plants contain a wide range of secondary substances like phenols, quinones, tannins, essential oils, flavonoids, alkaloids, sterols and saponins (Tripathi et al. 2005). Natural plant products have numerous potential antifungal and antibacterial compounds, such as acetaldehyde, benzaldehyde, benzyl alcohol, ethanol, methyl salicylate, ethyl benzoate, ethyl formate, hexanal, (E)-2 hexenal, lipoxygenases, jasmonates, allicin, glucosinolates, isothiocyanates, etc. (Utama et al. 2002; Tripathi and Dubey 2003; Palou et al. 2008). The use of natural volatile compounds to control postharvest diseases is gaining popularity day by day. In this method, the fruits can be fumigated with the volatile, or it could be incorporated into sprays or packaging (Table 9.2).

Botrytis rot (*Botrytis cinerea*) and Rhizopus rot (*Rhizopus stolonifer)* of strawberry and raspberry could be controlled with acetaldehyde vapour (Prasad and Stadelbacher 1973). The efficacy of volatile fungi toxicants like acetaldehyde, benzyl alcohol, benzaldehyde, cinnamaldehyde, ethanol, nerolidol and 2-nonanone was tested to protect citrus fruits against green mould fungus (Utama et al. 2002). Citral is an active compound in citrus fruit and suppressed germination of spores and growth of mycelium of *P. digitatum* spores (Klieber et al. 2002) and controlled spoilage caused by *P. digitatum* (Fisher and Phillips 2008). In addition, the citral production in the flavedo tissue of citrus fruit acted as a defence mechanism against *P. digitatum* infection (Rodov et al. 1999). The postharvest management of green mould fungus *P. digitatum* either after natural infection or artificial inoculation of grapefruit was found effective by using jasmonates (jasmonic acid and methyl

TABLE 9.2
Natural Compounds Isolated from Plants Tested Against Postharvest Pathogens

Crop	Pathogen	Plant products	Reference
Citrus	*Penicillium digitatum*	Benzaldehyde, cinnamaldehyde	Wilson and Wisniewski (1989)
		Acetaldehyde vapour	Prasad and Stadelbacher (1973)
		Citral	Fisher and Phillips (2008)
		Thymol, menthol	Jafarpour and Fatemi (2012)
		7-geranoxy coumarin	Agnioni et al. (1998)
		Jasmonic acid, methyl jasmonate nerolidol, 2-nonanone	Droby et al. (1999)
	P. italicum	7-geranoxy coumarin	Agnioni et al. (1998)
		Kaempferol	Tripathi et al. (2002)
	Geotrichum citri-aurantii	Heptanol, Decanol, Geraniol, Citronellol, Citral	Suprapta et al. (1997); Klieber et al. (2002); Zhou et al. (2014)
Raspberry	*Botrytis cinerea, Rhizopus stolonifer*	Acetaldehyde vapour	Prasad and Stadelbacher (1973)
	A. alternata, B. cinerea, C. gloeosporioides	hexan-1-ol, benzaldehyde, 2-nonanone, (Z)-hex-3-en-1-ol, (E)-hex-2-enal	Vaughn et al. (1993)
Strawberry	*B. cinerea, R. Stolonifer*	Acetaldehyde vapour	Prasad and Stadelbacher (1973)
	A. alternata, B. cinerea, C. gloeosporioides	hexan-1-ol, benzaldehyde, 2-nonanone, (Z)-hex-3-en-1-ol, (E)-hex-2-enal	Vaughn et al. (1993)
	C. acutatum, B. Cinerea	(E)-hex-2-enalin	Arroyo et al. (2007)
	Fruit decay	Thymol, eugenol, menthol	Wang et al. (2007)
Peach	*M. fructicola and B. Cinerea*	Benzaldehyde, methyl salicylate, ethyl benzoate	Wilson et al. (1987)
Grape	*B. cinerea*	(E)-hex-2-enalin	Arroyo et al. 2007
Blackberry	*B. cinerea*	(E)-hex-2-enalin	Arroyo et al. 2007
Blueberry	Fruit decay	p-Cymene, linalool, carvacrol, anethole, perillaldehyde	Wang et al. 2008
	P. digitatum	Carvacrol, cinnamaldehyde	Sun et al. 2014
Apricot	Rhizopus rot	Acetaldehyde, benzaldehyde, cinnamaldehyde	Abd Alla et al. 2008

jasmonate) (Droby et al. 1999). The fruits treated with jasmonates also effectively decreased chilling injury after storing in cold storage (Droby et al. 1999). Agnioni et al. (1998) isolated a bioactive compound from the flavedo tissue of grapefruit. The compound was identified as 7-geranoxy coumarin which showed antifungal activity against moulds such as *P. italicum* and *P. digitatum* under in vitro and in vivo conditions.

9.5 Mode of Action of Botanicals

9.5.1 Mode of Action of Plant Extracts

A large number of bioactive compounds are present in the extracts of plant and have antimicrobial activity to suppress growth of pathogens in different modes of action (Carson et al. 2002). Droby et al. (1999) found that jasmonates induced resistance responses in grapefruit and also suppressed development of *P. digitatum* in grapefruit (Droby et al. 1999). Jasmonates also played a vital role in the form of signal molecules in plant defence responses against pathogen attack (Tripathi and Dubey 2003). Essential oils play a role of defence mechanism in plants against plant pathogens. There is synergism between their different components, like application of essential oil and bioactive compounds simultaneously inhibit/suppress the development of resistant in fungus. For example, methanol extracts of *W. somnifera* and *Acacia seyal* controlled *P. digitatum* (green mould) due to the stimulatory effect on mechanism of the host defence (Lanciotti et al. 2004; Mekib et al. 2007). The defence mechanisms like synthesis of cell wall could serve as a physical and biological barrier to infecting pathogens or by increasing total soluble phenolic compound concentration in peels of orange (Mekib et al. 2007). These phenolic compounds have a role to change membrane functionality of pathogens (Lanciotti et al. 2004).

9.5.2 Mechanism of Volatiles and Essential Oils

Hexanol in different concentrations is found to be effective against fungal growth. To determine the effectiveness of volatile compounds against fungal growth, the actual vapour pressure is more important than the concentration of the compounds (Caccioni et al. 1997). The potency of hexane basically depends on its vapour pressure (effective concentration), its initial doses and the evaluated fungi. The antifungal property of essential oil is mainly based on the proportion of the individual share of bioactive compounds (Monzote et al. 2009). Almenar et al. (2007) reported that potato dextrose agar medium absorbed hexanol and checked growth of *A. alternata* and *C. acutatum* fungi at the rate of

3.0 and 3.3 μL, respectively. Hence, they hypothesized that if media can absorb hexanol, then fresh produce may also absorb a small quantity of volatile compounds, which can increase shelf life of produce by extruding/forcing out it. The saturation status of volatiles seems to also affect their efficacy; it was reported that saturated compounds are less reactive than unsaturated compounds. Hamilton-Kemp et al. (1992) and Andersen et al. (1994) reported that unsaturated aldehyde [(E)-hex-2-enal)] was found comparatively more efficacious against hyphal growth of *B. cinerea* and *A. alternata* than the saturated aldehyde (hexanal). Similarly, Arroyo et al. (2007) also reported that aldehydes [(E)-hex-2-enal and hexanol] compounds are less saturated than alcohols and esters and have a greater effect on fungal growth inhibition. There are some contradictory reports also, as among tested alcohols, (E)-hex-2-en-1-ol, and (Z)-hex-3-en-1-ol are less saturated than hexan-1-ol, and they showed that lower antifungal activity to inhibit mycelial growth of *C. acutatum* (Arroyo et al. 2007) might be due to *C. acutatum* showing a different reaction against volatiles as compared to *B. cinerea* and *A. alternata* (Almenar et al. 2007). In a transmission electron microscopic study of conidial cells of *C. acutatum* treated with (E)-hex-2-enal, the cell components were highly disorganized and cell wall and plasma membrane were disrupted, and the result ultimately showed lysis of organelles and cell death (Arroyo et al. 2007). However, Hamilton-Kemp et al. (1992) reported that aldehydes suppressed growth of *C. acutatum* and *B. cinerea* hyphae, but terpenes did not inhibit the growth. However, it was suggested that during interpretation more caution is required because culture conditions may be variable and the strain of test fungi may affect the findings. Inouye et al. (2001) reported that permeability of cell membrane as well as its interaction with volatile compounds is vital elements in the effectiveness of bioactive compounds. Change in the cell membrane permeability and degeneration of the ion gradients adversely affect vital cell processes and lead to cell death (Wang et al. 2007).

Kulakiotu et al. (2004) demonstrated that cell protoplast secretion without initial cell wall disruption in mycelia of *B. cinerea* caused rotting of grape, cv. Isabella fruits, when exposed to volatiles showing deformation of cell wall. But in untreated fruits, hyphae were quite healthy and conidiophores possess ample conidia. Thymol and carvacrol botanicals also demonstrated antifungal properties against grape yeasts in red wines by damaging the membrane, leakage of cytoplasmic content and ultimately inhibiting biosynthesis of ergosterol (Chavan and Tupe 2014). These volatile compounds have a different antifungal activity due to their different modes of action. Sivakumar et al. (2009) observed that the chalcones, which are aromatic ketones present in a few plant species, disrupted cell wall formation in fungus and cell lysis due to inhibition of cell wall polysaccharide 1, 3-beta-D-glucan synthesis. Further cell division may also be interrupted due to inhibition of the conversion of tubulin into microtubules (Rozmer and Perjesi 2016). Aldehydes are another group of volatile compounds which inhibit the cell division of fungus by reacting with and inactivating sulfhydryl. Some aldehydes like cinnamaldehyde, citral and perillaldehyde are good electron acceptors. They form a charge transfer complex with electron donors present in fungal cells. Thus they interfere with fungal metabolism (Kurita et al. 1979). Volatile compounds with α,β-unsaturated carbonyl groups like enones and enals often react with nucleophilus in fungi through the Michael reaction to create chemical modifications. They inhibit fungal growth of various postharvest fungal pathogens like *B. cinerea*, *C. acutatum*, *C. fragariae*, *C. gloeosporioides*, *Fusarium oxysporum* and *Phomopsis* sp. (Babu et al. 2006). Nevertheless, the reachability of α,β-unsaturated carbonyl groups to bulky nucleophile biomolecules have an important role in their activity against fungi (Wedge et al. 2000). Carvacrol mitochondrial electron-transfer respiration systems in mammals were inhibited by essential oil components of *C. ambrosioide* (Monzote et al. 2009). The phenol carvacrol has demonstrated antibacterial activity because of its ability to distribute into membranes. It also causes breakdown of ion gradients and increases the passive permeability of the cell membrane. It modulates certain Ca^{2+}-permeable transient receptor potential channels and inhibits sarcoplasmic reticulum Ca^{2+} ATPase. It activates ryanodine receptors in skeletal muscle, thereby influencing intracellular calcium homeostasis.

9.5.3 Mechanism of Fruit Resistance to Fungal Attack Treated by Volatiles and Essential Oils

Volatile compounds of plant origin have a significant role in self-defence of fruits against postharvest pathogens. Perez et al. (1999) reported that by a decrease of 25% in release of volatile compounds i.e. hexa 2-enal content from strawberry induces latent infectiont during strawberry devepment results in activation of lantent infections and cause extensive damage. Six and nine-carbon volatile compounds are produced in plant tissues against wounding in lipoxygenase (LOX) and hydroperoxide lyase (HPL) pathways. The products of both LOX and HPL pathways have a role in aroma biosynthesis. They also have antifungal and antibacterial properties and have a role in host-pathogen interactions. The (E)-hex-2-enal application in *Arabidopsis thaliana* activated various self-defence genes; i.e., lipoxygenase 2 and allene oxide synthase, lignified leaves and accumulated pathogenesis-related proteins such as 3 transcripts and camalexin, and they showed that the volatile treatments stimulate a mechanism of wound-repair and ultimately act as a physical barrier to direct penetration of pathogen (Neri et al. 2007). Wang et al. (2007) reported that external application of essential oils improved fruit resistance against fungal infection by increasing antiproliferative activity and free radical scavenging capacity in the fruits of strawberries (Kulakiotu et al. 2004). They demonstrated that application of grape (cv. Isabella) volatiles in kiwifruit against *B. cinerea*-induced mechanisms of disease resistance in the host, since plant volatiles inhibit the growth of fungus and are produced rapidly through the lipoxygenase pathway (in response to wounding). Essential oils also affect pathogens directly by contacting either in media or as volatiles during storage, which leads to the secretion of cell protoplasts without previous cell wall disruption. The grape (cv. Isabella) volatiles used for treatment the kiwifruits showed deformation of cell walls of hypha in *B. cinerea* (Kulakiotu et al. 2004).

Advantages of application of essential oils

- The antimicrobial activity of the vapour phase of essential oils is most effective when applied as fumigants in small fruits in which watery sanitation is not suitable (Wang et al. 2007).
- Essential oil vapours may be applied easily on fruits in storage and make them more attractive than the dipping method.
- Essential oils also limit the spread of pathogens by suppressing production of spores or decreasing inoculum load on fruit/vegetable surfaces or in the storage.

However, additional studies are needed into the role of volatile compounds in germination of spores and infectivity of pathogens, which will help in development of new postharvest disease control techniques.

9.6 Management of Postharvest Diseases of Fruits and Vegetables

Strawberry fruits inoculated with (E)-hex-2-enalin completely inhibited development of *C. acutatum* after 5 days (Arroyo et al. 2007) and also significantly reduced *B. cinerea* disease symptoms in blackberry, strawberry and grapes in in-vivo conditions (Archbold et al. 1997). Vaughn et al. (1993) tested 15 volatiles from strawberries and red raspberries during ripening. They reported that five compounds (hexan-1-ol, benzaldehyde, 2-nonanone, (Z)-hex-3-en-1-ol and (E)-hex-2-enal) completely inhibited growth of *A. alternata*, *B. cinerea* and *C. gloeosporioides* on strawberry and raspberry fruits. Similarly, in another study, thymol, eugenol and menthol reduced decay of strawberry fruits. Thymol was found most effective in reducing decay of berry as compared to eugenol and menthol (Wang et al. 2007). P-cymene, linalool, carvacrol, anethole and perillaldehyde compounds potently suppressed blueberry mould growth. Among them, cinnamic acid and cinnamaldehyde were less effective in suppressing blueberry mould growth (Wang et al. 2008) (Eucalyptus and cinnamon essential oil improved the quality of fruits and decreased the intensity of fruit decay (Tzortzakis 2007), and cinnamon bark essential oil reduced penetration, development and the number of infected fruits by both the fungi i.e., *C. acutatum* and *B. cinerea* at concentrations above 76.5 µL/L of air (Duduk et al. 2015). Thyme essential oil showed more antifungal activity and also reduced development of pathogens at above 15.3 µL/L of air. A higher concentration at 153 µL/L of air of thyme essential oil inhibited development of *C. acutatum* on inoculated fruits (Duduk et al. 2015). Moreover, strawberry spoilage incited by *B. Cinerea* and *R. stolonifer* was managed by volatiles of *T. vulgaris* (Bhaskara Reddy et al. 1998; Nabigol and Morshedi 2011) and they showed differences between thyme genotypes related to the chemistry and efficacy.

Efficacy of essential oil against different organisms causing postharvest spoilage in grapes was studied by several researchers. The essential oils from clove, thyme, and massoialactone significantly decreased necrotic lesions on leaves of grape incited by *B. cinerea* (Walter et al. 2001). Thymol or eugenol (volatile from clove) reduced the spoilage and loss of sensory quality of grape under CA storage (Valero et al. 2006). Spore and mycelia morphology was changed by infusing chitosan packages with *O. vulgare* essential oil. It inhibited the growth of *R. stolonifer* and *A. niger*, and germination of spores on inoculated grapes during storage (Santos et al. 2006). The essential oils of *Z. officinale*, *O. sanctum* and *P. persica* effectively managed storage rot of grapes under in vivo trials (Table 9.3), and treated grapes showed improvement of shelf-life up to 6 days (Tripathi et al. 2008).

9.6.1 Integrated Postharvest Disease Management

A combination of sesame oil (at 2.0%) and NaCl (1.0%) increased maximum shelf life 33.56 days of Nagpur mandarin fruits at room temperature from 18.78°C to 35.7°C and relative humidity from 22.69% to 44.32% (Thakre et al. 2008). Another example to improve fruit shelf life is application of plant extract with benzyl adenine, which was found to increase the shelf-life of orange (Bhardwaj et al. 2010). Bhardwaj and Sen (2003) reported the effect of neem leaf extracts indifferent concentrations on storage quality of mandarin (Nagpur Santra) and they found that neem leaf extract of 20% concentration significantly improved quality and retained higher ascorbic acid content (27.17 mg/100 mL juice) during storage. The antifungal activity of lemon (essential) oil was increased when it was applied with chitosan against artificially induced *B. cinerea* infection (inoculated with a spore suspension) in strawberries in vitro as well as in cold storage (Perdones et al. 2012)

9.7 Effect of Botanicals on Quality of Produce

Strawberry fruits treated with thymol and eugenol botanicals enhanced shelf life by increasing the fruit free radical scavenging capacity and thus increasing resistance against decay and deterioration of quality (Wang et al. 2007). Fruits treated with essential oil retained a larger quantity of sugars, flavonoids, anthocyanins, organic acid and phenolic compound and decreased fruit spoilage due to increasing the phenolic content in strawberries, which ultimately led to increase of oxygen radical absorbance capacity (Wang et al. 2007). Furthermore, they reported that antioxidant activity was increased by carvacrol, anethole and perillaldehyde. These botanicals also increased total anthocyanins, phenolics and hydroxyl radical (•OH) scavenging capacity in fruit tissues of blueberry. Chitosan coating and carvacrol and cinnamaldehyde maintained firmness and effectively extended shelf life of blueberry fruits (Sun et al. 2014). Similarly, fruit treatment with perillaldehyde, linalool, cinnamaldehyde, cinnamic acid, anethole and carvacrol essential oils increased antioxidant capacity in raspberries. Among these, perillaldehyde was found to be most effective (Jin et al. 2012). Grape

TABLE 9.3

Common Botanicals Used for Management of Postharvest Diseases of Horticultural Produce

Sr. No.	Fruit Crop	Name of Disease	Causal Organism	Plant Used	Plant Parts Used	Reference
1	Banana	Anthracnose	*Colletotrichum musae*	Lemon grass, *Cymbopogan flexuosus*, *Ocimum basilium*	Oil	Anthony et al. (2003)
				Allium sativum, *Copaifera langsdorfii*, *Cinnamomum zeylanicum* and *Eugenia caryophyllata*	Essential oil	Negreiros et al. (2013); Singh and Tripathi (2015)
				C. citrates	Essential oil	Jagana et al. (2018)
		Crown rot	*C. musae*, *Fusarium* spp.	*Ceyloncitronella*, *C. flexuosus*, *Ocimum* sp.	Oil	Anthony et al. (2003)
				Cinnamon	Plant extract	Win et al. (2007)
2	Mango	Anthracnose	*C. gloeosporiodes*	Lemon grass	Oil	Salomone et al. (2008)
				Annona squamosa, *Azadirachta indica*, *Vernonia amygdalina*	Aqueous extract	Onyeani and Amusa (2015)
				Datura stramonium, *Adhatoda schimperiana*, *Eucalyptus globulus*	*Methanol extracts*	Alemu et al. (2014)
				Basel oil, orange oil, lemon oil, mustard oil	Essential oil	Abd-Alla and Haggag (2013)
				Eupatorium odoratum (Siam weed), *Nerium oleander* (Oleander)	Aqueous extract (10%)	Dubey et al. (2007); Manasa et al. (2018)
		Fusarium rot	*Fusarium oxysporum*	*Indica*, *Allium cepa*, *O. sanctum*	Aqueous extract (10%)	Singh and Jain (2007)
		Stem end rot	*Botryodiplodia theobromae*	Geranium, mint, palmarosa, thyme oils	Essential oil	Tripathi and Shukla (2009)
				C. sinensis	Essential oil	Sharma and Tripathi (2006)
			Aspergillus niger	*C. sinensis*	Essential oil	Sharma and Tripathi (2006)
3	Citrus	Blue and green moulds	*Penicillium* spp. *Penicillium digitatum*	*C. zeylanicum*	Essential oil	Kouassi et al. (2012)
		Green mould		*Parastrephia lepidophylla*	Aqueous extract	Sayago et al. (2012)
				Thymus capitatus, *T. vulgaris*	Essential oil	Arras and Usai (2001); Fatemi et al. (2011)
				Sonchus oleraceus, *Borago officinalis*, *Sanguisorba minor*	Methanol extract	Gatto et al. (2011)
				Cymbopogon sp., *Lantana* sp.	Aqueous extract	Abd-El-Khair and Hafez (2006)
				Zataria multiflora	Essential oil	Solaimani et al. (2009)
				Mentha spicata, *Lippia scaberrima*	Essential oil	Du Plooy et al. (2009)
				A. sativum	Aqueous and ethanol extracts	Obagwu and Korsten (2003)
				Halimium umbellatum	Aqueous extract	Askarane et al. (2012)
				Withania somnifera, *Acacia seyal*	Methanol extract	Mekib et al. (2007)
				W. somnifera, *A. seyal*	Methanol extract	Tayel et al. (2009)

TABLE 9.3 *(Continued)*

Common Botanicals Used for Management of Postharvest Diseases of Horticultural Produce

Sr. No.	Fruit Crop	Name of Disease	Causal Organism	Plant Used	Plant Parts Used	Reference
				Bubonium imbricatum	Essential oil	Alilou et al. (2008)
				Citrus sp.	Essential oil	Badawy et al. (2011)
		Blue mould rot	*P. italicum*	*T. vulgaris*	Essential oil	Fatemi et al. (2011)
				Sonchus oleraceus, Borago officinalis, Sanguisorba minor	Methanol extract	Gatto et al. (2011)
				Zataria multiflora	Essential oil	Solaimani et al. (2009)
				Chrysanthemum sp.	Essential oil	Chebli et al. (2003)
				M. arvensis, Zingiber officinale, O. canum	Essential oil	Tripathi et al. (2005)
				A. nilotica	Aqueous extract	Tripathi et al. (2002)
				Ageratum conyzoides	Essential oil	Dixit et al. (1995)
				Simmondsia chinensis	Oil emulsion	Ahmed et al. (2007)
		Sour rot	*G. citri-aurantii*	*Cistus villosus,Halimium antiatatlanticum,H. umbellatum*	Aqueous and methanol extract	Talibi et al. (2012); Askarane et al. (2012)
				Thymus sp.	Essential oil	Liu et al. (2009)
				Chrysanthemum	Essential oil	Chebli et al. (2003)
				Halimium umbellatum	Aqueous and Methanol extract	Askarane et al. (2012); Talibi et al. (2012)
4	Apple	Brown rot	*Sclerotinia fructigena*	*Mentha* sp.	Aqueous leaf extract	Chauhan et al. (2008)
		Blue mould rot	*P. expansum*	*A. indica*	Leaf extract	Singh and Geeta (2007)
		Grey and blue moulds	*B. cinerea, P. expansum*	*E. globulus, thyme*	Essential oil	Latif (2016)
		Blue mould, Ulocladium rot, Alternaria rot	*P. expansum, Ulocladium chartarum, Alternaria mali*	*C. sinensis*	Essential oil	Sharma and Tripathi (2006)
5	Grape	Soft rot	*Penicillium* sp., *P. expansum*	*Aloe vera*	Gel	Parthasarathy et al. (2016)
		Grey rot	*B. cinerea*	*Mentha* sp., *O. sanctum, P. persica, Z. officinale*	Essential oil	Tripathi et al. (2008)
			Cladosporium cladosporioides, P. chrysogenum	*C. sinensis*	Essential oil	Sharma and Tripathi (2006)
			Aspergillus niger, R. stolonifer	*Origanum vulgare L.*	Essential oil	Santos et al. (2012)
6	Pear	Fruit decay	*B. cinerea, Glomerella cingulata, Monilinia fructigena, P. expansum*	Tea tree	Oil	Jobling (2000)
		Sclerotinia rot	*Sclerotinia sclerotiorum*	Garlic	Oil	Ziedan and Farrag (2008)

(Continued)

TABLE 9.3 *(Continued)*

Common Botanicals Used for Management of Postharvest Diseases of Horticultural Produce

Sr. No.	Fruit Crop	Name of Disease	Causal Organism	Plant Used	Plant Parts Used	Reference
7	Guava	Fruit canker, Anthracnose, fruit rot, dark mould, dairy mould	*Pestalosia psidii*, *Gloesporium psidii*, *Rhizoctonia solani*, *Fusarium* sp., *A. alternata*, *Geotrichumcandidum*	*A. indica*	Essential oil	Nongmaithem (2014)
8	Pomegranate	Soft rot	*Rhizopus arrhizus*, *R. stolonifer*	Peppermint, sweet basil	Oil	Sayed et al. (2008)
9	Papaya	Anthracnose	*C. gloeosporioides*	Thyme, Mexican lime, lemongrass, cinnamon, castor oil	Essential oil	Bosquez-Molina et al. (2010); Ali et al. (2015); Maqbool et al. (2011)
				Ginger	Oil and extract	Ali et al. (2016)
		Rhizopus rot	*R. stolonifer*	*Thyme, Mexican lime*	*Essential oils*	Bosquez-Molina et al. (2010)
10	Litchi	Fruit rots	*A. flavus*, *A. nidulans*, *A. niger*, *Botryodiplodia theobromae*, *Cladosporium* spp.	*Piper nigrum*, *Solanum nigrum*	Leaf extract	Mohamed et al. (1996)
11	Cherry	Rhizopus rot	*R. stolonifer*	Peppermint, sweet basil	Oil	Sayed et al. (2008)
		Brown rot	*M. fructicola*	Peppermint, sweet basil	Oil	Sayed et al. (2008)
12	Peach	Brown rot	*M. fructicola*	Peppermint, sweet basil	Oil	Sayed et al. (2008)
13	Jackfruit	Rhizopus rot	*R. artocarpi*	Peppermint, sweet basil	Oil	Sayed et al. (2008)
14.	Fig	Fruit rot	*Rhizopus* sp.	Peppermint, sweet basil	Oil	Sayed et al. (2008)
15	Aonla	fruit rot	*A. niger*	*Garlic (A. Sativum)*	Extract	Patel et al. (2008)
16	Strawberry		*R. stolonifer*, *P. digitatum*, *A. niger*, *B. cinerea*	*T. danensis* and *T. carmanicus*	Essential oil	Bhaskara Reddy et al. (1998); Nabigol and Morshedi (2011)
		Fruit decay	–	Eucalyptus, cinnamon, lemon grass	Essential oil	Tzortzakis (2007); Tzortzakis and Economakis (2007)
		Rhizopus rot, grey mould rot	*R. stolonifer*, *B. cinerea*	*T. vulgaris*	Essential oil	Nabigol and Morshedi (2011)
17	Peach		*R. stolonifer*, *M. fructicola*	*Agonis flexuosa* and *O. basilicum*	Essential oil	Ziedan and Farrag (2008)
18.	Apricot	Brown rot, grey mould rot	*M. fructicola*, *B. cinerea*	*T. vulgaris*, *Eugenia caryophyllata*, *C. zeylanicum*, *Carum copticum*	Essential oils	Hassani et al. (2011)
19	Tomato	Early blight and wilt	*A. solani*, *F. oxysporium*	Neem, Chinaberry *(Melia azadrach)*	Leaf extract	Siddiqui et al. (2004)
			Cladosporium fulvum, *B. cinerea*, *A. alternata*	*C. sinensis*	Essential oil	Sharma and Tripathi (2006)
		Fruit decay	–	Eucalyptus, cinnamon	Essential oil	Tzortzakis (2007)
		Black rot	*A. alternata*	Ginger oil formulation	Oil	Helal and Abdeldaiem (2009)
				Z. multiflora	Essential oil	Mahmoudi et al. (2012)

TABLE 9.3 ***(Continued)***

Common Botanicals Used for Management of Postharvest Diseases of Horticultural Produce

Sr. No.	Fruit Crop	Name of Disease	Causal Organism	Plant Used	Plant Parts Used	Reference
20	Cassava	Postharvest deterioration	*A. niger*, *B. theobromae*, *F. solani*, *P. oxalicum*	*A. indica* leaves, *Aframomum melegueta* seeds	Ethanolic and water extractions	Okigbo et al. (2009)
21	Carrot	Crater rot, Sclerotinia rot	*Rhizoctonia carotae*, *Sclerotinia sclerotiorum*		Essential oil	Ziedan and Farrag (2008)
22	Chilli	Anthracnose	*C. acutatum*	Garlic	Aqueous extract	Alves et al. (2015)
23	Yam		Rot-causing fungi	*A. indica*, *Chromoleana odorata*	–	Bhardwaj et al. (2010)
			A. niger, *A. flavus*, *Fusarium oxysporum*, *Rhizopus stolonifer*, *B. theobromae*, *P. chrysogenum*	*O. gratissimum*, *Afromamum melegueta*	Aqueous extract	Okigbo and Ogbonnaya (2006)
24	Bitter gourd		*Myrothecium roridum*, *Ulocladium* sp.	*C. sinensis*	Essential oil	Sharma and Tripathi (2006)

berries treated with eugenol or thymol essential oils packed in a modified atmosphere package showed minimum weight loss, slight changing of skin colour and also decreased ripening as well as minimized decay during storage (Valero et al. 2006). Sánchez-González et al. (2011) reported that chitosan-coated packages of grapes either alone or in combination with bergamot oil controlled respiration rate, water loss and also showed antifungal activity during storage which led to ultimately increasing the shelf life and quality of the berries. Chitosan packages infused with *O. vulgare* essential oil also retained the quality and sensory value of grape berries during storage (Santos et al. 2012).

9.8 Phytotoxicity, Off-Flavour and Off-Odour of Botanicals on Fresh Produce

There are many reports that the volatiles used against postharvest diseases have detrimental effects on quality and flavour of fruits. Among these materials, i.e., (Z)-hex-3-en-1-ol and (E)-hex-2-enal caused extensive fruit necrosis, while weight loss and other side effects were reported in (E)-hex-2-enal (Fallik et al. 1998), phytotoxicity was developed in strawberry fruits, and quality deteriorated due to increased exposure to volatiles (Arroyo et al. 2007). The phytotoxicity of (E)-hex-2-enal has also been reported in pear fruits (Neri et al. 2006), beans (Croft et al. 1993) and sliced apples (Corbo et al. 2000). It also caused extensive tissue necrosis in strawberry (Vaughan et al. 1993). Volatiles 2-nonanone, hexan-1-ol and benzaldehyde moderately harmed strawberry fruit and while extensive tissue necrosis was observed in (Z)-hex-3-en-1-ol in treated fruits. Walter et al. (2001) reported extra water loss from detached branches of grape in the laboratory, but not in the field unveiled to but not detected in field in unveiled condition when exposed to thyme. Also, incorporation of bergamot oil in chitosan coating showed browning in grapes during storage (Sánchez-González et al. 2011). However, Bhaskara Reddy et al. (1998) did not visualize any phytotoxic symptoms after 4 days of fruit treatment to essential oils of *T. vulgaris*. Natural flavour and palatability are very specific to each fruit even though they vary with different varieties of each fruit crop. It is essential to maintain natural flavour and taste of the fruit in every treatment, especially in case of volatiles and oils. Neri et al. (2007) reported off-odour development in peach and nectarine fruits treated with (E)-hex-2-enal described as a 'green'(leafy) and 'butyric' aromas. However, off-odours decreased in the fruits during ripening and nothing was perceived in ripened fruits. Off-odours were not observed in peach and nectar wine fruits treated with carvacrolor citral. Similarly, off-flavours and off-odours were completely absent in dates treated with citrus essential oils (Aloui et al. 2014) and also in cherries treated with eugenol, thymol or menthol (Serrano et al. 2005). However, cherries treated with eucalyptol generated off-flavours. Prasad and Stadelbacher (1974) reported that acetaldehyde-treated strawberries did not show any off-flavour at 1.0% concentration, but off-flavour was developed at 4.0% concentration. They reported that the majority of the essential oils applied at the minimum inhibitory concentrations did not leave off-flavours or off-odours on intact fruits, particularly the fruits that finished their ripening process during storage. However, extensive studies should be carried out, because of a great deal of variation between fruit species and cultivars is found which affects the development of off-flavours or off-odours. It is also important to test each horticultural produce item individually under different climatic conditions. In one study, Neri et al. (2007) demonstrated that stone fruits were comparatively more sensitive to (E)-hex-2-enal injury than pome fruits. The off-flavour recorded might be due to the fact that absorption of (E)-hex-2-enal was greater in stone fruits than pome fruits.

9.9 Application Methods of Botanicals and Criteria for Selecting a Good Product

Various methods are used to evaluate efficacy of plant extract/ botanicals against plant pathogens in in-vitro conditions such as diffusion test (agar well diffusion, agar disk diffusion, poison food technique, bio autography) and dilution methods (agar dilution, broth microdilution assay, broth macrodilution assay). However, the essential oils are applied to fruit coatings primarily to retain moisture (Du Plooy et al. 2009). This method later gained popularity when oil of *Simmondsia chinensis* (jojoba oil) was coated on 'Valencia' oranges and fruit quality was maintained for up to 60 days (Ahmed et al. 2007). The benefit of applying coatings of essential oils is to have close contact between the oils and fruit surfaces, and it allows exposing each fruit to similar concentrations of inhibitor for a longer duration (Du Plooy et al. 2009).

Another method of application is the dipping method, where the fruits are dipped in the desired concentration of botanical solution to control postharvest disease. The essential oil of Shiraz thyme was found to be effective against *P. digitatum* causing green mould rot in citrus only when it was applied using the dip method (Solaimani et al. 2009). However, the dipping method showed better results for managing green mould rot disease than the spray method. Similarly, in another example, guava fruits were dipped in the aqueous extracts of plant extracts, such as *Ocimum sanctum*, *Vinca rosea*, *Azadirachta indica*, *A. barbadensis*, *Withania somnifera* and *A. indica* karnel, *Allium sativum* bulb at 10% concentration to control Pestalotiopsis fruit rot disease caused by *Pestalotiopsis palmarum* (Meena et al. 2009).

Botanicals have properties as postharvest fungi toxicants if they are properly applied. To be effective during in vivo application, botanical products must meet conditions such as being efficacious even after short-duration treatments, have no negative effect on quality parameters like acidity, flavour and aroma, and be effective at low doses (Tripathi and Dubey 2003). The cost of application should be low and economically feasible, and the product should be easily available. The method of application should be standardized and should be easy.

9.10 Conclusion

Many scientists have worked on the efficacy of plant extracts, essential oils and volatile compounds against many postharvest diseases of horticultural produce in in-vivo conditions after harvest. Among different microbes, fungi and bacteria are mainly responsible for spoilage. Very few reports exist on the management of bacterial population on the surfaces of fruits and vegetables through using essential oils. It is expected that more work will be done on different aspects of botanical application, including dosage and method of application. There are ambiguous findings on botanicals like positivity of volatiles and essential oils under in vitro conditions on the media against fungi, but several researchers obtained variations in their efficacy and minimum inhibitory concentrations under in vivo conditions. Most of the variations in results recorded under in vivo conditions were due to the vapour pressure of volatile compounds. Further, it has been noticed that the efficacy of multi-compound essential oils is even more fugacious due to many factors like genetic variation of botanicals, environmental conditions and synergistic effects of multiple compounds present in plants, and their proportions. Major bioactive compounds of essential oils play an important role in antifungal and antibacterial activity. However, minor compounds present in the mixture may have synergistic effects to improve efficacy. Hence, there is an urgent need of future studies to establish the function of a single natural bioactive compound or mixture in plant-pathogen interactions. Bioactive compounds such as(E)-hex-2-enal and hexanal are found to be the most effective volatiles. Thyme is the most effective essential oil to control fungal diseases of both fruits and vegetables under in vitro conditions as well and postharvest treatment of farm produce before storage. It has been well documented that it has potential to apply on the fruits for enhancing shelf-life and also retain quality of the produce. It can be used to replace postharvest fungicides or controlled atmosphere storage.

There is a need for research on active packaging and improved formulation techniques to prolong activity, reduce volatility and improve surface coverage. Several researchers have already reported the toxicity of volatiles and essential oils after longer periods of exposure. Information on how we can best eliminate toxicity is currently needed. Future research is needed to study the effects of volatiles and essential oils on mass production of fungal spores, their effects on quality, structure and integrity of cell wall, and their role in plant self-defence mechanism and shelf life. Extensive research is necessary indifferent places and with different varieties of the fruit crop in various storage conditions to identify and minimize the causes of adverse effects on fruit quality by volatiles. Development of plant-based bioformulation is a new area of research which is recently the most preferred aspect. Researchers can easily utilize these formulations for plant protection in the field of postharvest disease management during storage and transportation, extending the shelf-life of fresh fruits and vegetables and retaining the quality of the produce. Protocol of edible coatings using oils can be standardized for various fruits and vegetables, nano-emulsion based delivery of plant materials as well as aromatic extracts for disease management. The strategies should be easily adoptable by farmers so that they can apply them to their produce and ultimately see greater profit. Simultaneously, the consumer will benefit from nutritionally rich quality fruits and vegetables at their doorstep.

REFERENCES

Abd-El-Khair H, Hafez O (2006) Effects of aqueous extracts of some medicinal plants as controlling the green mould diseases and improvement of stored "Washington" novel orange quality. Appl Sci Res 2: 664–674.

Abd Alla MA, El-Sayed H, Zeidan H, El-Mohamedy S (2008) Control of Rhizopus rot disease of apricot fruits (*Prunus armeniaca*) by some plant volatiles aldehydes. J Agric Biol Sci 45(4): 424–433.

Abd-Alla MA, Haggag WM (2013) Use of some plant essential oils as postharvest botanical fungicides in the management of anthracnose disease of mango fruits (Mangifera indica L.) caused by Colletotrichum gloeosporioides (Penz). Int J Agric Forestry 3(1): 1–6.

Agnioni A, Cabras P, Dhallewin G, Piris F, Reniero F, Schirra M (1998) Synthesis and activity of 7-geranoxy coumarin against *Penicillium* spp. in citrus fruits. Phytochemistry 47: 1521–1525.

Ahmed DM, EL Shami S, EL Mallah MH (2007) Jojoba oil as a novel coating for exported Valencia orange fruit. Part I: The use of trans (isomerized) jojoba oil. Am Eurasian J Agric Environ Sci 2: 173–181.

Alemu K, Ayalew A, Weldetsadi KJ (2014) Evaluation of antifungal activity of botanicals for postharvest management of mango anthracnose *Colletotrichum gloeosporioides*. Int J Life Sci 8(1): 1–6.

Ali A, Tan WP, Mustafa MA (2015) Application of lemongrass oil in vapour phase for the effective control of anthracnose of "sekaki"papaya. J Appl Microbiol 118: 1456–1464.

Ali A, Hei GK, Keat YW (2016) Efficacy of ginger oil and extract combined with gum arabic on anthracnose and quality of papaya fruit during cold storage. J Food Sci Technol 53(3): 1435–1444.

Alilou H, Akssira M, Idrissi Hassani LM, Chebli B, El-Hakmoui A, Fouad M, Rachida R, Herminio B, Amparo MB (2008) Chemical composition and antifungal activity of *Bubonium imbricatum* volatile oil. Phytopathol Mediterr 47(7): 3–10.

Almenar E, Auras R, Rubino M, Harte B (2007) A new technique to prevent the main post harvest diseases in berries during storage: Inclusion complexes β-cyclodextrin-hexanal. Int J Food Microbiol 118: 164–172.

Aloui H, Khwaldia K, Licciardello F, Mazzaglia A, Muratore G, Hamdi M, Restucciad C (2014) Efficacy of the combined application of chitosan and Locust Bean Gum with different citrus essential oils to control postharvest spoilage caused by *Aspergillus flavus* in dates. Int J Food Microbiol 170: 21–28.

Alves KF, Laranjeira D, Camara MPS, Camara CAG, Michereff SJ (2015) Efficacy of plant extracts for anthracnose controlling bell pepper fruits under controlled conditions. Horti Brasil 33(3): 332–338.

Andersen RA, Hamilton-Kemp TR, Hildebrand DF, McCracken CT Jr, Collins RW, Fleming PD (1994) Structure-antifungal activity relationships among volatile C6 and C9 aliphatic aldehydes, ketones, and alcohols. J Agric Food Chem 42: 1563–1568.

Anthony S, Abeywickrama K, Shanthi WW (2003) The effect of spraying essential oils of *Cymbopogon nardus*, *Cymbopogon flexuosus* and *Ocimum basilicum* on postharvest diseases and storage life of Embul banana. J Hortic Sci Biotechnol 78(6): 780–785.

Arras G, Usai M (2001) Fungitoxic activity of 12 essential oils against four postharvest pathogens: Chemical analysis of *Thymus capitatus* oil and its effect in subatmospheric conditions. J Food Prot 64: 1025–1029.

Archbold DD, Hamilton-Kemp TR, Barth MM, Langlois BE (1997) Identifying natural volatile compounds that control gray mold (*Botrytis cinerea*) during postharvest storage of strawberry, blackberry, and grape. J Agric Food Chem 45: 4032–4037.

Arroyo FT, Moreno J, Daza P, Boianova L, Romero F (2007) Antifungal activity of strawberry fruit volatile compounds against *Colletotrichum acutatum*. J Agric Food Chem 55: 5701–5707.

Askarane L, Talibi I, Boubaker H, Boudyach E, Msanda F, Saadi B, Serghini M, Ait B, Aoumar A (2012) Use of Moroccan medicinal plant extracts as botanical fungicide against citrus blue mould. Lett Appl Microbiol 56: 37–43.

Babu KS, Li XC, Jacob MR, Zhang Q, Khan S, Ferreia D, Clark AM (2006) Synthesis, antifungal activity, and structure-activity relationships of coruscanone analogs. J Med Chem 49: 7877–7886.

Badawy FMI, Sallem MAN, Ibrahim A, Asram (2011) Efficacy of some essential oils on controlling green mold of orange and their effects on postharvest quality parameters. Plant Pathol J 10: 168–174.

Bhardwaj RL, Dhashora LK, Mukherjee S (2010) Effect of plant extract and benzyl adenine on shelf life of orange (*Citrus reticulate* Blanco). Adv Dev Res 1: 32–37.

Bhardwaj RL, Sen NL (2003) Zero energy cool-chamber storage of Mandarin (*Citrus reticulata* Blanco) cv. 'Nagpur Santra'. J Food Sci Technol (Mysore) 40(6): 669–672.

Bhaskara Reddy MV, Angers P, Gosselin A, Arul J (1998) Characterization and use of essential oil from *Thymus vulgaris* against *Botrytis cinerea* and *Rhizopus stolonifer* in strawberry fruits. Phytochemistry 47: 1515–1520.

Bosquez-Molina E, Ronquillo-de Jesús E, Bautista-Banos S, Verde-Calvoa JR, Morales-Lopez J (2010) Inhibitory effect of essential oils against *Colletotrichum gloeosporioides* and *Rhizopus stolonifer* in stored papaya fruit and their possible application in coatings. Postharvest Biol Technol 57: 132–137.

Caccioni DRL, Gardini F, Lanciotti R, Guerzoni ME (1997) Antifungal activity of natural volatile compounds in relation to their vapor pressure. Sci Aliment 17: 21–34.

Carson CF, Mee BJ, Riley TV (2002) Mechanism of action of *Melaleuca alternifolia* (tea tree) oil on *Staphylococcus aureus* determined by time-kill, lysis, leakage and salt tolerance assays and electron microscopy. Antimicrob Agents Chemother 46(6): 1914–1920.

Chavan PS, Tupe SG (2014) Antifungal activity and mechanism of action of carvacrol and thymol against vineyard and wine spoilage yeasts. Food Control 46: 115–120.

Chebli B, Achouri M, Idrissi Hassani L, Hamamouchi M (2003) Chemical composition and antifungal activity of essential oils of seven Moroccan Labiatae against *Botrytis cinerea* Pers: Fr. J. Ethnopharmaco 189: 165–169.

Corbo MR, Lanciotti R, Gardini F, Sinigaglia M, Guerzoni ME (2000) Effects of hexanal, trans-2-hexenal and storage temperature on shelf life of fresh sliced apples. J Agric Food Chem 48: 2401–2408.

Cowan MM (1999) Plant products as antimicrobial agents. Clinical Microbiol Rev 12: 564–582.

Croft KPC, Juttner F, Slusarenko AJ (1993) Volatile products of the lipoxygenase pathway evolved from *Phaseolus vulgaris* (L.) leaves inoculated with *Pseudomonas syringae*pv. *phaseolicola*. Plant Physiol 101: 13–24.

Dixit S, Chandra H, Tiwari R, Dixit V (1995) Development of botanical fungicides against blue mould f mandarins. J Stored Prod Res 31: 165–172.

Droby S, Porat R, Cohen L, Weiss B, Shapiro B, Philosoph Hadas S, Meir S (1999) Suppressing green mold decay in grapefruit with postharvest jasmonate application. J Am Sci 124: 184–188.

Dubey RK, Kumar R, Jaya, Dubey NK (2007) Evaluation of *Eupatorium cannabium* L. oil in enhancement of shelf life of mango fruits from fungal rotting. World J Microbiol Biotechnol 23: 467–473.

Duduk N, Markovic T, Vasic M, Duduk B, Vico I, Obradovic A (2015) Antifungal activity of three essential oils against *Colletotrichum acutatum*, the causal agent of strawberry anthracnose. J Essent Oil Bear Pl 18: 529–537.

du Plooy W, Thierry R, Sandra C (2009) Essential oil amended coatings as alternatives to synthetic fungicides in citrus postharvest management. Postharvest Biol Technol 53: 117–122.

Eloff JN (1998a) Which extractants should be used for the screening and isolation of antimicrobial components from plants? J Ethnopharmacol 60: 1–8.

Eloff JN (1998b) A sensitive and quick microplate method to determine the minimum inhibitory concentration of plant extracts for bacteria. Planta Medica 64: 711–713.

EL-Kamali HH, EL-Amir MY (2010) Antibacterial activity and phytochemical screening of ethanolicextracts obtained from Sudanese Medicinal Plant. Cur Res J Biological Sci 2(2): 143–147

Fallik E, Archbold DD, Hamilton-Kemp TR, Clements AM, Collins RW, Barth MM (1998) (E)-2-Hexenal both stimulates and inhibits *Botrytis cinerea* growth in vitro and on strawberry fruit in vivo. J Am Soc Hortic Sci 123: 875–881.

Fatemi S, Jatarpour M, Eghbalsaied S (2011) Study the effect of *Thymus vulgaris* and hot water treatment on storage life of (*Citrus sinensis* cv. *valencia)*. J Med Plants Res 6: 968–971.

Fisher K, Phillips C (2008) Potential antimicrobial of essential oils in food is citrus the answer. Trends Food Sci Technol 19: 156–164.

Gatto MA, Ippolito A, Linsalata V, Cascarano NA, Nigro F, Vanadia S, Di Venere D (2011) Activity of extracts from wild edible herbs against postharvest fungal diseases of fruit and vegetables. Postharvest Biol Technol 61: 71–82.

GreenRJ (2004) Antioxidant activity of peanut plant tissues. Master's thesis, North Carolina State University.

Gurjar MS, Ali S, Akhtar M, Singh KM (2012) Efficacy of plant extracts in plant disease management. Agric Sci 3(3): 425–433.

Hamilton-Kemp TR, McCracken CT, Loughrin JH, Andersen RA, Hildebrand DF (1992) Effects of some natural volatile compounds on the pathogenic fungi *A. alternata* and *Botrytis cinerea*. J Chem Ecol 18: 1083–1091.

Harmala P, Vuorela H, Tornquist K, Hiltunen R (1992) Choice of solvent in the extraction of *Angelica archangelica* roots with reference to calcium blocking activity. Planta Medica 58: 176–183.

Helal IMM, Abdeldaiem MH (2009) Inhibition of green and blue rots in orange fruits by using clove and thyme essential oils treated by gamma radiation. Arab J Nucl Sci App 42: 257–268.

Inouye S, Takizawa T, Yamaguchi H (2001) Antibacterial activity of essential oils and their major constituents against respiratory tract pathogens by gaseous contact. J Antimicrobiol Chemother 47: 565–573.

Jafarpour M, Fatemi S (2012) Post harvest treatments on shelf life of sweet orange Valencia. J Med Plants Res 6: 2117–2124.

Jagana D, Hegde YR, Lella R (2018) Bioefficacy of essential oils and plant oils for the management of banana anthracnose-a major post-harvest disease. Int J Cur Microbiol Appl Sci 7(04): 2359–2365.

Jin P, Wang SY, Gao H, Chen H, Zheng Y, Wang CY (2012) Effect of cultural system and essential oil treatment on antioxidant capacity in raspberries. Food Chem 132: 399–405.

Jobling J (2000) Essential oils: A new idea for postharvest disease control. Good Fruit and Vegetables Magazine11.50. Available via: http://www.postharvest.com.au/gfvoils.

Klieber A, Scott E, Wuryatmo E (2002) Effect of method of application of antifungal efficacy of citral against postharvest spoilage fungi of citrus in culture. Aust Plant Pathol 31: 329–332.

Koparde AA, Doijad RCS and Magdum CS (2019) Natural Products in Drug Discovery. DOI:10.57772/intechopen.82860

Kouassi KHS, Bajji M, Jijakli H (2012) The control of postharvest blue and green molds of citrus in relation with essential oil-wax formulations, adherence and viscosity. Postharvest Biol Technol 73: 122–128.

Kulakiotu EK, Thanassoulopoulos CC, Sfakiotakis EM (2004) Postharvest biological control of *Botrytis cinerea* on kiwifruit by volatiles of 'Isabella' grapes. Phytopathology 94: 1280–1285.

Kurita N, Miyaji M, Kurane R, Takahara Y, Ichimura K (1979) Antifungal activity and molecular orbital energies of aldehyde compounds from oils of higher plants. Agric Biol Chem 43(11): 2365.

Lanciotti R, Gianotti A, Patrignani F, Belleti N, Guerzoni ME, Gardini F (2004) Use of natural aroma compounds to improve shelf-life and safety of minimally processed fruits. Trends Food Sci Technol 15: 201–208.

Latif F (2016) Postharvest application of some essential oils for controlling gray and blue moulds of apple fruits. Plant Pathol J 15(1): 5–10.

Liu D, Wang L, Li Y, Li H, Yu TND, Zheng X (2009) Antifungal activity of thyme oil against *Geotrichum citri-aurantic* in vitro and in vivo. J Appl Microbiol 51: 30–35.

Mahmoudi E, Ahmadi A, Naderi D (2012) Effect of *Zataria multiflora* essential oil on *Alternaria alternata* in vitro and in an assay on tomato fruits. J Plant Dis Protec 119: 53–58.

Manasa B, Jagadeesh SL, Thammaiah N, Sandhyarani N, Gangadharappa PM, Jagadeesha RC, Netravati (2018) Evaluation of fungicides, bioagents and botanicals on postharvest disease, shelf life and physicochemical properties of 'Alphonso' mango. J Pharmaco Phytochem 7(4): 1883–1888.

Maqbool M, Ali A, Alderson PG, Mohamed MTM, Siddiqui Y, Zahid N (2011) Postharvest application of gum arabic and essential oils for controlling anthracnose and quality of banana and papaya during cold storage. Postharvest Biol Technol 62: 71–76.

Mekib S, Regnier T, Korstem L (2007) Control of *Penicillium digitatum* on citrus fruit using plant extracts and study of their mode of action. Phytoparasitica 35: 264–276.

Meena OP, Gdara SL, Rathore GS, Pal V (2009) Eco-friendly management of postharvest fruit rot of guava caused by *Pestalotiopsis palmarum*. J Mycol Plant Pathol 39(3): 445–448.

Mishra KK, Kaur D, Sahu AK, Panik R, Kashyap, Mishra SP, Dutta S (2020) Medicinal plants having antifungal properties. DOI: 10.57772?intech open.90674.open access peer - reviewed chapter.

Mohamed S, Saka S, EL-Sharkawy SH, Ali AM, Muid S (1996) Antimycotic screening of 58 Malaysian plants against plant pathogens. Pesticide Sci 47(3): 259–264.

Monzote L, Werner S, Katrin S, Lars G (2009) Toxic effects of carvacrol, caryophyllene oxide, and ascaridole from essential oil of *Chenopodium ambrosioides* on mitochondria. Toxicol Appl Pharmacol 240(3): 337–347.

NabigolA, Morshedi H (2011) Evaluation of the antifungal activity of the Iranian thyme essential oils on the postharvest pathogens of strawberry fruits. Afr J Biotechnol 10: 9864–9869.

Ncube N, Afolayan SAJ, Okoh AI (2008) Assessment techniques of antimicrobial properties of natural compounds of plant origin: Current methods and future trends. Afri J Biotechnol 7: 1797–1806.

Negreiros RJZ, Salomão LCC, Perreira OL, Cecon PR, Siqueira DL (2013) Controle da antracnose na pós-colheita de bananas-"prata" com produtosalternativos aos agrotóxicos convencionais. Revista Brasil Fruti 35(1): 51–58.

Neri F, Mari M, Brigati S (2006) Control de *Penicillium expansum* by plant volatile compounds. Plant Pathol 55: 100–105.

Neri F, Mari M, Brigati S, Bertolini P (2007) Fungicidal activity of plant volatile compounds for controlling *Monilinia laxa* in stone fruit. Plant Dis 91: 30–35.

Nongmaithem N (2014) Control of post-harvest fungal diseases of guava by essential oil of *Azadirachta indica*. Indian J Hill Farm 27: 238–246.

Obagwu J, Korsten L (2003) Control of citrus green and blue molds with garlic extracts. Eur J Plant Pathol 109: 221–225.

Okigbo RN, Ogbonnaya OU (2006) Antifungal effects of two tropical plants extracts (*Ocimum gratissimum* and *Afromaomum melegueta*) on postharvest yam (*Dioscorea* spp.) rot. Afr J Biotechnol 5(9): 727–731.

Okigbo RN, Putheti R, Achusi CT (2009) Post-harvest deterioration of cassava and its control using extracts of *Azadirachta indica* and *Afromonium meleguata*. E J Chem 6(4): 1274–1280.

Onyeani CA, Amusa NA (2015) Incidence and severity of anthracnose in mango fruits and its control with plant extracts in South West Nigeria. Int J Agric Res 10(1): 33–43.

Palou L, Simlanick JL, Droby S (2008) Alternatives to conventional fungicides for the control of citrus green and blue moulds. Stewart Postharvest Rev 4: 1–16.

Parekh J, Jadeja D, Chanda S (2005) Efficacy of aqueous and methanol extracts of some medicinal plants for potential antibacterial activity. Turkish J Biol 29: 203–210.

Parthasarathy S, Rajalakshmi J, Narayanan P, Prabakar K (2016) Botanicals in eco-friendly post-harvest disease management. Innov Farm 1(3): 67–71.

Patel DS, Nath K, Patel RL, Patel S (2008) Management of foot rot of Anola caused by *Aspergillus niger*. Indian J Mycol Pl Pathol 38: 651–658.

Perdones A, Sánchez-González L, Chiralt A, Vargas M (2012) Effect of chitosan–lemon essential oil coatings on storage-keeping quality of strawberry. Postharvest Biol Technol 70: 32–41.

Perez AG, Sanz C, Olias R, Olias JM (1999) Lipoxygenase and hydroperoxide lyase activities in ripening strawberry fruits. J Agric Food Chem 47: 249–253.

Prasad K, Stadelbacher GJ (1973) Control of postharvest decay of fresh raspberries by acetaldehyde vapour. Plant Dis 57: 795–797.

Prasad K, Stadelbacher GJ (1974) Effect of acetaldehyde vapor on postharvest decay and market quality of fresh strawberries. Phytopathology 64: 948–951.

Rodov V, Ben-Tehstua S, Fang DQ, Kim JJ, Ashknazi R (1999) Preformed antifungal compounds of lemon fruit: Citral and its relation to disease resistance. J Agric Food Chem 43: 1057–1061.

Rozmer Z, Perjesi P (2016) Naturally occurring chalcones and their biological activities. Phytochem Rev 15: 87–120.

Salie F, Eagles PFK, Lens HMJ (1996) Preliminary antimicrobial screening of four South African *Asteraceae* species. J Ethnopharmacol 52: 27–33.

Salomone A, Scaritto G, Sacco A, Cabras G, Angioni A (2008) Inhibitory effects of the main compounds of oregano essential oil against some pathogenic fungi. Modern fungicide and antifungal compounds V. Deutsche Phytomediz Gesellschaft 44: 345–360.

Sánchez-Gonzalez L, Pastor C, Vargas M, Chiralt A, González-Martínez C, Chafer M (2011) Effect of hydroxypropyl methylcellulose and chitosan coatings with and without bergamot essential oil on quality and safety of cold-stored grapes. Postharvest Biol Technol 60: 57–63.

Santos NST, Aguiar AJAA, de Oliveira CEV, de Sales CV, Silva SM, Silva RS, Stamford TCM, de Souza EL (2012) Efficacy of the application of a coating composed of chitosan and *Origanum vulgare* L. essential oil to control *Rhizopus stolonifer* and *Aspergillus niger* in grapes (*Vitis labrusca* L.). Food Microbiol 32: 345–353.

Santos RP, Nunes EP, Nascimento RF, Santiago GMP, Menezes GHA, Silveira ER, Pessoa ODL (2006) Chemical composition and larvicidal activity of the essential oils of *Cordia leucomalloides* and *Cordia curassavica* from the northeast of Brazil. J Braz Chem Soc 17: 1027–1030.

Sayago JE, Ordonez RM, Kovacevich LN, Torres S, Isla MI (2012) Antifungal activity of extracts of extremophile plants from Argentine Peru to control post diseases of citrus fruit. Postharvest Biol Technol 65: 39–43.

Sayed EL, Hussian Z, Farag ES (2008) Fumigation of peach fruits with essential oils to control post-harvest decay. Res J Agric Bio Sci 4: 512–519.

Serrano M, Martinez-Romero D, Castillo S, Guille F, Valero D (2005) The use of natural antifungal compounds improves the beneficial effect of MAP in sweet cherry storage. Innov Food Sci Emerg Technol 6: 115–123.

Sharma N, Tripathi A (2006) Fungitoxicity of the essential oil of *Citrus sinensis* on postharvest pathogens. World J Microbiol Biotechnol 22(6): 587–593.

Sharma RN, Maharshi RP, Gaur RB (2009) Bio efficacy of phytoextracts in controlling postharvestblue mould rot (*Penicillium italicum)* of Kinnow (*Citrus delicosa*) fruits. J Mycol Plant Pathol 39(3): 480–483.

Siddiqui BS, Afshan F, Gulzar T, Hanif M (2004) Tetracyclic triterpenoid from the leaves of *Azadirachta indica*. Phytochemistry 65: 2363–2367.

Singh M, Jain KL (2007) Management of postharvest Fusarium rot of tomato fruits. J Mycol Plant Pathol 37(3): 449–450.

Singh MP, Jain RM (2007) Management of *Fusarium* rot of mango fruit. Hortic Adv 5: 102–112.

Singh R, Sinthi P (2015) *Cinnamomum zeylanicum* essential oil in the management of anthracnose of banana fruits. JIPBS 2(3): 290–299.

Singh YP, Geeta S (2007) Efficacy of leaf extracts and essential oils of some plant species against *Penicilluim expansum* rot of apples. Ann Plant Protec Sci 15(1): 135–139.

Sivakumar PM, Kumar TM, Doble M (2009) Antifungal activity, mechanism and QSAR studies on chalcones. Chem Biol Drug Des 74: 68–79.

Solaimani B, Ramezani S, Rahimi M, Saharkhiz MJ (2009) Biological control of postharvest disease caused by *Penicillium digitutum* and *P. italicum* on stored citrus fruits by Shiraz thyme essential oil. Adv Environ Biol 3: 249–254.

Sun X, Narciso J, Wang Z, Ference C, Bai J, Zhou K (2014) Effects of chitosan-essential oil coatings on safety and quality of fresh blueberries. J Food Sci 79: 955–960.

Suprapta DN, Arai K, Iwai H (1997) Effects of volatile compounds on arthrospore germination and mycelial growth of *Geotrichum candidum* citrus race. Mycoscience 38: 31–35.

Talibi I, Askarne L, Boubaker H, Boudyach EH, Msanda F, Saadi B, Ait Ben Oumar A (2012) Antifungal activity of Moroccan medicinal plants against citrus sour rot agent *Geotrichum candidum*. Lett Appl Microbiol 55: 155–161.

Tayel A, El-Baaz A, Salem M, El-Hadary M (2009) Potential application of pomegranate peel extract for the control of citrus green mould. J Plant Dis Plant Prot 6: 252–256.

Thakre S, Chopde N, Khobragade H, Golliwar VJ, Warahade A (2008) Effect of postharvest treatments on the storage behaviour and quality for Nagpur mandarin fruit. Orissa J Hort 36(2): 9–15.

Tiwari RKS, Chandravanshi SS, Ojha BM (2005) Efficacy of extracts of medicinal plant species on growth of *Sclerotium rolfsii* root rot in tomato. J Mycol Plant Pathol 34: 461–464.

Tripathi P, Dubey NK, Banerji R, Tiwari RKS, Chandravanshi SS, Ojha BM (2005) Efficacy of extracts of medicinal plant species on growth of *Sclerotium rolfsii* root rot in tomato. J Mycol Plant Pathol 34: 461–464.

Tripathi P, Dubey M (2003) Exploitation of natural products as an alternative strategy to control postharvest fungal rotting of fruits and vegetables. Postharvest Biol Technol 32: 235–245.

Tripathi P, Dubey NK, Pandey VB (2002) Kaempferol: The antifungal principle of *Acacia nilotica* Linn. Del J Indian Bot Soc 81: 51–54.

Tripathi P, Dubey NK, Shukla AK (2008) Use of some essential oils as post-harvest botanical fungicides in the management of grey mold of grapes caused by *Botrytis cinerea*. World J Microbiol Biotechnol 24: 39–46.

Tripathi P, Shukla A (2009) Application of essential oils for postharvest control of stem end rot of mango fruits during storage. Int J Postharvest Technol 1(4): 405–415.

Tzortzakis NG (2007) Maintaining postharvest quality of fresh produce with volatile compounds. Innov Food Sci Emerg Technol 8: 111–116.

Tzortzakis NG, Economakis CD (2007) Antifungal activity of lemongrass (*Cympopogon citratus* L.) essential oil against key postharvest pathogens. Innov Food Sci Emerg Technol 8: 253–258.

Utama IMS, Wills RBH, Ben-yehoshua S, Kuek C (2002) In vitro efficacy of plant volatiles for inhibiting the growth of fruit and vegetable decay microorganisms. J Agric Food Chem 50: 6371–6377.

Valero D, Valverde JM, Martınez-Romero D, Guillen F, Castillo S, Serrano M (2006) The combination of modified atmosphere packaging with eugenol or thymol to maintain quality, safety and functional properties of table grapes. Postharvest Biol Technol 41: 317–327.

Vaughn SF, Spencer GF, Shasha BS (1993) Volatile compounds from raspberry and strawberry fruit inhibits postharvest decay fungi. J Food Sci 58: 793–796.

Vlietinck AJ, Mvukiyumwami J 1995. Screening of hundred Rwandese Medicinal Plants for antimicrobial and antiviral properties. Journal of Ethno pharmacology. 46(1): 31–47.

Walter M, Jaspers MV, Eade K, Frampton CM, Stewart A (2001) Control of *Botrytis cinerea* in grape using thyme oil. Australas Plant Pathol 30: 21–25.

Wang CY, Wang SY, Chen C (2008) Increasing antioxidant activity and reducing decay of blueberries by essential oils. J Agric Food Chem 56: 3587–3592.

Wang CY, Wang SY, Yin JJ, Parry J, Yu LL (2007) Enhancing antioxidant, antiproliferation, and free radical scavenging activities in strawberries with essential oils. J Agric Food Chem 55: 6527–6532.

Wedge DE, Galindo JCG, Macias FA (2000) Fungicidal activity of natural and synthetic sesquiterpene lactone analogs. Phytochem 53: 747–757.

Wilson CL, Franklin JD, Otto B (1987). Fruit volatiles inhibitory to Monilinia fructicola and Botrytis cinerea. Plant Disease 71(4): 316–319.

Wilson CL, Wisniewski ME (1989) Biological control of postharvest diseases of fruits and vegetables: an emerging technology. Annu Rev Phytopathol 27: 425–441.

Win NKK, Jitareerat P Kanlayanarat S, Sangchole S (2007) Effects of Cinnamon extract ,chitosancoating, hot water treatment and their combinations on crown rot disease and quality of banana fruit. Postharvest Biol. Technol 45 (3): 333–340

Zhou H, Tao N, Jia L (2014) Antifungal activity of citral, octanal and a-terpineol against *Geotrichum citriaurantii*. Food Contr 37: 277–283.

Ziedan, EHE, Farrag ESH (2008) Fumigation of peach fruits with essential oils to control postharvest decay. Res J Agri Biol Sci 4(5): 512–515.

10

Edible Coatings for Quality and Disease Management of Fruits and Vegetables

Shruti Sethi, Alka Joshi, Swarajya Laxmi Nayak and Ram Roshan Sharma
Division of Food Science and Postharvest Technology, ICAR-Indian Agricultural Research Institute, New Delhi, India

CONTENTS

10.1 Introduction

Fruits and vegetables continue to respire and undergo metabolic changes until they are consumed. Improper handling of fruits and vegetables is one of the most significant contributors to total food production loss. Quality deterioration and pathogen attack on fresh produce may occur during harvesting, storage, handling and marketing until consumption. According to the United Nations Food and Agricultural Organization estimates, one-third of the food produced globally is lost after harvest in the postharvest handling chain (Gastavsson et al. 2011), with major losses inflicted due to diseases caused by pathogens. Incidence of physical injury, prevalence of humid conditions during storage and inherent high water activity and nutrient content of the respiring fruits and vegetables make the conditions conducive for attack by pathogens. In many fruits, ethylene plays a major role in enhancing susceptibility to postharvest diseases. *Botrytis cinerea* is the second most important fungal plant pathogen on the basis of scientific and economic importance (Dean et al. 2012). In fruits and vegetables, postharvest infection is not mainly because of bacteria but due to the presence of fungi, such as *Colletotrichum*, *Penicillium*, *Fusarium*, *Botrytis*, *Alternaria* and *Rhizopus* (Mendez and Mondino 1999). Control of these microorganisms by synthetic chemicals/fungicides has been the only commercial method that has long prevailed. Consumer demand for products free of synthetic chemicals has created a need for development of alternative solutions for pathogen control after harvest and during storage. Efforts to retard respiration rate, control senescence and minimize microbial infections and the subsequent loss in quality and decay incidence have led to development of various pre- and postharvest interventions. Various disease management strategies can be exercised on the fresh produce during storage such as bioagents, controlled atmosphere storage, ozonation, etc., as suggested by various workers (Ippolito and Nigro 2000; Teles et al. 2014; Nayak et al. 2019b). Manipulation of fruit ripening processes that can greatly retard postharvest infection may be achieved by retardation of ethylene production/action, controlled/modified atmospheres, application of hormones, bio-control agents or natural extracts (Madhav et al. 2018; Crisosto et al. 2002; Calvo-Garrido et al. 2013; Nayak et al. 2019c). One such alternative is the application of edible coatings on the fruit/vegetable surface. This chapter provides a general overview of the effect of developed edible coatings on mitigating decay and deteriorative quality changes of whole or minimally processed fruits and vegetables.

10.2 Historical Use of Edible Coatings

Edible coatings, also termed as 'environment friendly coatings' due to their biodegradability, are thin layers of food-grade edible substances, applied onto surfaces of fruits or vegetables as a substitute to the naturally occurring waxy coatings (Ansorena et al. 2018), that are safe to use (Dhall 2013). They have been used for a long time to control the migration of moisture, enhance aesthetic appeal and prolong shelf life of products. Dipping of fruits in a wax solution is one of the oldest methods, practiced since the twelfth century, especially in China, where waxing of citrus fruits was done to retard desiccation. The trend continued during the nineteenth century, wherein these edible coatings were used to prevent oxidation and rancidity of dried nuts during storage by application of sucrose coating. Commercial application of paraffin waxes as edible coatings for pears and apples started in the 1930s, followed by the use of sucrose fatty acid esters (SFAEs), Tal Pro-long, Semperfresh™, corn zein proteins in the 1980s to coat fruits and vegetables for retention of weight, colour and firmness during storage (Dhall 2013).

10.3 Application of Coatings

The application of edible coatings on the commodity surface can be accomplished by either dipping the fruit/vegetable/fresh-cut commodity in coating solution, spraying of solution, or brushing the solution on the surface of the commodity. Upon drying, the coatings form a semipermeable membrane around the surface and create a modified microclimate that forms a barrier for oxygen and moisture exchange and the movement of solutes (Tapia-Blácido et al. 2018). These coatings are edible and biodegradable in nature. Traditionally, utilization of edible coatings mainly focused on reducing water loss and retarding senescence through selective permeability to gases. However, the novel formulations of edible coatings being developed now allow the incorporation and controlled release of nutraceuticals, natural antimicrobial agents and antioxidants through different fabrication techniques (Gutiérrez 2018).

Edible coatings are classified according to the structural composition of the biopolymers used. Basic constituents of edible coatings are the hydrocolloids (polysaccharides or proteins), lipids (waxes, acetylated monoglycerides or fatty acids) and resins (Figure 10.1) with water or aqueous ethanol as solvent. The biopolymers have different film-forming mechanisms, such as covalent bonding (e.g. disulphide bonds and cross-linking), electrostatic forces and hydrophobic or ionic interactions. Polysaccharides have good film-forming ability with good mechanical properties, but they are poor barriers for water vapour transfer, whereas high hydrophobicity of lipid polymers makes them efficient barriers to moisture migration, while protein coatings provide mechanical stability. Appropriately formulated coatings can be utilized for senescence control, stable nutritional quality and safety enhancement. Composite/blended coatings contain a combination of the above biopolymers that are developed to extract the advantageous properties of each component for better functionality. To improve the mechanical and handling properties of the coatings, various other minor ingredients such as plasticizers, emulsifiers or surfactants maybe incorporated into the formulations. Plasticizers such as propylene glycol, glycerol and sorbitol help prevent brittleness, while emulsifiers/surfactants improve the emulsion stability. Better uniformity of an edible layer on cut surfaces of fruits and vegetables can be achieved by adding surfactants to the coating solution to reduce surface tension. These coating formulations can be directly used on food surfaces or they may be incorporated with food-grade additives such as antimicrobial,

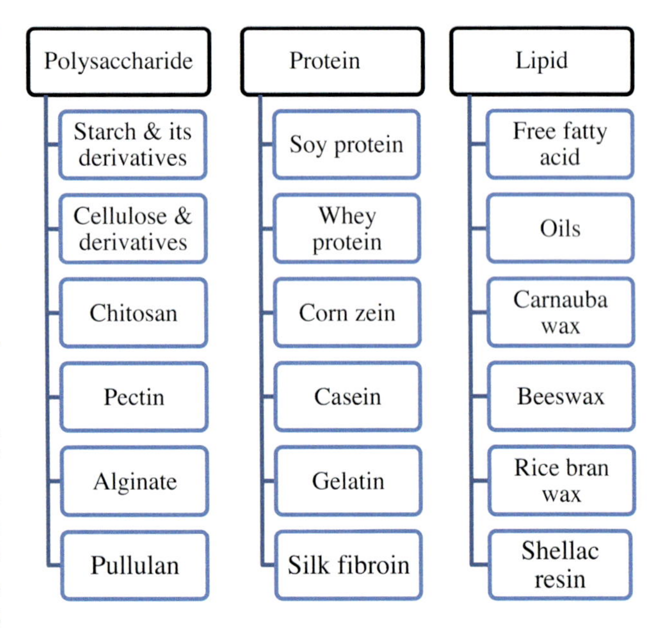

FIGURE 10.1 Classification of edible coatings.

nutraceuticals, antioxidants and flavourings. Care should be taken to include only GRAS additives and only those ingredients which are non-allergenic; e.g. a certain segment of population maybe allergic to wheat protein. Coatings maybe applied as a single- or multi-layered system to offer enhanced protection to the commodity. Also, use of materials of animal origin such as chitosan, collagen and gelatin may be avoided, if need be, owing to religious and food restrictions. Therefore, in such cases care should be taken to highlight the ingredient on the food label. Applying lipophilic films on cut produce also poses a challenge, as adhesion between the lipid layer on the wet surface is weak. This problem can be overcome by using a two-layer approach such as coating first with an alginate cover cross-linked with calcium followed by the hydrophobic acetylated monoglyceride (Wong et al. 1994).

10.4 Properties of Edible Coatings

The foremost requirement for any edible coating is that it should be constituted of GRAS substances, since it is consumed along with the commodity. The coatings should be flavourless, colourless, odourless and transparent, and should not impart any undesirable effect on the produce. Edible coatings should have good mechanical properties and gas and moisture barrier properties. Some coatings provide a gloss to the fruit/vegetable surface and contribute to improvement in the product appearance. The above mentioned characteristics are ultimately governed by the molecular makeup of the coatings. Reduced oxygen or excessive build-up of carbon dioxide in the microclimate around the fruit/vegetable as a result of the coating is to be prevented to avoid anaerobic respiration and spoilage of the commodity. Therefore, the thickness of edible coating applied on any food surface should usually be 300 μm (Pavlath and Orts 2009). Also, under high relative humidity, the coatings should be stable and should not disintegrate. The popularity of edible coatings may be further enhanced by developing economically better counterparts.

10.5 Advantages and Disadvantages of Edible Coatings

Edible coatings improve the quality of food products by retarding the deterioration processes. The application of these coatings helps to improve the physical strength of the fruit/vegetable and enhance their cosmetic appeal by imparting gloss to the surface. They can also protect food products from oxidative damage, moisture absorption/desorption, pathogen contamination and other degradative chemical reactions. Moreover, edible coatings can be incorporated with functional additives such as antioxidants, antimicrobials, colours, flavours, etc. that can further improve the shelf stability, nutritive property, sensory acceptability and microbial quality of the commodity (Nayak et al. 2019a). Nonetheless, edible coatings also have certain disadvantages. A thick layer of coating material can prohibit oxygen exchange and cause off-flavour development. Some edible coatings are hygroscopic in nature, which may result in increased microbial growth.

10.6 Role of Edible Coatings in Fruits and Vegetables

10.6.1 Weight Loss

Loss in weight postharvest is an important characteristic which affects the freshness of the produce. It results from respiration or transpiration processes. The water vapour permeability rate and thickness of the coating applied are important factors in governing the rate of weight loss. Different biopolymers show varied moisture migration rates owing to their biochemical nature. Kumar et al. (2018a) compared different coatings for maintaining quality of 'Santa Rosa' plum fruits during storage and observed that shellac-based coating could effectively retard the loss in weight of the plums due to its hydrophobicity. Similar results were also quoted by Zhou et al. (2008) in pears cv. Huanghua treated with shellac resin. The carboxymethyl cellulose (CMC) and Semperfresh™ coated fruits showed greater reduction in weight because of the higher water vapour transmission rates of these polysaccharide-based coatings. As mentioned earlier, coatings maybe applied as single layer or multi-layered. Velickova et al. (2013), while working on coated strawberry cv. Camarosa, found that composite coating consisting of beeswax-chitosan-beeswax resulted in reduced weight loss (~15–20% lower than the control) during storage at 20°C and 35–40% RH. Scanning electron micrographs of fruits coated with different edible coatings have shown filling up of the open stomatal pores on the fruit surface resulting in arresting of metabolic activity (Kumar et al. 2017b, 2018a). Coatings can also prevent surface dehydration in the case of fresh-cut produce. Composite coatings comprising hydrophobic moieties have also shown better retention of moisture from cut produce (Olivas et al. 2003; Wong et al. 1994).

10.6.2 Decay

Coatings may have varied influence on decay control on the basis of the biopolymer applied. Biopolymers provide a barrier around the produce that prevents pathogen entry. Various workers have demonstrated the impact of edible coatings in reduction of decay of the coated commodity. The stand-alone application of chitosan to control decay has been extensively studied by various workers on different horticultural crops such as strawberry, raspberry, plum and banana (Han et al. 2004; Wang and Gao 2013; Kumar et al. 2017b; Maqbool et al. 2010). A three-layer composite coating of beeswax-chitosan-beeswax was also successful in controlling fungal infection of strawberry cv. Camarosa (Velickova et al. 2013). El-Anany et al. (2009) observed a positive influence of edible coatings comprising jojoba wax, soybean gum, Arabic gum and glycerol on decay control of Anna apple. Carnauba wax (4.5%) has also demonstrated significant decrease in the brown rot incidence in plum and nectarine fruits (Gonçalves et al. 2010).

10.6.3 Firmness

Firmness of fruits is an important kinaesthetic quality parameter for selection of fresh produce. Loss of firmness is usually associated with cell wall breakdown and loss of turgor pressure. Edible coatings impart structural rigidity and preserve the firmness of the commodity by decreasing moisture loss and restricting the activity of pectin-degrading enzymes that cause softening owing to their oxygen barrier property. Coatings maybe used alone or in conjunction with other compounds to retain the firmness of fresh or minimally processed fruits/vegetables. The efficacy of coatings in firmness retention is further governed by the time of application (Kumar et al. 2018c), permeability of the coating materials (Zhou et al. 2008; Kumar et al. 2018a) and concentration of the coating solution applied (Malmiri et al. 2011; Valero et al. 2013). Commodities coated immediately after harvest tend to retain better firmness due to early retardation of metabolic activities in comparison to those that are coated after transportation. As mentioned previously, polysaccharide coatings are hydrophilic. Therefore, in comparison to lipid-based coatings they are less effective in retarding the respiratory processes that cause softening of the commodity. Also, increasing concentration of the biopolymer in the coating formulation has shown better control of turgor pressure, resulting in firmer commodity.

10.6.4 Skin Colour and Gloss

Colour is an important sensory attribute influencing consumer acceptability. Bright colour and gloss of the fresh produce that attracts the consumer can be well retained during storage by application of surface coatings. Improvement in glossiness and retardation of change of skin colour in climacteric fruits can be accomplished due to the barrier properties of the coatings. Kumar and co-workers (2017b, 2018a) have extensively studied the effect of edible coatings on plum peel colour during storage. Colour changes were delayed in the plum fruits by application of chitosan and shellac based coatings with a simultaneous inhibition of anthocyanin synthesis. Effectiveness of Semperfresh™ coating in improving luminosity in sweet cherry has been reported by Yaman and Bayoundurlc (2002). A synergistic effect of low temperature storage and concentration of coating solution on skin colour retention has been observed by Kumar et al. (2017a) and Valero et al. (2013) on coated plum fruits. Arnon et al. (2014) developed a polysaccharide-based edible bilayer coating comprising CMC and chitosan and found it to be equally as effective as the commercial polyethylene wax in enhancing citrus fruit gloss. In minimally processed apple wedges, CMC and aloe vera gel based coatings have proved to have better control of discolouration when used in conjunction with browning inhibitors rather than when used alone (Kumar et al. 2018b).

10.6.5 Respiration Rate and Ethylene Production

Studies suggest that the gaseous exchange between the fresh commodity and the external environment occurs either via diffusion through open pores or by permeation through skin of fruits (Amarante et al. 2001; Bai et al. 2002). Scanning electron micrographs of coated fruits clearly show that the edible coatings result in blocking of these open pores (Kumar et al. 2017b, 2018a). This is manifested by a reduction in the respiration rate and other metabolic activities that further suppresses the ethylene production. Kumar et al. (2018a) have demonstrated that coating of fruit immediately after harvest can effectively retard the ethylene evolution rate and the metabolic processes, which manifests in better retention of overall fruit quality. Previously, Perez-Gago et al. (2003) demonstrated that the impact of coating on the ethylene production relied on the inherent physiological quality of the produce and type of coating materials. HPMC coating resulted in decrease in C_2H_4 production of apple, while the same coating had no effect on ethylene production in plum fruits. Valero et al. (2013) reported suppression of climacteric peak in plum fruits by application of alginate coating, with higher concentration being more effective.

10.6.6 Soluble Solid Content, Titratable Acidity, Ascorbic Acid, Phenolics and Antioxidant Activity

Soluble solids content, titratable acidity, ascorbic acid, phenolics and antioxidant activity are important quality parameters related to the maturity and ripening of fruits and vegetables. Changes in these parameters in harvested produce can be controlled by the application of edible coatings that suppress the enzymatic activities and the overall metabolic processes due to the barrier properties. Breakdown of the complex starch to simple sugars during ripening results in the increase in the soluble solid content during storage. Slow increase in the soluble solid content as a result of coating has been observed by various workers (Zhou et al. 2008; Kumar et al. 2017b). Lipid-based coatings tend to be more effective in retarding the changes due to their lower gas permeability and better retardation of metabolic processes. Utilization of organic acids during continued respiratory metabolism in detached fleshy fruits results in decline of total acidity during their postharvest storage. Valero et al. (2013) have reported alginate coatings at varied concentrations (1–3%) to delay acidity loss in four plum cultivars.

Coatings have the potential to reduce the respiration rate of produce that contributes to maintaining the quality characteristics. Semperfresh™ was effective in retarding ascorbic acid loss for both ambient (30°C; 40–50% RH) and cold stored (0°C; 95–98% RH) sweet cherry fruits (Yaman and Bayoundurlc 2002). The low oxygen permeability of the edible coatings regulates the ascorbic acid oxidase and phenoloxidase enzymes that are responsible for greater retention of ascorbic acid in coated fruits. Similar observations were recorded by Zhou et al. (2008) in shellac coated pears. They also observed better retention of soluble solids and titratable acidity due to inhibition of the respiratory rates and other metabolic activities in pears during storage.

Antioxidant activity and phenolics have shown to decrease during prolonged storage. The suppression of phenylalanine ammonialyase (PAL) activity due to the coatings is responsible for the phenolic accumulation in fruits (Meng et al. 2008). Kumar et al. (2018d) and Wang and Gao (2013) have reported that irrespective of the type of coating applied, the total antioxidant activity was significantly higher in coated fruits.

10.6.7 Lipid Peroxidation/ Malondialdehyde Content

The increment in malondialdehyde (MDA) content is a measure of oxidative stress and damage of cell membranes that results from oxidation of polyunsaturated fatty acids. Edible coatings with their ability to develop a microclimate around the fruit/vegetable surface are able to inhibit the lipid peroxidation and the resultant rise in MDA content. Eum et al. (2009) and Kumar et al. (2017b) observed a slow rate of increase in production of malondialdehyde in Versasheen and chitosan coated plum fruits, respectively, during storage as compared to control. Chitosan + nano-silicon dioxide edible coating was found to delay the increase of MDA content in jujube (Yu et al., 2012) indicating the reduced free radical to the cytoplasmic membrane.

10.6.8 Enzyme Activity

Enzymes are responsible for the degradative changes occurring in respiring commodities. Coatings form a physical barrier around the commodity, leading to reduction of surrounding oxygen and subsequently reduction in the enzyme activity (Velickova et al. 2013). Chitosan at 2% has shown to retard the PME activity for coated fruits by about 44% in comparison to non-coated fruits (Kumar et al. 2017b). In a similar study, Kumar et al. (2018a) observed an 82–86% reduction in PME activity in plums coated with three different coatings, with maximum reduction obtained with shellac-based coating applied to fruits on the farm rather than after transportation. This suggests that prompt application of surface coatings on fresh produce after harvest can help curb the metabolic and enzyme activities which further helps in effectively retaining the quality attributes.

10.7 Active Edible Coatings Amended with Functional Additives

Edible coatings can be formulated using different biopolymers and amended with functional components such as texturizers, flavourings, colourants, anti-browning agents, antioxidants, nutraceuticals and probiotics. Calcium dips have been used to impart freshness and prolong shelf-life of whole or fresh-cut produce, while organic acids such as citric, ascorbic and oxalic acid have been incorporated to control excessive browning, especially in fresh-cut commodities. Likewise, antioxidants such as ferulic acid, N-acetylcysteine and glutathione have been used in edible coatings to prevent oxidation reactions in fresh-cut food matrices. Incorporation of nutraceuticals and probiotics in surface coatings can improve the nutritive value and confer health benefits to the coated produce.

10.8 Antimicrobial Edible Coatings

Chitosan and aloe vera are the only biopolymer-based edible coatings that inherently possess antimicrobial activity against a broad range of microorganisms.

10.8.1 Chitosan

Chitosan is known to inhibit postharvest decay of fruits and vegetables. It is produced commercially after deacetylation of chitin obtained from the exoskeletons of crustaceans such as lobsters, crabs and shrimps. Its non-toxic nature, edibility, biodegradability, ability to form films and excellent potential for preservation of fresh produce make it a popular candidate for use as an edible coating. It exhibits antimicrobial activity against a broad spectrum of spoilage microbes including fungi, bacteria and yeasts. It has been widely used for postharvest decay control in various horticultural crops including citrus, plums, apples, mango, berries, lettuce, carrots and tomatoes. In 2014, the EU regulation 2014/563 approved chitosan chloride as a plant protection product, to be used in plant disease management. Reduced pathogen growth in strawberry along with enhanced expression of defence-related enzymes was observed upon treatment with chitosan by Landi et al. (2014). Jin et al. (2016) have concluded that sanitizer wash, chitosan coating formulation containing acidulants (lactic, levulinic and acetic) along with modified atmospheres can be more effective to attain better log reduction in ginseng roots than their individual application, due to additive effects.

10.8.2 Aloe Vera

Aqueous extracts obtained from the leaves of the aloe vera plant can be used to coat whole or fresh-cut horticultural products. Aloe vera coating is basically used for arresting physiological processes of fruits after harvest, but it is also known to possess antimicrobial activity. In contrast to chitosan coatings that are of animal origin, the gel obtained from leaves of aloe vera plants can be accepted and exploited more for its antimicrobial potential. Saks and Barkai-Golan (1995) demonstrated significant inhibitory effect of aloe vera gel against *P. digitatum* under in vitro conditions and against green mould artificially inoculated on grapefruits. Jhalegar et al. (2014) observed reduced incidence of blue and green moulds infection on Kinnow mandarins by application of an aloe vera-based coating with a positive influence on fruit quality attributes under cold storage. A 50% reduction in *Botrytis cinerea* was observed in blueberry coated with composite edible coating comprising chitosan (0.5%), glycerol (0.5%), Tween 80 (0.1%) and aloe vera (0.5%) stored at 50°C for 10 days (Vieira et al. 2016). Kumar et al. (2018c) have shown aloe vera gel mixed with anti-browning agents to possess pronounced antimicrobial properties in comparison to aloe vera gel alone when applied on minimally processed apple wedges.

10.9 Edible Coatings Amended with Antimicrobial Compounds for Disease Management

Edible coatings can be employed as matrices to impregnate antimicrobial compounds to improve their effectiveness on specific microorganisms. To increase the efficacy of other edible coatings for antimicrobial effect, various antimicrobial compounds can be incorporated into these edible coatings. These can be categorized into three groups:

- Generally recognized as safe (GRAS) compounds: Acids and their salts (e.g. benzoates, propionates, sorbates, carbonates), methyl and ethyl parabens and their salts.
- Natural plant extracts (e.g. lemongrass, rosemary, cinnamon, citral, garlic, vanillin, cinnamaldehyde, grape seed extracts) and essential oils.
- Microbial antagonists/biological control agents (e.g. bacteria, yeasts).

These active edible coatings usually cause direct inhibition of pathogens and/or result in induction of defence mechanisms in the host tissues (Landi et al. 2014). Since these additives are going to be incorporated into coatings that are edible, they should be approved under food regulation, should be GRAS substances and should have no toxicological effects upon consumption. Table 10.1 shows examples of antimicrobial edible coatings developed against major disease-causing microorganisms of fruits and vegetables.

10.9.1 Plant Extracts

Plant extracts exhibit excellent antimicrobial properties and their incorporation into coatings can be a natural substitute for synthetic antibiotics. Several reports have shown the ability of composite films containing these natural extracts to control foodborne pathogens. Aitboulahsen et al. (2018) have demonstrated that gelatin-based edible coating incorporated with *Mentha* essential oil at 1% significantly inhibited total mesophilic microflora, moulds and yeasts in strawberry. The presence of oxygenated monoterpenes of mentha oil contributes to the antifungal and antibacterial properties of the oil (Ait-Ouazzou et al. 2012). Similarly, extracts of pomegranate seeds have shown good antibacterial action against the bacterial pathogens *Bacillus cereus* and *Staphylococcus aureus* when used in conjunction with mint extract in chitosan-PVA films (Kanatt et al. 2012). Coatings formulated with 2% sodium alginate and 1% grape seed extract exhibit excellent antifungal activity for preserving the grape fruit quality up to 15 days in cold storage (Aloui et al. 2014). A composite coating composed of sodium alginate + glycerol + sunflower oil with lemongrass can extend the shelf life of fresh-cut pineapple for 16 days at 10°C, with less than 10^6 CFU/g of sample, which is considered the acceptable limit as per the Institute of Food Science and Technology (Bierhals et al. 2011).

10.9.2 Antimicrobial Nanoparticles

The antimicrobial effect of nanoparticles may be due to the interaction between the positively charged polymer or metal ions and the charged cell membrane (Xing et al. 2019). Among inorganic nanoparticles, photocatalytic nature of titanium dioxide is responsible for its antimicrobial effect. Xu et al. (2017) observed that the composite coating incorporating graphene oxide, chitosan and titanium dioxide nanoparticles in the ratio 1:20:4 exhibited maximum inhibition of *A. niger* and *B. subtilis*, which induce cell membrane breakdown. Andrade et al. (2014) also showed that the ß-cyclodextrin-coated Ag NPs reduced more than 99% *Escherichia coli* and *Pseudomonas aeruginosa* population. Shankar et al. (2015) suggested the enhancement of antimicrobial property of chitosan film against *E.coli* and *Listeria monocytogenes* by incorporation of sulphur nanoparticles in horticulture matrices.

TABLE 10.1

Antimicrobial Edible Coatings for Target Microorganisms

S. No.	Disease	Causal Organism	Fruit/Vegetable	Antimicrobial Coating	Effectiveness Achieved	Reference
1.	Anthracnose	*Colletotrichum gloeosporioides*	Mango cv. Namdokmai Sithong	*Zingiber cassumunar* essential oil +HPMC	Decreased disease severity	Sothornvit and Klangmuang (2014)
2.	Wilt	*Fusarium oxysporum*	Cherry tomato	Aloe vera + pregelatinized corn starch + glycerol	Mycelium growth inhibition was found 65.16% with almost 4 times reduction in growth rate than control	Ortega-Toro et al. (2017)
3.	Rot	*Penicillium digitatum* and *P. italicum*	Lemon cv. Fino	Wax + thymol	68% and 47% fungal growth inhibition against *P. italicum* and *P. digitatum*, respectively, at 100 µL/L addition	Pérez-Alfonso et al. (2012)
4.	Alternaria rot	*Alternaria alternata*	Avocado (cv. 'Hass' and 'Gem)	CMC+ ethanolic extract of moringa leaves	42.9% inhibition against *A. alternata*	Tesfay et al. (2017)
5.	Grey mould	*Botrytis cinerea*	Kiwifruit	Chitosan hydrochloride	100% inhibition under in vitro conditions	Fortunati et al. (2017)
6.	Soft rot	*Rhizopus stolonifer*	Tomato	Chitosan+ bee wax+ lime essential oil	100% inhibition under in vitro trials	Ramos- García et al. (2012)

10.9.3 Bacteriocins

Bacteriocins are proteinaceous biopreservatives having a broad antimicrobial spectrum (Galvez et al. 2007). Acceptability of bacteriocins such as nisin and natamycin in preservation is well documented. Bacteriocins have the ability to penetrate the bacterial cell causing leakage of cytoplasmic contents and disturbing the membrane potential. Along with pomegranate and grape seed extract they have been used for strawberry. All combinations with 1.5% chitosan were protective against yeast and mould growth; however, seed extracts were effective against aerobic mesophilic bacteria at a 1% level along with 1.5% chitosan (Duran et al. 2016). A 4-log reduction in microbial load was observed by Narsaiah et al. (2015) in minimally processed papaya coated with 2% alginate and 20% bacteriocins. The coated papaya could be stored well up to 3 weeks under low temperature.

10.9.4 Bio-control Agents

Wickerhamomyces anomalus yeast, having Qualified Presumption of Safety (QPS) status by the European Food Safety Authority (EFSA) is a well-known biological control agent against different postharvest phytopathogenic fungi. Sodium alginate and locust bean gum coatings enriched with *W. anomalus* yeast were effective for exercising >73% control of green mould caused by *Penicillium digitatum* in inoculated fruits of 'Valencia' oranges (Aloui et al. 2015).

10.9.5 Enzymes

The lactoperoxidase enzyme system has a broad antimicrobial spectrum showing bactericidal (Gram negative) and a bacteriostatic effect (Gram positive) (Seacheol et al. 2005). Its efficiency against virus and yeast has also been documented. Lactoperoxidase generates hypothiocyanite ($OSCN^-$) and hypothiocyanate acid (HOSCN), both antimicrobial compounds that oxidize the sulfhydryl group of microbial enzymes (Martínez-Camacho et al. 2010). Edible coating based on chitosan and 1% lactoperosidase was found to be effective in controlling disease incidence in 'Kent' mangoes (Cisse et al. 2015).

10.9.6 Silk Fibroin

Silk fibroin, a structural beta-sheet protein obtained from silkworms, can also be used to modulate gas diffusion to retain freshness of the produce. It has successfully been tried for enhancement of storage life of strawberry and banana. A multi-layered coating can further increase the beta-sheet content and can possibly down-regulate microbial growth (Marelli et al. 2016).

10.9.7 Sorbic Acid and Potassium Benzoate

The preservative action of class II preservatives, sorbates and potassium benzoate is well documented. They are easily soluble and can be incorporated into the coating matrix. Inclusion of these antimicrobials into chitosan-based edible coating (chitosan 1.5 g/100 g + sodium benzoate 0.05 g/100 g + potassium sorbate 0.05 g/100 g + glycerol 0.5 g/100 g) has been shown to extend the shelf life of strawberries up to 15 days at 4°C (Treviño-Garza et al. 2015).

10.10 Regulatory Status

As their name suggests, edible coatings form an integral part of the commodity and are ingested along with the food. Thus, their formulation should conform to food regulations, and all ingredients should be food-grade, toxicologically safe and used within prescribed limits. In accordance with the European Directive (1998), the permitted ingredients that can be incorporated into edible coating formulations are karaya and Arabic gum, pectins, shellac, carnauba wax, candelilla wax, beeswax, lecithin, polysorbates and fatty acids and their salts. The FDA has also approved additives such as polydextrose, morpholine, SFAEs, sorbitan monostearate, castor oil and cocoa butter to be included in protective coatings applied to fresh horticultural produce (Vargas et al. 2008). Further, the premises/facility should strictly comply with good manufacturing practices. Food Safety and Standards Authority of India (FSSAI) permits the use of beeswax (white and yellow) or carnauba wax or shellac wax at levels not exceeding good manufacturing practices to coat fresh fruits under proper label declaration as provided in Regulation 2.4.5 (44) of FSS (Packaging and Labelling) regulations (FSSAI 2011). Several workers (Sánchez-González et al. 2011; Divya et al. 2018) have reported oral toxicity, cytotoxicity and skin allergies of certain plant extracts and nanoparticle formulations of edible coatings. Hence, utmost precaution needs to be taken before formulating a coating that will be consumed along with the commodity.

REFERENCES

Aitboulahsen M, Zantar S, Laglaoui A, Chairi H, Arakrak A, Bakkali M, Zerrouk MH (2018) Gelatin-based edible coating combined with *Mentha pulegium* essential oil as bioactive packaging for strawberries. J Food Quality 1–7. https://doi.org/10.1155/2018/8408915.

Ait-Ouazzou A, Lorán S, Arakrak A, Laglaoui A, Rota C, Herrera A, Pagán R, Conchello P (2012) Evaluation of the chemical composition and antimicrobial activity of *Mentha pulegium*, *Juniperus phoenicea*, and *Cyperus longus* essential oils from Morocco. Food Res Inter 45:313–319.

Aloui H, Khwaldia K, Sanchez-González L, Muneret L, Jeandel C, Hamdi M, Desobry S (2014) Alginate coatings containing grapefruit essential oil or grapefruit seed extract for grapes preservation. Int J Food Sci Technol 49: 952–959.

Aloui H, Licciardello F, Khwaldia K, Hamdi M, Restuccia C (2015) Physical properties and antifungal activity of bioactive films containing *Wickerhamomyces anomalus* killer yeast and their application for preservation of oranges and control of postharvest green mold caused by *Penicillium digitatum*. Int J Food Microbiol 200: 22–30.

Amarante C, Banks NH, Ganesh S (2001) Relationship between character of skin cover of coated pears and permeance to water vapour and gases. Postharvest Biol Technol 21:291–301.

Andrade PF, de Faria AF, da Silva DS, Bonacin JA, Gonçalves MDC (2014) Structural and morphological investigations of β-cyclodextrin-coated silver nanoparticles. Colloids Surf B 118: 289–297.

Ansorena MR, Pereda M, Marcovich NE (2018) Edible films. In: Gutiérrez TJ (Ed.). Polymers for food applications. Cham: Springer. pp. 5–24.

Arnon H, Zaitsev Y, Porat R, Poverenov E (2014) Effects of carboxymethyl cellulose and chitosan bilayer edible coating on postharvest quality of citrus fruit. Postharvest Biol Technol 87:21–26.

Bai JH, Baldwin EA, Hagenmaeir RH (2002) Alternative to shellac coatings provides comparable gloss, internal gas modification and quality for 'delicious' apple fruit. Hortic Sci 37:559–563.

Bierhals VS, Chiumarelli M, Hubinger MD (2011) Effect of cassava starch coating on quality and shelf life of fresh-cut pineapple (*Ananas comosus* L Merril cv "P'erola"). J Food Sci 76: 62–72.

Calvo-Garrido C, Viñas I, Elmer PAG, Usall J, Teixidò N (2013) Suppression of *Botrytis cinerea* on nectrotic grapevine tissues by early season applications of natural products and bio-control agents. Pest Manag Sci 70: 595–602.

Cisse M, Polidori J, Montet D, Loiseau G, Ducamp-Collin MN (2015) Preservation of mango quality by using functional chitosan-lactoperoxidase systems coatings. Postharvest Biol Technol 101:10–14.

Crisosto CH, Garner D, Crisosto G (2002) Carbon dioxide-enriched atmospheres during cold storage limit losses from Botrytis but accelerate rachis browning of 'Redglobe' table grapes. Postharvest Biol Technol 26: 181–189.

Dean R, van Kan JAL, Pretorius ZA, Hammond-Kosack KE, Di Pietro A, Spanu PD, Rudd JJ, Dickman M, Kahmann R, Ellis J, Foster GD (2012) The top 10 fungal pathogens in molecular plant pathology. Mol Plant Pathol 13: 414–430.

Dhall RK (2013) Advances in edible coatings for fresh fruits and vegetables: A review. Crit Rev Food Sci Nutr 53: 435–450.

Divya K, Smitha V, Jisha MS (2018) Antifungal, antioxidant and cytotoxic activities of chitosan nanoparticles and its use as an edible coating on vegetables. Int J Biol Macromol 114: 572–577.

Duran M, Aday MS, Zorba NND, Temizkan R, Buyukcan MB, Caner C (2016) Potential of antimicrobial active packaging containing natamycin, nisin, pomegranate and grape seed extract in chitosan coating to extend shelf-life of fresh strawberry. Food Bioprod Proc 98:354–363.

El-Anany AM, Hassan GFA, Ali FMR (2009) Effect of edible coatings on the shelf life and quality of 'Anna' apple (*Malus domestica* Borkh.) during cold storage. J Food Technol 7: 5–11.

Eum HL, Hwang DK, Linke M, Lee SK (2009) Influence of edible coating on quality of plum (*Prunus salicina* Lindl. cv. 'Sapphire'). Euro Food Res Technol 29: 427–434.

European Directive. European Parliament and Council Directive N 98/72/EC. (1998). On food additive other than colors and sweeteners. Available from: http://ec.europa.eu/food/fs/sfp/additflavor/flav11en.pdf.

Fortunati E, Giovanale G, Luzi F, Mazzaglia A, Kenny J, Torre L, Balestra G (2017) Effective postharvest preservation of kiwifruit and romaine lettuce with a chitosan hydrochloride coating. Coatings 7: 196.

FSSAI (2011) Food Safety and Standards Authority of India, Ministry of Health and Family Welfare, The Gazette of India: Extraordinary. Part III: Sec. 4.

Galvez A, Abriouel H, Lopez RL, Omar NP (2007) Bacteriocin based strategies for biopreservation. Int J Food Microbiol 120: 51–70.

Gastavsson J, Cederberg C, Sonesson U (2011) Global Food Losses and Food Waste. Food and Agriculture Organization (FAO) of the United Nations, Rome.

Gonçalves FP, Martins MC, Silva JS Jr, Lourenço SA, Amorim L (2010) Postharvest control of brown rot and *Rhizopus* rot in plums and nectarines using carnauba wax. Postharvest Biol Technol 58: 211–217.

Gutiérrez TJ (2018) Active and intelligent films made from starchy sources/blackberry pulp. J Polym Environ 26: 2374–2391.

Han C, Zhao Y, Leonard SW, Traber MG (2004). Edible coatings to improve storability and enhance nutritional value of fresh and frozen strawberries (*Fragaria* × *ananassa*) and raspberries (*Rubus ideaus*). Postharvest Biol Technol 33: 67–78.

Ippolito A, Nigro F (2000) Impact of pre-harvest application of biological control agents on postharvest diseases of fresh fruits and vegetables. Crop Prot 19: 715–723.

Jhalegar J, Sharma RR, Singh D (2014) Antifungal efficacy of botanicals against major postharvest pathogens of Kinnow mandarin and their use to maintain postharvest quality. Fruits 69: 223–237.

Jin TZ, Huang M, Niemira BA, Cheng L (2016) Shelf life extension of fresh ginseng roots using sanitiser washing, edible antimicrobial coating and modified atmosphere packaging. Int J Food Sci Technol 51: 2132–2139.

Kanatt SR, Rao MS, Chawla SP, Sharma A (2012) Active chitosan–polyvinyl alcohol films with natural extracts. Food Hydrocoll 29: 290–297.

Kumar P, Sethi S, Sharma RR (2017a) Combined effect of edible coatings and low temperature on plum fruit quality. Inter J Curr Microbiol Appl Sci 6: 4210–4218.

Kumar P, Sethi S, Sharma RR (2018b) Inhibition of browning in fresh-cut apple wedges through edible coatings and anti-browning agents. Indian J Hortic 75: 517–522.

Kumar P, Sethi S, Sharma RR, Singh S, Varghese E (2018c) Improving the shelf life of fresh-cut 'Royal Delicious' apple with edible coatings and anti-browning agents. J Food Sci Technol 55: 3767–3778.

Kumar P, Sethi S, Sharma RR, Srivastav M, Singh D, Varghese E (2018d) Edible coatings influence the cold-storage life and quality of 'Santa Rosa' plum (*Prunus salicina* Lindell). J Food Sci Technol. https://doi.org/10.1007/s13197-018-3130-1

Kumar P, Sethi S, Sharma RR, Verghese E (2018a) Influence of edible coatings on physiological and biochemical attributes of Japanese plum (*Prunus salicina* Lindell) cv. Santa Rosa Fruits 73 (1): 31–38.

Kumar P, Sethi S, Sharma RR, Srivastav M, Varghese E (2017b) Effect of chitosan coating on postharvest life and quality of plum during storage at low temperature. Sci Hortic 226: 104–109.

Landi L, Feliziani E, Romanazzi G (2014) Expression of defense genes in strawberry fruit treated with different resistance inducers. J Agric Food Chem 62: 3047–3056.

Madhav JV, Sethi S, Sharma RR, Nagaraja A (2018) Impact of salicylic acid treatments on storage quality of guava fruits cv. Lalit during storage Int J Curr Microbiol App Sci 7(9): 2390–2397.

Malmiri J, Osman A, Tan CP (2011) Development of an edible coating based on chitosan-glycerol to delay 'Berangan' banana (*Musa sapientum* cv. Berangan) ripening process. Int Food Res J 18: 989–997.

Maqbool M, Ali A, Ramachandran S, Smith DR, Alderson PG (2010) Control of postharvest anthracnose of banana using a new edible composite coating. Crop Protec 29 (10): 1136–1141.

Marelli B, Brenckle MA, Kaplan DL, Omenetto FG (2016) Silk fibroin as edible coating for perishable food preservation. Sci Reports 6. https://doi.org/10.1038/srep25263.

Martínez-Camacho AP, Cortez-Rocha MO, Ezquerra-Brauer JM, GracianoVerdugo AZ, Rodriguez-Félix F, Castillo-Ortega MM, Yépiz-Gómez MS, Plascencia-Jatomea M (2010) Chitosan composite films: thermal, structural, mechanical and antifungal properties. Carbohydrate Polym 82: 305–315.

Mendez S, Mondino P (1999). Control biologico postcosecha en Uruguay. Hortic Int 7: 29–36.

Meng X, Li B, Liu J, Tian S (2008) Physiological responses and quality attributes of table grape fruits to chitosan pre-harvest spray and postharvest coating during storage. Food Chem 106:501–508.

Narsaiah K, Wilson RA, Gokul K, Mandge HM, Jha SN, Bhadwal S, Anurag RK, Malik RK, Vig S (2015) Effect of bacteriocin-incorporated alginate coating on shelf-life of minimally processed papaya (*Carica papaya* L.). Postharvest Biol Technol 100: 212–218.

Nayak SL, Sethi S, Sharma RR, Prajapati U (2019a) Active edible coatings for fresh fruits and vegetables. In: Gutierrez TG (Ed.). Polymers for agri-food applications. Springer Nature, Switzerland, pp. 417–432.

Nayak SL, Sethi S, Sharma RR, Sharma RM, Singh S, Singh D (2019b) Aqueous ozone controls decay and maintains quality attributes of strawberry (*Fragaria* × *ananassa* Duch.) J Food Sci Technol. https://doi.org/10.1007/s13197-019-04063-3.

Nayak SL, Sethi S, Sharma RR, Singh D, Singh S (2019c) Improved control on decay and postharvest quality deterioration of strawberry (*Fragaria* x *ananassa* Duch.) by microbial antagonists Indian J Hortic 76: 502–507.

Olivas GI, Rodriguez JJ, Barbosa-Canovas GV (2003) Edible coatings composed of methylcellulose, stearic acid, and additives to preserve quality of pear wedges. J Food Proc Preserv 27: 299–320.

Ortega-Toro R, Collazo-Bigliardi S, Roselló J, Santamarina P, Chiralt A (2017) Antifungal starch-based edible films containing *Aloe vera*. Food Hydrocolloids 72:1–10.

Pavlath AE, Orts W (2009) Edible films and coatings: Why, what, and how? In: Embuscado ME, Huber KC (Eds.). Edible films and coatings for food applications. Springer Science + Business Media. https://doi.org/10.1007/978-0-387-92824-1_1.

Pérez-Alfonso CO, Martínez-Romero D, Zapata PJ, Serrano M, Valero D, Castillo S (2012) The effects of essential oils carvacrol and thymol on growth of *Penicillium digitatum* and *P. italicum* involved in lemon decay. Int J Food Microbiol 158:101–106.

Perez-Gago MB, Rojas C, Del Rio MA (2003) Effect of hydroxypropyl methylcellulose-lipid edible composite coatings on plum (cv. Autumn Giant) quality during storage. J Food Sci 68:879–883.

Ramos-García M, Bosquez-Molina E, Hernández-Romano J, Zavala G, Eduardo Terrés, Tejacal IA, Barrera-Necha L, Hernández-López M, Bautista-Baños S (2012) Use of chitosan-based edible coatings in combination with other natural compounds, to control Rhizopus stolonifer and *Escherichia coli* DH5α in fresh tomatoes. Crop Protec 38:1–6.

Saks Y, Barkai-Golan R (1995) *Aloe vera* gel activity against plant pathogenic fungi. Postharv Biol Technol 6: 159–165.

Sánchez-González L, Vargas M, González-Martínez C, Chiralt A, Cháfer M (2011) Use of essential oils in bioactive edible coatings: A review. Food Eng Rev 3: 1–16.

Seacheol M, Harris LJ, Krochta JM (2005) Antimicrobial effects of lactoferrin, lysozyme and the lactoperoxidase system and edible whey protein films incorporating the lactoperoxidase system against *Salmonella enterica* and *Escherichia coli* O157:H7. J Food Sci 70: 332–338.

Shankar S, Teng X, Li G, Rhim JW (2015) Preparation, characterization, and antimicrobial activity of gelatin/ZnO nanocomposite films. Food Hydrocoll 45: 264–271.

Sothornvit R, Klangmuang P (2014) Active edible coating to maintain the quality of fresh mango. In: V International Conference Postharvest Unlimited 1079, pp. 473–480.

Tapia-Blácido DR, Maniglia BC, Tosi MM (2018) Transport phenomena in edible films. In: Gutiérrez TJ (Ed.). Polymers for food applications, Springer, Cham, Switzerland, pp. 149–19.

Teles CS, Benedetti BC, Gubler WD, Crisosto CH (2014) Prestorage application of high carbon dioxide combined with controlled atmosphere storage as a dual approach to control *Botrytis cinerea* in organic 'Flame Seedless' and 'Crimson Seedless' table grapes. Postharvest Biol Technol 89: 32–39.

Tesfay SZ, Magwaza LS, Mbili N, Mditshwa A (2017) Carboxyl methylcellulose (CMC) containing moringa plant extracts as new postharvest organic edible coating for Avocado (Persea americana Mill.) fruit. Sci Hortic 226: 201–207.

Treviño-Garza MZ, García S, del Socorro Flores-González M, Arévalo-Niño K (2015) Edible active coatings based on pectin, pullulan, and chitosan increase quality and shelf life of strawberries (*Fragaria ananassa*). J Food Sci 80: 1823–1830.

Valero D, Mula-diaz HM, Zapata PJ (2013) Effect of alginate edible coating on preserving fruit quality in four plum cultivars during postharvest storage. Postharvest Biol Technol 77:1–6.

Vargas M, Pastor C, Chiralt A, McClements DJ, Gonzalez-Martinez C (2008) Recent advances in edible coatings for fresh and minimally processed fruits. Critical Rev Food Sci Nutr 48: 496–511.

Velickova E, Winkelhausen E, Kuzmanova S, Alves BD (2013) Impact of chitosan-beeswax edible coatings on the quality of fresh strawberries (*Fragaria ananassa* cv *Camarosa*) under commercial storage conditions. Food Sci Technol 52: 80–92.

Vieira JM, Flores-Lopez ML, de Rodrıguez DJ, Sousa MC, Vicente AA, Martins JT (2016) Effect of chitosan–aloe vera coating on postharvest quality of blueberry (*Vaccinium corymbosum*) fruit. Postharvest Biol Technol 116: 88–97.

Wang SY, Gao H (2013) Effect of chitosan-based edible coating on antioxidants, antioxidant enzyme system, and postharvest fruit quality of strawberries (*Fragaria* x *aranassa* Duch.). Food Sci Technol 52: 71–79.

Wong DWS, Tillin, SJ, Hudson JS, Pavlath AE (1994) Gas exchange in cut apples with bilayer coatings. J Agric Food Chem 42: 2278–2285.

Xing Y, Li W, Wang Q, Li X, Xu Q, Guo X, Bi X, Liu X, Shui Y, Lin H, Yang H (2019) Antimicrobial nanoparticles incorporated in edible coatings and films for the preservation of fruits and vegetables. Molecules 24:1695.

Xu W, Xie W, Huang X, Chen X, Huang N, Wang X, Liu J (2017) The graphene oxide and chitosan biopolymer loads TiO_2 for antibacterial and preservative research. Food Chem 221: 267–277.

Yaman O, Bayoundurlc L (2002) Effect of an edible coating and cold storage on shelf life and quality of cherries. Lebensm Wiss Technol 35: 146–150.

Yu Y, Zhang S, Ren Y, Li H, Zhang X, Di J (2012) Jujube preservation using chitosan film with nano-silicon dioxide. J Food Eng 113: 408–414.

Zhou R, Mo Y, Li Y (2008) Quality and internal characteristics of Huanghua pears (*Pyrus pyrifolia* Nakai, cv. Huanhhua) treated with different kinds of coating during storage. Postharvest Biol Technol 49:171–179.

11

Mycotoxins and Their Impacts on Human, Plant and Animal Health

Deeba Kamil, Shiv Pratap Choudhary, Rubin Debbarma, T. Prameeladevi and R. Sudeep Toppo
Division of Plant Pathology, ICAR-Indian Agricultural Research Institute, New Delhi, India

CONTENTS

11.1 Introduction

Many fungal species produce toxic chemical compounds known as mycotoxins, which are hazardous to human beings and animals and can cause disease and death. These are low molecular weight metabolites largely produced by Ascomycetous group of fungi such as *Aspergillus*, *Fusarium* and *Penicillium* (Liew and Redzwan 2018). When the toxic fungus grows on animals and produces disease it is called mycosis, while toxic fungus, when it comes in contact by dietary, respiratory, dermal and other exposures and produces disease, is called mycotoxicosis. Mycotoxins that make anoxious impact on human and animal health are known as mycotoxicosis, and the study of mycotoxicosis is called mycotoxicology (Bennett and Klich 2003). Some fungal species that produce mycotoxins which inhibit the growth of bacteria (like penicillin) are called antibiotics. The term mycotoxin was coined in 1962 as the result of Turkey X disease in which approximately 100,000 turkey poults died in England due to mycotoxins (Blount 1961; Forgacs 1962; Maggon et al. 1977). During the study of Turkey X disease, scientists observed that peanut seeds were also contaminated with a flatoxins secreted by fungus *Aspergillus flavus*, and researchers found that the secondary metabolites produced by many moulds might be poisonous (Richard 2007). Later, nearly 400 mycotoxins was identified; among them a flatoxins, ochratoxins, zearalenone, fumonisins and trichothecenes have great significance with respect to human health (Ates et al. 2013). Mycotoxins produced by fungus are not all toxic; the toxicity primarily depends on the quantity and period of dosage, type of mycotoxin and the physiological and nutritional status.

On the basis of food toxicity, mycotoxins are grouped in five major groups: aflatoxin, ochratoxin, fumonisin, deoxynivalenol/nivalenol and zearalenone. Based on mycotoxins found in food, the fungi have been divided in two groups: field fungi, which attack the crop and fungal mycotoxins that attack before harvest and storage fungi, which occur only after harvest. Mycotoxins can be produced in favourable conditions such as high temperature, high moisture content and unhygienic conditions during transportation and storage. These lethal mycotoxins generally occur in animal feeds, crops and animal products. The highest level of mycotoxins can be found in concentrated animal feed materials. The fungi contaminate another host such as plants and animals after establishment of infection and mycotoxins in their existing host. They transfer to the other growing crop by asexual spores, i.e. conidia, through wind or insects. Due to their harmful effects on plants and animals, it is necessary to remove or manage the development and production of mycotoxins. Many strategies have been developed by scientists to manage mycotoxins production. Methodologies include breeding resistant varieties, chemical control and cultural control. Resistant plants provide an effective and environmentally friendly approach to manage mycotoxin and fungi which produce mycotoxin.

11.2 Impact of Mycotoxins on Agriculture

Agriculture is the backbone of Indian economy because 60–70% people depend on agriculture for their basic needs or employment. Mostly every crop is infected with numerous types of diseases which are spread by fungi, bacteria and viruses. Nearly 25% of crops are affected with fungi worldwide. Mycotoxins are produced by fungi after infection on crops and most cereal crops are infected with fungal infections (Table 11.1). Maize is one of the highly susceptible crops, whereas rice is least susceptible to mycotoxins. The fungi associated with crops reduce the quality and quantity and also spread different types of diseases to human and animals. All these crops are main food and feed crops for human and animals; therefore, it is very necessary to study about the different mycotoxins and their effects on crops (Table 11.1).

11.3 Major Mycotoxins

Bennett (1987) defined that mycotoxins are produced by fungi naturally when introduced in low concentrations into animals and plants that induce a toxic response. Mycotoxins are mainly toxic secondary metabolites which directly or indirectly affect

TABLE 11.1

Important Mycotoxins and Their Effects on Plants

Mycotoxin	Range of Doses	Affected Plants	Effects on Plants
4-deoxynivalenol (DON)	2–50 μgmL^{-1}	Wheat, maize	Root and shoot inhibition, plantlet growth reduction
Aflatoxin	0.83–10 μgmL^{-1}	Barley, garden cress, maize, sorghum, wheat	Germination and seedling viability
Citrinin (CTN)	10^{-3} to 10^{-4} M	Bean, cotton, sorghum	Wilting symptoms appear
Cytochalasin (CB)	10^{-3} M	*Bromus tectorum*	Reduction in coleoptile elongation
Diacetoxyscirpenol (DAS)	1–10 $\mu g{\cdot}mL^{-1}$	Barley, wheat, sorghum	Reduction in seedling viability
Fumonisins (FB1)	1–1000 μM	Soya bean, maize, tomato, jimsonweed	Necrosis, wilting, inhibition in radicle elongation, height and biomass reduction
Fusaric acid (FA)	10–5 M	Tomato, *Arabidopsis thaliana*	Wilting symptoms, noxious effects like modification in cell growth, mitochondrial activity and membrane permeability
Moniliformin (MON)	10–100 μgmL^{-1}	Wheat, maize, duckweed	Leaf mass and plant growth decreased
Nivalenol (NIV)	48–100 μgmL^{-1}	Wheat	Root and shoot inhibition
Ochratoxin A (OTA)	50–200 μM	*Arabidopsis thaliana*	Formation of leaf lesions, blocking of root elongation
Patulin (PAT)	25–100 μgmL^{-1}	Maize, wheat	Radicle emergence, intermodal elongation, floret number, seed weight and seed number decrease
Penicillic acid (PA)	250–500 $\mu g{\cdot}mL^{-1}$	Lettuce, corn	Growth inhibition of seedlings and main root
Pyrenophoric acid	10^{-3} M	Cheatgrass	Reduction in coleoptile length
Pyrichalasin H	1 μgmL^{-1}	Rice	Curling of the shoot
T-2 toxin	1–6.7 μgmL^{-1}	Wheat, maize	Decrease in root, shoot and leaf growth rate
Tenuazonic acid (TA)	100–200 $\mu g{\cdot}mL^{-1}$	*Daturainnoxia*, green gram, lettuce, rice, rye, wheat	Leaf browning to chlorotic spots, growth inhibition of germinating seeds and root growth
Zearalenone/F-2 toxin (ZEA)	5 μgmL^{-1}	Maize	Inhibitory effect on root and shoot elongation and fresh mass accumulation of germinating embryos
Zygosporin D	1 nmol/plant	Rice	Reduction of the second leaf sheath length

Source: Ismaiel and Papenbrock (2015).

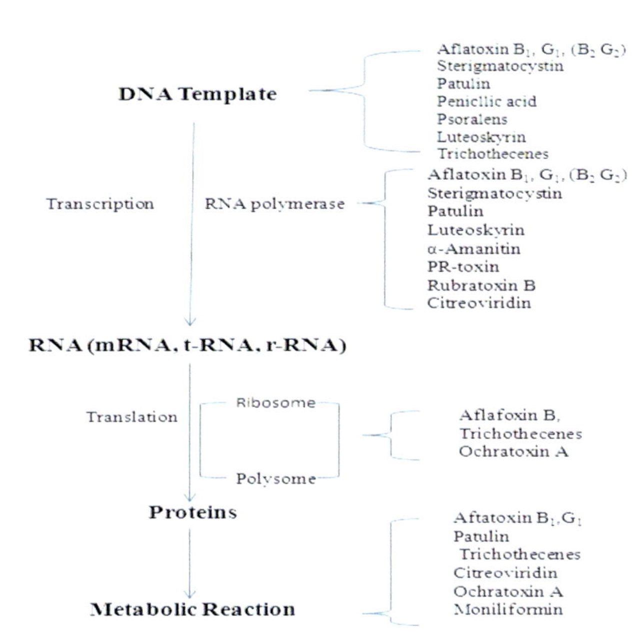

FIGURE 11.1 Mycotoxins affecting sites in RNA and protein synthesis (Kiessling 1986).

proteins (specific site of RNA) of humans, plants and animals (Figure 11.1). Nearly 25% of crops are affected by fungi which produce lethal mycotoxins worldwide (Bryden 2007). Among them, some mycotoxins contaminate most of the cereal grain and leguminous crops; namely, aflatoxins, penicillic acid, fumonisins, fusaric acid, fusarin C, citrinin, patulin, ochratoxin, deoxynivalenol and zearalenone. These major mycotoxins are mainly produced by Ascomycetous fungal genera; namely, *Aspergillus* and *Penicillium* (Ismaiel and Papenbrock 2015). Many cereal crops like barley, sorghum and wheat can contain mycotoxins.

11.3.1 Aflatoxins

Aspergillus species are known to produce aflatoxin, which is extremely potent toxin compound to animal and human health. Aflatoxins are found in highly humid and warm regions of the world and contaminate a number of crops. B aflatoxin (B1 and B2) is the main mycotoxin, which is produced by *A. flavus* and is highly toxigenic after cyclopiazonic acid which is extremely toxigenic and also produced by *A. flavus*. *A. parasiticus* produces G1 and G2 aflatoxin of G group along with B aflatoxins. In 1960, aflatoxins were first identified after a serious epidemic of Turkey X disease which was found to be associated with diet meal of peanuts in Britain. *A. flavus* grows throughout the world and is found in food crops such as peanuts, maize, and cotton seed in humid and warm regions. *A. parasiticus* has comparatively limited geographical distribution compared to *A. flavus* and generally infects peanuts.

11.3.2 Mode of Action of Aflatoxin

Aflatoxin inhibits chlorophyll a, chlorophyll b and protochlorophyllide biosynthesis (Sinha and Kumari 1990). The level of protein and nucleic acid is also suppressed by aflatoxin in germinating seeds (Sinha and Kumari 1990). In addition, aflatoxin inhibits the activity of chromatin-bound DNA-dependent polymerase. In the absence or unavailability of m-RNA, aflatoxin inhibits protein synthesis, whereas DNA synthesis is inhibited by binding to DNA during the replication process (Tripathi and Misra 1983). Aflatoxins impair the cell cycle and promote apoptosis in cultured cells and lead to apoptosis by the accumulation of DNA damage in cells (Fadl-Allah et al. 2011).

11.3.3 Ochratoxin

Aspergillus and *Penicillium* are the two main filamentous fungi known to produce ochratoxin. Ochratoxin is more complex to study than aflatoxin and was discovered in 1965 (Van der Merwe et al. 1965). Ochratoxin is divided into three groups, A, B and C (Heussner and Bingle 2015). Among these groups, ochratoxin A (OTA) is dominant and applicable fungal toxin which is chlorinated (Heussner and Bingle 2015). Ochratoxin B (OTB) and ochratoxin C (OTC) are non-chlorinated and grouped in ethyl esters, which are comparatively less toxic than ochratoxin A (Bayman and Baker 2006). Crops grown in temperate climates provide ideal conditions for the production of OTA. In tropical climates, OTA affects mainly coffee and grapes, whereas cereals are the main contaminant in temperate climates. *Penicillium verrucosum* produces OTA in cereal crops in cool temperate climates, while *Aspergillus* produces OTA in warm climates. When ochratoxin is produced by *A. ochraceus*, the required climatic conditions are 31°C, pH 3–10 and minimum water activity of 0.8, while 20°C, pH 6–7 and a minimum water activity of 0.86 are required by *P. verrucosum* (Reddy and Bhoola 2010).

11.3.4 Mode of Action of Ochratoxin

Ringot et al. (2006) reported that OTA toxicity involves suppression of protein synthesis, prohibition of mitochondrial respiration, DNA damage, disruption of calcium homeostasis and interference in the metabolic system. Wang et al. (2011) studied the phytotoxic mechanisms of OTA during gene and protein expression on *A. thaliana*. Some researchers summarized about OTA phytotoxicity as OTA is contact to *A. thaliana* leaves resulting the hypersensitive response. OTA affects protein synthesis by changing in expression of genes and also triggers the membrane transport system. Salicylic acid, ethylene and jasmonic acid are the main mediator molecules causing phytotoxicity of OTA. Wang et al. (2014) showed that OTA activates a defence response through enhancing the transcript level of enzymes to maintain the total glutathione content.

11.3.5 Deoxynivalenol

Deoxynivalenol (DON), also called vomitoxin, is mainly produced by fungus *Fusarium graminearum*, but in many geographical regions it is also produced by fungus *F. culmorum* (Richard 2000).

11.3.6 Mode of Action of Deoxynivalenol

DON mycotoxins play a major role in disease development. DON and 3-acetyldeoxynivalenol (3-ADON) inhibit seed germination and root development (Wakulinski 1989). DON is important for colonization of *F. graminearum* into the plant system. The fungus is allowed to spread because protein synthesis and development of the host enzyme is inhibited by DON (Snijders 1994).

11.3.7 T-2 Toxin

The T-2 mycotoxin comes under the Type A non-macrocyclic trichothecene group of mycotoxins. *F. sporotrichioides* is a primary fungus which produces T-2 toxin along with some closely related HT-2 toxins and diacetoxyscirpenol mycotoxins of the same chemical class (CAST 2003).

11.3.8 Mode of Action of T-2 Toxin

T-2 and HT-2 toxins are produced by *F. sporotrichioides*, known as a poor plant pathogen (Miller 1994). In germinating wheat seedlings, the dose of 1 µg·mL^{-1} of T-2 toxin fully affected root and shoot growth (Wakulinski 1989). In trichothecenes, the T-2 toxin tested as highly toxic for animal cells comparative to other trichothecenes (Thompson and Wannemacher 1986; Terse et al. 1993). Rahman et al. (1993) described that the concentration of T-2 toxins moderate cell division during prophase and metaphase. Mostly T-2 toxin inhibits protein, DNA and RNA synthesis. It was also reported that these toxins decrease antibody levels, immune-globulins and some hormones during cytokinesis.

11.3.9 Moniliformin

Moniliformin (MON) is a K/Na or sodium salt of 1-hydroxycyclobut-1-ene-3,4-dione produced by many *Fusarium* spp. but particularly *F. proliferatum* on the corn kernel (Vesonder et al. 1992).

11.3.10 Mode of Action of Moniliformin

The hostile effect of MON can be seen on developing leaf, root and seedlings of wheat and corn. (Wakulinski 1989). Van Asch et al. (1992) reported that MON significantly reduced fresh mass accumulation at a concentration of 10 µg·mL^{-1} and 100 µg·mL^{-1}.

11.3.11 Fumonisin

Fumonisin comes under non-fluorescent mycotoxins grouped in FB1, FB2 and FB3, which are primarily produced by *Fusarium verticillioides* and *F. proliferatum*. The strains of fumonisin have variable toxin-producing ability (CAST 2003). The group of fumonisin mainly affects corn but it is also found in sorghum and rice. The disease mouldy corn toxicosis was first reported in the horses fed with mainly maize from the United States (1891), and later reported in Argentina, China, Egypt, South Africa and Australia.

11.3.12 Mode of Action of Fumonisin

Fumonisin mainly causes physiological injury and sudden death in plants. Janardhan et al. (2011) reported that seeds of difficult pulse crops decreased their germinating ability when soaked in toxic fungal strain. Endosperm production was also inhibited by FB1 because it interferes metabolically with germination. When concentrated dependent FB1 treated with maize callus cells, it revealed alterations in cell ultrastructure (Van Asch 1990). FB1 was also found to decrease chlorophyll content (Abbas et al. 1993).

11.3.13 Fusarin and Fusaric Acid

Fusaric acid (FA) is a derivative of picolinic acid, which was first isolated from *Fusarium heterosporium* (Yabuta et al. 1934). This fusaric acid was first reported as involvement of toxins in pathogenesis of tomato wilt disease caused by *F. oxysporum* f. sp. *lycopersici* (Gäumann 1957). Many members of the *Gibberella fujikuroi* species complex were found to produce fusarin C on maize, soya bean, wheat, rye and barley. Several *Fusarium* species also produce fusarin C, fusaric acid and fumonisins FB1, FB2 and FB3.

11.3.14 Mode of Action of Fusarins and Fusaric Acid

The action mechanism is still not fully understood. It is likely to inhibit the enzyme that converts dopamine to norepinephrine, i.e. dopamine beta-hydroxylase. It also inhibits DNA synthesis and cell proliferation. Fusaric acid also acts as an inhibitor for quorum sensing. FA changes membrane permeability and decreases mitochondrial activity, oxygen uptake, ATP synthesis and root growth inhibition (Bouizgarne et al. 2006). The *Fusarium* wilt index is positively correlated with concentration of FA.

11.3.15 Penicillic Acid

The penicillic acid (PA) mycotoxin is mainly produced by fungus *Penicillium puberulum*, *P. roqueforti* mould and *Aspergillus flavus*. Penicillic mycotoxin was first isolated and named by Bainier but it was first synthesized by Ralph Raphael in 1947 during World War II. It is also the major product of acid degradation of penicillin. PA has also been isolated from many *Penicillium* species, especially *P. aurantiogriseum*, *P. canescens*, *P. Polonicum* and *P. Radicicola* (Ismaiel and Papenbrock 2015).

11.3.16 Mode of Action of Penicillic Acid

The PA mycotoxins activated when it interacts with the sulphydryl -residues in enzymes and is readily inactivated by thiols enzyme (Ciegler 1972). It has been found that PA inhibits the thiol enzymes, i.e. alcohol dehydrogenase and lactic dehydrogenase (Ashoor and Chu 1973). PA was isolated in infected corn when its quantities reached more than 2% (Ciegler and Kurtzman 1970). PA toxins increase during low temperature storage conditions and become serious for storage products (Kurtzman and Ciegler 1970; Keromnes and Thouvenot 1985).

11.4 Methods for Detection of Mycotoxins

Mycotoxins are chemical compounds or secondary metabolites that are produced or artificially manufactured by some important fungal species such as Aspergillus, *Fusarium* and *Penicillium*. Mycotoxins are poisonous compounds generally available in food and feeds that pose some serious health issues for humans, animals and plants. The following methods are used for the detection of mycotoxins from different sources.

11.4.1 Conventional Methods

Gas chromatography (GC), high performance liquid chromatography (HPLC) with mass spectrometry and some other techniques such as fluorescence or ultraviolet (UV) detection, thin layer chromatography (TLC) and enzyme-linked immunosorbent assay (ELISA) are some of the traditional procedures used in detection of mycotoxins. Overall, chromatography is the main method which is mostly used for mycotoxin detection in food and feed. Currently several methods and techniques have been developed for the mycotoxin detection, but the first mycotoxin was detected by TLC, and several laboratories routinely use this technique. For the rapid and visual detection, TLC is currently used for some mycotoxins such as AFM1, AFG1, AFG2, FB1 and OTA authorizing a qualitative assessment or, if coupled with instrumental densitometry, enabling a semi-quantitative assessment (Filazi et al. 2010; Shephard 2011; Li et al. 2017). Currently, mycotoxin identification from food and feed is focusing on applications that are rapid, inexpensive and easy to use and are capable of identifying several mycotoxins in single run with high sensitivity. To complete all these needs for identification of mycotoxins, different chromatographic techniques have been developed, such as HPLC with UV, diode array, fluorescence or mass spectrometry detectors, and ultra-HPLC with compact column packing material. For the analysis of volatile mycotoxins, several techniques are used, such as GC with electron capture (ECD), flame ionization (FID) or mass spectrometry detectors (Pereira et al. 2014). Before detection, all these chromatographic techniques require some additional steps such as preparation of chemicals, experienced operators and expensive instruments. All of these methods are laboratory-based and so are not appropriate for quick on-site detection. Therefore, an advanced conventional technique, ELISA, has been developed and is used widely for the analysis of mycotoxins. Direct assay, direct competitive assay and indirect competitive assay with competitive direct assay are some approaches of ELISA that have been most commonly used for mycotoxin detection (Goryacheva et al. 2007). Mycotoxins associated with different food media like AFs, DON, FB, OTA, T-2 toxin and ZEA were detected by commercial and fast screening kits of ELISA technique. These methods are less time consuming, more easily accessible, inexpensive and reliable techniques for analysis compared to other mycotoxin detectors (Goryacheva et al. 2007). Some of these techniques require wide preparation before analysis and take time to complete the process of mycotoxin identification. Therefore, some advanced devices, procedures or techniques have also been established for analysis of mycotoxins.

11.4.2 Advanced Methods

To ensure food and feed safety in the food industry, swift identification of mycotoxins is important because they directly harm human and animal health. Various works have been done to develop advanced approaches for swift mycotoxin detection. In the challenge to identify mycotoxins, some highly sensitive and specific techniques or devices, rapid analysis with low cost and transportability approaches have been developed. Furthermore, some more advanced methods or techniques like use of surface plasm on resonance, multiplex lateral flow, surface-enhanced Raman scattering, microarray chips and biosensors with nanoparticles have been developed. Among them, recently the thin film biosensor device is the best technique which is being used in current research for mycotoxin detection. Compared to other techniques, the biosensor has some advantages, such as it takes less time with swift detection, is highly sensitive, reliable, easy to prepare and cost effective. Vidal et al. (2013) studied electrochemical affinity biosensors for different mycotoxin detection. In nanoscience, some nanoparticles such as gold, silver, metal oxides and quantum dots are being broadly used in biosensors to enhance the detection ability of mycotoxins because these nanoparticles have some remarkable electronic, thermal and mechanical properties which increase their stability, sensitivity and selectivity to mycotoxins. Some biosensors have been studied, including biomedical immunochemical biosensors (monoclonal and polyclonal antibodies are used), optical biosensors, electrochemical biosensors, enzyme-based biosensors (Baeumner 2003), electrode-based biosensors (Tudorache and Bala 2007), whole cell-based biosensors (Baeumner 2003), whole organism-based biosensors, amperometric biosensors, acoustic biosensors, potentiometric biosensors, optical biosensors and colorimetric biosensors (Rana et al. 2010). Chaudhary and Castle (2011) have studied the use of nanotechnology for mycotoxin analysis in food and feed. They have used nano-sized biosensors in food labelling because during the storage and transport of food material, it is necessary to monitor the condition of packed food materials. During the storage and transportation food materials which come in contact with microorganisms by rough handling of transporter, nanosensors start releasing some food preservatives. Some advanced approaches have also been developed using DNA sensors like electrochemical DNA sensing and gold nanoparticle-based DNA chips by using nanoparticles (Im et al. 2006). By use of magnetic nanoparticles, Fernandez-Baldo et al. (2011) detected OTA mycotoxins. The advancement of modified biosensors is that they detect mycotoxins at a very low concentration.

11.4.3 On-Site Detection Methods

Moon et al. (2013) have developed some strip-based sensors, because these strips are cost-effective and small in size, and due to this, these can be used in field and lab for rapid on-site detection of mycotoxin. Burmistrova et al. (2014) developed a membrane-based flow-through immunoassay approach for multi-detection of mycotoxins. This technique helps in multi-detection of OTA, ZEN and FB1 mycotoxins a tone time and on-site. The advantage of this technique is that an unskilled person can perform it and there is no need for transportation of samples from field to lab. Samples sent from one state to another for identification and analysis often takes a long time, but with the development of this technique, it is possible to identify mycotoxins on-site.

Xu et al. (2012) studied a new approach in mycotoxin detection, i.e. colloidal gold immune chromatographic assay, which is economical, reliable, swift and sensitive to quantification and modification in analysis. Colloidal gold is used in labelling that can trace the labelled target. Currently, for immunological examination or detection, immune colloidal gold strips being used for multiple testing of mycotoxins. Microarray approaches and biochips are also popular for the identification of mycotoxins. Some other chip technologies have also been developed such as gene chip, in which target gene or oligonucleotide probe is fixed on chips, and protein chips, in which peptides or proteins are fixed on the chips. Among them, protein chips are currently used, which include chip microarray preparation, sample preparation, biological molecular reaction and signal detection and analysis. A method for quantitative analysis of AFB1 was established by Song et al. (2011) which is based on the microsphere array technology platform in which monoclonal anti-aflatoxin B1 is used.

Vicam's test strips are currently available commercially and are being used for swift, exact, on-spot existence of different mycotoxins at one time. These strips give reliable results in minimum time in the field and/or laboratory and do not require expensive chemicals and equipment. During the analysis, the test strips require just a dip into the samples and the results can be read. The advantage of these strips is that they are reliable, provide rapid detection and are easily available in the market. Because of reliability and rapidity, generally these strip assays have been well known in the medical field, but these methods have now been developed for mycotoxin detection.

11.5 Management Strategies and Their Impact on Mycotoxigenic Fungi and Associated Mycotoxins

The major food and feed crops, such as rice, maize, wheat, soya bean, sorghum and groundnut, are affected by mycotoxins, although several other crops are also affected. It is important to manage mycotoxigenic fungi and their mycotoxins for sustainable development of food and feed production because food and feed are basic needs of humans and animals. Many strategies for management of fungi and mycotoxin contamination exist. Among them, approaches include use of resistant varieties, use of organic material, chemical control and changes in production applications and reduction in plant stress. Use of resistant plant varieties is an effective and environmentally friendly strategy to control mycotoxigenic fungi and mycotoxins. Therefore, the resistant cultivars and landraces are important and currently used in the large-scale cultivation and by local producers. Due to some restrictions, commercial incorporation may be slow and complex. Management of mycotoxins by chemicals has been successfully used for major cereal crops, but for other important crops no method has been established. Due to harmful effects of chemicals on humans and animals, we are focusing on management of mycotoxins by biological control. Aflatoxins produced by *A. flavus* harms a large group of cereal crops, so researchers are currently focusing on biological control approaches for aflatoxins. Many African countries have developed commercial biological control products with promising results. In recent times, cultivation practices have been performed under different environmental conditions by which disease appearance and mycotoxin production is reducing. Advanced pre- and post-harvest methods can help in managing mycotoxins, but use of integrated methods may also provide more effective management of mycotoxins in crops. Unfortunately, farmers and producers have a little knowledge about fungi attacking the crops and host plants to manage mycotoxigenic fungi.

11.6 Future Directions

Mycotoxins are expected to become a greater concern for the feed and food supply chains in the future. Focused research on mycotoxin detection will help to manage them and reduce mycotoxin-related illnesses in food and feed supply. Mycotoxins should be treated as an emerging concern and we should try to keep them under control to provide low risk to the consumer. A number of tools to control their precursor fungi should be developed to save lives of people.

REFERENCES

Abbas HK, Gelderblom WCA, Cawood ME, Shier WT (1993) Biological activities of fumonisins, mycotoxins from *Fusarium moniliforme*, in jimson weed (*Datura stramonium* L.) and mammalian cell cultures. Toxicon 31: 345–353.

Ashoor SH, Chu FS (1973) Inhibition of alcohol and lactic dehydrogenases by patulin and penicillic acid in vitro. Food Cosmet Toxicol 11:617–624.

Ates E, Mittendorf K, Stroka J, Senyuva H (2013) Determination of *Fusarium* mycotoxins in wheat, maize and animal feed using on-line clean-up with high resolution mass spectrometry. Food Addit Contam 30:156–165.

Baeumner AJ (2003) Biosensors for environmental pollutants and food contaminants. Anal Bioanal Chem 377:434–445.

Bayman P, Baker JL (2006) Ochratoxins: Aglobal perspective. Mycopathologia 162:215–223.

Bennett JW (1987) Mycotoxins, mycotoxicoses, mycotoxicology and mycopathologia. Mycopathologia 100:3–5.

Bennett J W, Klich M (2003) Mycotoxins. Clin Microbiol Rev 16(3): 497–516.

Blount WP (1961) Turkey "X" disease. Turkeys 9(52): 55–58.

Bouizgarne B, El-Maarouf-Bouteau H, Frankart C, Reboutier D, Madiona K, Pennarun AM (2006) Early physiological responses of *Arabidopsis thaliana* cells to fusaric acid: Toxic and signaling effects. New Phytol 169:209–218.

Bryden WL (2007) Mycotoxins in the food chain: Human health implications. Asia Pacific J Clin Nutr 16:95–101.

Burmistrova NA, Rusanova T Yu, Yurasov NA, GoryachevaI Yu, Saeger SD (2014) Multi-detection of mycotoxins by membrane based flow-through immunoassay. Food Control 46:462–469.

CAST (2003) Mycotoxins Risks in Plant, Animal and Human Systems, Task Force Report, No.139. Council for Agricultural Science and Technology, Ames, Iowa, pp. 1–191.

Chaudhary Q, Castle L (2011) Food applications of nanotechnologies: An overview of opportunities and challenges for developing countries. Trends Food Sci Tech 22:95–603.

Ciegler A, Kurtzman CP (1970) Penicillic acid production by blue-eye fungi on various agricultural commodities. Appl Microbiol 20:761–764.

Ciegler A (1972) Bioproduction of ochratoxin A and penicillic acid by members of the *Aspergillus ochraceus* group. Can J Microbiol 18:631–636.

Fadl-Allah EM, Mahmoud MAH, Abd El-Twab MH, Helmey RK (2011) Aflatoxin B1 induces chromosomal aberrations and 5S rDNA alterations in durum wheat. J Assoc Arab Univ Basic Appl Sci 10:8–14.

Fernandez-Baldo MA, Bertolino FA, Fernandez G, Messina GA, Sanz MI, Raba J (2011) Determination of ochratoxin A in apples contaminated with *Aspergillus ochraceus* by using a microfluidic competitive immunosensor with magnetic nanoparticles. Analyst 136:2756–2762.

Filazi A, Ince S, Temamo GF (2010) Survey of the occurrence of aflatoxin M1 in cheeses produced by dairy ewe's milk in Urfa City, Turkey. Ankara Üniv Vet Fak Derg 57: 197–199.

Forgacs J (1962) Mycotoxicoses: The neglected diseases. Feedstuffs 34:124–134.

Gäumann E (1957) Fusaric acid as a wilt toxin. Phytopathology 47: 342–357.

Goryacheva IY, Saeger SD, Eremin SA, Peteghem CV (2007) Immunochemical methods for rapid mycotoxin detection: Evolution from single to multiple analyte screening: A review. Food Addit Contam 24:1169–1183.

Heussner AH, Bingle LE (2015) Comparative ochratoxin toxicity: A review of the available data. Toxins (Basel) 7:4253–4282.

Im Y, Vasquez RP, Lee C, Myung N, Penner R, Yun M (2006) Single metal and conducting polymer nanowires sensors for chemical and DNA detections. J Phys Conf Ser 38:61–64.

Ismaiel AA, Papenbrock J (2015) Mycotoxins: Producing fungi and mechanisms of phytotoxicity. Agriculture 5:492–537.

Janardhan A, Subramanyam D, Praveen A, Reddi M, Narasimha G (2011) Aflatoxin impacts on germinating seeds. Ann Biol Res 2(2): 180–188.

Keromnes J, Thouvenot D (1985) Role of penicillic acid in the phytotoxicity of *Penicillium cyclopium* and *Penicillium canescens* to the germination of corn seeds. Appl Environ Microbiol 49:660–663.

Kiessling KH (1986) Biochemical mechanism of action of mycotoxins. Pure Appl Chem 58(2):327–338.

Kurtzman CP, Ciegler A (1970) Mycotoxin from a blue-eye mold of corn. Appl Microbiol 20:204–207.

Li Q, Lu Z, Tan X, Xiao X, Wang P, Wu L, Shao K, Yin W, Han H (2017) Ultrasensitive detection of aflatoxin B1 by SERS aptasensor based on exonuclease-assisted recycling amplification. Biosens Bioelectron 97: 59–64.

Liew WPP, Redzwan MS (2018) Mycotoxin: Its impact on gut health and microbiota. Front Cell Infect Microbiol 8:60.

Maggon KK, Gupta SK, Venkitasubramanian TA (1977) Biosynthesis of aflatoxins. Bacteriol Rev 41:822–855.

Miller JD (1994) Epidemiology of *Fusarium* ear diseases in cereals. In: Miller JD, Trenholm HL (Eds.). Mycotoxins in Grain. Compounds Other Than Aflatoxin. Eagan Press, St. Paul, MN, pp. 19–36.

Moon J, Kim G, Lee SJ (2013) Development of nanogold-based lateral flow immunoassay for the detection of ochratoxin A in buffer systems. Nanosci Nanotechnol 13:7245–7249.

Pereira VL, Fernandes JO, Cunha SC (2014) Mycotoxins in cereals and related foodstuffs: A review on occurrence and recent methods of analysis. Trends Food Sci Technol 36:96–136.

Rahman MF, Bilgrami KS, Masood A (1993) Cytotoxic effects of DON and T-2 toxin on plantcells. Mycopathologia 124:95–97.

Rana JS, Jindal J, Beniwal V, Chhokar V (2010) Utility biosensors for applications in agriculture–A review. J Am Sci 6:353–375.

Reddy L, Bhoola K (2010) Ochratoxins-food contaminants: Impact on human health. Toxins (Basel) 2:771–779.

Richard JL (2007) Some major mycotoxins and their mycotoxicoses—An overview. Int J Food Microbiol 119:3–10.

Richard JL (Ed.) (2000) Mycotoxins—An overview. Romer Labs' Guide to Mycotoxins, 1: pp. 1–48.

Ringot D, Chango A, Schneider Y, Larondelle Y (2006) Toxicokinetics and toxicodynamics of ochratoxin A: An update. Chem Biol Interact 159:18–46.

Shephard GS (2011) Chromatographic separation techniques for determination of mycotoxins in food and feed. In: Saeger S (Ed.). Determining Mycotoxins and Mycotoxigenic Fungi in Food and Feed. Woodhead Publishing Limited, Philadelphia, PA, pp. 71–89.

Sinha KK, Kumari P (1990) Some physiological abnormalities induced by aflatoxin B1 in mungseeds (*Vigna radiata* var. Pusa Baishakhi). Mycopathologica 110:77–79.

Snijders CHA (1994) Breeding for resistance to *Fusarium* in wheat and maize. In: Miller JD, Trenholm HL (Eds.). Mycotoxins in Grain. Compounds Other Than Aflatoxin. Eagan Press, St. Paul, MN, pp. 37–58.

Song HJ, Liu SY, Ma HR et al. (2011) The quantitative analysis of AFB1 by using the microsphere array technology and indirect competition theory. Chin Agr Sci Bull 27(26): 144–150.

Terse PS, Madhyastha MS, Zurovac O, Stringfellow D, Marquardt RR, Kemppainen BW (1993) Comparison of in vitro and in vivo biological activity of mycotoxins. Toxicon 31:913–919.

Thompson WL, Wannemacher RW (1986) Structure-function relationship of 12,13-epoxy trichothecene mycotoxins in cell culture: Comparison to whole animal lethality. Toxicon 24:985–994.

Tripathi RK, Misra RS (1983) Mechanism of inhibition of maize seed germination by aflatoxin BI. In: Bilgrami KS, Prasad T, Sinha KK (Eds.). Mycotoxins in Food and Feed. Allied Press, Bhagalpur, India, pp. 129–141.

Tudorache M, Bala C (2007) Biosensors based on screen printing technology, and their applications in environmental and food analysis. J Anal Bioanal Chem 388:565–578.

Van Asch MAJ, Rijkenberg FHF, Coutinho TA (1992) Phytotoxicity of fumonisin B1, moniliformin and T-2 toxin in corn callus cultures. Phytopathology 82:1330–1332.

Van Asch MAJ (1990) Studies on the Resistance of Wheat and Maize to Fungal Pathogenesis. PhD thesis, University of Natal, Pietermaritzburg, South Africa.

Van Der Merwe KJ, Steyne PS, Fourie L, Scott DB, Theron JJ (1965) Ochratoxin A, a toxic metabolite produced by *Aspergillus ochraceus*. Nature 205:1112–1113.

Vesonder RF, Labeda DP, Peterson RE (1992) Phytotoxicactivity of selected water-soluble metabolites of *Fusarium* against *Lemna minor* L. (Duckweed). Mycopathologia 118:185–189.

Vidal JC, Bonel L, Ezquerra A, Hernandez S, Bertolin JR, Cubel C, Castillo JR (2013) Electrochemical affinity biosensors for detection of mycotoxins: A review. Biosens Bioelectron 49:146–158.

Wakulinski W (1989) Phytotoxicity of the secondary metabolites of fungi causing wheat head fusariosis (head blight). Acta Physiol Plant 11:301–306.

Wang Y, Peng X, Xu W, Luo Y, Zhao W (2011) Transcript and protein profiling analysis of OTA-induced cell death reveals the regulation of the toxicity response process in *Arabidopsis thaliana*. J Exp Bot 29:1–17.

Wang Y, Zhao W, Hao J, Xu W, Luo Y, Wu W (2014) Changes in biosynthesis and metabolism of glutathione upon ochratoxin A stress in *Arabidopsis thaliana*. Plant Physiol Biochem 79:10–18.

Xu F, Weijun K, Meihua Y, Zhen O (2012) Latest advancement for detection methods of mycotoxins in traditional Chinese Medicine. Mode Tradit. Chin Med Mater Med 14(5):1944–1952.

Yabuta T, Kambe T, Hayashi T. (1934) Biochemistry of 'bakanae' fungus of rice. J Agric Chem Soc Jpn 10: 1059–1068.

12

Characterization and Detection of Mycotoxins from Fruits and Vegetables

V. Koteswara Rao,[1] Vaibhavi C. Dhuri,[1] N. Vishwambar[2] and T. Bhagyashree[1]
[1]*Biochemical Sciences Division, CSIR-National Chemical Laboratory, Pune, Maharashtra, India*
[2]*Academy of Scientific and Innovative Research (AcSIR), New Delhi, India*

CONTENTS

12.1 Introduction

Fungi are a large diverse group of heterotrophic organisms, which are saprobes in the soil and on decomposing organic matter. Fungi infect plants, animals and humans and cause acute and chronic disease. Increases in the prevalence of disease-causing fungi are widely recognized globally for their pathogenicity and toxigenicity. Mycotoxins are a chemically diverse group of toxic secondary metabolites produced by filamentous fungi. *Aspergillus*, *Penicillium*, *Alternaria* and *Fusarium* produce chemically diverse mycotoxins in food grains, spices, fruits and vegetables under warm and humid conditions. The contamination of horticultural produce with fungal toxins is a major threat to global food safety. Fungal species accompanied by the deterioration of various fruits and vegetables are *Mucor indicus*, *M. amphibiorum*, *M. racemosus* and *M. hiemalis*; *Rhizopus* species (*R. stolonifer*, *R. nigricans* and *R. oligosporus*); *Candida albicans*, *A. fumigatus*, *A. niger*, *A. flavus*, *P. oxalicum* and *P. chrysogenum*, *F. accuminatum*, *F. oxysporum*, *F. eqiuseti*, *F. moniliforme*, *F. solani* and *F. dimerum*. Patulin (PAT), aflatoxins (AFTs), ochratoxin

A (OTA), fumonisin (FUM), zearalenone (ZEA) and nivalenol (NIV), deoxynivalenol (DON) and *Alternaria* toxins are major contaminants of fruits and vegetables. Most mycotoxins are chemically diverse, thermostable and difficult to remove during food processing operations. The Joint Scientific Expert Committee for Food and Agriculture (JECFA) with the World Health Organization (WHO) and the Food and Agriculture Organization of the United Nations (FAO) has indicated the health risk of mycotoxins. The adverse health effects include acute toxicity, which is the rapid onset from a single exposure, chronic toxicity, which is slow with delayed onset due to long-term exposure that can be hepatotoxic, neurotoxic, genotoxic, nephrotoxic, immunotoxic or carcinogenic in nature. Dietary exposure of mycotoxins can cause irreversible tissue damage through several biochemical mechanisms leading to cancer (WHO 2018).

Microbial activities pose a serious threat to food safety, of which mycotoxin contamination is the main concern (Koteswara Rao et al. 2015). The detection of these toxins' contamination and their source of pathogenic strains are crucial tasks in food safety which further correlate with human health. Hence, quick and accurate detection methodologies can help in the efficient spotting of contamination from foods and feed commodities. On the other hand, intake of the contaminated foodstuffs, carryover of the mycotoxins to animal products or from introduction to air and dust contaminated with toxins can lead to the exposure of various mycotoxins. About 300–400 mycotoxins have been identified present in livestock that are of concern to human health and have received widespread attention (Binder et al. 2007). Today, contamination of agricultural and horticultural produce is one of the main concerns to the humans and livestock (FAO 2019). Processing of this produce to meet the safety and quality needs for food is still a challenge for the food industry. The detection of these toxins' contamination and their source of pathogenic strain is the critical means in detecting and identifying the problems related to public health and food safety, and only rapid, sensitive and efficient detection technologies can enable users to make accurate assessments of the risks or contaminations of foods and feed commodities. Children and infants are another sensitive target of mycotoxin effects, especially by fruits. ZEA, OTA, aflatoxins, fumonisins (FUM-FB1, FB2 and FB3), trichothecenes (TRI)both A and B type toxins, namely DON (deoxynivalenol), NIV, T2, HT2, moniliformin (MON), enniatins (ENN), fusaric acid (FSA) and diacetoxyscirpenol (DAS) have been reported toxic to animals and humans. Several apple products including apple juice, baby food and mixed juice from stores in Tunisia revealed patulin (PAT) levels from 0 to 167 μg/L. These high levels of PAT in baby food can lead to inhibition of DNA, RNA and protein synthesis as well as rupturing of the plasma membrane due to the severity of the toxicity. The consumption of mycotoxin-affected fruit or food by children causes diverse effects, including stunted growth, malnutrition, etc. Aflatoxin-8, 9-epoxide binds with DNA with much affinity to form aflatoxin-N7-guanine, which results in the transversion of guanine to thymine (T) mutation further affect the cell cycle by affecting the P53 gene. Ergotism is another example of nephrotoxicity caused by ergot alkaloids, a derivative of citrinin (Bhat et al. 2010).

12.2 Fungi and Their Mycotoxins in Fruits and Vegetables

12.2.1 Banana

Banana (*Musa acuminata*),an important commercial crop of genus *Musa*, has evolved in Southeast Asia, including India as one of the main countries of origin. India contributes around 16% of the world's overall fruit production. It is cultivated as the fourth major staple food crop after rice, wheat and maize (FAS/USDA 2020). India is one of the highest banana-producing countries in the world at about 29 million tons per year, followed by China, the Philippines and Africa. About 1000 varieties of banana are cultivated worldwide. Musa (Banana Dwarf Cavendish) (45% of the global banana market) is commercially important due its high production value and resistance to adverse environmental conditions. Banana is used for various types of food servings such as to flavour muffins, cakes and bread. Banana is rich in carbohydrates, fibre, vitamin B_6, ascorbic acid and minerals that may help fight against atherosclerosis, which is responsible for heart attack and stroke. Banana is one of the most important fruits in consideration of food safety and community health (FAO, 2019).

Banana may be affected by various fungal infections during its growth and development. Species of *Fusarium*, *Aspergillus*, *Alterneria*, *Penicillium*, *Cylindrocarpon*, *Acremonium*, *Musicillium*, *Rhizoctonia* and *Drechslera* are responsible for various diseases of banana such as Panama disease (*Fusarium* wilt), root rot, Septoria brown spot, Cordana leaf spot, pitting disease, cigar end rot, and diamond spot, tip end rot, leaf spot, anthracnose and sigatoka. The fungal genera not only cause disease in banana but also lead to the production of toxic metabolites, which cause various diseases to humans and animals. Mycotoxins such as DON, ZEA, FB1, MON, OTA, AFs and TA are the major toxins found in banana. TRI is one of the most diverse groups of mycotoxins that lead to various acute and chronic diseases which are mainly produced by *Fusarium* species. However, *Stachybotrys*, *Acremonium*, *Myrothecium*, *Cylindrocarpon*, *Trichoderma*, *Dendrodochium* and *Trichothecium* are also reported as TRI producers (Ahmad and Yu 2017). The toxigenic *F. verticilloides* is found mainly in infected bananas and is also detected in various food samples collected throughout the world. *Fusarium oxysporum* f. sp. *cubense* (Foc) exists as different pathogenic races such as *Foc* races 1, 2, 3 and 4 to all the banana cultivars. Moreover, Foc, races 1 and 4 are of considerable concern (Zhang et al. 2018). The Foc mycelia directly penetrate the xylem vessels through the epidermal cell wall. The Foc race 4 of *Fusarium* wilt in banana was reported in India. Foc and host articulately recognize the relationship between the plant and microbe. Plant also developed resistance (R) genes, which are resistance-specific pathogens, and plant-produced antimicrobial compounds that are also pathogen-toxic. The biotic response includes the creation of networks and the regulation of plant protection/immune response. The molecular mechanisms of infection of host defence using OMICS approaches including genome editing, sequencing and proteome profile help in breeding programmes to develop resistance varieties of crops to encounter different pathogenic and toxigenic fungi.

12.2.2 Tomato

Tomato (*Solanum lycopersicum*) originated in South America and belongs to the family *Solanaceae*. It is one of the worldwide cultivated vegetable crops in area (around 4.4 million ha) and production is about 115 million metric tons. Due to its economic importance, cultivation areas and production increase day by day. China is the top tomato-producing country in the world and contributes about 31.8% of global production. Regular consumption of tomato has many benefits: it lowers the risk of osteoporosis, liver and cardiovascular diseases and various types of cancers because it contains lycopene, bioflavonoids and antioxidants, which act as anticancer agents. Tomato is more prone to diseases caused by several pathogenic fungi, bacteria, viruses and nematodes, which result in the reduction in yield that leads to severe economic losses (Worku and Sahe 2018). There are many fungal species that infect and cause various diseases in tomato; *Fusarium* wilt caused by *F. oxysporum* f. sp. *lycopersici* is one of the most devastating diseases of tomato (Srinivas et al. 2019). The toxigenic fungal genera infect tomato and produce different types of mycotoxins such as AFTs, sterigmatocystin, CIT, PAT, fumitremorgen, OTA, tenuazonicacid, ZEA and tricotesenler.

The species *F. oxysporum* f. sp. *lycopersici* (FOL) causes huge loss to vegetable crops. About 70 forms (special forms [f. sp.]) have been distributed; among these special forms FOL affects the tomato crops, causing *Fusarium* wilt. Mitogen-activated protein kinase (MAPK) and cyclic adenosine monophosphate (cMAP) are mainly responsible for the virulence of the FOL. Furthermore, cAMP-PKA and MAPK biosynthetic genes in FOL might regulate the infection process. In addition to the virulence genes, biosynthetic pathway genes of FOL, proteomics and transcriptomics are more useful for plant-pathogen interactions. OMICS approaches will help unrealize the possible virulence in tomato cultivars to assemble the gapless genome of FOL. This intuition to the succession of the constructional disease management system will offer new techniques for breeding resistance in order to resolve the huge losses of crops (Dong and Ronald 2019). Several FOL genes were also identified in 8 kb of a region of the chromosome, and the proteins released during host infection are xylem encoded cells (SIX1 and SIX 2) (Schmidt et al. 2016). The xylem of the eight fungal proteins of diseased plant in the similar region of chromosome. A top-down and bottom-up (shotgun) proteomic approach uses a high-throughput workflow to analyze the complex mixtures. These proteomic data can also be useful to complement to genomic data which would identify key and novel proteins. The use of advanced analytical equipment is centred around analysis by matrix-assisted laser desorption ionization (MALDI-TOF-MS), 2D gels and mass spectrometry. The MALDI-TOF-MS workflow also quickly analyses the protein sequences of intact protein targets of protein or peptide characterization.

12.2.3 Pineapple

Pineapple (*Ananas comosus* var. *comosus*) is an important perennial crop in tropical and subtropical regions. It belongs to the family *Bromeliaceae*. It is acknowledged as the "queen" of fruits due to its excellent taste and flavour, and also contributes to about 20% of the world's tropical fruit production. It can be consumed as canned pineapple slices, chips, juice, salads, syrup, alcohol, citric acid and puree. Pineapple contains important nutrients including sugars, potassium, vitamin C, calcium, fibre, vitamin B_1, vitamin B_6, copper and low amounts of fats and cholesterol. Pineapple plays an important role as an anti-inflammatory, antioxidant agent, for use in throat infection, arthritis, recovery from injuries and surgery (Rathnavelu et al. 2016). The enzymes obtained from pineapple are used to treat rheumatoid arthritis, diabetic ulcers and also improve blood circulation. Pineapple is a rich source of nutrients, but due to its acidic pH (4.5–6.5) it is more susceptible to various fungal infections such as *Aspergillus*, *Fusarium*, *Rhizopus*, *Penicillium* and *Candida*. Fusariosis is the most severe disease, caused by *Fusarium*, which infects all parts of the plant, mainly the fruit and stem apices. However, the fungal species that leads to this disease is ambiguous, and also causes fruitlet core rot. *Fusarium* has the ability to spread and completely colonize inside of perennial crops with only few detectable symptoms of infection. *Fusarium* species which infect pineapple fruit are *F. ananatum*, *F. concentricum*, *F. fujikuroi*, *F. guttiforme*, *F. incarnatum*, *F. oxysporum*, *F. polyphialidicum*, *F. proliferatum*, *F. temperatum* and *F. verticillioides*. However, FB, beauvericin (BEA) and MON are the mycotoxins produced by these species and can be detected in pineapple fruit. Based on in vitro and in planta mycotoxin synthesis comparison of the GFSC, moniliformin MON, BEA, FUM, fusaproliferin, fusarins and fusaric acid, TRI and ZEN are the major contaminants found in pineapple that affect human and animal health.

12.2.4 Grape

Grape (*Vitis vinifera L.*) is a worldwide grown, economically important crop belonging to the genus *Vitis*, family *Vitaceae*. Maharashtra is India's largest producer of grapes (Thompson seedless variety) and Nashik, also known as the wine capital of India, Karnataka, is the second largest producer of grapes in India(Bangalore Blue, Thomson seedless) followed by Tamil Nadu, Andhra Pradesh, Mizoram, Punjab, Haryana, Madhya Pradesh, Jammu and Kashmir and Himachal Pradesh (Patil et al. 2020). The various varieties of grapes can be used as a table fruit, and for making wine, raisins, jam, jelly and juice. It is a rich source of vitamin K, calcium, magnesium, manganese, potassium and phosphorus. Italy is one of the top grape producing countries, with an annual production of about 8.2 million metric tonnes, followed by France, the United States, Spain, Peru, South Africa, China and India. Low temperature and high humid conditions lead to various types of fungal infections in grapes which affect shoot, stem, leaves and fruits and ultimately lead to severe loss in yield. Fungi that infect grapes and cause numerous diseases and fungal species responsible for disease are depicted in Table 12.1.

TABLE 12.1

Effects of Different Fungal Species on Fruits and Vegetables

Name of the Fungi	Disease	Crop	Toxin
Fusarium oxysporum f. sp. cubense	*Fusarium* wilt	Banana	DON
Alternaria alternata	*Alternaria* rot	Banana, peach	FB1
Botrytis cinerea	Septoria	Peach	OTA
Penicillium expansum	Blue mould rot	Peach	PAT
Penicillium digitatum	Brown spot	Orange, banana	MON
Cylindrocarpon	Cordana	Banana	OTA
Acremonium	Leaf spot	Banana	AFs
Musicillium	Pitting disease	Banana	TA
Rhizoctonia	Cigar end rot	Banana	AME
Drechslera	Diamond spot	Banana	ALT
Botryodiplodia theobromae	Tip end rot	Banana	TEA
Colletotrichum musae	Anthracnose	Strawberry, banana	MON, ZEN
Pseudocercospora musicola	Sigatoka	Banana	PAT
A. alternata	Alternaria brown spot	Mandarin, orange, lemon, melon, apple, berry	ATX-I, -II, -III, ALT, AME
A. clavitus	Alternaria brown spot	Apple, cherry	PAT
A. amstelodami	Alternaria brown spot	Apple, cherry	AOH
A. longivesica	Alternaria brown spot	Apple, cherry	PAT
A. terreus	Brown rot, blossom blight	Apricot	PAT
A. alternata	Alternariosis	Grape	PAT
Aspergillus niger	Berry rots	Grape	OTA
C. herbarum	Grape black rot	Grape	CIT
Penicillium spp.	Angular leaf spot	Grape	PAT
Rhizopus stolonifer	Black rot	Grape	CIT
R. arrhizus	Bot canker	Grape	CIT
Fusarium equiseti	Fusariosis	Grape	PAT
F. moniliform	Fusariosis	Grape	CIT
F. scirpi	Fusariosis	Grape	CIT
F. solani	Fusariosis	Grape	PAT
Plasmopara viticola	Downy mildew	Grape	CIT
Gloeosporium ampelophagum	Anthracnose	Grape	PAT
Guignardia bidwellii	Black rot	Grape	PAT
Septoria melanosa	Brown spot	Grape	OTA
Cercospora sessilis	Cercosporose	Grape	OTA
Uncinula necator	Powdery mildew	Grape	PAT
B. cinerea	Gray rot	Grape	PAT
Pseudopezizatracheiphia	Rotbrenner	Grape	PAT
Eutypalata	Trunk and arm dieback	Grape	CIT
Phomopsis viticola	Trunk and arm dieback	Grape	OTA
Botryosphaeria sp.	Trunk and arm dieback	Grape	OTA
Coniella diplodiella	White rot	Grape	OTA
Armillaria mellea	*Armillaria* root rot	Blueberry	AOH, AME
A. gallica	*Armillaria* root rot	Blueberry	AOH, AME
Phytophthora cinnamomi	*Phytophthora* root rot	Blueberry	TEN
A. tenuissima	*Alternaria* fruit rot	Blueberry	AFB1
Monilinia vaccinia-corymbosi	Botrytis blight	Blueberry	AFB2
Colletotrichum acutatum	Ripe rot	Blueberry	AFG1, AFG2
Phomopsis vacinii	Twig blight	Blueberry	OTA
Sclerotium rolfsii	Southern blight	Tomato	PAT
Verticillium dahlia	*Verticillium* wilt	Tomato	AFTs
Colletotrichum spp.	Anthracnose	Tomato	STC
Phytophthora spp.	Buckeye rot	Tomato	CIT
A. solani	Early blight	Tomato	FTM (fumitremorgen)
Septoria lycopersici	Septoria leaf spot	Tomato	OTA
A. tomatophila	Early blight	Tomato	Tenuazonic acid and ZEA

TABLE 12.1 *(Continued)*

Effects of Different Fungal Species on Fruits and Vegetables

Name of the Fungi	Disease	Crop	Toxin
F. oxysporum f. sp. *lycopersici*	*Fusarium* wilt	Tomato	Tricotesenler
F. guttiforme	Fusariosis	Pineapple	FB
F. concentricum	Fruitlet core rot	Pineapple	BEA
F. fujikuroi	Fusariosis	Pineapple	MON
F. ananatum	Fruitlet core rot	Pineapple	FUM
F. incarnatum	Leaf spot	Pineapple	Fusaproliferin
F. oxysporum	Pineapple fruit rot	Pineapple	Fusarins
F. polyphialidicum	Pineapple fruit rot	Pineapple	TRI
F. proliferatum	Pineapple fruit rot	Pineapple	ZEN
F. temperatum	Leaf spot	Pineapple	GFSC
F. verticillioides	Leaf spot	Pineapple	Fusaric acid
A. alternata	Red spot	Apricot	AFTs
A. nominus	Black spot	Apricot	AFTs, PAT
P.verrucosum	Ripe fruit rot	Apricot	OTA
P. Nordium	Ripe fruit rot	Apricot	OTA
A. ochraceous	Brown rot	Apricot	OTA
A. niger	Brown rot	Apricot	OTA
M.fructicola	Brown rot	Apricot	PAT
Paecilomyces spp.	Brown rot	Apricot	PAT
Zasmidium citrigriseum	Greasy spot	Orange	Alternariol
Diaporthecitri	Melanose	Orange	AFTs
Elsinoëfawcettii	Scab	Orange	PAT
Phyllosticta citricarpa	Black spot	Orange	ZEN
Colletotrichum acutatum	Postbloom fruit drop	Orange	
B. cinerea	Graymould	Strawberry	AFTs, CIT
P. cactorum	Leather rot	Strawberry	PAT
C. truncatum	Anthracnose, fruit rot	Capsicum	TRI, ZEA, CIT
C. gloeosporioides	Anthracnose, fruit rot	Capsicum	OTA, PAT
Cercospora abelmoschi	*Cercospora* leaf spot	Capsicum	AFB1
F. oxysporum f. sp. *vasinfectum*	*Fusarium* wilt	Capsicum	Gliotoxin
Erysiphe cichoracearum	Powdery mildew	Capsicum	DON
Pythium spp., *Rhizoctonia* spp.	Damping off	Capsicum	CIT
Curvularia	Stalk rot	Cauliflower	Nivalenol
Aspergillus	Black rot	Cauliflower	OTA
Helminthosporium	Clubroot	Cauliflower	FUM
Pyricularia	Downy mildew	Cauliflower	ZEA
Trichothecium	Stem rot	Cauliflower	AFTs
F. crookwellense	Fusariumdry rot	Potato	DAS
F. avenaceum	Fusariumdry rot	Potato	ZEA (F-2)
F. culmorum	Fusariumdry rot	Potato	DON
F. equiseti	Fusariumdry rot	Potato	3-AcDON
F. acuminatum	Potato dry rot	Potato	MON

12.2.5 Peach

Peach (*Prunus persica* L.) is a small climacteric fruit belonging to the family *Rosaceae* which grows in dry, continental or temperate climates. Its nectar is used as baby food for children and infants. More than 2000 varieties of peaches are available today. Peaches are high in vitamins, fibre and minerals and have a large amount of potassium, phosphorus, vitamin B and also provitamin A and vitamin C. Consumption of peaches helps to develop resistance against infections and helps in eliminating harmful free radicals that may cause cancer. Peach can help to control obesity-related diseases like diabetes and metabolic syndrome. Peaches are reported to have anti-inflammatory properties which help in reducing cholesterol levels associated with cardiovascular disease. Peach and its products are more susceptible to contamination by *Penicillium expansum*, *Alternaria alternata* and *Botrytis cinerea*. However, *P. expansum* is the major agent causing blue mould rot of various fruits including peach. *Botrytis cinerea* causes peach spoilage and is reported to be pathogenic to more than 200 species of fruits. *Alternaria* cause *Alternaria* rot and are involved in spoilage of stored fruits during refrigeration, as they have the ability to grow at low temperatures (Abata et al. 2017). Peach is highly susceptible to mould invasion and mycotoxin contamination. Peaches are majorly contaminated with OTA, and PAT is mainly produced in rotten parts of the fruit.

12.2.6 Blueberry

Blueberry (*Vaccinium* section *Cyanococcus*) is one of the most nutritious fruits, belonging to the family *Ericaceae*, and has antioxidant, anti-diabetes, anti-obesity, anticancer, anti-osteoporosis, anti-neurodegenerative and anti-inflammatory properties. It also helps to maintain bone strength, mental health. Blueberry is rich in anthocyanins, which improve night vision and reduce risk of muscular degeneration and heart diseases. Several fungi infect blueberries and cause a variety of diseases. Fungal pathogens can attack blueberry fruit during preharvesting and postharvesting.

Fungal spoilage associated with blueberries is species of *Botryosphaeria*, *Fusicoccum*, *Dothiorella*, *Phomopsis*, *Cylindrocladium*, *Curvularia*, *Phoma*, *Phytophthora*, *Penicillium*, *Aspergillus*, *Alternaria*, *Botrytis*, *Colletotrichum*, *Phomopsis*, *Monilinia*, *Rhizoctonia*, *Fusarium*, *Pestalotiopsis*, *Pucciniastrum*, *Dothichiza*, *Nigrospora* and *Rhizopus*. The fungal species that produce mysterious toxins in blueberries are alternariol (AOH), alternariol monomethyl ether (AME), tentoxin (TEN), AFB1, AFB2, AFG1, AFG2, OTA.

12.2.7 Apricot

Apricot (*Prunus armeniaca* L.) is a stone fruit which is found in the Mediterranean climate and matures during early summer. Apricot is an excellent source of iron, potassium and vitamin A. India produces 18,000 tons of apricots every year. Turkey is one of the world's largest apricot producers, with an annual production of 20.6%, and export was 77.8% world production (ERS, 2018). Several studies were performed on the natural occurrence of mycotoxins in apricot which showed the contamination of AFT, OTA, PAT, and Alternaria toxins. Among these, AFTs are produced by *A. flavus*, *A. parasiticus* and *A. nominus*, while OTA is produced by *P. verrucosum*, *P. Nordium*, *A. ochraceous* and *A. carbonarius*, and PAT is produced by *Aspergillus terreus*, *Byssochlamys*, *Paecilomyces* and *Penicillium* species. AFTs and OTA are observed in apricot-derived products, which are used by people in their daily diets.

12.2.8 Orange

Oranges are one of the most important commercially demanded fruits in India, taking about 40% of the total Indian land, producing various delicious species like *Citrus sinensis*, *Citrus reticulata* and *Citrus aurantifolia*, commonly known as sweet orange, mandarin and acid lime, respectively. Oranges – especially Nagpur oranges – are exported to China and European countries. Orange is a vital fruit consumed in the population across the world. Problems like infections by several bacterial and fungal species lead to the production of several toxic secondary metabolites including PAT, alternariol, AFTs, etc. The prevalence of the mycotoxins in fruit also leads to several harmful effects on infants and children. Multi-occurrence of mycotoxins in the fruit or food products can lead to a wide array of cancers in humans. The range of mycotoxins causes critical reduction in male reproductive efficiency, including epimutations and infertility (El Khoury et al. 2019).

12.2.9 Strawberry

Strawberry is one of the favourite fruits of children as well as adults in the summer season. It is exported to around 111 countries. Highly demanded seasonal fruits, strawberries are cultivated and procured in a systematic way. However, many infections occur through various means, leading to the production of severe toxic secondary metabolites like those of AFTs, PAT, CIT, etc.

12.2.10 Capsicum

Capsicum is the genus of flowering plants belonging to the family *Solanaceae*, native to Latin America. It contains about 38 species. Most of the varieties of pepper are pungent; those that are non-pungent are consumed as vegetables. The global market for capsicum was US $ 8686.7 million in 2018 and is predicted to reach upto US $12240 million by 2025 (Costa et al. 2019). It is rich in proteins, carbohydrates, minerals, vitamin C, vitamin B_6, cryptoxanthin and some other antioxidant molecules, which play an important role against various types of cancers, cardiovascular diseases, stroke, atherosclerosis and cataracts. It also helps in treatment of arthritis, rheumatism, pulmonary diseases, skin rashes, oedema, haemorrhoids and leprosy, and it is also used as a food supplement and additive. Capsicum extract has a range of biological properties, which include antioxidant, antidiabetic, chemoprotective, gastroprotective and several metabolic syndromes. The major genera of fungi that infect capsicum are *Alternaria*, *Aspergillus*, *Fusarium*, *Penicillium*, *Cladosporium*, *Colletotrichum* and *Rhizopus*. Among *Aspergillus*, *Fusarium* and *Penicillium* are the potent mycotoxin producers, which are able to produce AFTs, OTA, GT, TRI, FUM, ZEA, PAT and CIT during pre- and postharvesting conditions of storage, drying and transportation. These mycotoxins can lead to a number of acute and chronic diseases or even death in humans and animals.

12.2.11 Okra

Okra (*Abelmoschus esculentus*) belongs to the family *Malvaceae*, originated in Ethiopia and spread in various parts of the world such as North Africa, the Mediterranean, Arabia, and India. It is an economically important crop cultivated in tropical and subtropical areas of various countries including India. India is one of the top okra producers (67.1%). Okra is a rich source of fats, fibre, proteins, vitamins, folate, mucilage and minerals. Mucilage has medicinal applications in detoxification of toxic compounds in the liver. The seeds of okra are a rich source of oil (20–40%) which contains nutritionally important polyunsaturated fatty acid and linoleic acid; the protein component contains a large amount of essential amino acids as compared to soya bean and other plant proteins. Its soluble fibre plays an important role in lowering serum cholesterol level and reduces the risk of heart disease. Okra is well known for its high antioxidant properties and some potential effects in human diseases like type II diabetes, digestive disorders, genitourinary disorders, spermatorrhoea and various types of cancer. Besides its nutritional and medical applications, it can be used in pharmaceutical and herbal formulations in various industries to cure diseases. Fungal infection in okra leads to several diseases such as *Cercospora* leaf spot (*Cercospora abelmoschi*), *Fusarium* wilt (*Fusarium oxysporum* f. sp. *Vasinfectum*), powdery mildew (*Erysiphe cichoracearum*), damping off (*Pythium* spp., *Rhizoctonia* spp.) and associated *Alternaria* and *Aspergillus* species that lead to severe loss in yield (Singh et al. 2018).

12.2.12 Cauliflower

Cauliflower (*Brassica oleracea*) is a well-known cash crop which is a member of *Brassicaceae* family. It has various medicinal properties as an anticarcinogenic agent and anti-neutralizing agent. Cauliflower is rich in vitamins and minerals and also has tremendous nutritious significance such as low calorific value, no cholesterol and low-fat content. It also acts a good source of vitamin K, thiamine, riboflavin, niacin, manganese phosphorus, fibre, vitamin B_6, folate, pantothenic acid, potassium and manganese. It helps reduce risk of cancer and brain and heart problems (Szalay, 2018) and also helps to prevent oxidative stress, stomach disorders, several respiratory problems and immune disorders. Several diseases affect cauliflower, like stalk rot, black rot, clubroot, downy mildew, stem rot and verticillium wilt. India ranks second for the production of cauliflower with about 29.9% global production (Egel et al. 2018). Cauliflower is an essential crop that is more prone to fungal infections mainly by *Aspergillus*, *Penicillium*, *Fusarium*, *Alternaria*, *Curvularia*, *Helminthosporium*, *Rhizoctonia*, *Pyricularia* and *Trichothecium*. These fungal genera cause various types of diseases and also lead to mycotoxin production that impairs the viability and quality of cauliflower. Cauliflower is likely to be contaminated by a variety of fungal species that leads to extensive loss of yield. The mycotoxigenic fungal species isolated from infected cauliflower have been successfully identified and their toxigenic potential has also been determined, which shows the TA, alternariol, alternariol monomethyl ether, altenuene and tentoxin as the major mycotoxins produced by *Alternaria* species.

12.3 Identification and Detection of Mycotoxigenic Fungi

Fungal identification and differentiation of closely related species are very difficult; however, predict the progression of species by metabolite profiling is very important for precise identification. The metabolic profiling can be targeted by use of metabolic pathway genes respective to specific metabolite. Importance in secondary metabolite production includes mycotoxins, its chemistry and biology remains a challenging task (Raja et al. 2017). Morphology alone gives misidentification due to diversity and ecology of the species of *Aspergillus*, *Fusarium* and *Penicillium* as well as other species. Phenotypic differentiation of most of the fungal species/strains has been performed based on morphological characterization in different culture medium. In addition, the appearance and authentication of these food- and feed-borne fungal strains is typically complicated. Precise identification of toxigenic fungi is a predefined criterion for efficient execution of the various diseases, including population inheritance studies. The use of precise nucleic acid-based techniques, such as polymerase chain reaction (PCR), has smoothened the approach for molecular phylogeny of fungal species to promote food safety and security. Furthermore, in microbial taxonomy and molecular phylogeny, a number of PCR assays have been developed for the fungal species producing mycotoxins. However, secondary metabolite-based classification also plays an important role for modern taxonomists for the natural classification and its evolutionary criteria of the fungal species.

Traditional identification of fungi including mycotoxigenic *Aspergillus* species further authenticated with molecular methods, and DNA/RNA sequencing may attribute to a greater extent in identification, depending on the aim of the study. Genomic sequencing is important in the description of new species – not only those derived from food but also from environmental sources. Fungal internal transcribed spacer (ITS) is the first authorized DNA barcoding marker from a precise gene used for the identification of fungi. Due to the limitation of ITS for better segregation of closely related species/species complex in the fungal genome, various other protein coding region/genes such as translation elongation factor 1-α (TEF1α), calmodulin (CaM), beta-tubulin (BenA) and DNA-directed RNA polymerase are commonly used to explain phylogenetic relationships among fungi (Amol et al. 2019). Currently, the mini-chromosome maintenance supermolecule (MCM7) has been used as an outstanding marker for inferring each higher and lower level phyletic relationships. Furthermore, molecular markers are extensively used to study high genotypic diversity and variability in the species complex. These markers and/or probes are also used to determine the species diversity and differentiate novel species. Identification and detection of toxigenic fungi in fruits and vegetables is the main task for researchers because of their genomic variability and complexity. Many fungi cause several severe diseases and also are able to produce different types of mycotoxins that affect fruit quality as well as productivity and result in harmful effects to consumers. A molecular approach is one of the best ways to overcome this problem because it reduces the time and cost and gives accurate identification at the species level. PCR-based detection of toxigenic fungi is considered a good alternative due to its specificity and sensitivity, particularly when multi-copy sequences are used to design species specific primers. Other molecular approaches such as amplified fragment length polymorphism (AFLP) and random amplified polymorphic DNA (RAPD) help in developing specific marker sequences that are used for species-specific primer designing that allow identification and characterization of toxigenic fungi. PCR-ITS-RFLP analysis is a powerful tool used for molecular identification and characterization of fungi from spoiled and healthy pineapple fruits, using different endonucleases such as SduI, HinfI, HhaI, NlaIII, HaeIII and RsaI would differentiates *Fusarium oxysporum* f. sp. *cubense* in banana from other races. The morphological, vegetative compatibility group (VCG) analysis, volatile production and PCR using Foc tropical race (TR) specific primers give accurate identification of Foc in banana. Inter simple sequence repeats (ISSRs) provide the characterization and polyphyletic nature of the Fol. Thus, ISSR molecular typing and PCR are potent tools for complete analysis of genetic diversity and toxigenic potential of *F. oxysporum* in banana. The ISSR analysis used to study the diversity at species level specified that there is a wide genetic variability among the strains, whereas ITS-based phylogenetic analysis showed four lineages of fungal pathogens. Further analysis confirmed that there is intra-species variation, nucleotide insertions and deletions in the ITS region which is clade specific. The pathogenic strain also showed toxigenic potential which is determined by targeting their metabolic pathway genes specific to *fum1* for detection of FUM production which clearly indicated that the strains infecting tomato crops have potential to produce FUM (Nirmaladevi et al. 2016). Databases of genomes and genetic markers used as sources for molecular barcodes are being created, and the fungal world is in progress to be unveiled with the help of bioinformatic tools.

The fruits and vegetables contaminated by various diverse fungal species are known to produce numerous secondary metabolites which are toxic in nature. Toxins play diverse roles, which kill the cells and promote infection. Furthermore, other chemotypes interfere with programmed cell death which leads to necrosis and pathogenesis. These toxic molecules or metabolites are derived from specific biosynthetic pathways present within the gene clusters, and some of these clusters transfer between the species (Villafana et al. 2019). Conserved region/homology within a toxin biosynthetic gene from different species of fungi was used to detect the presence of the toxigenic species highly specific for the particular toxin which also was reported to cause disease to the plants and animals (Okoth et al. 2018). The molecular detection of these genes can be done by targeting their metabolic pathway genes respective to toxins which are specific for genus and species. Furthermore, the multiple toxin-producing genes can also be detected by using multiplex PCR in a single run. This technique is not only used for the detection of multi-toxins/toxin-producing fungi but also useful for detection of plant pathogenic fungi in horticultural crops. In fungi, biosynthetic pathway genes are located in clusters in different loci and relative expression of respective genes depends on the coexisting gene in the cluster. The advanced tools and techniques used to quantitate the expression of genes can be evaluated using real-time quantitative PCR (qPCR) (Abdullah et al. 2018). The production and biosynthesis of mycotoxins is a highly complex process, which is influenced by environmental factors and needs to be understood at the transcriptional level.

Mycotoxin production in the food chain can have severe effects on the health of humans and animals; therefore, rapid detection of mycotoxins is important to ensure food safety and quality. The appraisal of mycotoxin production levels and toxigenic fungal species is important to evaluate food quality and improvement of executive approach. Traditional methods augment the cost and analysis period. Molecular methods are a better substitute to predictable methods for recognition of toxic fungi both qualitatively and quantitatively (Nwachukwu et al. 2019a). Furthermore, comparatively monoplex PCR, multiplex PCR (mPCR) based screening of toxin chemotypes reduce the cost of analysis and is important for co-occurrence of mycotoxins in various food and feed systems. mPCR assay is a sensitive, rapid and reliable tool that simultaneously detects and differentiates mycotoxin-producing and non-producing strains. Moreover, amplification of species-specific, structural or regulatory genes involved in toxin biosynthesis issued for toxin detection (Nwachukwu et al. 2019). Furthermore, PCR-based detection of AFB1, OTA, ZEA, FUM and PAT works by targeting key gene involved in their biosynthesis for toxigenic fungal species. Also, in PAT biosynthesis, 15 genes were recognized in the genome of *A. clavatus*. Recently, *P. griseofulvum* and *P. expansum* have also been detected as PAT producers. The CIT biosynthesis includes non-reduction polyketides of polyfunctional proteins (nrPKS), which include putative domain methyltransferase (MT), acyl transferase (AT), acyl carrier protein (ACP) and ketosynthase (KS). Of the TR biosynthesis involved in metabolite export (*terG*, *terJ*), *terA* and *terR* are the key genes responsible for TR production used as a key gene for detection.

Gene-specific PCR assays are widely used, sensitive methods to identify essential genes involved in mycotoxin biosynthetic pathways as well as to find out the toxigenic potential of fungi. The FUM biosynthetic pathway has been well studied in *Fusarium* species, whereas the *fum1* gene is particularly used to study the FUM producing fungi. Recently, the structure of the BEA biosynthetic gene cluster in *F. proliferatum* has been discovered and genes responsible for the biosynthesis of fusarins and fusaric acid by *F. verticillioides*. This includes TRI5 and TRI13 genes from the TRI biosynthetic cluster, PKS4 and PKS13 genes from ZEA. However, *F. graminearum* TRI13 gene is responsible for NIV production and was amplified using a DNA template extracted from pineapple skin samples. Although it is difficult to purify *F. graminearum* isolates from this plant material, the PCR-based approach for identification of biosynthetic pathway genes to predict the presence of mycotoxins can be used. It is very sensitive, even for complex matrices such as uncultured strains present in the plant material. The potential mycotoxin producing fungal species in grapes are *A. niger*, *A. carbonarius*, *Fusarium* spp. and *Trichothecium roseum* which produce OTA, AFB2, MPA, CPA, TRI, FB1 and ZEN. The detection of OTA was carried out by targeting its biosynthetic pathway gene such as polyketide synthase (PKS) which is involved in the initial steps of the pathway and non-ribosomal peptide synthetase (NRPS) which plays an important role in the formation of the peptide that links the polyketide and phenylalanine which is the key step in OTA biosynthetic pathway.

12.3.1 Analytical-Based Detection of Mycotoxins in Fruits and Vegetables

Globally, several researchers are actively engaged in developing methods ranging from conventional thin-layer chromatography (TLC) densitometer to advanced immunosensors for detecting mycotoxins in food and other samples. However, chromatographic methods like HPLC, HPTLC, GC, liquid chromatography with tandem mass spectrometry(LC-MS/MS), fluorescence spectrophotometry, frontier infrared spectroscopy, fluorometer, Fourier-transform infrared spectroscopy (FTIR), radioimmunoassay (RIA), ELISA, lateral flow devices (Immunodipstick), surface plasm on resonance (SPR) and electro-chemical immunosensors are routinely used for the detection and quantification of mycotoxins in agricultural food crops (Li et al. 2012). The analytical detection of these mycotoxins has become a widely used tool because of the involvement of more specific and sensitive methods such as high performance liquid chromatography with fluorescence detector (HPLC/FLD), LC/MS/MS and ultra-high pressure liquid chromatography (UHPLC), which have more advantages than the traditional ones, such as smaller sample requirement, multiple samples can be processed in a single run, reduced time, rapid diagnosis and high sensitivity. A gas chromatography coupled with tandem mass spectrometry (MS/MS) and dispersive liquid-liquid microextraction (DLLME) able to detect various toxins (PCA, MPA, CPA, FB1, ZEN, OTA, AFB2, DON, 3-AcDON, 15-AcDON, NEO, DAS, NIV, ZON, α-ZOL, β-ZOL, α-ZAL, β-ZAL, FUS X, T-2, HT-2 and PAT) in fruit juice (Carballo et al. 2018).

To date, monoclonal/polyclonal antibodies are used for the detection of several toxins; however, cross-reactivity is another challenging task. Detection of mycotoxins in various cereal-based foods using immunoassay techniques has increased in the last decade, including field-based immunochromatographic test strips carried out on the basis of interaction among the toxins, and antibody-coated nanomaterials. Lateral-flow devices (LFDs) have evolved and are commercially used for the detection of mycotoxigenic species and their toxins in agricultural produce. Quantitative detection of mycotoxins in food and feedstuffs was improved by the advancements in portable photometric strip readers. Moreover, in recent years aptamer-based diagnostic methods have also been developed by several researchers targeting AFB1, FB1 and OTA, but no attempts were made to adapt developed technologies to prepare point-of-care (POC) diagnostic platforms. The detection of these mysterious toxins by sensitive and field-deployable rapid POC diagnostics using aptamers through the application of novel nanobiotechnological approaches is more feasible than with conventional methods. However, LC-MS/MS yielded efficient reproducibility for the detection of mycotoxins from apple puree used as infant food (Malachová et al. 2014). Another approach for the detection of mycotoxins reported the use of mid-infrared/attenuated total reflection (ATR) measurements and principal component analysis (PCA) followed by cluster analysis classification. Systematic preparation of anti-CIT monoclonal antibody (McAb) 2B9, CIT-BSA and McAb 2B9 conjugates was carried out for reliable quantification within a short time.

12.3.2 Aptamers in Detection of Mycotoxins in Horticultural Produce

Aptamers (Apt) are third-generation molecular probes. "Apt" is short oligonucleotide or peptide molecules classified into DNA/RNA/XNA with specificity and affinity towards the target molecule. They can be readily used for the detection of the predetermined molecule and can fold into three-dimensional structures, helping in efficient binding to the targets. Apt are lower molecular weight and size (less than 10 kDa) than the antibodies, so they can be effectively substituted with the aptamers. This upcoming synthetic material is prepared through systematic evolution of ligands by exponential enrichment (SELEX) (Sefah et al. 2010). To date, several SELEX methods have been employed including quantum dots, reduced graphene oxide-SELEX (rGO-SELEX), capture-SELEX, immuno-SELEX, high affinity resins, etc. At one time, one generates several billion copies of the selected aptamer. Animals or high-end in vitro animal cell culture facilities are not needed. Hence, the cost and labour are cheaper. Moreover, the specificity and sensitivity of the aptamers are on par with any given antibodies. In fact, aptamers have many advantages over antibodies. Fungal toxins being non-protein organic compounds, generating specific and sensitive antibodies is always a difficult job with several constraints. However, aptamers can easily be synthesized in a laboratory with simple PCR and million copies can be generated in limited time. It is cost-effective when compared to the current probes such as antibodies and sensors. These are versatile and minimize the cross-reactions that can lead to false results. To date, no detection methods have been developed based on aptamers for fungal toxins. Hence, APT opens up new technology solutions for rapid and robust detection of fungal toxins in agricultural produce. These toxins can be detected by sensitive and field-deployable rapid POC diagnostics using aptamers through application of novel nanobiotechnological approaches. Since the probes and molecules used in the development of these platforms can be easily synthesized in the lab and the materials are also available in lower prices, the cost for each testing would come down considerably. These are versatile and minimize the cross-reactions that can lead to false results. In the present scenario, FDA-approved aflatoxin kits such as AflaTest (VICAM Co.), Agriscreen (NEOGEN Corp.), AflaCup 10 (Romer Labs), AflaCup 20 (Romer Labs) and EZ-Screen (MedTox) are available in the market. So, POC is accessible to all types of food manufacturing industries.

These aptamers specifically recognize and bind to their targets. They are alternative to antibody-based systems, and have several advantages, particularly thermal and chemical stability, practical synthesis and increased binding affinity. These unique characteristics of aptamers have great potential for detection of biomolecules. These novel properties make aptamers the material of choice for highly sensitive biosensing platforms (aptasensors), conjugated with nanomaterials with highly specific recognition abilities and a unique way to analyze target analytes in food and feed samples.

12.3.3 Nanobiotechnology in Mycotoxin Detection

Nanotechnology is the most studied branch of science, with tremendous application and great scientific interest as it is a bridge between bulk materials and atomic or molecular structures that are small in size and have a high surface area (Kumar et al. 2018). Magnetic nanoparticles (NPs) have gained wide attention due to their super-paramagnetic behaviour, and they show strong magnetic dipole-dipole attraction and ferromagnetic behaviour. Super paramagnetic iron oxide NPs with modified surface chemistry can be used for numerous in vivo applications such as MRI contrast enhancement, tissue repair, DNA nanotechnology and immunoassay, detoxification of biological fluids, hyperthermia, drug delivery and cell separation. Also, these NPs are being used in the electrical, biological, textile and pharmaceutical industries, and in paints, dyes, preparation of nanobiocomposites and chemistry. They are also used in the development of biosensors, biomedicine and nanobiotechnology, drug delivery, medical diagnostic tools, cancer treatment, as antimicrobial agents and for detection and separation of a pathogen. Furthermore, iron oxide NPs adsorb fluorescently labelled DNA oligonucleotide with the help of backbone phosphate, and quench the fluorescence. Also, iron oxide NPs attached to DNA can detect arsenate by using phosphate backbone instead of the base pairs.

There has been a tremendous increase in the use of NPs for the purpose of biosensor development in the recent years. Nanostructures such as NPs, nanowires and nanorods are used as promising materials for construction of biosensors helping in specificity of the sensors. Implementation of nanomaterials in the development of nano-biosensors and their use for the detection of mycotoxins in food and feed is highly essential (Sertova and Ignatova 2019). Rather than only aptamer alone, aptamer-conjugated with the nanoparticles bring high local densities. Also, NPs provide a larger surface area, helping in the attachment of the biomolecules like enzymes, aptamers, antibodies, etc. Various nanomaterials including magnetic NPs, metal oxide NPs and the noble metal NPs are used in a wide range of selection processes of the different analytes.

They possess size as small as 1–100 nm, having properties like the electronic, chemical as well as optical differing among the bulk material, eventually endorsing their use in the preparation of biosensors. The noble metal possesses exclusive physicochemical properties which are highly beneficial for the development of nano-size biosensors. Along with the several noble metal NPs, many biocompatible and optically active metals including gold as an efficient candidate are used on a large scale. For an even broader methodology, nano-silver has also been used for sensing of mycotoxins using the aptamer (Sojinrin et al. 2019).

12.3.4 Quantum Dots

Quantum dots (QDs) are nanosized (2–10 nm) man-made semiconductors which are used in the electron transport based on electrical current variation. They possess electronic and optical properties which make them different from bigger size particles due to quantum mechanics. QDs are now used extensively in the preparation of FRET-based fluorescence biosensors for detection of antigens, mycotoxins, antibodies, etc. They also have the advantage of a narrow emission band which helps in permitting the detection of mycotoxins, antibodies, DNAs, etc. QDs hold properties like size-tuneable spectra, luminescence, high stability, high photo-resistance and higher quantum yields, which make them efficient tools for FRET-based detection. QDs can be effectively moulded into required shapes so as to coat with the biomaterials for the development of biosensors. Moreover, the specificity of the aptamer is not disturbed due to the conjugation of QDs with the aptamer, hence is very reliable during studies. Also being more photostable and 10–20 times brighter compared to the common dyes and less prone to destruction, helping in long-term and efficient cellular tracking processes. They also possess a higher signal-to-noise ratio than organic dyes. After the excitation energy exceeds the band gap, absorption of the photons is carried out using QDs, followed by shift of the electrons from the ground state to the excited state (Liu et al. 2019).

12.3.5 Carbon-Based Nanoparticles

Carbon based NPs are of great interest and are employed in the fabrication of various sensing platforms. Carbon NPs are a good choice for biosensors due to the characteristic of chemical inertness. Carbon NPs include graphene oxide used in various detection studies of mycotoxins.

12.3.6 Graphene Oxide

Graphene oxide (GO) forms the π-π stacking interaction, van der Waals force, also hydrogen bonds with the ssDNA, which helps in specific detection methods. GO ideally consists of two important components. The first is the fluorophore-labelled aptamers and the second is the water-soluble GO sheets, where the first is used as the probe and the second as the substrate, which is why it can be used for the process of FRET. When there is no target present, the aptamer is adsorbed onto the GO sheet using the π-π stacking interaction among the aromatic rings of aptamer-exposed nucleotides and the sheet giving no fluorescence emission, whereas in the presence of the target molecule, the target molecule binds to the aptamer and this complex is released from the surface emitting fluorescence (Wang et al. 2019). GO can be efficiently used for processes like cell labelling and protein detection. It has larger surface areas, better photoluminescence, surface modifications and optical, thermal, structural and electronic properties including the quantum Hall effect or fluorescence quenching ability. Also, it is highly sensitive, specific, selective and lower cost which helps in the detection of various molecules like proteins, DNAs, RNAs, metal ions, pollutants, mycotoxins, etc.

12.3.7 Semiconductor Quantum Dots

Quantum dots (QDs) reported interesting and novel characteristics and have been studied in both fundamental research and practical application. In the past two decades, significant advances have been made in the synthesis of colloidal semiconductor (QDs), especially II–VI compounds such as cadmium sulphide (CdS), cadmium selenide (CdSe) and cadmium telluride (CdTe), which are very stable. Also, it is quite possible to distinguish different kinds of antigen with CdTe QDs at different sizes. These highly visible-luminescent nanomaterials are very promising for various applications in optoelectronics and biological labelling (Medintz et al. 2006). These QDs are of very good crystallinity, with a spherical shape and strongly emit in the green-to-red spectral region (500–700 nm) with a luminescence quantum yield (LQY) of 30–85%. Extremely light-emitting QDs have a wide variety of research applications. CdTe/CdS QDs combined with nanoparticles allow various targeted pathogens and influenza H5N1 viruses to be identified. Gold nanoparticles (GNPs) have received special attention because of their unique properties such as colorimetric, conductivity and nonlinear optical properties and simple functionality with biological recognition components.

They hold great promise for biological and medicinal applications. AuNPs functionalized with oligonucleotides (ONT-AuNPs) have emerged as a kind of novel nanomaterial for diagnosis and therapy, taking advantage of the highly efficient fluorescence quenching properties of AuNPs for proximately fluorescent dyes through energy-transfer processes.

12.3.8 Fluorescence Resonance Energy Transfer

Fluorescence resonance energy transfer (FRET) is a physical phenomenon being used more and more in biomedical research and drug discovery today. FRET relies on the distance-dependent transfer of energy from a donor molecule (chromophore) to an acceptor molecule, in which the energy is subsequently transferred. It occurs over greater than interatomic distances, without any conversion to thermal energy, and molecular collision. The transfer of energy leads to a reduction in the donor's fluorescence intensity and excited state lifetime, and an increase in the acceptor's emission intensity. Considering the advantages of the FRET-based assay for the detection of these mysterious toxins in food and feeds, we wish to develop a nontechnology-based platform utilizing the properties of FRET between QDs and AuNPs. The interactions between ssDNA, nanomaterial and the target were combined to develop a sensitive and selective method for fluorescence quenching-based detection platform. In rGO-SELEX, the reduced graphene oxide interacts with the nucleotides and helps in the selection process. This is an immobilization-free process and hence lesser time consumption, whereas the capture-SELEX involves immobilization of the nucleotide sequences leading to the specific selection of the ssDNA against the target. However, the process of using magnetic beads for the selection is another method usually involving the principle of affinity chromatography, with several rounds eventually producing highly specific aptamers for detection (Kumar et al. 2018). Immuno-SELEX includes preparation of the anti-target antibodies and the use of sandwich ELISA principle for better efficacy and authentication of the aptamer. In another study, Brevetoxin-2-bovine serum albumin (BTX-2-BSA) was coated on a microtitre well plate for the selection of specific aptamers binding to the target toxin BTX-2. A study was performed by immobilizing the kanamycin on the beads used for affinity column chromatography. Here, the FAM-labelled aptamer was used for the colorimetric assay for the detection of kanamycin.

12.3.9 Magnetic Nanoparticles

Magnetic nanoparticles (MNPs) are biocompatible and their diversely fulfilling applications make them an efficient molecule for sensing purposes (Becheva et al. 2019). MNP surfaces can be easily chemically modified with desired functional groups and can be further used as the detecting probe for the detection of several biomolecules. Their larger surface areas indeed help in the sufficient enhancement of the kinetics of the detection process leading to improvement in the efficiency of immobilization. Also, MNPs can be used in the fields of diagnostics, medicines, food safety, etc. due to unique characteristics. Commonly used MNPs include iron oxide NPs for the detection of mycotoxins from food samples, protein purification, as well as immobilization of enzymes.

12.3.10 Electrochemical Sensors

Electrochemical sensors are basically made up of the electrodes which possess highly sensible and efficient modified surfaces with the help of electrical current variations for the detection of mycotoxins (Santos et al. 2019). Electrochemical impedance spectrometry allows the sensitive monitoring of changes in conductivity/resistance or charging capacity of an electrochemical interface. The bio-electrochemical property serves as a transduction component in electrochemical biosensors. The electrochemical reaction causes various changes in the signals for conductance, further measured using potentiometry, conductometry or amperometry. Using electrochemical sensors, the interaction between the target molecule and the signal emitted by the bio-component is measured. This particular method possesses higher selectivity and specificity, a higher signal-to-noise ratio, a lower price as well as high feasibility, which makes it one of the most trusted techniques. Its usual requirements are the reference, auxiliary and a working electrode. The effective performance of a working electrode can be improved by metals like gold, silver or mercury.

12.3.11 Colorimetry

A very well-known and simple method for the detection of any common analysis is the method of colorimetry. Measurement of the colour changes due to the interaction between the target molecule and the aptamer helps to select the right selection and modifications. The colorimetric method is quite simple, easy to operate, and is lower in price as well as highly efficient, which draws a great deal of attention to its use in mycotoxin detection. Use of NPs results in its aggregation in the presence of the target molecule, and the changes in the localized surface plasmon resonance (LSPR) signal due to this aggregation can be also viewed by the naked eye, making it more reliable. The gold NPsare the colorimetric indicators used in the DNA-based detection of the analyte due to various characteristics like the distance-dependent optical properties and higher extinction coefficients. The aggregation alters the LSPR effect and results in the red shift of the UV-Vis adsorption spectrum.

12.3.12 Labelled Free

Label-free aptasensors are of great interest now due to their properties like simple experimental procedures, which include the gain of direct information about the interaction of the target with the aptamer aligned molecule by measuring properties like refractive index, mass and electrical resistivity produced by this interaction. They help in measuring the real-time quantification of the biomolecular interaction kinetics.

The impedimetric sensors used in label-free detection allows efficient detection of mycotoxins from food and feed due to several properties including portability, sensitivity, easy usage as well as miniaturization. Use of various electro transducers like indium tin oxide (ITO) coated glass, gold electrodes or screen-printed carbon electrodes (SPCEs) are being used for detection mycotoxins like OTA. These label-free biosensors need the stable and relatively lower resistive electro-transducer surface. Electro-polymerization can be used as another method for improvement of the electrical conductivity on the irregularly shaped surfaces of the electrode with the help of dyes, including thionine for the production of the stable redox-active coatings on its surface.

12.4 Conclusion

The diverse and polymorphic complexion of fungi contaminate various agricultural and horticultural produce is known to cause several cancers. Mycotoxin contaminations emerge in the food chain due to mould infection of agricultural commodities. Horticultural produce such as banana, tomato, pineapple, grape, peach, blueberry, apricot, orange, capsicum, strawberry, okra and cauliflower are nutritionally important crops, which are contaminated with both toxigenic and pathogenic fungi that produce a variety of mycotoxins into their products. The development of rapid and early detection of toxigenic fungi is important to ensure food security and food safety. The detection of mycotoxins by sensitive and field-deployable rapid POC diagnostics using aptamers through the application of novel nanobiotechnological approaches is more feasible than conventional methods. Nanostructures such as nanoparticles, nanowires and nanorods are promising materials for construction of biosensors helping in specificity and selectivity of the sensors. The electrochemical, calorimeter and fluorescence and FRET-based detection systems are more feasible for the high throughputs (HTs) for various toxins. In addition, molecular characterization of toxigenic fungi and toxins and genetic diversity is useful to develop disease resistance varieties through breeding programs.

12.5 Future Directions

Mycotoxin contamination is inevitable in horticultural crops and food grains. The current research focus is on the development of precise methods for early diagnosis of plant pathogenic fungi to replace the traditional methods by producing various biothreat toxins. In addition, the development of biomarkers for the plant breeding programme for the development of disease-resistant crops is imperative for managing the toxigenic and pathogens fungi. Since food safety is the essential target for improving human well-being, the regulation of the tainting of mycotoxins in food and feed is of utmost importance. Moreover, the complete elimination of toxins is impossible, hence forth, various scientific communities have been made biological management of mycotoxin, which are more beneficial, simple and eco-friendly over physical and chemical agents.

REFERENCES

Abata LK, Viera W, Paz IA, Flores FJ (2017) First report of Alternaria rot caused by *Alternaria alternata* on peach in Ecuador. Plant Dis 100: 2323.

Abdullah AS, Turo C, Moffat CS, Lopez-Ruiz FJ, Gibberd MR, Hamblin J, et al (2018) Real-time PCR for diagnosing and quantifying co-infection by two globally distributed fungal pathogens of wheat. Front Plant Sci 9: 1086.

Ahmad A, Jae-Hyuk Y (2017) Occurrence, toxicity, and analysis of major mycotoxins in food. Int J Environ Res Public Health 14: 632.

Amol MS, Ramu V, Vishwambar N, Rajkamal K, Prabla K, Santhakumari B, et al (2019) Morphological and molecular characterization of *Penicillium rubens* sp.nov isolated from poultry feed. Indian Phytopath. DOI: 10.1007/s42360-019-00165-2.

Becheva ZR, Gabrovska KI, Godjevargova TI, Zvereva EA (2019) Aflatoxin B1 determination in peanuts by magnetic nanoparticle–based immunofluorescence assay. Food Anal Methods 12: 1456–1465.

Bhat R, Rai RV, Karim AA (2010) Mycotoxins in food and feed: Present status and future concerns. Compr Rev Food Sci Food Saf 9: 57–81.

Binder EM, Tan LM, Chin LJ, Handl J, Richard J (2007) Worldwide occurrence of mycotoxins in commodities, feeds and feed ingredients. Anim Feed Sci Technol 137: 265–282.

Carballo D, Pinheiro-Fernandes-Vieira P, Font G, Berrada H, Ferrer E (2018) Dietary exposure to mycotoxins through fruits juice consumption. Rev Toxicol 35: 2–6.

Costa J, Rodriguez R, Garcia-Cela E, Medina A, Magan N, Lima N, Santos C (2019) Overview of fungi and mycotoxin contamination in capsicum pepper and in its derivatives. Toxins 11: 1–16.

Dong OX, Ronald PC (2019) Genetic engineering for disease resistance in plants: Recent progress and future perspectives. Plant Physiol 180(1): 26–38.

Economic Research Service (ERS), USDA, Fruit and Tree Nut Data - Exports/Imports (2018).

Egel D, Foster R, Maynard E, Weller S, Babadoost M, Nair A, Rivard C, Kennelly M, Hausbedk M, Szendra Z, Hutchinson B, Orshinsky A, Eaton T, Welty C, Miller S (2018) Midwest Vegetable Production Guide for Commercial Growers.

El Khoury D, Fayjaloun S, Nassar M, Sahakian J, Aad PY (2019) Updates on the effect of mycotoxins on male reproductive efficiency in mammals. Toxins 11(9): 515.

FAO (2019) Banana Facts and Figures. http://www.fao.org/economic/est/estcommodities/bananas/bananafacts/en/#.XKzikZhKg2w.

Foreign Agricultural Service/USDA. Office of global analysis (2019). https://ipad.fas.usda.gov.

Koteswara Rao V, Girisham S, Reddy SM (2015) Inhibitory effect of essential oils on growth and ochratoxin-A production by *Penicillium* species. Res J Microbiol 10: 222–229.

Kumar A, Gupta K, Dixit S Mishra K, Srivastava S (2019) A review on positive and negative impacts of nanotechnology in agriculture. Int J Environ Sci Technol 16:2175–2184.

Kumar, AYVV, Renuka RM, Venktatramana M, Poda S (2018) Development of a FRET-based fluorescence aptasensor for the detection of aflatoxin B1 in contaminated food grain samples. RSC Advances 8(19): 10465–10473.

Li YM, Liu X, Lin Z (2012) Recent developments and applications of surface plasmon resonance biosensors for the detection of mycotoxins in foodstuffs. Food Chem 13: 1549–1554.

Liu L, Tanveer Z I, Jiang K, Huang Q, Zhang J, Wu Y, Han Z (2019) Label-free fluorescent aptasensor for ochratoxin–A detection based on CdTe quantum dots and (*N*-Methyl-4-pyridyl) porphyrin. Toxins 11(8):447.

Malachová A, Sulyok M, Beltrán E, Berthiller F, Krska R (2014) Optimization and validation of a quantitative liquid chromatography-tandem mass spectrometric method covering 295 bacterial and fungal metabolites including all regulated mycotoxins in four model food matrices. J Chromatogr A 1362:145–156.

Nirmaladevi D, Venkataramana M, Rakesh KS, Uppalapati SR, Gupta VK, Yli-Mattila T, et al. (2016) Molecular phylogeny, pathogenicity and toxigenicity of *Fusarium oxysporum* f. sp. *lycopersici*. Sci Rep. DOI: 10.1038/srep21367.

Nwachukwu IN, Amadi ES, Ogwo UC, Umeh SI, Opurum CC, Chinakwe EC, et al. (2019) Molecular screening of fungal isolates of palm oil from south eastern Nigeria for aflatoxin and ochratoxin biosynthetic genes using multiplex polymerase chain reaction (mPCR). MRJI 28: 1–9.

Okoth S, Boevre DM, Vidal A, Mavungu JDD, Landschoot S, Kyallo M, et al. (2018) Genetic and toxigenic variability within *Aspergillus flavus* population isolated from maize in two diverse environments in Kenya. Front Microbiol 9: 57.

Patil S, Shinde M, Prashant R, Kadoo N, Upadhyaya A, Gupta VS (2020) Comparative proteomics unravels the differences in salt stress response of own rooted and 110R grafted Thompson seedless grapevines. Proteome Res 19 (2): 583–599.

Raja HA, Miller AN, Pearce CJ, Oberlies NH (2017) Fungal identification using molecular tools: A primer for the natural products research community. J Natural Products 80(3): 756–770.

Rathnavelu V, Alitheen NB, Sohila S, Kanagesan S, Ramesh R (2016) Potential role of bromelain in clinical and therapeutic applications. Biomed Rep 5: 283–288.

Santos AO, Vaz A, Rodrigues P, Veloso ACA, Venancio A, Peres AM (2019) Thin film sensor devices for mycotoxins detection in foods: Applications and challenges. Chemosensors 7: 3.

Schmidt SM, Lukasiewicz J, Farrer R, van Dam P, Bertoldo C, Rep M (2016) Comparative genomics of *Fusarium oxysporum f. sp.* melonis reveals the secreted protein recognized by the Fom-2 resistance gene in melon. The New phytologist *209*: 307–318.

Sefah K, Shangguan D, Xiong X, O'Donoghue MB, Tan W (2010) Development of DNA aptamers using cell-selex. Nat Protoc 5(6):1169–1185.

Sertova N, Ignatova M (2019) Detection of mycotoxins through nanobiosensors based on nanostructured materials: A review. Int J Biosci 14(5): 106–115.

Singh P, Orbach MJ, Cotty PJ (2018) *Aspergillus texensis*: A novel aflatoxin producer with morphology from the United States. Toxins 10:513.

Sojinrin T, Liu K, Wang K, Cui D, Byrne HJ, Curtin JF, Tian F (2019) Developing gold nanoparticles - conjugated aflatoxin B1 antifungal strips. Int J Molec Sci 20: 6260.

Srinivas C, Devi D, Murthy K, Chakrabhavi DM, Lakshmeesha, Singh B, Kalagatur N, Niranjana S, Hashem A, Alqarawi A, Allah E, Nayaka S, Srivastava R (2019) *Fusarium oxysporum* f. sp. *lycopersici* causal agent of vascular wilt disease of tomato: Biology to diversity - A review. Saudi J Biol Sci 26:1315–1324.

Szalay J (2018) Cauliflower: Health Benefits & Nutrition Facts. LiveScience.

Villafana RT, Ramdass AC, Rampersad SN (2019) Selection of *Fusarium* trichothecene toxin genes for molecular detection depends on TRI gene cluster organization and gene function. Toxins 11(1): 36.

Wang X, Xiaoyi G, Jiale H, Xiaochen H, Yunchao L, Xiaohong L, Louzhen F, Hua-Zhong Y (2019) Systematic truncating of aptamers to create high-performance graphene oxide (GO)-based aptasensors for the multiplex detection of mycotoxins. Analyst 144: 3826–3835.

Worku M, Sahe S (2018) Review on Disease management practice of tomato wilt caused *Fusarium oxysporum* in case of Ethiopia. J Plant Pathol Microbiol 9:11.

World Health Organization (2018) Monitoring health for the SDGs, sustainable development goals. World Health Organization, Geneva, Licence: CC BY-NC-SA 3.0 IGO.

Zhang L, Yuan T, Wang Y, et al (2018) Identification and evaluation of resistance to *Fusarium oxysporum* f. sp. *cubense* tropical race 4 in *Musa acuminata* Pahang. Euphytica 214:106.

13

Application and Monitoring of Health Hazard and Codex Alimentation for Product and Commodity

Mahender Singh Yadav
National Research Centre for Integrated Pest Management (ICAR), Pusa Campus, New Delhi, India

CONTENTS

13.1 Introduction

Food is important to life; thus, food safety should be a basic right. Billions of individuals in the world are in danger of unsafe food. Traditionally, documented human tragedies and economic disasters due to badly contaminated food have occur red because of intentional or unintentional personal conduct and governmental failure to safeguard food quality and safety. To confirm food safety and to stop foodborne diseases, speedy and correct detection of infective agents is important. Understanding the risks of health hazards will facilitate the United States to require action to avoid or mitigate these risks. The Codex Alimentarius has massive impact on the thinking of food producers and processors, as well as on the end users, the consumers. Government agencies should enforce food safety laws to safeguard public and individual health. Medical suppliers should alert to stop foodborne diseases and ensure disease free safe diet. The intimate collaboration between all the stakeholders can ultimately guarantee food safety within the 21st century. The food producers, distributors, handlers and vendors bear primary responsibility, while consumers should stay alert and literate.

13.2 Health Hazards

Health hazards are chemical, physical or biological factors in our surroundings that may have negative impacts on our short-term or semi-permanent health. Understanding the risks of those hazards will facilitate the United States to require action to avoid or mitigate these risks. For a typical food safety hazard, the extent of the harmful effects of the hazards on the health of the buyer is established by risk analysis and hazard analysis. Risk analysis is sometimes conducted by a national food or health administrative unit that addresses public health issues concerning a specific food safety hazard related to the given sector of the food trade. Risk assessment, risk management and risk communication are important components of risk analysis (https://www.who.int/foodsafety/risk-analysis/risk-management/en/ [29 August 2020]). The primary objective of risk analysis is to ascertain a national food safety objective for a hazard. The food safety objectives are often look thoughtfully for a long time, because the most acceptable level from the hazard in a varied food. Once a food safety objective for a hazard has been established by risk analysis, it should be considered throughout the hazard analysis step (the first of seven principles, below) for hazard analysis and critical control point (HACCP).

13.3 HACCP Codex Alimentarius

HACCP Codex Alimentarius is a global specification supporting the preventive management of food questions of safety. It assists the Codex Alimentarius Commission within the implementation of the Joint FAO/WHO Food Standards Programme, coverage to the Director-Generals of Food and Agriculture Organization of the United Nations and United Nations agency. This international forum includes scientists, technical specialists, government regulators and international client and trade organizations. Governments use codex texts as a part of their national food safety guidelines and they are used by industrial partners to specify the grade and quality of consignments. The HACCP system may be used in any stages of an organic phenomenon, including food production and preparation processes as well as packaging, distribution, etc. HACCP is recognized internationally as a logical tool for adapting outdated methods of review to a contemporary, science-based food safety system.

13.4 Hazard Analysis and Critical Control Points

HACCP was enforced in 1997 and modified food safety methodology to be a science-based rather than a standard "sight, smell and touch" review. It is mostly preventive and science based. The system consists of identification of hazards and controlling them prior to consumption. Effective hygiene practices will certainly reduce the hazards leading to production of safe foods. HACCP is Associated with internationally recognized system to reduce the chance of safety hazards in food. It focuses on identifying and preventing hazards which will render food unsafe. It additionally focuses on review activities on the important areas of food safety and can guarantee a scientific basis for controls in operations within the food trade. It is a recognized systematic and preventive approach that addresses biological, chemical and physical hazards through anticipation and interference, instead of through product review and testing. HACCP emphasized management of the method as upstream within the process system as attainable by utilizing operator management and for continued observation techniques at important management points. It requires a long-term commitment by trade management in every food premises and vessel. HACCP may be applied throughout the organic phenomenon from the producer to client. Where attainable, the full food process and distribution system should be evaluated for attainable HACCP application.

13.4.1 The Importance of HACCP

- It provides enhanced food safety and reduced risk of foodborne diseases.
- It provides greater confidence to customers.
- It provides reduction in production prices through reduced wastage.
- It facilitates compliance with statutory necessities.

13.4.2 The Seven Principles of HACCP

- Conductance of hazard analysis
- Identification of critical control points
- Establishment of essential limits
- Monitoring of essential management points
- Establishment of corrective actions
- Establishment of record
- Establishment of verification procedures

13.4.3 HACCP Principle 1 for Development of an HACCP Plan

- Hazard identification
- Hazard analysis
- Identification of management for the hazard

The above activities need to be performed by a group of suitably qualified employees at a food company that processes products with which the hazard is associated. Successful execution of the hazard analysis activity as a part of an HACCP plan development depends on the availability of a broad range of scientific information relating to the hazard. A vital source of this information is the report from risk assessment conducted on the hazard. HACCP is an internationally recognized system for reducing the chance of safety hazards in food. Full compliance with good management practices (GMP) and good hygiene practices (GHP) within whole production and handling operations are assumed to be in place when introducing HACCP. Adequate resources must be available for its implementation, and maintenance and education and training of all personnel should be conducted.

13.5 Food Safety

The key concerns in food safety include foodborne pathogens, pesticides residue, chemical contaminants, food allergens and nutritional as well as physical hazards. For both regulatory bodies and industry, foodborne pathogens are still the most likely hazards, and diseases caused by microbiologically contaminated food are on the rise in both developing and developed countries. Since the trade in food commodities has expanded globally, food safety can no longer be considered a domestic issue. With the advent of the World Trade Organization (WTO) regime, standards, guidelines and recommendations of Codex Alimentations can be considered as a benchmark for international trade. Appropriate quality control measures during food processing can reduce losses and stimulate trade in food products. Food safety is an issue that affects the lives of consumers both economically and in terms of public health. However, lack of institutional coordination, the shortage of technical skill and equipment, the absence of responsible monitoring systems, lack of awareness in safety and quality issues on the part of food handlers in the organized and unorganized sectors, the increasing incidence of foodborne diseases and the emergence of newer variants of pathogens are several issues which require considerable attention.

Quality aspects of processed food products have to be nurtured by the whole someness of not only of the product but also of the raw material. In HACCP, hazard analysis activity can be equated to the risk assessment component of risk analysis.

13.5.1 Food Safety Needs

Food manufactures should ensure that their products are safe and do not make people sick. Firms should have a food safety arrangement and observe the food safety laws. Manufacturers are accountable for manufacturing foods that area safe and reliable. The food sector additionally has its own internal monitoring systems. Businesses that keep their management systems in good order are less frequently inspected. The government checks that these needs are met. In recent years, there have been numerous scandals involving deceitful food. As an example, a product may contain ingredients that are forbidden, or a product may be put forward as a different product than expected. Like other meats being put forward as beef, or honey bulked up with sugar syrups. These deceptions mislead shoppers. Incorrectly labelled food can endanger public health as a result of the precise ingredients not being clear or being falsified. It should contain substances that are helpful in nursing.

13.5.2 Food Safety Laws within the Commodities Act

All the foundations of the General Food Law Regulation are place in Dutch laws. Food and food products may not endanger the health or safety of shoppers. The Commodities Act lays down rules for foods and alternative products. This includes rules on the hygienic preparation and labelling of foods. Besides the overall Food Law, European Economic Community(EEC)Member States area unit operating to push food safety through:

- The Codex Alimentarius, a global agreement involving the EU and 186 countries. All EU food standards and laws area unit supported the Codex Alimentarius (FAO / WHO, 2018).
- The European Food Safety Authority (EFSA), Associate in Nursing Freelance Analysis Institute advises the EU Commission on the risks related to food safety and animal health.

13.5.3 The HACCP Food Safety Setup for Businesses

Involved food should meet national safety and hygiene necessities so as to safeguard client health. Businesses are needed by law to formulate a food safety setup in line with the HACCP system. This describes any hazards which will occur in their operations, and what the business will do to correct any issues. For instance, they will take measures concerning:

- Personal hygiene of employees.
- Product hygiene throughout the transportation of raw materials (such as milk containers).
- The means by which food is treated or processed (such as exploitation clean machines to slice bread).
- Packaging. High quality packaging material used
- Storage in clean and dark facilities.

13.5.4 Reporting Unsafe Food Products

It is competent for traces of undesirable substances like hepatotoxic mould or chemical residue to occur in food for animal or human consumption. Harmful microorganisms, such as Eubacteria, may contaminate foods throughout the producing method. It may be harmful for human health if these substances exceed safety limits.

13.5.5 Tracing Unsafe Food Products

Food makers should be able to trace the origin of the ingredients they use and record the destination of their product. If necessary, they will recall unsafe product and destroy it.

13.6 Export Inspection and Certification System

The government of India set up the Export Inspection Council of India (ECI) under Section 3 of the Export (Quality and Inspection) Act of 1963 as an apex body for the sound development of trade quality and shipment inspection. The most system of export review and certification being followed by EIC that is consignment wise review, in method internal control, self-certification and Food Safety Management Systems Base Certification (FSMSC). FSMSC is based on international standards together with HACCP, GMP and GHP. The Codex HACCP and food-hygiene standards are adopted by the Bureau of Indian Standards (BIS), the national standards body in the republic of India. The Food Safety and Standards Authority of India (FSSAI) has been established under the Food Safety and Standards Act of 2006 that could be a consolidating statute associated with food safety and regulation in the Republic of India. FSSAI is part of the Bureau of the Ministry of Health and Family Welfare of the Government of India. The Sanitary and Phytosanitary (SPS) measures are outlined as any measures applied at intervals the territory for

- To guard living being from risk arising from entry, spread of diseases, malady-carrying organisms or malady inflicting organisms.
- To guard living being from risks arising from additives, contaminants, toxins or malady-inflicting organisms in food, beverages or foodstuffs.

The reference created to Codex food safety standards in WTO agreement on the SPS measures agreement implies that Codex has comprehensive implications for breakdown trade disputes.

Food process units are being inspired to adopt these systems on a voluntary basis. Risk analysis was approved by Codex Alimentarius Commission (CAC) in 1991–1993, and since then Codex has developed many vital text-like definitions of terms associated with risk analysis statements of principle about the role of Food Safety Risk Assessment (1997). Application of biotechnology to food processing and production of raw food materials is presently under the scrutiny of the Codex Alimentarius Commission (CAC). The Commission frequently examines new ideas and systems related to food safety and thereby the protection of consumers against potential health hazards.

13.7 Codex Alimentations of Product and Trade Goods

United Nations Food, Agricultural and World Health Organizations (FAO/WHO) established the Codex Alimentarius in 1963. It is a set of internationally adopted food standards and connected texts conferred in a uniform manner (FAO of UN, 2016). These food standards and connected texts aim at protecting consumer health and creating certain honest practices within the food trade. The Codex Alimentarius has become the worldwide reference for customers, food producers and processors, national food agencies and international food trade. The Codex Alimentarius has had a huge impact on the thinking of food producers and processors, as well as on the end users, the consumers. Its influence extends to each continent and its contribution to the protection of public health and honest practices within the food trade is immeasurable. The publication is meant to guide and promote the elaboration and institution of definitions and necessities for foods to assets in their harmonization and in doing therefore to facilitate international trade. The Codex Alimentarius includes standards for all the principal foods, whether or not processed, semi processed or raw, for distribution to the buyer. It includes provisions associated with food hygiene, food additives, residues of pesticides and veterinary medication, contaminants, labelling and presentation, strategies of study and sampling and import and export review and certification. Employees selected by UN agency and UN agency, set at UN agency headquarters in Rome, run the Codex Secretariat. The Codex Alimentarius Commission has been supported in its work by the currently universally accepted principle that people have the right to expect their food to be safe, of good quality and appropriate for consumption. Creating standards that protect customers, guarantee honest practices in food sales and facilitate trade may involve the collaboration of specialists in various food-connected scientific disciplines. Codex Alimentarius Standards: Codex Alimentarius CAC has developed a suggested international code to follow. General principles of food hygiene lay a firm foundation for ensuring food hygiene. They follow the food chain from primary production through to the final buyer, highlighting the key hygiene controls at every stage. These controls square measure internationally recognized as essential to ensuring the protection and suitableness of food for human consumption (CAC and FAO / WHO Food Standards Programme, 1999). One example of the numerous standards established by the Codex Alimentarius is maximum residue limits (MRLs) for residues of pesticides in foods. The strategies of study and sampling for chemical residue in commodities are normal standards instituted by the Codex Alimentarius. For several years FAO/WHO and Joint Meeting on Pesticide Residues have created internationally acclaimed knowledge wide used the government trade and analysis centres. Publications ensuing from conferences and connected activities square measure acclaimed international references. International norms for commercial import of food Commodities, 7200 food standards developed by codex square measure contained beside labelling and additive standards and different recommendations in 19 volumes. The cluster known as "commodity standards" is the most important variety of specific standards within the Codex Alimentarius. These square measure the trade goods standards. The commodities are as follows.

- Cereals, legumes and derived merchandise, as well as vegetable proteins.
- Fat, oils and connected merchandise.
- Fish and fish items.
- Fresh fruits and vegetables.
- Processed and fast-frozen fruits and vegetables.
- Fruit juices.
- Meat and meat merchandise as well as soup and milk, and milk merchandise.
- Sugars, cocoa merchandise, chocolate etc. products.

Most standards in the Codex Alimentarius take a number of years to develop (FAO / WHO, 2018). The Codex Alimentarius' commission secretariat arranges the preparation of projected draft standards and circulates drafts to government members for comment. Comments are reviewed by the subsidiary body allotted responsibility for development of projected draft. Once the commission adopts a customary law, it is supplementary to Codex Alimentarius. At present, compliance for chemical MRLs is established for 200 active ingredients in each specific commodity. The Codex Alimentarius International Standards provide pointers for quality and fairness of the international food trade.

13.8 Conclusion

Food is essential to life; hence, food safety is a basic human right. Foodborne pathogens, pesticide residues, chemical contaminants, food allergens and nutritional as well as physical hazards are key determinants in food safety. For both regulatory bodies and industry, foodborne pathogens are still the most likely hazards. Codex Alimentarius food standards and connected texts are effective in protecting consumer health and ensuring fair practices in the food trade. HACCP Codex Alimentarius specifications should be followed for preventive management of food safety issues. Adequate resources must be made available for implementation, maintenance, education and training of all personnel associated with commodity trade. The intimate collaboration between all the stakeholders will ultimately ensure food safety in the 21stcentury

REFERENCES

Codex Alimentarius Commission and the FAO/WHO Food Standards Programme. (1999). Basic Texts on Food Hygiene, 2nd ed., FAO, Rome Italy.

FAO/WHO. (2018). Codex Alimentarius: Understanding Codex, 5th ed. FAO/WHO, Rome, Italy.

FAO/WHO (2005). Food Safety Risk Analysis. Part I: An overview and framework manual, provisional ed. WHO/FAO of United Nations at Rome, Italy, June.

Food and Agricultural Organization of United Nations. (2016). Procedural Manual of the Codex Alimentarius Commission, 25th ed. FAO/WHO Food Standards Programme. Food and Agricultural Organization of United Nations, Rome, Italy.

Satin M (2008) Food alert: The ultimate sourcebook for food safety. 2nd ed. Facts on File Inc., New York.

World Health Organization. Risk Management (2020). https://www.who.int/foodsafety/risk-analysis/risk-management/en/ [29 August 2020].

14

Recent Advances in Postharvest Handling and Management of Citrus Diseases

J.U. Vinay,[1] Jahagirdar Shamarao,[1] V. Devappa,[2] S.A. Ashtaputre[1] and Dinesh Singh[3]
[1]Department of Plant Pathology, University of Agricultural Sciences, Dharwad, Karnataka, India
[2]Department of Plant Pathology, College of Horticulture, University of Horticultural Sciences Campus, Bengaluru, India
[3]Division of Plant Pathology, Indian Agricultural Research Institute, New Delhi, India

CONTENTS

14.1 Introduction

In modern agriculture, the horticultural sector plays an important role in providing livelihood security to farmers in India. According to production data of the National Horticultural Board, India stands as the second largest producer of fruits in the world with 684.66 lakh tonnes from an area of 61.01 lakh hectares. Citrus ranked as the third largest fruit industry after banana and mango in India (Anonymous 2012). Citrus (*Citrus* spp.) is a non-climacteric and unique berry-like fruit, which is a woody and evergreen perennial plant. Citrus fruits probably rank third among the subtropical fruits of the world. Citrus represents the third most important group of fruits in India (Das et al. 2018). In India, the major citrus cultivated regions are Tamil Nadu, Madhya Pradesh, Punjab, Karnataka, Assam, Orissa, Maharashtra and Uttar Pradesh. The postharvest losses in horticultural commodities occur in different stages *viz*., during production or harvesting, postharvest handling, storage, processing and transportation. The major commercial citrus fruit in India is mandarin orange, followed by sweet orange, lemon and acid lime. *Citrus* spp. are vulnerable to more than 150 diseases that are continuously emerging and causing damage in terms of production and also affecting the citrus industry. The losses caused by postharvest diseases account to 20–40%under poorly managed situations. The supply chain period in case of citrus is 3months, which is a crucial period to protect the produce from postharvest decay. Postharvest loss causes economic damage in production, harvesting, packing, marketing and transportation. Postharvest fungal pathogens have a prolonged dormant (quiescent) stage after infection, and at the time of storage become active, resulting in development of disease symptoms. The key diseases are caused by *Diplodia*, *Phomopsis* and *Alternaria* stem-end rots which become established in necrotic tissue on the button surface (Table 14.1). These fungal pathogens are major contaminants in the packinghouse. As per study of Nanda et al. (2010), the total postharvest loss of citrus fruits is 6.38%. Among this loss, 4.84% was at farm level and 1.54% was in storage. The total postharvest loss reported in mandarins was 3.00% in Sikkim and 6.15% in Assam. The Indian Institute of Horticultural Research, Bangalore (Karnataka) reported the postharvest losses in Coorg mandarin. Nearly 3.50% of the total market arrivals were discarded due to postharvest diseases. The list of postharvest diseases of citrus are presented in Table 14.1.

TABLE 14.1

Postharvest Diseases of Citrus and Their Causal Agents

Sl. No.	Disease	Pathogen(s)
1.	Citrus gummosis/ brown rot	*Phytophthora parasitica, P. citrophthora, P. nicotianae*
2.	Blue mould	*Penicillium italicum*
3.	Green fruit mould	*Penicillium digitatum*
4.	Greasy spot	*Mycosphaerella citri*
5.	Anthracnose	*Colletotrichum gloeosporioides*
6.	*Alternaria* brown spot	*Alternaria alternata, A. citri*
7.	Citrus black spot	*Guignardia citricarpa*
8.	Diplodia stem-end rot	*Diplodia natalensis* (Syn: *Lasiodiplodia theobromae*)
9.	Phomopsis stem-end rot or citrus melanose	*Diaporthe citri* (Anamorph: *Phomopsis citri)*
10.	Sour rot	*Geotrichum candidum* or *Geotrichum citri-aurantii* (Teliomorph: *Galactomyces citri-aurantii*)
	Aspergillus rot	*Aspergillus niger*
11.	Canker	*Xanthomonas citri* subsp. *citri*

14.2 Major Postharvest Diseases and Their Management

14.2.1 Citrus Gummosis/Brown Rot

Citrus gummosis is caused by three important species such as *P. parasitica*, *P. citrophthora* and *P. palmivora*. Trunk gummosis and root rot are the most destructive forms of the *Phytophthora* group of diseases and after the pandemic outbreak of the nineteenth century and the consequent widespread use of resistant rootstocks, they are regarded as endemic diseases in all citrus-growing areas of the world. The citrus belt of the Vidarbha and Marathwada regions of Maharashtra, Punjab, Madhya Pradesh, Andhra Pradesh and northeastern states of India were surveyed to assess the impact of *Phytophthora* diseases. Two mating types (A1 and A2) of *Phytophthora parasitica* (nicotianae) have been recorded from the orchards of Nagpur mandarin in Nagpur district, Maharashtra (Naqvi 2002). *P. parasitica* Dastur has been a common species associated with citrus disease in Assam and northwest India. In Madhya Pradesh, adjoining to Vidarbha region of Maharashtra, around 20–50% of Nagpur mandarin plants were found to be severely declined due to *P. parasitica* and *P. citrophthora* along with *P. palmivora*. In Andhra Pradesh, 20–100% of the acid lime plantation was severely affected by *P. parasitica* along with *P. citrophthora* and *P. palmivora*. In Punjab state, 10–80% plants of *C. sinensis* and 10–100% plants of kinnow mandarin showed symptoms of diseases caused by *P. parasitica*, *P. citrophthora* and *P. palmivora* due to excessive flood irrigation (Naqvi 2002).

14.2.1.1 Symptoms

The symptoms caused by gummosis are characterized by damping-off of seedlings in the nursery bed, which is a widespread problem in the citrus industry and frequently occurs in citrus orchards where phytosanitary conditions are difficult

FIGURE 14.1 Symptoms of green mould rot (a) and blue mould rot (c) on kinnow fruits and conidiophores and phialides arrangement of *Penicillium digitatum* (b) and *Penicillium italicum* (d).

to maintain. More than 20% of seedling mortality has been observed in central India due to *Phytophthora* infection. Foot rot and gummosis are serious diseases caused by *Phytophthora* spp. Infected bark remains firm, with small cracks through which abundant gum exudation occurs. The gum disappears after heavy rains but remains persistent on the trunk under dry conditions. Lesions spread around the circumference of the trunk, slowly girdling the tree (Figure 14.1). Pale green leaves with yellow veins is a typical symptom of severe infection of *Phytophthora*. Brown rot epidemics are usually restricted to areas where rainfall coincides with the early stages of fruit maturity. *Phytophthora* causes the following common diseases in citrus cultivars. The brown spot stage has been characterized as light brown and leathery. In humid conditions, white mycelium is grown on the surface of rind. Fruits near the ground become infected and rotten. Most of the infected fruits soon abscise but harvested fruits may not show symptoms until they have been kept in storage for a few days. In storage, infected fruits have a characteristic pungent and rancid odour. All the three associated pathogens are involved in causing gummosis but *P. palmivora* has been the predominant in causing brown rot stage (Naqvi 2002).

14.2.1.2 Causative Agents and Epidemiology

At least ten species of *Phytophthora* have been reported to attack citrus in the world, but the predominant species are *P. citrophthora* and *P. parasitica* or *P. nicotianae*. The latter is the dominant species which is usually associated with root rot while *P. citrophthora* is frequent in old plantings and is commonly associated with trunk gummosis. Both *P. citrophthora* and *P. nicotianae* are polyphagous in nature. Other species such as *Phytophthora citricola*, *Phytophthora cactorum*, *Phytophthora hibernalis* and *Phytophthora syringae* are found occasionally. The last two species are found during winter months solely because of their low-temperature requirements.

Sporangia produced in the most superficial soil layer (0–30 cm depth), on contact with air, are the main source of inoculum. Natural infections are most frequently caused by zoospores and occasionally by direct or indirect germination of sporangia through a germ tube or by releasing zoospores, respectively. Production and germination of sporangia is influenced by temperature and soil water potential. Their dissemination is mostly by water splash and occasionally by wind, within water droplets. The zoospores are motile and can swim short distances by flagellar movement or can be carried over longer distances by soil water. They swim towards roots, as they are attracted by root exudates, and encyst upon contact, germinate and penetrate fruits, leaves, shoots and green twigs directly (Naqvi 2002). Both *P. citrophthora* and *P. nicotianae* are soil-borne pathogens. They complete their life cycles in the soil. Under dry desert conditions, infection by these two pathogens has been noticed only on roots and lower bark tissue at the crown of the tree. In humid, wet areas, these pathogens can infect fruits and upper trunk tissue. *P. nicotianae* has a higher optimum temperature for growth (85–90°F) than *P. citrophthora* (75–80°F). Both these fungus-like organisms produce motile, swimming zoospores when soils are saturated from irrigation or rainfall. These spores are attracted to root tips. After attachment, they germinate and infect cortical root tissue. Infected feeder root cortical tissue become rotten and dark in appearance.

Nutritional depletion and light also stimulate sporangial production from mycelium. Indirect germination of sporangia to produce zoospores requires free water which is stimulated by a drop in temperature. Under moist conditions, sporangia may also germinate directly by growth of germ tubes but the correlation between soil saturation and severity of Phytophthora root rot suggests that indirect germination is more important in the root disease cycle. Chlamydospore production by *P. nicotianae* occurs under unfavourable conditions for fungal growth, i.e. nutrient depletion, and low oxygen levels and temperatures (59–64°F). Water requirements for germination of chlamydospores are similar to those for sporangia. Chlamydospores of *P. nicotianae* appear to become dormant below 59°F. Hence, temperature of 82–90°F is used to stimulate germination. The requirements for oospore germination are thought to be nearly identical to those of chlamydospores. Oospore maturation appears to be an important factor in germinability of *Phytophthora* spp. This ecological feature explains sudden epidemic explosions of brown rot following persistent rainfall. Neither of the two species forms sporangia on the gummy cankers at the base of the trunk. It is known that *P. citrophthora* does not reproduce sexually, and *P. nicotianae* reproduces sexually only occasionally, since only A1 mating type is found in most citrus orchards (Naqvi 2002).

14.2.1.3 Disease Management

Most citrus trees are budded onto a rootstock. Most of these rootstocks are more resistant to *Phytophthora* diseases than the scions (tops). Hence, it is necessary to keep the bud union at least 4–6 inches above the soil line at planting. The "Budwood Certification Programme", including surveillance, indexing and identification of mother trees resistant to *Phytophthora*, should be strengthened for raising disease-free planting material. The marked *Phytophthora* resistant trees should be multiplied and distributed to nurserymen for producing *Phytophthora* resistant seedlings. Strict plant quarantine measures should be enforced to check the movement of contaminated budwood and planting material in the new planting sites. Contacting low-hanging fruits to the soil should be avoided, and if already contacted to soil, those fruits should be kept as a separate lot. Trees must be well pruned in the manner that branches hang low enough. Exposure of seeds to hot water at 48.9°C for 4–10 min can eradicate seed-borne infection. In general, water should not be allowed to stand around the crown of citrus during irrigation. Brown rot disease can be reduced by providing proper ventilation under the tree canopy to shorten durations of high relative humidity and wetness required for inoculum production. Electrolysed chlorine injected into citrus micro-irrigation systems effectively killed propagules of *Phytophthora* (Savita and Nagpal 2012). Moderately resistant rootstocks such as *Citrus macrophylla*, Cleopatra mandarin and Troyer citrange are recommended for cultivation. Over-irrigation resulting in soil saturation is detrimental to normal feeder root development even in the absence of *Phytophthora*. It is important to dig a planting hole as deep as necessary to obtain good drainage. Control of *Phytophthora* diseases in citrus has been enhanced by the availability of highly active, systemic fungicides specific for control of *Phytophthora*. These chemicals are applied to the soil and the crown and trunk areas of affected trees. Disease spread can be controlled by removing diseased bark and a buffer strip of healthy, light brown to greenish bark around the margins of the infection. Preventive measures include selection of a proper site with adequate drainage. Use of resistant rootstocks and avoiding contact of water with the tree trunk by adopting the ring method of irrigation is equally effective. Application of aureofungin 46.15% w/v. SP at300 (1%) g or mL per acre has been recommended. Wound dressing of Bordeaux paste (1.0%) amended with mustard oil from August to December is very effective in managing the disease spread (Thakre et al. 2017). Management of brown rot can be easily achieved through pre-harvest applications of copper and fosetyl Al (Aliette). These fungicides are applied between August and September and Aliette has a 30-daywaiting period after application to harvest the fruits.

14.2.2 Blue and Green Mould Rot

Blue and green moulds are common postharvest diseases of citrus which are caused by *Penicillium italicum* and *P. digitatum*, respectively. The major problem of these pathogens is due to their spores (conidia) which appear as mould and become airborne. All kinds of citrus fruit are susceptible to these two diseases. Among them, *P. digitatum* alone is responsible for approximately 90%of postharvest losses.

14.2.2.1 Symptoms

The initial symptom of blue and green mould is the appearance of a soft, watery and slightly discoloured spot (Figure 14.1a and c). These spots become enlarged to 1–2 inches. Later, white mycelium appears on the surface, and when the growing pathogen is about 1 inch in diameter, blue (*P. italicum*) or olive green (*P. digitatum*) spores are produced at the centre. The entire fruit surface is rapidly covered with the blue or olive-green spores. Typical terpenous odour in the surrounding infected area is characteristic of disease. Pathogens produce ethylene in sufficient quantities which results in the rapid senescence of adjacent fruits. In both diseases, fruits rapidly spoil and collapse. At lower relative humidity, the fruit may shrink and become mummified. In the case of green mould, the pathogen penetrates fruit rind through the wounds. Symptoms begin as a water-soaked area at the fruit surface, followed by growth of colourless mycelium and finally sporulation occurs (green colour). In the case of blue mould, the pathogen penetrates through the uninjured peel and can easily spread from one fruit to surrounding fruits (Pallottino et al. 2012).

14.2.2.2 Causative Agents and Epidemiology

The genus *Penicillium* was first described by John H.F. Link in 1809. The species level identification of the genus *Penicillium* was based on the morphological characteristics and branching of the conidiophores (Figure 14.1b and d). The pathogen produces brush-like conidiophores, which bear conidia. Conidiophores are simple or branched and are ended with flask-shaped phialides on which conidia arise. The characteristics of both the pathogens are presented hereunder.

P. digitatum	**P. italicum**
Causes citrus green mould of citrus.	Causes citrus blue mould of citrus.
Colonies are plane and grow rapidly on malt extract agar (MEA) and potato dextrose agar (PDA) but poorly on synthetic media.	Colonies grow rapidly on MEA and PDA. At 25°C, colonies grow restrictedly on Czapek agar.
Colonies are olive green colour and the reverse of the plate is colourless to cream yellow or pale dull brown.	Colonies are plane, heavy sporulating, blue coloured and often appear as granular due to the presence of bundles of conidiophores and conidial heads. The reverse is colourless to grey to yellow-brown.
Colony texture is velutinous with no exudate droplets.	Colony texture is velutinous to fasciculate, crustose, with exudates absent or very limited.
The odour produced is strong due to volatile metabolites (limonene, valencene, ethylene, ethyl alcohol, ethyl acetate, or methyl acetate).	The odour is produced due to volatile metabolites (ethyl acetate, isopentanol, linalool, isobutanol, 1-octene, ethyl butanoate, 1-nonene, styrene or citronellene).
The conidial apparatus is very fragile and tends to break up into many cellular elements.	The conidial apparatus consists of asymmetric penicilli bearing tangled chains of conidia.
Conidiophores are terverticillate, borne from subsurface or aerial hyphae, irregularly branched and consist of short stipes with few metulae and branches that terminate in whorls of three to six phialides.	Conidiophores originate from the substratum or occasionally from superficial hyphae and are terverticillate, hyaline, branches appressed with stipes and metulae more or less cylindrical, smooth-walled, bearing three to six phialides each.
Phialides are often solitary, cylindrical with a short neck.	Phialides are slender, cylindrical with short but distinct necks.
Conidia are smooth-walled, ellipsoidal to cylindrical, variable in size, but mostly 3.5–8.0 × 3.0–4.0 µm.	Conidia are cylindrical at first, but often become elliptical or subglobose, smooth, 4.0–5.0 × 2.5–3.5 µm in size, greenish, smooth-walled.
Penicillium digitatum is the first phytopathogenic *Penicillium* species whose complete genome has been entirely sequenced.	Colourless to light brown sclerotia, measuring 200–500 µm, have been occasionally observed in fresh isolates.

Source: Pallottino et al. (2012).

The fungi that cause blue mould and green mould both belong to the genus *Penicillium*. Spores of the fungi are produced in chains in massive quantities on infected fruit. The fungi survive in the orchard as spores. Infection occurs by airborne spores entering the fruit rind through injuries. The infection and sporulation cycle can occur many times during the season in the packinghouse. Blue and green mould develop most rapidly at around 75°F; on the other hand, rot is virtually stopped at a temperature of 34°F. Both pathogens prefer cooler temperatures for infection. However, *Penicillium digitatum* is most destructive and causes major postharvest decay. *Penicillium italicum* has the ability to grow rapidly at temperatures below 10°C and is more common in cold storage. The *Penicillium* species are identified by the size of conidia, length of the stalks that hold the spores and the colour of the spore mass on their hosts.

Blue mould becomes more dangerous in storage because it rapidly spreads in the box and directly infects healthy fruits regardless of injury. *Penicillium italicum* is nesting-type pathogen which produces enzymes that help in softening of the adjacent fruit and causes infection. *P. digitatum* is not able to spread by nesting but infects fruits through wounds. Hence, if single fruit is infected, it remains as such without contaminating adjacent fruit. Green mould is quite common in India and grows slowly at lower temperatures but grows rapidly at higher temperatures (25–30°C) (Pallottino et al. 2012).

14.2.2.3 Disease Management

Careful picking, handling, packing and storage of fruits minimizes injuries to the fruit rind and the risk of blue or green mould development. Sanitary packinghouse practices, including the use of disinfectants such as chlorine or other materials, help in reducing the concentration of fungal spores capable of causing infection. Disinfection is achieved by spraying sodium hypochlorite solution (100–150 ppm) (Rouissi et al. 2009). Peroxyacetic acid (PAA) (a strong oxidizer formed from hydrogen peroxide and acetic acid) is effective against green mould. Dipping of fruits with PAA (800 µg/mL) for 1 min was found effective to reduce the green mould incidence on lemon. Hot water dip treatment at 56°C for 20 s has inhibited the *P. digitatum* spore germination. The short hot water brushing treatment (60°C for 20 s) reduced the green mould disease on citrus fruits (Sapers 2001). Perez et al. (2005) reported that intermittent curing treatment of two cycles (18 h at 38°C) completely inhibited the *P. italicum* on mandarin stored under ambient conditions. Carbonate and bicarbonate salts are most effective antifungal agents which reduced *Penicillium* rots on citrus fruit after immersion treatment in 3.00% (wt/vol.) of these salt solutions at high temperature.

Aspire™ (*Candida oleophila*), Pantovital™ (*Pantoea agglomerans*) and Biosave™ (*Pseudomonas syringae*) are registered in the United States and Spain as commercial biopesticides for the control of green and blue mould diseases of citrus fruits. Leelasuphakul et al. (2008) reported the antifungal efficacy of *Bacillus subtilis* strain against *Penicillium digitatum* through inhibition of spore germination by the action of water-soluble antibiotic secondary metabolites, proteins, enzymes and volatile compound production. Metabolites of *Lactobacillus plantarum* IMAU10014, *viz.*, 3-phenyllactic acid and allyl phenylacetate, showed antifungal activities against green mould (Wang et al. 2012). Metabolites of *Streptomyces* sp. RO3 (molecular masses >2000 Da) showed fungicidal action against green mould disease (Maldonado et al. 2010). Treatment with yeasts such as *Wickerhamomyces anomalus*, *Metschnikowia pulcherrima* and *Aspergillus pullulans* increased the accumulation of peroxidase and superoxide dismutase in mandarins which resulted in reducing severity of blue mould. Fruit dipping in *Pichia membranefaciens* (1×10^8 CFU/L) had lowered disease incidence (Luo et al. 2012). Fruit immersion treatments of citrus with *B. subtilis* strains, Bs 167 and COB5Y1 were most effective against *Penicillium*moulds with 78.75% and

75.00% of disease inhibition (Tariang et al. 2019). Pre-harvest spray of organic elicitors like salicylic acid (8 mM) and methyl jasmonate (3 mM) to orange plants proved to be effective in reducing colony/lesion diameter, wound rotting and spore mass density of *P. digitatum* in contrast to postharvest treatment (Iqbal et al. 2012). Chitosan combined with salicylic acid had better treatment of green mould than these isolated compounds without compromising the quality of fruit. Extracts of garlic, neem, *Withania somnifera*, *Acacia seyal*, mustard and radish showed antifungal activity against *P. digitatum*.

Fungicides are commonly used as dips or sprays to prevent *Penicillium* rots. Thiabendazole is an approved fungicide in Australia for postharvest dipping treatments against *Penicillium* moulds of citrus.

Dipping fruitsin thiabendazole (30 s) followed by rinsing within 24 h of harvest effectively manages these moulds. Postharvest application of selected fungicides can help in delaying the development of blue and green mould, especially in combination with immediate cooling of fruits after packing. Storage rots can be avoided by careful handling during harvesting to minimize cuts, scratches and bruises. Treatment of fruits with Bavistin (1000 ppm), maintenance of optimum temperature range and relative humidity and exclusion of ethylene during transport can reduce postharvest losses. The fungicide fludioxonil has curative efficacy against *P. digitatum* on Tarocco orange fruits (Schirra et al. 2005). Benomyl is used as a pre-harvest spray in South Africa and many other citrus-growing countries to prevent *Penicillium* rots. Many researchers have developed the commercial E-Nose that is capable of differentiating between non-infected and infected lemons and oranges with *P. digitatum* through early recognition of volatile organic compounds which are produced during pathogen fruit interaction (Pallottino et al. 2012). Maintenance of 100 ppm of ammonia in the storage atmosphere for 9–10 h can control the blue and green mould of orange and lemon fruit. Exposure of Valencia oranges and Eureka lemons continuously to the ozone (1.00 ppm vol/vol) at 10°C in export container for 2 weeks resulted in the delay of incidence of blue and green mould (Palou et al. 2001).

Evaluation of different ratios of clay-chitosan nanocomposite (CCNC) against *P. digitatum* on 'Valencia Late' sweet orange revealed that complete inhibition of the pathogen was noticed at 20 µg/mL for CCNC (1:0.5) through collapsing and irregular branching of hyphae in the apical part (Youssef and Hashim 2020). Silver nanoparticles (10.00 nm size) showed antifungal activity against *P. digitatum* of citrus fruits. Copper nanoparticles (48.00 nm) completely inhibited the growth of green mould of 'Valencia Late' sweet orange at 20 µg/ml concentration in both *in vitro* and *in vivo* conditions (Youssef et al. 2017). Application of TiO_2 nanoparticles (7.00 nm) on polypropylene film coated lemon fruits effectively controlled the *Penicillium* mould.

14.2.3 Greasy Spot

The period from late spring to early summer is when the majority of greasy spot infections occur. Greasy spot is most problematic on fresh grapefruit but it should be a concern for all citrus species, while rind blotch symptom is the greatest concern for fresh fruit. Significant leaf loss caused by greasy spot can reduce the ability of trees of any type to photosynthesize. Over time, trees can be weakened and are no longer able to support large numbers of fruits.

14.2.3.1 Symptoms

Swelling on lower leaf surfaces and yellow mottle appears at the corresponding point on upper surface. Swollen tissue starts to collapse and turns brown and eventually black in colour. Infection causes premature leaf drop, which occurs mostly in winter and early spring. On leaves, symptoms generally appear in the early winter but can take longer if the temperature is cold. Initially, the leaves have a mottled appearance, especially on the lower surface. Brown patches form within the mottle, especially on the undersides of the leaves, and remain surrounded by a yellow halo. As the lesions develop, they become visible on both sides of the leaves and have the appearance of greasy spots. If infection is severe, leaves can drop before the brown lesions occur. On fruits, lesions are frequently referred to as rind blotch. Rind blotch is composed of a large number of black pinpoint lesions between the oil glands. Light infections can lead to diffuse blotches, but more severe infection scan greatly discolour the rind. The cells surrounding the lesions can retain their chlorophyll for longer and can fail to colour properly. Occasionally, larger blotchy, slightly sunken areas occur on grape fruit which are termed pink pitting (Timmer and Gottwald 2000).

14.2.3.2 Causative Agents and Epidemiology

Mycosphaerella citri can be readily isolated from leaf or fruit lesions but care must be taken to maintain aseptic conditions and to use very small pieces of infected mesophyll. The fungus forms dark, grey-green colonies on most common cultural media. *M. citri* grows slowly and colonies reach about 2 cm in diameter after 3 weeks of growth. Conidia are sometimes formed in culture, especially when isolations are first made. Conidia measure 2–3.5 µm × 10–70 µm are multiseptated, pale olive-brown, cylindrical to slightly bent and verrucose. Pseudothecia on decomposing leaves are aggregated, have papillate ostioles, and measure up to 90 µm in diameter. Ascospores are slightly fusiform, have a single septum, and often contain two oil globules in each cell. They are hyaline and measure 2–3 × 6–12 µm (Timmer and Gottwald 2000).

The major source of inoculum for greasy spot is wind-dispersed sexual spores (ascospores) produced during periods of wetness on decomposing fallen leaves. The disease cycle of this pathogen is unusual because most of the infection occurs on mature leaves. In areas with high temperature and high rainfall, the disease cycle continues year-round and infection may occur at any time. Growth of *M. citri* in the leaf mesophyll is very slow, and many penetrations are required for macroscopic symptoms to develop. The pathogen produces ethylene which eventually induces premature defoliation (Graham et al. 1984).

14.2.3.3 Disease Management

Copper fungicides directly kill germinating ascospores and epiphytic mycelium and prevent infections. Spray zineb 75% WP at 600–800 g in 300–400 L of water/acre. Foliar fertilizers, especially those containing heavy metals such as zinc, manganese and iron, are quite effective for greasy spot control if applied at sufficiently high rates (Timmer and Zitko 1995).

14.2.4 Anthracnose

Anthracnose of sweet orange caused by *Colletotrichum gloeosporioides* has been reported as a severe epidemic disease in pre-harvest conditions in several orchards of Sicily, one of the most important orange production areas for Europe. Nowadays, this emerging pathogen strongly reduces the amount of marketable fruits worldwide and fungicide applications are needed. *C. gloeosporioides* (Penz.) Penz. & Sacc. represents a serious threat for worldwide orange production in all production stages, inducing various symptoms. *C. gloeosporioides* was reported as causal agent of postharvest anthracnose but also as causal agent of pre-harvest symptoms such as wither-tip on twigs, tear-stain and stem-end rot on fruit (Timmer et al. 2000).

14.2.4.1 Symptoms

On leaves, common symptoms are characterized as more or less circular spots, light tan in colour with a prominent purple margin that at a later phase of infection will show the fruiting bodies of the fungus (tiny dispersed black flecks). Tissues injured by various environmental factors (such as mesophyll collapse or heavy infestations of spider mites) are more susceptible to anthracnose colonization (Figure 14.2a). On fruits, anthracnose usually occurs only on fruits that have been injured by other agents, such as sunburn, chemical burn, pest damage, bruising or extended storage periods. The lesions are brown-to-black spots of 1.5 mm or greater diameter (Figure 14.2b). The decay is usually firm and dry but if deep enough can soften the fruit. If kept under humid conditions, the spore masses are pink to salmon but if kept dry, the spores appear brown-to-black. On ethylene degreened fruits, lesions are flat and silver in colour with a leathery texture. On degreened fruits, much of the rind is affected. The lesions will eventually become brown to grey black leading to soft rot (Timmer et al. 2000).

14.2.4.2 Causative Agents and Epidemiology

Anthracnose is caused by *Colletotrichum gloeosporioides*. The pathogen has septate mycelia, asexual fruiting body is the acervulus and setae are present. Once the spores germinate, they form a resting structure that allows them to remain dormant until an injury occurs. Cool weather (temp 20°C) responsible for development of disease in plants. Long periods of high relative humidity >80% with mists increase the disease severity. The primary source of inoculum is dormant mycelia, and the secondary source is conidia produced in acervulus. Disease is favoured by warm weather at temperature of 30–32°C, RH 80–85%, cloudy weather and presence of susceptible hosts (Timmer et al. 2000).

14.2.4.3 Disease Management

Management is by pruning infected leaves, twigs and fruits and burying them outside the garden. Fruit bagging at fruit-set stage (late October to early November) is a good practice. Spray fungicides like propineb, tebuconazole or prochloraz at early phase of infection. Collect affected leaves and burn them. Avoid excess nitrogenous fertilizer application. Irrigation during summer is required. Chemicals such as carbendazim at 1.5g/L and benomyl at 1g/L can be used. Spraying of pyraclostrobin fungicide gave protection to both pre-harvest and postharvest disease incidence (Piccirillo et al. 2018).

14.2.5 *Alternaria* Brown Spot or Black Rot

Alternaria species cause four distinct diseases in citrus, such as *Alternaria* brown spot of tangerines and their hybrids, *Alternaria* leaf spot of rough lemon, *Alternaria* black rot/fruit rot and Mancha foliar on lime. The disease occurs occasionally in lemons and oranges. *Alternaria* rot is primarily a problem in storage but it sometimes occurs in the orchard where it can cause premature fruit drop. In other parts of the world, the disease has been reported on citrus fruits other than navels and lemons. *Alternaria* fruit rot is the most important in areas where citrus is processed for juice because of juice contamination by masses of black fungal mycelium found in the interior of the infected fruits.

14.2.5.1 Symptoms

Fruits infected with *Alternaria* may turn light in colour several weeks before the colour break in healthy fruits. Some infected fruits may prematurely drop while others may remain on the tree. Infected fruits may appear normal. The simple method

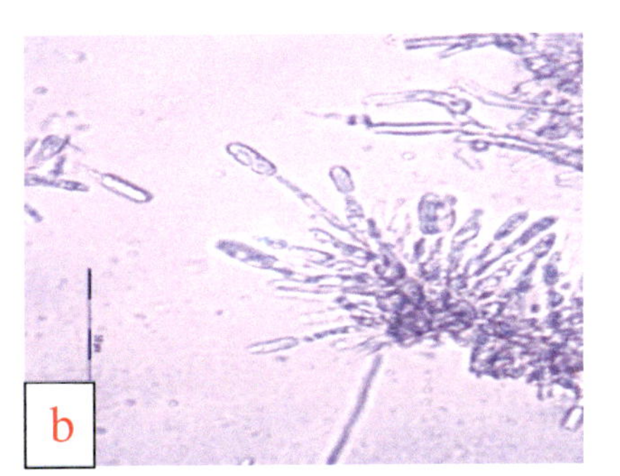

FIGURE 14.2 Symptom of anthracnose disease on fruits (a) and morphological structure of *Colletotrichum gloeosporioides* showing acervuli, setae and conidia (b).

of diagnosis is to cut the fruit in half, exposing the stylar or "blossom" end and the central cavity. Diseased fruits have a brown to blackish discolouration at the "blossom" end. The discolouration and decay may be restricted to the blossom end or may be extended deep into the central cavity. Observations of infected tissue in the laboratory reveal the presence of *Alternaria*. Fungal spores are unique in appearance. There may be little or no external evidence of infection. In lemons, the disease is most common during storage. Splitting, caused by environmental factors, often predisposes oranges to infection by *Alternaria*. Large navel fruits may split and drop in the fall during hot and dry weather. The incidence of splitting is higher in sunburnt fruits, and in trees stressed by drought and frost injury.

14.2.5.2 Causative Agents and Epidemiology

The spores, called conidia, are the primary source of infection during flowering. The microscopic, diagnostic features of *Alternaria* black core rot are as follows: Conidia are produced solitarily or in chains, ranging in shape from ellipsoidal to obclavate and are dark blackish-brown in colour. Transverse and longitudinal septa range in numbers from 1to6, and up to 3, respectively. Conidia are beakless when ellipsoidal or with a cylindrical beak, 3–4 µm in diameter and up to 13 µm long (Timmer et al. 2000).

Alternaria is an active saprophyte which grows on dead citrus tissues during wet weather. Airborne conidia are produced during these periods. These spores germinate and the fungus establishes itself in the button or stylar end of the fruit. Entrance into the fruit is facilitated when splits or growth cracks occur. The fungus grows into the central core of the fruit and causes a black decay. *Alternaria* produces large numbers of conidia in infected fruits. Release of conidia from sporulating brown spot lesions is triggered by rainfall or by sudden changes in relative humidity. Spores are dispersed by wind currents and are eventually deposited on the surface of susceptible tissues. With dew, the following night, the conidia germinate and eventually infect the leaves or fruits. Penetration of the leaf can occur directly or through stomata. The fruits may dry and become black and mummy-like in appearance. This fruit becomes one of the survival mechanisms for the fungus (Timmer et al. 2000).

14.2.5.3 Disease Management

Avoidance of overhead irrigation and use of under-tree irrigation systems will reduce the disease severity. Wider spacing and skirting of trees allows better ventilation and tends to reduce the disease severity. Avoidance of excess irrigation and nitrogen fertilization has been recommended for some time to avoid production of large amounts of susceptible tissues. Foliar fungicide applications are usually necessary to produce fruits with good external quality in areas where Alternaria brown spot is common. Depending on the climate in different areas, from 3 to 8 applications may be needed. Azoxystrobin and pyraclostrobin are generally more effective in managing the disease. Commercial control of *Alternaria* black core rot can be achieved through the use of systemic fungicides such as tebuconazole (Horizon, 25% EW) and difenoconazole (Score, 25% EC). Tebuconazole (0.021%) as a pre-harvest spray can control *Alternaria* leaf spot in the field and postharvest *Alternaria* decay in tangelos.

14.2.6 Citrus Black Spot

Citrus black spot (CBS) affects all citrus varieties, and all sweet oranges, grapefruit and lemons are highly susceptible to black spot. This fungus causes severe lesions on the rind of fruit, significantly decreasing fruit quality and marketability as fresh produce. Symptomatic fruit is not acceptable in the fresh markets. In December 2016, CBS-like symptoms were observed on Nagpur mandarin (*Citrus reticulata*), 'Mosambi' sweet orange, and Cutter Valencia (*C. sinensis*) trees in two orchards situated at Mohpa and Kachimet localities of Nagpur district, Maharashtra state. Later symptoms were also noticed on Nagpur mandarin and Mosambi fruits in the fresh fruit market of Nagpur (also known as "Orange City") (Das et al. 2018).

14.2.6.1 Symptoms

Lower fruits often have more symptoms but they do not cause internal decay. Symptoms will most likely appear about a month before harvest on the sunny side of the tree. The four symptom types are hard spot, cracked spot, false melanose and virulent spot. A hard spot is a sunken lesion with black margins and a grey centre. Fruiting bodies that contain fungal spores are produced within the grey centre. These lesions occur prior to harvest on the side of the fruits exposed to sunlight. A virulent spot is a spreading sunken necrotic (dead) lesion without defined borders. The colour varies from brown to black to brick-red. These lesions usually occur towards the end of the fruit growing season. False melanose consists of small, black necrotic lesions that are less than one millimetre in diameter. These symptoms often appear on young green fruits. Freckle spot occurs on mature fruits, postharvest, and often while in storage. These spots are round, depressed and reddish-brown with a red to brown border. They range in size from 1 to 3 mm and are an indication of a heavy infestation.

14.2.6.2 Causative Agents and Epidemiology

Ascomata of *Phyllosticta citricarpa,* previously known as *Guignardia citricarpa*, are formed on fallen, decomposing leaves. Perithecia are aggregated, globose, non-papillate, and about 100–175 µm in diameter (Kotz 1981). Asci are cylindrical, clavate, and each contain eight spores. Ascospores are 4.5 × 6.5 µm wide by 12.5–16 µm long, hyaline, non-septate, multiguttulate and swollen in the centre. A colourless appendage occurs at each end. Pycnidia are found in abundance on dead fallen leaves and are also produced on fruits and peduncles. They are dark brown to black and 115–190 µm in diameter. Conidia are obovate to elliptical, hyaline, non-septate, multiguttulate with a colourless appendage and are 5.5–7.0 µm wide by 8.0–10.5 µm long. In culture, colonies of *G. citricarpa* are dark brown to black and the mycelium is thick and prostrate.

The primary inoculum is from leaf litter and ascospores which are ejected when leaf litter is wet. Spores move with wind current. An additional source of inoculums is lesions on infected fruits, leaves and branches (conidia). Pathogen survives on leaves, leaf litter branches, fruits and peduncles. Dark brown or black pycnidia, structures that produce conidia, are formed on fruits, pedicles and leaves. They are also abundant on dead leaves. Conidia are not wind-borne but may reach susceptible fruits by rain splash. However, in climates with frequent summer rains, conidia play a larger role in the epidemic. Infections are latent until the fruit becomes fully grown or mature. At this point, the fungus may grow further into the rind, producing black spot symptoms months after infection, often near or after harvest. Symptom development is increased in high light intensity, intensifying temperatures, drought, and low tree vigour.

14.2.6.3 Disease Management

Sour orange (*Citrus aurantium*) is one of the few species of citrus that is resistant to black spot. Grapefruit (*C. paradisi*) and lemons (*C. limon*) are the most susceptible whereas some mandarins (*C. reticulata*) are more tolerant (Kotz 2000). Sanitation in orchards including collection and burning of fallen leaves should be followed. Spraying of 5% urea on fallen leaves litter reduces diseases incidence. A number of fungicides such as copper products, dithiocarbamates, benzimidazoles and strobilurins are effective against black spot. Pre-harvest sprays of benzimidazole fungicides are effective in preventing or delaying symptom expression during transport or storage. Treatment with guazatine or imazalil, hot water or waxing decreased the viability of the pathogen in black spot lesions (Korf et al. 2001).

14.2.7 Diplodia Stem-End Rot

This is an important postharvest disease in humid and warm regions. The disease is not common in arid areas. In India, the disease is most common in Central India (Nagpur mandarins) than in arid parts of Punjab (kinnows). In the literature, causal organism of citrus diplodia stem-end rot is described as *Diplodia natalensis* and its synonyms are *Lasiodiplodia theobromae* and *Botryodiplodia theobromae*. The term Alternaria has frequently been used in the literature to differentiate it from other citrus stem-end rots.

14.2.7.1 Symptoms

The pathogen infects citrus fruits from the button (calyx or disk) at the stem end of the fruit. Infection proceeds through the fruit core rapidly, resulting in development of soft brown-to-black decay symptoms at both ends of the fruit. The pathogen is grown unevenly on the rind and produces finger-like projections with black-to-brown discolourations at the lesion margin (Figure 14.3). The pathogen also infects through injuries, resulting in development of a soft brown-to-black lesion that enlarges very rapidly. Decay develops rapidly during and after excessive fruit ethylene degreening treatment and can be observed in fruit at the packing house.

14.2.7.2 Causative Agents and Epidemiology

Lasiodiplodia theobromae grows rapidly on potato dextrose agar (PDA) medium at 20–30°C with a white to light or dark grey aerial mycelia and later produces abundant black

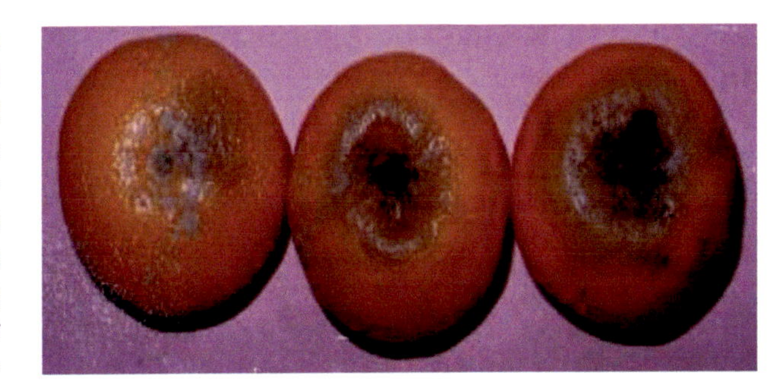

FIGURE 14.3 Symptoms of stem-end rot in kinnow fruits treated caused by *Lasiodiplodia theobromae*.

pigmentation which can be clearly seen from the reserve side of the culture plate. In PDA, concentrations of dextrose in the medium affect the mycelial growth, pigmentation and pycnidium productions. In higher dextrose concentrations, the pigment production is greater, but less pycnidium formation is noticed (Alan et al. 2001). Pycnidia are sub-globose to globose (300–700 μm in diameter) and single or grouped in a stroma and papillate. Pathogen produces pycnidia on dead citrus twigs. Conidiophores are simple, hyaline, occasionally septate, rarely branched, cylindrical and arising from the inner layers of cells lining the pycnidial cavity. Conidia are formed inside the pycnidia. Young conidial cells are hyaline, granular and have no septum. Matured conidia are striated and have one septum i.e., two-celled. The two-celled conidia survive over seasons on citrus trees and can cause latent infections on citrus fruits during winter and spring seasons. Perithecia are globose, papillate and capable of producing ascospores. Ascospores are single celled, hyaline and elliptical to ellipsoidal in shape. Ascospores become airborne and responsible for long-distance dispersal to newly planted citrus trees. Usually, pathogen does not sporulate on infected fruits. The fruit decay does not spread from infected fruits to healthy ones in packed containers.

The pathogen is facultative parasite, i.e., it completes its life cycle on dead twigs in groves through growth and sporulation in the deadwood including bark and twigs of citrus trees. The pathogen can survive over seasons on deadwoods. Rainfall water carries the spores from deadwood to the surfaces of immature fruits. The pathogen becomes latent and does not cause any fruit decay before harvest. Infection is developed after harvest under high temperatures and relative humidity. Optimal temperature for pathogen mycelial growth is 30°C. Spores remain latent until favourable conditions (*viz.*, high temperatures, injuries, during ethylene degreening, over maturity, when calyx dies and abscises) prevail. The pathogen causes infection both internally and externally (Zhang and Swingle 2005).

14.2.7.3 Disease Management

Hot water treatment (53°C for 5 min) has the ability to eradicate quiescent infection of the pathogen in oranges. The development of disease is completely inhibited at 10°C. Deadwoods of trees should be removed and burned outside the orchard. The fruits should be harvested at the right stage of maturity and avoid over-maturity fruit harvesting. Application of 2,4-D should be done in the packhouse to maintain green calyx on the fruit. Refrigeration of fruits immediately after packing helps to control the disease. Application of fungicides before degreening of fruits is a more effective approach for control of stem-end rot than later applications on the packingline. Preharvest spray of benzimidazole fungicides, *viz.* benomyl and thiophanate methyl, was effective for the control of disease (Zhang and Timmer 2007). Thiabendazole (TBZ) and imazalil are the effective fungicides for management of the disease. The fruit immersion treatment for 60 s in thiabendazole (2000 mg/L) at pH 5 and 20°C can effectively manage the disease. Immediate postharvest (4–12 h after harvest) application of fludioxonil (1000–6000 ppm) as drench or with wax to the fruits was effective against diplodiastem-end rot. Application of azoxystrobin fungicide also managed the disease. The combined application of yeast (*Candida saitoana*) with glycolchitosan (0.2%) reduced the incidence of stem-end rot on Valencia oranges. Hot water dip (52°C for 5 min) treatment of fruits with 10.00% ethyl alcohol prior to degreening reduced the disease in Hamlin sweet oranges. The water dip treatment or drenching with *Sec*-butylamine (2 AB) (1.00%) was found to be effective against stem-end rot. Phosphate salt is also effective against stem-end rot.

14.2.8 Phomopsis Stem-End Rot or Citrus Melanose

The disease affects all the citrus cultivars, but it is most severe in grapefruit and lemons. The pathogen causes two distinct diseases on citrus species. The perfect stage of the pathogen causes melanose (lesions on fruit and foliage), and the imperfect stage causes *Phomopsis* stem-end rot (postharvest disease). It most commonly occurs in citrus worldwide. The disease is occurring in larger proportions and reduces marketable fruit yield. The disease is more active in cooler seasons.

14.2.8.1 Symptoms

The initial symptom of *Phomopsis* stem-end rot is development of softness on the stem end of the fruit with no initial discolouration. Further, colour starts to develop and eventually turn to tan-brown and finally black. The infected tissue is slightly shrivelled that forms ridge between the rotten and healthy tissue. The pulp becomes mushy without discolouration. Later development of unpleasant rancid odour is noticed. The melanose symptoms of the disease are small spots or scab-like lesions which are referred as tear-drop, mud-cake and star melanose.

14.2.8.2 Causative Agents and Epidemiology

Diaporthecitri is the causal agent for this disease which produces hyaline two-celled ascospores within each cell, and two oil droplets or guttulae are prominent. Ascospores are slightly constricted at the septum. Ascospores are formed inside the perithecium which are usually flask-shaped sexual fruiting bodies. Perithecia are circular in shape and flattened at the base with long black coloured beaks that taper outwards. The beaks ranged from 200 to 800 μm long, 40 to 60 μm wide, and the base of the perithecium ranges from 125 to 160 μm in diameter (Wolf 1926). The conidial stage of the pathogen is the most important in the disease cycle. *Phomopsiscitri* produces pycnidia, which are scattered on substratum. Pycnidia are ovoid, dark coloured, thick-walled with erumpent. Pycnidia are formed on deadwood and rotting fruit. The pathogen produces two kinds of conidia i.e., alpha and beta conidia. The alpha conidia are hyaline, single-celled and contain two guttulae. They are the primary means of dissemination of spore for short distances. The beta conidia are long, slender rod-shaped spores, hooked at one end, and produced in older pycnidia. The beta conidia are not able to germinate (Wolf 1926).

Phomopsis has similar mode of infection and life cycle to *Diplodia*. The pathogen remains latent in the button tissue and infects the fruit when the calyx abscises. Further, infection develops after the harvest when calyx loosening occurs. The pathogen enters the fruit through natural openings. The pathogen infects young leaves and fruits of certain citrus species when the tissue is grown and expands during extended periods of rainy and humid weather. *D. citri* is mainly a saprophyte that survives on deadwoods and derives energy from both living and dead wood. The fruiting bodies, perithecia and pycnidia, are only produced on dead or dying twigs and infected fruit. Perithecia are rarely formed and the conidia produced inside the pycnidia are considered a primary source of inoculum. Ascospores are discharged in a forcible manner, and this is the main reason for long-distance dispersal of the disease.

Infection process is started when the ascospores (*Diaporthe* stage) or conidia (*Phomopsis* stage) land on the leaf tissue. In dry conditions (temperatures <17°C and >35°C), the spores die and infection does not occur. It is well known that the pathogen requires 10–24 h of moisture for the spore germination. After a spore landing on susceptible tissue, 10–12 h of free available moisture is required for infection at temperature of 25°C, but at 15°C of temperature, 18–24 h of wetness are necessary for infection. Thus extended wet periods resulting from afternoon rain showers plus dew periods in May and June coupled with warmer temperatures during these months create favourable weather for infection (Kuhara 1999).

14.2.8.3 Disease Management

The disease can be prevented by adopting suitable cultural practices that promote the vigorous development of trees and by removing the deadwoods or dried twigs from the tree. Avoid predisposed conditions that will result in abscission of the button, *viz.*, degreening of fruit with ethylene, injuries, over-maturity, high temperatures during transport of packed fruit and prolonged storage. Application of potassium sorbate and potassium phosphate salts at 20°C temperature reduced disease development (Cerioni et al. 2013). Protectant sprays of copper fungicides at suitable times are widely recommended for melanose control. Foliar applications of pyraclostrobin fungicide to the spring flush growth of citrus trees provide control of melanose disease (Mondal et al. 2007). Aspire, a yeast *Candida oleophila* (which competes with the pathogen for nutrients released by injuries), is a promising biological control agent that is commercially registered for control of postharvest rots of citrus.

14.2.9 Sour Rot

The disease is reported from the majority of citrus growing areas of the world and infects all varieties of citrus, including tangerine (*Citrus reticulata*), orange (*C. sinensis*), grapefruit (*C. paradisi*) and lemon (*C. limon*). *Geotrichum* infection requires more extensive damage on fruit. Over-matured fruits are more susceptible to this disease. Sour rot is major problem in warmer regions of India where the citrus fruits are handled on the bare ground and where the packaging is done on paddy straw. Among citrus fruits, Nagpur mandarin is infected more in the March to April harvest season, but disease usually does not occur on crops harvested in November to December because of low temperatures.

14.2.9.1 Symptoms

Initially, extensive water-soaked lesions are formed on the fruits and further lesions become light brown to yellow in colour. Further, the mycelium is developed on the fruit which bears large amount of conidia. Then, fruit becomes a softened, stinking and semi-solid mass due to the action of extracellular enzymes of the pathogen (Figure 14.4a). Further, infected tissues of fruit rapidly degrade. Infection spreads quickly inside the package, and corrugated cartons may also lose strength as they get wet.

14.2.9.2 Causative Agents and Epidemiology

Geotrichum citri-aurantii, the causal agent sour rot, conidia is cylindrical to oval shaped (Figure 14.4b). Mycelium is white in colour and has prominent dichotomous branching. Ascospores are formed in pairings with isolates of the opposite mating type in the laboratory, and production of ascospores has been noticed in decaying lemon fruit in the field (Butler et al. 1988).

The pathogen is present in the soil of most citrus producing areas. Spores spread to the fruit through dust and rain splashing on the low hanging fruits. Fruit is infected through injuries that are caused by insects (false codling moth, fruit fly and fruit-sucking moths) and by snap-picking fruits. The disease should be considered a serious threat to the citrus industry, particularly in the warm and humid autumn season. Firmness of citrus fruits predisposes the attack of rotting fungi. Poor orchard sanitation, fruit damage during harvest and uncut branches touching the soil surface in citrus orchards predisposes the pathogen infection and spread. Higher nitrogen application and excessive irrigation reduces the fruit firmness and increases the susceptibility to mechanical and pathogen damage (Olmo et al. 2000). The sour odour developed in the advanced stages of the disease attracts vinegar flies, which help to spread the disease. Sour rot develops rapidly at 28–30°C temperature. Disease spreads in packed cartons from infected fruit to healthy fruit, and its development is stimulated by the presence of green mould spores. Temperature <10°C is known to suppress the pathogen development.

14.2.9.3 Disease Management

Good harvesting practices should be used. Avoid mechanical injury to the fruits while harvesting. Avoid packing of fruits that are contacted with soil. Fruits must be harvested at ideal maturity stage and always avoid over-mature fruits. Immediately after harvesting, fruits should be packed and refrigerated. Sanitation of the packhouse is essential to prevent vinegar flies. Avoid fruits touching the soil surface with proper handling of trees. Ferraz et al. (2016) reported that *Rhodotorula minuta*, *Candida azyma* and *Aureobasidium pullulans* showed antifungal activity against the sour rot pathogen through deformation of pathogen hyphae. Fruit immersion treatment with potassium sorbate or sodium bicarbonate at 1.00% (wt/vol) for 30 s at 25°C reduced sour rot development. Triazole fungicides, propiconazole and cyproconazole are highly effective against sour rot disease (McKay et al. 2007). Cecropin A-Melittin hybrid peptide BP21 can effectively control sour rot infections of citrus fruits. Tea saponin also inhibited mycelial growth and spore germination of the pathogen individually and in combination with imazalil and prochloraz. The best ratio of tea saponin and prochloraz or imazalil to control *G. candidum* was 8:2. Guazatine is the widely used fungicide for control of sour rot. Application of water suspension of yeast (*Debaryomyces hansenii*) cells to the wounds on the fruit surface prior to inoculation or infection reduced disease incidence by 80–90% in orange fruits (Mehrotra et al. 1998). Aspire (*Candida oleophila*) is also highly effective against sour rot disease. Bui-fumigant fungus *Muscodor albus*

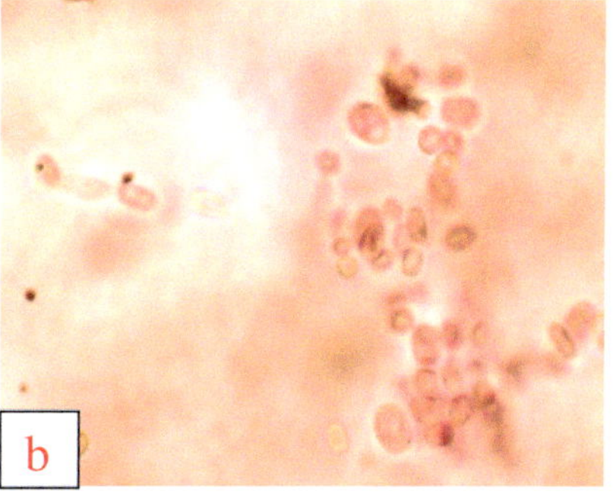

FIGURE 14.4 Soft watery decay with distinct margin between decayed and healthy tissue (a) and conidia of *Geotrichum citri-aurantii* (b).

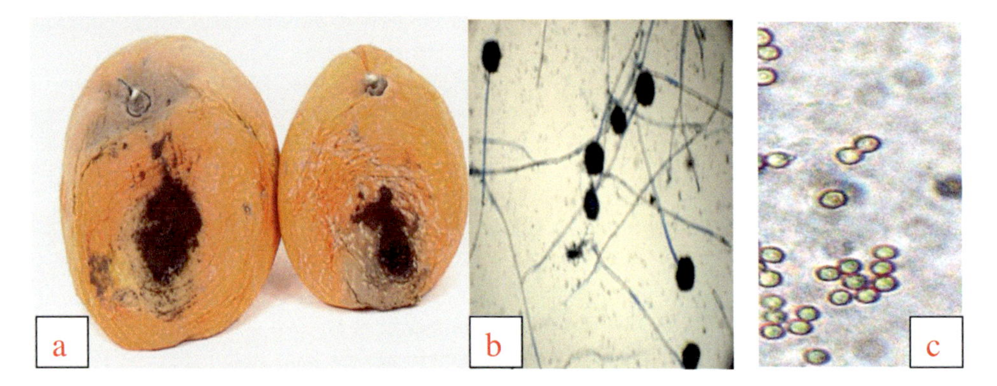

FIGURE 14.5 Symptoms of Aspergillus rot on kinnow fruits caused by *Aspergillus niger* (a) and mycelium, conidiophores and globose vesicle. Metulae and phialides cover the entire vesicle and condia (b & c)

(produces the volatiles compounds) was used to fumigate storage rooms of lemons to control the sour rot pathogen (Mercier and Smilanick 2005).

14.2.10 Aspergillus Rot

Aspergillus sp. is one of the predominant pathogens associated with the spoilage of citrus fruits. Black mould caused by *Aspergillus niger* results in postharvest spoilage in of citrus fruits and spoilage of acid lime at field.

14.2.10.1 Symptoms

Aspergillus infected fruits are covered with black mycelial growth which produces conidial heads which is referred to as black mould. Further, fruits become rotted and even adjacent fruits are infected, as the spores contaminate the whole lot (Figure 14.5a).

14.2.10.2 Causative Agents and Epidemiology

Aspergillus niger hyphae are branched and septate (Figure 14.5b). Conidia are greenish, globose to sub-globose with rough surface and occur in chains of two or more. Conidiophores are prominent under microscopes with short circular to semicircular shape; the spores are dark and oval in shape. *Aspergillus niger* is parasitic fungi that penetrate citrus fruit tissue through micro-wounds and bruises. High temperatures and high relative humidity favour the disease development. The most favourable temperature for disease development is 25–40°C. The pathogen causes rapid decay and spreads very quickly at 30–35°C. *Aspergillus* does not grow below 15°C, and at refrigerated conditions there is no growth at all. The fungus is very temperature specific.

14.2.10.3 Disease Management

Fruits stored under refrigerated conditions (8–10°C) are free from decay caused by *A. niger.* The strain of commonly used yeast in wineries, *Sacchromyces cerevisiae*, was found to inhibit the growth of *A. niger* in Nagpur mandarins and acid limes (Naqvi 1998). The black mould disease can be reduced by following management practices recommended for other storage rots.

REFERENCES

Alan MS, Begum MF, Sarkar MR, Islam MR, Alam MS (2001) Effect of temperature, light and media on growth, sporulation, formation of pigments and pycnidia of *Botryodiplodia theobromae*. Pak J Biol Sci 4:1224–1227.

Anonymous (2012) Annual Report. Citrus Research Station (2012) Dr. Y.S.R. Horticultural University, Tirupati. pp. 125–128.

Butler EE, Fogle D, Miranda M (1988) *Galactomyces citri-aurantii* a newly found teleomorph of *Geotrichum citri-aurantii* –the cause of sour rot of citrus fruit. Mycotaxon 33:197–212.

Cerioni L, Sepulveda M, Rubio-Ames Z, Volentini SI, Rodriguez-Montelongo L, Smilanick J, Ramallo J, Rapisarda VA (2013) Control of lemon postharvest diseases by low-toxicity salts combined with hydrogen peroxide and heat. Postharvest Biol Technol 83: 17–21.

Das AK, Nerkar S, Kumar A (2018) First report of *Phyllosticta citricarpa* causing citrus black spot on *Citrus sinensis* and *C. reticulata* in India. Plant Dis 23: 56–61.

Ferraz LP, Cunha T, Silva AC, Kupper KC (2016) Bio-control ability and putative mode of action of yeasts against *Geotrichum citri-aurantii* in citrus fruit. Microbiol Res 188: 72–79.

Graham JH, Whiteside JO, Barmore CR (1984) Ethylene production by *Mycosphaerellacitri* and greasy spot infected citrus leaves. Phytopathology 74: 817.

Iqbal Z, Singh Z, Khangura R, Saeed A (2012) Management of citrus blue and green moulds through application of organic elicitors. Australas Plant Pathol 41: 69–77.

Korf HJG, Schutte GC, Kotz TJM (2001) Effect of packhouse procedures on the viability of *Phyllosticta citricarpa*, anamorph of the citrus black spot pathogen. African Plant Protec 7(2): 103–109.

Kotz TJM (1981) Epidemiology and control of citrus black spot in South Africa. Plant Dis 65(12):945–950.

Kotz TJM (2000) Black spot. In: Timmer LW,Garnsey SM, Graham JH (Eds.). Compendium of Citrus Diseases. APS Press, St. Paul, MN, pp. 23–25.

Kuhara S (1999) The application of the epidemiologic simulation model "MELAN" to control citrus melanose caused by *Diaporthe citri* (Faw.) Wolf. Food Fertilizer Technol Center Exten Bull 8.

Leelasuphakul W, Hemmanee P, Chuenchitt S (2008) Growth inhibitory properties of *Bacillus subtilis* strains and their metabolites against the green mold pathogen (*Penicillium digitatum* Sacc.) of citrus fruit. Postharvest Biol Technol 48: 113–121.

Luo Y, Zeng K, Jian M (2012) Control of blue and green mold decay of citrus fruit by *Pichia membranefaciens* and induction of defense responses. Sci Hortic 135: 120–127.

Maldonado MC, Orosco CE, Gordillo MA, Navarro AR (2010) *In vivo* and *in vitro* antagonism of *Streptomyces* sp. RO3 against *Penicillium digitatum* and *Geotrichum candidum*. Afr J Microbiol Res 4: 2451–2456.

McKay AH, Forster H, Adaskaveg JE (2007) Sensitivity of isolates of *Geotrichum citri-aurantii*, the causal pathogen of sour rot in citrus, to DMI fungicides. Phytopathology 97: S74.

Mehrotra NK, Sharma N, Nigam M, Ghosh R (1998) Biological control of sour-rot of citrus fruits by yeast. Proc Nat Acad Sci India Section B Biol Sci 68(2): 133–139.

Mercier J, Smilanick, JL (2005) Control of green mold and sour rot of stored lemon by biofumigation with *Muscodor albus*. Biol Control 32: 401–407.

Mondal SN, Vicent A, Reis RF, Timmer LW (2007) Efficacy of pre and post inoculation application of fungicides to expanding young citrus leaves for control of melanose, scab, and Alternaria brown spot. Plant Dis 91: 1600–1606.

Nanda SK, Vishwakarma RK, Bathla HVL, Anil R, Chandra P (2010) Harvest and postharvest losses of major crops and livestock produce in India. All India Coordinated Research Project on Post Harvest Technology (ICAR), Ludhiana.

Naqvi SAMH (2002) Fungal Diseases of Citrus–Diagnosis and Management. Technical Bulletin No. 5, NRC for Citrus. Nagpur, India, 61 p.

Naqvi SAMH (1998) *Phytophthora* diseases of citrus in India and their integrated management. National Symposium on Perspectives in Integrated Plant Disease Management. Indian Society of Plant Pathologists, Ludhiana. NRC for Citrus, Nagpur and CICR, Nagpur at NRC for Citrus, Nagpur. 13–14 Feb. 2002 (Abstract) pp. 21–22.

Olmo M, Nadas A, Garcia GM (2000) Nondestructive methods to evaluate maturity level of 160 oranges. J Food Sci 65: 365–369.

Pallottino F, Costa C, Antonucci F, Strano MC, Calandra M, Solaini S, Menesatti P (2012) Electronic nose application for determination of *Penicillium digitatum* in Valencia oranges. J Sci Food Agricul 92 (9): 2008–2012.

Palou L, Smilanick L, Crisosto CH, Mansour M (2001) Effect of gaseous ozone exposure on the development of green and blue molds on cold stored citrus fruit. Plant Dis 85: 632–638.

Perez AG, Luaces P, Olmo M, Sanz C, Garcia JM (2005) Effect of intermittent curing on mandarin quality. J Food Sci 70: 64–68.

Piccirillo G, Carrieri, R, Giancarlo P, Antonino A, Ernesto L, Dolores F (2018) *In vitro* and *in vivo* activity of QoI fungicides against *Colletotrichum gloeosporioides* causing fruit anthracnose in *Citrus sinensis*. Sci Hortic 236(16): 90–95.

Rouissi W, Sanzani SM, Ligorio A, Khamis Y, Yaseen T, Cherif M, D'Onghia AM, Ippolito A (2009) Application of salts and natural substances to reduce incidence of *Penicillium* rot on Maltaise and Valencia Late oranges in Tunisia. In: Proc. of the 10th Arab Congress of Pl Protec, Arab J Pl Protec; 26–30 October 2009; Beirut, Lebanon, p. E–71.

Sapers GM (2001) Efficacy of washing and sanitizing methods for disinfection of fresh fruits and vegetable products. Food Tech Biotech 39: 301–311.

Savita GSV, Nagpal A (2012) Citrus diseases caused by *Phytophthora* species. GERF Bull Biosci 3(1):18–27.

Schirra M, D'Aquino S, Palma A, Marceddu S, Angioni A, Cabras P, Scherm B, Migheli Q (2005) Residue level, persistence, and storage performances of citrus fruit treated with fludioxonil. J Agric Food Chem 53: 6718–6724.

Tariang J, Majumder D, Firake DM (2019) Bio-control management of postharvest *Penicillium* rot disease of Khasi Mandarin (*Citrus reticulata* Blanco) oranges using native *B. subtilis* in Meghalaya, India. Int J Pure App Biosci 7(2): 292–302.

Thakre B, Uttam S, Gour CL (2017) Use of suitable fungicides for the control of gummosis caused by (*Phytophthora* sp.) on Nagpur Mandarin in Satpura Plateau of Madhya Pradesh, India. Int J Curr Microbiol App Sci 6(7): 2395–2400.

Timmer LW, Garnsey SM, Graham JH (2000) Compendium of citrus disease. 2nd ed. The American Phytopathological Society, St, Paul, MN.

Timmer LW, Gottwald TR (2000) Greasy spot and similar diseases. In: Timmer LW, Garnsey SM, Graham JH (Eds.). Compendium of Citrus Diseases. American Phytopathological Society, St. Paul, MN, pp. 25–28.

Timmer LW, Zitko SE (1995) Evaluation of nutritional products and fungicides for control of citrus greasy spot. Proc Fla State Hortic Soc 108:83–87.

Wang HK, Yan YH, Wang JM, Zhang HP, Qi W (2012) Production and characterization of antifungal compounds produced by *Lactobacillus plantarum* IMAU10014. PLOS ONE 7: e29452.

Wolf FA (1926) The perfect stage of the fungus which causes melanose of citrus. J Agri Res 33(7):621–625.

Youssef K, Hashim AF (2020) Inhibitory effect of clay/chitosan nanocomposite against *Penicillium digitatum* on citrus and its possible mode of action. Jordan J Biolog Sci 13; in press.

Youssef K, Hashim AF, Margarita R, Alghuthaymi MA, Abd-Elsalam KA (2017) Antifungal efficacy of chemicallyproduced copper nanoparticles against *Penicilliumdigitatum* and *Fusarium solani* on citrus fruit. Philip Agric Sci 100: 69–78.

Zhang J, Swingle P (2005) Effects of curing on green mold and stem-end rot of citrus fruit and its potential application under Florida packing system. Plan Dis 89: 834–840.

Zhang J, Timmer, LW (2007) Pre-harvest application of fungicides for postharvest disease control on early season tangerine by hybrids in Florida. Crop Prot 26: 886–893.

15

Postharvest Diseases of Mango and Their Management

P.K. Shukla, Tahseen Fatima, Baradevanal Gundappa and Nidhi Kumari
Central Institute for Subtropical Horticulture (ICAR), Lucknow, Uttar Pradesh, India

CONTENTS

15.1 Introduction

Mango (*Mangifera indica* L.) is an important fruit crop of tropical as well as subtropical regions of the world (Alemu et al. 2014a). The place of origin of mango is accepted to lie between India and Burma (De Candolle 1904). The great variety in taste and flavour and the medicinal and nutritional properties of mango have given it great value among the consumers worldwide (Diedhiou et al. 2007). Mango ranks fifth in area (56.81 million hectares) and production (50.64 million tonnes) among leading fruits grown worldwide (FAO 2017). India shares about 56% in global production and has the number one position in area as well as production, followed by Thailand, Mexico, China, Pakistan, Indonesia, Philippines, Nigeria, Egypt, Brazil, Haiti, etc. (Swart 2010). It is grown in almost all parts of India with a production of 21.82 million tonnes mango fruits from an area of 2.26 million hectares (NHB Database 2017–2018).

There are about 1700 varieties of mango in the world, but only about three dozen are grown commercially. In India, Dashehari, Chausa, Langra, Kesar, Alphonso, Bagganpallu, Neelum, Fazli and Totapuri are the most appreciated mango cultivars among mango lovers. Unripe mango fruit cultivars are utilized in making pickle, chutney, amchur, murabha, aamras, etc., while cultivars like Totapuri and Alphonso are used in various types of drinks (Kalita 2014).

Mango crops have been reported to suffer from a large number of diseases at various growth, development and production stages. Various pre-harvest and postharvest diseases of mango are responsible for the immense loss to plant growth, qualitative aspects of fruits and ultimately the yield (Diedhiou et al. 2007; Prakash 2004; Prakash et al. 2011; Shukla et al. 2018a, 2018b). Among those, important pre-harvest diseases are powdery mildew, wilt, anthracnose, blossom blight, dieback/twig blight, malformation and sooty mould; postharvest diseases include anthracnose and several other types of rot. The pre-harvest diseases are directly or indirectly responsible for yield and quality of fruits. In this chapter, we discuss various postharvest diseases of mango and different measures to manage them.

15.2 Major Postharvest Diseases

The postharvest losses in mango have been reported to be up to 17.36% (Haggag, 2010). Mango fruitlets used to suffer from infection of anthracnose, and powdery mildew and infected fruitlets drop down. Mango fruits remain comparatively resistant to infection during developmental stages, however the susceptibility of fruits to infection gradually increases towards maturity and they become highly susceptible during the period between harvest and consumption because of chemical changes during ripening. Mango fruits, being highly perishable, are marketed and consumed as early as possible after harvest. The mango crop is the host of many pathogens, among which fungi are the major agents throughout the world (Diedhiou et al. 2007). Bacterial canker is also a severe constraint in humid regions. The major postharvest diseases of mango are discussed here.

15.2.1 Anthracnose

Anthracnose is considered the most serious postharvest disease of mango worldwide. *Colletotrichum gloeosporioides* (Penz. and Sacc.) was identified as the causal organism of this disease. The fungus produces symptoms on every tender part of the tree including fruitlets and mature fruits. It is reported to be a key constraint almost worldwide, particularly in areas where rain and high humidity prevail during fruit development and maturity periods (Akem 2006; Haggag 2010; Shukla et al. 2016, 2017), and therefore is a major limitation in export trade expansion (Sangeetha and Rawal 2008). Fruits at all growth stages are prone to the disease and produce typical disease symptoms (Yenjit et al. 2004). Sunken necrotic lesions on the fruit surface and mummified fruits are typical symptoms (Rivera-Vargas et al. 2006). The rotting of mature fruits caused during postharvest handling, ripening and transport is the greatest matter of concern, which reduces the number of marketable fruits resulting in huge losses. Postharvest rot is directly associated with pre-harvest care when symptomless and quiescent infection is caused during fruit development but causes decay and rotting of fruit after harvest during ripening (Haggag 2010).

15.2.1.1 Symptoms

The symptoms of anthracnose include blossom blight – typical shot hole spots on leaves and spots or lesions of different sizes on fruits. Infection on the panicles appears as brown to black streaks on peduncles and spots on flowers which may later cover the entire panicle; under favourable conditions and whole panicle becomes blighted. There is no chance of fruit set on such panicles. Leaf symptoms appear on young leaves as minute reddish brown round spots during heavy dew periods or blight during rainy season. Infected young fruits develop black spots and later shrivel and finally drop off. The disease in harvested fruits appears in the form of dark lesions during the ripening stage. Within a few days, disease progresses rapidly, and the entire fruit becomes rotten in severe cases (Figure 15.1). The symptoms on shoulder browning or sap flow injury affected fruits may appear in a tearstain pattern,

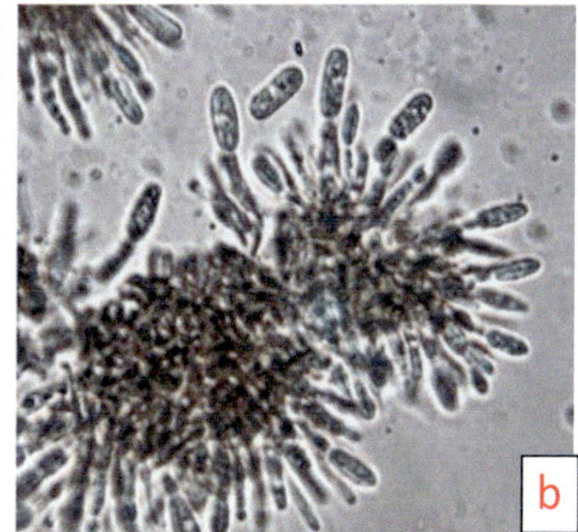

FIGURE 15.1 Symptoms of anthracnose disease of mango (a) and morphological structure of *Colletotrichum gloeosporioides* showing acervuli, setae and conidia (b).

from stem end portion down (Arauz 2000; Giblin et al. 2010; Shukla et al. 2016). Initially the infection remains confined to the epicarp, however within a short period of time it proceeds into pulp. Rotten fruits may also have sporulation of fungus on epicarp (Akem 2006).

15.2.1.2 Disease Cycle and Epidemiology

Environmental parameters *viz.*, optimum temperature, high relative humidity, rainfall for a longer period of time or dew at night, especially during initiation and development of new flushes and fruit maturity greatly influence the extent of infection and damage. The infection is mostly aggravated by temperature between 20°C and 30°C (Arauz 2000). Regular rains during panicle development and fruit maturity result in severe infection. Availability of free water on fruit surface as well as high humidity for more than 10 hours is necessary for conidial germination, which is the primary and secondary inoculum source of infection. However, storage temperature that reaches up to 40°C during ripening can cause severe rotting (Prakash 2004; Prakash et al. 2011).

15.2.2 Stem-End Rot

Stem-end rot is another economically serious postharvest disease of mango fruits worldwide (Ni et al. 2012). The disease incidence is more severe in orchards having older trees, and if fruits are subjected to prolonged storage (Cooke et al. 2009). Several fungi *viz.*, *Lasiodiplodia theobromae* (Pat.) Griffon & Maubl., *Aspergillus niger* van Tieghem, *Phomopsis mangiferae* S. Ahmad or *Dithiorella dominicana* Petr. & Cif. and *Colletotrichum gloeosporioides* Penz. have been isolated from diseased fruits (Cooke et al. 2009). Among these, *B. theobromae* has been considered the most important wound/secondary pathogen and is a saprophyte, capable of living on different hosts or substrates at relatively high temperatures. It has also been associated with postharvest rots of a wide range of vegetables.

15.2.2.1 Symptoms

The pathogen remains dormant on the unripe fruits; as the ripening process begins, fruits start rotting near the stem end and become brown-grey, soft, and rotting begins which gradually covers the entire fruit surface (Govender, 2005; Prakash, 2004; Prakash et al. 2011). Injuries to fruits may lead to lesion formation even away from the stem end (Figure 15.2). Adjacent healthy fruits also get infected after coming into physical contact with these decaying fruits. The symptoms may vary according to the causal agent. *P. mangiferae* results in formation of firmer lesions with defined margins which spread comparatively more slowly than those caused by other stem-end rot fungi. The fruiting bodies formed by *P. mangiferae* are dark and pinhead-sized, whereas *C. gloeosporioides* produces pink spore masses (Prakash 2004; Cooke et al. 2009; Prakash et al. 2011).

FIGURE 15.2 Symptoms of stem-end rot disease of mango caused by several fungi.

15.2.2.2 Disease Cycle and Epidemiology

The fungi remain within the branches and twigs, causing latent infection. The endophytic hyphae colonizes the inflorescence and reaches the stem end several weeks after flowering but does not extend into fruit until harvesting (Govender 2005; Cooke et al. 2009). The stem-end rot-causing fungi may also harbour in soil and fallen leaves, where the fruit also gets infection from after harvest (Cooke et al. 2009). The pathogen is suspected to invade the mango trees either through natural openings or wounds and further causes latent infection in the fruit by entering through stem ends. The conidia are dispersed and released with free water under high humidity. Conidia present in rainwater and fruiting structures of the pathogen growing on fallen leaves are the sources of inoculum. The pathogen can overwinter anywhere on tree or in soil in the form of black pycnidia and perithecia.

15.3 Bacterial Black Spot (Canker)

In general, bacterial black spot is not considered an important postharvest disease of mango fruits; however, when fruits are severely infected by *Xanthomonas citri* pv. *mangiferae-indicae* (Patel, Moniz & Kulkarni) Robbs, Ribeiro & Kimura may suffer when high humidity is high during storage. The disease appears as light-coloured water-soaked spots which later turn into dark star-shaped craters. Bacterial infection is confirmed by oozing of gummy substance from cracks caused due to infection. The tissue injured due to bacterial infection also allows other saprophytic fungi to grow. Its incidence in new orchards can be avoided by planting bacteria-free grafts. Spray of copper-based bactericides has also been recommended during the rainy season to avoid pre-harvest infection in fruits.

15.4 Other Fruit Rots

After harvest, rotting of mango fruits is common and may be due to anthracnose and stem-end rot or infection caused by other pathogenic fungi or infestation by other saprophytic fungi through injury (Prakash et al. 2011). Such rots can be avoided by proper sorting and removing the injured fruits from a healthy lot. Hot water treatment (HWT), mentioned for anthracnose, may also help in managing such rots (Mansour et al. 2006). Significant decrease in mango fruit rot after treatment with essential oils of *Thymus vulgaris* (1000 µL/L), *Artemisia persica* (1000 µL/L) and *Rosmarinus officinalis* (500 µL/L) has been reported by Javadpour et al. (2018). The description of various types of rots is briefly mentioned below.

15.4.1 Black Rot

Black rot is caused by *Aspergillus niger* V. Tiegh, *A. variecolor* (Berk & Br.) Thorn & Raper, *A. nidulans* (Eidorn) Wint, *A. fumigatus* Fres., *A. flavus* Link, *A. chevalieri* (Mang.) Thorn. & Church. After harvest, fungus invades the fruits through wounds or cut ends and infection starts in the form of greyish to pale brown soft, sunken spots which later coalesce into dark brown to black lesions and spread rapidly. Profuse fungal growth develops later over these lesions (Figure 15.3). The disease development intensifies at 30–36°C. Careful handling at all stages to avoid mechanical injury and sap burn damage to the fruits can significantly reduce incidence of black rot.

15.4.2 Botryosphaeria Rot

Botryosphaeria rot is caused by *Botryosphaeria ribis* Gross. & Dugg. Initially the fruit tissue around the spot softens, followed by brown discoloration of affected skin and darkening of the flesh beneath. Dark, sunken, oval to round or sometimes elliptical lesions with minute black bodies and white to grey mouldy growth over it in advance stages accompanied with shallow decay of flesh beneath are formed. Variable symptoms are produced by *B. ribis* and therefore sometimes confused with stem-end rot. Hence, association of *B. ribis* must be confirmed by culturing and microscopic examination. In order to minimize the severity of fruit rot caused by *Botryosphaeria*, fruits should be harvested along with a 2–3 cm stalk and sanitation should be maintained in orchards.

15.4.3 Brown Rot

Brown rot is caused by *Lasiodiplodia theobromae* (Pat.) Griffon & Moubl. The presence of dark brown to black necrotic patches without any defined margin near the stalk or anywhere on a wounded fruit is a characteristic symptom. A heavily infected fruit turns dark, brittle, light weighted with disintegrated and off-flavoured internal pulp. To reduce the incidence of brown rot, fruits should remain free of bruising/wounding during picking and subsequent handling.

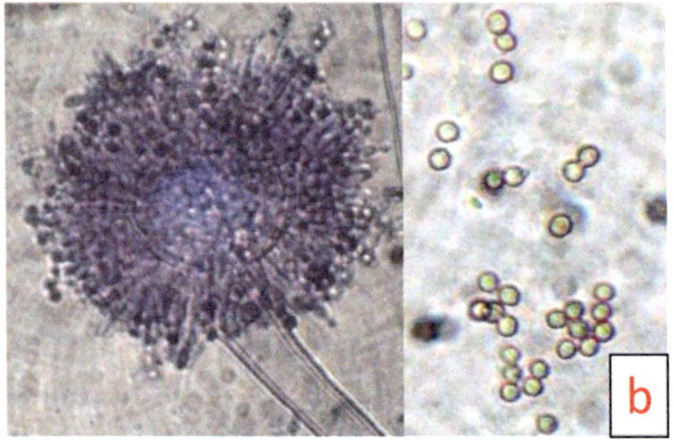

FIGURE 15.3 Symptoms of black rot disease of mango (a) and morphological structure of *Asprgillus niger* having condiophore, vesicles and conidia (b).

15.4.4 Stem End Soft Rot

Ceratocystis paradoxa (Dade) Moreau is the causal agent of this disease. Sometimes this rot causes severe postharvest losses. Pale, soft and watery irregular lesions develop near the stem end which later darkens due to spore formation and subsequently covers the entire fruit. Rotting is accompanied by a rancid odour. By proper handling to prevent fruit injuries, avoiding packing of wet fruits and following proper hygienic practices, incidence of stem end soft rot can be reduced.

15.4.5 Pestalotiopsis Rot

Pestalotiopsis versicolor (Speg.) Stey, *P. mangiferae* (P. Henn.) Stey is the causal agent of this disease. The symptoms are characterized as small brown spots which slowly increase in size and later turn into dark brown spots with greyish white centres and black dots of acervuli. As the disease advances to the stalk end, the fruits drop off the trees.

15.4.6 Charcoal Rot

The causal agent of charcoal rot is *Macrophomina phaseolina* Maubl. Ashby which usually infects the mango fruits through cut ends. Black irregular necrotic leathery patches usually originating from the distal end of the fruits are the characteristic symptoms of this disease. Rotting of fruits and brown discoloration of tissue generally take place from the lower side of the fruit. As the infection advances under humid conditions, the entire fruit surface as well as the internal pulp turns dark brown and completely rots. Incidence of charcoal rot can be reduced by maintaining field hygiene and removing diseased fruits as well as careful handling of harvested fruits to avoid soil contamination and bruising.

15.4.7 Alternaria Rot

Alternaria tenussima (Fr.) Wiltshire and *A. alternata* (Fr: Fr.) Keissl. are the causal agents of this disease. Initially, water-soaked circular and sub-circular spots appear on the fruits, which later become black with sunken centres and then enlarge in size. Dark brown spores may grow on lesions under humid conditions. Reddish patches on the flesh below the lesions are clearly visible when the epicarp is removed.

15.4.8 Phoma Rot

This disease is caused by *Phoma mangiferae* P. Henn. Its infection appears as brown to black spots on the epicarp of fruits. The outer peel remains hard but the internal portion rots quickly. In advanced stages, dark pycnidia are formed over such rotten tissues.

15.4.9 Macrophoma Rot

This disease is caused by *Macrophoma mangiferae* Hingorani & Sharma on mango fruits. Water-soaked, brownish and circular lesions appear on infected fruits which later become irregular in shape, covering the entire fruit surface and get covered by deep brown mould.

15.4.10 Rhizopus Rot

Rhizopus arrhizus Fisch and *R. oryzae* Went & Gerlings commonly cause soft decay in mango fruits which emits a very peculiar and unpleasant foul odour. Initially, there is development of a profuse coarse white area on the fruits, which later turns to black spore heads on the fruit surface.

15.4.11 Cladosporium Rot

This is caused by *Cladosporium herbarum* (Pers.) Link, *C. cladosporioides* (Fres) de Vries. Lesions covered with a white mould are associated with injury on fruits.

15.4.12 Fusarium Rot

The losses due to Fusarium rot are quite common in the rainy season. The causal agent of this disease is *Fusarium oxysporum* Schld. Like most of the postharvest rots, infection in this case also takes place through stem end or other injuries. There is appearance of large, dark-brown, irregular water-soaked areas on the fruits. Later there is a profuse growth of dull pink fungal colonies on fruits. Such fruits also emit a putrescent odour.

15.4.13 Rhizoctonia Rot

Rhizoctonia rot is caused by *Rhizoctonia solani* Kuhn, a fungus which thrives well under warm, wet conditions. Development of dark colour lesions takes place on the surface of fruits. The lesion development is followed by hard brown rotting on the upper portion with a clear line of distinction between healthy and affected tissue rotting with abundant sclerotial bodies of the fungus. In the advanced stages, the fungus also penetrates inside the fruits which are close to the ground.

15.4.14 Mucor Rot

Mucor rot is caused by *Mucor subtilissimus* Oud, a wound parasite. In this disease, white mould develops at the stem end or at any place on the injured fruit surface. Yellow colour sporangia usually grow on the water-soaked lesions. Infection is escalated with fruit fly infestation.

15.4.15 Grey Mould Decay

Botrytis cinerea Pers. is the causal pathogen of this disease. A light brown firm decay is later covered by grey or brown fungal growth containing mass of spores. The pathogen spreads readily by contact with adjacent fruits, giving rise to larger nests of diseased fruit in packed containers.

15.4.16 Blue Mould

Blue mould is caused by *Penicillium expansum* link ex Grey. A soft brown decay with white powdery spore masses, which become blue when mature, of the skin and flesh develops at the stem end or at the wound, which later on may spread over the entire fruit surface.

15.4.17 Hendersonia Rot

Pale brown circular lesions with cracked centre caused by *Hendersonia creberrima* Sydow and Butler appears on the fruit. Decayed tissue becomes soft and moist.

15.4.18 Watery Soft Rot

Sclerotinia sclerotiorum Lib. de Bary-caused rot becomes severe during the rainy season under subtropical conditions. The infected tissue becomes soft and ooze comes out. White coloured fungal growth may also appear on fruits.

15.4.19 Sclerotium Fruit Rot

This rot is caused by *Sclerotium rolfsii* Sacc., which is more severe in warm, moist conditions and common in mango fruits harvested at relatively late stages and subsequently heaped on the soil. The sclerotia and hyphal strands surviving in the soil act as the source of inoculum. In the early stages, the lesions have sharply defined margins. Later, rotting becomes intrusive.

15.4.20 Yeasty Rot

Yeasty rot is caused by *Saccharomyces* spp. The infection of yeast takes place through wounds. It is characterized by oozing of juice and a fermented smell. To reduce further infection of the disease, fruits should be stored at low temperature and infected fruits should promptly be discarded and destroyed. Strict field hygiene should be maintained with crop rotation if possible. Harvested fruits should be kept in clean places to avoid the contamination.

15.5 Factors Affecting Postharvest Disease Development

The yield and quality of fruits depends upon many factors, but much can be improved by adopting good horticultural practices. Some factors like microclimate, chemicals used for pest management and nutrition of crops, organic inputs applied, round-the-year pest management, harvesting methods and stage of fruits at harvest are responsible for fruit quality parameters and their susceptibility to diseases (Huong 2008; Lalel et al. 2003; Léchaudel and Joas 2006; Prakash 2004). Major factors associated with disease incidence in mango are been presented here.

15.5.1 Light

Sunlight is one of the most important needs of plants for proper growth, development and fruiting. The importance of light is mostly unnoticed in young orchards, but old orchards may become senile due to low light penetration within the canopy. Therefore, plants should be trained and pruned from the start, and intense pruning should be done in old orchards for maximum interception of sunlight. Pruning enhances the amount of light interception in the canopy, which enhances growth and productivity (Ahmad et al. 2006). It is also helpful in modifying the microclimate conditions against crop pests and diseases.

15.5.2 Rains

Rains play extremely important role in the development of initial inoculums in orchards and also facilitate pathogen invasion. Mango cv. Dashehari is an early maturing variety which is mostly harvested before the onset of monsoon rains and hardly has any issue of fruit rotting during storage and ripening. As the monsoon season proceeds, infection takes place and symptoms appear on fruits ripening on the tree, and severe postharvest rotting takes place in the orchard or during storage and transportation (Shukla et al. 2016, 2017).

15.5.3 Temperature

After the hot summer period, the temperature falls to optimum with the onset of monsoon season and infection of fungi takes place in mature fruits. The pathogens remain latent until the ripening process begins and set severe rotting during ripening. The fruits can be protected by harvesting at proper maturity and storage at low temperature to reduce metabolic and biochemical reactions (Crane et al. 2009). The low temperature for storage of mature mango fruits is recommended between 10°C and 13°C (Johnson and Hofman 2009; Kader and Mitcham 2008).

15.5.4 Microclimate

The environment within the canopy plays an extremely important role in population buildup of insect pests and pathogens, and ultimately to the extent of losses. Canopies favouring free flow of air and penetration of light create a microclimate against the majority of pests. The irrigation system also plays an important role. Drip irrigation is best, followed by basin irrigation, and the worst is flood irrigation. Development of high humidity within the canopy and large areas of wetness for longer periods enhances the incidence of pests and diseases.

15.5.5 Cultural Practices

Every activity which contributes towards minimizing the incidence of pests and/or diseases and promoting plant growth is important. Inoculum buildup of several pathogens can be reduced by ploughing, weed management, pruning of severely infected twigs and removal of fallen leaves, twigs and fruits.

15.5.6 Pest Management Practices

Harvested fruits are safer in orchards where year-round integrated pest management practices are followed. Management of sucking pests results in the least development of sooty moulds and shoulder browning (Shukla et al. 2016, 2018a). Pre-harvest spray of fungicides 1 month prior to harvest also reduces postharvest disease development.

15.5.7 Nutrition

The health and physiological status of trees also play an important role in the production of better quality fruits. Balanced crop nutrition is helpful in inducing tolerance to pathogenic infection. Application of organic matter as a source of nutrition also promotes development of bio-control agents in soil which indirectly reduce inoculum development in orchards.

15.5.8 Harvesting Method, Stage and Postharvest Handling

Fruits should be harvested at proper maturity and without causing any injury to avoid increased chances of infection and rotting. The best practice is to pick the fruits with 2–3 cm stem-end portions by hand or with the help of secateurs, keeping those gently in plastic carets, followed by de-sapping, hot water treatment, cooling, drying and packing into CF boxes.

15.6 Management Strategies for Postharvest Diseases

Postharvest management means handling of fruits after harvest to prolong disease-free shelf life and to maintain their look. Proper postharvest management is necessary to reduce postharvest losses. Care must be taken at every stage, including harvesting, de-sapping, grading, packaging, storage and transport. Non-seriousness of the majority of mango cultivators towards postharvest disease management is responsible for huge losses. Farmers need to understand the importance of postharvest handling and know that their responsibility does not end until the product reaches the consumer.

Prevention is a better and more effective tool of postharvest disease management. Chemicals have mainly been used to prevent and manage postharvest diseases; however, the residue results in environmental contamination and therefore poses potential health hazards (Alemu et al. 2014). Therefore, it is necessary to adopt non-chemical methods instead of chemical ones, particularly those applied during 1 month before harvest, for postharvest disease management. Management of postharvest diseases using economically viable, non-chemical methods is necessary. This can be achieved by adopting integrated disease management (IDM) strategies (Prakash 2004; Ploetz 2009). However, chemicals are an important component of disease management in IDM, when integrated at the correct time for protection of crop at critical stages. Management strategy against postharvest diseases can be grouped into three: anthracnose, stem-end rot and other diseases.

15.6.1 Management of Anthracnose Disease

Postharvest management of this disease can be achieved by applying year-round integrated orchard management at critical phenological stages (Ploetz 2009). Inoculum buildup in the orchard can be minimized by taking care of each new flush. The first critical stage is after fruit set and during the early development phase of fruits. At this stage, fruits are highly susceptible to infection under high relative humidity. Use of fungicides at this stage ensures crop protection as well as no residue remaining in mature fruits. An ideal anthracnose management strategy includes both pre- and postharvest management practices.

15.6.1.1 Pre-harvest Management

15.6.1.1.1 Cultural Practices

The pre-harvest application of fertilizers, manures, growth regulators and canopy management influences the fruit physiology, chemical composition and extent of infection in fruits (Prakash 2004). Well-managed orchards with over 30% light penetration and free wind blow have minimum inoculum buildup. Selection of an area for establishment of a new orchard should be done keeping in mind the monsoon period. It should not be during fruit development and maturity phase (Arauz 2000).

15.6.1.1.2 Grow Resistant Varieties

Plant resistance is the best and environmentally safest approach to plant disease management. However, resistance of commercial cultivars of mango has not been consistent at different locations against anthracnose. All commercial cultivars are susceptible to disease but susceptibility varies considerably among them (Akem 2006). Thus, area-specific performance of cultivars can provide the opportunity for selection of tolerant ones for the establishment of new orchards.

15.6.1.1.3 Use of Chemicals, Bio-control Agents and Botanicals

Since *C. gloeosporioides* remains active on trees and on fallen fruits or leaves throughout the year except during winters, inoculum is mostly available to cause fresh infection on tender tissue under favourable conditions. It is better to reduce the inoculums buildup. Excellent disease control has been achieved using the fungicides dithiocarbamate, mancozeb and febran in the field (Akem 2006). Benzimidazoles, tinidazole and prochloraz are the most commonly used fungicides for mango having after-infection activity. Benomyl is applied along with protectants to impede the development of a resistant pathogen population (Arauz 2000). Carbendazim (0.1%), thiophanate methyl (0.1%), myclobutanil (0.1%), metalaxyl + mancozeb (0.2%), tricyclazole (0.1%) and difenoconazole (0.1%) have also been found effective (Agrios 2005; Shukla et al. 2018a); however, none of these fungicides should be sprayed during the 30 days before harvesting, to avoid residual toxicity. Spray of mancozeb in mango orchards reduced anthracnose incidence on leaves and thus was found helpful in reducing inoculum buildup (Islam et al. 2016).

Comparative efficacy of fungicides like carbendazim, tricyclazole, azoxystrobin, thiophanate methyl at 0.1% and zineb at 0.2%, and botanicals (*Eupatorium odoratum* and *Nerium oleander*) and bioagents (*Trichoderma viride* and *T. harzianum*)at 5% was evaluated against anthracnose at different fruit growth stages [pre-flowering stage (December), peanut stage (February) and marble stage (March)] cv. Alphonso for 3 years under field conditions. Azoxystrobin was found highly effective in managing the disease. Significant and consistent reduction in disease was recorded during the first year (20.71%), second year (22.33%) and third year (24.54%) in leaf anthracnose. The rest of the test fungicides, bioagents and botanicals were also effective in combating the disease and had a significant difference as compared to control. The treatments were effective in reducing the inoculum buildup, which ultimately reduced latent infection in fruits and postharvest rotting (Manasa et al. 2018).

Efficacy of thiophanate methyl, azoxystrobin, myclobutanil and mancozeb spray was tested against fruit anthracnose in the southern part of Senegal, where heavy rains during fruit ripening cause severe losses. During 2 years' experimentation, 73.2 and 80% of mangoes ripened free from disease –those treated with thiophanate methyl, and 46.6 and 60% treated with azoxystrobin, and untreated fruits had 100% infection (Diedhiou et al. 2014).

15.6.1.2 Postharvest Management

The postharvest anthracnose management measures are applied to reduce the level of infection on the fruits. Keeping in mind the set standards regarding fruit quality and fungicide residue in fruits, no chemical treatment should be applied at this stage, and if possible safe approaches should be used, as discussed here.

15.6.1.2.1 Biological Control

Not much research has been carried out for the management of disease using micro-organisms. *Bacillus licheniformis*, a gram-positive bacterium, has been reported to reduce anthracnose incidence in South Africa (Govender et al. 2005). Significant reduction in anthracnose of fruits during ripening was also reported by Senghor et al. (2007). Gram-negative bacteria and some other amendments have also been employed by a few workers for biological management of mango anthracnose (Vivekananthana et al. 2004). Similarly, application of some bacterial and yeast isolates was found effective in reducing the severity of anthracnose on fruits artificially inoculated with *C. gloeosporioides* (Kefialew and Ayalew 2008). Those investigations were successful only by inoculating bio-control agents before the pathogens but not when pathogen was inoculated before application of bio-control agent. At the time of harvest, fruits mostly have symptomless infection; therefore, application of bacterial bioagents could not be popularized.

Two bioproducts (*Bacillus pumilus* QST 2808 and *B. subtilis* QST 713 of Bayer) were compared with azoxystrobin and thiophanate methyl for efficacy against anthracnose in artificially inoculated fruits, and both bioagents were found at par with fungicides (Diallo et al. 2017).

15.6.1.2.2 Hot Water Treatment

Hot water treatment (HWT), which includes dipping of fruits in hot water with or without fungicides at 50–55°C for up to 5 minutes (Alemu 2014a, 2014b) just after harvesting, is an age-old and successful practice that can alone efficiently reduce anthracnose development on fruits (Uddin et al. 2018). HWT is an effective, easy to apply and simple practice still used in many mango producing countries. However, the effectiveness of HWT in reducing anthracnose development is greatly influenced by the initial infection level as well as storage conditions. In the case of possible severe incidence of the postharvest fruit rot, the potency of HWT can be greatly enhanced by combining low doses of fungicides (Kefialew and Ayalew 2008; Kapse et al. 2009). For example, Singh (2011) applied hot water (45°C or 50°C), hot water (45°C) and sodium bicarbonate, hot water (45°C) and propiconazole dip for 5 minutes each on fruits artificially inoculated with *C. gloeosporioides* and recorded 82.13–96.44% reduction in disease development over control. HWT of mango fruits combined with fungicides such as carbendazim, prochloraz, benomyl, imazalil and thiophanate methyl was found to be effective in reducing postharvest anthracnose severity (Shukla et al. 2017; Uddin et al. 2018). HWT should not be delayed beyond 2 days after harvesting, and the treated fruits should be transferred to ambient cool water after treatment, followed by air drying before packing.

15.6.1.2.3 Use of Plant Extracts

Because of increased awareness regarding the harmful effects on environment and health hazards posed by fungicides, use of plant extracts and essential oils has gained popularity in crop production schedules due to their apparently safe nature and antimicrobial activities. These extracts are composed of secondary metabolite mixtures (Gottlieb et al. 2002). A few workers studied the potential of different products of plant origin and essential oil *viz.*, castor oil, eucalyptus oil, lemongrass oil, garlic bulb extract, *Azadirachta indica*, *Zingiber officinale*, *Lantana* and *Curcuma longa* leaf extract in suppressing anthracnose development and postharvest rots of mango fruits (Singh et al. 2003; Duamkhanmanee 2008; Alemu et al. 2014b). Singh (2011) evaluated extracts of *Argemone mexicana* L. *Azadirachta indica A.* Juss, *Bombax ceiba* L., *Cannabis sativa* L., *Cassia fistula* Herbb. Ex Oliver, *Citrus aurantifolia* L (Christm.) Hiroe, *Hyptis suaveolens* Poit., *Murraya koenigii* Spreng., *Piper methysticum* Forst. f, *Tabernaemontana divaricata* G. Don on fruits artificially inoculated with *C. gloeosporioides*, and 54.1–89.4% reduction in disease development was recorded over control. Besides the anthracnose control, these extracts also positively influence mango's storage quality by retaining total soluble solids and sugar contents better.

15.6.1.2.4 Chemical Control

Thiabendazole, benomyl, mancozeb, ferbam, prochloraz and imazalil were found effective in postharvest management of anthracnose (Akem 2006; Govender and Korsten 2006; Alemu 2014). Singh (2011) evaluated sodium bicarbonate (5%), boric acid (5%), carbendazim (500 ppm), thiophanate methyl (500 ppm), propiconazole (500 ppm) and hexaconazole (500 ppm) against anthracnose under laboratory conditions and a 74.0–89.5% reduction in disease on fruits was recorded.

15.6.2 Management of Stem-End Rot Disease

15.6.2.1 Pre-harvest Management

Disease incidence can be managed by reducing inoculum load and infection in twigs by adopting year-round orchard management. Latent infection in branches and twigs can be reduced by pruning, orchard sanitation and fungicide spraying.

Practices *viz.*, resistant or tolerant cultivar planting, avoiding injuries which are the entry points of pathogen and fungicide sprays can contribute in reducing the deposition of pre-harvest inoculum. Application of *Bacillus licheniformis* and covering fruits with polyethylene or paper bags was also found effective in reducing the incidence of stem-end rot (CABI 2005).

15.6.2.2 Postharvest Management

Irradiation, HWT and controlled atmosphere with regulated O_2 and CO_2 concentration coupled with low temperature during storage contribute greatly to minimizing postharvest diseases in mango (Govender 2005; Cooke et al. 2009). A warm water dip containing *B. licheniformis* and low dose of prochloraz effectively controlled fruit rots (CABI 2005; Cooke et al. 2009). Thiophanate methyl, carbendazim, tebuconazole + trifloxystrobin and azoxystrobin were found effective in managing postharvest stem-end rot (Khanzada et al. 2005; Syed et al. 2014). Tecto (thiabendazole) 1.8 mL/L alone and/or in combination with Sportak (prochloraz) 0.5 ML/L and carbendazim (450 mg/L) was highly effective in managing stem-end rot of mango (Amin et al. 2011). Singh (2011) evaluated sodium bicarbonate (5%), boric acid (5%), carbendazim (500 ppm), thiophanate methyl (500 ppm), propiconazole (500 ppm) and hexaconazole (500 ppm) against stem-end rot under laboratory conditions and recorded 76.9–92.3% reduction in disease on fruits. Singh (2011) also evaluated extracts of *Argemone mexicana* L. *Azadirachta indica A*. Juss, *Bombax ceiba* L., *Cannabis sativa* L., *Cassia fistula* Herbb. Ex Oliver, *Citrus aurantifolia* L (Christm.) Hiroe, *Hyptis suaveolens* Poit., *Murraya koenigii* Spreng., *Piper methysticum* Forst. f, *Tabernaemontana divaricata* G. Don on fruits artificially inoculated with *Botryodiplodia theobromae* and recorded 47.9–87.0% reduction in disease development over control. Singh (2011) applied hot water (45°C or 50°C), hot water (45°C) and sodium bicarbonate, hot water (45°C) and propiconazole dip for 5 minutes each on fruits artificially inoculated with *B. theobromae* and observed 83.79–96.41% reduction in disease development over control.

15.6.3 Management of Other Diseases

A large number of pathogens have been isolated from rotting fruits, and the diseases caused by those have been reproduced. However, most of these fungi infested the fruits due to careless harvesting and postharvest handling or storage of fruits. Since the associated fungi remain in soil as well as air, their contact with fruits mostly enables them to enter the fruit through the stem end portion or physical injuries. Such infection/infestation can be minimized if fruits are carefully picked, handled and treated as mentioned for anthracnose and stem-end rot.

15.7 Conclusion

Postharvest diseases of mango cause severe losses worldwide. The mature green fruits are mostly symptom free but during ripening and storage, symptoms appear and result in considerable losses. To manage such diseases, knowledge and understanding of the pathogen's biology, the conditions aggravating disease development and the economics, efficacy and market acceptability of management practices is necessary. Integrated disease management, including both chemical-based pre-harvest and non-chemical-based postharvest approaches must be utilized to avoid residual toxicity and to maintain quality standards of fruits.

15.8 Future Directions

The future research should be focused on the use of nontoxic compounds, bio-control agents and further refinement of HWT, particularly variety specific. Newer molecules should also be evaluated as pre-harvest treatment for postharvest disease management. It is not easy to cover the whole fruit surface during spray; therefore, spray techniques should also be revisited and new-generation spreader-stickers should also be evaluated. Traditional picking of fruits causes injury to 10–25% of fruits and prevailing mango harvesting is labour oriented and increases the cost of harvesting. Therefore, work should also be done towards practically applicable solutions of fruit harvesting. Farmers should also be sensitized about the importance of safe harvesting and careful postharvest disease management.

REFERENCES

Agrios GN (2005) Plant Pathology 5th Edition, Elsevier Academic Press, Amsterdam, 26–27398–401

Ahmad S, Chatha ZA, Nasir MA, Aziz A, Virk NA, Khan AR (2006) Effect of pruning on the yield and quality of Kinnow fruit. J Agric Soc Sci 2: 51–53.

Akem CN (2006) Mango anthracnose disease: Present status and future research priorities. JPP 5(3): 266–273.

Alemu K (2014) Dynamics and management of major postharvest fungal diseases of mango fruits. J Bio Agric Healthcare 27:13–21.

Alemu K, Ayalew A, Woldetsadic K (2014a) Effect of aqueous extracts of some medicinal plants in controlling anthracnose disease and improving postharvest quality of mango fruit. Pers Gulf Crop Prot 3(3): 84–92.

Alemu K, Ayalew A, Woldetsadic K (2014b) Evaluation of antifungal activity of botanicals for postharvest management of mango anthracnose (*Colletotrichum gloeosporioides*). Int J Life Sci 8(1):1–6.

Amin M, Malik AU, Khan AS, Javed N (2011) Potential of fungicides and plant activator for postharvest disease management in mangoes. Int J Agric Biol 13: 671–676.

Arauz LF (2000) Mango anthracnose: Economic impact and current options for integrated management. Plant Dis 84(6): 600–611.

CABI (2005). Crop Protection Compendium. 2005 edn. CABI, Wallingford, UK.

Cooke T, Persley D, House S (2009) Diseases of fruit crops in Australia. CSIRO Publishing, Australia, pp. 157–173.

Crane JH, Salazar-Garcia S, Lin TS, Queiroz PAC Shue ZH (2009) Crop production: Management. In: Litz RE (Ed.). The mango: Botany, production and uses. 2nd edn. CABI, Wallingford, UK, pp. 471.

De Candolle A (1904). Origin of cultivated plants. K. Paul, Trench, Trubner, pp. 468.

Diallo Y, Diédhiou, Bush ME, Nita M, Baudoin A (2017) Biofungicides: An efficient an alternative control strategy against mango anthracnose in Senegal. Int J Adv Res 5(6): 990–995.

Diedhiou PM, Diallo Y, Faye R, Mbengue AA, Sene A (2014) Efficacy of different fungicides against mango anthracnose in Senegalese Soudanian agroclimate. AJPS 5(15): 2224–2229.

Diedhiou PM, Mbaye N, Dramé A, Samb PI (2007) Alteration of postharvest diseases of mango *Mangifera indica* through production practices and climatic factors. Afr J Biotech 6(9): 1087–1094.

Duamkhanmanee R (2008) Natural Eos from lemon grass (*Cymbopogon citrates*) to control postharvest anthracnose of mango fruit. Int J Biotech 10(1): 104–108.

FAO (2017) Mango Production. In: FAOSTAT online database. http://www.fao.org/default.htm.

Giblin FR, Coates LM and Irwin JAG (2010) Pathogenic diversity of avocado and mango isolates of *Colletotrichum gloeosporioides* causing anthracnose and pepper spot in Australia. J Australas Pl Pathol 39: 50–62.

Gottlieb OR, Borin MR, Brito NR (2002) Integration of ethno-botany and photochemistry: Dream or reality? Photochemistry 60: 145–152.

Govender V, Korsten L, Sivakumar D (2005) Semi-commercial evaluation of *Bacillus licheniformis* to control mango postharvest diseases in South Africa. Postharvest Biol Technol 38: 57–65.

Govender V (2005) Evaluation of biological control systems for the control of mango postharvest diseases. MSc thesis. University of Pretoria, South Africa, 126 pp.

Govender V, Korsten L (2006) Evaluations of different formulations of *Bacillus licheniformis* in mango pack house trials. Biol Control 37: 237–242.

Haggag WM (2010) Mango diseases in Egypt. Agric Biol J North Am 1(3): 285–289.

Huong PT (2008) Some initial results of improvement of neglected mango orchards in Coc Lac hamlet, Yen Chau district, Son La Province. J Sci Devel Hanoi Univ Agric Vietnam 2: 105–109.

Islam MM, Hassan ME, Hossain SMM, Hasan MM, Islam NB (2016) Prevalence and management of some major diseases of mango at Dijapur district in Bangladesh. Int J Expt Agric 6(2):1–11.

Javadpour SA, Golestani S, Rastegar MM, Dastjer (2018) Postharvest control of *Aspergillus niger* in mangos by means of essential oil. Adv Horti Sci 32(3): 389–398.

Johnson GI, Hofman PJ (2009) Postharvest technology and quality treatment. In: Litz RE (Ed.). The Mango: Botany, Production and Uses. 2nd edn. CABI, Wallingford, UK, pp. 537–571.

Kader A, Mitcham B (2008) Optimum procedures for ripening mangoes. In: Fruit ripening and ethylene management. Univ Calif Postharvest Technology Research Centre, pp. 47–48.

Kalita P (2014) An overview on Mangifera Indica: Importance and its various pharmacological action. Pharma Tutor 2(12): 72–76.

Kapse BM, Pawar VN, Sakhale BK (2009) Postharvest disease management in mango (*Mangifera indica* L.) cv. Kesar Acta Hortic 820: 493–503.

Kefialew Y, Ayalew A (2008) Postharvest biological control of anthracnose (*Colletotrichum gloeosporioides*) on mango (*Mangifera indica*). Postharvest Biol Technol 50: 8–11.

Khanzada MA, Lodhi M, Shahzad S (2005) Chemical control of *Lasiodiplodia theobromae,* the causal agent of mango decline in Sindh. Pak J Bot 37: 1023–1030.

Lalel HJD, Singh Z, Tan SC (2003) The role of ethylene in mango fruit aroma volatiles biosynthesis. J Hortic Sci Biotechnol 78: 485–449.

Léchaudel M, Joas J (2006). Quality and maturation of mango fruits of cv. 'Cogshall' in relation to harvest date and carbon supply. Australian Agri 57: 419–426.

Manasa B, Jagadeesh SL, Thammaiah N. Jagadeesh RC, Gangadharappa PM, Nethravathi (2018) Pre-harvest application of azoxystrobin minimized anthracnose of mango (cv. Alphonso) both at field and postharvest level enhancing yield and quality of fruits. J Pharmacol Phytochem 7(3): 2962–2967.

Mansour FS, Abd-El-Aziz SA, Helal GA (2006). Effect of fruit heat treatment in three mango varieties on incidence of postharvest fungal disease. J Pl Pathol 88(2): 141–148.

NHB Database (2017) National Horticulture Board, Department of Horticulture and Cooperation, Government of India, Aristo Printing Press, New Delhi. http://www.nhb.gov.in. NHB Database 2017–2018.

Ni H-F, Yang H-R, Chen R, Lio R-F, Hung TH (2012) New Botryosphaeriaceae fruit rot of mango in Taiwan: identification and pathogenicity. Botanical Stud 53: 467–478.

Ploetz RC (2009) Management of the most important pre- and post-harvest disease. University of Florida, TREC, Homestead, Department of Plant Pathology, Homestead, FL.

Prakash O (2004) Diseases and disorders of mango and their management. Dis Fruits Veg 1: 511–620.

Prakash O, Misra AK, Shukla, PK (2011) Postharvest diseases of mango and their management. Proc Global Conference on Augmenting Production and Utilization of Mango: Biotic and Abiotic Stresses, ICAR-CISH, Lucknow, 21–24 June 2011, pp. 137–144.

Rivera-Vargas LI, Lugo-Noel Y, McGovern RJ, Seijo T, Davis MJ (2006) Occurrence and distribution of *Colletotrichum* spp. on mango *(Mangifera indica* L.) in Puerto Rico and Florida, USA. Pl Pathol J 5: 191–198.

Sangeetha CG, Rawal RD (2008) Nutritional studies of *Colletotrichum gloeosporioides* (Penz.) Penz. and Sacc. the incitant of mango anthracnose. World J Agri Sci 4(6): 717–720.

Senghor A, Liang WJ, Ho WC (2007) Integrated control of *Colletotrichum gloeosporioides* on mango fruit in Taiwan by the combination of *Bacillus subtilis* and fruit bagging. Biocon Sci Technol 17(8): 865–870.

Shukla PK, Adak T, Baradevenal G (2017) Anthracnose disease dynamics of mango orchards in relation to humid thermal index under subtropical climatic condition. J Agrometeorol 19(1): 56–61.

Shukla PK, Savita V, Fatima T, Mishra R, Misra AK, Bajpai A, Baradevenal G, Muthukumar M (2018b) First report on wilt disease of mango caused by *Ceratocystis fimbriata* in Uttar Pradesh. Indian Phytopathol 71(1): 135–142.

Shukla, PK, Adak T, Misra AK, Singh A (2016) Appraisal of shoulder browning disease of mango (*Mangifera indica* L.) in subtropical region of India. J Mycol Plant Pathol 46(1): 38–46.

Shukla, PK, Bhattacharjee AK, Dixit A (2018a) Efficacy of difenoconazole against shoulder browning disease of mango (*Mangifera indica* L.) and its residue analysis for safety evaluation in mango fruit. Indian Phytopathol 71(1): 147–151.

Shukla PK, Gundappa (2017) Management of post-harvest anthracnose of mango by pre- and post-harvest treatments. In: Int Conf on Bioresource and Stress Management, State Insti Agri Manag, Durgapura, Jaipur, Rajasthan, 08–11 November 2017.

Singh P (2011) Integrated management of storage anthracnose of mango. J Mycol Plant Pathol 41(1): 63.

Singh D, Thakur RK, Singh D (2003) Effect of pre harvest sprays of fungicides and calcium nitrate on postharvest rot of Kinnow in low temperature storage. Plant Dis Res 18: 9–11.

Swart G (2010) Epidemiology and control of important postharvest diseases in mangoes in South Africa. Southern African Society for Plant Pathology. Midrand, South Africa.

Syed NR Mansha N, Khaskheli MA, Khanzada MA, Lodhi AM (2014) Chemical Control of Stem end rot of mango caused by *Lasiodiplodia theobromae*. Pak J Phytopathol 26(2): 201–206.

Uddin MN, Shefat SHT, Afroz M, Moon NJ (2018). Management of anthracnose disease of mango caused by *Colletotrichum gloeosporioi*des: A review. Acta Sci Agric 2(10): 169–177.

Vivekananthana R, Ravia M, Saravanakumara D, Kumar N, Prakasama V, Samiyappana R (2004) Microbially induced defense related proteins against postharvest anthracnose infection in mango. Crop Protec 23: 1061–1067.

Yenjit P, Intanoo W, Chamswarng C, Siripanich J, Intana W (2004) Use of promising bacterial strains for controlling anthracnose on leaf and fruit of mango caused by *Colletotrichum gloeosporioides*. Walailak J Sci Techn 1(2): 56–69.

16

Postharvest Diseases of Banana and Their Management

Allada Snehalatharani,[1] V. Devappa,[2] and C.G. Sangeetha[2]
[1]Horticultural Research Station, Kovvur, Dr. YSRHU, Andhra Pradesh, India
[2]Department of Plant Pathology, College of Horticulture, UHS Campus, Bengaluru, India

CONTENTS

16.1 Introduction

Banana (*Musa* sp.) is a commercially important fruit crop in India. The fruit is available year-round with a number of dessert and cooking type varieties and is available at affordable prices throughout the year; it is also a crop of good export value. India leads the world in banana production, occupying nearly 30% of 102.02 million tons produced worldwide. In India, the crop is cultivated in 883.8 thousand ha, producing 30807.5 thousand MT with a productivity of 34.9 MT/ha (Horticultural Statistics at a Glance, 2018). Major states contributing to the country's banana production are Tamil Nadu, Maharashtra, Andhra Pradesh, Gujarat and Karnataka.

The fruit is nutritionally important for its carbohydrates, vitamins (B complex), and minerals like potassium, calcium, phosphorus and magnesium. Banana fruit is easily digestible and does not contain fat or cholesterol. Hence it is a good source for baby food. Continuous intake of fruit lessens health problems like heart disease, high blood pressure, arthritis, ulcer, gastroenteritis and kidney disorders.

Banana value can be improved with many added products like chips, jam, jelly, banana puree, juice, wine, halwa, etc.

The pseudo stem after removing the leaf sheaths is used as a vegetable. Culinary bananas contain high starch and their composition resembles that of potato (Padam et al. 2014). Fibre from the plant is used to make craft items such as bags, pots and wall hangers. It is also used in preparation of rope and paper. In addition, leaves are used as healthy eating plates.

Favourable climatic conditions for crop growth are temperatures between 15°C and 35°C and a relative humidity of 75–85%. The crop grows well in tropical humid lowlands from sea level to 2000 m above mean sea level. Banana grows well in rich and fertile soil at a pH range of 6.5–7.5. Fertile soil with adequate moisture and well-drained conditions are necessary for good growth. Banana cultivation is not advisable under too acidic and too alkaline soil conditions. Soil should be rich in organic material, macro elements and micro elements. Commercial triploid bananas are of two broad categories: dessert, with AAA or AAB genome, and cooking with ABB genome. Cooking bananas with high starch at maturity are used as vegetables. Important cultivars of banana are Dwarf Cavendish, Robusta, Monthan, Poovan, Nendran, Red banana, Rasthali, Karpurvalli, Grand Naine, etc.

Banana fruit is a highly perishable commodity with postharvest losses estimated up to 25–30% (Kachhwaha et al. 1991). The climacteric fruit nature of banana leads to attack of numerous pathogens from the time of harvest to consumption. Many primary and secondary pathogens attack banana and deteriorate fruit quality and thereby postharvest shelf life. Crown rot, anthracnose, cigar-end rot, ripe rot, stem-end rot and black end are some of the important postharvest diseases of banana. However, crown rot is the major factor reducing crop profits after harvest (Krauss and Johanson 2000; Lassois et al. 2011). In addition to pre-harvest disease management, understanding major postharvest diseases of banana, their etiology, mode of spread and management certainly improves the shelf life and fruit quality of banana.

16.2 Major Postharvest Diseases of Banana

16.2.1 Crown Rot

Crown rot is one of the important yield loss factors in banana from the time of harvest to storage to final marketing of the produce. It is the common and frequently occurring postharvest disease on cut hands after de-handling of the bunch for transport or storage. The disease incidence ranges from 4–98% across the world. Fruits are more susceptible to crown rot during the rainy season when compared to other seasons, and yield losses up to 86% were recorded in some countries (Alvindia et al. 2000).

16.2.1.1 Symptoms

The infection starts from the cut portion of the hand, otherwise called the crown. Infection occurs on the pad of tissue severed when the hands are cut from the bunch. Crown rot initiates as a dark brown or black rot and spreads through the crown. It later penetrates the pedicels of individual fingers. In many cases, crown rot is limited to the crown region of the hand, but under severe situations, infection moves to the pedicels of the fingers (Figure 16.1). In advanced conditions, mycelium with white or grey or pink colour and fruiting bodies of the fungus are visible on the rotted crown or stalks of fingers or fingers themselves (Lassois et al. 2010). Infected fingers may fall from the weakened crown in cases of severe infection. Rotting then enters the pulp itself and the entire fruit is lost.

Disease symptoms on crown or infection by the pathogen is not visible when bananas are de-handled during harvest or during packing and storage. The symptoms are visible after a few days and when fruits are in transport. Crown rot infection progresses rapidly after ripening of the fruit. Initial crown rot infection is superficial, and with disease progression it enters into the inner tissues. Occasionally, infection starts with peduncles and enters the fruits. The disease deteriorates fruit quality, induces early ripening of the fruit as ethylene is produced by stressed tissues, necrotic fruits and fungal mycelium (Ewané et al. 2012).

16.2.1.2 Causal Organism

Crown rot is a complex disease, and many fungal pathogens are associated with it. The fungal organisms associated with the disease are *Musicillium theobromae* (Turconi) Zare & W. Gams, *Colletotrichum musae* (Berk. & Curt.) Arx., *C. gloeosporioides* Penz., *Ceratocystis paradoxa* (Dade) Moreau, *Lasiodiplodia theobromae* (Pat.) Griffon and Maubl,

FIGURE 16.1 Crown rot of banana.

Nigrospora sphaerica (Sacc.) Mason, *Cladosporium* sp., *Acremonium* sp., *Penicillium* sp. and *Aspergillus* sp., as well as several *Fusarium* spp., including *F. pallidoroseum* (Cooke) Sacc., *F. semitectum* Berk. & Ravenel, *F. verticillioides* Sacc., *F. sporotrichoides* Sherb, *F. oxysporum* Schlect, *F. solani* Mart. Sacc. And *Fusarium musae* Van Hove et al. sp.nov.) (Lassois et al. 2010; Kamel et al. 2016; Molnar et al. 2018). Daniel et al. (2018) reported *Phellinus noxius* (Corner) and *Botryodiplodia* sp. as causal organisms of crown rot disease in Côte d'Ivoire. *Colletotrichum musae* is the most commonly associated pathogen and is more aggressive than *C. gloeosporioides* (Bele et al. 2018).

F. musae is also found as a causal organism of crown rot disease. *F. musae* is a population within *F. verticilloides* that can infect banana. Even though both species are morphologically similar, they differ in host specificity. *F. verticillioides* infect maize, rice, banana whereas *F. musae* can only infect banana fruits. *F. musae* was found associated with bananas of Latin America, the Canary Islands, and the Philippines. The same pathogen was not associated with bananas of African origin. It was reported that the pathogenicity of *F. musae* strains is more than *F. verticillioides* strains. Fumonisin, the mycotoxin, was not produced in *F. musae* strains, as fumonisin biosynthesis gene cluster was not detected.

16.2.1.3 Disease Cycle

Infected flowers are the main source of infection. However, decaying leaves and leftover banana stalks in the field, and contaminated water used for cleaning bunches also aid in fungal inoculum transfer. Crown rot pathogens infect the de-handled crowns after harvest. However, symptoms of pathogen infection are visible later during storage or long-distance transport. Crown rot fungi are saprophytic on dead banana leaves, bracts, discarded fruits and stems. Pathogen inoculum in the field settle by two means: airborne fungal spores that settle on bunch and waterborne fungal spores that are splashed on to the fruit by rain or irrigation water. The flowers and the last bracts are the main sources of inoculum. Wounds caused by cutting off the hands is accessible by the fungal spores, and the spores germinate in response to favourable conditions. The infection then deepens based on the fruit's physiological state, length of time before ripening and environmental conditions.

Crown rot pathogens *Fusarium* sp. or *Colletotrichum* sp. harbour in flower parts and their remnants and then initiate the infection process. The bract of last bunch also plays an important role in infection by *Colletotrichum* sp. The primary inoculum first enters flower parts and the bract of last bunch through rain splash conidia, transmission by insects or through the aerial dissemination of ascospores. Infection by primary inoculum leads to sporulation on the young flowers and on the bracts of last bunch which then provides the secondary inoculum (de Lapeyre de Bellaire and Mourichon 1997).

Dessert bananas are more susceptible to the disease than cooking bananas. Within the genome group also, varieties differ in susceptibility to the disease. Many pre-harvest factors such as geographical variations, seasonal variations, age of the bunch, source-sink ratio, biotic and abiotic factors affect crown rot infection on fruits. Cultivation areas and seasonal variations influence fruit susceptibility. Bananas from low altitude regions are more easily prone to crown rot infection than fruits from high altitude regions. Ewané et al. (2013) reported that rainy season, in particular some harvest days in rainy season, influences crown rot disease. Krauss and Johanson (2000) also reported that incidence of crown rot is greater during rainy season, and losses up to 10% have been recorded in bananas coming from the Windward Islands. The state of the banana fruit during harvest is of utmost importance as it influences host response to pathogen attack and finally crown rot infection.

Other factors such as source-sink ratio during the vegetative and filling stage also influence crown rot infection. Removal of banana hands from the bunch decreases total sink and thereby crown rot infection rate. Fruit age is linearly correlated to crown rot infection and has been demonstrated in Guadeloupe. Older fruits are more susceptible to the disease (Ewané et al. 2012). Crown rot infection is also affected by the presence of phenolic compounds in the fruits. Waxy layer on the leaf, components of cell wall such as cuticle, suberin and the cell wall itself are the physical barriers that prevent infection by crown rot fungi. These contain the phenolic compounds lignin, cross-linking hydroxy cinnamic acid, etc. (Ewané et al. 2012). Phenolic compounds like dopamine and cyanidin-related compounds, lignin and ferulic acid, phytoalexins such as phenylphenalenones, irenolone and emenolone are found in higher concentrations in resistant fruits after wounding or inoculation with *C. musae* (Figure 16.2).

There is variation in the gene expression pattern of crown rot-susceptible and crown rot-resistant bananas. Experiments based on cDNA-amplified fragment length polymorphism (AFLP) and a real-time reverse-transcription polymerase chain reaction showed variation in gene expression of susceptible and resistant banana cultivars. Two of the signal transduction genes, three of the proteolytic machinery genes, two pathogenesis-related protein 14 genes, one CCR4-associated factor protein and one cellulose synthase gene were significant in resistant bananas. In particular, cellulose synthase gene and dopamine-β-monooxygenase gene played a very important role in susceptibility of fruits to crown rot. The cellulose synthase gene when expressed excessively led to susceptibility of the fruits. The catecholamine pathway component dopamine-β-monooxygenase is found related to banana responses to crown rot disease (Lassois et al. 2011).

Some researchers have raised the concern of *F. musae* infecting banana as a threat to human beings and view the importance of the disease from the public health point of view (Triest et al. 2015; Triest and Hendrickx 2016; Triest et al. 2016). *F. musae* pathogen is isolated from banana fruits only, unlike *F. verticillioides*. Infection by opportunistic human pathogen *F. musae* seems to be from crown rot disease of banana even though the link is not yet clear. However, the only known environmental habitat of *F. musae* is the banana fruit and it was found that the patients with *F. musae* infection are from non-banana producing countries. Hence, it was suspected that latent infection of *F. musae* in crown rot infected banana may be the reason for human infection when the human patient is in proximity to the *F. musae* of banana fruits.

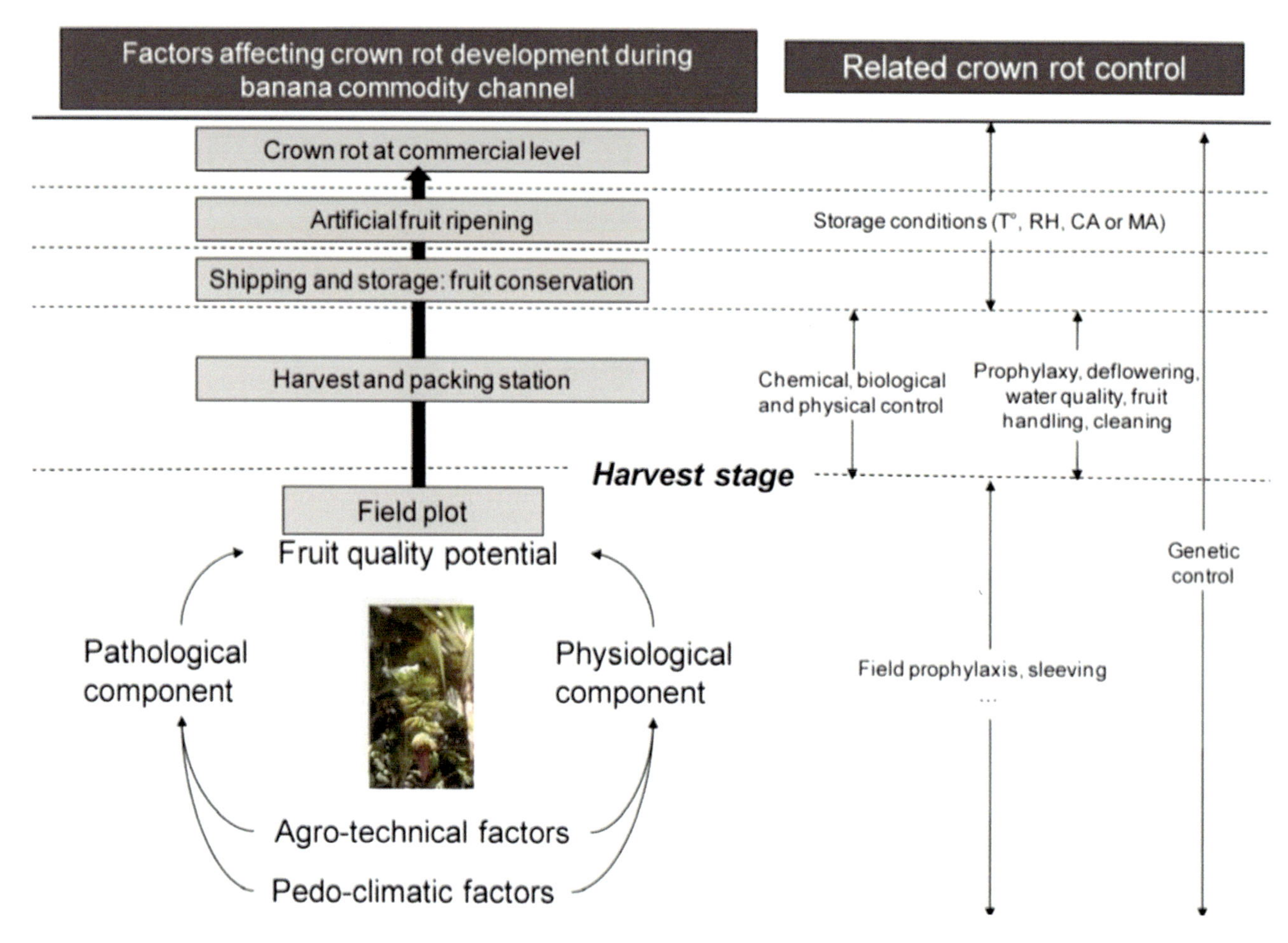

FIGURE 16.2 Key factors concerning crown rot development and related control methods.

16.2.2 Anthracnose

Anthracnose is another major postharvest disease and accounts for nearly 40% losses (Coelhoet al. 2010). It forms brown sunken spots on the peel as the fruit ripens. The spots increase in size, coalesce and form extensive areas of sunken, brown-black tissue (Figure 16.3). Orange-coloured spore masses may develop under favourable conditions. The pulp remains unaffected until the fruit becomes over-ripe. The disease usually starts in the field itself as quiescent infections on the fruit. However, successful entry of the pathogen is influenced by the accumulation of phytoalexins during fruit ripening. Symptoms only appear on overripe fruits. It reduces the quality and nutritive value of the fruits and renders fruits unfit for consumption and marketing.

16.2.2.1 Causal Organism

Colletotrichum musae (Berk. & Curt.) Arx. causes anthracnose in banana. It produces conidia and conidiophores on acervuli. Fungus produces spores on senescing banana tissue, leaves, bracts, discarded fruits and fruit stems. Spores dislodged by water reach the fruits in the field by rain splash or irrigation water. *C. musae* grows well at temperatures from 27°C to 30°C. Conidia germinate and form appressoria within 24–48 h. After infecting the fruit, *C. musae* remains in latent phase until the initiation of ripening. Infection also takes place directly through wounds which stimulate conidial germination and mycelial growth. Conidia are hyaline, without any septum, ellipsoid in shape, and the spore size ranges from 10 to 18 µm and 5 to 9 µm (average of 14.5–6.9 µm) (Unnithan and Thammaiah 2017).

FIGURE 16.3 Symptoms of banana anthracnose.

Dessert cultivars are more susceptible to the disease than the cooking cultivars. In the dessert cultivars, Gros Michel is less susceptible than the Cavendish group of cultivars. Anthracnose causes significant wastage in domestic markets, particularly in developing countries, where the damage level is higher and storage conditions are not controlled.

16.2.3 Cigar End Rot

Cigar end rot is an important pre-harvest as well as postharvest fungal disease. This disease is of economic importance in Central and Western Africa. It also occurs in the Canary Islands, Egypt, India, Iran, South Africa and South America. The causal organisms are *Musicillium theobromae* (Turconi) Zare & Gams, *Verticillium theobromae* Mason & Hughes (Amani 2006; Amani 2008) and *Trachysphaera fructigena* Tabor & Bunting. The pathogen of cigar end rot enters the banana finger through the flower, causing dry rot of immature banana fingers. The pathogen infects the fruit immediately after its emergence.

The diseased portion remains attached to the finger and looks like the ash of a cigar. It damages the fruits in the field (pre-harvest conditions) and fruits in transportation and storage (post-harvest conditions). *M. theobromae* is the widely distributed and associated pathogen with cigar end rot disease of banana when compared to *T. fructigena* (Masudi and Bonjar 2012).

16.2.3.1 Symptoms

The field symptoms appear as localized necrosis at the tip end of the fruit, followed by darkening and wrinkling of the skin. Because of the disease infection, fruits ripen early without maturity. Powdery greyish conidia appear on the wrinkled black portion of the tip under favourable atmospheric conditions. It leads to a burned tip of the fruit generally called the "cigar end." One or all fingers on a hand may be affected by this disease (Figures 16.4 and 16.5). The first symptoms are localized wrinkling and darkening of the peel at the tip of the fruit. The darkened area is seen as a black band with a narrow chlorotic region between infected and healthy tissues (Gul et al. 2018).

FIGURE 16.4 Symptom of cigar end rot.

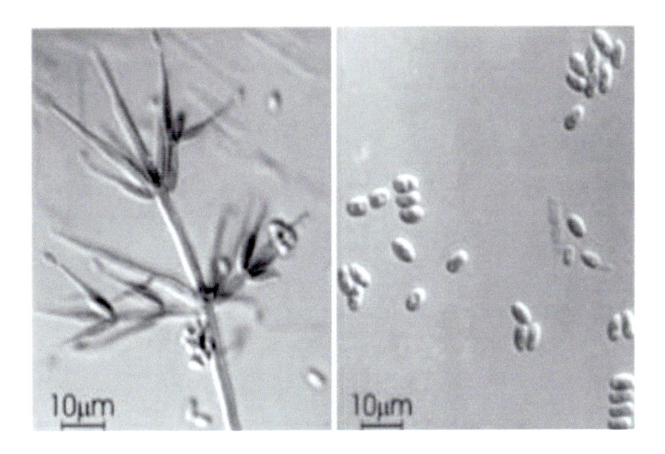

FIGURE 16.5 Conidia and conidiophores of *Musicillium theobromae*.

16.2.4 Fungal Scald

The disease superficially resembles anthracnose. It appears on fruits which are in transit under controlled atmospheric conditions for more than 14 days. Scald develops at the fruit tips touching polythene liners in the presence of moisture. Reddish brown sunken spots form on green fruit near the fingertips and on the bottom fruits of the carton. The spots increase in size during ripening. *Colletotrichum musae* initiates scald which is followed by secondary infection with *Fusarium pallidosporeum* (Jones 2000).

16.2.5 Stem End Rot

If the pedicel or stem of green banana fruit is injured, a rot similar to crown rot often develops. In case of stem end rot, the decay starts from the cut end when individual fingers are cut from the bunch and marketed as a single fruit rather than in clusters or hands. *Colletotrichum musae* and *Chalara paradoxa* are the causal organisms of stem end rot and the disease is not important now, as the handling procedures have been changed (Jones 2000).

16.2.6 Main Stalk Rot

Main stalk rot is the decay of the peduncle when bananas are exported as a bunch. Spreading from the proximal end, the rot has a characteristic odour and often engulfs the lower hands, causing finger drop. *Chalara paradoxa*, *Lasiodiplodia theobromae* and *Colletotrichum musae* are implicated in main stalk rot (Jones 2000).

16.2.7 Botryodiplodia Finger Rot

This is a soft rot that advances from the tip of the fruit below the flower remnants. The pulp is converted to a black mass and the entire fruit can decay (Figures 16.6 and 16.7). The skin of the fruit becomes black and wrinkled, ripens prematurely and is encrusted with pycnidia. Grey to black woolly mycelium appears on the surface of the fruit under conditions of high humidity. Fully mature fruit is more susceptible to infection, and affected clusters ripen earlier.

FIGURE 16.6 Disease score for finger rot.

FIGURE 16.7 Disease score for fruit rot.

The disease is caused by *Lasiodiplodia theobromae*. Pycnidia are black, flask shaped with short necks and 250–300 μm in diameter. They contain short conidiophores accompanied by paraphyses. Conidia are exuded as single hyaline cells which mature to two-celled, brown, unconstructed longitudinally striated spores. Wind and water disseminate the conidia. Generally, finger rot initiates at the distal end of the finger and through wounds (Jones 2000).

16.2.8 Squirter

This is a disease of single fruit. It is caused by *Nigrospora sphaerica* (Sacc.) Mason. The pathogen enters through cut pedicel and advances into the fruit as it ripens. The infected fruit pulp becomes liquid mass and may squirt from the end of the fruit if squeezed. External symptoms become visible as a bluish tan discoloration of the peel as the fruit ripens. Conidia of *N. sphaerica* in culture are black when mature and measure 18 × 15 μm. They form on dead vegetation, including grasses and banana trash. As it is of a single fruit infection, it is of little importance because of changing handling techniques (Jones 2000).

16.2.9 Fuzzy Pedicel

The disease is characterized by fluffy grey to white mycelium occupying fruit pedicels under stored conditions. Tarnowski et al. (2010) isolated two pathogens, *Sporothrix* (72%) and *Fusarium* (6%) from the infected pedicels. Internal transcribed spacer (ITS) region and β-tubulin genes of *Sporothrix* were amplified and the sequence data from the two genes when aligned showed a close relationship to environmental *Ophiostoma/Sporothrix* clad involving *Sporothrix stylites*, *S. humicola* and *S. pallida*. *Fusarium proliferatum*, *F. pseudocircinatum*, *F. sacchari* and *F. verticillioides* and unnamed taxa in the *F. incarnatum-equiseti* were identified based on the amplification and alignment of gene sequence of EF1α gene. Disease development was inhibited at 14°C but not at 25°C.

16.3 Disease Management Strategies

The major pathogen of postharvest losses in banana is *C. musae*. Adopting the management practices to control *C. musae* also helps in managing the other postharvest pathogens. The following management strategies have been developed for postharvest diseases of banana.

16.3.1 Managing the Ripening Process

Green bananas pass through three physiological stages: the pre-climacteric or pre-ripening phase, the climacteric or ripening phase and senescence. Stage of ripening is closely related to the resistance of banana to postharvest diseases. The pre-climacteric phase of the fruit retains much of the resistance whereas the fruit in the ripening phase and senescence phase are susceptible to the postharvest diseases.

16.3.2 Season

Season influences both incidence and severity of postharvest diseases. Anthracnose incidence and severity is high in rainy season when compared to winter season and summer season (Jagana et al. 2017)

16.3.3 Avoidance

The amount of inoculum affects the level of crown rot. Clean cultivation and taking care of pre-harvest diseases in the banana orchard helps to reduce the inoculum levels of postharvest pathogens.

16.3.4 Sanitation

The majority of the postharvest fungal species are saprophytes that can infect fruits under unfavourable conditions. They usually harbour senescent portions of the plant, particularly dead and decomposed leaves. Old and drooping leaves of the plant harbour the fungal inoculum and may lead to severe contaminations by the pathogens. Floral parts of banana are another important site for pathogens, in particular *C. musae* and *Fusarium* species. Hence, removing and destroying dried flower parts and maintaining field sanitation are important practices for removing pathogen inoculum from the field.

16.3.5 De-Handling Technique

Separation of hands from the bunch after harvest is known as the de-handling technique and is important for long distance transport of the fruits. However, it may lead to infection by the pathogens if not properly taken care of.

16.3.6 Bunch Sleeving

Bunch sleeving with a perforated plastic film helps to protect bunches from fungal contamination.

16.3.7 Modified Atmosphere

Modified atmospheric conditions prolong the green life of fruit and prevent premature ripening during transit and storage. Reducing the oxygen level to 3–7%, increasing carbon dioxide to 10–13% and absorbing ethylene coupled with refrigerated conditions extend the green life of the banana fruits and thereby decrease crown rot initiation.

16.3.8 Physical Control

The following physical treatments are useful in managing postharvest diseases of banana.

- Hot water treatment at 45°C for 20 min within 15–20 min after de-handling or 48°C for 8–10 min reduces the incidence of crown rot (Mirshekari et al. 2011; Fernandes et al. 2017).
- UV irradiations also show inhibition of postharvest pathogens (Bokhari et al. 2013). Of the three radiations, UV-C irradiation is most effective in managing postharvest diseases if given for 45 min and above.

16.3.9 Chemical Treatments

The following chemical treatments are useful for controlling postharvest diseases of banana.

- Chemical fungicides are the best control method of postharvest diseases. Use of chemicals in postharvest disease management dates back to the 1960s with the introduction of the systemic fungicides thiabendazole and benomyl. The benzimidazoles act as antimitotic chemicals. Imazalil (500 ppm) and bitertanol (1 g/L) are the other fungicides that inhibit ergosterol biosynthesis and are used in disease management.
- Chemical treatment generally involves fruit dip in the required concentration of chemical solution, chemical spray of the fruits, etc. However, the fungicide should be applied uniformly to fruit and care should be taken while treating with chemicals for their efficacy. Diedhiou et al. (2014) reported that imazalil is very effective when the fruits are dipped in the solution for 5 min, which reduced disease incidence, maintained quality of the fruit and thereby increased shelf life of the fruit.
- Fungicides, benlate (benomyl) (1000 ppm), ascorbic acid (0.1%), thiabendazole, chlorothalonil (1.5 mL/L), triazoles, strobilurin and imidazoles are used to control postharvest diseases of banana.
- Propiconazole (1 mL/L) and carbendazim (0.5 g/L) successfully controlled the pathogen *Lasiodiplodia theobromae* causing fruit rot both in field conditions as well as the postharvest diseases of banana (Nath et al. 2015).
- An edible film forming polymers such as potassium sorbate and sodium benzoate are inhibitory to *C. musae.*
- Chitosan is generally used as preservative for fruits and vegetables. It forms a thin film coating around the produce and protects the fruit from pathogen infection. It protects the produce by three mechanisms. The compound itself has antifungal properties which manages postharvest pathogens. It enhances host immunity to pathogens by targeting the defence system, and finally it changes the micro-atmospheric conditions of the fruit by preventing transpiration loss. This combined action of the chitosan increases shelf life and quality of the fruit (Jinasena et al. 2011).

16.3.10 Biological Control

- Biological control using yeast, bacteria or fungi is another alternative for postharvest disease management of banana. Individual biological control agents or the consortium of organisms are used in postharvest disease management.
- Conidia and culture filtrates of *T. asperellum* are found to suppress growth of *Fusarium oxysporum* and *Colletotrichum musae* under *in vitro* conditions (Adebesin et al. 2009).
- Consortium of three bioagents (*Bacillus subtilis*, *Pseudomonas fluorescens* and *Trichoderma harzianum*) arrest the mycelial growth of *C. musae* (Hedge 2018).
- Postharvest application with *B. amyloliquefaciens* DGA14 along with hot water treatment is effective and comparable with fungicide-treated banana (Alvinda 2013).
- Khleekorn et al. (2015) reported that two organisms, *Pantoea agglomerans* and *Enterobacter* sp., were found effective in controlling *C. musae*.
- *T. viride*, *Metarhizium anisopliae*, *P. fluorescens* and *B. thuringiensis* and their consortia are found useful to suppress *C. musae* (Dutta and Bora 2017). Ather et al. (2018) reported 40% reduction in crown rot disease development with yeast (*Saccharomyces cerevisiae*).

16.3.11 Plant Extracts

- Application of essential oils from plants is another attractive method for controlling postharvest diseases. Essential oils are classified as safe and would therefore be more acceptable to consumers. Essential oils are biologically active in their vapor phase. In vapour phase, they may act as fumigants and thereby manage the postharvest pathogens. As the essential oils are more complex with multicomponents, the chances of pathogen resistance to essential oils is low, which makes them a suitable alternative.

- The essential oils of *Cinnamomum zeylanicum*, *Azadiractha indica* and *Mentha arvensis* show 100% activity against the postharvest pathogen. Essential oil from *C. zeylanicum* acts as fungi static at 100 ppm and fungicidal at hypertoxic concentration of 200 ppm (Singh and Tripathi 2015).
- Abd-Alla et al. (2014) reported that cinnamon, thyme and bitter and sweet almond oils are some safe alternatives in protecting bananas from postharvest pathogens.
- Clove oil at 0.5% and eucalyptus oil at 2% concentration also control *C. musae* (Hedge 2018; El Zahaby et al. 2018).
- Latex from *Carica papaya* (papaya) also reduces postharvest pathogens. The activity of chitinase enzyme in latex is responsible for the conidial digestion process. The de-handled crowns should be treated with the latex after 1 hour of separation from the bunch. This provides ample time for the exudates to evaporate. The latex acts as a physical barrier for the entry of the postharvest pathogens (Indrakeerthi and Adikaram 2011).
- Plant extracts such as *Acacia albida* and *Prosopis julifera* at 20% concentration along with hot water treatment at 50°C reduces anthracnose incidence and severity and results in an increase in marketable fruit yield (Bazie et al. 2014).
- Plant extract of *Solanum torvum* also reduces crown rot disease and is on par with fungicide benomyl (0.1%) (Thangavelu et al. 2015). The extract also increases the shelf life of the fruitsby 16–20 days.

16.3.12 Novel Molecule: Melatonin

Li et al. (2019) reported melatonin as an alternative to chemicals in postharvest disease management. Exogenous application of melatonin reduces the incidence of anthracnose and delay fruit senescence. External spray with melatonin changed the gene expression pattern of 339 genes in banana. The upregulated genes are found to be involved in the processes like signal transduction, cell wall formation, secondary metabolism, volatile compound synthesis and response to stress. The signalling pathways that showed changes after melatonin treatment are auxin-mediated signalling pathway, ethylene-mediated signalling pathway and mitogen-activated protein kinase-mediated signalling pathway.

16.4 Conclusion and Future Directions

Even though India is the largest producer of bananas, 30% of crop loss is due to postharvest pathogens. Crown rot, anthracnose and cigar end rot are the major postharvest diseases in banana. Various postharvest diseases were discussed in detail with respect to etiology, disease cycle, mode of spread and available management methods. Both latent/quiescent infection and wound infection was reported in banana postharvest pathogens which emphasizes pre- as well as postharvest fruit management. Control strategies, including physical, chemical, biological and modified atmospheric practices that are being followed at present for postharvest management in banana were compiled and presented.

Management of postharvest diseases of banana is a difficult task and there is a great need for improvement. Identification and evaluation of new chemical molecules that are safe for banana, exploration of effective alternative methods such as physical, biological and the methods that enhance host resistance to the postharvest pathogens need to be thoroughly investigated for postharvest disease management. Effective epiphytes or endophytes need to be explored in addition to clearly differentiating the pathogenic and non-pathogenic forms of the organisms. Research may be on biochemical mechanisms that induce host resistance to the pathogens or use of biotechnological approaches in obtaining disease resistance. Integrated disease management methods combining cultural, physical, biological and GRAS chemicals along with pre-harvest management of the crop need to be developed for postharvest diseases of banana.

REFERENCES

Abd-Alla MA, El-Gamal NG, El-Mougy NS, Abdel-Kader MM (2014) Postharvest treatments for controlling crown rot disease of Williams banana fruits (*Musa acuminata* L.) in Egypt. Plant Pathol Quara 4(1): 1–12.

Adebesin AA, Odebode CA, Ayodele AM (2009) Control of postharvest rots of banana fruits by conidia and culture filtrates of *Trichoderma asperellum*. J Plant Protec Res 49(3): DOI: 10.2478/v10045-009-0049-6.

Alvinda DG (2013) Improving control of crown rot disease and quality of pesticide-free banana fruit by combining *Bacillus amyloliquefaciens* DGA14 and hot water treatment. Eur J Plant Pathol 136:183–191.

Alvindia DG, Kobayashi T, Yaguchi Y, Natsuaki KT (2000) Symptoms and the associated fungi of postharvest diseases on non-chemical bananas imported from the Philippines. Jpn J Trop Agric 44:87–93.

Ather M, Waris M, Azhar M, Ahmed S, Ahmed M, Basharat M, Mohsin M (2018) Efficacy of *Saccharomyces cerevisiae* to control crown rot of banana caused by *Fusarium semitectum*. Pak J Phytopathol 30(01): 11–17.

Bazie S, Ayalew A Woldetsadik K (2014) Integrated management of postharvest banana anthracnose (*Colletotrichum musae*) through plant extracts and hot water treatment. Crop Prot 66:14–18.

Bele L, Kouamé DK, Atta HD (2018) Sensitivity of Colletotrichum species responsible for banana anthracnose disease to some fungicides used in postharvest treatments in Côte d'Ivoire. Int J Environ Agric Biotechnol 3(2): 537–542.

Bokhari NA, Siddiqui I, Parveen K, Siddique I, Rizwana H, Soliman DAW (2013) Management of anthracnose of banana by UV irradiation. J Animal Plant Sci 23(4): 1211–1214.

Coelho AFS, Dias MSC, Rodrigues MLM, Leal PAM (2010) Controlepós-colheita da antracnose da banana 'Prata- Anã' tratada com fungicidas e mantida sob refrigeração. Ciência e Agrotecnologia, Lavras. 34(4):1004–1018.

Daniel KK, Yeyeh TMN Hortense AD (2018) Control of fungal isolates responsible for postharvest crown rot of banana (Musa sp. cavendish subgroup, cv. Grande Naine) by three fungicides in Côte d'Ivoire. Int J Curr Res 10(4):68404–68408.

De Lapeyre de Bellaire L, Mourichon X (1997) The pattern of fungal contamination of the banana bunch during its development and potential influence on incidence of crown rot and anthracnose diseases. Plant Pathol 46:481–486.

Diedhiou PM, Zakari AH, Mbaye N, Faye R, Samb PI (2014) Control methods for post-harvest diseases of banana (*Musa sinensis*) produced in Senegal. Int J Sci Environ Technol 3(5): 1648–1656.

Dutta J, Bora LC (2017) Pathogens associated with micropropagated banana plantlets and their management with microbial bioagents. Int J Curr Microbiol Appl Sci 6(7): 1673–1686.

El Zahaby HM, Maswada HF, Ziedan EH, Zoeir EHAER (2018) Safe integrated control of postharvest rot diseases on banana fruit. Plant Arch 18(2): 1345–1351.

Ewané CA, Chillet M, Castelan F, Brostaux Y, Lassois L, Ngando JE, Hubert O, Chilin-Charles Y, Lepoivre P, De Bellaire LL (2013) Impact of the extension of black leaf streak disease on banana susceptibility to post-harvest diseases. Fruits 68: 351–365.

Ewané CA, Lassoisi L, Brostaux Y, Lepoivrei P, de Lapeyre de Bellaire L (2013) The susceptibility of bananas to crown rot disease is influenced by geographical and seasonal effects. Can J Plant Pathol 35(1): 27–36.

Ewané CA, Lepoivre P, de Lapeyre L, de Bellaire LL, Lassois L (2012) Involvement of phenolic compounds in the susceptibility of bananas to crown rot. A review. Biotechnol Agron Soc Environ 16(3): 393–404.

Fernandes MB, Mizobutsi EH, e Silva LM, Ribeiro RCF, Rodrigues MLM (2017) Hydrothermal treatment in the management of anthracnose in 'Prata-Anã' banana produced in the semiarid region of Minas Gerais, Brazil. Revista Brasileira de Fructicultura. DOI: http://dx.doi.org/10.1590/0100-29452018871.

Gul N, Khanzada AM, Khanzada MA, Rajput AQ, Shah GS, Sahito OM (2018) Responses of banana cigar end rot pathogen to chemical fungicides. Plant Prot 02(02): 35–44.

Hedge YR (2018) Novel approaches for management of anthracnose of banana caused by *Colletotrichum musae*. J Plant Physiol Pathol. DOI: 10.4172/2329-955X-C2-017.

Anonymous (2018) Horticultural Statistics at a Glance. Horticulture Statistics Division, Department of Agriculture, Cooperation & Farmers Welfare, Ministry of Agriculture & Farmers Welfare, Government of India.

Indrakeerthi SRP, Adikaram NKB (2011) Control of crown rot of banana using *Carica papaya* latex. J Natl Sci Foundation Sri Lanka 39(2): 155–162.

Jagana D, Hegde YR, Rajasekhar L (2017) Post-harvest diseases of banana (*Musa paradisiaca* L.)- A survey and pathological investigations. Int J Pure App Biosci 5(5): 706–714.

Jinasena D, Pathirathna P, Wickramarachchi S, Marasinghe E (2011) Use of chitosan to control anthracnose on "Embul" banana. Int Conf Asia Agric Animal IPCBEE, Vol.13, IACSIT Press, Singapore.

Jones DR (2000) Diseases of banana, abaca and enset. CABI, Wallingford, UK.

Kachhwaha M, Chile A, Khare, Mehta A, Mehta P (1991) A new fruit rot disease of banana. Indian Phytopath 43: 211.

Kamel MAM, Cortesi P, Saracchi M (2016) Etiological agents of crown rot of organic bananas in Dominican Republic. Postharvest Biol Technol 120: 112–120.

Khleekorn S, McGovern R Wongrueng S (2015) Control of the banana anthracnose pathogen using antagonistic microorganisms. Int J Agric Technol 11(4): 965–973.

Krauss U, Johanson A (2000) Recent advances in the control of crown rot of banana in the Windward Islands. Crop Prot 19: 151–160.

Lassois L, Frettinger P, de Bellaire L, Lepoivre P, Jijakli H (2011) Identification of genes involved in the response of banana to crown rot disease. Mol Plant Microge Interact 24(1): 143–153.

Lassois L, Jijakli MH, Chillet M, de Bellaire LL (2010) Crown rot of bananas: Pre-harvest factors involved in postharvest disease development and integrated control methods. Plant Dis 94(6):648–658.

Li T, Wu O, Zhu H, Zhou Y, Jiang Y, Gao H, Yun Z (2019) Comparative transcriptomic and metabolic analysis reveals the effect of melatonin on delaying anthracnose incidence upon postharvest banana fruit peel. BMC Plant Biol 19: 289–304.

Masudi S, Bonjar GHS (2012) Fulfillment of Koch's postulates for in vitro pathogenicity of *Musicilliumtheobromae* (Turconi) Zare & Gams as the cause of banana cigar end rot disease. J Plant Protec Res 52: 410–414.

Mirshekari A, Ding P, Kadir J, Ghazali HM (2011) Effect of hot water dip treatment on postharvest anthracnose of banana var. Berangan. African J Agric Res 7(1): 6–10.

Molnar O, Bartok T, Szecsi A (2015) Occurrence of *Fusarium verticillioides* and *Fusarium musae* on banana fruits marketed in Hungary. Acta Microbiol Immunol Hung 62: 109–119.

Nath K, Solanky KU, Bala M (2015) Management of banana (*Musa Paradisiaca*) fruit rot diseases using fungicides. J Plant Pathol Microbiol 6: 8.

Padam BS, Tin HS, Chye FY, Abdullah MI (2014) Banana byproducts: An underutilized renewable food biomass with great potential. J Food Sci Technol 51(12): 3527–3545.

Singh R, Tripathi P (2015) Cinnamomum zeylanicum essential oil in the management of anthracnose of banana fruits. JIPBS 2(3): 290–299.

Tarnowski TL, Pérez-Martínez JM, Ploetz RC (2010) Fuzzy pedicel: A new post-harvest disease of banana. Plant Dis 94: 621–627.

Thangavelu R, Sundararaju P, Sathiamoorthy S (2015) Management of anthracnose disease of banana caused by *Colletotrichum musae* using plant extracts. J Hort Sci Biotechnol 79(4): 664–668.

Triest D, Pierard D, De Cremer K, Hendrickx M (2016) *Fusarium musae* infected banana fruits as potential source of human fusariosis: may occur more frequently than we might think and hypotheses about infection. Commun Integr Biol 9: e1162934.

Triest D, Hendrickx M (2016) Postharvest disease of banana caused by *Fusarium musae:* A public health concern? PLOS Pathog 12(11): e1005940.

Triest D, Stubbe D, De Cremer K, Pierard D, Detandt M, Hendrickx M (2015) Banana infecting fungus, *Fusarium musae*, is also an opportunistic human pathogen: are bananas potential carriers and source of fusariosis? Mycologia 107: 46–53.

Unnithan RR, Thammaiah N (2017) Isolation, identification and proving the pathogenicity of banana anthracnose pathogen *Colletotrichum musae*. Int J Plant Prot 10: 399–403.

17

Postharvest Diseases of Grapes and Their Management

Raghavendra Achari[1] and V. Devappa[2]
[1]College of Horticulture, Munirabad (Koppal), Karnataka, India
[2]College of Horticulture, UHS (B) Campus, Bengaluru, Karnataka, India

CONTENTS

17.1 Introduction

Grape has special importance among the major horticultural crops in view of its value added as raisins. The European Union is a major destination for export of table grape from India. In India, about 95% of total grape production comes from Maharashtra and Karnataka. Grapes are mainly produced for fresh consumption (71%), raisins (27%), winemaking (1.5%) and fresh juice (1.5%) (Adsule et al. 2012; Sharma et al. 2018). The perishable nature of grapes results in considerable loss both during harvesting and after harvest. Postharvest loss during transportation and storage is mainly attributed to the physical handling of produce and damage caused by diseases and pests. At this time, progress in addressing postharvest diseases and pests has not made the desired progress compared to progress in addressing physical handling issues leading to postharvest loss. It is difficult to assess exactly the quantitative postharvest loss due to disease, and loss in the nutritional value and quality of food is generally overlooked. Association of Chamber of Commerce and Industry of India's study states that 30% of India's fresh produce is rendered unfit for consumption as a result of spoilage after harvesting (Anonymous 2013). Worldwide, postharvest losses as rots caused by microorganisms have been estimated up to 50% of the harvested crop. In India, overall postharvest loss in grape is 8.30%, which includes loss during sorting and grading (3.21%), transportation (1.93%) and farm level storage (5.54%) (Nanda et al. 2012). The present postharvest loss statistics reveals that India is losing about 2,23,000 tonnes of grapes annually. Beyond the visible loss, the economic loss is much higher than estimated. Loss occurring in extremely perishable grapes (table purpose) is mainly during preparation, harvest, packing, storing, transport and distribution (Sharma et al. 2018). During storage or long-distance transportation, spoilage due to moulds that develops in transit or when carried from the field is considerable.

Under a Network project, a study was conducted in Andhra Pradesh to assess the postharvest losses on Thompson seedless grapes. The study documented that loss in the domestic market was about 7.96%, of which the major loss was at the retail market level (4.56%) followed by field loss of 3.40%, whereas loss in the export market was further split into loss at field level (8%) and loss at cold storage level (11.13%), amounting to a total loss of 19.13% (Anonymous 2003). The study also documented postharvest loss in Karnataka on Thompson seedless variety at the local market amounting for 14.40%, which further include 7.31% loss at field level, 4.24% loss during transportation to wholesale market and 2.85% during retail level disposal. Ladania et al. (2005) conducted a study during 2000–2001 in major grape growing regions of Maharashtra, i.e. Nashik and Sangli districts, contributing 70% total grape production of state. Thompson seedless, Sonaka, Tas-e-Ganesh and Sharad seedless (black) are the major grape varieties, occupying an area of about 90%. For assessing the loss, wholesale and the retail level grape markets at Mumbai, Pune and Nagpur were selected for the study. It has been documented that the total loss aggregated was 19–30.9%, which includes farm level loss of 1.00–1.25%, wholesale level loss of 5.5–8.65% and retailer level 12.25–16%. The observations were specific for grapes packed in boxes, whereas for those grapes carried in bamboo baskets to Mumbai market the loss was 24.19%, which included shattering of bunches, rupturing and rotting of berries. The value of loss has been assessed as Rs.434 crore.

To assess the postharvest losses in grapes at different levels, several studies were conducted. Postharvest loss is less if produce is disposed of in local retail markets (14.4%) and more for sale in distant markets (21.3%) (Sreenivas Murthy et al. 2004). Similarly, between wholesale and retail markets, the postharvest loss varied from 17.75 to 24.65% (Ladania et al. 2005). The same has been reiterated by a recent study conducted under AICRP Fruits by ICAR-NRC Grapes, Pune, which revealed a postharvest loss of 16.69–26.30%. Thereby we can ascertain that around 25% of what is produced is being lost as postharvest loss.

17.2 Major Postharvest Diseases

The grapes are soft and delicate, and their high water content makes them vulnerable for postharvest diseases caused by several fungi including grey mould rot, anthracnose, blue rot, Botrytis rot, Fusarium rot and soft rot. In India, the major postharvest diseases of grapes include black rot and Rhizopus rot. The major fungal pathogens associated with postharvest diseases of grapes are *Aspergillus niger*, *A. ochraceus*, *A. terreus*, *A. flavus*, *Alternaria alternata*, *Botrytis cinerea* (grey mould), *Colletotrichum gloeosporioides*, *Cladosporium* sp., *Mucor* sp., *Penicillium expansum* (blue mould), *P. funiculosum* (blue mould), *Phomopsis viticola*, *Rhizopus stolonifer* and *Lasioplodia theobromae* (Barkai-Golan 2001). These are reported as either on-field infection or infection after harvest. *Botrytis cinerea* is a major pathogen present across the world and especially seen where table grapes are harvested late and exposed to more rainfall along with high relative humidity. Postharvest fungal infection not only reduces the quality of grapes but is also known to pose serious health hazards by metabolites produced by associating pathogens like mycotoxins. Ochratoxin A is an important mycotoxin produced by *P. verucosum* and *A. ochraceous* in both fresh and wine grapes followed by "aflatoxins" known to be produced by *A. flavus* and *A. parasiticus*. Brown skin spot, a new postharvest disease caused by *Cadophora luteo-olivacea*, has been reported on Shine Muscat, a Japanese grape cultivar (Nakaune et al. 2015).

17.2.1 Botrytis Bunch Rot

Botrytis rot, which is caused by *Botrytis cinerea* Pers., is the most predominant and most damaging postharvest disease worldwide. The grey mould caused by *B. cinerea* has been studied elaborately as a postharvest disease of table grapes among the good number of postharvest diseases reported on grapes. Due to its common prevalence, other postharvest diseases caused by other fungi are often overlapped with grey mould disease. Being a high sugar disease, Botrytis infection can occur at the bloom stage, but symptoms appear only close to the harvesting stage and thereby can easily carry to the postharvest stage. On berries, *Botrytis* infection develops early symptoms of water-soaked spots coupled with skin of the berries slipping off from the pulp easily, and hence it is also identified as "slip skin problem". The infected berries change colour and grey-coloured fungus growth appears (Figure 17.1) (Anonymous 2020). Spores of botrytis spread easily to other berries. If dry and hot conditions prevail after infection of berries, Botrytis may lead to shrivelling of berries that later turn mummified.

17.2.2 Aspergillus Rot

Grape berry rot caused by *Aspergillus* spp. is commonly observed after sugar development and close to maturity of berries. *Aspergillus* spp. is believed to be a secondary pathogen, but it can directly infect berries which are intact to vines that are not previously damaged or infected by other pathogens (Figure 17.2) (Anonymous 2020). The pathogen can cause active infection when water is present on skin of berries during relatively warm temperatures (20–30°C). Upon infection the berries turn tan or brown, and are covered with brown to black coloured spores masses quickly.

FIGURE 17.1 Botrytis bunch rot caused by *Botrytis cinerea*.

FIGURE 17.2 Aspergillus rot.

17.2.3 Bitter Rot

Bitter rot is mainly observed when berries ripen completely, and the pathogen associated is *Greeneria uvicola* (Berk. & M. A. Curtis) Punith. Even though the pathogen infects at the flowering stage, symptoms will not be observed until berries mature. Berry colour changes to brown with the appearance of spore bodies arranged in concentric rings initially, and as the disease progress it covers the entire berries (Figure 17.3) (Anonymous 2020). Affected berries turn soft, with poor pedicel attachment leading to shattering of infected berries. Optimum infection occurs at 28–30°C.

FIGURE 17.3 Bitter rot symptoms on berries caused by *Greeneria uvicola*.

FIGURE 17.4 Symptoms of Botryosphaeria bunch rot.

17.2.4 Botryosphaeria Bunch Rot

Botryosphaeria species are known to cause berry or bunch rot which is usually observed when the crops near maturity. Even though disease initiation is slow in the beginning, disease later progresses very quickly. Initial symptoms on berries look similar to that of grey mould with change of colour of water-soaked berries (Figure 17.4) (Anonymous 2020). As the disease progresses, small black pimples appear on the berries before they shrivel, and eventually berries drop off from the peduncle. The common association of trunk disease has been observed in plants where Botryosphaeria causing bunch rot infection is observed. Fungi conforming to seven species of the Botryosphaeriaceae, *Diplodia seriata*, *Neofusicoccum parvum*, *N. luteum*, *Dothiorella viticola*, *Lasiodiplodia theobromae*, *D. mutila* and *Botryosphaeria dothidea*, were isolated (Nicola et al. 2010).

17.2.5 Black Rot

Black rot of grapes is one of the symptoms of infection by *Guignardia bidwellii* (Ellis) Viala & Ravaz [anamorphic state, *Phyllosticta ampelicida* (Engleman) Van der Aa]. It is more serious and becomes destructive under conditions of high relative humidity coupled with warm weather. It can infect almost all parts of the vine including foliage, fruits, stems, shoots and tendrils, and it appears more severely on fruits. On leaves, there is an appearance of angular to circular spots with reddish brown colour and coalesce giving reddish brown blotchy appearance. Black-coloured pycnidial fruiting bodies are observed on the inside margin of spots and predominantly observed on young leaves. On green berries, small circular spots with brown rings appear initially and later turn tan. It spreads quickly, covering 50% of berry and later the entire berry turns black followed by mummification. On mummified berries, black pycnidial fruiting bodies are also observed.

High temperature and high humidity during storage is most favourable for the disease. The pathogen is known to enter through hairline cracks on the skin of berries which are caused by either improper water management or physical damage during postharvest stage. The pulp of berries becomes soft and watery upon infection by the pathogen.

17.2.6 Rhizopus Rot (*Rhizopus* sp.)

This disease is characterized by rapid production of a coarse grey mat of mycelium under favourable warm and moist conditions. The disease is caused by *Rhizopus stolonifer* Vuillemin, predominantly observed in tight bunches, and on injured berries if not trimmed off during sorting and packing, the disease becomes severe under storage conditions. Hence, trimming infected or injured berries at harvest or while grading and packing will considerably reduce the disease during storage in spite of favourable weather conditions.

17.3 Disease Management Strategies

Grape is no exception to postharvest losses caused by either natural senescence or association of pathogenic and/or saprophytic microbes, which are known to greatly affect both quality and quantity of produce. Under field conditions, colonization of grape berries by fungus can occur usually at beginning of sugar accumulation/colour development stage (Chulze et al. 2006). To reduce the postharvest spoilage of grape berries and to enhance shelf life and market value, it is important to manage postharvest diseases. A number of techniques are in use to address the causes of postharvest loss including disease. The use of synthetic fungicides, physical measures, biological agents and plant extracts will have both individual and collective efficacy in managing the postharvest diseases of grapes, with distinct pros and cons. Physical and chemical methods are often limited by their cons. Hence, plant-based products have been overrated for their pros in managing postharvest diseases (Hassani et al. 2012). Hazard Analysis Critical Control Points (HACCP) compliance must be considered before designing strategies for the management of postharvest diseases in grapes by following Good Agricultural Practices (GAPs) and Good Manufacturing Practices (GMPs) as per the requirements of market. Postharvest handling and storage conditions need extra attention.

17.3.1 Principles of Management

Principles of management of major postharvest diseases of grapes are mainly grouped into preventive type and curative type. The preventive measures include (i) avoidance of the pathogen-cultural practices; (ii) host resistance –resistant varieties; (iii) exclusion – quarantines and sorting/grading; (iv) eradication/eliminating or reducing inoculum-sanitation; (v) protection/prevention/chemical or biological or physical treatments – cold temperature. Curative methods include therapy by physical or chemical treatments.

The integrated management of postharvest diseases of grapes begins with selected pre-harvest operations, especially selection of fungicides near to harvest, and harvesting operations including handling of berries during and after harvest. Further, depending on the quality of produce and targeted market, postharvest practices need to be designed to reduce postharvest loss besides increasing shelf life for better market value. Less frequently, some physical practices are used including hot water treatment, heat treatment, chemicals used either singly or along with fungicides, biological agents and plant-based formulations (Sukatta et al. 2008; Tripathi et al. 2008; Hassani et al. 2012). Practices for the management of postharvest diseases of grape including physical and chemical methods are limited for commercialization due to the disadvantages associated with them. Hence, plant-based biologically active products have drawn the attention of researchers to use them to protect highly perishable fruits and vegetables. The properties of antifungal, antioxidant, antibacterial and biosafety associated with botanicals and essential oils is gaining importance in plant protection. Further, plant-based natural products including botanical extracts and essential oils, plant defence molecules such as chitosan and salicylic acid, either used separately or used along with storage conditions, approaches such as modified storage atmosphere and controlled atmosphere have been studied for efficacy in reducing postharvest loss in fruit crops including table grapes (Shahi et al. 2003; Valero et al. 2006). However, deploying controlled atmosphere in grapes is still debated for its merits and demerits. Lichter et al. (2006), in a review published on the control of postharvest loss in grapes, emphasized chemical methods followed by physical methods of management when compared to safer use of biological agents or plant-based products. In an another review published (Romanazzi et al. 2012) on management of postharvest disease of grapes, grey mould disease of fresh grapes by the use of safer and effective biological control methods as an alternative to use of hazardous chemical measures. However, they restricted the study to grey mould disease and excluded almost all other postharvest diseases of grapes, did not emphasize physical and chemical methods of management of postharvest diseases of grapes.

Recently, it was found that calcium supplementation will help in maintaining the turgidity of cell wall besides ensuring the quality of highly perishable vegetables and fruits. Hence, integrating the different components of disease management makes our efforts for the management of postharvest diseases of fruits, especially in grapes, more sustainable, practically useful and consistent in the long run to all the stakeholders of the production and supply chain.

17.3.2 Physical or Chemical Methods

Murthy et al. (2014) documented the practice of bunch cleaning at 15 days before harvest of the crop for dispersion to local markets, leading a postharvest loss of about 3.40%. Important postharvest techniques, especially refrigerated transportation, modified atmosphere storage such as cold storage and storage under low pressure conditions practices aiming at managing postharvest diseases by avoiding further spread are mainly either preventive or curative in action. In an effort to eliminate the microbial load on the fruits, some physical sterilization methods like exposure to UV-C, UV-B and UV-A have been used, mainly to reduce decaying of fruits and vegetables, and thereby enhancing shelf life (Qadri et al. 2020). Candir et al. (2011) reported that prepackaging treatment of fresh grapes in polyethylene bags with hot water (50°C and 55°C) has reduced the rate of decay due to fungal pathogens in transportation and storage. Disinfectants and carbonate and bicarbonate salts are promising in controlling grey mould disease when treated to fresh berries of grapes. Ammonium bicarbonate when applied at 500 mM was found effective among the different bicarbonate salt solutions analyzed. Similarly, SO_2 fumigation for a relatively longer period (2–6 weeks) and repeated treatments (since SO_2 is contact action) at regular intervals was found promising in reducing postharvest fungal diseases, but it is known to cause berry injuries and affect the taste of berries (Crisosto et al. 2002).

Eradication of pathogens using mild to strong oxidizing agents like chlorine dioxide, ozone or peroxide is also promising. For inhibition of spore germination of *B. cinerea*, low doses of ethanol treatment (10% and 20%) with potassium sorbet (0.5% and 1.0%) for 10 seconds was found effective, whereas 30% and 40% ethanol completely inhibited spore germination (Karabulut et al. 2004, 2005).

17.3.3 Fungicides and Plant Defence Molecules

Fungicides as postharvest treatment can be applied directly as sprays, dips, fumigants, waxes and coatings and indirectly as treated wraps and box liners. The concentration of SO_2 in the boxes is crucial and above 10 mg/kg of fresh fruit known to affect fruits. The production of SO_2 is mainly influenced by humidity inside boxes. Use of synthetic fungicides for the management of postharvest fungal pathogens is a common practices but is limited recently because of associated residue problems. Microbial and plant products or metabolites are known to have fungicidal properties. Chitosan is one such molecule which directly has fungicidal action and also activation of host defence against plant pathogens, especially postharvest pathogens. The efficacy of chitosan enhanced when dissolved it in different acids to activate its antimicrobial and eliciting properties (Romanazzi et al. 2009). Use of UV-C treatments and chitosan is reported to have synergistic efficacy in reducing both blue and grey mould of fresh grapes (Romanazzi et al. 2006).

17.3.4 Botanicals

Many natural compounds have been documented with considerable antifungal activity (Dubey and Jalal 2013). Plant-based volatile compounds which are used mainly to flavour and season foods are also known to reduce microbial load. The plant-based volatile compounds act as protectants and antimicrobials and are less harmful to mammalians and friendly to the environment, which makes them the best alternatives to synthetic fungicides. Plants represent a good source of natural compounds having antifungal properties which can potentially replace the use of chemical fungicides. Many plant species produce volatile substances and essential oils known to have antifungal or antimicrobial property and can be used as preservatives, especially during the postharvest stage. Plant extracts, pyrethrum, rotenone, neem and essential oils are five major types of plant-based products that sustained stiff competition from synthetic fungicides in managing plant diseases including postharvest diseases.

Essential oils and plant defence molecules like salicylic acid and chitosan have been found effective when used with physical methods of controlled atmosphere (like combination of high CO_2 and low O_2 levels) and modified storage atmosphere found promising in managing the postharvest diseases of grapes (Valero et al. 2006). Chitosan in combination with grapefruit seed extract (GSE) (Xu et al. 2007) has been found promising to control postharvest decay and to ensure quality of grapes.

Aloe vera gel is an edible antimicrobial coating for retaining moisture, texture and to control respiratory rate of fresh grapes (Martinez-Romero et al. 2006). Specifically, it enhanced about 35 days of storability at 1°C along with retention of major quality parameters (Valverde et al. 2005). Cinnamon and clove oils have shown their efficacy in reducing postharvest rotting of grapes caused by major postharvest rotting fungi such as *Colletotrichum*, *Lasiodiplodia*, *Phomopsis*, *Aspergillus*, *Alternaria* and *Rhizopus* in table grapes (Lopez-Malo et al. 2007; Sukatta et al. 2008). Postharvest treatment of grape with *Ocimum sanctum*, *Zingiber officinale* or *Prunus persica* essential oils delayed onset of rotting of table grapes to 8, 10 and 9 days, respectively, as against 4 days in control (Tripathi et al. 2008). Similarly, eugenol or thymol treatment to grape cluster delayed onset of decaying to 56 days at storage temperature of 1°C. Spray treatment on grape bunches with *Thymus vulgaris* and *Satureja hortensis* oils reduced grey mould disease severity under 0°C storage conditions (Abdollahi et al. 2010). In an experiment, grapes about 1 kg were treated with *Artemisia nilagirica* (200 μL) and *Cymbopogon citratus* oils (300 μL), the shelf life of grapes has been found enhanced by up to 10 days by protecting them from microbial decaying (Sonker et al. 2015, 2014). Guillen et al. (2007) reported that modified atmosphere packaging and treatment of eugenol-thymol-carvacrol essential oils collectively reduced rotting of berries by 30% (control 37% and treated 7% rotting) after 56 days of cold storage at 1°C.

17.3.5 Biological Control Using Microbes

Bio-control agents such as, fungi, bacteria and yeast with antagonistic properties are promising for controlling the postharvest diseases in grapes. Postharvest biological control agents used during the postharvest stage have been studied extensively in many fruit crops, including grapes. By introducing natural enemies of the pathogen, normal growth or activity of pathogen can be restricted. These include *Muscodor albus* (Gabler et al. 2006), *Metschnikowia fructicola*, *Metschnikowia pulcherrima* (Karabulut et al. 2003), *Hanseniaspora uvarum* (Liu et al. 2010) against *B. cinerea*, *Cryptococcus laurentii* and *Aureobasidium pullulans* against black Aspergilli (Dimakopoulou et al. 2005). The *Muscodor albus* formulation is used at 5 or 20 g per kilogram of grape cluster at 15°C for 7 days or with 5 or 10 g/kg of grapes at 0.5°C for 28 days against postharvest grey mould. Its performance has been observed in −0.5 to 1°C during commercial storage, 2–5°C during transportation and 15–20°C during marketing (Mercier and Smilanick 2003). Antibiotics produced by *Trichoderrna* spp. exhibited antifungal properties against *Sclerotinia sclerotiorum*, *Botrytis cinerea*, *Corticium rolfsii* and other major plant pathogenic fungi.

17.4 Future Directions

Postharvest loss is not an exception in grape, a perishable fruit. Loss at the postharvest level should be deemed as double loss over and above value of the commodity because of exhaust of inputs of production with the same value. Considerable postharvest loss reduces the availability of fruits for consumption, in turn effecting nutritional security. The present review mainly emphasis the loss caused by postharvest pathogens, its cause, and strategies to manage loss by integrating physical, chemical, biological including botanical components which are useful for addressing the common problems of postharvest handling of fresh grapes. The integrated approaches for management of postharvest diseases of table grapes, starting from practices on standing crop, preferably as preventive measures can only address the problem better over either use components of disease management individually. As postharvest diseases pose challenges near to harvest or after harvest, selection of components of management is limited by residue and effect on quality and shelf life. Hence, a combination approach will be practical. Storage at low temperature (0°C) helps in delaying the ripening process besides maintaining quality. The postharvest handling aspects are not as prominent as in other fruit crops. Even though exploring botanical and biological tools outscores chemical synthetic fungicides, not so prominent use has been recorded on a commercial scale. Safer storage methods can be further developed that address the challenges of postharvest diseases of grapes.

REFERENCES

Abdollahi A, Hassani A, Ghosta Y, Bernousi L, Meshkatalsadat MH (2010) Study on the potential use of essential oils for decay control and quality preservation of 'Tabarzeh' table grape. J Plant Prot Res 50:45–52.

Adsule PG, Sharma AK, Upadhyay A, Sawant IS, Jogaih S, Upadhyay AK, Yadav DS (2012) Grape research in India: A review. Progr Hortic 44(2):180–193.

Anonymous (2003) Retrieved May 31, 2013, from http://nhb.gov.in/ horticulture%20crops/grape/grape1.htm

Anonymous (2013) Huge postharvest losses for India. (http://www.fruitnet.com/asiafruit/article/159139/huge-postharvest-losses-for-india)

Anonymous (2020) Know your grapevine bunch rots. (https://www.agric.wa.gov.au/summer/know-your-grapevine-bunch-rots)

Barkai-Golan R (2001) Soft fruits and berries: Grapes. In: Barkai-Golan R (Ed.), Postharvest Diseases of Fruits and Vegetables Development and Control. Elsevier, Amsterdam, pp. 315–316.

Candir E, Kamiloglu O, Ozdemir AE, Celebi S, Coskun H, Ars M, Alkan S (2011) Alternative postharvest treatments to control decay of table grapes during storage. J Appl Bot Food Qual 84:72–75.

Chulze SN, Magnoli CE, Dalcero AM (2006) Occurrence of ochratoxin A in wine and ochra-toxigenic mycroflora in grapes and dried vine fruits in South America. Int J Food Microbiol 111: S5–S9.

Crisosto CH, Palou L, Garner D, Armson DA (2002) Concentration by time product and gas penetration after marine container fumigation of table grapes with reduced doses of sulphur dioxide. Hortic Tech 12:241–245.

Dimakopoulou M, Tjamos SE, Tjamos EC, Antoniou PP (2005) Chemical and biological control of sour rot caused by black aspergilla in the grapevine variety Agiorgitico of Korinth region. In: International Workshop on Ochratoxin A in Grapes and Wines: Prevention and Control, Marsala (TP), p. 77.

Dubey SR, Jalal AS (2013) Species and variety detection of fruits and vegetables from images. Int J Appl Pattern Recognit 1(1):108–126.

Gabler FM, Fassel R, Mercier J, Smilanick JL (2006) Influence of temperature, inoculation interval and dosage on biofumigation with *Muscodor albus* to control postharvest gray mold on grapes. Plant Dis 90:1019–1025.

Guillen F, Zapata PJ, Martinez-Romero D, Castillo S, Serrano M, Valero D (2007) Improvement of the overall quality of table grapes stored under modified atmosphere packaging in combination with natural antimicrobial compounds. J Food Sci 72:S185–S190.

Hassani A, Fathi Z, Ghosta Y, Abdollahi A, Meshkatalsasat MH, Marandi RJ (2012) Evaluation of plant essential oils for control of postharvest brown and gray mold rots on apricot. J Food Safety 32: 94–101.

Karabulut OA, Mlikota GF, Mansour M, Smilanick JL (2004) Postharvest ethanol and hot water treatments of table grapes to control gray mold. Postharvest Biol Technol 34: 169–177.

Karabulut OA, Romanazzi G, Smilanick JL, Lichter A (2005) Postharvest ethanol and potassium sorbate treatment of table grapes to control gray mold. Postharvest Biol Technol 37: 129–134.

Karabulut OA, Smilalanick JL, Gabler FM, Mansour M, Droby S (2003) Near harvest application of *Metschnikowia fructicola*, ethanol and sodium bicarbonate to control postharvest diseases of grape in central California. Plant Dis 87:1384–1389.

Ladania MS, Wanjari V, Mahalle B (2005) Marketing of grapes and raisins and postharvest losses of fresh grapes in Maharashtra. Indian J Agric Res 39(3): 167–176.

Lichter A, Gabler FM, Smilanick JL (2006) Control of spoilage in table grape: Review. Stewart Postharvest Rev 6:1–10.

Liu HM, Guo JM, Cheng YJ, Luo L, Liu P, Wang BQ, Deng BX, Long CA (2010) Control of grey mould of grape by *Hanseniaspora uvarum* and effects on postharvest quality parameter. Ann Microbiol 60:31–35.

Lopez-Malo A, Barreto-Valdivieso J, Palou E, Mart FS (2007) *Aspergillus flavus* growth response to cinnamon extract and sodium benzoate mixture. Food Control 18:1358–1362.

Martinez-Romero D, Alburquerque N, Valverde JM, Guillan F, Castillo S, Velero D, Serrano M (2006) Postharvest sweet cherry quality and safety maintenance by *Aloe vera* treatment: A new edible coating. Postharvest Biol Technol 39: 93–100.

Mercier J, Smilanick JL (2003) Control of green mold and sour rot of lemon and gray mold rot of grapes by fumigation with *Muscodor albus* (Abstr.). Phytopathology 93: S61.

Murthy MRK, Reddy GP, Rao KH (2014) Retail marketing of fruits and vegetables in India: A case study on export of grapes from Andhra Pradesh, India. Euro J Logist Purch Supply Chain Manag 2(1): 62–70.

Nakaune R, Tatsuki M, Matsumoto, Ikoma Y (2015) First report of a new postharvest disease of grape caused by *Cadophora luteoolivacea.* J Gen Plant Pathol 82: 116–119.

Nanda SK, Vishwakarma RK, Bathla HVL, Rai A, Chandra P (2012) Estimation of quantitative harvest and postharvest losses of major agricultural produce in India. In: AICRP on Post Harvest Technology, CIPHET, Ludhiana, India.

Nicola W, Gavin A, Steel C, Raman H, Savocchhia S (2010). Botryosphaeriaceae associated with bunch rot of grapes in South Eastern Australia. International Council on Grapevine Trunk Diseases, The Mediterannean Phytopathological Union (Eds.). Phytopathologia Mediterannea, Firence, p. 106.

Qadri R, Azam M, Khan I, Yang Y, Ejaz S, AkramM T, Khan A (2020) Conventional and modern technologies for the management of post-harvest diseases. In: Ul Haq I, Ijaz S (Eds.). Plant Disease Management Strategies for Sustainable Agriculture through Traditional and Modern Approaches. Sustainability in Plant and Crop Protection. Vol 13. Springer, Cham. https://doi.org/10.1007/978-3-030-35955-3_7.

Romanazzi G, Gabler FM, Margosan DA, Macky BE, Smilanick JL (2009) Effect of chitosan dissolved in different acids on its ability to control postharvest gray mold of table grapes. Phytopathology 99:1028–1036.

Romanazzi G, Gabler FM, Smilanick JL (2006) Pre-harvest chitosan and postharvest UV-C irradiation treatments suppress gray mold of table grapes. Plant Dis 90: 445–450.

Romanazzi G, Lichter A, Gabler FM, Smilanick JL (2012) Recent advances on the use of natural and safe alternatives to conventional methods to control postharvest gray mold of table grapes. Postharvest Biol Technol 63: 141–147.

Shahi SK, Patra M, Shukla AC, Dikshit A (2003) Use of essential oil as botanical pesticide against postharvest spoilage in *Malus pumilo* fruits. Bio-control 48: 223–232.

Sharma AK, Sawant SD, Somkuwar RG, Naik S (2018) Postharvest losses in grapes: Indian status. Technical Report, ICAR NRC Grapes, Pune, India. DOI:10.13140/RG.2.2.17999.89761/1.

Sonker N, Pandey AK, Singh P (2015) Efficiency of *Artemisia nilagirica* (Clarke) Pamp essential oil as a mycotoxicant against postharvest mycobiota of table grapes. J Sci Food Agri 95:1932–1939.

Sonker N, Pandey AK, Singh P, Tripathi NN (2014) Assessment of *Cymbopogon citratus* (DC.) stapf. essential oil as herbal preservatives based on antifungal, antiaflatoxin and antiochratoxin activities and in vivo efficacy during storage. J Food Sci 79: M628–M634.

Sreenivas Murthy D, Gajanan TN, Sudha M (2004) Postharvest loss and its impact on marketing cost, margin and efficiency: A study on grapes in Karnataka. Indian J Agric Econ 59(4):772–786.

Sukatta U, Haruthaithanasan V, Chantarapanont D, Suppakul P (2008) Antifungal activity of clove and cinnamon oil and their synergistic against postharvest decay fungi of grape *in vitro.* Kasetsart J (Natural Sci) 42:169–174.

Tripathi P, Dubey NK, Shukla AK (2008) Use of essential oils as postharvest botanical fungicides in the management of grey mould of grapes caused by *Botrytis cinerea.* World J Microbiol Biotechnol 24: 39–46.

Valero D, Valverde JM, Martinez-Romero D, Guillen S, Castillo S, Serrano M (2006) The combination of modified atmosphere packaging with eugenol or thymol to maintain quality and safety and functional properties of table grapes. Postharvest Biol Technol 42: 222–227.

Valverde MJ, Valero D, Romero MD, Guillen F, Castillo S, Serrano M (2005) Novel edible coating based on *Aloe vera* gel to maintain table grape quality and safety. J Agri Food Chem 53: 7807–7813.

Xu WT, Huang KL, Guo F, Qu W, Yang JJ, Liang ZH, Luo YB (2007) Postharvest grapefruit seed extract and chitosan treatment of table grapes to control *Botrytis cineria.* Postharvest Biol Technol 46: 86–94.

18

*Postharvest Diseases of Guava (*Psidium guajava *L.) and Their Management*

Manju Sharma
Amity Institute of Biotechnology, Amity University Haryana, Manesar, Gurugram, India

CONTENTS

18.1 Introduction

The origin of guava (*Psidium guajava* L.) is obscure. Its origin is said to be South America and the West Indies but it is frequently grown in other parts of the tropics and subtropics including India, where it is considered an important hardy crop grown in neglected soils. India is known for large production of guava fruits, followed by countries like Pakistan, Mexico and Brazil (Singh 2011).

Australia, Bangladesh, Colombia, Cuba, Egypt, Indonesia, Malaysia, Sudan, Thailand, Venezuela, Vietnam, the United States and Puerto Rico contribute substantially to guava production. Guava is prone to be attacked by a wide range of diseases, from root to crown and fruits due to a wide variation in climatic conditions. A number of plant pathogens of fungal, bacterial, algal and nematode origin have been reported to cause various types of diseases. As many as 177 pathogens have been reported on guava to date, of which 167 are fungal, 3 bacterial, 3 algal, 3 nematodes and 1 epiphyte (Figure 18.1) (Misra AK 2004, 2005). Before harvesting the fruits, pathogens show various manifestations viz., anthracnose, damping off, die back, leaf spot, canker, rots, rust, red rust, scab, sooty mould, stylar-end rot and wilt. Guava is usually grown in two seasons, rainy season and winter. Winter season crops provide better quality fruits than rainy season crops.

The chances of viral infections in the form of disease are rare. Many fungi prefer an acidic (pH 2.5–6) environment to grow and develop, whereas neutral conditions are preferred by

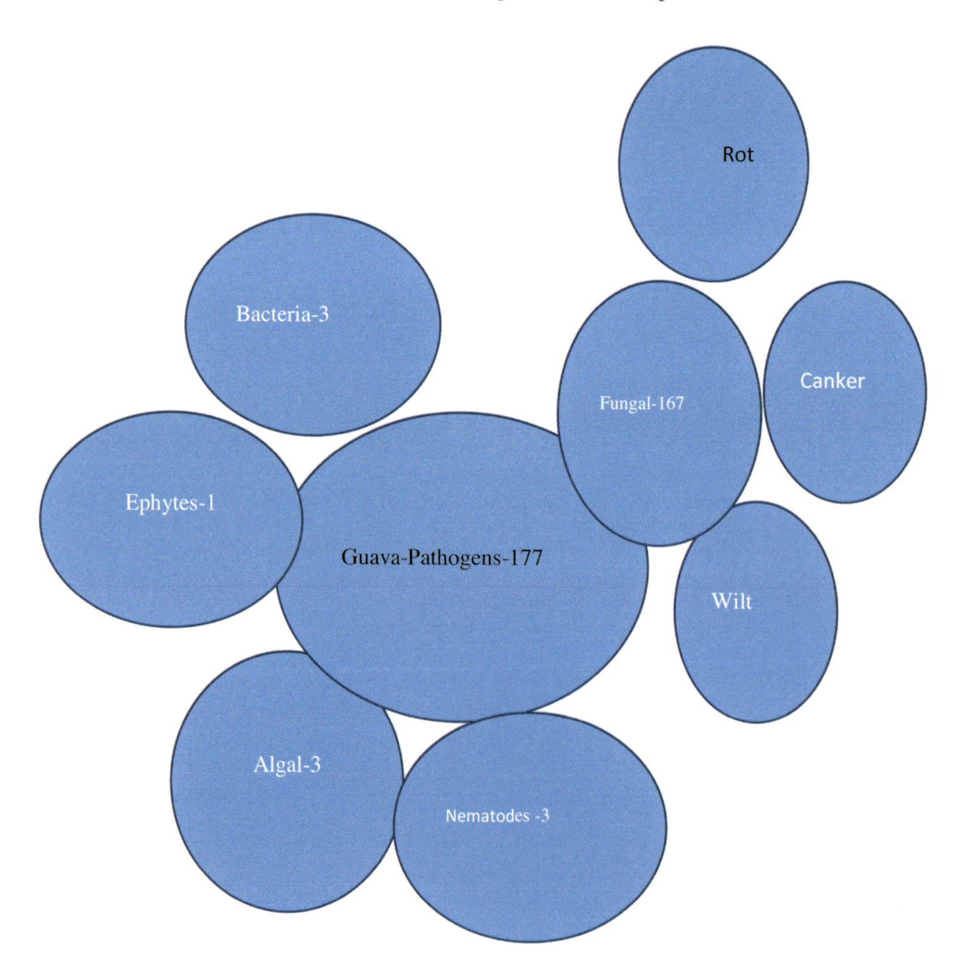

FIGURE 18.1 Pathogens of guava that cause diseases.

bacteria to flourish. However, few species can tolerate a pH below 4.5. Therefore, too acidic nature of fruits does not allow bacteria to cause infection (Thompson 2010).

In addition, nutrition and growth regulators and environment conditions are the abiotic aspects of deterioration of fruits. Environmental factors that exceed the optimum ranges induce incidence of physiological disorders such as abnormal growth patterns of fruits (Burzo et al. 2001) due to disruption in physiological processes during the pre- and postharvest periods. Abnormal temperature, fluctuating moisture, light, nutrient, hindrance in growth regulator supply and accumulation of harmful gases in the vicinity of fruit storage, harvest and marketing or during growth in fields are responsible for spoilage of fruits. The symptoms that appear due to physiological disorders seem disease-like, but alteration in environmental conditions is the only preventive measure to this problem.

Major loss to guava fruits is mainly due to field, transport and storage injuries so they become prone to fungal infections. Postharvest fungal attack spoils fruit during transit, storage and transportation, which affects the quality and marketability of damaged fruit, leading to economic loss. The presence of high levels of biochemicals and their low pH values cause fruits to perish due to fungal infection (Singh and Sharma 2007). Guava production is favoured if flowering and fruiting occur in the dry season, where the anthracnose, stylar-end rot are not allowed to infect fruits. The extent of fruit senescence determines the incidence of disease, which doubles with the ripening of fruits (Srivastava and Lal 2009). Fruits are perishable and unavoidably prone to numerous diseases, and require judicious preventive measures right in the orchards. Hence, the cumulative information regarding epidemiological studies and disease management shall be discussed in length on the basis of research work published by various workers. Pre-harvest disease management treatments with safer chemicals including Integrated Pest Management (IPM) and irradiation alone or combined with cold storage, growth retardants, and coating comes as good option for postharvest strategy to control considerable postharvest losses. Innovative harvesting can also help in minimizing the problem of postharvest damage.

18.2 Postharvest System of Guava

Judicious and careful handling of guava fruits is required to get good market price. Postharvest system of guava includes many steps from harvesting to marketing to be taken to minimize losses (Figure 18.2)

18.2.1 Harvesting

Timely harvesting of fruits helps in mitigating significant postharvest losses in fruit production. The maturity period is always directly related to the climacteric condition of the growing areas. The change in fruit colour from dark green to pale green or yellowish green, specific gravity, total soluble solids, acidity, etc. determine the maturity of fruit and are indicators of its quality and storage life. For better shelf life and long-distance travel, specific gravity between 1.00 and 1.02 is considered the best. Preferably guavas are hand-picked and harvested at 2–3 day intervals during rainy season, while a 4–5 day gap is required to harvest winter fruits. Fruit harvesting with the stalk along with one or two leaves is desirable to reduce friction while keeping in field containers. It is better not to mix immature, infected, damaged and diseased fruits for long life of healthy fruits. Traditionally fruits are hand-picked, but a notched stick with a bag is often used for harvesting. During the process of harvesting, falling of fruits on ground causes bruises and cracks and leads to compromised quality of produce. Mechanical injury of fruits and packaging them with healthy fruits makes them more vulnerable for pathogen entry, thereby leading to rots during different operations. Fruits should be plucked at proper maturity, keeping distance of transport in consideration to obtain optimum quality and shelf life to minimize the losses. The general practice of leaving plucked fruits on the ground covered by leaves results in latent infections, which in turn leads to rots during storage and ripening operations. Therefore, use of a tarpaulin, thick sheet of newspaper and mats are recommended with farm level disinfection if required.

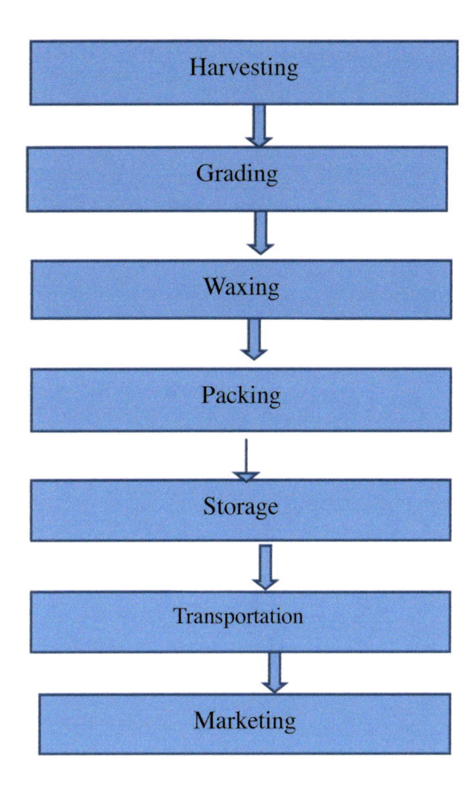

FIGURE 18.2 Steps in the postharvest system.

18.2.2 Grading, Waxing and Packing

Farmers practice very informal sorting and grading by removing highly damaged, unattractive and small fruits. Generally, guava fruits are sent to local markets packed in gunny bags, cloth bags or baskets, but for distant markets, baskets or boxes are used. To protect the fruits against thrust and jerks during handling and transportation, plastic crates are sometimes used. The most advisable popular method of the fruit packaging today is cell pack in CFB boxes.

Coating of guava with edible wax is known as waxing, which is an effective method to maintain freshness of guava fruits for a long period. Wax coating can reduce physical loss in weight and maintains colour, texture and aroma. It also minimizes spoilage during storage and transportation of fruits. Since guava is edible item, the standard quality government agency approved wax should be used to avoid health hazards (Gerg 2019).

18.2.3 Storage

On the basis of harvesting season of guava, the shelf life of guava fruits varies form 6–9 days to 2–4 days in winter and rainy season, respectively, at ambient temperature but it can be extended up to 14 days with defined ventilation conditions. The shelf life can be increased by lowering the temperature upto 10°C or below. The low temperature does not suit immature and ripe guava fruits, although hard, green and mature fruits without colour break can stay longer.

18.2.4 Transportation

The transport of fruits in countries like India is done via trucks, which is not suitable for delicate perishable commodities like guava. Globally, the quality of the produce is of paramount importance. Therefore, introduction of reefer vans specially designed with the least possible thrusts and having differential temperature systems could provide some relief from damage during long transportation. The use of palletization of CFB boxes can also contribute in minimizing the loss.

18.2.5 Marketing

After proper handling of guava produce, it should reach the market in a timely fashion to avoid a glut situation resulting in crashing of price. Appropriate market information systems and infrastructure are needed for movement of fruits. Strengthening of processing industry infrastructure may lead to immediate significant use of guava produce and indirectly contribute to save spoilage of this perishable item.

18.3 Postharvest Diseases of Guava

As of now, 177 postharvest diseases have been reported in guava, of which 167 are caused by fungal pathogens, and of these 91 invade fruits and deteriorate their biochemical constitution (Misra AK 2007). Environmental factors viz., temperature and relative humidity affect the growth of postharvest fungi (Figure 18.1).

18.4 Postharvest Fungal Diseases

18.4.1 Fruit Canker

18.4.1.1 Symptoms

Appearance of rust or brown coloured necrotic spots on green fruits is the initial indication of fruit canker which becomes advanced with increased lesions, and infection penetrates into the pulp (Arya 1993; Aulakh et al. 1998). Fruit canker is a common postharvest diseases of guava and rarely affects the leaves (Mathew 2010; Rao et al. 2012). In the beginning the spots are unbroken, circular and scabby, 2–4 mm in diameter, on the affected area but later they tear open the epidermis in a circinate manner. *Pestalotia psidii* is major postharvest fungi next to *Rhizopus stolonifer* (Srivastavaand Lal 2009). The temperature of 25–35°C along with high humidity was reported amicable for maximum disease development (Arya 1993).

18.4.1.2 Causal Organism

Pestalotiopsis is a genus of ascomycete fungi known for invading plant species. It is diverse group of species primarily classified on the basis of conidial pigmentation, size, septation and appendages (CABI Biosciences 2001; Nag Rag 1993; Lisa et al. 2006). *Pestalotiopsis* spores have mostly four pigmented median cells with two to four apical appendages arising as tubular extensions from the apical cell and a centric basal appendage (Jeewon et al. 2002). It is considered to be a complex genus which cannot be classified to the species level due to swift changes in growth rate, morphology of conidia and fruiting bodies within species (Kaushik et al. 1972).

18.4.1.3 Disease Cycle and Epidemiology

Dormant mycelium is the primary source of infection. A *kajji* bug (*Helopeltis antonii*) punctures the young fruit sucking juice and facilitates the entry of pathogen. The conidia are considered the secondary source of windborne infection.

18.4.2 Fruit Rot

Guava fruits are soft and pulpy due to the presence of high water content in tissues which makes them more vulnerable to rot. On the basis of symptoms appearing on fruits, Lim and Manicom (2003) classified fruit rots into dry and soft rot. Shallow and superficial necrotic lesions in the form of either small spots or patches were classified as dry rots.

Dry rots are mostly caused by *Cladosporium* sp., *Diplodia theobromae*, *Guignardia* sp., *Macrophoma* sp. and *Macrophomina phaseolina*. Brown water-soaked areas that expand and extend into the fruit mesocarp or into the seed cavity are called soft rots. The fungal pathogens that cause soft rots include *Aspergillus niger*, *R. stolonifer*, *Mucor heamalis*, *Fusarium solani*, etc. Different fungal strains for guava rot have been identified in different parts of the world and reviewed by Fatima (2019) with varying symptoms. Fruit rots can be observed at varying times: (i) on mature and ripening of fruit prior to harvest (ii) at harvest and (iii) after harvest during transportation and storage.

18.4.3 Anthracnose

18.4.3.1 Symptoms

The small brown lesions on the fruit surface later turn into bigger and sunken patches. High humidity favours accumulation of a salmon-coloured conidial mass in the centre part of the lesions (Figure 18.3). The disease proliferates after fruit ripening due to some changes such as increments in sugar content which is a great temptation for pathogens as base food

FIGURE 18.3 Anthracnose disease in guava.

material, a decrease in natural antifungal compounds and phytoalexins, lowering of phenolics that are potent to render resistance against pathogens, and rupturing of fruit tissues that facilitates the entry of pathogen (Prusky 1996).

18.4.3.2 Causal Organism

Colletotrichum gloeosporioides is the major cause of severe production loss in guava (Lim and Manicom 2003). The conidia of *Colletotrichum gloeosporioides* are normally a septate, however development of septa occur at the time of germination due to mitotic division (O'Connell et al. 1993). The germ tube directly enters via penetration and pegs from the appressoria develop in different positions over the conidia and show varied shapes such as lobed, spherical and irregular (Moraes et al. 2013).

18.4.3.3 Disease Cycle and Epidemiology

Colletotrichum species enters guava by three possible pathways: through natural openings or directly via appressorium and/or through wounds (Jeffries et al. 1990). Most of *Colletotrichum* species prefer the presence of appressorium to directly penetrate the host tissues, while in some other species the appressorium is optional or not required. For colonization, the pathogen may utilize two distinct strategies, i.e., intracellular hemibiotrophy or subcuticular intramural settlement, or a combination of both (Bailey and Jeger 1992). Immediately after penetration, formation of spherical infection vesicle takes place to develop hyphae from this vesicle and subsequently colonize other host cells quickly. Initially, biotrophical growth of *C. gloeosporioides* was observed without inducing any symptoms or changes in the necrotrophic phase to induce symptoms and cause decay of the host cells (O'Connell et al. 2000). *In vitro* studies proved that infection begins with the development of appressoria from conidia within 6 h time of launching on fruits, and infection vesicles establish hyphae in epidermal and parenchymatous cells after 120 h (Moraes et al. 2013).

18.4.4 Guignardia Fruit Rot

18.4.4.1 Symptoms

On maturation, the guava fruit becomes more susceptible to rot which manifests with the appearance of dark brown water-soaked spots (Lal et al. 1980). The circular, purple-to-brown lesions (0.5–1.0 cm) spread over all surfaces and turn black and shrunken with a high degree of severity. According to Soares-Colletti et al. (2015), temperature above 10°C and below 35°C favours black spot development.

18.4.4.2 Causal Organism

Guignardia psidii Ullasa & Rawal belongs to family *Botryosphaeriaceae*. The fungus is dense, submerged, brown-to-black mycelium with septate hyphae. Ascocarps were perithecial and unilocular with prominent long necks; the stromatic ascocarp walls are composed of multiple cell layers, and are deeply impregnated with pigments on the outside. Conidia are unicellular, hyaline, obovate, 6–12 (7.5) × 5–8 μm, having a gelatinous envelope and apical appendage as outgrowth. Appendages are hyaline, smooth, tubular, and 3.0–4.5 × 0.5 μm. Single ascospores bearing single conidia developed in ascigerous states inform about the homothallic nature of the fungus. The ascigerous state has been identified as *G. psidii* and the anamorph as *Phyllosticta psidiicola* (González and Rondón 2005).

18.4.4.3 Disease Cycle and Epidemiology

Guignardia fruit rot is very common and serious in Hawaii in the case if fruits that are not harvested in a timely manner and left to over-ripen on the tree or on the ground. The moist weather favours the production of ascospores which keep releasing throughout spring and summer to cause continuous primary infection on fruits. Optimum growth of black rot fungus requires warm weather, while cool weather diminishes growth. Drizzle, fog or rain for 2–3 days is favourable for infection (Ries 1999). The conidial spore can germinate within 10–15 h in rainwater droplets and transfer spores to different plant parts, especially susceptible young leaves, as well as to over-ripened fruits.

18.4.5 Phomopsis Fruit Rot

18.4.5.1 Symptoms

The estimated loss due to this rot is 60%. The affected area of fruits shows many dark brown spots which turn dark black with time (Figure 18.4). A minimum temperature required for this fungus growth is approximately 10°C; near 25°C is considered optimum while 35°C observed maximum to limit its growth.

18.4.5.2 Causal Organism

The disease is caused by *Phomopsis destructum* Rao, Agarwal & Saksena. *Phomopsis* sp. belongs to the genus *Ascomycete* and member of family *Valsaceae*. The fungus has thin-walled, septate, branched and hyaline mycelium. Pycnidia are dark coloured, ovoid, leathery to carbonaceous and thick walled, being 300–600 μm diameter. Two types of spores are produced by this fungus. stylospores are cylindrical, long and bent at one side like the hook of walking stick, and measures 11–30 × 1.5–2.0 μm. Pycnidiospores are comparatively small, hyaline and fusiform at 2.8–8.0 × 2.0–2.8 μm.

FIGURE 18.4 Symptom of Phomopsis fruit rot of guava.

18.4.5.3 Disease Cycle and Epidemiology

The disease appears in December and spreads quickly during January to March. Initially, it affects the fruit near the stalk, but infection can spread all over in amicable environmental conditions (Misra 2005). In severe cases, the fruit pulp becomes soft, brown and drops from the plant.

18.4.6 *Pestalotiaolivacea* Rot

18.4.6.1 Symptoms

Brownish-coloured watery lesions initiate infection on fruits, which later change to russet-coloured spots. Black pinhead-like acervuli start appearing on white fluffy growth of mycelium after a week (Maharachchlkumbura et al. 2011).

18.4.6.2 Causal Organism

The disease is caused by *Pestalotia olivacea* Guba. The conidial characters can be utilized to distinguish taxa as per the recent available molecular data. The genera have much complex taxonomy but remain ambiguous or debatable. Pestalotiopsis is the member of the family *Amphisphaeriaceae* and molecular studies reveal its monophyletic origin (Jeewon et al. 2004). The size of conidia is a good morphological marker to divide species into diverse groups. In most cases, the length and width are considered good taxonomic markers due to stability on various growth media and further generations.

The white colony of fungi appears with hyaline branched, septate 2.5- to 3.5-µm wide hyphae. The conidia of variable colour have four to five cells, having dark brown middle cells with thick wall and septa. The middle cell is surrounded by two hyaline and conical end cells on either side. Conidia size is 2.5–32.5 × 4.5–6.0 µm. A short pedicel of 23–25 µm long is threadlike with pointed ends keeping setulae 2–3 cells on either side of the conidia attached to hyaline cell (Maharachchlkumbura et al. 2011).

18.4.7 Diplodia Soft Watery Rot

18.4.7.1 Symptoms

The brownish discolouration usually starts from the stem end of the fruit and slowly spreads downwards in an irregular wavy manner to cover the entire surface of the fruit. As infection advances, numerous small pycnidia are produced that cover the entire surface of the fruit. Finally, soft and watery rot is caused by the pathogen. Pycnidia production in *B. theobromae* takes place at 25°C, followed by conidial germination maximum at 30°C (Patel and Pathak 1995). It has been observed that storage of fruits at 10–15°C prevents them from decay for 10 days.

18.4.7.2 Causal Organism

Botryodiplodia theobromae Pat. is the fungus belongs to genus mold. There is absence of a defined sexual state, thus it is a part of the fungi imperfecti.

18.4.7.3 Disease Cycleand Epidemiology

At relative humidity of 100%, the highest spore germination is noticed which favours the spread of disease (Patel and Pathak 1995). The entry point of the fungus is the stem end of the fruit. Wind or rain facilitates spores to reach out to the stem ends of very small fruits growing on trees. The scars at the stem end allow the fungus to enter the fruit once it reaches it. Humid areas are more prone to this serious disease.

18.4.8 Aspergillus Soft Rot

18.4.8.1 Symptoms

Aspergillus awamori mainly causes postharvest decay, which starts affecting unripe fruits and results in 10–15%fruit loss. Small, circular, water-soaked spots start exhibiting on fruit skin, which get bigger and turn russet-brown colour with time. Development of the disease is immensely favoured by high humidity and temperature. However, very young, green fruits are safe from the attack of soft rot disease. Softening of the diseased lesion in middle leads to development of reddish brown to clove brown coloured conidial heads surrounded by a white circular ring mixed with citron yellow coloured fungus mycelia.

18.4.8.2 Causal Organism

Several species of *Aspergillus* have been identified which cause guava fruit soft rot, viz. *Aspergillus awamori*, *A. wentii* and/or *A. niger*. *Aspergillus* is characterized with the presence of conidia and is also known as a conidial fungus, which denotes the asexual state of fungi. A few, however, are known to have a teleomorph (sexual state) in the Ascomycota. All members of the genus *Aspergillus* are members of the Ascomycota which has been proved with DNA evidence.

18.4.8.3 Disease Cycle and Epidemiology

The entire fruit surface shows black mouldy growth due to rapid spread of disease. The fruit rot completely covers the fruit within 10–12 days, and emission of a fermented odour from soft pulpy tissue indicates severity of disease (Mathew 2010; Rao et al. 2012; Amadi et al. 2014; Salau et al. 2015).

18.4.9 Rhizopus Rot

18.4.9.1 Symptoms

Soft rot affects green or fully ripe fruits in the orchard. Oily and water-soaked lesions appear initially during the early disease development stage, with distinct sunken lesion margins. The fruit flesh rapidly reduces to a semi-solid state in a few days with extending lesions. The epidermis generally does not show any damage.

18.4.9.2 Causal Organism

Rhizopus stolonifer (Fr.) Lind has white hyphae, which enters through the surface of the root and gives rise to a large number of brown-black sporangiophores (34 μm diam. × 1000–3500 μm length) to support a sporangium (100–350 μm diam.). The sporangium produces sporangiospores (4–11 μm diam.) that are ovoid, single celled and brown in colour. Sporangiospores are considered primary inoculums and release passively on breakdown of the outer layer of the sporangium. Sexual recombination is rare and occurs between the mycelium of two compatible strains on contact. Progametangia fuse to develop to ganitangia and produce thick-walled zygospores, which germinate into sporangiophores with a single sporangium at the terminal end. *R. stolonifer* prefer smooth wounds to bruised/crushed tissue wounds, but evidence is needed to prove this.

18.4.9.3 Disease Cycle and Epidemiology

Insects and fruit flies frequently visit the orchards and play major role in spreading infection but have not been fully investigated. Lesions are the point of infection, which facilitates aerial hyphae development that swiftly extends over the lesions and covers them with a sparse white to grey mycelium. Initially, white coloured sporangiophores and sporangia develop densely at the point of infection but later turn black in colour. Even after infection, fruits do not fall from trees until harvested manually or follow the course of natural maturation to dislodge. Mucor-rotted fruits can be distinguished from *Rhizopus* rot due to the presence of abundant yellow mycelia and sporangia, while *Rhizopus* rot is characterized with sparse aerial mycelium with dark grey to black sporangiophore and sporangia on fruit surface (Ooka 1980).

18.5 Minor Postharvest Diseases

18.5.1 Drechslera Fruit Rot

Causal organism: *Drechslera halodes* f. sp. *destrcutum*.

Fruits show rounded, small, dark brownish spots with light yellow margins. The advancement of disease is manifested by increase in size of spots that leads to soft rot of fruit, which transudes an unpleasant odour (Lal et al. 1980). This dematiaceous (dark-walled) mould belongs to the fungi imperfecti due to lack of known sexual state.

18.5.2 Dry Rot or Charcoal Rot

Causal organism: *Macrophomina* sp.

Dark black coloured and nearly circular necrotic regions appear on the fruit. Slowly, when infection reaches the pulp, the colour changes to black (Singh and Thakur 2005).

18.5.3 Beltrania Rot

Causal organism: *Beltrania rhombica*.

The disease appears on ripened fruits in the form of water-soaked spots which makes the fruit black within 6–7 days due to deposition of conidial mass on the fruit surface (Fatima 2019). It is mitosporic fungus, which has not yet been extensively explored for structure and spread. It is air-borne and requires special media to grow.

18.5.4 Myrothecium Soft Rot

Causal organism: *Myrothecium roridum*.

This fungus infects fruit in the pre-harvest stage but affects the fruit much later after harvest. The slightly sunken, olive-grey spots that appear on the fruit are small and circular and later increase in size. Affected fruit is covered quickly with white woolly mycelium with viscid spore mass having dark green colour (Singh and Thakur 2005). Dissemination of spores takes place at the time of plucking.

18.5.5 Physalospora Fruit Rot

Causal organism: *Physalospora psidii*.

The disease starts catching fruits from the stylar end with gradual creeping towards the stem end region. Lesions that appear in the beginning are circular, brown and water-soaked, and later changes to dark brown.

18.5.6 Mucor Rot

Causal organism: *Mucor haemalis* Wehmer.

Initiation of water-soaked lesions, within a week slowly covers the entire fruit with yellowish, hairy mass of fungal fruiting bodies along mycelia. Yeasty odour emits from diseased fruits. Wounding of fruit is necessary to cause infection; therefore, caution at the time of postharvest handling can be very useful to save the fruits from this wound parasite of guava fruit (Kunimoto et al. 1977).

18.5.7 Alternaria Rot

Causal organism: The disease is caused by *Alternaria alternata*.

Alternariol, tenuazonic acid, altenuene and alternariol monomethyl ether are the mycotoxins secreted during *in vitro* culture of *A. alternata* associated with rot (Ammar and El-Naggar 2014). The rot has been reported by several researchers (Naureen et al. 2009; Srivastava and Lal 2009; Ammar and El-Naggar 2014; Salau et al. 2015; Zahra 2016).

18.6 Management of Postharvest Diseases

In addition to getting bumper crop produce, ignorance about not keeping the harvest safe from postharvest pathogens may ruin the supply chain. Fruits must be harvested carefully to avoid physical damage which can provide the way to primary infection. Careful postharvest operations are also imperative to maintain the integrity of natural barriers of fruit intact and ward off infection by secondary pathogens. Initially, the use of chemicals was the only option to get rid of postharvest diseases (PHDs) but with the developments in science and technology bio-control has taken up a prime seat.

The use of certain chemicals and chemical fungicides has lost its relevance due to environmental concerns as well as their nature of disintegration after sometime, leaving crops vulnerable to pathogen attack. Residual toxicities, non-biodegradability and the nature of adding to pollution brought the use of synthetic fungicides at halt. The incidence of the Anar butterfly can be controlled by the "remove and destroy" policy of affected fruits (Vikaspedia). Keeping fruits covered with polythene bags on reaching 5cm is the best option of avoidance (Misra and Gupta 2005). With the advancement of technology, the integrated approach has been developed combining different control measures. The use of bio-inoculants and essential oils started taking front seat as curative and preventive measures to check postharvest losses. At postharvest stage, bacteria, yeasts and fungi are in use as the main antagonistic organisms to deal with pathogens. Druvefors (2004) explained that the use of one living organism to control another is the central theme of biological control. The main controlling strategies of antagonists include production of lytic enzymes, creating competition for nutrients and space, the induction of resistance and antibiosis (Nunes 2012; Spadaro and Droby 2016).

Treatment of guava fruits with carbendazim and benomyl inhibited radial growth of *P. psidii* (Rao et al. 2012). Since insect wounds predispose fruit to infection, spray the young fruits after pollination with a suitable systemic insecticide (dimethoate, 2 mL/L) will take care of the infection. Spread of the disease can be checked by three or four sprayings with Bordeaux mixture 1.0% or copper oxychloride 0.2% at 15-day intervals. Summer irrigation and nutritional management reduce disease incidence. The homoeopathic drugs potassium iodide and arsenic oxide completely inhibited spore germination. They can also inhibit growth of the pathogen.

Nowadays natural plant- and animal-derived compounds are also part of biological control agents along with antagonistic microorganisms. Biological controls have been coming up as an alternative to synthetic fungicide treatments and achieving success using antagonistic microorganisms to handle both pre-harvest and postharvest diseases (Janisiewicz and Korsten 2002). Fravel (2005) and Chanchaichaovivat et al. (2007) have tested numerous microbial antagonists on different fruits and vegetables as pathogen control measures.

The use of essential oils as antifungal agents has been well documented in the literature (Pitarokili et al. 1999; Meepagal et al. 2003; Antunes and Cavaco 2010). Biologically active essential oils have potential as alternative and environmentally friendly options for disease management. Various essential oils were observed to be effective against a wide range of fungi due to their antifungal activity. Various botanicals and bio-control agents such as yeast and *Saccharomyces cerevisiae* have been used to control guava canker. Garlic extract has been used *in vitro* as a botanical and exhibited upto 30% inhibition in pathogens growth, while bio-control agents restricted growth upto 40%. It has been concluded for the study that the botanical garlic bulb extract, bio-control agent, *S. cerevisae* and fungicides like mancozeb and mancozeb + carbendazim, were effective in the management of guava canker disease (Surwade 2013). Timely exposure of green fruits of guava with ethanol and acetic acid vapours for 2 h has been identified as an effective safety measure to reduce microbial load and provide improved shelf life.

The efficacy of neem essential oil has been evaluated against six postharvest fungi (PHFs) viz. *Pestalosia psidii*, *Gloesporium psidii*, *Rhizoctonia solani*, *Fusarium* sp., *Alternaria alternata* and *Geotrichum candidum* and found effective when used proportionally to its concentration in controlling wilt. It is difficult to control the radial growth of the fungus with application of 1% oil, but an increase of upto 5–10% in concentration was observed to inhibit fungal activity moderately against all the postharvest pathogens. The most significant reduction in mycelium growth was noticed at 50% concentration of neem oil (Nongmaithem 2014).

Aspergillus niger strain AN-17, *Trichoderma harzianum* and *Penicillium citrinum* were also used as bioinoculents to control PHDs of guava. An option of intercropping with *Curcuma domestica* or *Tagetes erecta*, use of resistant rootstock, separate basin irrigation and avoidance of tillage during July to November were seen as good alternatives for managing growth of microorganisms. A combination of two or more practices for disease management can be used successfully to control guava wilt. Standardization of bioagent application time and development of techniques for easy and cheap availability of bioagents can bring positive change in postharvest handling (Misra 2003).

Five yeast strains (*Metschnikowia lunata* Y-1209, *Pichia anomala* Moh 93, *Pichia guilliermondii* Moh 10, *P. anomala* Moh 104 and *Lipomyces tetrasporus* Y-115) were evaluated for their antagonistic property against *Botryodiplodia theobromae* which causes Diplodia rot of guava. *In vitro* studies have revealed that both strains of *P. anomala* were most effective to control pathogen due accumulation of extracellular matrices around the pathogen hyphae at the time of direct interaction between *B. theobromae* and *P. anomala* Moh 93. The antagonistic yeast penetrates the hyphae of *B. theobromae* and destroys the cells. *In vivo* both strains of Moh 93 and Moh 104 reduced the incidences of disease by 39.1% and 50.0%, respectively. Use of specific strains of *P. anomala* is emphatically recommended as a safe and effective bio-control measure against Diplodia rot of guava (Mohamed and Saad 2009). The potential of different strains of the yeast *P. anomala* as an effective safeguard against Diplodia rot of guava fruit has been proved. Use of two other antagonistic yeasts (*Candida* sp. and *Rhodotorula* sp.) was individually observed to be very effective in controlling the predominant pathogens *P. expansum* and *P. psidii* (Deeba et al. 2008). Sometimes using a combination of a recommended fungicide in low concentrations and other physical treatments with these biological control agents can further improve their efficacy to control pathogens. Precolonization of the fruit surface with the antagonists just before harvesting reduces the chance of infliction of wounds with pathogen (Janisiewicz and Korsten 2002). Secondary infection in guava fruit can be prevented through this approach.

High temperature and humid climatic conditions in India make guava fruits more prone to rot caused by various fungi. To avoid fruit rot, cautious handling procedures should be followed to minimize physical damage and thereby diminish occurrence of infection. Effective cold-chain management can considerably diminish the development of fruit rot efficaciously, as storage at low temperature suppresses fungal pathogens activities (Prusky et al. 2000). Storage of fruits at temperatures below a critical limit may cause chilling damage, which further enhances the susceptibility of fruit to decay. The onset of disease can be supressed for sometime by subjecting fruits to postharvest treatment that causes delay in ripening.

Treatment of guava fruits with 1-MCP has been reported to reduce the decay incidence in guava fruit (Singh and Pal 2009). On the other hand, the potential reduction in fruit delay has been noticed by exposing mature unripe guava fruits to gamma irradiation (0.25–0.5 kGy) (Barkai-Golan and Follett 2017; Singh and Pal 2009). Storage of tropical fruits for long time necessitates treatment with a potent fungicide to maximize the life of produce in storage. High intrinsic resistance to infection and unfavourable environmental conditions make the use of fungicides most effective and do not allow pathogens to grow (Eckert and Ogawa 1985). Hence, postharvest fungicide treatment and optimum storage conditions can be combined to diminish the fruit rot incidence in guava.

For good postharvest health of fruits, pre-harvest care is critical. Good orchard management practices and pre-harvest fungicide sprays are essential to lessen latent infections when fruits reach maturity. In most countries, the choice of fungicide type is subject to the climatic conditions of the specific area, with approved treatment doses (Singh 2009). According to Omayma et al. (2010), treatment with calcium chloride and lemon grass fumigation together is effective to retard *Rhizopus* rot establishment.

The application of phosphites [phosphite K (40% P_2O_5 and 30% K_2O) and phosphite-Ca (10.7% P_2O_5, 3.89% Ca, and 0.5% B)] including carbendazim as reference, acetyl salicylic acid, calcium chloride ($CaCl_2$), 1-methylcyclopropene (1-MCP) and hot water (HW) applied singly or in combination control anthracnose (*Colletotrichum simmondsii*) during the postharvest stage. The effectiveness of treatment was proved with reduction in diameter of anthracnose lesion and number of lesions without compromising fruit quality (pH, fresh weight loss, total soluble solids and acidity) when treated with HW at 47°C for 20 min. The 1-MCP treatment alone was least effective but when combined with phosphite, it effectively alleviated the anthracnose effect on fruits (Cruz et al. 2015).

18.7 Future Directions

The prime issues that take priority are quality of fresh fruit, safety, nutrition and flavour after harvesting. Environmental issues are of concern and should be heeded. Adoption of a triple bottom-line concept, viz. social, economic and environmental aspects, can help to provide more balanced handling of postharvest problems in guava (Watkins and Akman 2005). Guava fruit is highly perishable in nature and infested by many species of fruit fly, which makes it vulnerable to fruit rots. It also cannot tolerate phytosanitary treatments, being sensitive to high and low temperatures. Application of edible coatings (Watkins 2008) and approval of using 1-MCP and DCA storage are some options that have revolutionized postharvest handling to some extent. The replacement of synthetic fungicides viz. imazalil, thiabendazole, pyrimethanil, and fludioxonil by more resilient and eco-friendly postharvest disease management practices such as utilization of natural compounds and bioinoculent/biological control. The combination of two or more techniques is being cited as a good option for fruit safety. Infrastructural constraints cause a great loss to growers due to deterioration in quality of the produce and directly affecting availability of fruit at the consumer level. Exploration of doses, time of treatment and different growing and postharvest conditions to determine the efficacies of various bio-control agents is underway against a range of pathogens.

REFERENCES

Amadi JE, Nwaokike P, Olahan, GS, Garuba T (2014) Isolation and identification of fungi involved in the postharvest spoilage of guava (*Psidium guajava*) in Awka metropolis. Int J Eng Appl Sci 4(10): 7–12.

Ammar MI, El-Naggar MA (2014) Screening and characterization of fungi and their associated mycotoxins in some fruit crops. Int J Adv Res 2(4): 1216–1227.

Antunes MDC, Cavaco AM (2010) The use of essential oils for postharvest decay control. A review. Flavour Fragr J 25: 351–366.

Arya A (1993) Tropical Fruits – Diseases and Pests. Kalyani Publishers, New Delhi, p. 82.

Aulakh KS, Sokhi SS, Ratan GS (1998) Postharvest diseases of tropical and subtropical fruits. In: Neeta S, Mashkoor AM (Eds.). Postharvest Diseases of Horticulture Perishables. International Book Distributing Co., Lucknow, India, pp. 42–76.

Bailey JA, Jeger MJ (1992) *Colletotrichum:*Biology, Pathology and Control. CABI, Wallingford, p. 388.

Barkai-Golan R, Follet PF (2017) Irradiation for Quality Improvement, Microbial Safety and Phytosanitation of Fresh Produce. Academic Press, San Diego, CA, p. 302.

Burzo I, Delian E, Craciun C (2001) Ultrastructural changes induced by the physiological disorders in fruits and vegetables. Acta Hortic 553: 255–256.

CABI Biosciences database (2001). Published online.

Chanchaichaovivat A, Ruenwongsa P, Panijpan B (2007) Screening and identification of yeast strains from fruits and vegetables: Potential for biological control of postharvest chilli anthracnose (*Colletotrichum capsici*). Biol Cont 42: 326–335.

Corata U De (2019) Improving the shelf-life and quality of fresh and minimally-processed fruits and vegetables for a modern food industry: A comprehensive critical review from the traditional technologies into the most promising advancements. Crit Rev Food Sci Nutr 60(6): 940–975.

Cruz AF, Medeiros NL, Benedet GL (2015) Control of postharvest anthracnose infection in guava (*Psidium guajava*) fruits with phosphites, calcium chloride, acetyl salicylic acid, hot water, and 1-MCP. Hortic Environ Biotechnol 56: 330–340.

Deeba K, Lal AA, Bashyal BM, Gupta P (2008) Ecofriendly management of the postharvest diseases of guava (*Psidium guajava* L.). Crop Res (Hisar) 35: 131–134.

Druvefors U (2004) Yeast bio-control of grain spoilage mold. Doctoral dissertation, Swedish University of Agricultural Sciences. Retrieved January 17, 2005, from Epsilon dissertations and graduate. Theses archive website: <http://dissepsilon.slu.se/archive/00000552/

Eckert JM, Ogawa JM (1985) The chemical control of postharvest diseases: Subtropical and tropical fruit. Annu Rev Phytopathol 23: 421–454.

Fatima S (2019) Introduction to major postharvest diseases of guava. J Drug Deliv Therapeut 9(4): 591–593.

Fravel DR (2005) Commercialization and implementation of biocontrol. Annu Rev Phytopathol 43: 337–359.

Gerg N (2019) Handbook on Postharvest Management of Guava. Central Institute for Subtropical Horticulture, Lucknow, India.

González MS, Rondón A (2005) First report of Guignardia psidii, an Ascigerous state of Phyllosticta psidiicola, causing fruit rot on guava in Venezuela bio-control of postharvest disease (Botryodiplodia theobromae) of guava (Psidium guajava L.) by the application of yeast strains. http://krishikosh.egranth.ac.in/handle/1/5810007118.

Janisiewicz WJ, Korsten L (2002) Biological control of postharvest diseases of fruits. Annu Rev Phytopathol 40: 411–441.

Jeewon R, Liew ECY, Hyde KD (2002) Phylogenetic relationships of *Pestalotiopsis* and allied genera inferred from ribosomal DNA sequences and morphological characters. Mol Phylogen Evol 25: 378–392.

Jeewon R, Liew ECY, Hyde KD (2004) Phylogenetic evaluation of species nomenclature of *Pestalotiopsis* in relation to host association. Fungal Divers 17: 39–55.

Jeffries P, Dodd, JC, Jeger MJ, Pumbley RA (1990) The biology and control of Colletotrichum species on tropical fruit crops. Plant Pathology. London, 39;343–366.

Kaushik CD, Thakur DP, Chand JN (1972) Parasitism and control of *Pestalotia psidii* causing cankerous disease of ripe guava fruits. Indian Phytopathol 25: 61–64.

Kunimoto RK, Ito PJ, Do WH (1977) Mucor rot of guava fruits caused bv *Murrn hiemalis*. Trop Agric (Trinidad) 54(2): 185–IR7.

Lal B, Rai RN, Arya A, Tiwari DK (1980) A new soft rot of guava. Natl Acad Sci Lett 3: 259–260.

Lim TK, Manicom BQ (2003) Diseases of guava. In: Ploetz RC Diseases of Tropical Fruit Crops, CABI, Oxfordshire, pp. 275–290.

Lisa MK, Maile EV, Francis TZ (2006) Identification and characterization of *Pestalotiopsis* spp. causing scab disease of guava, *Psidium guajava*, in Hawaii. Plant Dis 90(1): 16–23.

Maharachchlkumbura SM, Guo L, Chukeatirote E, Ali B, Kevin H (2011) Pestalotiopsis—morphology, phylogeny, biochemistry and diversity. Fungal Diversity 50(1): 167–187.

Mathew S (2010) The prevalence of fungi on the post harvested guava (*Psidium guajava* L.) in Aksum. Int J Pharmac Sci Res 1(10): 145–149.

Meepagal KM, Kuhajek JM, Sturtz GD, David Wedge E (2003) Vulgarone B, the antifungal constituent in the steam-distilled fraction of *Artemisia douglasiana*. J Chem Ecol 29: 1771–1780.

Misra AK (2003) Guava diseases–their symptoms, cause and management. In: Naqvi SAMH (Ed.). Diseases of Fruits and Vegetables: Diagnosis and Management, Vol. 2, Kluwer Academic Publishers, the Netherlands, pp. 81–119.

Misra AK (2004) Guava diseases – their symptoms, causes and management. In Book: Diseases of Fruits and Vegetables, Diagnosis and managment-Volume-II; Naqvi SAMH (Ed.). Kluwer Academic Publishers, pp. 81–119. www.springeronline; www.ebooks.kulweronline.com

Misra AK (2004a) Guava diseases – their symptoms, causes and management. In: Naqvi SAMH (Ed.). Diseases of Fruits and Vegetables, Volume II, Springer, Dordrecht.

Misra AK (2004b) Important diseases of guava in India with special reference to wilt. Naqvi SAMH (Ed.). Diseases of Fruits and Vegetables, Volume II, Kluwer Academic Publishers, the Netherlands (eBook), pp. 81–119.

Misra AK (2005). Important diseases of guava in India with special reference to wilt. Souvenir. 1st Int Guava Symp., December 5–8, CISH, Lucknow, pp 75–90.

Misra AK (2007) Present status of important diseases of guava in India with special reference to wilt. Acta Horticulturae 735: 507–523.

Misra AK, Gupta VK (2005) Faster and efficient multiplication of bioagents through use of polythene sheet for management of guava wilt. Intl Conf on Plastic Culture and Precision Farming, Nov. 17–21, 2005, New Delhi, p. 329.

Mohamed H, Saad A (2009) The bio-control of postharvest disease (Botryodiplodia theobromae) of guava (*Psidium guajava* L.) by the application of yeast strains. Postharvest Biol Technol 53: 123–130.

Moraes SRG, Tanaka FAO, Nelson M (2013) Histopathology of *Colletotrichum gloeosporioides* on guava fruits (*Psidium guajava* L.). Rev Bras Frutic Jaboticabal – SP 35(2): 657–664.

Nag Rag TR (1993) Coelomycetous Anamorphs with Appendage-Bearing Conidia. Mycologue Publications, Waterloo, Ontario, Canada.

Naureen F, Humaira B, Viqar S, Jehan A, Syed E-H (2009) Prevalence of postharvest rot of vegetables and fruits in Karachi, Pakistan. Pak J Bot 41(6): 3185–3190.

Nongmaithem N (2014) Control of postharvest fungal diseases of guava by essential oil of *Azadirachta indica*. Indian J Hill Farm 27(1): 135–139.

Nunes C (2012) Biological control of postharvest diseases of fruit. Eur J Plant Pathol 133: 181–196.

O'Connell RJ, Uronu AB, Waksman G, Nash C, Keon JPR, Bailey JA (1993) Hemibiotrophic infection of *Pisum sativum* by *Colletotrichum truncatum*. Plant Pathol London, 42: 774–783.

Omayma MI, Eman AA, Abd-Allah ASE, El-Naggar MAA (2010) Influence of some post-harvest treatments on guava fruits. Agric and Biol J North Am 1(16): 1309–1318.

Ooka, J1 1980. Guava fruit rot caused by Rhizopus stolonifer in Hawaii. Pi. Dis., 64 (4): 412–413

Patel KD, Pathak VN (1995) Development of Botrydiplodia rot of guava fruits in relation to temperature and humidity. Indian Phytopathol 48: 86–89.

Pitarokili D, Tzakou O, Couladis M, Verykokidou E (1999) Composition and antifungal activities of essential oils of *Salivia pomifera* subsp. *calycena* growing wild in Greeces. J Esential Oil Res 11: 655–659.

Prusky D (1996) Pathogen quiescence in postharvest diseases. Annu Rev Phytopathol 34: 413–434.

Prusky D, Freeman S, Dickman MB (2000) *Colletotrichum*: Host specificity, pathology and host-pathogen interaction. APS Press, St. Paul, MN, p. 393.

Rao A, Abhilasha K Lal, Sobita S, Chandra S, Singh R, Singh L (2012). Post-harvest fungal diseases of guava: A brief review. Management of canker (*Pestalotia psidii*) disease of guava (*Psidium guajava* L.). Ann Plant Protec Sci 20(2): 383–385.

Ries SM (1999) Reports on Plant Diseases: Black Rot of Grape. Integrated Pest Management at the University of Illinois. 2010. http://ipm.illinois.edu/diseases/series700/rpd703/

Salau IA, Shehu K, Kasarawa AB, Sambo S, Shahida AA (2015) Fungi associated with postharvest rot of commonly consumed fruits in Sokoto Metropolis Nigeria. J Adv Bot Zool 3(3): 1–4.

Singh SP (2009) Prospective and retrospective approaches to post harvest management quality of fresh Guava (*Psidium guavaja L.*) fruit in the supply chain. Fresh Produce 4(special issue1): 36–48.

Singh SP (2011) Guava (Psidium guajava L.). In: Yahia EM (Ed.). Postharvest Biology and Technology of Tropical and Subtropical Fruits, Woodhead Publishing, Sawston, Cambridge, pp. 213–246e.

Singh SP, Pal RK (2009) Ionizing radiation treatment to improve postharvest life and maintain quality of fresh guava fruit. Radiation Phy Chem 78: 135–140.

Singh D, Sharma RR (2007) Postharvest diseases of fruit and vegetables and their management. In: Prasad D. (Ed.). Sustainable Pest Management, Daya Publishing House, New Delhi, India.

Soares-Colletti AR, Fischer IH, Afonseca Lourenco S (2015) The effects of temperature and wetness duration on the development of *Guignardia psidii* in guava fruit naturally infected. Australas Plant Pathol 44(4): 413–418.

Spadaro D, Droby S (2016) Development of bio-control products for postharvest diseases of fruit: The importance of elucidating the mechanisms of action of yeast antagonists. Trends Food Sci Technol 47: 39–49.

Srivastava R, Lal A (2009) Incidence of post-harvest fungal pathogens in guava and banana in Allahabad. J Hortic Sci 4(1): 85–89.

Surwade KA (2013) Studies on canker disease of guava (*Psidium guajava*) caused by *Pestalotiopsi psidii*. Thesis, Mahatma Phule Krishi Vidyapeeth Rahuri.

Thompson AK (2010) Postharvest chemical and physical deterioration of fruit and vegetables. Chemical Deterioration and Physical Instability of Food and Beverages. A volume in Woodhead Publishing Series in Food Science, Technology and Nutrition. pp. 483–518

Watkins CB (2008) Postharvest ripening regulation and innovations in storage technology. Acta Hortic 796: 51–58.

Watkins CB, Ekman JH (2005) How postharvest technologies affect quality. In: Ben Yehpshua S (Ed.). Environmentally Friendly Technologies for Agricultural Produce Quality, 1st edn, CRC Press, Roca Raton, FL, pp. 437–481.

Zahra IE-G (2016). Isolation and identification of fungi associated with fruits sold in local markets. Int J Res Studies Biosci 4(11): 61–64.

19

Postharvest Diseases of Litchi and Their Management

Vinod Kumar,[1] and Ajit Kumar Dubedi Anal[2]
[1]*ICAR-National Research Centre on Litchi, Mushahari, Muzaffarpur, Bihar, India*
[2]*Amity Institute of Microbial Technology, Amity University, Uttar Pradesh, India*

CONTENTS

19.1 Introduction

Litchi (*Litchi chinensis* Sonn.) is an economically important fruit crop that belongs to the family *Sapindaceae* and is commercially cultivated in subtropical and tropical regions of the world. Postharvest diseases generally occur during the period from harvest to consumption. The postharvest diseases reduce the value of fruits because during the time that these fruits pass from field to consumers they become infected. Losses of postharvest disease caused by infection of several microorganisms may reach approximately 50% (Jiang et al. 2001) through transportation and storage (Swarts and Anderson 1980). During the supply of litchi fruits, postharvest disease loss was observed in the range of 35–44% due to decay (Kumar et al. 2016b). Postharvest decay in litchi is caused by *Alternaria alternata*, which has also been reported from Australia (Johnson et al. 2002) and Pakistan (Alam et al. 2017). Fungal pathogens such as *Alternaria* sp., *Aspergillus* sp., *Penicillium* sp., *Rhizopus* sp., *Colletotrichum* sp. and *Botryodiplodia* sp. attach to litchi fruits through injuries that occur at the time of transportation and the managing process (Jiang et al. 2003; Scott et al. 1982). Postharvest loss due to fungal infection varied at different stages from farmer field to consumer. In India, the highest percent of postharvest losses was found to be due to *A. alternata* (86.7%) followed by *C. gloeosporioides* (7.2%), *A. flavus* (3.9%) and *A. niger* (2.2%) in different storage conditions during two consecutive years (Kumar et al. 2016b). Earlier, Prasad and Bilgrami (1974) reported that the most common pathogens responsible for postharvest diseases are *Aspergillus*, *Cylindrocarpon*, *Botryodiplodia* and *Colletotrichum*. *C. gloeosporioides* is the major pathogen which responsible for postharvest decay or losses in China (Liu et al. 2006). In Africa, the *Phomopsis*, *Pestalotiopsis*, *Penicillium*, *Alternaria*, *Botryosphaeria* and *Fusarium* sp. are leading fungal pathogens responsible for decay in litchi (de Jager et al. 2003). Some *Penicillium* spp. have also been reported to cause postharvest losses in litchi; among them,

Penicillium expansum is the main pathogenic species (Jacobs and Korsten 2004).

From the harvesting process to reaching the market and purchasing by consumer, pathogen may infect the fruit at any stage. During the above process cracks may develop and provide an easy pathway for entry of pathogens, resulting in disease developing and starting decay of fruits (Sivakumar et al. 2005).

SO_2 fumigants are often used in industry for commercially grown litchi to deal with fruit decay issues (Swarts 1985). However, consumer preferences, pesticide residue regulations and environmental concerns have threatened the use of chemicals to manage postharvest diseases in fruits (Droby et al. 1991). Therefore, some new techniques and new methodologies can be developed that may be helpful in the management of postharvest disease of litchi. With the help of some bio-control agents including fungi and bacteria there may be an emerging and promising alternative to control postharvest diseases in litchi (Sivakumar et al. 2010; Liu et al. 2013). Bio-control methods are eco-friendly and nontoxic for humans and can also be easily produced at the commercial level. This chapter gives an account of postharvest diseases of litchi, their symptoms and aetiology, and various options for managing them.

19.2 Postharvest Diseases

A wide range of pathogens including bacteria, fungi and viruses attack litchi fruits during postharvest stage and actively grow in fruits. Postharvest diseases in litchi fruit are mainly caused by microscopic fungi in litchi resulting in decay or rotting. This is a serious problem for litchi production. Numerous postharvest diseases of litchi have been reported from India (Prasad and Bilgrami 1973). Several fungal pathogens, such as *Alternaria*, *Colletotrichum*, *Botryodiplodia*, *Aspergillus*, *Fusarium* and *Penicillium* sp, have been reported from India (Prasad and Bilgrami 1973; Kumar et al. 2016b), and due to improper handling become the reason for postharvest fruit diseases. A mean pathological loss of 23.2% in 2012 and 17.9% in 2013 has been reported in the supply chain of litchi in India by Kumar et al. (2016b), the highest being at the retail level. Postharvest diseases of litchi are given in Table 19.1, and symptoms and aetiology of some important diseases are described in the next section.

19.2.1 Symptoms and Aetiology of Postharvest Diseases

19.2.1.1 Alternaria Rot

Alternaria rot in litchi is caused by *Alternaria alternata*. Primary rotting symptoms of Alternaria rot on fruit surface are light to dark coloured water-soaked lesions 2–6 mm in diameter.

These lesions enlarge progressively and coalesce, covering the whole fruit and leading to rot disease. With disease advancement, the decayed portions of fruit become depressed and the rot starts to penetrate deeply into the pulp. After disease development, infected fruit starts to ferment and emits a foul odour. Sometimes the outer surface of fruit cracks and

TABLE 19.1

Major Postharvest Fungal Pathogens of Litchi Fruit

Sr. No.	Postharvest Disease	Pathogen	Reference(s)
1.	Alternaria rot	*Alternaria* sp.	Coates et al. (1994); Scott et al. (1982); Kumar et al. (2016b)
2.	Aspergillus rot	*Aspergillus* spp.	Roth (1963); Scott et al. (1982); Prasad and Bilgrami (1973)
3.	Botryodiplodia rot (Stem end rot)	*Lasiodiplodia theobromae* (synonyms: *Botryodiplodia theobromae* and *Diplodia natalensis*)	Coates et al. (1994); Roth(1963); Prasad and Bilgrami (1973)
4.	Botrytis rot	*Botrytis cinerea*	Chandel and Chauhan (2017)
5.	Cladosporium rot	*Cladosporium* sp.	Scott et al. (1982)
6.	Anthracnose	*Colletotrichum* spp.	Coates et al. (1994); Scott et al. (1982); Prasad and Bilgrami (1973)
7.	Curvularia rot	*Curvularia* sp.	Roth (1963); Prasad and Bilgrami (1973)
	Cylindrocarpon rot	*Cylindrocarpon tonkinense*	Prasad and Bilgrami (1973)
8.	Dothiorella rot	*Dothiorella* sp.	Coates et al. (1994)
9.	Fusarium rot	*Fusarium* spp.	Roth (1963); Prasad and Bilgrami (1973)
10.	Geotrichum rot (Sour rot)	*Geotrichum candidum*, *G. ludwigii*	Tandon and Tandon (1975); Wild (1992); Tsai and Hsieh (1998)
11.	Blue mould rot	*Penicillium* spp.	Scott et al. (1982); Prasad and Bilgrami (1973)
12.	Peronophythora rot	*Peronophythora litchii*	Qu et al. (2001); Jiang et al. (2001); Vien et al. (2001)
13.	Pestalotia rot	*Pestalotiopsis* sp.	Prasad and Bilgrami (1973)
14.	Phoma rot	*Phoma* sp.	Coates et al. (1994)
15.	Phomopsis rot	*Phomopsis* sp.	Coates et al. (1994); Scott et al. (1982)
16.	Rhizopus rot	*Rhizopus* sp.	Roth (1963); Scott et al. (1982)

FIGURE 19.1 Some pathogens associated with litchi fruit decay/rot in storage (left to right): *Colletotrichum gloeosporiodis*, *Alternaria alternata* and *Aspergillus niger*.

comes in contact with fungi which cover the entire surface with dark mycelium (Figure 19.1 & Figure 19.2).

19.2.1.2 Anthracnose

The causal organism of anthracnose disease of litchi fruit is *Colletotrichum gloeosporioides*. The rotting symptoms are slightly depressed black dots that are initially small in size with circular shape and appear on rind of fruits. These dots later coalesce and form a large spot with irregular shape. After pathogen establishment (5–6 days later) acervuli (dot-like fungal bodies) start to appear on infected area. These acervuli are small asexual fruit bodies found on the outer surface of the host plant which have deep multicellular setae (Figure 19.1).

19.2.1.3 Aspergillus Rot

Aspergillus rot is caused by three species of *Aspergillus* as given below.

- ***Aspergillus niger*:** The rotting symptoms of *Aspergillus niger* on litchi fruit first appear on the stalk end of the fruit, which shows light brown lesions encircling the stalks and later change to dark brown. After 4–5 days, distinct black conidial heads develop on these lesions (Figure 19.1).
- ***Aspergillus flavus*:** Litchi fruits infected with *Aspergillus flavus* first attack fruit margins, causing brownish discoloration which later turns velvety and becomes slowly depressed. After the disease has been established (6–8 days), fruiting bodies (conidial mass) appear and cover the entire or most of the fruit. After complete disease development or in advanced stages of disease, the infected portion looks dirty and becomes yellowish green and powdery in appearance.
- ***Aspergillus verisicolor*:** In the beginning of disease, the infected part of fruit shows circular patches that subsequently coalesce and become irregular. These disease symptoms start to change into velvety appearance and finally become brownish green in colour. In the advanced stage (after 6–8 days), some round, green cleistothecial bodies are also seen.

FIGURE 19.2 Symptoms developed on litchi fruits infected by *A. Alternata*, stored at low temperature in cold storage.

19.2.1.4 Geotrichum Rot (Sour Rot)

Geotrichum rot of litchi fruit is caused by *Geotrichum candidum* and was first observed on Allahabad (Jamaluddin et al. 1975). Symptoms start as a soft area on the fruit which turns soft and loose, emitting a foul odour. White waxy growth of the fungus appears on the rotten surface. Injury on the fruit surface facilitates infection.

19.2.1.5 Pestalotia Rot

Initially Pestalotia rot symptoms appear as light brown discoloured spots but after 5–6 days convert into russet coloured. As the disease develops, or with age, these spots start to enlarge and become depressed in the centre. After establishment of the pathogen, small and single fruit acervuli bodies of gregarious black colour appeared on infected regions with viscid spore masses.

19.2.1.6 Botryodiplodia Rot

Botryodiplodia rot of litchi is caused by *Botryodiplodia theobromae*. The disease symptoms first appeared as water-soaked lesions and brownish discolouration on the rind. It progresses irregularly on either side but more often towards the stalk end of the fruit. In general, rotting of the pulp is comparatively faster than that of the rind. Under humid conditions, light grey cottony mycelia are observed over the infected tissues. Pycnidia are seen later as minute bodies.

19.2.1.7 Cylindrocarpon Rot

Postharvest rot of litchi due to *Cylindrocarpon tonkinense* Bugn was reported by Prasad and Bilgrami (1972). When the pathogen first attacks the fruit, disease symptoms of Cylindrocarpon rot appear as water-soaked lesions in large size. After infection, the pathogen progressively enters deep into the pulp and the fruit starts to crack. Finally, these cracks become completely covered with thick mycelium of pathogen.

19.2.1.8 Peronophythora Rot

Fruit has white to creamy semitranslucent flesh and red colour in pericarp, but it is easily subjected to the infection of *Peronophythora litchii* Chen, and market value is lost. It is a major disease of litchi in Southern China. The fungus is a facultative necrotrophic plant pathogen. Infected flower of litchi with peronophythora become brown and covered with whitish masses of sporangia and sporangiophores. The pathogen is polycyclic in nature, resulting in the dispersal of inoculums over an extended time period and a wide area during the season. Young and ripe fruits, pedicels and leaves are also attacked, their tissues turning brown and dying, especially in periods of heavy rain (Hall 1989).

19.2.1.9 Botrytis Rot

Botrytis rot was observed on fruits of litchi from Himachal Pradesh, India. Greyish fungal mass entirely covers the stem end of the litchi fruit. Small spots appear on the surface of mature fruits. Grey mycelia and conidiophores develop on the diseased fruits and a severity of 20.5% was reported by Chandel and Chauhan (2017). The disease is caused by *Botrytis cinerea* Pers. (Fr.).

19.3 Conventional Strategies for Postharvest Disease Prevention and Control

19.3.1 Pre-harvest Treatment

The treat of litchi fruits using aureofungin fungicide, 500 ppm, was reported effective to control the fruit rot (Prasad and Bilgrami 1975). Huang and Scott (1985) reported that fruit infected by some fungal pathogens that cause rotting disease can be controlled by dipping the fruits into ambient temperature and then in hot benomyl (0.05–0.2%) for 2–16 min. These fruit samples can be supplied in the market after air drying and packaging in PVC bags.

19.3.2 Heat Treatment

This is a common method of controlling rotting disease in litchi and other fruits causes by several pathogens after postharvest. Heat works in two ways; primarily it may completely kill the pathogen, and secondly it can delay the development of pathogens. This method is applied in the market or in industries to maintain fruit quality and shelf life. Hot water is the most common and efficient method compared to hot air because it is less injurious for fruit health. To control postharvest disease, this method is also used with the combination of fungicide. 'Tai So' and 'Wai Chee' litchi cultivars are stored at 5°C for 4 weeks and then the quality can be maintained (retain the look or disease control) by vapour heat treatment methods at 45°C core temperature for 42 min (Jacobi et al. 1993). However, vapour heat treatment was reported to be cultivar dependent, and Taiwanese litchi cultivars were found to respond well to vapour heat treatment, suggesting that these cultivars were more heat tolerant. The polyphenol oxidase and peroxidase activity response to discoloration of fruit pericarp can be controlled by hot water treatments at 45°C for 5–10 min (Souza et al. 2010).

19.3.3 Radiation Treatment

The incidence of postharvest disease of litchi is also reduced by the use of ionizing radiation. Ionizing radiation, particularly gamma radiation, with low temperature storage is recommended for postharvest storage of litchi. Use of radiation is another substitute for sulphur dioxide (SO_2) fumigation for short-time storage of litchi. Fruit rot incidence caused by *Colletotrichum* sp. can reduced with the dose of 75 and 300 Gy irradiation and fruit quality is also maintained during storage at 5°C for 3weeks (Mc Lauchlan et al. 1992). Storage for short duration with irradiation up to 1 kGy dose with low temperature is also a good substitute for SO_2 fumigation. However, this treatment does not help in extending the storage life beyond 16 days (Hangantileke et al. 1993).

19.3.4 Chemical Control

Fungicides can be used for postharvest disease control. Use of fungicide totally depends on target pathogen, type of fungicide and time of pathogen attack and application of fungicide. Generally, fungicides are recommended for pathogens that attack fruits before harvest. The fungicides are generally applied in repetition on growing season/crops except systemic fungicides which are applied with a strategic plan. Generally fungicide in postharvest is used to control infection or remove infection already present on the host, and works to defend against pathogens during transportation activities. The fungicide should also be effective during the dormant stage of pathogens. Systemic fungicides are generally used for this purpose, but after fungicides are applied, it is not known the extent of their penetration. If the infection is already in the host or is found after harvest in this condition, fungicide can be used to suspend the pathogen growth. How successful fungicides are in doing this depends largely on the extent to which infection has developed at the time of fungicide application and how effectively the fungicide penetrates the host tissue.

In this sense, most of the "fungicides" used postharvest are actually fungistatic rather than fungicidal in their action under normal usage. Commonly, benzimidazole (e.g. benomyl and thiabendazole) and the triazole (e.g. prochloraz and imazalil) fungicides are used as spray or dip. *Penicillium* and *Colletotrichum* postharvest pathogens are generally controlled by use of the benzimidazole group of fungicides. Several fumigants such as ammonia, carbon dioxide, and ozone are used to control postharvest disease, but sulphur dioxide is particularly used to control grey mould disease of litchi. Many countries are using fruit wraps or box liners impregnated with the biphenyl fungicide to control *Penicillium* rot.

Benomyl, iprodione, prochloraz, imazalil and thiabendazole fungicides are evaluated to control postharvest disease in litchi (Scott et al. 1982; Botha et al. 1988). Swart (2009a) reported that dipping of litchi fruit in 3% hydrochloric acid for 120 s and either 405 ppm prochloraz or 300 ppm fludioxonil is the best method to treat postharvest decay and colour retention in litchi. Brown et al. (1984) reported that dipping of infected fruit in 0.125–0.25 g/l prochloraz solution at 25°C for 5 s to 5 min controls postharvest decay. In another study, postharvest decay in cv. Madras was reduced by treating fruits with SO_2+ prochloraz 0.05% and packing in everfresh bags (Swarts 1990). Litchi fruit cv. Mauritius decay was controlled by treating with Nustar (flusilazol), Punch C (250 g flusilazol + 125 g carbendazim) and Punch X (125 g flusilazol + 250 g carbendazim). However, flusilazol caused a browning problem in the fruits. Schutte et al (1990) used two slow-releasing paper sheets impregnated with sodium metabisulphite placed in the outer edge of litchi fruits in the polythene bags for controlling 100% decay with only 2% browning. In cv. Madras of lichi, excellent decay and browning control was achieved with a dip in 1% Semperfresh (a sucrose ester preparation) + 600 ppm benomyl + 250 ppm prochloraz in combination with one foam sheet (as buffer) and one slow-release sheet impregnated with sodium metabisulphite and storage at 3.5°C. Fruit rot during storage caused by *G. candidum* was found to be controlled by dipping fruits for 10 min in nickel chloride (0.1%), nickel sulphate (0.1%) or captan (0.2%) and then storing them at 3.8–5.5°C. However, only nickel chloride (0.1%) proved superior in maintaining the bright colour of the fruits (Yadav et al. 1984). In Australia, treatment of fruits with hot benomyl at 52°C for 2 min and air dried prior to PVC wrapping and storage at 5°C was found useful in controlling the postharvest rots of litchi by *Cladosporium* and *Fusarium* spp. (Sittigul et al. 1994). Dipping fruits in water at 55°C for 30 s followed by a dip in either 3% HCl or 10% oxalic acid for 2 min also showed good control of postharvest decay and retention of pericarp colour (Swart 2009b). For litchi anthracnose (*Colletotrichum* spp.), 120 mg/L chlorine dioxide (ClO_2) solution treatment was effective in inhibiting the disease and improving the quality of litchi fruits (Wu et al. 2011). Litchi fruits that were dipped in hot benomyl (0.1%) at 48°C for 1–3 min and 50°C for 1–2 min managed the postharvest diseases with the most acceptable fruit appearance and with least skin colour loss (Wong et al. 1991).

Because of worldwide consumer concerns, regulation changes and as per government guidelines or environment rules, fungicides cannot be used continuously in the future. Due to the nontoxic nature, bio-control agents are environmental friendly and a good alternative to control of decay pathogens. The crude extract of *Bacillus subtilis* is good source which effectively controls postharvest pathogens (Jiang et al. 2001).

19.3.5 Sulphur Fumigation

India, South Africa, Thailand, Israel and Madagascar effectively use SO_2 to treat postharvest disease of litchi. Paull et al. (1995) reported that browning of fruit pericarp is reduced by treatment with hot water, low temperature and use of SO_2 fumigation alone or in combination with an acid dip. Fruits are usually treated on-farm by either burning sulphur powder or fumigating directly with SO_2 gas. In this process, fruits are placed in a closed chamber where 50–100 g sulphur per m^3 of air space is burned for 20–30 min (Kumar 2018). A major limitation with the use of SO_2 is the health risk associated with the sulphur residues remaining in the edible aril of sulphated litchi fruit. High relative humidity influences the movement and absorption of SO_2 negatively in dried fruit, leading to lower SO_2 residues. The lower the humidity and higher the temperature, the higher the residue recorded. The practice of leaving fruit overnight before fumigation leads to increased residue levels. Care should be taken to keep the time lapse between harvest and fumigation as short as possible (Lemmer and Kruger 2001). Browning is prevented since SO_2 results in the formation of colourless quinine–sulphite complexes and inhibits polyphenol oxidase activity, preventing the formation of quinines, which rapidly polymerize to form brown pigments (Macheix et al. 1990). Studies on using SO_2 sheets on the bottom of cartons of commercially fumigated fruits showed that these sheets can significantly reduce the incidence of fungal infection when stored under export simulation conditions (30 days at 1°C) followed by a shelf life phase of 12 days at 13°C. The sheets did not increase the sulphur residues in the aril of the fruit (Schoeman et al. 2003; Schoeman et al. 2009).

19.4 Emerging Technologies for Postharvest Disease Control

Heavy economic losses and a growing concern over food safety awareness and environment pollution have become the driving forces for developing alternative postharvest treatments that are environmentally friendly, safe and economically acceptable. Emerging technologies that have been studied in recent years include natural fungicides (fruit coatings) (Duvenhage 1993; Zhang and Quantick 1997), biological control agents (Korsten et al. 1993; Kumar et al. 2016b), alternative postharvest dip treatments (Sivakumar et al. 2005) and controlled atmosphere (Kumar 2018) and modified atmosphere packaging (Jiang and Fu 1999; Sivakumar and Korsten 2006). Harvesting fruits during the early morning hours (4.00–8.00 AM), avoiding mechanical injury, prompt pre-cooling (temperature 4°C RH 85–90%) and maintenance of optimum temperature and relative humidity during transportation help to prevent fruit rot in litchi (Kumar 2018).

19.4.1 Modified Atmosphere Packaging

Modified atmosphere packaging (MAP) is most important technique which is helpful in the reduction of postharvest losses. In this technique, the inner environment is different from that of normal air, usually consisting of reduced oxygen and increased carbon dioxide concentrations combined with a low temperature, which increase the product's shelf life. MAP technology has advantages for farm and industrial purposes because it is low cost and easy to implement and handle (Flores et al. 2004). The successful use of MAP is based on the specific permeation properties of polymer films to O_2 and CO_2 to generate atmospheres that are suitable for the postharvest life of many horticultural commodities (Persis et al. 2000). In MAP packaging, it is essential that the fruits are almost 100% disease free, since the pre-sorting of fruits before sale is not practicable in the large-scale marketing chain. Postharvest pathogens can act as inoculums and contaminate surrounding fruits in the packaging. In litchi fruit, this technology reduced postharvest diseases and maintained a high humid environment for the retention of fruit at low temperatures (De Reuck et al. 2009).

19.4.2 Biological Control

In the last few years, it has significantly come into notice that bio-control agents are used against antimicrobial activities to control the postharvest disease of litchi. Several microorganisms can be isolated from the surfaces of fruits, and these isolates screen against inhibition of pathogens that cause postharvest diseases. In the pathogen-antagonist interaction, several activities are involved, such as site exclusion, nutrient and space competition and antibiotic production, which suppress the disease development. Generally, it was noticed that antimicrobial agents work significantly when applied before the infection established. The litchi fruit pathogen *Peronophythora litchii* was controlled by the *Bacillus subtilis*. Besides postharvest disease control, it is also helpful in storage of fruits at 5°C for 30 days to increase shelf life (Jiang et al. 2001). Kumar et al. (2016a) reported that treatment with a novel isolate of *Bacillus subtilis* NRCL BS-01 significantly reduced fruit decay as well as percent disease index. The efficacy of NRCL BS-01 and chitosan in controlling fruit rots and enhancing shelf life was on par with carbendazim treatment. The *Candida tropicalis* YZ27 is an antagonistic yeast reported to work against postharvest decay and to maintain quality of litchi cv. Bombai (Zhimo et al. 2018). Antagonist yeast works significantly against microorganisms by rapid colonization on the fruit surface, resulting delayed decay and reduced disease incidence and severity at 28 ± 2°C, 78 ± 1% RH for 6days as compared to control fruit. Additionally the uses of C. *tropicalis* YZ27 also helps to prevent pericarp browning and anthocyanin pigments with weight loss of fresh fruit (Zhimo et al. 2018).

19.4.3 Natural Fungicides

Several plants and microorganism naturally produce some biochemical compounds which have antifungal and bacterial properties that can be used in postharvest diseases. Chitosan is a natural modified biochemical compound derived by deacetylation of chitin, a major component of crustacean shells (No and Meyers 1995). Chitosan is water-insoluble but soluble in weak organic acid solutions. Chitosan has antimicrobial activity against some fungi and bacteria that have the potential to reduce postharvest losses and maintain preservation of fruit (Sagoo et al. 2002). Chitosan coating is mostly used today in preservation because it changes the internal environment without causing anaerobic respiration and is more selectively permeable to O_2 than CO_2 (Bai et al. 1988). Zhang and Quantick (1997) reported that different concentrations (1% and 2% dissolved in 2% glutamic acid) of chitosan for coating can change into secondary metabolites that can be useful for preservation. Similarly, Jiang et al. (2005) also observed that the concentration of chitosan (2% in 5% acetic acid) increased the PPO activity and delayed the decrease in anthocyanin content. Thus, chitosan might form a protective barrier on the fruit surface, reducing weight loss and browning. One of the most beneficial effects of chitosan coating is control of postharvest losses during storage of litchi fruit (Jiang et al. 2005; Zhang and Quantick 1997).

19.5 Conclusions

Several fungal pathogens are responsible for postharvest disease in litchi. Some of these are produced after harvest during transport, and some infect before harvest and remain dormant until favourable conditions arrive for disease development. Generally, the pre-harvest factors may directly and indirectly responsible for the postharvest disease development, even in the case of infections initiated after harvest. Postharvest disease causes fruit yield loss in litchi, and these losses can be reduced or controlled by several techniques such as use of fumigation, hot water treatment, modification in transport and packing techniques and use of some bio-control agents. While developing strategies for postharvest disease control, it is imperative to take a step back and consider the production and postharvest handling systems in their entirety. Conventional methods may also play a major role in reducing or controlling postharvest disease because these techniques are eco-friendly and less harmful to humans and animals. On the other hand, decreasing the use of chemical in the agricultural sector or horticulture is a good initiative, to be replaced by development of new strategies like MAP, use of bio-control agents, and natural fungicides derived from plants or polymers from animal origins.

REFERENCES

Alam MW, Gleason ML, Amin M, Ali S, Fiaz M, Rehman A, Ahmed R, Khan AS (2017) First report of *Alternaria alternata* causing postharvest fruit rot of lychee in Pakistan. Plant Dis 101: 1041.

Bai RK, Huang MY, Jiang YY (1988) Selective permeability of chitosan-acetic acid complex membrane and chitosan-polymer complex membrane for oxygen and carbon dioxide. Polym Bull 20: 83–88.

Botha T, Schutte GC, Lonsdale JH, Kotze JM (1988) Swammegeassosieër met na-oesbederf by lietsjies en die beheer van na-oessiektes en verbruining met swamdoders en poli-etileenverpakking. S Afr Litchi Growers Assoc Yearb 1: 20–22.

Brown BI, Scot KJ, Mayer DG (1984) Control of ripe fruit rot of guava, lychee and custard apple by postharvest prochloraz dip. Singapore J Primary Indus 12: 40–44.

Chandel S, Chauhan P (2017) Botrytis rot on *Litchi chinensis*, a new record from Himachal Pradesh, India. Int J Curr Microbiol App Sci 6(8): 1315–1317.

Coates L, Johnson GI, Sardsud U, Cooke AW (1994) Postharvest diseases of lychee in Australia, and their control. ACIAR Proc 58: 68–69.

de Jager ES, Wehner FC, Korsten L (2003) Fungal post-harvest pathogens of litchi fruit in South Africa. S Afr Litchi Growers Assoc Yearb 15: 24–32.

De Reuck K, Sivakumar D, Korsten L (2009) Integrated application of 1-methylcyclopropene and modified atmosphere packaging to improve quality retention of litchi cultivars during storage. Postharvest Biol Technol 52(1): 71–77.

Droby S, Chalutz E, Wilson CL (1991) Antagonistic microorganisms as biological control agents of postharvest diseases of fruits and vegetables. Postharvest News Inf 2: 169–173.

Duvenhage JA (1993) Control of postharvest decay and browning of litchi fruit by sodium metabisulphite and low pH dips. S Afr Litchi Growers Assoc Yearb 5: 31–32.

Flores FB, Martínez-Madrid MC, Amor MB, Pech JC, Latche A, Romojaro F (2004) Modified atmosphere packaging confers additional chilling tolerance on ethylene inhibited cantaloupe Charentais melon fruit. Eur Food Res Technol 219: 431–435.

Hall G (1989) *Peronophthora litchii*. IMI Descriptions of Fungi and Bacteria, 1989 No. 98 Sheet 974, CABI, Wallingford.

Hangantileke SG, Noomhorn A, Upadhyay IP, Srinivas-Rao M (1993) Effect of irradiation and storage temperature on the shelf life and quality of Thai lychee. Postharvest Handling of Tropical Fruits, ACIAR Conference Proceedings, Chiang Mai, Thailand 50: 352–354.

Huang PY, Scott KJ (1985) Control of rotting and browning of litchi fruits after harvest at ambient temperature in China. Trop Hortic 62: 2–4.

Jacobi KK, Wong LS, Janet EG (1993) Lychee (*Litchi chinensis* Sonn.) fruit quality following vapour heat treatment and cold storage. Postharvest Biol Technol 3: 111–119.

Jacobs R, Korsten L (2004) Preliminary identification of *Penicillium* species isolated through the litchi export chain from South Africa to distribution centres in the Netherlands and United Kingdom. S Afr Litchi Growers Assoc Yearb 16: 34–39.

Jamaluddin M, Tandon P, Tandon RN (1975) Rot of fruits of litchi (*Litchi chinensis*) in marketing processes. Indian Phytopathol 28: 530–531.

Jiang Y, Li J, Jiang W. (2005) Effects of chitosan coating on shelf life of cold-stored litchi fruit at ambient temperature. LWT - Food Sci. Tec. 38:757–61

Jiang YM, Fu JR (1999) Biochemical and physiological changes involved in browning of litchi fruit caused by water loss. J Hortic Sci Biotech 7: 43–45.

Jiang YM, Yao L, Lichter A, Li J (2003) Postharvest biology and technology of litchi fruit. Food Agric Environ 2: 76–81.

Jiang YM, Zhu, XR, Li YB (2001) Postharvest control of litchi fruit rot by *Bacillus subtilis*. LWT Food Sci Technol 34: 430–436.

Johnson GI, Cooke AW, Sardsud U (2002) Postharvest disease control in lychee. Acta Hortic 575: 705–715.

Korsten L, De Villiers EE, De Jager ES, Van Harmelen MWS, Heitmann A (1993) Biological control of litchi fruit diseases. S Afr Litchi Growers Assoc Yearb 5: 36–40.

Kumar V, Anal AKD, Nath V (2018) Bio-control fitness of an indigenous *Trichoderma viride*, isolate NRCL T-01 against *Fusarium solani* and *Alternaria alternata* causing diseases in litchi (*Litchi chinensis*). Int J Curr Microbiol Appl Sci 7(3): 2647–2662.

Kumar V, Purbey SK, Alemwati P, Anal AKD, Nath V (2016a) Effect of some fructoplane antagonists and postharvest dip treatments on litchi fruit rots and shelf life. Int J Trop Agric 34(2): 333–343.

Kumar V, Purbey SK, Anal AKD (2016b) Losses in litchi at various stages of supply chain and changes in fruit quality parameters. Crop Prot 79: 97–104.

Kumar V (2018) Alternaria diseases of litchi. NRCL TB-13, ICAR-National Research Centre on Litchi, Muzaffarpur, Bihar, India.

Lemmer D, Kruger FJ (2001) Identification and quantification of the factors influencing sulphur dioxide residue levels in South African Export litchi fruit. Acta Hortic 558: 331–337.

Liu A, Chen W, Li X (2006) Developments of anthracnose on harvested litchi fruits and the effects of the disease on storage of the fruits. Acta Phyto Sin 33(4): 351–356.

Liu J, Sui Y, Wisniewski M, Droby S, Liu Y (2013) Review: Utilization of antagonistic yeasts to manage postharvest fungal diseases of fruit. Int J Food Microbiol 167: 153–160.

Macheix JJ, Fleuriet A, Billot J (1990) Fruit Phenolics. CRC Press, Boca Raton, FL.

Mc Lauchlan RL, Mitchell GE, Johnson GI, Nottingham SM, Hammerton KM (1992) Effect of disinfection dose of irradiation on the physiology of Taiso Lychee. Postharvest Biol Technol 1: 273–281.

No HK, Meyers SP (1995) Preparation and characterization of chitin and chitosan- a review. J Aquat Food Prod T 4(2): 27–52.

Paull RE, Reyes MEQ, Reyes MU (1995) Litchi and rambutan insect disinfection treatments to minimize induced pericarp browning. Postharvest Biol Technol 6: 139–148.

Persis E, Dvir O, Feygenberg O, Ben Aire R, Ackerman M, Lichter A (2000) Production of acetaldehyde and ethanol during maturation and modified atmosphere storage of litchi fruit. Postharvest Biol Technol 26: 157–165.

Prasad SS, Bilgrami RS (1972) Unrecorded *Cylindrocarpon* rot of litchi from Muzaffarpur, India. Indian Phytopathol 25: 459–461.

Prasad SS, Bilgrami RS (1973) Investigation on diseases of litchi. III. Fruit rots and their control by postharvest treatments. Indian Phytopathol 26: 523–527.

Prasad SS, Bilgrami RS (1974) Investigations on diseases of litchi: VI. Postharvest diseases of fruit in India. Plant Dis Rep 58: 1134–1136.

Prasad SS, Bilgrami RS (1975) Investigations on diseases of litchi: VII. Antibiotics for control of Aspergillus fruit rot. Proc Natl Acad Sci India B, 45: 40–42.

Qu HX, Sun GZ, Jiang YM, Zhu XR (2001) Induction of several pathogenesis-related proteins in litchi after inoculation with *Peronophythora litchii*. Acta Hortic 558: 439–442.

Roth G (1963) Postharvest decay of litchi fruit. Citrus and Subtropical Fruit Research Institute, Department of Agriculture, South Africa, Technical Communications 11: 1–16.

Sagoo S, Board R, Roller S (2002) Chitosan inhibits growth of spoilage micro-organisms in chilled pork products. Food Microbiol 19(2–3): 175–182.

Schoeman MH, Botha FA, Kruger FJ (2009) Evaluation of a ‘one sheet fits all carton types’ SO_2 sheet on HLH Mauritius and McLean’s Red litchi fruit. S Afr Litchi Growers Assoc Yearb 21: 36–39.

Schoeman MH, Lemmer D, Kruger FJ (2003) Preliminary evaluation of SO_2 sheets to prevent postharvest diseases of South African export litchi fruit. S Afr Litchi Growers Assoc Yearb 15: 6–9.

Schutte GC, Botha T, Kotze JM (1990) Postharvest control of decay and browning of litchi fruit by fungicide dips and paper sheets impregnated with sodium metabisulphite. S Afr Litchi Growers Assoc Yearb 3: 10–14.

Scott KJ, Brown BI, Chaplin GR, Wilcox ME, Bain JM (1982) The control of rotting and browning of litchi fruit by hot benomyl and plastic film. Sci Hortic 16: 253–262.

Sittigul C, Sardsud U, Sardsud V, Chaiwangsri T (1994) Effects of fruit maturity at harvest on disease development in lychee during storage. *Proc Post-harvest Handling of Tropical Fruits.* In: Champ BR, Highley E, Johnson GI (Eds.). Australian Centre for International Agricultural Research, pp. 9–14.

Sivakumar D, Korsten L (2006) Evaluation of the integrated application of two types of modified atmosphere packaging and hot water treatments on quality retention in the litchi cultivar McLean's Red. J Hortic Sci Biotechnol 81: 639–644.

Sivakumar D, Regnier T, Demoz B, Korsten L (2005) Effect of postharvest treatments on overall quality retention in litchi fruit during low temperature storage. J Hortic Sci Biotechnol 80: 32–38.

Sivakumar D, Terry LA, Korsten L (2010) An overview on litchi fruit quality and alternative postharvest treatments to replace sulfur dioxide fumigation. Food Rev Int 26: 162–188.

Souza AV, de Vieites RL, Kohatsu DS, Lima CPP (2010) Thermal treatment in lychee colour maintenance. Rev Bras Frutic 32: 67–73.

Swart SH (2009a) Evaluation of prochloraz and fludioxonil fungicides, alone and in combination to control postharvest decay on litchi fruit. S Afr Litchi Growers Assoc Yearb 21: 46–49.

Swart SH (2009b) Evaluation of prochloraz and fludioxonil fungicides, applied before and after acid dip treatments, to control postharvest decay and prevent pericarp browning on litchi fruit. S Afr Litchi Growers Assoc Yearb 21: 51–57.

Swarts DH (1985) Sulfur content of fumigated South African litchi fruit. Subtropica 6: 18–20.

Swarts DH (1990) The postharvest treatment of Madras Litchis. S Afr Litchi Growers Assoc Yearb 3: 21–22.

Swarts DH, Anderson T (1980) Chemical control of mold growth on litchi during storage and sea shipment. Inf Bull Citrus Subtrop Res Inst 98: 13–15.

Tandon MP, Tandon RN (1975) Rot of fruits of litchi (*Litchi chinensis*) in marketing processes. Indian Phytopathol 28: 530–531.

Tsai JN, Hsieh WH (1998). Occurrence of litchi sour rot and characteristics of the pathogens *Geotrichum candidum* and *G. ludwigii*. Plant Pathol Bull 7: 10–18.

Vien NV, Benyon FHL, Trung HM et al. (2001) First record of *Peronophythora litchii* on litchi fruit in Vietnam. Austra Plant Pathol 30: 287–288.

Wild BL (1992) Variations in sensitivity of isolates of *Geotrichum candidum* to the fungicide guazatine. Australas J Plant Pathol 21: 13–15.

Wong LS, Jacobi KK, Giles JE (1991) The influence of hot benomyl dips on the appearance of cool stored lychee (*Litchi chinensis* Sonn.). Sci Hortic 46: 245–251.

Wu B, Li X, Hu H, Liu A, Chen W (2011) Effect of chlorine dioxide on the control of postharvest diseases and quality of litchi fruit. Afr J Biotechnol 10(32): 6030–6039.

Yadav GR, Prasad B, Upadhayay J (1984) Effect of postharvest treatment on storage rot of litchi fruits. Progr Hortic 16: 351–352.

Zhang DL, Quantick PC (1997) Effects of chitosan coating on enzymatic browning and decay during postharvest storage of litchi (*Litchi chinensis* Sonn.) fruit. Postharvest Biol Technol 12: 195–202.

Zhimo VY, Saha J, Singh B, Chakraborty I (2018) Role of antagonistic yeast *Candida tropicalis* YZ27 on postharvest life and quality of litchi cv. Bombai. Curr Sci 114(5): 1100–1105.

20

Postharvest Diseases of Papaya and Their Management

Atul Kumar,[1] Jai Prakash Singh,[2] Pooja Singh[3] and Shaily Javeria[1]
[1]Division of Seed Science and Technology, ICAR-Indian Agricultural Research Institute, New Delhi, India
[2]Fruits and Horticultural Technology, ICAR-Indian Agricultural Research Institute (ICAR), New Delhi, India
[3]Department of Botany, DDU Gorakhpur University, Gorakhpur, Utter Pradesh, India

CONTENTS

20.1 Introduction

Papaya (*Carica papaya* L.) is a plant that produces fruits throughout the year. It needs a smaller area for the tree, comes to fruiting in 1 year, is easy to harvest, and next to banana it provides more income/ha. It has a great deal of medicinal and nutritional value. The dried latex of its immature fruits forms papain which is used in the manufacture of chewing gum, meat tenderizing, and in cosmetics. Papaya is most popular because of its (i) short duration, (ii) high yield, (iii) more income, (iv) high palatability and (v) continuous bearing. It is a rapidly growing, hollow stem, short lived, perennial, herbaceous plant. It is grown largely in Orissa, Andhra Pradesh, Karnataka, West Bengal, Assam, Tamil Nadu, Kerala, Madhya Pradesh, Bihar and Gujarat. It is highly sensitive to frost and waterlogging.

Papaya can be consumed at ripe and unripe stages (Mendoza 2007). It is rich in vitamins C, A and the presence of many polyphenols, making it a good source of antioxidants in the diet (Lim et al. 2007; Adetuyi et al. 2008). The fruit also contains proteolytic enzymes, viz. chymopapain and papain, regarded helpful for digestion, and it has many industrial applications. Because of these properties, papaya is a rich source of fat-free nutrition and balanced diet for all humans. The major part of the harvest is loose during transportation and handling, which leads about 30% losses (Ravindran et al. 2007). In papaya fruit, postharvest losses are significantly high due to fungal infections. Rotting in fruits of papaya may occur by various fungi (Sawant and Gawai 2011). Postharvest pathogens like *Phomopsis*, *Botryodiplodia theobromae* Pat., *Colletotrichum gloeosporioides* (Penz.) Sacc., *Pestalotiopsis*, *Stemphylium*, *Fusarium*, *Alternaria* and *Aspergillus* attack fruits and cause a great deal of damage to fruit production and quality (Chowdhury et al. 2014). In Bangladesh, some postharvest diseases of papaya viz. Fusarium rot, Anthracnose, Rhizopus rot, Aspergillus rot, stem-end rot and Penicillium rot were recorded (Hamim et al. 2014). Postharvest losses are caused by both external and internal factors. Helal et al. (2018) reported 19 species of fungi associated with postharvest diseases of papaya.

20.1.1 External Factors

Factors which lead to postharvest losses are as follows.

20.1.1.1 Mechanical Injury

Fresh vegetables and fruits are highly sensitive to mechanical injury due to their tenderness and high moisture content. Poor handling and improper and unsuitable packing during transportation are the cause of breaking, bruising, impact wounding, cutting and other forms of injury in fresh vegetables and fruits.

20.1.1.2 Parasitic Diseases

A major cause of postharvest losses in vegetables and fruits is the invasion by bacteria, fungi, insects, viruses and nematodes. Fresh produce is readily attacked by microorganisms that spread rapidly due to the absence their own defence mechanisms in the tissues and the ample amount of moisture and nutrients which bear their growth. It is an increasingly difficult task to control postharvest decay because while consumer concern for food safety is increasing, the number of safe pesticides available is rapidly declining.

20.1.2 Internal Factors

Physiological deterioration of vegetable and fruit tissues continues after harvest, and towards their physiological activity. Physiological disorders occur with significant low or high temperature injury, mineral deficiency or undesirable environmental conditions, like high humidity. Physiological deterioration may also occur naturally owing to enzymatic activity, leading to over-senescence and ripeness (Choudhary 2006).

20.2 Postharvest Diseases of Papaya

Most of the pathogens that infect papaya are fungal, although bacteria and insects are also reported during postharvest. Some important postharvest diseases of papaya are described below.

20.2.1 Anthracnose

20.2.1.1 Symptoms

Small spots appear first on the skin as brown superficial discolorations and gradually develop into circular, minutely sunken areas, 1–3 cm in diameter. Generally, on ripe mature fruits they appear water soaked. Gradually, the lesions coalesce on the margins of these spots and sparse mycelial growth often appears. Under humid conditions, encrustation of salmon-pink spores arranged in a concentric pattern also may develop on the surface of many older spots. Fruits later turn dirty brown and rot. Rot on papaya fruit is also initiated at the pre-harvest stage since the pathogen is generally present on other plant parts such as stem and petiole. Infection in the very early stage of fruit growth results in deformation and mummification, and at the mature stage in soft rot. Other types of symptoms on very young fruit at about 5–0 days old show white growth of fungus covering the entire fruit surface, while slightly bigger fruits show symptoms similar to cigar end disease of banana. On fruits, white concentric rings of spores over the infected part are observed (Figure 20.1). Later, the colour changes to pink and then becomes dark and finally turns black. *C. capsici* develop stem-end rot of premature fruits. Water-soaked dark

FIGURE 20.1 Symptoms of anthracnose disease of papaya.

brown lesions are formed, which later show cracking, resulting in weakening of pedicel, and finally the premature fruits drop. *C. acutatum* cause rotting of the taper end.

20.2.1.2 Causal Organisms

Colletotrichum caricae, *C. papayae*, *C. gloeosporioides*, *C. capsici*, *C. acutatum* and *Gloeosporium caricae*, *G. papaya*.

20.2.2 Stem-End Rot or Botryodiplodia Fruit Rot

20.2.2.1 Symptoms

The disease occurs on both green and ripe fruits. The development of disease is faster on half-ripe and ripe fruits than on green fruits. In the initial stages, the dark green water-soaked spot appears first. With disease development, the affected area becomes shriveled, dark brown and marked into concentric zone with numerous pycnidia. These spots are mainly surrounded by water-soaked dark green area. The infection may involve from one-third to half of the whole surface of the fruit. When the stalk of the fruit is infected, the fruit drops down.

20.2.2.2 Causal Organism

Botryodiplodia theobromae Pat.

20.2.3 Macrophomina Fruit Rot

20.2.3.1 Symptoms

The rotting is generally observed on half-grown to mature fruits. Tiny water-soaked spots appear as circular specks on the surface of the fruit. The spots enlarge, become sunken and the pathogen advances deep into the tissues of the fruit, causing rotting and disintegration. The fruits become dark brown to black with loss of moisture. The affected tissues harden and become dotted with small sclerotia of the pathogen. When cut open, the pulp has a brownish black colour and has dark mycelial growth of the fungus (Figure 20.2). The fruit stalk, when infected, results in the fall of fruit.

20.2.3.2 Causal Organism

Macrophomina phaseoli (Maubl.) Ashby

20.2.4 Phomopsis Fruit Rot

20.2.4.1 Symptoms

On the surface of the diseased fruit water-soaked spots develop, which increase in size towards maturity. The entire area becomes pulpy and soft and gives the appearance of soft rot. The rotted portion becomes dark brown to black in colour and cracks at a late stage. The part surrounding the diseased area of the fruit develops water soaking and the rotted portion becomes raised and turns white. The fruits can get infected at all stages of their growth until ripening.

20.2.4.2 Causal Organism

Phomopsis caricae papaya Petrak and Cif.

20.2.5 Rhizopus or Watery Fruit Rot

20.2.5.1 Symptoms

Very watery fruit rot, or Rhizopus rot, appears in the form of water-soaked spots on the skin of fruits, which enlarge quickly involving the large area of the fruit. The flesh becomes watery and soft but not discoloured otherwise. In the later stages, the

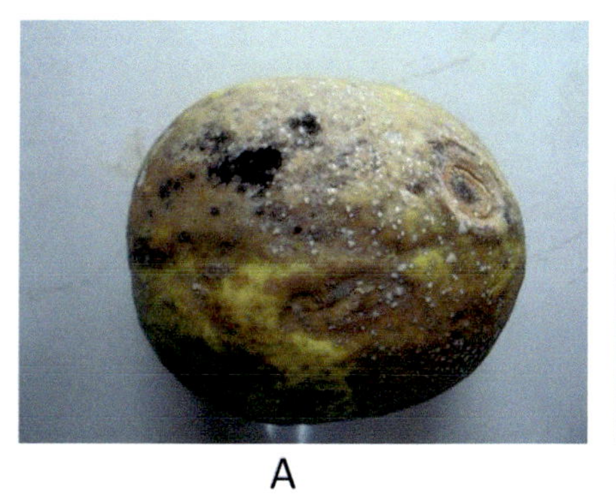

A

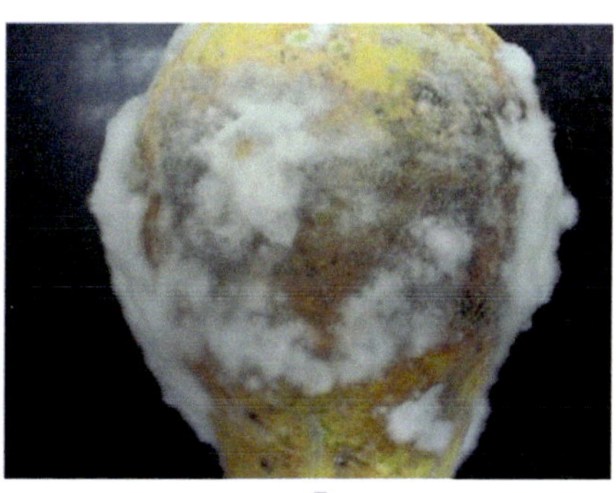

B

FIGURE 20.2 Symptoms of Macrophomina fruit rot of papaya. (A) Initial stage. (B). Mature stage of infection.

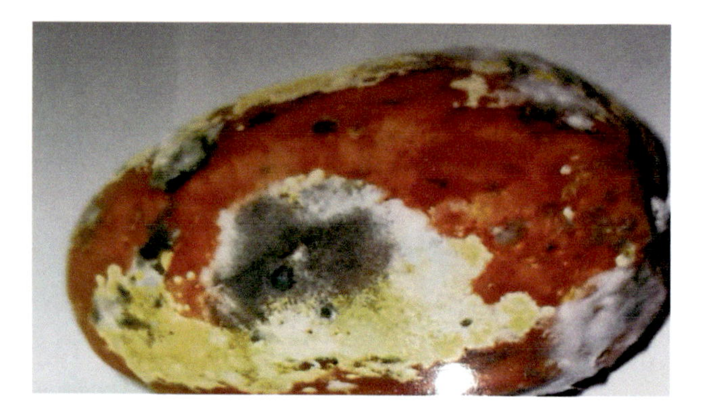

FIGURE 20.3 Symptoms of Rhizopus rot and other fungal infection on papaya fruits during storage.

FIGURE 20.4 Symptoms of fusarium rot disease of papaya.

surface of the collapsing fruit strands is covered by coarse grey or white mould (Figure 20.3).

The rot swiftly destroys the complete fruit and rapidly spreads to other fruits. The affected fruits emit a foul odour. Green papaya is also affected by *R. stolonifer*.

20.2.5.2 Causal Organism

Rhizopus stolonifer (Syn. *Rhizopus nigricans*).

20.2.6 Fusarium Fruit Rot

20.2.6.1 Symptoms

Large and slightly depressed lesions are formed due to fusarial rot. The lesions appear as water-soaked spots which later develop into soft depressed areas with scanty fungal growth.

20.2.6.2 Causal Organism

Fusarium equiseti Corda.

20.2.7 Fusarium Soft Rot

20.2.7.1 Symptoms

The disease initiates with water-soaked lesions, and shortly circular depressed spot develop which increase rapidly. Under moist conditions, white mycelial fungus growth appears on the surface of the fruit. The whole fruit rots within a week and a fermented odour emits from the soft pulpy tissue of the fruit.

20.2.7.2 Causal Organism

Fusarium moniliforme Sheldon.

20.2.8 Fusarium White Rot

20.2.8.1 Symptoms

White coloured rotting on the lower half of the fruit due to *F. pallidoroseum* (cooke) Sacc. was recorded from Lucknow (Figure 20.4).

20.2.8.2 Causal Organism

Fusarium pallidoroseum (Cooke) Sacc.

20.2.9 Ceratocystis Black Rot

20.2.9.1 Symptoms

The black rot disease appears as brown water-soaked appearance on the infected tissue, which gradually covers the whole fruit. A black crust composed of dark spores of the fungus develops on the infected surface within 6–7 days of infection. Rotted fruits become soft and pulpy with advancement of the disease.

20.2.9.2 Causal Organism

Ceratocystis paradoxa Dade.

20.2.10 Hyalodendron Soft Rot

20.2.10.1 Symptoms

The soft rot disease appears white to light brown coloured small irregular necrotic lesions over the ripe fruit. The infection is severe in the case of ripe fruits and results into 25–30% loss during storage and marketing. Artificial inoculation of the fungus shows disease symptoms in unripe fruits as well but the severity of infection is much less. The disease is highly favoured by damp conditions of storage.

20.2.10.2 Causal Organism

Hyalodendran sp.

20.3 Management of Postharvest Diseases

For managing and controlling postharvest diseases of fruits, a number of treatment methods have been employed. These methods may be chemical, biological control, physical or by use of botanicals. Although no management strategy is completely reliable and may be modified according to numerous factors, viz., climate, place, maturity, fruit variety, transportation, packaging, handling, etc., a unified approach encompassing various techniques is shown to be beneficial for some fruits (Singh 2010). Some of these treatment methods are described here.

20.3.1 Physical Treatments

Some physical methods, viz. forced hot air, heat treatment by water, gamma and ultraviolet irradiation, low and high pressure, low-high temperature, modified atmosphere packaging (MAP) and synthetic non-edible/edible fruit coatings, have been used for protecting the fruit at the time of postharvest phase. Postharvest heat treatments are non-polluting and are used for disease control and insect disinfestation in fresh horticultural crops (Lurie 1998). Hot water dip treatment method at 44°C controls the decay of papaya fruits due to *C. gloeosporioides* (Chan et al. 1996). Hot water spray treatment method for 2–5 min at 50–60°C has been suggested for the preservation of papaya fruits (Couey et al. 1984). Hot water treatment method at 49°C for20 min and then immediately washing the fruit with cold water minimizes the loss of papaya fruits due to soft rots infected by *Rhizopus*. The method of heat treatment has been used for the removal of fruit flies on papaya. The U.S. Department of Agriculture allows treatment of papaya with dry or vapour heat, forced hot air to prevent the spread of fruit fly in different areas of the United States. However, the above-mentioned treatments are not effective to manage diseases (Nishijima 1992). For controlling market fungal pathogens, forced hot air (48.5°C for 3–4 h) treatment of papaya with low relative humidity have been reported by Lay-Yee et al (1998) and Perez-Carillo et al. (2003). UV-C exposure for 2–4 min was efficacious in managing sporulation and anthracnose fungi conidial growth on papaya fruits (Cia et al. 2007). Gamma rays are capable of penetrating fruits and destroying pathogens in established lesions deep in host tissue. Gamma radiation at1kGy safeguarded anthracnose of papaya fruits caused by *Colletotrichum gloeosporioides* (Zhao et al. 1996; Cia et al. 2007).

For fruit preservation, variation in storage pressure has also been used. Low temperatures with low humidity levels decreased development of decay in papaya fruits (Paull et al. 1997). Modified atmosphere (MA) storage or controlled atmosphere (CA) storage has a vast application in expanding the shelf life of harvested fruits (Lazan et al. 1993). These methods are basically focussed on maintaining (i) low oxygen tension, (ii) low temperature, (iii) low ethylene concentration around fruits and (iv) adequate humidity. Chen and Paull (1986) found reduced chilling injury symptoms in papaya fruit with application of a solution of Sta-Fresh 7051 wax (1:70 v/v) and preserved/stored at 2°C for 14 days and 10°C for 24 days. Some edible fruit coatings, viz., Samp-Fresh and Cryovac, are available for preservation of papaya fruits (Yusof and Salleh 1992). Currently, MAP is widely used for papayas. "Sunrise" and "Kapoho" varieties of papaya fruit individually sealed in high density polyethylene (HDPE) have fewer chilling injury symptoms than unsealed fruit (Chen and Paull 1986). When hot water, thiabendazole dip is used with shrink wrapping along with various films like Dupont 50-EHC, Dupont75-EHC, Cryovac D-955 or Cryovac MPD-2055, there was 90% reduction in postharvest loss after 14 days of storage (Chen and Paull 1986). Fruits covered with eggs or first Instar larvae of oriental fruit fly (*Daucus dorsalis* Hende) wrapped in Cryovac D-955 and stored at 24–25°C showed a decreased in number of insects that survived after 96 h (Shetty et al. 1989). Seal packaging of "Backcross Solo" papaya in three layers of low-density polyethylene (0.0125 mm) followed by storage at 24–25°C for 2.5 weeks reduced the ripening of peel colour and fruits reduced the increase in total titratable acidity (Lazan et al. 1993). The internal gas concentration delayed ripening and expanded the fruit shelf life after applying a cellulose base film to papaya. MAP (using 0.05 mm shrinkable polyethylene film) at 15°C delayed loss of firmness in papaya fruit "Exotica" (Lazan et al. 1993). Sunrise papaya stored in an insecticidal atmosphere (0.17–0.35% O_2, balanced N_2) for upto 5days at 20°C had less loss of firmness than the control, and no apparent internal or external injury was found (Yahia et al. 2000). MAP (at 10°C) along with low-density polyethylene film packaging showed superior maintenance of papaya fruits during storage (Gonzalez-Aguilar et al. 2002).

A combination of hydrothermal treatment at 48°C along with calcium chloride solution (1% w/v) was able to preserve postharvest papaya for up to 20 days (Ayon-Reyna et al. 2017). Terao et al. (2019) reported successful management of stem-end rot of papaya fruit with combined use of ozonated water and short exposure at high temperature of 60°C. In spite of adequate research, no commercial utility of MA/CA storage has been reported for papaya. Controlled studies are still needed to establish the prospective application and ideal atmosphere for management of papaya fruits during storage.

20.3.2 Chemical Methods

Currently, for fruit storage lots of soft/synthetic chemicals, viz. colloidal sulphur, antibiotics, inorganic copper compounds, phenolic compounds, calcium compounds, sulphur compounds, dithiocarbamates and various fungicides, are used. Although the relevance of chemicals as storage in the current perspective is not positive related the environment and consumer health, the following account compiles the related work on papaya fruit.

Khare and Dhingra (1974) obtained favourable results in managing *Phomopsis* rot of papaya by spraying Captan, Benlate, Borax, Dithane M-45 and Diflotan applied in a range of 1–10% solution. Dickman et al. (1983) observed that certain organophosphorous pesticides may reduce papaya anthracnose. Ogbadu and Otaru (1986) found effectiveness of dipping treatment of 5% derivative benzoic acid (i.e. 1-chloro-2,4 dinitrobenzene) treatment against rot of papaya fruits caused by *Aspergillus flavus*. According to Tewari et al. (1988), Bavistin at 1000 ppm was efficient against Thielaviopsis rot of papaya and caused 95% control of disease. Further, Aureofungin, Bavistin and Benlate (500-ppm solutions) were also able to reduce conidial germination of *Fusarium semitectum* on papaya fruit surfaces during storage (Prasad et al. 1988). Sharma and Khan (1989) recorded better management of *Ulocladium chartarum* infection of papaya with Rovral and Bavistin. Nishijima et al. (1990) recorded that Mancozeb managed Rhizopus soft rot of papaya fruits. Furthermore, anthracnose of papaya was managed by using postharvest treatment of 1000 ppm of Propiconazole or Prochloraz as a contact pesticide (Sepiah 1993). Treatment with Prochloraz markedly decreased papaya stem-end rot disease (Lay-Yee et al. 1998). Oliviera (2002) found that after Benomyl treatment at 500 ppm a reduction in fruit decay occurred. Sivakumar et al. (2002) described the efficient control of papaya anthracnose by sodium bicarbonate (2%) and ammonium carbonate (3%) in solution as contact pesticide. Ammonium carbonate (3%) subsumed into a formulation of wax successfully reduced anthracnose disease by70% for 21days at 13.5°C and 95% relative humidity. Lichanporn and Kanlavanarat (2006) used various organic acids viz. citric, ascorbic acid (2–5%) as a dipping procedure for postharvest controlling of papaya disease. Chitosan coating steeped with fungicides was reported to control postharvest losses of papaya fruits (Bautista-Banos et al. 2003; Hewajulige et al. 2007). Papaya fruits treated with ethylene suppressor, i.e. 1-methyl cyclopropane, resulted in improved shelf life (Manenoi et al. 2007). According to Aleryani-Raqeeb et al. (2008), the use of calcium chloride (2%) with chitosan coatings on fruits extended the storage life of papaya upto 33 days.

Lata et al. (2018) showed that salicylic acid (1 and 2 mM), calcium chloride (1 and 2%) and nitric oxide (1 and 2 mM) solution in water enhanced the shelf life of papaya at the surrounding conditions with less disease incidence. Papaya treated with phosphate-Ca significantly reduced anthracnose disease incidence in Sunrise Solo papaya and maintained physicochemical properties of stored papaya (Lopes et al. 2018).

20.3.3 Biological Control

Biological control is the novel approach for management of postharvest diseases. It includes microorganism application to control fruit pathogens. Several natural epiphytes constitute natural antagonists that interact with pathogens in a negative way. A successful bio-control agent has many modes of action and has a great potential to take over existing strategies of disease control. Several bio-control agents have been successfully tested on many fruits during postharvest phase. Some bio-control agents have been commercialized viz. Biosave (*Pseudomonas syringae*, in the United States), Shemer (*Metchnikowia fructicola*, in Israel), Biocoat (*Candida saitoano*, in the United States) and Aspire (*Candida oleophila*, in the United States and Israel). In many countries, this method is not used commercially on a large scale but today a great deal of investigation is needed on low-cost use of biological means. Some bio-control agents have also worked against papaya storage rots. Paraffin wax formulation of *Candida oleophila* along with sodium bicarbonate (2%) was used by Gamagae et al. (2004) for managing storage of papaya anthracnose of fruits for 2 weeks at 13.5°C at 90% relative humidity. In in vivo studies, Sánchez et al. (2005) reported that *Bacillus* spp. inhibits the growth of *Penicillium* infection on papaya fruits and reduced blue mould rotting of fruit. Capdeville et al. (2007) showed bio-control potential of *Cryptococcus magnus* for-managing anthracnose of papaya fruit; antagonist activity was found to significantly decrease the disease incidence caused by *C. gloeosporioides* at a concentration of 10^7–10^8 cells/mL 48 h prior to inoculation. Emulsions of *Pseudomonas aeruginosa* and *Burkholderia cepacia* were efficient in bio-controlling *C. gloeosporioides* on papaya fruit (Rahman et al. 2007). Wijeratna et al. (2008) reported biological control activity of *Trichoderma harzianum* against *C. gloeosporioides*, causing stem-end rot and anthracnose of papaya in vitro. Shi et al. (2010) recorded that papaya anthracnose has been effectively controlled by *Pseudomonas putida* isolated from pericarp of papaya up to 54% during storage. Hernandez-Montiel et al. (2018) examined excellent antagonistic capacity of *Debaryomyces hansenii* towards C. *gloeosporioides* causing papaya anthracnose. Over the previous 25 years, bio-control has appeared as an efficacious strategy to control major postharvest fruit diseases, but research into biological control of postharvest decay is in its early stage. Although few bio-control agents viz. Aspire (the USA) and Biosave 100 are accessible in the market, commercial use of it on a large scale is still missing. Due to its high cost, developing countries like India cannot use it, as most of the farmers are in small and marginal lands with a very small source of income. Further studies regarding the effects of these biological agents with same pathological environment of plants and on human health have still to be worked out.

20.3.4 Use of Botanicals

Plants contain a variety of secondary metabolites (terpenoids, alkaloids, glycosides, phenols and tannins), many of which exhibit fungicidal nature (Graingeand Ahmed 1988). Botanicals in the form of aqueous or organic extracts, essential oils, plant powders and purified active compounds have reportedly been used for postharvest preservation of several tropical as well as temperate fruits (Ippolito and Nigro 2003; Tripathi and Dubey 2004; Burt 2004).

20.3.4.1 Plant Extracts/Natural Compounds

Polyamines (putrescine, spermidine and spermine) were found by Purwoko et al. (1998) to delay papaya fruit ripening, thereby increasing shelf life up to 2–6 days when applied as 1% contact pesticide. Ribeiro and Bedendo (1999) reported an inhibitory effect of peppermint, castor bean and pepper extracts on mycelia growth of causal agent of papaya anthracnose; all extracts were effective above 200 mg/g, causing reduced conidium production in the range of 41–84%. Aqueous extract of *Pithecellobium dulce* exhibited acute antifungal activity against *Fusarium* sp. isolated from infected papaya in vitro (Bautista-Banos et al. 2002). Methyl jasmonates and MAP (low temperature of 10–13°C and 80% RH) reduces decay and maintains postharvest quality of papaya (Gonzalez-Aguilar et al. 2002). Chitosan (1.5%) coating on papaya fruit applied before *C. gloeosporioides* inoculation reduced anthracnose disease by 80% (Bautista-Banos et al. 2003). Silva et al. (2006) observed in vitro antifungal activity of ethanolic extract of *Plectrantus barbatus* against *Glomerella cingulata* causing stem-end rot in papaya fruits. Hernandez-Albiter et al. (2007) found promising in vitro antifungal effect of crude plant extract of *Cestrumnocturnum* and *Annonacherimola* against *C. gloeosporioides* obtained from diseased papaya. Pramod et al. (2007) reported effectiveness of papaya latex (5%) on inhibition of spore germination of several postharvest pathogens of papaya. Postharvest dipping in Borneol solution for few minutes could extend shelf life of papaya up to 7 days without decline in fruit quality (Venkateshan et al. 2010). Ali et al. (2014) found promising results from ethanolic extract of propolis (EEP) in combination with gum arabic (GA) for controlling papaya anthracnose. Coating of papaya fruits with 50% *Aloe vera* gel effectively preserved quality attributes of fruits during storage up to 12 days (Mendy et al. 2019).

20.3.4.2 Essential Oils

The essential oils or volatile oils produced by plants are aromatic, oily liquids obtained from plant organs–flower bud, seed, leaf, twig, bark, wood and root and rhizomes. In many cases, these are biologically active, endowed with antimicrobial, allelopathic, antioxidant and bioregulatory properties. The fungicidal nature of oils is basically due to their components which are mainly monoterpenes and sesquiterpenes which are hydrocarbons with general formula (C5H8)n and their derivatives. Until recently, essential oils have been used as food flavourings due to their flavour and fragrance, but nowadays essential oils and their pure components are gaining increasing interest because of their safe status, wide acceptance by consumers and their exploitation for multipurpose uses (Cowan 1999). Essential oils can widely betappedas antimicrobials to preserve perishables. Their use is an innovative and significant alternative to synthetic fungicides in maintenance of organoleptic parameters, nutrition of fruits and controlling microbial spoilage of highly perishable fruits, but it is yet to be commercialized. Antifungal activity of essential oil of *Cinnamomum zeylanicum* and *Syzygium aromaticum*on *C. gloeosporioides* has been reported by Barrera-Necha et al. (2008). In vivo application (dip treatment of 1% oil) of both oils preserved papaya with minimum quality changes for more than 10 days. Singh (2009) reported absolute fungi toxicity of oil of *Mentha arvensis* (at 500 ppm) against *Aspergillus flavus* and *Fusarium moniliforme*, two dominant postharvest pathogens of papaya. The oil exhibited prophylactic activity, protected stored papaya at 800 ppm and increased shelf life up to 2 days more than control and pesticide-treated fruits. Overall, the oil-treated fruits were superior in acceptability, nutritional status and organoleptic attributes to pesticide-treated fruits after 7days of storage. Bosquez-Molina et al. (2010) used a fruit coating of Mesquite gum impregnated with *Citrus aurantifolia* and *Thymus vulgaris* oil (1–5%); there was 50% reduction in fruit decay. More work needs to be done regarding application of essential oils in the form of formulations, integrated fruit coatings, treated wraps, dip emulsions and their commercialization on a large scale.

Oil of *Lippiasidoides* along with carboxymethyl cellulose coating preserved quality of commercial papaya fruit for 9 days at normal temperatures (Zillo et al. 2018). Papaya fruits treated with a combination of essential oils and paraffin (rosemary and pepper) reduced anthracnose incidence in papaya fruits (Dias et al. 2020).

20.4 Conclusion

India has achieved a record horticultural production in the last few years, but the per capita consumption of vegetables and fruits is still below the level prescribed by Indian Council of Medical Research nutritional guidelines. Fruits and vegetables are still found mostly on the plates of wealthy people and are too expensive for poorer people. The reason is the postharvest losses incurred due to abiotic and biotic stresses every year. Improper transit, storage, handling and overall management tend to create huge losses, sometimes amounting to greater than 60% of produce. A lower supply than demand tends to raise the commodity's cost, which ultimately leads to lack of availability of fruits to many people. An approximate annual postharvest perishable loss in India is nearly identical to the amount which is consumed by the whole of the United Kingdom. Thus, to our agro-based economy, postharvest losses of perishables are a great setback. Modern techniques, viz. use of radiation, edible coatings and MAP, can safely be suggested and could be used for extending the shelf life of many fruits. Use of biological control agents commercially on a large scale as tools for postharvest management has still many issues to resolve, especially in developing countries like India, where there is no regulatory control on bioagents and their formulations. Taking due care of the ecology and the environment bioagents should be given top priority.

REFERENCES

Adetuyi FO, Akinadewo LT, Osmosuli SV, Ajala, L (2008) Antinutrient and antioxidant quality of waxed and unwaxed *Carica papaya* fruit stored at different temperatures. Afr J Biotechnol **7**: 2920–2924.

Aleryani-Raqeeb A, Mahmud TMM, Syed Omar SR, Mohamed Zaki AR (2008) Effect of calcium infiltration and chitosan coating of storage life and quality characteristics during storage of papaya (*Carica papaya*). Int J Agric Res 3: 296–306.

Ali A, Cheong CK, Zahid N (2014) Composite effect of propolis and gum arabic to control postharvest anthracnose and maintain quality of papaya during storage. Int J Agri Biol 16(6): 1117–1122.

Ayón-Reyna LE, González-Robles A, Rendón-Maldonado JG, Báez-Flores ME, López-López ME, Vega-García MO (2017) Application of a hydrothermal-calcium chloride treatment to inhibit postharvest anthracnose development in papaya. Postharvest Biol Technol 124: 85–90.

Barrera-Necha LL, Bautista-Banos S (2008) Efficacy of essential oils on the conidial germination, growth of *Colletotrichum gloeosporioides* (Penz.) Penz. and Sacc. and control of postharvest diseases in papaya (*Carica papaya* L.). Plant Pathol J **7**: 174–178.

Bautista-Banos S, Hernandez- Lopez M, Bosquez-Molina E, Wilson CL (2003) Effect of chitosan and plant extracts on growth of *Colletotrichum gloeosporioides*, anthracnose level and quality of papaya fruit. Crop Prot 22: 1087–1092.

Bautista-Baños S, Hernández-López M, Barrera-Necha LL (2002) Antifungal screening of some plants of the state of Morelos,

Mexico against four fungal postharvest pathogens of fruits and vegetables. Rev Mexicana de Fitopatol 18: 36–41.

Bosquez-Molina E, Ronquillo-de Jesús E, Bautista-Baños S, Verde-Calvo JR, Morales-López J (2010) Inhibitory effect of essential oils against *Colletotrichum gloeosporioides* and *Rhizopus stolonifer* in stored papaya fruit and their possible application in coatings. Postharvest Biol Technol 57(2): 132–137.

Burt S (2004) Essential oils: Their antibacterial properties and potential applications in food – A review. Int J Food Microbiol 94: 223–253.

Capdeville G, de Souza MT Jr, Santos-Carlos A et al. (2007) Selection and testing of epiphytic yeasts to control anthacnose in post-harvest of papaya fruit. Sci Hortic 111: 179–185.

Chan HT Jr, Maindonald JM, Laidlaw WG, Seltenrich M (1996) ACC Oxidase in papaya sections after heat treatment. J Food Sci 61: 1182–1186.

Chen NM, Paull RE (1986) Development and prevention of chilling injury in papaya fruit. J Am Soc Hortic Sci 111(4): 639–643.

Choudhary ML (2006) Recent Developments in Reducing Postharvest Losses in the Asia Pacific Region. Reports of APO seminar Reduction of Postharvest Losses of Fruit and Vegetables, India, 5–11 October 2004, and Marketing and Food Safety: Challenges in Postharvest Management of Agricultural/Horticultural Products in Islamic Republic of Iran, 23–28 July 2005, Rolle RS (Ed.). Rome, Italy.

Chowdhury SM, Sultana N, Mostofa G, Kundu B, Rashid M (2014) Postharvest diseases of selected fruits in the whole sale market of Dhaka. Bangladesh J Plant Pathol 30(1&2): 13–16.

Cia P, Pascholati SF, Benato EA, Camili EC, Santos-Carlos A (2007) Effects of gamma and UV-C irradiation on the postharvest control of papaya anthracnose. Postharvest Biol Technol 43: 366–373.

Couey HM, Alvarez AM, Nelson MG (1984) Comparison of hotwater spray and immersion treatments for control of postharvest decay of papaya. Plant Dis 68: 436–437.

Cowan MM (1999) Plant products as antimicrobial agents. Clin Microbiol Rev 12: 564–582.

Dias BL, Costa PF, Dakin MS et al. (2020) Control of papaya fruits anthracnose by essential oils of medicinal plants associated to different coatings. J Med Plant Res 14(6): 239–246.

Dickman MB, Patil SS, Kolatktukudy PE (1983) Effects of organophosphorous pesticides on cutinase activity and infection of papayas by *Colletotrichum gloeosporioides*. Phytopathology 8: 1209–1214.

Gamagae SU, Kumar S, Wijesundera RLC (2004) Evaluation of post-harvest application of sodium bi carbonate incorporated wax formulation and *Candida oleophila* for the control of anthracnose of papaya. Crop Prot 23: 575–579.

Gonzalez-Aguilar GA, Buta J, Wang CY (2002) Methyl jasmonates and modified atmosphere packaging reduces decay and maintains post-harvest quality of papaya "Sunrise". Postharvest Biol Technol 28: 361–370.

Grainge M, Ahmed S (1988) Handbook of Plants with Pest Control Properties. Wiley, New York, 407 pp.

Hamim I, Alam MZ, Ali MA, Ashrafuzzaman M (2014) Incidence of post fungal diseases of ripe papaya in Mymensingh. J Bangladesh Agric Univ 12(1): 25–28.

Helal BR, Hosen S, Shamsi S (2018) Mycoflora associated with postharvest disease of papaya (*Carica papaya* L.) and their pathogenic potentiality. Bangladesh J Bot 47 (3): 389–395.

Hernández-Albíter RC, Barrera-Necha LL, Bautista-Baños S, Bravo-Luna L (2007) Antifungal potential of crude plant extracts on conidial germination of two isolates of *Colletotrichum gloeosporioides* (Penz.) Penz. and Sacc. Mexican J Phytopathol 25(2): 180–185.

Hernandez-Montiel LG, Gutierrez-Perez ED, Murillo-Amador B et al. (2018) Mechanisms employed by *Debaryomyces hansenii* in biological control of anthracnose disease on papaya fruit. Postharvest Biol Technol 139: 31–37.

Hewajulige IGN, Sivakumar D, Sultanbawa Y, Wilson WRS, Wijesundera RLC (2007) Effect of chitosan coating on the control of anthracnose and overall quality retention of papaya (*Carica papaya* L.) during storage. Acta Hort 740: 245–250.

Ippolito A, Nigro F (2003) Natural antimicrobials in post-harvest storage of fresh fruit and vegetables. In Roller S (Ed.). Natural Antimicrobials for Minimal Processing of Food. CRC Press, Boca Raton, FL, pp. 201–234.

Khare MN, Dhingra OD (1974) *In vitro* and *in vivo* testing of fungicides against *Phomopsis caricae papaya* causing fruit rot of papaya (*Carica papaya* L.). JNKVV Res J 8: 258–259.

Lata D, Aftab MA, Homa F, Ahmad MS, Siddiqui MW (2018) Effect of eco-safe compounds on postharvest quality preservation of papaya (*Carica papaya* L.). Acta Physiol Plantarum 40(1): 8.

Lay-Yee M, Clare GK, Petry RJ, Fullerton RA, Gunson A (1998) Quality and disease incidence of Waimanalo Solo papaya following forced air heat treatments. Hortic Sci 33: 878–880.

Lazan H, Ali ZM, Selamat MK (1993) The underlying biochemistry of the effect of modified atmosphere and storage temperature on firmness decrease in papaya. Acta Hortic 343:141–147.

Lichanporn I, Kanlavanarat S (2006) Effect of organic acid and modified atmosphere conditions on quality of Shredded green papaya (*Carica papaya* L.). Acta Hortic 712: 729–734.

Lim YY, Lim TT, Tee JJ (2007) Antioxidant properties of several tropical fruits: A comprehensive study. Food Chem 103: 1003–1008.

Lopes LF, Cruz AF, Barreto MLDA, Vasconcelos TMMD, Blum LEB (2018) Postharvest treatment with Ca-phosphite reduces anthracnose without altering papaya fruit quality. J Horti Sci Biotechnol 93(3): 272–278.

Lurie S (1998) Postharvest heat treatments. Postharvest Biol Technol 14(3): 257–269.

Manenoi A, Ruth E, Bayogan V, Thumbdee S, Paull R (2007) Utility of 1-methyl cyclopropene as a papaya postharvest treatment. Postharvest Biol Technol 44: 55–62.

Mendoza EMT (2007). Development of functional foods in the Philippines. Food Sci Technol Res 13: 179–186.

Mendy TK, Misran A, Mahmud TMM, Ismail SI (2019) Application of *Aloe vera* coating delays ripening and extend the shelf life of papaya fruit. Sci Hortic 246: 769–776.

Nishijima KA (1992) Effect of forced hot air treatment of papaya fruit on fruit quality and incidence of postharvest diseases. Plant Dis 76: 723–727.

Nishijima WT, Ebersole S, Fernandez JA (1990) Factors influencing development of postharvest incidence of *Rhizopus* soft rots of papaya. Acta Hortic 269: 495–202.

Ogbadu LJ, Otaru EB (1986) Influence of structural variation in selected benzoic acid derivatives on their effectiveness at controlling *Aspergillus flavus* in fruits. Microbiol Lett 32: 65–68.

Oliviera JJ (2002) Benomyl residues in papaya (*Carica papaya* L.). Pesticidas Ecotoxicol Meio Ambiente Curitiba 22: 51–58.

Paull RE, Nishijima W, Reyes M, Cavaletto C (1997) A review of postharvest handling and losses during marketing of papaya (*Carica papaya* L.). Postharvest Biol Technol 11: 165–179.

Perez-Carillo E, Yahia EM, Ariza R, Vega M, Cornejo G (2003) Forced hot air treatment at low relative humidity is effective in reducing chilling injury and decay development in papaya fruit. Acta Hortic 604: 697–702.

Pramod G, Swami AP, Srnivas P (2007) Post-harvest diseases of papaya fruits in Coimbatore markets. Ann Plant Prot Sci 15: 33–35.

Prasad J, Pathak VN, Pathak AK (1988) Efficacy of certain fungicides to inhibit conidial germination of *Fusarium semitectum* causing fruit rot of papaya (*Carica papaya*) during storage. Indian J Mycol Pl Pathol 18: 277.

Purwoko BS, Kermayanti N, Susanto S, Nasution MZ (1998) Effect of polyamines on quality changes in papaya and mango fruits. Acta Hortic 464: 510–512.

Rahman MA, Kadir J, Mahmud TMM, Rahman RA, Begum MM (2007) Screening of antagonistic bacteria for bio-control activities on *Colletotrichum gloeosporioides*in papaya. Asian J Plant Sci 6: 12–20.

Ravindran C, Kohli A, Srinivas M (2007) Fruit production in India. Chron Hortic 42: 21–26.

Ribeiro LF, Bedendo IP (1999) Efeitoinibitório de extratos vegetaissobre *Colletotrichum gloeosporioides* - agente causal da podridão de frutos de mamoeiro. Sci Agric 56: 1267–1271.

Sánchez JAR, Rodríguez-Cervantes CH, Vázquez-Arce MA, Calderón-Santoyo M (2005) Bio-control of postharvest fungi in papaya using *Bacillus* sp. IFT Annual Meeting, July 15–20. New Orleans, Louisiana. http://ift.confex.com/direct/ift/2005/techprogram/paper_29802.htm (accessed on 20.1.10).

Sawant SG, Gawai DU (2011) Effect of fungal infections on nutritional value of papaya fruits. Curr Bot 2(1): 43–44.

Sepiah M (1993) Efficacy of propiconazole against fungi causing post-harvest disease on Eksotika papaya. In: Proceedings of International Postharvest Conference on Handling Tropical Fruits, Chingmai, Thailand, p. 53.

Sharma N, Khan AM (1989) Chemical control of post-harvest diseases of papaya fruit caused by *Ulocladium chartatum*. Indian Bot Reptr 8: 65–66.

Shetty KK, Klowden MJ, Jang EB, Kochan WJ (1989) Individual shrink wrapping: A technique for fruit fly disinfestations in tropical fruits. Hortic Sci 24: 317–319.

Shi J, Liu A, Li X, Feng S, Chen W (2010) Identification of endophytic bacterial strain MGP1 selected from papaya and its bio-control effects on pathogens infecting harvested papaya fruit. J Science Food Agric 90(2): 227–232.

Silva MB, Silva CA, Viana LAS, Brasileiro BG, Jamal CM (2006) Potential use of *Plectrantus barbatus* ethanolic extract to control phytopathogenic fungi. Rev Bras Pl Med Botucatu 8:78–79.

Singh, P (2010) Advances in control of post-harvest diseases of papaya fruit. A review. Agric Rev 31(3): 194–202.

Singh P (2009) Studies on postharvest deterioration of papaya fruit and its management by plant volatile. PhD thesis. D.D.U. Gorakhpur University, Gorakhpur, India.

Sivakumar D, Hewarathgamagae NK, Wilson WSS, Wijesundera RLC (2002) Effect of ammonium carbonate and sodium bicarbonate on anthracnose of papaya. Phytoparasitica 30: 486–492.

Terao D, de Lima Nechet K, Frighetto RTS, Sasaki FFC (2019) Ozonated water combined with heat treatment to control the stem-end rot of papaya. Sci Horti 257(6): 108722.

Tewari DK, Srivastava RC, Katiyar N, Lal B (1988) *Thielaviopsis* rot of papaya. Indian Phytopathol 41: 491–492.

Tripathi P, Dubey NK (2004) Exploitation of natural products as an alternative strategy to control postharvest fungal rotting of fruits and vegetables. Postharvest Biol Technol 32: 235–245.

Venkateshan S, Sudhagar R, Kamala KS, Manivannan K (2010) Acta Hortic 851: 541–544.

Wijeratnama SW, Dharmatilakaa Y, Weerasinghe D (2008) Conf on Int Res on Food Security, Natural Resource Management and Rural Development; University of Hohenheim, Germany, October7–9, 2008. http://www.tropentag.de/2008/abstracts/links/Wilson.

Yahia EM, Villagomez G, Juarez A (2000) In: Artcs F, Gil MI, Conesa MA (Eds.). Improving Postharvest Technologies of Fruits, Vegetables and Ornamentals. International Institute of Refrigeration, Paris, pp. 714–718.

Yusof S, Salleh M (1992) Physicochemical response of papaya to waxing. Acta Hortic 292: 223–230.

Zhao M, Moy J, Paull RE (1996) Effect of gamma–irradiation on ripening papaya pectin. Postharvest Biol Technol 8: 209–222.

Zillo RR, da Silva PPM, de Oliveira J, da Glória EM, Spoto MHF (2018) Carboxymethylcellulose coating associated with essential oil can increase papaya shelf life. Sci Hortic 239: 70–77.

21

Management Postharvest Diseases of Peach Fruits and Their Management

Monica Sharma[1] **and Amit Sharma**[2]
[1]*Department of Plant Pathology, Dr YS Parmar University of Horticulture and Forestry, Hamirpur, Himachal Pradesh, India*
[2]*Department of Basic Science, College of Horticulture and Forestry, Dr YS Parmar University of Horticulture and Forestry, Hamirpur, Himachal Pradesh, India*

CONTENTS

21.1 Introduction

The term "stone fruit" is a general term which describes fruits of plant species of genus *Prunus*. Peaches, nectarines, apricots, plums and cherries are the main stone fruits. The fruit consists of a mesocarp, which is a layer of fleshy, edible pulp that surrounds a relatively large, hard pit, i.e. "stone", which eventually shields and protects a seed. Stone fruits are vulnerable to many diseases due to their soft texture. The perishable nature of stone fruits predisposes them to many postharvest diseases. Postharvest diseases in stone fruits are caused by various fungi, bacteria and in some cases by viruses. The infection of fruits may either be pre-harvest or postharvest. In pre-harvest infection, the pathogen may enter through penetration of outer covering of fruit or through natural openings or through wounds on fruits. In postharvest infection, the cut ends of fruits are the usual entry point for pathogens. In addition, other methods of penetration are also involved. For example, *Colletotrichum* and *Sclerotinia* may also enter through direct penetration of the peel of fruits. In general, many pre-harvest and postharvest factors predispose stone fruits to various postharvest diseases (Table 21.1). If these factors are not maintained properly, the chances of losses due to postharvest diseases increase to large scale. Knowledge of mode of infection, predisposing factors to infection, disease dissemination, varietal susceptibility or resistance, etc. is useful to develop strategies for management of postharvest diseases.

Fruits are an ideal substrate for the development of pathogenic microorganisms during storage, which could lead to losses of product ranging from a minimum of 10–15% in developed countries compared to over 50% in developing countries. It has been estimated that in fresh fruits and vegetables, various factors including improper handling, inappropriate storage and transportation and infection by pathogens can result up to 45% postharvest losses (Siddique et al. 2018). Postharvest losses in peaches alone may reach upto 80% during transport and storage (Sestari et al. 2008). Many fungi and bacteria can cause postharvest spoilage of different stone fruits (Table 21.2).

TABLE 21.1

Factors Influencing Postharvest Infection

Pre-harvest Factors	Postharvest Factors
Cultivars	Maturity stage
Planting material	Methods of harvesting
Weather	Time of harvesting
Crop husbandry/cultural operations	Precooling
	Packaging and packaging materials
	Temperature during storage
	Transportation

21.2 Postharvest Diseases of Stone Fruits

21.2.1 Brown Rot

Brown rot is distributed worldwide and is an important disease of all stone fruits. However, brown rot disease is more prevalent in humid weather. The disease affects fruit in the orchard, during transportation and in shops while selling. The infection of brown rot is very rapid and results in rotting of stone fruits during short storage in the consumers' houses during consumption. Brown rot is among the serious diseases of peaches, nectarines, apricots and cherries. On plums and prunes, brown rot disease is not a serious disease; however, considerable losses occur sometimes in some regions.

21.2.1.1 Symptoms

The symptoms appear as circular and non-sunken spots covered with a tough leathery peel which remains intact over the decaying brown- to black-coloured flesh. In advanced stages, powdery spore masses of tan-grey colour cover the surface of affected areas in concentric rings. The complete fruit may be rotten within 3–4 days but the outer peel of fruit remains intact. Such fruits eventually shrivel and mummify (Figure 21.1).

TABLE 21.2

Important Postharvest Diseases of Stone Fruits

Sr. No.	Disease	Pathogen (s)	Host Crops
1.	Brown rot	*Monilinia fructicola*	Peaches, nectarines, plums, cherries, apricots
2.	Rhizopus rot	*Rhizopus stolonifer*	
3.	Grey mould rot	*Botrytis cinerea*	
4.	Black mould rot	*Aspergillus niger*	
5.	Blue mould rot	*Penicillium* sp.	
6.	Coryneum blight/scab	*Coryneum carpophilum*	
7.	Cladosporium rot	*Cladosporium herbarum*	
8.	Alternaria rot (green mould rot)	*Alternaria alternata*	Cherries, plums
9.	Anthracnose (bitter rot)	*Glomerella cingulata*	Peaches, nectarines, apricots
10.	Pink mould rot	*Trichothecim roseum*	
11.	Diplodia rot	*Diplodia natalensis*	
12.	Powdery mildew	*Sphaerotheca pannosa*	
13.	Sour rot	*Geotrichum candidum*	
14.	Gilbertella rot	*Gilbertella persicaria*	Peaches
15.	Bacterial spot/canker	*Xanthomonas arboricola* pv. *pruni*	Peaches, plum, nectarines, apricots

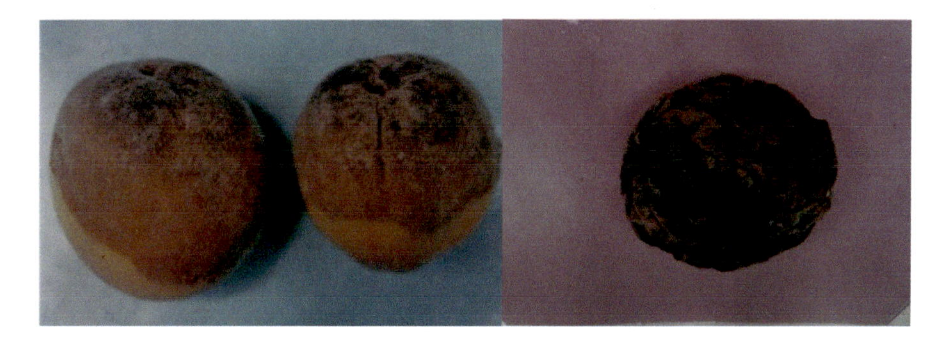

FIGURE 21.1 Symptoms of brown rot disease on peach fruits.

21.2.1.2 Causal Organism

Brown rot disease is caused by *Monilinia fructicola* (Winter) Honey and *Monilinia laxa* (Aderh. and Ruhland) Honey. *M. fructicola* occurs in all regions of the world where stone and pome fruits are grown. The mycelium of fungus produces conidiophores that are long, branched and bear chains of oval or lemon-shaped conidia. Conidia are the source of secondary infection. *Monilinia* is an ascomycetous fungus. The fungus produces apothecia on mummified fruit that have fallen to the ground. Apothecia are usually 5–20 mm in diameter and bear tubular-shaped asci. The asci bear ascospores which initiate new infections on susceptible hosts. Mycelium of the fungus secretes powerful enzymes that dissolve the middle lamellae of host cells, which results on softening of tissues.

21.2.1.3 Disease Cycle and Epidemiology

The pathogen mainly perpetuates in the form of dormant mycelium in mummified dried fruits. In addition, the fungi can also overwinter on twig cankers. During rainy periods, the apothecia are produced in mummified fruits that are fallen on ground. The apothecia have long stalk and bear asci and ascospores. The ascospores are released forcibly by a puffing action and initiate the disease on blossoms, twigs and young leaves. Conidia are produced in the mummified fruits attached to the trees and also from the dormant mycelium in the twigs and leaves. Spores are released from mummified fruits or twig cankers and infect blossoms, twigs, leaves and young fruits. The infection can also occur on maturing fruits later in the season in the presence on rainy period. Fruits also become infected in packinghouses or in markets. Spores of *M. fructicola* can germinate at0–32°C, with optimum at 15–26°C.

21.2.2 Grey Mould Rot

Grey mould rot occurs frequently on cherries, whereas it occurs on peaches, nectarines or apricots and other stone fruits when the fruits are stored for too long.

21.2.2.1 Symptoms

Initially, the symptoms appear as light-brown coloured lesions on the outer skins of fruits. The spot later becomes covered with a delicate growth of the fungus, and the underlying tissue becomes soft and darker brown in colour. As the infection progresses, the white- to grey-coloured growth of the fungus appears on the fruit surface. Ultimately, grey-brown coloured spores are produced abundantly on the spots. The underlying decaying tissues of the fruits are soft and watery and slip easily from the peel under pressure.

21.2.2.2 Causal Organism

Grey mould disease is mainly caused by *Botrytis cinerea* Pers. However, complexes of different species are also known to cause this disease (Walker et al. 2011; Rupp et al. 2017). Moreover, multiple gene lineages showed that *B. cinerea* is complex system of species and not a single organism (Fournier et al. 2005). The pathogen survives as sclerotia on plant debris and in the soil. Sclerotium geminates under favourable conditions and forms the mycelium. Mycelium forms conidiophore on which numerous conidia are formed (Figures 21.2 and 21.4). Conidiophores are branched in the upper part and each branch ends with a hemispherical or spherical swelling having spore-bearing projections minute called sterigmata. The sterigmata eventually bear conidia. The conidia are globose, ellipsoidal or egg-shaped, smooth, hyaline to pale brown, usually with a protuberant hilum.

21.2.2.3 Disease Cycle and Epidemiology

The fungus usually enters through the breaks or injuries on the peels of the fruits. The fungus may also enter through cut stems or calyx ends. The infection spreads rapidly under high humidity and at 22–25°C. However, the rotting can also proceed even at −1°C and infection can spread to adjacent fruits causing extensive nesting.

21.2.3 Rhizopus Rot

In India and other warmer areas, Rhizopus rot is the main serious postharvest disease of peaches, nectarines, apricots and cherries. The disease is of common occurrence in harvested fruits although sometimes it occurs in the orchard. The incidence of Rhizopus rot increases as the fruit ripens.

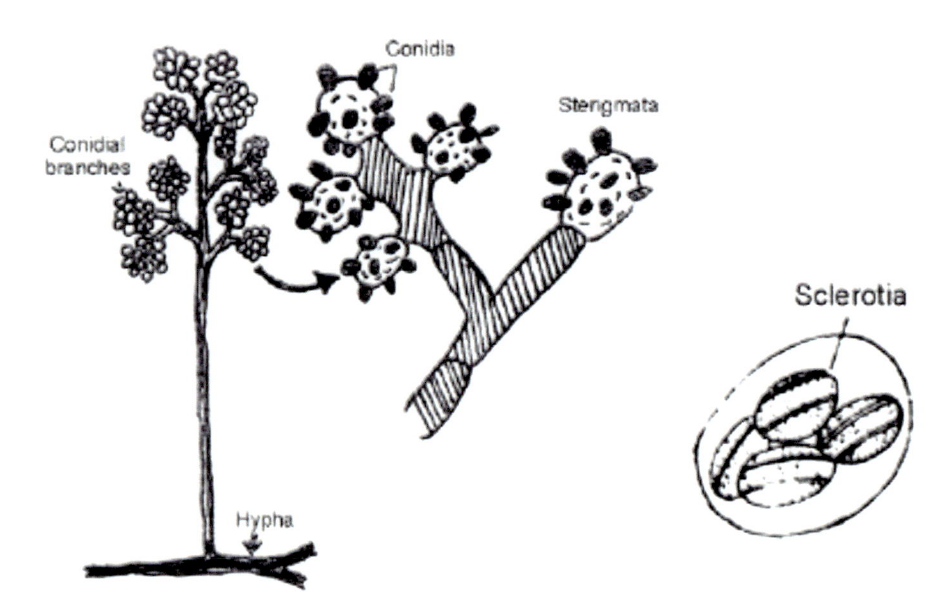

FIGURE 21.2 Conidiophore with conidia and sclerotia of *B. cinerea*.

21.2.3.1 Symptoms

The initial symptoms of the disease appear as a circular, tan area which is surrounded by an island of apparently healthy peel of fruit. Later, the peel of the affected areas becomes tan to brown in colour. The peel of the infected fruit easily slips from the decaying tissues, which produce a characteristic sour odour. The flesh of infected areas is tan, mushy and watery with a characteristic sour odour. As the infection progresses, white and cottony growth of the fungus appears and completely covers the entire affected area. The fungal mycelium soon produces the spore-producing fruiting bodies, which are initially white and eventually turn black (Figure 21.3). The temperature suitable for ripening of fruits also favours disease development and the fruit could completely rot in 48 hours.

21.2.3.2 Causal Organism

Rhizopus rot is caused by *Rhizopus stolonifer* Vuillemin and *R. arrhizus* Fisher. The mycelium of fungus produces many aerial stolons. The stolons form rhizoids at some points. One or more sporangiophores arise over the rhizoids. The sporangiophores are long, unbranched and bear sporangia. Sporangia are sac-like structures containing sporangiophores (Figure 21.3b). The central portion of sporangium is highly vacuolated columella, and the peripheral zone is the spore-bearing region containing sporangiospores. The sporangiospores are globose to oval and multinucleate. The sporangiophores are released by disintegration of the sporangial wall. *R. stolonifera* is heterothallic and needs the existence of two physiologically distinct compatible mycelia for sexual reproduction. By gametangial fusion, zygospore is formed. During the germination of zygospore, meiosis takes place and spores of the same type or a mixture of spores are formed. However, zygospore is rarely produced during the postharvest infection of stone fruits.

21.2.3.3 Disease Cycle and Epidemiology

Immature fruits are resistant, whereas mature fruits are extremely vulnerable to the disease. Infection occurs in injured or over-ripe fruits. The fungus can also directly penetrate the fruits. Recently, evidence was provided to demonstrate the capability of *R. stolonifer* to directly enter uninjured stone fruits, which were correlated with the secretion of esterase enzymes by the fungus (Baggio et al. 2016). The fungus is unable to grow at temperature less than 5°C. The most

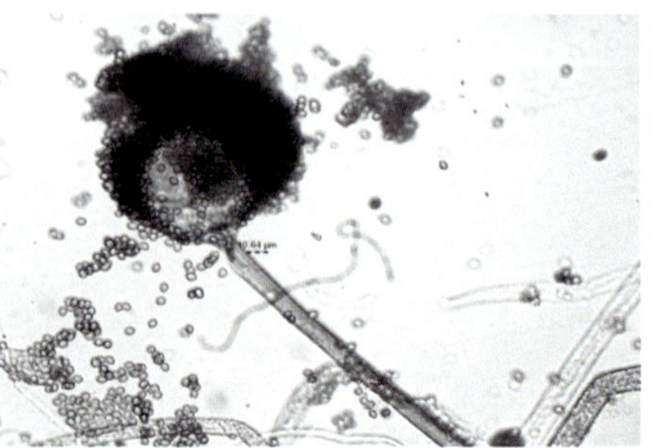

FIGURE 21.3 Symptoms of Rhizopus rot on peach fruits (a) and morphological structure of *Rhizopus stolonifer* having sporangiophores, sporangium and spores (b).

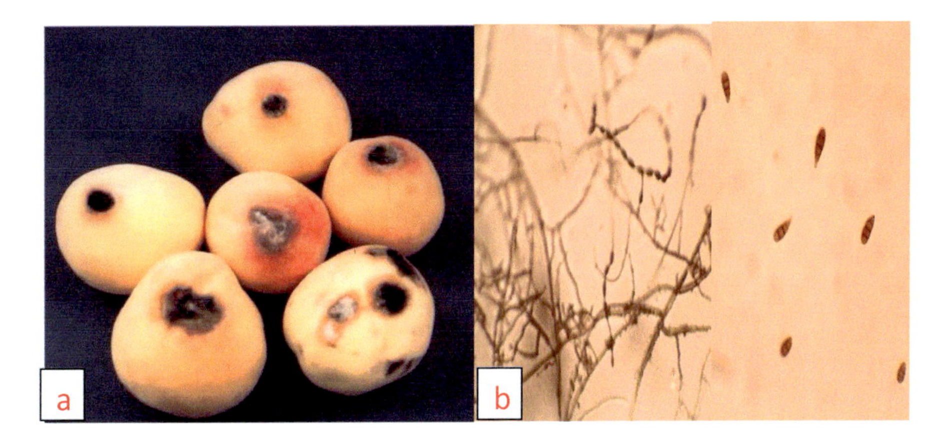

FIGURE 21.4 Symptoms of Alternaria rot (a) and mycelium and conidia of *Alternaria alternata* (b).

favourable temperature for spore germination and development of Rhizopus rot is 25–35°C.

21.2.4 Alternaria Rot

Alternaria rot of stone fruits has been documented from different regions of the world. The disease more commonly occurs on cherries and plums and is occasionally found on peaches, apricots and nectarines.

21.2.4.1 Symptoms

The symptoms appear as firm, dark brown to black and somewhat wet spots (Figure 21.4). The spots are covered with white mycelia growth of the fungus and spores of olivaceous green colour. The decayed tissue beneath the spot is cone shaped, with the apex of the cone extended towards the pit. The cone separates readily from the surrounding tissue and can be lifted out. The symptoms of Alternaria rot are similar to those of Cladosporium rot, but in the former, the surface mould is darker, the decayed tissue is more moist and dark and the enlargement of the lesions is more rapid.

21.2.4.2 Causal Organism

The disease is caused by different species of *Alternaria*, i.e. *Alternaria alternata* (Fr.) Keissl., *A. tenuis* Nees, *A. tenuissima* Samuel Paul Wiltshire. The conidia are beak-shaped with horizontal and vertical septation. Conidia of *A. alternata* are dark brown in colour and have short beaks. In contrast, conidia of *A. tenuissima* are smooth-walled, light brown to golden brown in colour, have long and tapering beaks, occur singly or in chains of 2–5 and have slight constriction at the transverse septa (Ellis et al. 1997; Simmons 2007; Pastor and Guarro 2008). On the other hand, conidia of *A. alternata* usually grow in long chains. The culture of *A. tenuissima* on natural medium forms concentric rings (Nasehi et al. 2012).

21.2.4.3 Disease Cycle and Epidemiology

The pathogen usually enters through breaks on the surface of the peel of fruits. The pathogen grows rapidly at 29°C. The decay of fruits is very slow during storage at 4°C.

21.2.5 Blue Mould Rot/Penicillium Rot

Blue mould rot disease of stone fruits is a very common occurrence in cherries. It appears on injured or over-ripe fruits of peaches, nectarines and apricots. However, this rot does not occur on freshly harvested and uninjured fruits. The blue rot also appears on fruits during storage at low temperature for very long duration.

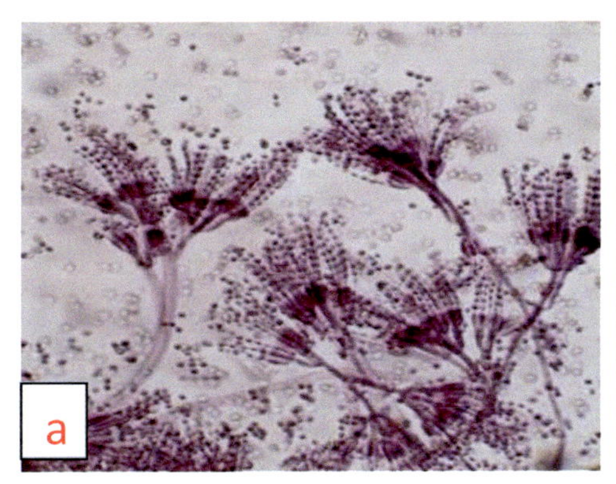

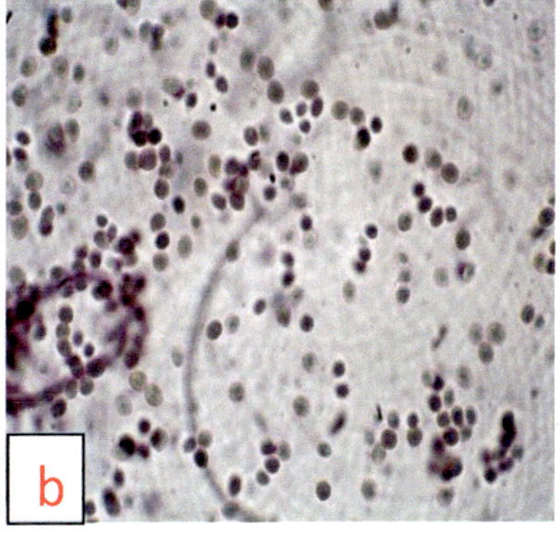

FIGURE 21.5 Conidiophores and their ornamentation, and phialides (a) and conidia (b) of important pathogenic fungi of peach.

21.2.5.1 Symptoms

The symptoms of blue rot appear as circular, flat, light brown/tan coloured spots on the surface of the fruits. Internally, the tissue is very soft and moist. With the progress in infection under humid conditions, the mould appears as white tufted cushions near the centre of spot and soon produces masses of bluish-green spores. The affected areas gradually enlarge. The pathogen rapidly infects adjoining healthy fruits by forming "nests" of decay. The decaying fruits emit a musty odour with a musty taste.

21.2.5.2 Causal Organism

The disease is mainly caused by *Penicillium expansum* Link. The fungus occurs throughout the world. It is psychrophilic fungus and can survive at low temperatures. However, in some cases, some other species of *Penicillium* may also be involved with the development of blue mould rot. The conidiophores bear phialides which are arranged in a brush-like manner and eventually bear conidia. The conidia are usually in chains, dry, smooth, elliptical, and "dull-green" in colour and are usually disseminated by air (Figure 21.5a, b).

21.2.5.3 Disease Cycle and Epidemiology

The conidia of *P. expansum* enter through wounds on the fruit surface and initiate infection (Torres et al. 2006). Generally, the handling of fruits during harvesting, packaging and processing results in creation of wounds through puncturing, rubbing, etc. which predisposes the fruits to entry of fungal conidia. Conidia can survive in soil, crop debris and tree bark. The conidia are airborne also and can infect the fruits at any stage of growth, harvesting, packaging, processing, transportation and storage. After gaining entry in the fruits through wounds, conidia germinate and form germ-tubes. The germ-tubes eventually grow to form hyphae which establish the infection and colonization of fungus occurs in infected stone fruits. The infected fruit cells are killed and infection spreads. A large number of conidia are produced by *Penicillium* spp. Conidia are widely disseminated through air and breaches on the surface of fruits. *Penicillium* is a wound parasite which usually gains entry through lenticles or cracks caused by changes in moisture or injuries during harvesting and packing. The fungus rarely infects the fruits through intact peel. Over mature fruits are more susceptible. The optimum growth of fungus occurs at 25°C and growth almost stops at 0°C. The conidia can germinate at 0°C and initiate the infection but the germination is greatly delayed and the infection proceeds very slowly.

21.2.6 Black Mould Rot/Aspergillus Rot

Black mould rarely occurs on cherries, peaches, nectarines, apricots and plums. It rarely occurs on freshly harvested fruits. This rot usually occurs when ripening of fruits occurs in warm periods followed by prolonged transit or storage.

21.2.6.1 Symptoms

The symptoms initiate as small, tan, slightly sunken spots on the fruit surface. Later, white fluffy growth of the fungus appears near the centre of the spots which produces abundant masses of spores. Initially, the spores are yellow but eventually turn jet black.

21.2.6.2 Causal Organism

Aspergillus niger Tiegh. is responsible for black mould rot disease of stone fruits. The colour of *A. niger* colonies is white or yellow and bears numerous dark coloured conidia. The fungal mycelium is transparent and septate. Conidiophores of fungus are long, usually 900–1600 µm. Each conidiophore has a typical foot cell and a terminal swollen bulbous cell called a "vesicle". The diameter of vesicles ranges from 40–60 µm. On vesicles, bottle-shaped structures called sterigmata or phialides arise all over the surface and produce chains of conidia at the tips (Figure 21.5f). Each globose vesicle bears brown-coloured metulae which eventually bear phialides. Hence, the arrangement of phialides on vesicles is biseriate. The phialides are conidiogenous cells and form conidia in a blastic basipetal succession through the process of conidiogenesis. The conidia are globose in shape and have a diameter of 3–5 µm (Debets et al. 1990).

21.2.6.3 Disease Cycle and Epidemiology

The fungus grows rapidly at high temperature, i.e. around 30–35°C, whereas no or little growth occurs below 7°C.

21.2.7 Bacterial Spot/Bacterial Canker

Bacterial spot or bacterial canker is an important disease of peaches, nectarines and apricots. The disease is especially important in the fruits grown in subtemperate or subtropical regions. This disease causes huge economic losses in stone fruits throughout the world (Stefani 2010). However, the disease occurs less often in temperate regions. Fruit quality and quantity are drastically reduced due to the disease. The affected fruits are prone to other fungal, bacterial and insect attack.

21.2.7.1 Symptoms

Initially, the disease appears as small, brown, circular lesions on maturing fruits on trees. With the enlargement of the spots, the centres become darker and sunken and the margins are water-soaked. With the advancement of the infection, numerous cracks appear on the fruits which may be either small or shallow. There is exudation of yellowish and gummy mass from the cracks. The exudate may also be produced in larger spots which later dry into hornlike projections.

21.2.7.2 Causal Organism

Bacterial spot or bacterial canker is caused by bacterium *Xanthomonas arboricola* pv. *pruni* (Smith). (Vauterin et al. 1995), earlier known as *Xanthomonas campestris* pv. *pruni* (Vauterin et al. 1995). The bacterium attacks the fruit crops of *Prunus* species only. The bacterium is an important pathogen of peach, nectarine and plum. The bacterium is also reported to cause disease on cherry, almond and apricot but causes less damage on these crops. *X. arboricola* pv. *pruni* is listed as a quarantine pathogen in many countries (Garita-Cambronero et al. 2018). The bacterium is motile, strictly aerobic, gram-negative, rod-shaped with round ends, non-spore forming and possesses a single polar flagellum, i.e. monotrichous. The pathogen occurs singly, occasionally in pairs and rarely in short chains.

21.2.7.3 Disease Cycle and Epidemiology

The bacterial pathogen overwinters in infected twigs. During spring, the pathogen is disseminated by wind and rain to leaves, blossoms and fruits. The bacterial pathogen initiates the infection on blossoms or the unbroken peel of the fruit. Severe infection of bacterium occurs at 21–30°C in the orchard. The bacterial growth slows down at temperature above 30°C. The disease should be managed in the orchard only. Postharvest treatments are not effective in managing bacterial canker of stone fruits.

21.3 Management of Postharvest Diseases of Stone Fruits

Management of postharvest diseases depends on understanding the nature of causal organisms, the conditions that promote the occurrence of diseases and the factors that affect their capacity to cause losses.

21.3.1 Sanitation and Proper Handling of Fruits

Usually, the inoculum available during the harvesting, grading, packing, storage and transportation is the main source of infection of postharvest diseases. Most of the postharvest fruit pathogens gain entry through wounds, bruises and injuries besides natural openings. Hence careful picking, handling, grading, packing, storing and transportation should be ensured to prevent or minimize mechanical injury of fruits during and after harvesting. Disinfection of baskets and boxes should be followed to avoid contamination. Sanitation practices in the orchard, packinghouse and during storage and transport should be followed. Destruction of mummified fruits in the orchard, cleaning of grading and packing facilities, and destruction of culled fruits near orchards and packinghouses will reduce the inoculum load and will eventually reduce the probability of fruit infection.

21.3.2 Proper Storage Conditions

Refrigeration after harvesting of fruits is the most commonly used method for managing the infection of postharvest diseases. In addition to temperature, the humidity and carbon dioxide concentration should also be controlled to increase the shelf life of fruits. The activity of postharvest fruit pathogens is slow at low temperatures. Moreover, the ripening process of fruits inhibits the synthesis of antimicrobial substances (Agrios 2005), hence slow-down of ripening process by maintaining temperature, humidity and carbon dioxide level will help in production of antimicrobial substances which will eventually help in reducing the occurrence of postharvest diseases. Harvesting of fruits should be avoided during wet weather. Harvesting of fruits should be done early in morning. The field heat should be removed before packing.

21.3.3 Physical Methods

Postharvest infections can be reduced by controlling temperature and humidity, modified atmosphere, ionizing radiations, good sanitation. After harvesting the fruits should be sterilized either by exposing to gamma radiation or hot water treatment or chemical dip treatments. Heat treatment of fruits before storage is also a recommended practice for management of postharvest disease of stone fruits. Singh et al. (2006) reported that hot water dip treatments (50°C for 3 min) reduced incidence of Rhizopus rot in peach fruits. Hot water dip or moist hot air exposure of fruits helped to reduce both surface infection and deep infections in pulp of fruit. Exposure of cherry fruits to hot air at 44°C for 114 minutes lowered the incidence of blue mould decay due to *P. expansum* (Wang et al. 2015). It was also observed that the activity of tissue-softening enzymes, i.e., polygalacturonase and pectinmethylesterase, was significantly inhibited by hot air treatment, whereas the activity of defence enzymes, i.e., superoxide dismutase, catalase, peroxidase, and polyphenoloxidase, chitinase and β-1,3-glucanase, was enhanced by hot air treatment.

21.3.4 Use of Fungicides

Fungicides are widely used for the management of postharvest diseases of stone fruits. Some of the fungicidal dip treatments reported to be effective for managing grey mould, brown mould, Rhizopus rot diseases in stone fruits are carbendazim, hexaconazole, iprodione andbitertanol (Parveen et al. 2013, 2016). Postharvest application of fungicides like fludioxonil, tebuconazole, azoxystrobin, fenhexamid, fenbuconazole, myclobutanil, cyprodinil and propiconazole was found to be effective in reducing the Rhizopus rot of peaches in Canada (Northover and Zhou 2002). "Reduced-risk" fungicides such as fenhexamid, fludioxonil and tebuconazole were highly effective in reducing the incidence of brown rot, grey mould, Rhizopus rot of peach, nectarine and plum caused by *Monilinia fructicola*, *Botrytis cinerea* and *Rhizopus stolonifera*, respectively (Förster et al. 2007). It was further observed that these fungicides were more effective when applied as protectants. Treatment of apricot, peach and nectarine fruits with fludioxonil was also effective in reducing the postharvest decay of stone fruits (D'Aquino et al. 2007).

Pre-harvest application of fungicides such as fenhexamid tebuconazole, boscalid + pyraclostrobin, difenoconazole and cyprodinil + fludioxonil, applied 1 week before harvest of fruits, lowered the occurrence of brown rot, blue mould and grey mould of peach, nectarine and apricot (D'Aquino et al. 2012). Pre-harvest application of four fungicides, viz. captan, iprodione, iminoctadine and tebuconazole, alone and in combination with postharvest application of *Trichoderma harzianum*, were evaluated at 5°C and 20°C against brown rot,

Rhizopus rot and blue mould rot of peach fruits (Pavanello et al. 2015). Tebuconazole application before harvest lowered the incidence of brown rot at 20°C. The postharvest application of *T. harzianum* alone did not show any benefit for the management of brown rot, but its combination with tebuconazole resulted in lowering the occurrence of brown rot, Rhizopus rot and blue mould rot diseases of peach during shelf life.

21.3.5 Use of Biological Control Agents

Biological control through microbial antagonists has been exploited as a substitutive strategy to manage postharvest diseases of stone fruits. For the management of postharvest diseases of peaches and nectarines, 103 isolates of yeast were evaluated, out of which 7 isolates were found to be effective to reduce the incidence and severity of blue mould and grey mould diseases caused by *Penicillium expansum* and *Botrytis cinerea*, respectively (Karabulut and Baykal 2003). Pre-harvest applications of *Epicoccum nigrum* on peach trees in orchards reduced the incidence of postharvest brown rot (Larena et al. 2005). Bio-control effects of *Bacillus subtilis* and induction of resistance by *B. cereus* AR156 for the reduction of Rhizopus rot of peach fruit was demonstrated (Wang et al. 2013a, 2013b). Culturable bacterial microflora of plums was isolated and *Curtobacterium*, *Pseudomonas*, *Microbacterium* and *Clavibacter* were found in higher frequency (Janisiewicza et al. 2013). Further, at the species level, *Curtobacterium flaccumfaciens*, *Clavibacter michiganensis*, *Microbacterium lacticum*, *Enterobacter intermedius* and *Chrysomonas luteola* were the most frequently isolated species. Among all the isolated bacterial flora of plum, *Pantonea agglomerans* and *Citrobacter freundii* were the most effective for the management of brown rot disease of plum, and grew very well on wounded fruits of plums. For the control of different postharvest diseases of peach, like Alternaria rot and grey mould rot, *Bacillus subtilis* JK-14, was demonstrated to be effective (Zhang et al. 2019). Further, it was showed that activity of the antioxidant enzymes, i.e., superoxide dismutase, peroxidase and catalase, increased after treatment with *B. subtilis* JK-14 in peach fruits which were challenge-inoculated with *Alternaria tenuis* and *Botrytis cinerea*.

21.3.6 Use of Biopesticides

Brown rot and anthracnose of peach were effectively managed under *in vitro* and *in vivo* conditions by essential oil of *Eucalyptus globulus*, *Cinnamomum camphora* and *Cymbopogum citrates* (Pancera et al. 2015). Essential oil of *Thymus vulgaris* was effective to decrease the mycelial growth of *R. stolonifer* and also lowered the incidence of Rhizopus rot in peach fruits (Taheri et al. 2018). The ethanolic extracts of avocado seeds were assessed for their inhibitory action against the mycelial growth of *M. fructicola* and were reported to have antifungal activity at different concentrations (Fagundes et al. 2018). Recently, antifungal potential of polyphenolic extract of orange peel was reported against three important postharvest fungal pathogens, namely *M. fructicola*, *B. cinerea* and *A. alternata* (Hernández et al. 2020). Further, two phenolic compounds, namely flavonoids (naringin, hesperidin and neohesperidin) and phenolic acids (ferulic and *p*-coumaric), were assessed for their antifungal potential, and it was found that phenolic acids had significantly greater potential to inhibit the mycelia growth of pathogenic fungi in synthetic medium.

21.3.7 Induction of Resistance

Pre-harvest and postharvest applications of resistance-inducing chemicals have been reported to be efficacious to reduce the occurrence of various postharvest diseases of stone fruits. Liu et al (2005) demonstrated that postharvest dip of peach fruits in benzo-(1,2,3)-thiadiazole-7-carbothioic acid (BTH) for 5 minutes resulted in enhanced resistance against. It was further demonstrated that the activity of defence-related enzymes, phenols and hydrogen peroxide increased significantly. The incidence and severity of blue mould rot caused by *P. expansum* was reduced with 0.5 mM salicylic acid treatment on sweet cherries (Chan and Tian 2006). Chitosan at 2 mg/mL was efficient in reducing the growth of mycelial, sporulation and spore-germination of *R. stolonifer* and also reduced the disease incidence and severity of Rhizopus rot disease of peach (Hernández-Lauzardo et al. 2010). The treatment of peach fruits with a synthetic cytokinin, i.e., 6-benzylaminopurine (BAP) resulted in 63% disease reduction of brown rot and also lowered the diameter of lesions in comparison to control (Zhang et al. 2015). Further, it was demonstrated that activities of defence-related enzymes, specifically polyphenol oxidase and peroxidase, increased, which resulted in triggering of defence responses in the host plant. Li et al. (2017) demonstrated that application of nitric oxide activates the expression of enzymes of phenylpropanoid pathway, i.e., phenylalanine ammonia-lyase, cinnamate-4-hydroxylase, 4-coumaroyl-CoA ligase, chalcone synthase and chalcone isomerase and induced resistance in peach fruits against brown rot caused by *M. fructicola*. Another study showed that the incidence of blue mould rot, caused by *P. expansum*, of peach fruits was reduced by chlorogenic acid treatment which induced resistance in plants by triggering signalling pathway of salicylic acid (Jiao et al. 2018).

21.3.8 Integrated Disease Management Practices

Integration of a yeast antagonist, i.e., *Candida oleophila* with hot water treatment (55°C) for 10 seconds and subsequently storage at 0°C was found to be effective for the management of grey mould rot and blue mould rot diseases of peaches (Karabulut and Baykal 2004). Four-time application of *Epicoccum nigrum*, alone and in integration with fungicides, was recommended as a disease management approach to reduce the postharvest decay of peach fruits and also for reducing fungicide treatments and residues (Larena et al. 2005). Combined postharvest treatment of peach cv. Shan-e-Punjab fruits with hot water treatment (50°C for 3 min) and bio-control agent (*Debaryomyces hansenii*) and packed in 1.0% perforated 200-gauge HDPE film and stored in cold storage saved the fruits synergistically from postharvest decay and maintained the quality of fruits for a longer period (Singh and Mandal 2006; Mandal et al. 2007). Combined application of salicylic acid at 0.05 mM with ultrasound resulted in reducing the disease incidence of blue mould rot of peach fruits

(Yang et al. 2011). Combination of pre-harvest spray application of fungicides such as fenhexamid tebuconazole, boscalid +pyraclostrobin, difenoconazole and cyprodinil + fludioxonil followed by 2 minutes postharvest dip in 2% sodium bicarbonate at 20°C reduced rotting of peach, nectarine and apricot fruits by postharvest diseases (D'Aquino et al. 2012). Singh (2005) and Singh and Sharma (2009) reported that *D. hansenii* and calcium chloride (2.0%) synergistically reduced Rhizopus rot of peaches during storage. Combined application of ultrasounds and potassium sorbate solution to plum fruits reduced the blue rot decay of plum fruits (Molinu et al. 2012). Wu et al. (2019) demonstrated that the combination of *Bacillus subtilis* with heat treatment of peach fruits resulted in significant reduction in the incidence of brown rot disease. In comparison to treatment of either *B. subtilis* or heat treatment alone, the integrated treatment significantly enhanced the activity of phenylalanine ammonia-lyase (PAL), β-1,3-glucanase, chitinase, polyphenol oxidase (PPO), catalase (CAT), peroxidase (PO) and superoxide dismutase (SOD) and induced resistance in peach fruit against pathogenic fungus.

21.4 Conclusions

Stone fruits are highly perishable fruits and are subject to many postharvest diseases during storage and transportation. The infections of stone fruits with various pathogens lead to quick rot and decay of stone fruits after harvesting. Disease management practices like fungicides, bio-control agents, physical methods, induced resistance, etc. are effective for the management of postharvest diseases of stone fruits.

21.5 Future Directions

Although a large number of studies have demonstrated the efficacy of postharvest treatments, only a few of them are currently applied under commercial conditions. Many treatments are developed under *in vitro* and have limitations such as economic cost, delays due to registration issues, or reduced efficacy compared to conventional fungicides. Research is needed to resolve these obstacles to make all these strategies available in the near future. Additionally, development of new molecular techniques for identification of pathogens for disease forecasting and risk management will lead to better management of postharvest diseases of stone fruits.

REFERENCES

Agrios GN (2005) Plant Pathology. 5th edition. Elsevier/Academic Press, Amsterdam, 952 pp.

Baggio JS, Goncalves FP, Lourenco SA, Tanaka FAO, Pascholati SF, Amorim L (2016) Direct penetration of *Rhizopus stolonifer* into stone fruits causing Rhizopus rot. Plant Pathol 65: 633–642.

Chan ZL, Tian SP (2006) Induction of H_2O_2-metabolizing enzymes and total protein synthesis by antagonistic yeast and salicylic acid in harvested sweet cherry fruit. Postharvest Biol Technol 39: 314–320.

D'Aquino S, Barberis A, Satta D, De Pau L, Schirra M (2012) Pre-harvest treatments with fungicides and postharvest dips in sodium bicarbonate to control postharvest decay in stone fruit. Commun Agric Appl Biol Sci 77(3): 197–205.

D'Aquino S, Schirra M, Palma A et al. (2007) Residue levels and storage responses of nectarines, apricots, and peaches after dip treatments with fludioxonil fungicide mixtures. J Agric Food Chem 55(3): 825–831.

Debets A, Holub E, Swart K, van den Broek H, Bos C (1990) An electrophoretic karyotype of *Aspergillusniger*. Mol Gen Genet 224: 1432–1874.

Ellis MB, Ellis JP, Hawksworth A (1997) Microfungi on Land Plants: An Identification Handbook. New Enlarged Edition. Richmond Publishing, Slough, 860 pp.

Fagundes MCP, Oliveira AF, Carvalho VL, Ramos JD, Santos VA, Rufini JCM (2018) Alternative control of plant pathogen fungi through ethanolic extracts of avocado seeds (*Persea americana* Mill.). Brazilian Arch Biol Technol 61: e18180052.

Förster H, Driever GF, Thompson DC, Adaskaveg JE (2007) Postharvest decay management for stone fruit crops in California using the "reduced-risk" fungicides fludioxonil and fenhexamid. Plant Dis 91: 209–215.

Fournier E, Giraud T, Albertini C, Brygoo Y (2005) Partition of the *Botrytis cinerea* complex in France using multiple gene genealogies. Mycologia 97(6): 1251–126.

Garita-Cambronero J, Palacio-Bielsa A, Cubero J (2018) *Xanthomonas arboricola* pv. *pruni*, causal agent of bacterial spot of stone fruits and almond: Its genomic and phenotypic characteristics in the *X. arboricola* species context. Mole Plant Pathol 19: 2053–2065.

Hernández A, Ruiz-Moyano S, Galván AI, Merchán AV, Nevado FP, Aranda E, Serradilla MJ, Córdoba MG, Martin A (2020) Anti-fungal activity of phenolic sweet orange peel extract for controlling fungi responsible for post-harvest fruit decay. Fungal Biol. DOI.10.1016/j.funbio.2020.05.005.

Hernández-Lauzardo A, Velázquez-del Valle M, Veranza-Castelán L, Melo-Giorgana G, Guerra-Sánchez M (2010) Effect of chitosan on three isolates of *Rhizopusstolonifer* obtained from peach, papaya and tomato. Fruits 65(4): 245–253.

Janisiewicza WJ, Jurick WM, Vico I, Peter KA, Buyer JS (2013) Culturable bacteria from plum fruit surfaces and their potential for controlling brown rot after harvest. Postharvest Biol Technol 76: 145–151.

Jiao W, Li X, Wang X, Cao J, Jiang W (2018) Chlorogenic acid induces resistance against *Penicillium expansum* in peach fruit by activating the salicylic acid signaling pathway. Food Chem 260: 274–282.

Karabulut OA, Baykal N (2003) Biological control of postharvest diseases of peaches and nectarines by yeasts. J Phytopathol 151(3): 130–134.

Karabulut OA, Baykal N (2004) Integrated control of postharvest diseases of peaches with a yeast antagonist, hot water and modified atmosphere packaging. Crop Protec 23(5): 431–435.

Larena I, Torres R, De Cala A, Liñána M, Melgarejoa P, Domenichini P et al. (2005) Biological control of postharvest brown rot (*Monilinia* spp.) of peaches by Weld applications of *Epicoccum nigrum*. Biol Control 32: 305–310.

Li G, Zhu S, Wu W, Zhang C, Peng Y, Wang Q, Shi J (2017) Exogenous nitric oxide induces disease resistance against

Monilinia fructicola through activating the phenylpropanoid pathway in peach fruit. J Sci Food Agric 97(9): 3030–3038.

Liu H, Jiang W, Bi Y, Luo Y (2005) Postharvest BTH treatment induces resistance of peach (*Prunus persica* L. cv. Jiubao) fruit to infection by *Penicillium expansum* and enhances activity of fruit defense mechanisms. Postharvest Biol Technol 35: 263–269.

Mandal G, Singh D, Sharma RR (2007) Effect of hot water treatment and bio control agent (*Debaryomyce shansenii*) on shelf life of peach. Indian J Hortic 64(1): 25–28.

Molinu MG, Pani G, Venditti T, Dore A, Ladu G, D'Hallewin G (2012) Alternative methods to control postharvest decay caused by *Penicillium expansum* in plums (*Prunus domestica* L.). Commun Agric Appl Biol Sci 77(4): 509–514.

Nasehi A, Kadir JB, Abidin MA Z, Wong M Y, Mahmodi F (2012) First report of *Alternaria tenuissima* causing leaf spot on eggplant in Malaysia. Plant Dis 96(8): 1226–1226.

Northover J, Zhou T (2002) Control of rhizopus rot of peaches with postharvest treatments of tebuconazole, fludioxonil, and *Pseudomonas syringae*. Can J Plant Pathol 24(2): 144–153.

Pancera MR, Conte RI, Silva SM, Sartori VC, Silva Ribeiro RT (2015) Strategic control of postharvest decay in peach caused by *Monilinia fructicola* and *Colletotrichum gloesporioides*. Braz J Appl Technol Agric Sci 8(1): 7–14.

Parveen S, Ganie AA, Wani AH (2013) *In vitro* efficacy of some fungicides on mycelial growth of *Alternaria alternata* and *Mucor piriformis*. Arch Phytopathol Plant Protec 46(10): 1230–1235.

Parveen S, Wani AH, Bhat MY, Koka JA, Wani FA (2016) Management of postharvest fungal rot of peach (*Prunus persica*) caused by *Rhizopus stolonifer* in Kashmir Valley, India. Plant Pathol Quora 6(1): 19–29.

Pastor FJ, Guarro J (2008) Alternaria infections: Laboratory diagnosis and relevant clinical features. Clinical Microbiol Infec 14(8): 734–746.

Pavanello EP, Brackmann A, Thewes FR, Venturini TL, Weber A, Blume E (2015) Postharvest biological control of brown rot in peaches after cold storage preceded by pre-harvest chemical control. Rev Ceres Viçosa 62(6): 539–545.

Rupp S, Weber RWS, Rieger D, Detzel P, Hahn M (2017) Spread of *Botrytis cinerea* strains with multiple fungicide resistance in German horticulture. Front Microbiol 7: 2075.

Sestari I, Giehl RFR, Pinto JAV, Brackmann A (2008) Condições de atmosfera control adaparpêssegos "Maciel" colhidosemdoisestádios de maturação. Ciência Rural 38: 1240–1245.

Siddique SS, Hardy GEStJ, Bayliss KL (2018) Advanced technologies for controlling postharvest diseases of fruit. Acta Hortic 1194: 193–200.

Simmons EG (2007). Alternaria: An Identification Manual (fully illustrated and with catalogue raisonné 1796-2007). CBS Fungal Biodiversity Centre, Utrecht, the Netherlands, pp. 500–502.

Singh D (2005) Interactive effect of *Debaryomyces hansenii* and calcium chloride to reduce Rhizopus rot of peaches. Indian J Mycol Pl Pathol 35(1): 118–121.

Singh D, Mandal G (2006) Improved control of spoilage with a combination of hot water immersion and *Debaryomyces hansenii* of peach fruit during storage. Indian Phytopathol 59(2): 168–173.

Singh D, Mandal G, Agarwal MK, Kumar P, Jain RK (2006) Effect of hot water dip treatments to reduce incidence of Rhizopus rot in peach fruits. Indian Phytopathol 59: 52–55.

Singh D, Sharma RR (2009) Post harvest behaviour of peaches (*Prunus persica*) pre-treated with antagonist *Debaryomyces hansenii* and calcium chloride. Indian J Agric Sci 79(9): 674–678.

Stefani E (2010) Economic significance and control of bacterial spot/canker of stone fruits caused by *Xanthomonas arboricola* pv. *pruni*. J Plant Pathol 92: 99–104.

Taheri P, Ndam LM, Fujii Y (2018) Alternative approach to management of Rhizopus rot of peach (*Prunus persica* L.) using the essential oil of *Thymus vulgaris* (L.). Mycosphere 9(3): 510–517.

Torres R, Teixidó N, Viñas I, Mari M, Casalini L, Giraud M, Usall J (2006) Efficacy of *Candida sake* CPA-1 formulation for controlling *Penicillium expansum* decay on pome fruit from different Mediterranean regions. J Food Protec 69(11): 2703–2711.

Vauterin L, Hoste B, Kersters K, Swings J (1995) Reclassification of *Xanthomonas*. Int J Syst Bacteriol 45: 472–489.

Walker AS, Gatier A, Confais J, Martinho D, Viaud M, Le Pêcheur P, et al. (2011) *Botrytis pseudocinerea*, a new cryptic species causing gray mold in French vineyards in sympatry with *Botrytis cinerea*. Phytopathology 101: 1433–1445.

Wang L, Jin P, Wang J, Gong H, Zhang S, Zheng Y (2015) Hot air treatment induces resistance against blue mold decay caused by *Penicillium expansum* in sweet cherry (*Prunus cerasus* L.) fruit. Sci Hortic 189: 74–80.

Wang XL, Wang J, Jin P, Zheng YH (2013a) Investigating the efficacy of *Bacillus subtilis* SM21 on controlling Rhizopus rot in peach fruit. Int J Food Microbiol 164: 141–147.

Wang XL, Xu F, Wang J, Jin P, Zheng YH (2013b) *Bacillus cereus* AR156 induces resistance against Rhizopus rot through priming of defense responses in peach fruit. Food Chem 136: 400–406.

Wu S, Zhen C, Wang K, Gao H (2019) Effects of *Bacillus subtilis* CF-3 VOCs combined with heat treatment on the control of *Monilinia fructicola* in peaches and *Colletotrichum gloeosporioides* in litchi fruit. J Food Sci 84(12): 3418–3428.

Yang ZF, Cao SF, Cai YT, Zheng YH (2011) Combination of salicylic acid and ultrasound to control postharvest blue mold caused by *Penicillium expansum* in peach fruit. Innov Emerg Technol 12: 310–314.

Zhang Y, Zeng L, Yang J, Zheng X, Yu T (2015) 6-Benzylaminopurine inhibits growth of *Monilinia fructicola* and induces defense-related mechanism in peach fruit. Food Chem 187: 210–217.

Zhang S, Zheng Q, Xu B, Liu J (2019) Identification of the fungal pathogens of postharvest disease on peach fruits and the control mechanisms of *Bacillus subtilis* JK-14. Toxins 11: 322.

22

Postharvest Fungal Diseases in Stone Fruits and Their Management

Efath Shahnaz, Ali Anwar and Saba Banday
Division of Plant Pathology, Sher-e-Kashmir University of Agricultural Sciences & Technology of Kashmir, Shalimar Campus, Srinagar, Jammu & Kashmir, India

CONTENTS

22.1 Introduction

The fresh fruit industry is highly vulnerable to losses caused by postharvest diseases, particularly the highly perishable fruits. The stone fruits have been ranked as cherries, nectarines, peaches, plums, and apricots in order of their decreasing susceptibility to postharvest decay (Eckert and Ogawa 1988). *Monilinia fructicola*, *Botrytis cinerea* and *Rhizopus stolonifer* are the primary decay pathogens in semi-arid regions, whereas *Geotrichum candidum* and *Gloeosporium gloeosporioides* are more commonly encountered in humid regions (Ogawa et al. 1961). Similarly, *Penicillium expansum* and *Alternaria alternata* are more commonly encountered on sweet cherries, while *Gilbertella persicaria* and *Mucor piriformis* frequently associated with decays of peaches and nectarines. The different species of *Monilinia* and *Botrytis cinerea* cause brown rot and grey mould, respectively, of almost all pome and stone fruits. However, the frequency and extent of damage caused by these pathogens on different fruits is considerably influenced by the host, type and virulence of pathogens and above all various environmental factors, particularly temperature and relative humidity.

22.2 Postharvest Diseases of Stone Fruits

22.2.1 Brown Rot

Brown rot is a significant postharvest disease of pome as well as stone fruits all over the world, causing significant losses each year. The causal pathogen, *Monilinia* Honey (anamorph *Monilia*) causes similar symptoms on almost each of the stone and pome fruits, most notably cankers, blossom and twig blight, and fruit rot. The most notable symptoms are produced on *Cerasus avium* (bird cherry), *C. vulgaris* (morello cherry), *Cydonia* spp. (quince), *Malus domestica* (apple), *M. pumila* (apple), *Prunus armeniaca* (apricot), *P. avium* (sweet cherry), *P. cerasus* (sour cherry), *P. domestica* (plum), *P. persica* (peach), *Prunus* spp. (stone fruit), and *Pyrus* spp. (pear) (Van Leeuwen et al. 2002; Hilber-Bodmer et al. 2012; Poniatowska et al. 2016).

22.2.1.1 *Symptoms*

The pathogen causes various symptoms depending on the site of infection, like leaf blights, twig blights, cankers of stem and brown fruit rots. On leaves, infection appears as slightly circular, brown areas that may turn necrotic and collapse, resulting in either a "shot hole" or death of the entire leaf. Infected blossoms are brown in colour and ultimately die (Figure 22.1a–c). The pathogen then invades the adjacent twig tissue from blossoms, where it results in sunken dead areas of bark delimited by sharp borders. In some cases, the pathogen causes brownish dieback on the leaf petioles, laminas, small fruits and fruit pedicels (Petróczy and Palkovics 2009).

The pathogen can attack the developing fruits at any stage. Initially the fruit shows brown, circular and firm lesions which later cover the whole fruit, resulting in rotting of fruit. This is the highly damaging stage and economically important. Under optimal conditions of temperature and relative humidity, trusses of mycelium and conidia may develop from the infected fruit skin in concentric rings (Byrde and Willetts 1977). The fruits thus infected may remain attached to the tree or fall to the ground. In the former case, they dry, shrivel and persist as mummies (Van Leeuwen et al. 2002). The mummified fruit may remain hanging on the tree or later fall from the tree onto the orchard floor where they survive the harsh winter buried completely or partially under the leaf litter or soil (Byrde and Willetts 1977).

22.2.1.2 *Causal Organism*

Monilinia fructicola (Winter) Honey; Synonym: *Sclerotinia fructicola* (Winter) Rehm; Anamorph: *Monilia fructicola* Batra is causal agent of brown rot disease of stone fruits (Figure 22.1a–c). There are about 30 species of the genus reported but only a few are of economic importance. The commonly encountered species in Europe are *M. laxa* and *M. fructigena*, while *M. fructicola* is found in some parts of Asia, North, Central and South America, Australia and New Zealand (Poniatowska et al. 2013). *Monilia polystroma* (anamorph), which closely resembles *M. fructigena*, is reported only from Japan (Van Leeuwen et al. 2002). Hu et al. (2011) reported *Monilia mumecola* and *M. yunnanensis* from China in addition to these four species. Yin et al. (2014) reported *M. mumecola*, *M. fructigena* and *M. polystroma* from the ripening fruit of *Prunus armeniaca* (apricot), whereas *M. laxa* was isolated from infected blossoms and blighted twigs of apricot (Van Leeuwen et al. 2002; Poniatowska et al. 2013). *M. fructicola* has been isolated from both blighted blossom and rotten fruit (Anonymous 2009).

22.2.1.3 *Disease Cycle and Epidemiology*

The mummified fruit or cankerous lesions or infected fruit stalks, scars and buds, consist of the conidia of the pathogen which function as primary sources of inoculum with the advent of favourable climactic conditions in the spring season. Primary infection causes blossom blight, which under favourable conditions leads to twig blight and branch canker. These areas result in the formation of secondary inoculum from which latent infection of immature green fruit takes place, resulting in pre-and postharvest brown rot on mature stone fruit (Schlagbauer and Holz 1989; Hong et al. 1998). The pre-harvest losses are less severe than postharvest stages, especially losses caused during storage and transit (Ogawa and English 1991; Guijarro et al. 2008). Under unfavourable environmental conditions, the pathogen remains latent in immature fruit until conditions are more favourable for disease development towards the end of growing season (De Cal et al. 2008).

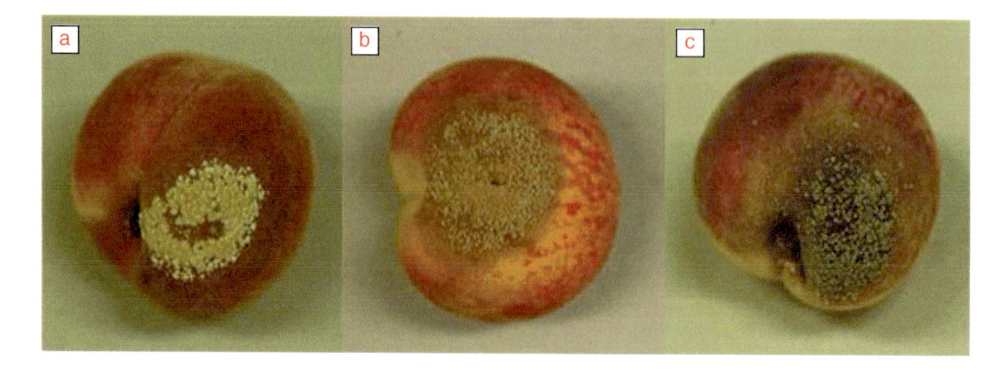

FIGURE 22.1 Peaches inoculated with (a) *Monilinia fructigena*, (b) *M. Fructicola* and (c) *M. Laxa*. (Anonymous 2009.)

The pathogen germinates over a wide temperature range of 0–35°C (Casals et al. 2010) and infection occurs easily by fruit-to-fruit contact (Michailides and Morgan, 1997) or through cracks and wounds in the fruit skin (Xu and Robinson 2000). The conidia are further spread by the activities of wind, water, insects, birds, and man (Byrde and Willetts 1977; Bannon et al. 2009). Van Leeuwen et al. (2002b) observed that fruits infected late in one season can add to primary inoculum of *M. fructigena* in the following season, and infection on trees may be caused by infected fruit present on the ground.

22.2.2 Grey Mould or Green Fruit Rot

22.2.2.1 Symptoms

Symptoms start as tiny light brown to dark brown lesions on the petals, styli, filaments as well as anthers of flowers. The spots and lesions soon cover the entire petals. In early stages, V-shaped necrotic lesions are often visible. On fruits, brown lesions appear which soon take on a watery appearance. Under moist conditions, woolly strands of superficial fungal growth and grey sporulation are visible on the diseased organs. In most cases symptoms are not present on twigs and spurs. Early infection causes the whole fruit to be rotten, whereas late infections result in misshapen and deformed fruit with areas of green and rotted tissues (Figure 22.2). Infection of young fruit can lead to mummification also (Beever and Elvidge 1986; Ferrada et al. 2016).

22.2.2.2 Causal Organism

Botrytis cinerea, *S. sclerotiorum*, *M. laxa* and *M. fructicola* have been found to be associated with the green or grey fruit rot of stone fruits (Wilson and Ogawa 1979). Ferrada et al. (2016) reported *B. cinerea* and *B. prunorum* as causal agents of blossom blight and postharvest decay on plum fruit. However, *B. cinerea* is the most important and the most consistently isolated pathogen from stone fruits showing grey mould symptoms. This pathogen has recently been classified as the most important plant pathogen (Dean et al. 2012), causing heavy annual economic losses (Weiberg et al. 2013). The pathogen infects peach, almond, apricot, plum and nectarine (Abata et al. 2016; Fourie and Holz 1995)

FIGURE 22.2 Grey mould on nectarine. (Image Number: 5538856 Mourad Louadfel, Homemade, Bugwood.org.)

22.2.2.3 Disease Cycle and Epidemiology

Green fruit is usually more resistant to infection than mature fruit (Fourie and Holz 1987). Infection can be through stomata or directly by the formation of minute infection pegs formed from the inner appressorium wall, where it appears as a wedge-like infection. In green apricot and almond fruits, infection occurs via the withered floral parts (Ogawa and English, 1960). Chlamydospores are short-term survival structures, whereas sclerotia are the main survival structures (Holz et al. 2007)

22.2.3 Rhizopus and Mucor Rot

22.2.3.1 Symptoms

Infection occurs through the stem end, the calyx and puncture wounds or wounds caused by insects. Lesions are circular, light brown, soft, and watery and are soon covered with masses of shiny erect sporangiophores that appear at breaks in the fruit skin or emerge through lenticels. Peaches and nectarines infected in cold storage sometimes show a narrow band of clear, water-soaked tissue at the lesion margin and usually give off a pleasant aroma (Figure 22.3a). The diseased internal tissue is very soft, watery, light brown in colour and can be easily separated from the healthy tissue. All three, viz. *Mucor*, *Rhizopus* and *Gilbertella* soft rot, are tan soft rots and look identical in the beginning but can be distinguished in the advanced stages. In the case of Mucor rot, erect white or yellowish, shiny sporangiophores with gray to black sporangia (in the laboratory) or brown sporangia (in the field) densely cover the decay lesions (Figure 22.3b). Under high relative humidity, the sporangia absorb water, the sporangial wall dissolves, and the whole sporangium becomes a "sporangial drop". At 0°C, Mucor rot develops very rapidly.

In the field, *Gilbertella* rot of stone fruit, caused by *Gilbertella persicaria* (Ed. Eddy) Hesseltine, resembles Rhizopus rot because of the dense black sporangia. However, the sporangiophores of *Gilbertella* are short and the sporangia remain wet ("sweat"). The sporangial wall splits in half on top but remains attached at the tip of the sporangiophores, and rot develops rapidly at 20–25°C but does not develop at 0°C.

22.2.3.2 Causal Organisms

Rhizopus arrhizus, *Mucor piriformis* and *Gilbertella persicaria* are important Mucoraceous fungi that cause postharvest storage losses in most of the perishable fruits (Ogawa

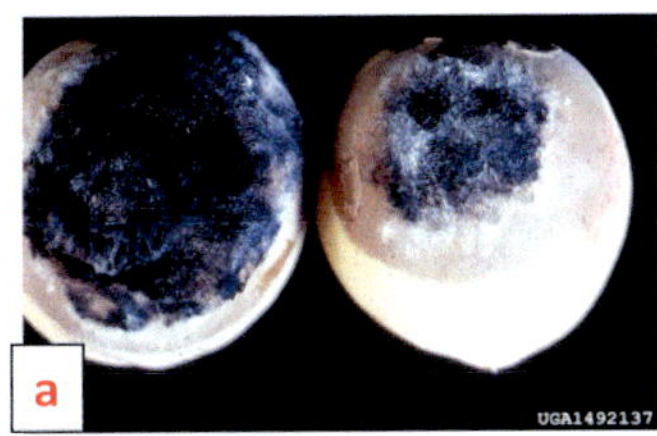

FIGURE 22.3 (a) Rhizopus rot on peach. (Image Number: 1492137 Plant Pathology, University of Georgia, Bugwood.org.) (b) Mucor rot on peach. (HORTGRO image.)

et al. 1971; Sholberg and Ogawa 1983). Rhizopus rot, caused by *Rhizopus stolonifer*, is among the most harmful pathogens causing postharvest diseases of stone fruits. The airborne spores of *R. stolonifer* are very common in the atmosphere and infection occurs mainly at wound sites and bruises caused during improper handling at harvesting, processing or packaging. The decaying apricots may sometimes show the presence of *R. circinans* as well as soft rots of almond and walnut caused by *Rhizopus oryzae* and *R. stolonifer*, respectively, in fruit shops in Jammu, India (Badyal 1991).

M. piriformis has recently been isolated from several stone fruits, especially at low temperatures (Bertr and Saulie-Carter 1980; Michailides 1984; Smith et al. 1979). However, in hot semi-arid regions it does not pose a major threat major because of low survivability of the fungus in the soil at high temperatures (Michailides and Spotts 1986). It was initially described by Fischer in 1892, was first reported as a cause of fruit decay in 1895 and from 1962 to 1990 several *Mucor* spp. including *M. circimelloides*, *M. racernosus*, *M. plumbeus* and *M. hiemalis* were isolated from stone fruit in the field and in cold storage in California, with *M. piriformis* being the most destructive pathogen in cold storage. *M. piriformis* was commonly isolated from infected Fuji apples, Asian pears, and plums in California (Michailides and Spotts 1990).

22.2.3.3 Disease Cycle and Epidemiology

Fallen fruit are infected by direct contact with infected soil or by contact spread which is accelerated by birds, rodents and insects. In addition, fall and winter rains wash spores from decayed fruit into the soil or splash them onto other fruit.

M. piriformis cannot compete effectively with other soil microbes at 20°C. It grows and sporulates on peach leaves fallen on the ground or on common weeds of pome and stone fruit orchards, such as chickweed (*Stellaria media* L.) and ryegrass (*Lolium perenne* L.), when temperatures are low and antagonistic microflora are limited. Other species of *Mucor* reported to cause decay of stone fruit can usually survive as sporangiospores or chlamydospores in soil and can withstand higher soil temperatures.

22.2.4 Penicillium Rot

*Penicillium*is an important opportunistic fungus that is capable of causing heavy losses when conditions are favourable. The pathogen infects nectarine, plum, peach, apricot and almond (Badyal 1991; Neshawy 1997; Brito et al. 2019; Louw and Korsten 2019).

22.2.4.1 Symptoms

Penicillium digitatum causes green mould, while *P. expansum*, *P. crustosum* and *P. solitum* cause blue mould on nectarine and plum (Figure 22.4). The infected tissues become soft and discoloured with sunken lesions and white mycelial growth. The pathogens are distinguishable only after sporulation. Wrinkling of fruit is evident with infection by *P. digitatum*. Sporulation is more profuse on nectarine than on plum (Louw and Korsten 2019).

FIGURE 22.4 *Penicillium* rot on plum. (HORTGRO image.)

22.2.4.2 Causal Organism

Penicillium expansum and *P. digitatum* are important pathogens responsible for heavy losses and mycotoxin production (Ceponis and Friedman 1957; Wells et al. 1994; Pitt and Hocking 2009; Ma et al. 2003; Snowdon 2010). Recently some other *Penicillium* species have also been reported as potential risk to nectarine and plum (Louw and Korsten 2016). Some of these are *Penicillium crustosum* on peach (Restuccia et al. 2006), *Penicillium chrysogenum* on black plum (Seigbe and Bankole 1996) and *Penicillium digitatum* on nectarine (Navarro et al. 2011).

22.2.5 Aspergillus Rot

22.2.5.1 Symptoms

The infected tissues are pale and water soaked, and the spores are green, powdery and easily released when mature (Figure 22.5). It has been observed that *A. flavus* occurs more frequently on almonds from orchards in an area with high average day and night temperatures than on almonds from orchards in an area with low average day and night temperatures and more in sun than in shade (Purcell et al. 1980).

22.2.5.2 Causal Organism

Purcell et al. (1980) reported *Aspergillus flavus* on almond, whereas Badyal (1991) has reported the presence of *A. ustus* as an important postharvest pathogen of almond. *A. flavus* has

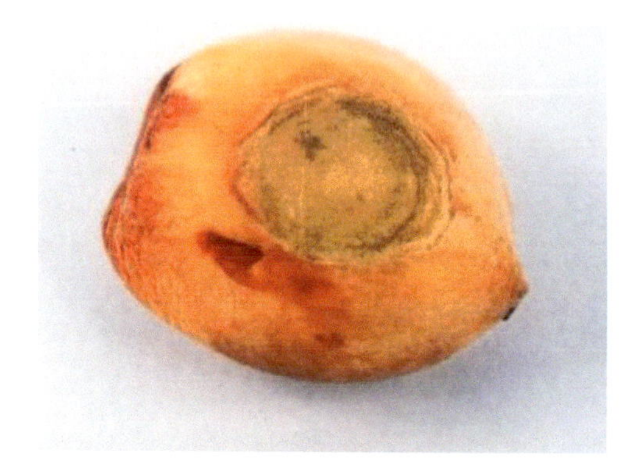

FIGURE 22.5 Aspergillus rot disease on peach fruits during storage.

also been reported in peach (Michailides and Thomidis 2007). Fungi associated with fruit rot of peach and plum were identified as A. *flavus*, *A. fumigatus*, *A. japanicus [A. japonicus]*, *A. niger*and *A. terreus* (Singh and Prashar 1989). The pathogen infects plum and almond fruits (Purcell et al. 1980; Bamba and Sumbali 1999).

22.2.6 Sour Rot

Sour rot of nectarineas well as peach is mainly linked with injured or bruised fruit and fruit having split pits (Adaskaveg and Crisosto 2006). The disease is caused by *Geotrichum candidum*. The fungus infects peach, nectarine and plum (Bamba and Sumbali 1999; Adaskaveg and Crisosto 2006).

22.2.6.1 Symptoms

The disease affects both ripe fruit and young and severely injured fruit. The initial symptoms are visible as brown and watery with soft decay. Under favourable conditions, a thin layer of white mycelial growth is clearly visible on the fruit surface (Wells 1977). The fruit decay may extend to the pit and cover the entire fruit (Burton and Wright 1969). Rotted fruit have a distinguishing odour, ranging from yeasty to vinegary. Under severe conditions, juice may stream from the lesion, resulting in skin disintegration and formation of furrows (Michailides et al. 2004). The inoculum can be present in orchard soils upto a depth of 10 cm and in packing houses. Nitidulid beetles and fruit flies can play a role in disease transmission (Yaghmour and Bostock 2012).

22.3 Management of Postharvest Diseases

Postharvest management of diseases of stone fruits can be achieved by adopting a multipronged strategy. Nabi et al. (2017) have integrated the various methods into four broad categories of physical, chemical, resistance and bio-control in various combinations. However, for the sake of simplicity, we shall elaborate them individually.

22.3.1 Chemical Methods

The compounds of calcium, silicon and borate, which directly inhibit the growth of pathogenic fungi, have successfully been used in controlling a number of postharvest diseases of stone fruits (Conway et al. 1991; Qin and Tian, 2005; Qin et al. 2007). A 20-min dip in peracetic acid (PAA) of pathogenic conidia at 250 μg/mL or ClO_2 at 10 μg/mL or for 5 min in PAA at 250 μg/mL completely inhibited their pathogenesis, and no symptoms of brown rot were observed in artificially inoculated and wounded nectarines and plums (Mari et al. 1999).

Cherry fruits treated with a pre-harvest iprodione application and postharvest dip in a suspension of *C. infirmo-miniatus* containing 0.5–1.5 × 10^8 CFU/mL were protected from brown rot disease. The disease was also reduced by modified atmosphere packaging (MAP) alone, and further protection was provided by *C. infirmo-miniatus*–MAP synergism. A remarkable reduction of 0.4% from 41.5% incidence of brown rot was obtained by combining pre-harvest iprodione and postharvest *C. infirmo-miniatus* treatments with MAP (Spotts et al. 1998). Cyprodinil, fenhexamid, fludioxonil and a mixture of boscalid plus pyraclostrobin are examples of reduced risk fungicides applied to stone fruits for controlling postharvest (Adaskaveg et al. 2005).

In order to avoid fungicide resistance, Beever and Elvidge (1986) suggested the following spray schedule: Apply triforine over the bloom period under high disease pressure, naphthalimide (Captan) after bloom under low disease pressure, a dicarboximide (vinclozolin) spray at pre-harvest, and a postharvest spray of benzimidazole (benomyl).

22.3.2 Physical Methods

Physical methods include elimination of all diseased parts (mummified parts, outbreaks, branches, fruit), collection of rotten fruit before and during harvest and of fallen fruit rots after harvest, burying them deeply or dumping them rather than allowing them to produce future apothecia, and increasing ventilation in the orchard (Villarino et al. 2010). Some other methods are proper handling of harvested fruit to prevent wounding, proper sanitation both at pre- and postharvest stages, rapid cooling after harvest, and storage at 0°C has to be followed (Sommer 1985; Osorio et al. 1993; Hong et al. 1998).

The shelf life and protection of fresh fruit may be extended by various preservation methods which aim at the removal of the pathogens or inhibition of their growth. They include storage under refrigerated conditions, modified and controlled atmosphere storage and hypobaric or low pressure storage (Kader and Ben-Yehoshua 2000; Gorny and Kader 1997; Tian et al. 2004).

Hot water immersion at 60°C for 60 sec is effective for plums for reduction of the incidence of brown rot from more than 80% among control fruit to less than 2%. This treatment reduced decay incidence from 100% to less than 5% on nectarine fruit stored at 20°C and from 73 to 28% on cold-stored fruit. Therefore, brief immersion in heated water can effectively manage postharvest brown rot of stone fruits, especially for the organic fruit industry (Karabulut et al. 2010).

22.3.3 Biological Control

Biological control with antagonistic yeasts has shown promising results (Wilson and Wisniewski 1989; Janisiewicz and Korsten 2002). Qing and Tian (2000) reported successful management of Rhizopus rot of nectarine fruits by *Pichia membranefaciens*. The antibiotic-producing bacteria, *Pseudomonas corrugata* and *P. cepacia*, significantly reduced rot when applied up to 12 h after inoculation (Smilanick et al. 1993). Recent studies by Francesco et al. (2020) have shown the capability of volatile organic compounds produced by *Aureobasidium pullulans* L1 and L8 strains to effectively reduce the incidence of brown rot incidence *Monilinia* spp.

An attempt was made to manage the brown rot of stone fruits by treatment of bio-control agents viz. the yeast *Rhodotorula kratochvilovae* strain LS11 and the commercial biofungicide Serenade Max® (based on *Bacillus subtilis* strain QST 713), either alone or along with a low dose of fungicides (25% of full dosage, integrated control) cyprodinil (30 days before harvest), and boscalid (7 days before harvest). The results showed that biological and integrated control based on two important treatments at 30and 7days before harvest are effective and safer control strategies for brown rot of stone fruits (Curtis et al. 2019).

22.3.4 Resistance

Droby et al. (2002) and Tian and Chan (2004) have attributed the resistance induction by biotic and abiotic factors for the management of postharvest diseases of stone fruits. Higher levels of resistance have been found in some cultivars such as 'Bolinha', 'Contender' and 'F8, 1-42' against *Monilinia* spp. Using a segregating peach population derived from crosses between 'Contender' and 'F8, 1-42', somequantitative trait loci (QTL) of *Monilinia* spp. resistance havebeen described (Garcia et al. 2013; Pacheco et al. 2014). Phenotypic data, after artificial inoculations in different peach cultivars, has revealed that Rich Lady was relatively susceptible, whereas Royal Glory was moderately resistant to *Monilinia* spp. (Papavasileiou et al. 2020).

REFERENCES

Abata LK, Izquierdo AR, Viera W, Flores FJ (2016) First report of Botrytis rot caused by Botrytis cinerea on peach in Ecuador. J Plant Pathol 98(3): 690.

Adaskaveg JE, Förster H, Gubler WD, Teviotdale BL, Thompson DF (2005) Reduced-risk fungicides help manage brown rot and other fungal disease of stone fruit. Cal Agricul 59: 109–114.

Adaskaveg, JE, Crisosto CH (2006) Sour rot control. Central Valley Postharvest newsletter 15: 2–5.

Anonymous (2009) Diagnostic protocols for regulated pests PM 7/18 (2). *Monilinia fructicola*. EPPO Bulletin 39: 337–343.

Badyal K (1991) New records of fungal species associated with rots of almond and walnut fruits. Indian J Mycol Plant Pathol 21(3): 295.

Bamba, R, Sumbali, G. 1999. Some new mycopathogens associated with postharvest decay of plum fruits (*Prunus domestica* L.) in India. Proceed Nat Acad Sci India. 69(3/4): 323–330.

Bannon F, Gort G, van Leeuwen G, Holb I, Jeger M (2009) Diurnal patterns in dispersal of *Monilinia fructigena* conidia in an apple orchard in relation to weather factors. Agr Forest Meteorol 149: 518–525.

Beever RE, Elvidge J (1986) Green fruit rot of apricot caused by *Botrytis cinerea* resistant to benzimidazole fungicides. New Zealand J Agri Res 29(2): 299–304.

Bertr PF, Saulie-Carter JL (1980) Mucor rot of pears and apples – Oregon St. Agric. Exp. Sta. Spec. Rept. 568: 21.

Brito, ACQ, Mello JF, Vieira JCB, Câmara MPS, Bezerra JDP, Souza-Motta CM, Machado AR. 2019. First Report of Penicillium expansum causing postharvest fruit rot on black plum (Prunus domestica) in Brazil. Plant Dis. https://doi.org/10.1094/PDIS-04-19-0889-PDN

Burton CL, Wright WR (1969) Sour rot of peaches on the market. Pl Dis Repor 53: 580–582.

Byrde RJW, Willetts HJ (1977) The brown rot fungi of fruit. Their biology and control. Pergamon Press, Oxford.

Casals C, Vinas I, Torres R, Griera C, Usall J (2010) Effect of temperature and water activity on in vitro germination of *Monilinia* spp. J App Microbiol 108: 47–54.

Ceponis MJ, Friedman BA (1957) Effect of bruising injury and storage temperature upon decay and discolouration of fresh, Idaho-grown Italian prunes on the New York City market. Plant Dis Repor 41: 491–492.

Conway WS, Sams CE, Abbott JA, Bruton BD (1991) Postharvest calcium treatment of apple fruit to provide broad-spectrum protection against postharvest pathogens. Plant Dis 75: 620–622.

Curtis F, Giuseppe I, Assunta R, Alberto RM, Succia PT, Raffaello C (2019) Integration of biological and chemical control of brown rot of stone fruits to reduce disease incidence on fruits and minimize fungicide residues in juice. Crop Prot 119: 158–165.

Dean R, Vankan JL, Pretorius ZA, Hammond KKE, Pietro AD, Spanu PD, Rudd JJ, Dickman M, Kahmann R, Ellis J, Foster GD (2012) The top 10 fungal pathogens in molecular plant pathology. Mol Plant Pathol 13: 414–430.

De Cal A, Gell I, Usall I, Vinas P, Melgarejo (2009) First report of brown rot caused by *Monilinia fructicola* in peach orchards in Ebro Valley, Spain Plant Dis 93: 763.

Droby S, Vinokur V, Weiss B, Cohen L, Daus A, Goldschmidt EE, Porat R (2002) Induction of resistance to *Penicillium digitatum* in grapefruit by the yeast bio-control agent *Candida oleophila*. Phytopathology 92: 393–399.

Eckert JW, Ogawa JM (1988) The chemical control of postharvest diseases: Deciduous fruits, berries, vegetables and root/tuber crops. Ann Rev Phytopathol 26: 433–469.

Ferrada EE, Latorre BA, Zoffoli JP Castillo A (2016) Identification and characterization of Botrytis blossom blight of Japanese plums caused by *Botrytis cinerea* and *B. prunorum* sp. Nov. in Chile. Phytopathology 106(2): 155–165.

Fourie JF, Holz G (1987) Infection and decay of stone fruit by *Botrytis cinerea*, *Monilinia laxa* and *Rhizopus stolonifer.* Phytophylactica 17: 179–181.

Fourie JF, Holz G (1995) Initial infection process by Botrytis cinerea on nectarine and plum fruit and the development of decay. Phytopathology 85(1): 82–87

Francesco AD, Foggia MD, Baraldi E (2020) *Aureobasidium pullulans* volatile organic compounds as alternative postharvest method to control brown rot of stone fruits. Food Microbiol 87: 103395.

Garcia, M., PJ, Parfitt DE, Bostock RM, Fresnedo-Ramirez J, Vazquez Lobo A, Ogundiwin EA, Gradziel TM, Crisosto CH (2013) Application of genomic and quantitative genetic tools to identify candidate resistance genes for brown rot resistance in peach. PLOS ONE 8.

Gorny JR, Kader AA (1997) Low oxygen and elevated carbon dioxide atmospheres inhibit ethylene biosythesis in preclimacteric and climacteric apple fruit. J Amer Soc Hortic Sci 122: 542–546.

Guijarro BP, Melgarejo R, Torres N, Lamarca J, Usall A, De Cal (2008) *Penicillium frequentans* population dynamics on peach fruits after its applications against brown rot in orchards. J App Microbiol 10: 659–671.

Hilber-Bodmer M, Knorst V, Smits THM, Patocchi A (2012) First report of Asian brown rot caused by *Moniliapolystroma* on apricot in Switzerland. Pl Dis 96(1): 146.

Holz G, Coertze S, Williamson B (2007) The ecology of Botrytis on plant surface. In: Elad Y, Williamson B, Tudzynski P,

Delen N (Eds.), Botrytis: Biology, Pathology and Control. Springer, Heidelberg, pp. 9–27.

Hong CX, Michailides TJ, Holtz BA (1998) Effects of wounding, inoculum density, and biological control agents on postharvest brown rot of stone fruits. Pl Dis 82: 1210–1216.

Hu MJ, Cox KD, Schnabel G, Luo CX (2011) *Monilinia* species causing brown rot of peach of China. PLOS ONE 6(9): e24990.

Janisiewicz WJ, Korsten L (2002) Biological control of postharvest diseases of fruits. Ann Rev Phytopathol 40: 411–441.

Kader AA, Ben-Yehoshua S (2000) Effects of super atmospheric oxygen levels on postharvest physiology and quality of fresh fruits and vegetables. Postharvest Biol Technol 20: 1–13.

Karabulut OA, Smilanick JL, Crisosto CH, Palou L (2010) Control of brown rot of stone fruits by brief heated water immersion treatments. Crop Prot 29(8): 903–906.

Louw JP, Korsten L (2016) Postharvest decay of nectarine and plum caused by *Penicillium*. Eur J Plant Pathol 146: 779–791.

Louw JP, Korsten L (2019) Penicillium rot of nectarine and plum. SA Fruit J Feb/Mar 2019. 58–59.

Mari M, Cembali T, Baraldi E, Casalini L (1999) Peracetic acid and chlorine dioxide for postharvest control of *Monilinia laxa* in stone fruits. Plant Dis 83(8): 773–776.

Ma Z, Luo Y, Michailides TJ (2003) Nested PCR assays for detection of *Monilinia fructicola* in stone fruit orchards and *Botryosphaeria dothidea* from pistachios in California. J Phytopathol 151: 312–322.

Michailides TJ, Morgan DP (1997) Influence of fruit-to-fruit contact on the susceptibility of French prune to infection by *Monilinia fructicola*. Plant Dis 81: 1416–1424.

Michailides TJ, Morgan DP, Day KR (2004) First report of sour rot of California peaches and nectarines caused by yeasts. Plant Dis 88: 222.

Michailides TJ, Spotts RA (1986) Factors affecting dispersal of *Mucor piriformis* in pear orchards and into the packinghouse. Plant Dis 70: 1060–1063.

Michailides TJ, Spotts RA (1990) Transmission of *Mucor piriformis* to fruit of *Prunuspersica* by *Carpophilus* spp. and *Drosophila melanogaster*. Plant Dis 74: 287–291.

Michailides T, Thomidis T (2007) First report of *Aspergillus flavus* causing fruit rots of peaches in Greece. Plant Pathol 56(2): 352

Michailides TJ (1984) Studies on the ecology and epidemiology of the postharvest fruit pathogen *Mucor pyriformis* Fischer. PhD thesis, University of California, Davis, 88 pp.

Nabi SU, Raja WH, Kumawat KL, Mir JI, Sharma OC, Singh DB, Sheikh MA (2017) Postharvest diseases of temperate fruits and their management strategies. A review. Ind J Pure App Biosci 5(3): 885–898.

Navarro D, Díaz-Mula HM, Guillén F, Zapata PJ, Castillo S, Serrano M, Valero D, Martínez-Romero D (2011) Reduction of nectarine decay caused by *Rhizopus stolonifer, Botrytis cinerea* and *Penicillium digitatum* with *Aloe vera* gel alone or with the addition of thymol. Int J Food Microbiol 151: 241–246.

Neshawy SM (1997) Efficacy of Candida oleophila strain 128 in preventing *Penicillium expansum* infection on apricot fruit. Acta Hortic 485: 141–148.

Ogawa JM, English H (1991) Diseases of Temperate Zone Tree Fruit and Nut Crops. Vol 3345. University of California (System). Division of Agriculture and Natural Resources.

Ogawa JM, English H (1960) Blossom blight and green fruit rot of almond, apricot and plum caused by *Botrytis cinerea*. Plant Dis Rep 44: 265–268.

Ogawa JM, Leonard S, Manji BT, Bose E, Moore CJ (1971) *Monilinia* and *Rhizopus* decay control during controlled ripening of freestone peaches for canning. J Food Sci 36: 331–334.

Ogawa JM, Lyda SD, Weber OJ (1961) 2,6-dichloro-4-nitroaniline effective against *Rhizopus* fruit rot of sweet cherries. Plant Dis Rep 45: 636–638.

Osorio JM, Adaskaveg JE, Ogawa JW (1993) Comparative efficacy and systemic activity of iprodione and the experimental anilide E-0858 for control of brown rot on peach fruit. Plant Dis 77: 1140–1143.

Pacheco I, Bassi D, Eduardo I, Ciacciulli A, Pirona R, Rossini L, Vecchietti A (2014) QTL mapping for brown rot *(Monilinia fructigena)* resistance in an intraspecific peach (*Prunus persica* L. Batsch) F1 progeny. Tree Genet Gen 10: 1223–1242.

Papavasileiou, A, Georgia T, Anastasios S, Martina S, Athanassios M, Karaoglanidis G (2020). Proteomic analysis upon peach fruit infection with *Monilinia fructicola* and *M. laxa* identify responses contributing to brown rot resistance. Sci Rep10: 7807.

Petróczy M, Palkovics L (2009) First report of *Monilia polystroma* on apple in Hungary. Eur J Plant Pathol 125: 343–347.

Pitt JI, Hocking AD (2009) Fungi and food spoilage. Springer Science + Business Media, New York.

Poniatowska AM, Michalecka J, Puławska (2016) Genetic diversity and pathogenicity of *Monilinia polystroma* - The new pathogen of cherries. Plant Pathol 65(5): 723–733.

Poniatowska A, Michalecka M, Bielenin A (2013) Characteristic of *Monilinia* spp. fungi causing brown rot of pome and stone fruits in Poland. Eur J Plant Pathol 135: 855–865.

Purcell SL, Douglas JP, Bruce EM (1980) Distribution of *Aspergillus flavus* and other fungi in several almond growing areas of California. Phytopathology 70(9): 926–929.

Qin GZ, Tian SP (2005). Enhancement of biological control activity by silicon and the possible mechanisms involved. Phytopathology 95: 69–75.

Qing F, Tian S (2000). Postharvest biological control of Rhizopus rot of nectarine fruits by *Pichia membranefaciens*. Plant Dis 84(11): 1212–1216.

Qin QZ, Tian SP, Chan ZL, Li BQ (2007) Crucial role of antioxidant proteins and hydrolytic enzymes in pathogenicity of Penicillium expansum: Analysis based on proteomic approach. Mol Cell Proteomics 6(3): 425–438.

Restuccia C, Giusino F, Licciardello F, Randazzo C, Caggia C, Muratore G (2006) Biological control of peach fungal pathogens by commercial products and indigenous yeasts. J Food Prot 69: 2465–2470.

Schlagbauer HE, Holz G (1989) Occurrence of latent Monilinia laxa infections on plums, peaches and apricots. Phytophylactica 21, 35–38.

Seigbe DA, Bankole SA (1996) Fungi associated with postharvest rot of black plum (*Vitex doniana*) in Nigeria. Mycopathologia 136: 109–114.

Sholberg PL, Ogawa JM (1983) Relation of postharvest decay fungi to slip-skin maceration disorder of dried French prunes. Phytopathology 73: 708–713.

Singh RS, Prashar M (1989) Postharvest spoilage of peach and plum fruits in north India due to *Aspergillus*. J Res 26(1): 62–64.

Smilanick JL, Denis-Arrue R, Bosch JR, Gonzalez AR, Henson D, Janisiewicz WJ (1993) Control of postharvest brown rot of nectarines and peaches by *Pseudomonas* species. Crop Prot 12(7): 513–520.

Smith WL Jr, Moline HE, Johnson KS (1979) Studies with *Mucor* species causing postharvest decay of fresh produce. Phytopathology 69: 865–869.

Snowdon AL (2010) A Colour Atlas of Postharvest Diseases and Disorders of Fruit and Vegetables Vol. 1. General Introduction & Fruits. Manson Publishing, London.

Sommer NF (1985) Role of controlled environments in suppression of postharvest diseases. Can J Plant Pathol 7: 331–339.

Spotts RA, Cervantes LA, Facteau TJ, Chand-Goyal T (1998) Control of brown rot and blue mold of sweet cherry with pre-harvest iprodione postharvest *Cryptococcus infirmo-miniatus*, and modified atmosphere packaging. Plant Dis 82(10): 1158–1160.

Tian SP (2007) Management of postharvest diseases in stone and pome fruit crops. In: Ciancio A, Mukerji KG (Eds.). General Concepts in Integrated Pest and Disease Management, Vol. 1. Springer, Dordrecht, pp. 131–147.

Tian SP, Chan ZL (2004) Potential of induced resistance in postharvest disease control of fruits and vegetables. Acta Phytopathol Sin 34: 385–394.

Tian SP, Jiang AL, Xu Y, Wang YS (2004). Responses of physiology and quality of sweet cherry fruit to different atmospheres in storage. Food Chem 87: 43–49.

van Leeuwen GCM, Baayen RP, Holb IJ, Jeger MJ (2002) Distinction of the Asiatic brown rot fungus *Monilia polystroma* sp. nov. from *M. fructigena*. Mycol Res 106: 444–451.

Villarino M, Melgarejo P, Usall J, Segarra J, De Cal A (2010) Primary inoculum sources of *Monilinia* spp. in Spanish peach orchards and their relative importance in brown rot. Plant Dis 94: 1048–1054.

Weiberg A, Ming W, Feng L, Hongwei Z, Zhihong Z, Isgouhi K, Hsien-Da H, Hailing J (2013) Fungal small RNA as suppress plant immunity by hijacking host RNA interference pathways. Science 342: 118–123.

Wells JM (1977) Sour rot of peaches caused by *Monilinia implicata* and *Geotrichum candidum*. Phytopathology 67: 404–408.

Wells JM, Butterfield JE, Ceponis MJ (1994) Diseases, physiological disorders, and injuries of plums marketed in metropolitan New York. Plant Dis 78: 642–644.

Wilson EE, Ogawa JM (1979) Fungal, Bacterial and Certain Nonparasitic Diseases of Fruit and Nut Crops in California. Division of Agricultural Sciences, University of California, Berkeley, 190 pp.

Wilson CL, Wisniewski ME (1989) Biological control of postharvest diseases of fruits and vegetables: An emerging technology. Annu Rev Phytopathol 27: 425–441.

Xu XM, Robinson JD (2000) Epidemiology of brown rot (*Monilinia fructigena*) on apple: Infection of fruits by conidia. Plant Pathol 49: 201–206.

Yaghmour MA, Bostock RM (2012) Biology and sources of inoculum of *Geotrichum candidum* causing sour rot of peach and nectarine fruit in California. Plant Dis 96(2): 204–210.

Yin YF, Chen SN, Cai ML, Li GQ, Luo CX (2014). First report of brown rot of apricot caused by *Monilia mumecola*. Plant Dis 98(5): 694–696.

23

Postharvest Diseases and Disorders of Apple: Perspectives for Integrated Management

K. P. Singh and T. Aravind
Department of Plant Pathology, College of Agriculture, G.B. Pant University of Agriculture, Pantnagar, Udham Singh Nagar District, Uttarakhand, India

CONTENTS

23.1 Introduction

Apple is one of the most important temperate fruits and is fourth among the most widely produced fruits in the world after, pear, orange, banana and grape. China is the leading producer of apple, followed by the United States, Turkey, Italy and India (FAOSTAT 2017). In India, apple is mostly grown in the states of Jammu and Kashmir, Himachal Pradesh, Uttarakhand and Arunachal Pradesh. The North Eastern Hills region, viz., the states of Arunachal Pradesh, Nagaland, Meghalaya, Manipur and Sikkim, also grows apple on a small scale on a small scale. Red, Royal and Golden Delicious are among the important apple varieties, accounting for 46, 21, and 16% of the total acreage, respectively. Delicious cultivars are famous for their quality and taste. Apple can be grown at altitudes 1,500–2,700 m above mean sea level in the Himalayan range, which experiences 1,000–1,500 hours of chilling (temperature remains at or below 7°C during the winter season). The temperature during the growing season is around 21–24°C. Fog and excessive rains (>100–125 cm annually) near the fruit maturity result in poor fruit development and quality due to non-uniform colour development and attack by fungal pathogens. The area under apple cultivation in India has increased tremendously from 241.88 thousand hectares in 2001–2002 to 306.0 thousand hectares in 2017–2018, and the production from 1158.4 thousand metric tonnes in 2001–2002 to 2371.0 thousand metric tonnes in 2017–2018. During 2013–2014, the productivity of apple at world level was 15.48, whereas it was only 6.14 in India (Bhat 2019).

In the past decades, production has greatly increased due to increase in the cultivated area, but the focus on plant protection measures to enhance productivity has not yet gained much momentum. The keeping qualities of a number of varieties of apple were studied by Bose (1969) in Uttarakhand. He observed that during storage, a large number of fungi (*Rhizopus nigricans*, *Pestalotia* sp., *Penicillium expansam*) caused rotting of the stored fruits. The rotting may start from any point where there is the slightest injury to the skin. The inner tissues of the affected area of the fruit become light brown in colour and watery. Spoilage of apples is a big problem in the North West Himalayan states, i.e., Jammu and Kashmir, Himachal Pradesh and Uttarakhand, aggravated by the absence of proper storage facilities and paucity of information on factors causing loss during storage. It has been internationally recognized that one of the most efficient and economically feasible means to enhance food production is to reduce postharvest loss. Due to their high water content and tenderness of tissues, these postharvest losses are greatest in fruits and vegetables. These losses are high even in areas having sophisticated technologies, where scientific harvesting, handling, transit, storage and antimicrobial treatments are adopted.

Major postharvest losses occur due to improper harvesting time, long storage and faulty transportation conditions which is the result of fluctuations in demand-supply chain. The tough skin of the apple provides protection from storage diseases, but wounds created during harvesting and handling pave the way for early decay. Scab is the most devastating of the postharvest diseases, which greatly impairs the marketability of the fruits. The initiation of scab is determined by the orchard conditions, but the severity is mostly influenced by the storage and transport conditions.

23.2 Postharvest Losses

Postharvest losses can exceed 50% of the harvested fruits in developing countries (Eckert and Ogawa 1988). It is estimated that about 20–25% of the harvested fruits are decayed by pathogens during postharvest handling even in developed countries (Nabi et al. 2017). As of now, over 90 fungal species have been reported to cause decay of apple during storage (Ilyas et al. 2007). Postharvest losses occur at harvesting, sorting, grading, packaging, transporting and storing stages; in apple, the highest losses were noticed during storage. The infection by the pathogenic bacteria and fungi can occur during the pre-harvest (under field conditions) and/or postharvest (during harvesting, handling, storage, transport and marketing) stages. In India, losses from postharvest diseases in fruits are reported to be between 10% and 40%. It has been estimated that due to improper storage and transport facilities, Indian farmers incur a loss of Rs 92,651 crores as postharvest losses.

23.3 Predisposing Factors for Postharvest Losses

The storage conditions of the harvested produce play a crucial role in postharvest losses. The most important factors include moisture, temperature and composition of the surrounding air, in particular the O_2/CO_2 ratio. High temperature during storage and transport favours pathogens, and if other conditions like high moisture are present it can lead to severe disease. According to China (2007), in order to minimize the loss of moisture and prevent loss of tissue turgidity and shrivelling, fresh fruits need high humidity levels (>95% relative humidity) in storage.

23.4 Origin and Development of Diseases

Apples can be stored for several months (up to 1 year) under controlled conditions (Thompson et al. 2018). Several scientific studies have been undertaken on the storage conditions and prolonging safe storage of apple (Janisiewicz and Korsten 2002; Konopacka and Plocharski 2004; Thompson et al. 2018). Postharvest fungal infections, however, still cause major shortfalls during storage and along the supply chain (FAO 2015). Depending on cultivar, season and geographical location, storage fungi can cause severe losses in apple. Most severe infection by fungal pathogens occurs when the storage moisture is in the range of 82.46–84.48%.

The postharvest diseases of apple can be divided into two categories:

1. Diseases initiated due to pre-harvest infection.
2. Diseases that develop due to postharvest infections.

The major pre-harvest pathogens of apple cause diseases such as brooks fruit spot, sooty blotch, fly speck, core rot and scab. These fungi require moisture, temperature and humid conditions for spore production, dispersal and fruit infection. Postharvest infections mostly occur through the injuries/wounds produced during harvesting, handling transport or storage or through damages caused by insect attack.

23.5 Postharvest Physiological Disorders

Other than the various postharvest diseases, certain physiological disorders caused by several abiotic factors also cause postharvest losses in apple. Important physiological disorders include bitter pit, brown heart, storage scald, water core, etc. (Westercamp and Monteils 2006).

23.5.1 Bitter Pit

Bitter pit is induced by calcium deficiency. When it occurs on the tree, it is known as tree pit, whereas when it occurs in storage, it is known as storage pit (Ferguson and Watkins 2011). Pits are often associated with the terminals of branched vascular bundles. The lesions are small, brown, dry, slightly bitter tasting, and about 3–5 mm in diameter. The affected fruits have dark spots of about ½ cm diameter, which occur on skin or in flesh or both. Internal lesions can occur anywhere in the tissue from the core line to the skin, but they are more common just below the skin. In badly infected fruit, they may coalesce into blotches. Initially the skin of the fruit over the affected area appears water soaked. Later, the spots become purple on blushed skin and dark green on less coloured areas. The skin over the pits gradually becomes brown or sometimes black. The cells are dead and turn brownish black. Large fruit and early harvesting increase the severity of bitter pit (Figure 23.1). The incidence of bitter pit can be minimized by avoiding excessive tree vigour, harvesting immature and large fruit, excessive thinning, high levels of potassium and magnesium, irregular water supply, and application of large amounts of nitrogenous fertilizer. Rapid cooling after harvest, controlled atmosphere (CA) storage, harvesting at optimum maturity, spraying with calcium chloride or calcium nitrate during apple growing season and drenching of fruits in 2% $CaCl_2$ solution before storage may reduce the occurrence.

FIGURE 23.1 Symptoms of bitter pit.

23.5.2 Brown Heart

This disorder in stored apples causes higher concentrations of CO_2 (>1%) and is caused by internal browning (Bhat and Khan 2018). This physiological disorder is mostly associated with over-mature and large fruits. It is characterized by the appearance of brown discoloration in the flesh, mostly initiating near or in the core of the fruit. Brown areas may include dry cavities developed due to desiccation, and have well-defined margins. Harvesting at optimal maturity and reduction in concentrations of CO_2 below 1% in CA storage minimizes the incidence.

23.5.3 Storage Scald

Storage scald usually appears on the yellow or green surface of the fruit. The skin turns faint brown and internal breakdown occurs rapidly. The skin is killed, and it readily peels off from the fruit with slight pressure. The flesh turns brown and breaks down into a pulp. The disorder is favoured by high nitrogen and low calcium concentrations in the fruit, harvesting of immature fruits and hot, dry weather before harvest. Irregular brown patches of dead skin become rough when the disease is severe and develop within 3–7 days upon warming of fruits followed by cold storage. Inadequate air circulation in packaging boxes or in storage rooms also promotes storage scald (Bhat and Khan 2018). Harvesting of the fruits at proper maturity and ensuring adequate ventilation in storage can help in managing this disorder.

23.5.4 Water Core

Water core is characterized by water-soaked regions in the flesh of fruit. These water-soaked areas are found near the core or on the entire apple (Bhat and Khan 2018). The external symptom is not very prominent. When cut open, the flesh of affected fruit has a watery or glossy appearance and has poor keeping quality (Figure 23.2). Intercellular air spaces in the

FIGURE 23.2 Symptoms of water core.

core and cortical tissues seem to be filled with sorbitol. This malady is associated with advancement in fruit maturity, low night temperatures prior to harvest, exposure of fruits to high temperatures, high levels of nitrogen, boron and low levels of calcium in the fruit. Large fruits are the most susceptible to infection. Severely affected fruits may smell (sweetish odour) and have a fermented taste. Since severely affected fruits are heavier than the normal ones, they can be segregated by flotation method by a mixture of water and alcohol. The disease increases as fruits become over-mature. The most effective way to reduce incidence is pre-harvest or postharvest spray of calcium and avoiding delayed harvests. Fruits should be stored in a ventilated atmosphere at the correct stage of maturity.

23.5.5 Shrivelling

Apple loses weight after harvesting by means of transpiration and respiration, resulting shrivelling of fruits. Shrivelling is caused by high temperatures with low relative humidity in the storage room. Shrivelling negatively influences the marketability and potential shelf life of apple fruits (Subedi and Gautam 2018). Intensity of shrivelling is high in Golden Delicious due to the presence of a thinner layer of natural wax on the fruit surface.

23.5.6 Sunburn

Sunburn is the result of high-intensity solar radiation. Affected areas become discoloured to white, brown, or tan colour. Mild injury results in superficial damage only, but prolonged exposure can damage the inner flesh (sunken dead areas) as well (Figure 23.3). Allowing a more vigorous growth of the trees, resulting in more foliage and branches for more shade can help in minimizing this disorder. Fruit should be moved quickly to shaded areas or packing houses to avoid injury on newly exposed surfaces.

23.5.7 Russeting

Russet is usually initiated early during fruit formation, about 30 days after anthesis, which coincides with the greatest increment of tangential growth, which occurs 23 days after anthesis based on equatorial cross-section measurements. Russeting occurs in various forms, ranging from the rough scabrous type, which covers full fruit, to light netted. Severe damage

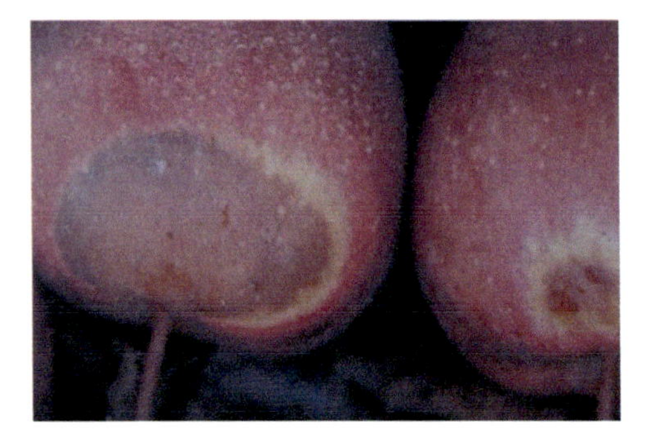

FIGURE 23.3 Symptoms of sunburn on apple fruits.

FIGURE 23.4 Symptoms of russeting.

produces cracks on the fruit and ultimately fruit splits. Under storage conditions, such fruit shrivels (Figure 23.4). A number of structural, physiological and environmental factors affect russet development. The incidence of russeting varies widely and it develops under conditions of high humidity, frequent rain, dew, frost or low temperature. Fruit enlarges during the night, which is influenced by temperature. Fruit cracking and russeting are both associated with disintegration or breakdown of the cuticular layer. Cultivars with thin cuticles or areas of fruit having thin cuticle are more subjected to russet formation. Apple exposed to the sun is more prone to russeting, and it rarely occurs on shaded fruit. Repetitive solar radiation enhances the polymerization of cuticular waxes into an amorphous matrix, which makes fruits susceptible to cracking. The incidence of fruit russet is also related to impaired foliage and greater exposure to weather conditions, especially at an early stage of fruitlet development. Chemical sprays, which are toxic to cells, are found to be the major factor causing russeting in apple fruits. Fungicide sprays such as Dodine, Kavach, Mancozeb, etc. at the green tip to petal fall stage results in russeting. Severe russeting occurs in areas with a poor water supply and in clayey soils. Imbalance of organic acids also causes russeting, and penetration of oxygen into the deeper layers of epidermis and hypodermis through the cracks which may initiate phellogen formation. Powdery mildew infections on blossoms and fruitlets are a direct cause of fine web-like russeting. Fungal infection, breaking of the epidermal hairs and mechanical injury to the fruit also causes russeting. Spray of Captan (0.3%) or calcium chloride (0.5%) 20 days before harvesting reduces russeting to some extent by improving the colour of the fruit (Thakur and Xu 2004).

23.5.8 Cork Spot

Affected fruits show dry, brownish, corky spots scattered in the flesh. The fruits may become dimpled or misshapen with vertical depressions. Initial symptoms appear as small, blushed areas on the skin of the fruit above the affected brown spot (Bhat and Khan 2018). Sometimes there is no external sign, but during storage, brown spots develop and extend throughout the flesh. Affected tissue is usually much harder than the healthy tissue. Development of external or internal cork depends on the variety of apple and the time when disease starts. The internal cork always affects the core area. As boron and calcium deficiencies are responsible for development of cork spot, pre- or postharvest spray of calcium can reduce this disorder.

23.5.9 Low Temperature Breakdown

This malady is associated with calcium deficiency; the affected tissues are more firm, moist and dark in colour. Pre-harvest spray and postharvest drenching with $CaCl_2$, harvesting at optimum maturity, rapid cooling and reducing storage duration reduce this disorder (Bhat and Khan 2018).

23.5.10 Replant Disease Complex

Apple replant disease is attributed to many biotic and abiotic edaphic factors and is characterized by suppressed growth of young trees in old orchards. Abiotic factors include nutritional deficiency, pH reaction and accumulation of phytotoxins. A number of plant pathogens belonging to fungi, bacteria, actinomycetes and nematodes cause replant disease complex. The complex disease occurs when apple is grown after apple in the same site wherein both abiotic and biotic factors affect the growth of new apple plant. It has no characteristic symptoms other than stunted growth of trees during the first few years of planting (especially during 2–4 years). However, the intensity of the disease depends on the quantity of the plant residues accumulated in each site. In spring season, the pronounced symptoms include the poor growth and light coloration of the leaves and fibrous and hairy nature of the lateral roots resulting in poor nutrient uptake (Thakur and Xu 2004).

23.5.11 Jonathan Spot

The disorder is also known as spy spot, physiological spot or senescent spot. The cause of this disorder is calcium deficiency. The spots may be sharply sunken, irregularly shaped or somewhat lobed area developed at the lenticels that is very dark brown or black (Bhat and Khan 2018). These spots penetrate the flesh somewhat below the surface of the skin. These discolorations can be removed by thinly peeling the fruit. The disorder can be reduced by harvesting apples at optimum maturity pre-harvest or postharvest spray of calcium, harvesting of fruits at optimum maturity, cooling and controlled storage temperatures reduce this disorder.

23.6 Diseases Caused by Pre-harvest Infection

A broad range of factors during the pre-harvest of fruits influences the incidence and severity of postharvest diseases in apple. The most important factors are climatic factors of the region (viz. temperature, humidity, rainfall, etc.), cultivar, and quality of planting material, cultural practices, plant protection measures adopted and the overall orchard maintenance. They can have both direct (survival and amount of inoculum, dispersal, etc.) and indirect (altering the host physiology to influence the host response to infection) influence on the disease occurrence. For example, cultural practices such as pruning of the infected parts reduce the initial inoculum of the pathogen, resulting in lower disease incidence. On the other hand, application of a large quantity of nitrogenous fertilizers predisposes the fruits to infection as a result of soft skins which are highly prone to injuries (Giraud and Bompeix 2012).

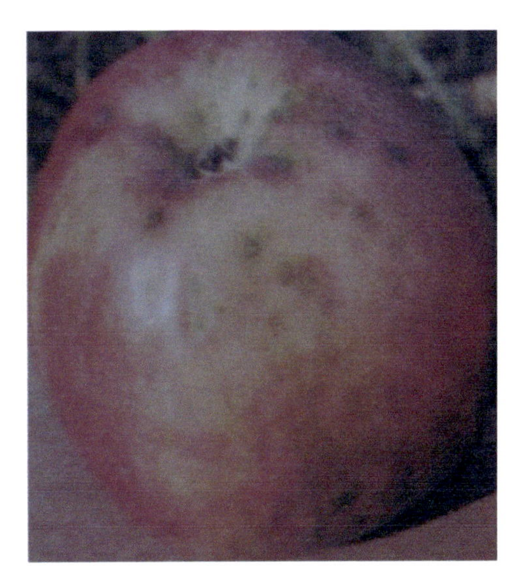

FIGURE 23.5 Symptoms of Brooks fruit spot on apple.

23.6.1 Brooks Fruit Spot

This disease is incited by *Mycosphaerella pomi*, and symptoms first appear on the surface of immature apple fruit as slightly sunken, irregular dark green lesions. The lesions turn dark red or purple with increase in fruit maturity. Severe infection results in pitted and cracked fruit (Figure 23.5). Apple scab spray schedule should be followed to manage Brooks fruit spot. The disease is more prevalent in storage. The spots coalesce, covering larger areas on the surface during storage.

23.6.2 Sooty Blotch

Sooty blotch caused by *Gloeodespomigena* sp. is found on mid-season and late maturing varieties of apple and increases during storage. On the mature fruits, the symptoms appear as shades of olive green. They have a sooty appearance and form discrete small circular colonies to slightly diffuse large colonies. These spots are not a permanent spot on the surface of fruit. The fungus overwinters as pycnidia and mycelium on infected parts of apple and other forest trees. During early summer or spring, the fungus gets dispersed in the form of conidia or chlamydospores. Though most conidia are liberated by early summer, the secondary spread occurs throughout the seasons with the help of mycelial fragments and chlamydospores. Frequent rains and low temperature are conducive for disease development (≤18°C) during the months of May and June. The disease is more prevalent in areas receiving less sunlight (Giraud and Bompeix 2012).

23.6.3 Fly Speck

The disease is caused by *Zygophiala jamaicensis* and *Mycrotheria* spp. and the symptoms are approximately described by the name. The disease is characterized by the black lesions containing 6–50 black superficial pseudothecia on the surface of fruits. The disease is more prevalent in areas receiving abundant rainfall and low wind speed. This disease appears

FIGURE 23.6 Symptoms of fly speck disease on apple fruit.

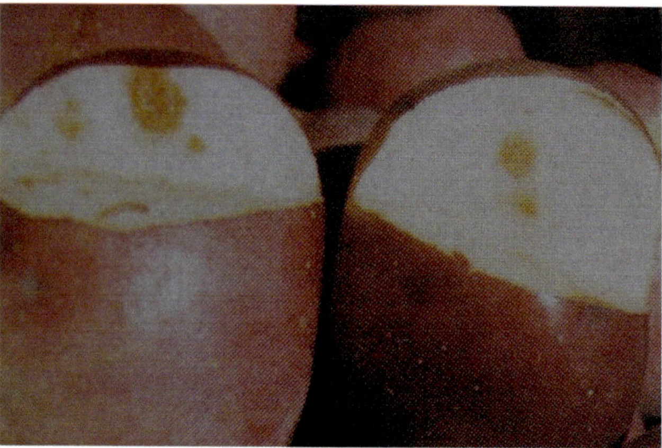

FIGURE 23.7 Symptoms of core rot disease in apple fruits.

during the rains and spreads rapidly through rain splash. The atmospheric temperature between 18°C and 30°C and high atmospheric humidity helps to increase the intensity of the disease. The small spots, which are light brown net-like in appearance, soon increase to 20 mm in diameter and become numerous (Figure 23.6). Since flyspeck and sooty blotch infection occur while the fruits are still on the trees, early adoption of prophylactic measures is essential. Both diseases can be controlled by (i) spraying with lime sulphur or zineb, (ii) thinning and (iii) dipping the picked fruits for 1 minute in bleaching powder (5%) or sodium chlorate (3%). Captan (0.2%) and Benlate (0.05%) are the most effective fungicides for the control of both the diseases of apple (Xiao and Kim 2010).

23.6.4 Core Rot

Core rot of apple has been described from Kumaon (Bose 1969). The disease is commonly found in apples having open pores at the calyx end. The most common fungus associated with the disease is *Alternaria alternata* (Fr.) Keissler. The spores splashed in by raindrops fall at the calyx end, and when such fruits are stored, the fungus grows within the core and penetrates inside the flesh. Typical rot symptoms of this disease are brown to black nearly round lesions, often centred on a skin break or weakened tissue. These lesions contain air pockets and appear corky and dry within the seed locules. The surface of the lesion turns black to dark brown, and in later stages the affected flesh turns black and the rotted tissues become spongy (Figure 23.7). The rot may be wet rot that develops rapidly in the fruit mesoderm or a dry slow rot. The pathogens colonize the senescing flower and later, when the fruit matures, cause fruit infection. The important factors determining the disease severity include the cultivar susceptibility, high relative humidity and warm temperature (Niem et al. 2007). Durrani (1976) recorded occurrence of Alternaria rot in several markets and found maximum disease incidence (12.62%) in March–April. He also screened chemicals in post-harvest dip treatment and found that none except Aureofungin could prevent the disease.

Another disease is incited by *Alternaria mali* Robt and *A. tenuis*. The symptoms appear more during September to December and more in refrigerated fruits than the fresh fruits. Typical rot symptoms of this disease include brown to black coloured oval to irregular lesions, often in the vicinity of weakened tissue or a skin breakage. These spots are firm, dry and shallow, and enlarge and coalesce to form irregular patches. The rotted tissues later turn black and become spongy. The spores are splashed in by raindrops falling at the calyx end. When such fruits are stored, the fungus grows into a sooty velvety colony within the core and gradually penetrates inside the flesh, causing a dry rot. It is very difficult to detect the affected fruits by external inspection in the store. Fruits with open pores at the stylar end should not be stored. While picking, fruits pointing upwards should be sorted out (Bose 1969).

23.6.5 Storage Scab

Another important postharvest disease in apple is storage scab. The symptoms of the scab are observed on all the above-ground plant parts including leaves, petioles, twigs and fruits under field conditions; the disease first appears on the under-surface of tender leaves that emerge in the spring. It is characterized by circular, dark olive green to brownish rough spots having velvety texture (Singh 2019). The lesions from the primary infection are scarce, with 1–2 lesions per leaf. The secondary infections result in more numerous spots. The entire surface of the leaves gets covered when there are numerous secondary infections, and the condition is termed sheet scab. The secondary infection occurs during late summer; the lesions are small and hence the condition is termed pinpoint scab (Singh et al. 2021). Infection that occurs just before harvest may be symptomless at picking yet develop into storage scab lesions after harvest. In the initial stages, the scabs appear as minute brown to black shiny spots which later coalesce to form larger lesions (Figure 23.8).

Singh et al. (2017) observed that scab development in storage was more pronounced in apples collected from poorly maintained orchards as compared to orchards adopting integrated plant protection measures. In Gangothri fruit valley, apple scab lesions appeared after 45 days of storage in all the cultivars. Golden Delicious fruits showed more scab as compared to other cultivars after 60 days of storage. The new scab lesions developed more in fruits having scab lesions at the different time period of storage. There was a gradual increase in the number

FIGURE 23.8 Symptoms of apple scab.

of scab lesions in all the fruits of Delicious cultivars. Compared to healthy fruits, the scabbed fruits showed pronounced shrivelling. Other pathogens found associated with scab lesion were identified as *Trichothecium roseum*, *Monilinia* spp., *Glomerella cingulata*, *Alternaria alternata*, *Penicillium* spp., *Aspergillus* spp. and *B. cinerea*. Pink mould rot caused by *T. roseum* is also associated with the scab lesions and penetrates the healthy fruits in storage (Singh et al. 2017). The symptoms appear as small brown spots, later increase in size, and are characterized by light centres with brown margins (Singh et al. 2021).

Fungicides are the primary means of controlling postharvest diseases, including that caused by *Venturia inaequalis* on apple. Several workers reported that storage scab was controlled by a spray schedule. The curative fungicides controlled storage scab under conditions of low inoculum potential but at moderate and high potential, the use of protective fungicides during growing season was necessary. Kaul (1985) reported that the following is the best control strategy of scab on storage fruit.

1. Control by orchard spray management to prevent late infections and storage scab.
2. Postharvest fungicidal treatment with systemic fungicides (carbendazim).
3. Temperature management influences the deterioration rate of harvested apple in general and development of fungal rots and storage scab in particular. Refrigerated storage at 32°F (0°C) is advocated.

Recently, Singh et al. (2017) reported that postharvest spray of carbendazim (0.05%), bitertanol (0.075%) and flusilazole (0.015%) were effective in controlling storage fungi, including scab. In protective spray programme, propineb (0.4%) protected fruits from storage rot even after 60 days. Penconazole (0.05%) and bitertanol (0.075%) were highly effective as pre-harvest sprays. Bio-control agents such as *P. expansum* and *B. cinerea* and botanicals like econeem, garlic extract, aonla leaf extract, etc., were also effective in managing the scab (Singh et al. 2017).

23.6.6 Phacidiopycnis Fruit Rot

The disease was first reported by Kim and Xiao (2006) from north central Washington State, and the pathogen has been named *Phacidiopycnis washingtonensi*. The disease is characterized primarily by calyx end rot and stem-end rot. The decayed area was light brown to brown and spongy. The symptoms were found to advance along the vascular region and the cut fruits exhibited U- or V-shaped decayed portions. On extended storage, the skin colour of the infected fruits turned dark brown to black. Pycnidia of the pathogen were visible on the fruit surface during later stages of infection.

23.7 Diseases Caused by Postharvest Infection

23.7.1 Blue Mould Rot

Blue mould, incited by *Penicillium expansum* Link., is one of the most common postharvest fungal diseases on apples worldwide (De Capdeville et al. 2003; Giraud and Bompeix 2012). The disease occurs wherever apples are grown or shipped. It is the most destructive disease of apple in transit and in storage conditions most prevalent in the North West Himalayan states. Symptoms originate usually from wounds, stems and invasions of core rot. Later in the season, when the fruits are weak, infection occurs through lenticels. The infection starts from a wound and blue fungal growth appears on the surface of fruits. The rot develops around any minor injury as water soaked, small pale to yellowish brown spots, initially shallow but subsequently deep and wide. The affected parts of the fruits become watery and turn soft, light or yellowish-brown in colour. With the progress of the disease, these watery spots enlarge, resulting in fruit rot. These rotted areas have very soft texture and fruiting bodies of the fungus can be seen on these areas. The mycotoxins produced by the fungus add to the losses caused by the disease.

23.7.2 Grey Mould Rot

Grey mould is caused by *Botrytis cinerea* Fr. It has been recorded as a major cause of spoilage in some areas. The disease spreads from the infected fruits to the healthy ones that come in contact with it (Giraud and Bompeix 2012). Hence, the disease can cause 20–60% losses in prolonged storage conditions. The visible symptoms include translucent, watery, light brown to dark brown spots. *Candida saitoana* is a potent bio-control agent against *B. cineria* as it can induce systemic resistance in the apple fruits (Ghaouth et al. 2003).

23.7.3 Pink Mould

The pink mould rot incited by *Trichothecium roseum* Link ex. Fr. is widely prevalent and very destructive. Delicious cultivar shows higher percentage of infection. The infection is both external and internal but sometimes internal only. The internal infection is evident based on pinkish fluffy growth of the fungus. The rot is usually found associated with scab. The rot appears as sunken, brown bands about 1/8 inch wide encircling the scab spots. White, fungal threads appear and under moist, warm conditions, pink spore masses are produced on the rotted fruits. The advanced stage of rot is characterized by chocolate-brown, sunken area with an irregular outline. The rotted areas remain firm and dry. The affected tissues have a bitter taste. The rot begins from any point of injury on the fruit surface in absence of apple scab. The symptoms are light brown lesions which radiate from the point of infection produce dark brown rot. The rotted flesh becomes musty. The seed cavities contain a pink spore mass. The rot affects all Delicious cultivars of apple. The disease develops slowly below 10°C and it is considerably checked at 32°C.

23.7.4 Bull's Eye Rot

Bull's eye rot (*Pezicula malicorticis*) is more commonly seen on apples from orchards with perennial canker problems on trees. Golden Delicious and Mashadi varieties of apple are highly susceptible to this disease under conditions in Jumla, Nepal (Subedi and Gautam 2018). The initial phase is observed as brown, depressed, circular spots in storage, which gives the name "bull's-eye" to the rot (Giraud and Bompeix 2012).

23.7.5 Bitter Rot

Bitter rot (*Collectotrichum gloeosporiodes*) is a fruit-rotting fungal disease of apple. Symptoms appear on the surface of apple directly exposed to the sun as small, circular brown lesions that change to sunken, dark brown lesions as they enlarge (Giraud and Bompeix 2012). Fruiting bodies visible to the naked eye after the lesion are 1 inch in diameter and are arranged in a concentric circle pattern at the centre of the lesion. Care must be taken during harvesting; fruits must be carefully handled while picking and transferring from bag to bin to avoid bruising.

23.7.6 Whiskers Rot

Whiskers rot is known to occur in all the apple-growing countries. It is caused by *Rhizopus arrhizus* Fischer. Affected fruits show a russet to Verona-brown discoloration of the skin which easily peels away from the underlying tissue. Tissues become soft and watery. The flesh becomes sour in taste without any marked undesirable odour. Occasionally, the infected fruits appear to be free from the fungal growth but on cutting, sparse mycelial growth with black sporangial heads is seen in the seed cavities. The temperature should be kept below 10°C for control of disease. Over-ripe fruits should not be stored.

23.7.7 Rot Due to Other Saprophytic Fungi

Saprophytes also occur occasionally in injured fruits during storage, viz., *Rhizopus nigricans*, *Aspergillus niger*, *A. flavus*, *Pestalotia* sp., etc. All the saprophytic fungi spread by contact and through dispersal of the spores, causing heavy damage. Fruits damaged during picking and storing or by insect, birds and wind fall at the time of maturity are very prone to rotting due to saprophytic fungi. Mucor rot (*Mucor piriformis*) can cause significant losses of fruit (Giraud and Bompeix 2012), but is generally not a major problem, when good harvest management and water sanitation practices at packing are implemented.

23.8 Management of Postharvest Diseases

Since the postharvest diseases of apple are caused due to preharvest and postharvest factors, effective control of postharvest diseases also depends on their management. Impregnated fruit wrappers are reported effective in reducing decay in fruits and prevent sporulation of some fungi. Wrappers not only provide cover to the fruit surface but provide safeguard from bruises and depressions which form easy pathways for decay-causing organisms. It is highly effective in large-scale shipment of wrapped fruits. The skin coating can improve the keeping quality of stored fruits by decreasing the water loss and retarding ripening, rotting and development of scald and other diseases (Kaul and Munjal 1982). Harvested apples are drenched with diphenylamine (DPA) to control superficial scald, a physiological disorder (Janisiewicz et al. 2005). Among them, sanitary conditions in the orchard are probably the most important, followed by different levels of susceptibility of cultivars (Jones et al. 1996; Spotts et al. 1999; Blazek et al. 2006). Many improper sanitation practices during the period of packaging, storage and transit to long distance lead to increase in the surface microflora population on fruits (De Roever 1998).

23.8.1 Cultural Practices

The main approach of cultural practices is to destroy the infected leaves in the pre-leaf fall stage. Sanitation practices such as ploughing leaves into the soil in autumn or early spring have been recommended. There was evidence that these practices did, indeed, reduced the amount of scab and other diseases, but they were not considered reliable enough to replace seasonal fungicide treatments. In New Zealand, removing of leaves by ploughing reduced the amount of scab to nearly two-thirds. Raking and burning of leaves and then ploughing them into the soil was also effective in reducing scab and premature leaf fall, but the treatment was considered too time-consuming for practical use in commercial apple production (MacHardy 1996). Orchard sanitation practices like regular pruning of trees, removal and destruction of fallen leaves and plant debris, etc., play an important role in pest management. Proper tree management and care such as chiselling out the dead portion of trees and wound dressing help in enhancing tree vitality. Other good orchard management practices include proper selection of the orchard site and ensuring that the graft union is well above the soil line.

High-yielding resistant cultivars/rootstocks should be selected for the control of economically important pests and diseases. Important cultural management options in apple include the following.

1. Use of locally adapted pest and disease resistant varieties. This eliminates the need for frequent use of fungicides to control diseases.

2. Maintaining good tree vigour through integrated nutrient management and regular pruning of the trees.
3. Field sanitation - this includes cleaning of orchards by regular removal of weeds, fallen leaves and pruned twigs to reduce the primary inoculum.
4. Reducing inoculum potential levels in the orchard which help to develop pre- and postharvest diseases by following the methods mentioned below.
   a. Destroying the fallen leaves by pulverizing them into soil to destroy the primary inoculum.
   b. Application of nitrogen in the orchard basins for fast decomposition of the debris and fallen leaves. Most of the orchards in hilly states of India have used 5 kg of urea per 100 litres of water (Singh and Kumar 1999).
   c. Collect and burn the fallen leaves, which could be easier in valley areas.
   d. Application of excessive nitrogenous fertilizers has been deemed to induce susceptibility to scab in apple cultivars even in combination with phosphorus. Induction of potassium fertilizer reduces the scab incidence. Thus, a balanced approach in the use of fertilizers without being excessive is recommended, which will also save undue expenditure.
   e. High density of vigorously growing trees in established orchards tends to prevent aeration. Consequently, it results in slow drying of leaf surface and hence long periods of leaf wetness favours infection. Such a situation needs to be corrected through planned pruning and cutting of the overlapping branches.
   f. There is no complete cure to the malady and orchardists alone can keep it in check through intelligent scientific management following cultural and organic control measures.
5. Regular pruning of the trees will help with better air circulation, light penetration and spray coverage. This will help in reducing the humidity within the crop canopy, thereby reducing many diseases.
6. Removal and destruction of the mummified and dried fruits on the tree.
7. Careful handling of the fruits during and after harvesting to reduce wounds and bruises.
8. Quick refrigeration of the fruits to below 9°C.

23.8.2 Harvesting and Handling

Harvesting at the correct physiological maturity is essential to ensure high quality fruits. When harvested early, fruit is more susceptible to many storage diseases apart from poor eating and storage qualities. Completely ripe fruits should be kept separately and storing them for a long time should be avoided because they will rapidly become mealy and too soft for sale. The incidence and severity of postharvest diseases varies among the cultivars. According to Ahmadi-Afzadi et al. (2013), the severity of bitter rot and blue mould caused by *C. gloeosporioides* and *P. expansum*, respectively, varied with the germplasm. They reported that the lesion diameter varied significantly among the late and early maturing cultivars. It was negatively correlated with harvest date and fruit firmness and positively correlated with fruit softening wherein all the three characteristics are genetically determined. The boxes that are used for storage should be dry and properly cleaned before use. This should ensure good aeration of the fruits to avoid postharvest losses.

Apples are very susceptible to mechanical damage and hence they need to be handled with utmost care. Harvested apples should not be allowed to drop down, and fruits should not be handled roughly by workers. After harvesting apples which are often stored in the field should be drenched with fungicide and/or calcium chloride before packaging for control of apple bitter pit during storage. Apples are mostly stored and transported in bulk boxes of 18–20 kg capacity. They also should not be overfilled, as the fruits get bruised when such boxes are stacked.

The innovative storage techniques, such as dynamic controlled atmosphere (DCA), ultra-low oxygen (ULO), extreme ULO, initial low oxygen stress(ILOS), etc., have been found beneficial in increasing the shelf life of apple fruits. Of these, DCA is the most promising. It is characterized by decreasing the oxygen rate to <0.7% until stress peak period and then pushing up again by 0.2%. These techniques can be used in combination with other management strategies (Giraud and Bompeix 2012).

23.8.3 Hot Water Treatment (Thermotherapy)

HWT is an eco-friendly approach for the effective management of the postharvest diseases in many fruits and vegetables (Kaul and Munjal 1980; Maxin et al. 2012b, 2012ba, 2014; Maxin et al. 2005; Wassermann et al. 2019). Dipping of apple fruits in water at 50°C for 5 minutes was highly effective to control fungal decay in apple fruits (Kaul and Munjal 1980). Recently, it has been proven that combination of HWT with bio-control agents and/or bioactive molecules are more effective for postharvest disease management than the individual approach (Conway et al. 2004; Spadaro et al. 2004). Store healthy apples reported beneficial microbes comprising 18 bacterial and 4 fungal species belonging to different taxa which were absent in diseased apples. Wassermann et al. (2019) reported that combined treatment of apple fruits with a biological control consortium consisting of *Pseudomonas paralactis* 6F3, *Pantoea vagans* 14E4 and *Bacillus amyloliquefaciens* 14C9 along with HWT was effective for the management of many postharvest diseases. HWT also has proven effective for the management of *Monilia fructigena* and *Phytophthora* spp. apart from *Neofabraea* spp. and superficial scald. Hence, HWT offers an effective means of non-chemical management strategy for postharvest disease management in apple and other fruits (Giraud and Bompeix 2012).

23.8.4 Microbial Biopesticides/Bioagents

The use of bio-control agents is an emerging alternative for fungicides in postharvest disease management. Beneficial fungi are effective only when humidity is high (60–80%) and so their usefulness is restricted to greenhouses or to regions, such as hilly areas where humidity is always very high during the summer season. On the other hand, the beneficial bacteria have a wider range of applications, as they are less sensitive to moisture. Among the known genus, *Trichoderma* (*T. harzianum* and *T. asperellum*) is one of the most commonly used bio-control agents. It is known to have tremendous potential as bio-control agent based on selection of the species on the basis of antagonistic traits as well as other behavioural traits. It is a very potential fungus for the control of many plant diseases, especially soil- and seed-borne diseases, as it either directly parasitizes the pathogen or competes with other microorganisms and pathogens for food requirements. Antagonists have the added advantages of non-persistence in the environment and non-toxicity to humans, and in many cases have proven to be on a par with the chemical fungicides. Phytotoxic compounds produced by *Trichoderma* spp. are directly antimicrobial in nature and show herbicidal properties. It produces a number of enzymes such as cellulases and hemicellulases, chitinases and glucanases. The enzymes, glucanases and chitinases are responsible for its fungitoxic activity. *Trichoderma* spp. controls *Phytophthora cactorum* on apple (Singh 2006; Singh et al. 2008). Application of *T. asperellum* controls apple canker. Application of *T. harzianum* and *T. asperellum* along with soil solarization has been found to protect apple plants from white root rot pathogen.

Sodium bicarbonate is a potent substance for postharvest management and has been found to be effective in combination with synthetic fungicides and certain bio-control agents for pome fruits (Palou et al. 2001; Smilanick et al. 1999; Wisniewski et al. 2001). Vero et al. (2013) isolated psychrotrophic yeasts from Antarctic soils and they have been observed as potential bio-control agents for the management of postharvest diseases of apple (blue and grey mould) during cold storage. There are three major approaches by which we can improve the effectiveness of the bio-control agents: (i) selection of more effective strains having ecological fitness, (ii) modifying the environment to enhance survival of bio-control agents and (iii) development of microbial consortia. Of these, use of microbial consortia is more popular as it enhances the spectrum of activity, efficiency and reliability and allows combining the traits of different bio-controls (Janisiewicz 1996). Giraud and Bompeix (2012) observed that certain antagonistic microorganisms (*Candida* spp.) were effective against storage fungi such as *P. expansum*, *B. cinerea* and *M. fructigenae*, and are allowed in European countries. The yeast *Pichia guilliermondi* strain M8 reduced the grey mould of apple (*B. cinerea*) to 20% as compared to control (45.3%) after storage at 1°C for 120 days. Pretreatment with the strain at 10^8 cells was found to induce the defence response in apple fruits which reduced the grey mould incidence. The yeast is also found to exhibit antagonistic microorganisms through direct attachment, competition for nutrients, secretion of hydrolytic enzymes, etc. (Zhang et al. 2011). Three more yeast species, viz., *Rhodotorula glutinis*, *Cryptococcus laurentii* and *Cryptococus infirmominiatus*, alone or in combination with thiabendazole, were also found to be beneficial in reducing postharvest disease of apple (Chand-Goyal and Spotts 1997).

Janisiewicz and Korsten (2002) have reported the use of different bacteria and yeast for postharvest disease management in apple. Many fungi producing volatile organic compounds are also found effective (Lee et al. 2009; Park et al. 2010; Stinson et al. 2003a, 2003b). *Oxyporus latemarginatus* EF069 isolated from healthy tissue of red pepper showed fumigation activity against postharvest apple decay (Lee et al. 2009). Mycofumigation with cultures of *Nodulisporium* sp. CF016 was proved effective against the blue and grey mould of apple (Park et al. 2010). These fungi are potential sources of new biomolecules for postharvest disease management.

23.8.5 Botanical Pesticides and Other Natural Compounds

Several compounds of natural origin have been found promising for the eco-friendly management of postharvest diseases of fruits and vegetables including the pome fruits.

Chitosan is another promising compound for postharvest disease management. The mode of action of chitosan can be broadly grouped into two – direct toxicity and elicitation of plant defence. Indirect toxicity, chitosan acts by membrane damage and electrostatic interaction, by acting on pathogen nucleic acids, metal chelation and by deposition on microbial surface. Chitosan also induces the synthesis of PR protein, defence-related enzymes and various other secondary metabolites. The beneficial effect of chitosan on postharvest disease management in apple has been reported by De Capdeville et al. (2003)

23.8.6 Chemical Control

Spoilage of apple is a big problem in India which is aggravated by the absence of proper storage facilities and paucity of information on factors causing losses during storage. Susceptibility of apple fruits to postharvest rotting increases after prolonged storage. Synthetic commercial fungicides have been mostly used to manage postharvest diseases of fruits. Pre-harvest sprays of fungicides can be resorted to reduce decay during storage conditions. Pre-harvest application of ziram (0.2%) has been shown to reduce the apple decay to less than 50% with a single spray (Coates et al. 1995). Iprodione has been used as pre-harvest spray for many years for the management of *Monilinia* infection in stone fruits. It is an effective postharvest fungus against *Rhizopus*, and *Alternaria* is increased when used in combination with oils/waxes reframe the sentence. The new class of Strobilurin fungicide is effective in controlling postharvest diseases in fruits including apple scab (Singh et al. 2017). Singh et al. (2017) reported that the flusilazole, bitertanol and carbendazim were the most effective fungicides for controlling postharvest diseases and other rotting fungi.

23.8.7 Inorganic Fungicides

Inorganic compounds (containing no carbon atoms) are used in conjunction with good cultural practices. The copper-based fungicidal schedule has been found to provide acceptable control of many postharvest diseases when sprayed multiple times. The inorganic fungicides popular across the globe include elemental sulphur, lime sulphur, Bordeaux mixture (copper sulphate, 2 kg, quick or unslaked lime, 2 kg and water, 200 litres), Bordeaux paste (copper sulphate, 1 kg + lime 3 kg + water, 9 litres + linseed oil, 1 litre), Chaubattia paste (copper carbonate, 800 g + red lead oxide, 800 g + raw linseed oil, 1 litre), cow dung paste (fresh cow dung, 1 kg + clay soil, 1 kg + copper sulphate, 10 kg), and fixed copper fungicides. Although cupric hydroxide alone and in combination with sulphur was found effective in scab and powdery mildew management, its use is limited as it causes russeting of fruits.

Use of sulphur in plant disease control is probably the oldest method and continues to be in maximum use even today. Elemental sulphur can be used in dust form or as wettable powder, the latter being more commonly used. It is prepared by fine grinding of the mineral. A major problem with the use of sulphur as a fungicide is that it is toxic to many beneficial insects, spiders and mites. Sulphur only provides about 1 week of protection and requires up to 18 applications per season to obtain satisfactory control.

Jain et al. (2010) reported that biochemical studies of cow urine revealed the presence of sodium, nitrogen, sulphur, vitamins, minerals and hormones and showed antimicrobial action against many clinical as well as plant pathogens (Jarald et al. 2008). Several workers reported the antimicrobial action of cow urine against a wide range of pathogenic microbes including phytopathogenic fungi (Akhter et al. 2006; Deshmukh et al. 2012; Ruchira et al. 2016) and is due to a lower pH, presence of inorganic reactive compounds like formaldehyde, ketones amines, and phenolic compounds like gallic, caffeic, ferulic acids during long-term storage (Naotoshi et al. 2007; Singh et al. 2012). Prasad et al. (2018) reported complete inhibition of a perfect stage of *Venturia inaequalis* (pseudothecial formation and ascospore productivity) during pink bud to petal fall stage of apple in the subsequent spring season. Singh et al. (2017) suggested that antagonists and plant products were highly effective in controlling postharvest losses of apple and increased the shelf life during storage. Sharma and Raj (2019) observed that cow urine-based botanical formulation was quite effective in reducing the microbial population during storage of fruit for a longer duration.

23.8.8 Cooling and Storage

As the apple fruits continue to respire even after harvest, it is essential to cool them immediately. The storage containers and rooms should be dried and thoroughly cleaned before storing. They can be disinfected using 0.25% sodium hypochlorite. Surfaces must be dried before storing. The temperature and humidity of the storage rooms must be checked frequently and be adjusted to the optimum level. The optimum storage temperature for storing apple is 30–40°F, with an optimum of 32°F and relative humidity of 90–95%. Since apples are moderately susceptible to freezing injury, sufficient care must be taken to keep the temperature 1–2 degrees above the freezing point. Relative humidity and saturation point encourages the growth of pathogens and hence should be avoided. Allowing the fruits to warm in the storage area after removing the chilling will help to prevent sweating of the fruits.

New storage techniques, such as ultra-low oxygen (1.5% O_2), initial low oxygen stress (0.5% O_2) ultralow oxygen, extreme ultra low oxygen (0.7–0.9% O_2), dynamic controlled atmosphere, consistent in decreasing the O_2 rate to <0.7% until the stress peak, and then push it up again by 0.2%, reduces scald incidence considerably, even on highly susceptible cultivars (Monteils and Westercamp 2007–2009). However, they may reappear during the sorting and packaging process. DCA seems to be the most promising technique for long term storage

23.9 Perspectives for Integrated Control

With the aim of reducing residues, an integrated strategy combining several techniques could be envisaged:

1. Overall sanitation, disinfecting bins and stores, good practices during handling and keeping harvest date for each variety.
2. In pre-harvest, maintaining only one chemical treatment in the orchard, 3 or 4 weeks before harvest and before rainfall, followed by alternative products (if available and authorized).
3. In postharvest, treatment with hot water followed by bio-control.
4. Cold storage under controlled atmospheric conditions.
5. Grading after regular checking and good practices in packaging.

With a better assessment of risks in orchard and storage, the combination of these techniques could achieve control of the main economically significant postharvest diseases and disorders and minimize the risk of fungicide residues in pome fruits.

REFERENCES

Ahmadi-Afzadi M, Tahir I, Nybom H (2013) Impact of harvesting time and fruit firmness on the tolerance to fungal storage diseases in an apple germplasm collection. Postharvest Biol Technol 82: 51–58.

Akhter N, Begum F, Alam S, Alam MS (2006) Inhibitory effect of different plant extract, cow dung and cow urine on conidial germination of *Bipolaris sorokiniana*. J Biol Sci 14: 87–92.

Bhat SA (2019) Trends and growth in area, production and productivity of apples in India from 2001-02 to 2017-18. Res Ambition 4: 13–23.

Blazek J, Kloutvorova J, Krelinova J (2006) Incidence of storage diseases on apples of selected cultivars and advanced selections grown with and without fungicide treatments. Hortic Sci 33: 87–94.

Bose SK (1969) Spoilage of apples in Kumaon during storage. Progr Hortic 1: 65–88.

Chand-Goyal T, Spotts RA (1997) Biological control of postharvest diseases of apple and pear under semi-commercial and commercial conditions using three saprophytic yeasts. Biol Control 10: 199–206.

China PR (2007) Management of postharvest diseases in stone and pome fruit crops. In: General Concepts in Integrated Pest and Disease Management, pp. 131–147.

Coates L, Cooke A, Parsley D, Beattie B, Wade N, Ridgeway R (1995) Postharvest diseases of horticultural produce. Volume 2. Tropical Fruit. DPI, Queensland.

Conway WS, Leverentz B, Janisiewicz WJ, Blodgett AB, Saftner RA, Camp MJ (2004) Integrating heat treatment, bio-control and sodium bicarbonate to reduce postharvest decay of apple caused by *Colletotrichum acutatum* and *Penicillium expansum*. Postharvest Biol Technol 34: 11–20.

De Capdeville G, Beer SV, Watkins CB, Wilson CL, Tedeschi LO, Aist JR (2003) Pre and postharvest harpin treatments of apples induce resistance to blue mold. Plant Dis 87: 39–44.

De Roever C (1998) Microbiological safety evaluations and recommendations on fresh produce. Food Control 9: 321–347.

Deshmukh SS, Rajgure SS, Ingole SP (2012) Antifungal activity of cow urine. IOSR J Pharm 2: 27–30.

Durrani MK (1976) Studies on Alternaria rot of apple. MSc thesis, University of Udaipur, Jobner.

Eckert JW, Ogawa JM (1988) The chemical control of postharvest disease: Deciduous fruits, berries, vegetables and root/tuber crops. Annu Rev Phytopathol 26: 433–469.

FAO (2015) Postharvest Losses along Value and Supply Chains in the Pacific Island Countries. Vol. 1. Food and Agriculture Organization, Rome, Italy, pp. 0–5.

FAOSTAT (2017) Food and Agriculture Organization of the United Nations. Food and Agriculture Organisation. Rome, Italy. Available at: http://www.fao.org/faostat/en/#data/QC.

Ferguson IB, Watkins CB (2011) Bitter pit in apple fruit. Hortic Rev 11: 289–355.

Ghaouth AE, Wilson CL, Wisniewski M (2003) Control of postharvest decay of apple fruit with *Candida saitoana* and induction of defense responses. Phytopathology 93: 344–348.

Giraud M, Bompeix G (2012) Postharvest diseases of pome fruits in Europe: Perspectives for integrated control. Integrated Plant Protection in Fruit Crops" Subgroup "Pome Fruit Diseases." IOBC-WPRS Bulletin 84: 257–263.

Ilyas MB, Ghazanfar MU, Khan MA, Khan CA, Bhatti MAR (2007) Postharvest Losses in Apple and Banana During Transport and Storage. Department of Plant Pathology, University of Agriculture, Faisalabad. [pdf] Available at: http://pakjas.com.pk/upload/96195.pdf.

Jain NK, Gupta YB, Garg R, Silawat N (2010) Efficacy of cow urine therapy on various cancer patients in Mandsaur district, India. A survey. Inter J Green Pharm 4: 29–35.

Janisiewicz W (1996) Ecological diversity, niche overlap, and coexistence of antagonists used in developing mixtures for bio-control of postharvest diseases of apples. Phytopathology 86: 473–479.

Janisiewicz WJ, Korsten L (2002) Biological control of postharvest diseases of fruits. Annu Rev Phytopathol 40: 411–441.

Janisiewicz WJ, Peterson DL, Yoder KS, Miller SS (2005) Experimental bin drenching system for testing bio-control agents to control postharvest decay of apples. Plant Dis 89: 487–490.

Jarald E, Edwin S, Tiwari V, Garg R, Toppo E (2008) Antioxidant and antimicrobial activities of cow urine. Global J Pharmacol 2: 20–22.

Jones AL, Ehret GR, Meyer MP, Shane WW (1996) Occurrence of bitter rot on apple in Michigan. Plant Dis 80: 1294–1297.

Kaul JL (1985) Storage scab and its management. In: Agarwala RK, Kaul JL, Bhardwaj LN (Eds), Principles and Concepts of Apple Scab Disease Management. H P Krishi Vishva Vidyalaya, Solan, pp. 96–98.

Kaul JL, Munjal R (1980) Effectiveness of hot water treatments in controlling rots of apple. Indian Phytopathol 33: 484–485.

Kaul JL, Munjal RL (1982) Fruit wrappers and skin coatings for control of postharvest decay of apple. Indian J Mycol Plant Pathol 12: 179–184.

Kim YK, Xiao CL (2006) A postharvest fruit rot in apple caused by *Phacidiopycnis washingtonensis*. Plant Dis 90: 1376–1381.

Konopacka D, Plocharski WJ (2004) Effect of storage conditions on the relationship between apple firmness and texture acceptability. Postharvest Biol Technol 32: 205–211.

Lee SO, Kim HY, Choi GJ, Lee HB, Jang KS, Choi YH, Kim JC (2009) Mycofumigation with *Oxyporus latemarginatus* EF069 for control of postharvest apple decay and *Rhizoctonia* root rot on moth orchid. J Appl Microbiol 106: 1213–1219.

MacHardy WE (1996) Apple Scab: Biology, Epidemiology and Management. Academic Press, APS, St. Paul, Minnesota, 545 pp.

Maxin P, Huyskens-Keil S, Klopp K, Ebert G (2005) Control of postharvest decay in organic grown apples by hot water treatment. Acta Hortic 682: 2153–2158.

Maxin P, Weber RWS, Lindhard PH, Williams M (2012a) Hot-water dipping of apples to control *Penicillium expansum*, *Neonectria galligena* and *Botrytis cinerea*: Effects of temperature on spore germination and fruit rots. Eur J Hortic Sci 77: 1–9.

Maxin P, Weber RW, Pedersen HL, Williams M (2012b) Control of a wide range of storage rots in naturally infected apples by hot-water dipping and rinsing. Postharvest Biol Technol 70: 25–31.

Maxin P, Williams M, Weber RWS (2014) Control of fungal storage rots of apples by hot-water treatments: Anorthern European perspective. Erwerbs-Obstbau 56(1): 25–34.

Monteils G, Westercamp P (2007–2009) Pommier Pink Lady® Cripps Pinkcov: Lutte contreles maladies de conservation à l'aide de très basses teneurs en oxygène. Fiches compterendus des essais 2007, 2008, 2009, Ctifl-CEFEL.

Nabi SU, Raja WH, Kumawat KL, Mir JI, Sharma OC, Singh DB, Sheikh MA (2017) Postharvest diseases of temperate fruits and their management strategies. A review. Int J Pure App Biosci 5: 885–898.

Naotoshi K, Osamu Y, Yoshihiko S, Fuminobu M, Masahiro Y, Yoshimitsu M (2007) Clinicopathological findings in peripartum dairy cows fed amino salts lowering the dietary cation-anion difference: Involvement of serum inorganic phosphorus, chloride and plasma estrogen concentrations in milk fever. Japanese J Vet Res 55: 3–12.

Niem J, Miyara I, Ettedgui Y, Reuveni M, Flaishman M, Prusky D (2007) Core rot development in red delicious apples is affected by susceptibility of the seed locule to *Alternariaalternata* colonization. Phytopathology 97: 1415–1421.

Palou L, Smilanick JL, Usall J, Vinas I (2001) Control of postharvest blue and green molds of oranges by hot water, sodium carbonate, and sodium bicarbonate. Plant Dis 85: 371–376.

Park MS, Ahn J, Choi GJ, Choi YH, Jang KS, Kim JC (2010) Potential of the volatile-producing Fungus *Nodulisporium* sp. CF016 for the control of postharvest diseases of apple. Plant Pathol J 26: 253–259.

Prasad RK, Singh KP, Gupta RK (2018) Pre leaf fall spray of chemical, cow urine and fungal antagonists on spring ascospore production of the apple scab pathogen, *Ventura inaequalis*. Int J Curr Microbiol Appl Sci 7: 575–586.

Ruchira T, Tewari AK, Brijesh B, Puspendra S, Megha P (2016) Role of cow urine in beekeeping and crop protection I Uttarakhand, India. Res J Recent Sci 5: 100–107.

Sharma K, Raj H (2019) Effect of postharvest treatments on surface microflora density of apple fruits. Plant Dis Res 34: 22–28.

Singh BK, Yadav KS, Verma A (2017) Impact of postharvest diseases and their management in fruit crops: an overview. J Bio Innov 6: 749–760.

Singh KP (2006) Investigation on development of bio-control measures for the management of saprophytic stage of apple scab pathogen *Venturia inaequalis* in Uttaranchal Himalayas. ICAR Report, New Delhi, 53 pp.

Singh KP (2019) Aerobiology, epidemiology and management strategies in apple scab: Science and its applications. Indian Phytopathol 72: 381–408.

Singh KP, Aravind T, Srivastava AM, Karibasappa CS (2021) Decision making tools for integrated disease management In: Singh KP, Jahagirdar S, Sarma BK (Eds.). Emerging trends in plant pathology. Springer Nature, Singapore

Singh KP, Kumar J (1999) Studies on ascospore maturity of *Venturia inaequalis,* the apple scab pathogen, in Central Himalayas of India. J Mycol Plant Pathol 29: 408–415.

Singh KP, Prasad D, Kumar J (2008) Devising an integrated apple disease management program through the use of antagonists, need based fungicides and advisory services in Uttarakhand hills. (abst.). Indian Phytopathol 61: 398.

Singh KP, Singh A, Prasad RK, Kumar J (2017) Postharvest application of fungicides, antagonists and plant products for controlling storage scab and rots of apple fruits. Indian Phytopathol 70: 315–321.

Singh UP, Maurya S, Singh A, Nath G, Singh M (2012) Antimicrobial efficacy, disease inhibition and phenolic acid-inducing potential of chloroform fraction of cow urine. Arch Phytopathology Plant Prot 45: 1546–1557.

Smilanick JL, Margosan DA, Mlikota F, Usall J, Michael IF (1999). Control of citrus green mold by carbonate and bicarbonate salts and the influence of commercial postharvest practices on their efficacy. Plant Dis 83: 139–145.

Spadaro D, Garibaldi A, Gullino ML (2004) Control of *Penicillium expansum* and *Botrytiscinerea* on apple combining a biocontrol agent with hot water dipping and acibenzolar-S-methyl, baking soda, or ethanol application. Postharvest Biol Technol 33: 141–151.

Spotts RA, Cervantes LA, Mielkee A (1999) Variability in postharvest decay among apple cultivars. Plant Dis 83: 1051–1054.

Stinson AM, Ezra D, Hess WM, Sears J, Strobel G (2003a) An endophytic *Gliocladium* sp. of *Eucryphia cordifolia* producing selective volatile antimicrobial compounds. Plant Sci 165: 913–922.

Stinson AM, Zidack NK, Strobel GA, Jacobsen BJ (2003b) Mycofumigation *Muscodoralbus* and *Muscodorroseus* for control of seedling diseases of sugar beet and Verticillium wilt of eggplant. Plant Dis 87: 1349–1354.

Subedi GD, Gautam DM (2018) Postharvest technology for apple commercialization in Nepal (Nepali). Horticulture Research Division. Lalitpur.

Thakur VS, Xu X (2004) Integrated apple orchard management. Impact analysis and transfer of technology. European Union international and National Agricultural Technology Project. 210 pp.

Thompson AK, Prang RK, Bancroft R, Puttongsiri T (2018) In: Thompson AK, Prange RK, Bancroft RD, Puttongsiri T (Eds.). Controlled Atmosphere Storage of Fruit and Vegetables. CABI, Wallingford.

Vero S, Garmendia G, Gonzalez MB, Bentancur O, Wisniewski M (2013) Evaluation of yeasts obtained from Antarctic soil samples as bio-control agents for the management of postharvest diseases of apple (*Malus* x *domestica*). FEMS Yeast Res 13: 189–199.

Wassermann B, Kusstatscher P, Berg G (2019) Microbiome response to hot water treatment and potential synergy with biological control on stored apples. Front Microbiol 10: 1–12.

Westercamp P, Monteils G (2006) Maîtrise de l'échaudure de prématurité: stress d'oxygène initial et autres méthodes alternatives. Infos-Ctifl 225: 24–27.

Wisniewski M, Wilson C, Ghaouth AE, Droby S (2001) Increasing the ability of the bio-control product, Aspire, to control postharvest diseases of apple and peach with the use of additives. Biological control of fungal and bacterial plant pathogens. IOBC WPRS Bull. 24: 157–160.

Zhang D, Spadaro D, Garibaldi A, Gullino ML (2011) Potential bio-control activity of a strain of *Pichia guilliermondii* against grey mold of apples and its possible modes of action. Biol Control 57: 193–201.

24

Postharvest Diseases of Strawberry and Their Management

Satish K. Sharma,[1] **Joginder Pal**[1] **and Anju Sharma**[2]
[1]*Department of Plant Pathology, Dr YS Parmar University of Horticulture and Forestry Nauni, Solan, Himachal Pradesh, India*
[2]*Department of Basic Sciences, Dr YS Parmar University of Horticulture and Forestry Nauni, Solan, Himachal Pradesh, India*

CONTENTS

24.1 Introduction

Strawberry (*Fragaria× ananassa* Duch.) is grown for its attractive berry fruits and is considered a delicacy fruit because of limited availability in the market for a short duration. The fruits are highly delicate and require utmost care from plucking to until final consumption. Any lapse in between may cause extensive damage to the produce due to mechanical injuries caused by rubbing of fruits against each other, desiccation, appearance of fungal and bacterial pathogens or saprophytes on the injured portion developing into rots and spoilage of the fruit in transit, storage or at retail outlets (Feliziani and Romanazzi 2016). The important fungal pathogens causing postharvest spoilage are *Botrytis cinerea*, *Penicillium* spp., *Mucor* spp., *Rhizopus stolonifer*, *Colletotrichum* spp., etc., which not only cause economic losses but also generate mycotoxins, thereby affecting human health (Pétriacq et al. 2018). These pathogens may infect strawberry fruits at the time of harvesting, handling or during storage. Losses caused by postharvest diseases are greater as compared to pathogens occurring under field conditions. Postharvest losses have been estimated to be around 30% every year despite sophisticated storage facilities being used in the last few years (Singh et al. 2017). Hence, in order to meet the demand of this fruit for growing population, it is important to control postharvest spoilage in addition to increasing production and productivity.

24.2 Major Postharvest Diseases

24.2.1 Grey Mould/Botrytis Fruit Rot

Grey mould is one of the most important postharvest diseases of strawberry; however the disease may also appear while the fruits are about to ripen but are still in the field. If the infection occurs at pre-harvest stage, then fruits are damaged in storage, transportation and at market outlets when the fruit reaches at over-ripened stage due to delayed consumption or utilization. Pre-harvest infection of grey mould pathogen may develop disease symptoms on petals, flower stalks, fruit caps, and on fruit.

24.2.1.1 Causal Organism

Grey mould of strawberry, also known as Botrytis grey mould or Botrytis fruit rot, is incited primarily by *Botrytis cinerea* (de Bary) Whetzel. Recently, the other species of *Botrytis*, viz. *B. caroliniana* Li & Schnabel, *B. fragariae* Plesken, Hahn & Weber, *B. mali* Heald and *Botryotinia ricini* (G.H. Godfrey) Whetzel (Amiri et al. 2016; Dowling et al. 2017) have been found associated with the strawberry fruit rots. The pathogen is necrotrophic in nature, and more than 200 plant species fruits are susceptible to grey mould rot, including fruits and vegetables (Choquer et al. 2007). The perfect stage of the pathogen is *Botryotinia fuckeliana*.

24.2.1.2 Symptoms

The initial infection initiates at the flowering stage of the strawberry plants where the depleting flower petals invite the infection of the pathogen and start harbouring fungus for future development on the fruits. Subsequently, the infection may spread to the healthy blossoms and may result in discoloration, browning extending further to pedicel. Later in the growing season the fungus produces spores which settle on the fruits and develop small dark indifferent spots where the fruit tissue has become soft as compared to nearby fruit tissue depicting unusual appearance. Consequently, heavy sporulation may appear at the point of infection, and the colour may change from white to grey giving fuzzy, greyish-coloured, soft growth on the berry. Later such infected fruits shrivel, dry and eventually form a "mummy" (Grout et al. 2019). The late infections on the mature fruit the symptoms develop as soft, light brown areas (Figure 24.1A) and with progress of the pathogen fruits dry out and are covered by a distinctive velvety grey, dusty-appearing fungal growth. Slowly the infected plant parts are coated with a grey fuzzy mass of spores followed by a soft rot (Figure 24.1B). Further, high relative humidity aggravates multiplication of the fungus, resulting in the development of white cottony mass of mycelium and spores. Sometimes with the infection on partially ripened or unripe fruits, lesion development may mis-shape the fruit, thereby rendering it unfit for marketing or may be destroyed before maturity and become tough and mummified (Williamson et al. 2007).

FIGURE 24.1 (A) Light brown discoloration of fruits. (B) Sporulation on rotting fruits.

24.2.1.3 Disease Cycle and Epidemiology

The fungus over summers on dead mummified fruits or decaying leaf tissues and plant debris. The fungus colonizes the necrotic tissue and thereafter remains quiescent in the base of the floral receptacle. During the spring, when the weather is cool and wet, fungal spores germinate which are disseminated to susceptible plant portions by wind and splashing rain or irrigation water (Beattie et al. 1989; Grout et al. 2019). These spores serve as the primary source of inoculum for blossom and fruit infection. Conidia of *B. cinerea* on the surface of necrotic flower parts germinate in the proximity of moisture.

The infection of blossom and mature fruits requires wet rainy conditions or heavy dew or overhead irrigation coupled with temperature range of 5–30°C; however, the most appropriate temperature for infection subsists in the range of 15–25°C. Infection requires 6 h of wetness, and the rate of infection approaches 90% when flowers or fruit remain wet for 24 h or more. Spore production accelerates with increasing duration of free moisture and occurs over a varied temperature range. Flower infection is the primary means through which fruits become infected. Flowers are susceptible once they have opened, but the vulnerability to infection increases exponentially after 2–3 days of flower opening. The fungus penetrates the petals, stamens, and pistils but not the sepals. One to several blossoms per cluster may become infected, and subsequently the infected blossoms often turn brown, wilt, and die; this phase is termed as blossom blight. The fungus enters immature or developing fruits through these individual flower infections where it remains quiescent until the fruit starts to ripen (Coates et al. 1995). Green fruits are substantially resistant to direct invasion of the pathogen. As the berries begin to ripen, the fungus becomes active and starts to colonize the fruit. Symptoms initiate as a discoloration typically at the calyx end (Figure 24.1A and B). The infected fruits shrivel, dry and convert to "mummy". Such mummified fruits become sacs of grey powdery fungal spores which are easily dispersed by wind and splashing rain to nearby plants and fruits (Turechek 2004).

24.2.1.4 Disease Management

- Certified or disease-free planting material should be used for new plantations. In disease-prone areas the disease-resistant or tolerant varieties should be grown; 'Camarosa', 'Carmine', and 'FL Radiance' and 'FL Elyana' are less susceptible.
- Avoid overwatering, sprinkler irrigation, or stagnation of water in the greenhouse/field.

- The botanical formulations comprising of bougainvillea (*Bougainvillea glabra*), artemisia (*Artemisia roxburghiana*), tulsi (*Ocimum sanctum*), darek (*Melia azedirach*), eucalyptus (*Eucalyptus globules*), lavender (*Lavandula angustifolia*) at different concentrations have been found effective in reducing the severity of the disease.
- Antagonistic fungi such as *Trichoderma harzianum*, *T. viride*, *Gliocladium roseum*, *Bacillus subtilis*, *Streptomyces griseoviridis* and *S. lydicus* have been suggested as alternatives to fungicides.
- Application of fungicides such as carbendazim 50%WP (0.05%), hexaconazole 5EC (0.03%) and Captan 75%WP (0.12%) have been reported effective against the disease.

24.2.2 Soft Rot

The disease has synonyms such as black rot, Rhizopus rot, or Mucor rot, and is most important under storage and transit as it spreads from damaged but infected fruits to the healthy fruits quickly (Feliziani and Romanazzi 2016) when the temperature is more than 5°C (Maas, 1998; Aliasgarian et al. 2013) as compared to grey mould rot, which requires little higher temperature. The pathogen has an extensive host range (Bautista-Baños et al. 2014).

24.2.2.1 Symptoms

First, discoloured, water-soaked spots appear on the infected fruits (Figure 24.2A–C) which enlarge rapidly, followed by enzymatic breakdown resulting in leakage of secondary metabolites of the pathogen in the form of putrefying juice. Prior to complete rotting of the fruit under high relative humidity, the fruits are covered with white mycelium and sporangiophores developing black, spherical sporangia, containing thousands of spores erupting to release a cloudy mass carrying millions of spores (Figure 24.2C) (Feliziani and Romanazzi 2016).

24.2.2.2 Causal Organism

The *Rhizopus stolonifer* (Ehrenb.) Vuill. and *Mucor* sp. appear similar to each other and it is difficult to distinguish the two under field conditions; however, using a hand lens and examining the mycelia on the infected fruit, *Mucor* sporangia have conspicuous film of viscous liquid on the surface, whereas *Rhizopus* sporangia have a dry appearance (Bautista-Baños et al. 2014).

24.2.2.3 Disease Cycle and Epidemiology

Rhizopus stolonifer is fast-growing fungus, particularly under high humid conditions. It colonizes host tissue at lower temperatures, usually from 5 to 10°C, than other postharvest fungi, such as *B. cinerea* which needs more than 15°C temperature for fruit colonization. *Rhizopus* spp. and *Mucor* spp. are found in soil, plant debris and in the air. Fruits touching the soil surface are more prone to soft rot; however, rain plays an important role for the spread of the spores from the infected to healthy plants or fruits. It has been observed that soft rot is more common on strawberry plants exposed to rain in the field or grown under plastic tunnels due to the presence of a continuous film of moisture. Zygospores serve as resting spores whereas sporangiospores are responsible for secondary spread of the disease. Mechanical injury to the fruits predisposes them to initiation of infection by *R. stolonifer* and *Mucor* spp., and subsequently they invade the fruits very rapidly. Further dissemination of the fungus is also facilitated by stolons that can adhere to the surfaces of the nearby healthy fruit by means of rhizoids. After harvest, infection and disease spread caused by *R. stolonifer* depends mainly on the storage temperature, as the minimum temperature for spore germination and growth is about 5°C; however, for *Mucor* spp. can grow and infect the fruit at a temperature of nearly 0°C as well as at 24°C. Both pathogens are closely associated with respect to their enzymatic activities of polygalacturonases, which degrade the middle lamellae of the infected cells. Polygalacturonases and other enzymes like xylanase, cellulase and amylase soften the

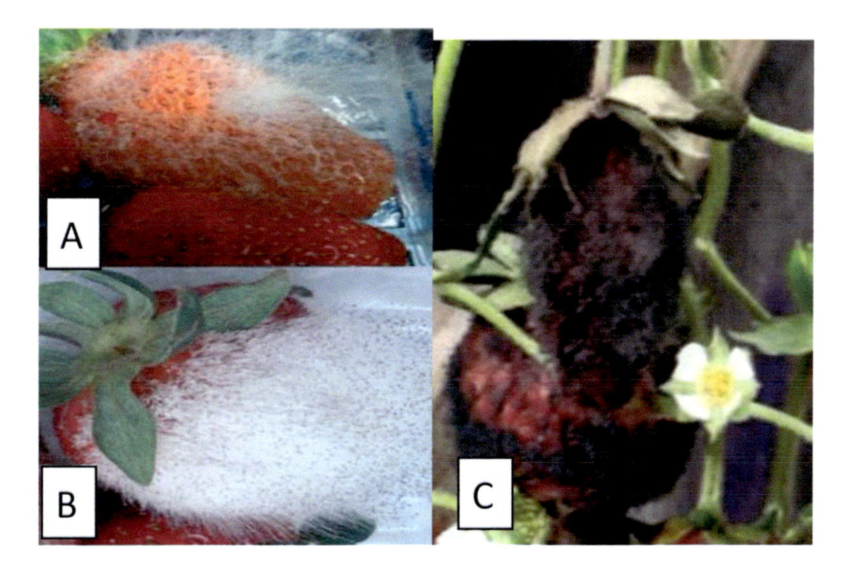

FIGURE 24.2 (A) *Rhizopusstolonifer.* (B) *Mucor* spp. (C) *Rhizopus.*

fruit tissues in a short time. Leakage of extract from the softened fruits often contributes to the dispersal of fungal conidia to the adjoining fruits for initiation of a new attack (Bautista-Baños et al. 2014).

24.2.2.4 Disease Management

- Destruction of infected fruits and crop refuse after harvesting of crop.
- Precooling of harvested fruits to below 8–10°C to avoid congenial atmosphere for the growth of the Rhizopus, followed by refrigerated transportation of precooled fruits.
- Antagonists such as *Aureobasidium pullulans*, *Candida oleophila*, *T. harzianum*, *Cryptococcus laurentii*, etc. have been found effective against the disease.
- Essential oils and plant extracts such as *L. angustifolia*, *Foeniculum vulgare*, *Cuminum cyminum*, *Urticadioica*, *Abies sibirica*, etc. have been found effective to control soft rot of strawberry.
- Broad-spectrum and synthetic fungicides such as boscalid, fludioxonil and fenhexamid can be used to control soft rot during the fruit ripening period.

24.2.3 Anthracnose

Anthracnose is one of the most destructive diseases of strawberry and affects the foliage, runners, crowns and most importantly, the fruit. The disease is caused by *Colletotrichums* spp., which can infect all parts of the plant such as roots, leaves, blossoms, twigs, and fruits, and causes symptoms such as crown root rot, defoliation, blossom blight and fruit rot. The pathogen is of quarantine importance in some European countries. In India the pathogen is endemic to the northeast, though it is not destructive until warm temperature and rainfall prevails during fruit set and harvest.

24.2.3.1 Symptoms

Colletotrichum spp. infects both immature fruits at pre-harvest stage and mature fruits during storage. Symptoms of anthracnose fruit rot appear as dark and sunken lesions (Figure 24.3A and B) on infected fruits which may become tan to dark brown in colour at later stage. During rainy or humid conditions, masses of fungal spores develop around the centre of spots in cream- to salmon-coloured spore masses encircling the fruit lesions. Spots often enlarge until the entire fruit is affected. Diseased fruits frequently become mummified (Ward and Hartman 2012). Under field conditions it is known to reduce plant stand and yield.

24.2.3.2 Causal Organism

Anthracnose of strawberry is caused by *Colletotrichum acutatum* Simmonds, *C. fragariae* Brooks and *C. gloeosporioides* (Penz.) Penz. and Sacc. (Howard et al. 1992; Smith and Black 1990). *C. acutatum* is the primary causal agent and renders severe economic losses to strawberry produce all over the world.

24.2.3.3 Disease Cycle and Epidemiology

The pathogen overwinters in contaminated soil and plant debris, particularly mummified fruits. The pathogen has a wide temperature range at which it can initiate infection; however, the optimum temperature for disease development is 27°C. The spring infections may remain latent for a long time until the weather becomes warm and wet, leading to rapid development of the disease. The secondary spread of the disease occurs through conidia during the entire growing season. The incubation latent period varies according to temperature, i.e. the higher the temperature (25°C) the shorter the time (2–3 days) and vice versa (7–11 days at 5°C). However, the most appropriate temperature is 22–26°C. In addition, an adequate period of surface wetness is required for infection.

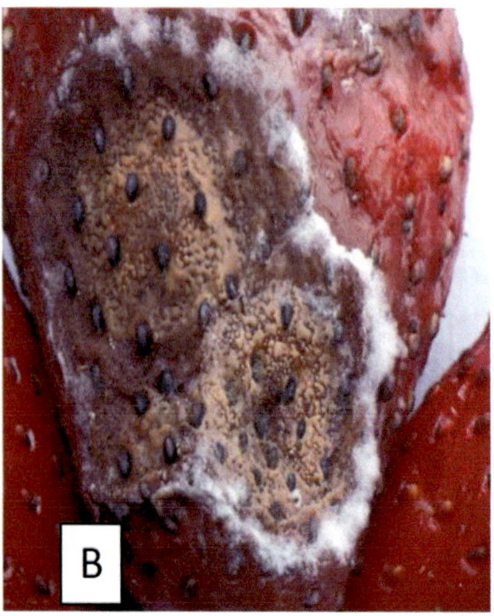

FIGURE 24.3 (A) Infected flowers possess brown spore masses. (B) Fruit lesions with mycelium and spore masses orange spore masses.

At 25–30°C, infection occurs in less than 24 h, but at lower temperatures a longer wetness period is required. Secondary spread is through splashing rain or by insects, animals and workers. Although *C. acutatum* has an extensive host range that includes many fruits, vegetables, and weed species, the strains of *C. acutatum* which are pathogenic on strawberries are relatively host specific in nature (Smith 1998; Smith 2008).

24.2.3.4 Disease Management

- Use of disease-free runners for fresh planting is important to avoid introduction of the disease through planting material. Disease-resistant cultivars such as 'Pelican' and 'Sweet Charlie' etc. can be grown in endemic disease situations.
- Use of drip irrigation rather than overhead irrigation and straw/grass mulch can minimize the rain or irrigation splashes and can prevent spread of the disease.
- Bioagents such as *T. viride* and *B. subtilis* QST 713) can also be utilized under low pressure disease situations.
- Under field conditions where the disease pressure is high, sprays of protective and systemic fungicides such as Captan50WP, quadric 2.08F, 80WP, or 80WDG, pyraclostrobin, azoxystrobin, trifloxystrobin and cyprodinil can be utilized for keeping the disease under control.

24.3 Other Minor Postharvest Decay

Penicillium fruit rot is predominantly caused by *Penicillium expansum*, and along with some other species of the genus *Penicillium* can cause a great deal of postharvest losses in strawberry fruit. Repetitive usage of agrochemicals against grey mould can increase the incidence of other postharvest rots caused by *Penicillium* spp., *Aspergillus* spp., *Alternaria* spp., *Geotrichum* spp. and *Cladosporium* spp. These postharvest diseases usually show negligible occurrence; however, their incidence has increased among cultivars that show resistance to grey mould.

24.4 Conclusion

A large number of fungal pathogens are known to initiate postharvest diseases in strawberry fruits. Some of these infect produce prior to harvest and then remain quiescent until conditions are favourable for disease development after harvest. Other pathogens infect produce during and after harvest through surface injuries. In order to develop strategies for postharvest disease control, it is imperative to take action and consider the production and postharvest handling systems as a serious concern. Many pre-harvest factors directly and indirectly influence the development of postharvest diseases, even in the case of infections after harvest. Traditionally, fungicides have played a key role in postharvest disease control of strawberry. However, trends towards reduced chemical usage in horticulture are forcing the progression of new strategies.

24.5 Future Directions

Postharvest diseases account for substantial reduction in strawberry fruit production, resulting in millions of dollars of loss every year. Over the past years, the use of physical treatments, synthetic fungicides, natural compounds and biocontrol agents have been investigated for the management of postharvest diseases of such perishable fruits; however, none of these techniques are robust which can be employed for the complete control of postharvest diseases of strawberry. Such approaches require a deep understanding of the epidemiology of the causal agents, fruit defence mechanisms against pathogens and understanding of host-pathogen interactions at the molecular level in order to develop the novel inherited disease control mechanisms in which the deployment of resistant cultivars can be a cornerstone. The possibility of integrating the different effective strategies to achieve a greater level of control of postharvest pathogens and to minimize or replace the use of synthetic fungicides needs to be explored in certain host-pathogen systems. The usefulness of integrating different strategies to provide better control of diseases and to obtain safe, disease and residue-free food products is the need of hour. Many details about grey mould of strawberries, which is a common and most important disease, are still poorly understood. Future research is necessary to characterize the genetic pathways and biochemical components that are involved in strawberry – *B. cinerea* interactions. Research on genetic modifications of strawberry that restrict postharvest infections could also be used for guiding conventional breeding efforts or developing new varieties once the market is ready for their acceptance.

REFERENCES

Aliasgarian S, Ghassemzadeh HR, Moghaddam M, Ghaffari H (2013) Mechanical damage of strawberry during harvest and postharvest operations. J World Appl Sci 22(7): 969–974.

Amiri A, Onofre RB, Peres NA (2016) First report of gray mould caused by *Botryotinia ricini Amphobotrys ricini* on strawberry in the United States. Plant Dis 100: 1007.

Bautista-Baños S, Bosquez-Molina E, Barrera-Necha LL (2014) *Rhizopus stolonifer* (Soft rot). In: Bautista-Baños S (Ed.). Postharvest Decay: Control Strategies. Elsevier, p. 383.

Beattie B, McGlasson WB, Wade NL (1989) Postharvest diseases of horticultural produce. Volume 7: Temperate Fruit. CSIRO, Melbourne.

Choquer M, Fournier E, Kunz C, Levis C, Pradier JM, Simon A, Viaud M (2007) *Botrytis cinerea* virulence factors: New insights into anecrotrophic and polyphagous pathogen. FEMS Microbiol Lett 277: 1–10.

Coates L, Cooke A, Persley D, Beattie B, Wade N, Ridgeway R (1995) Postharvest Diseases of Horticultural Produce. Volume 2: Tropical Fruit. DPI, Queensland.

Dowling ME, Hu MJ, Schnabel G (2017) Identification and characterization of *Botrytis fragariae* isolates on strawberry in the United States. Plant Dis 101: 1769–1773.

Feliziani E, Romanazzi G (2016) Postharvest decay of strawberry fruit: Etiology, epidemiology, and disease management. J Berry Res 6: 47–63.

Grout M, Daughtrey M, Mattson N (2019) Gray mould of greenhouse strawberries caused by *Botrytis cinerea*. eGRO Edible Alert. www.e-gro.org4:2.

Howard CM, Maas JL, Chandler CK, Albregts EE (1992) Anthracnose of strawberry caused by the *Colletotrichum* complex in Florida. Plant Dis76: 976–981.

Maas JL (1998) Compendium of Strawberry Diseases, 2nd edition. APS Press, St. Paul, MN, 138 pp.

Pétriacq P, López A, Luna E (2018) Fruit decay to diseases: Can induced resistance and priming help? Plants 7(4): 77.

Singh BK, Yadav KS, Verma A (2017) Impact of postharvest diseases and their management in fruit crops: An overview. J Bio Innov 6: 749–760.

Smith BJ (1998) Anthracnose Fruit Rot (Black Spot). In: Maas JL (Ed.). Compendium of Strawberry Diseases, 2nd edition. APS Press, St. Paul, MN, pp. 31–33.

Smith BJ (2008) Epidemiology and pathology of strawberry anthracnose: A North American perspective. Hortic Sci 43: 69–73.

Smith BJ, Black LL (1990) Morphological, cultural, and pathogenic variation among *Colletotrichum* species isolated from strawberry. Plant Dis 74: 69–76.

Turechek W (2004) Managing gray mould of strawberry. Cornell Blog. https://cpb-us-e1.wpmucdn.com/blogs.cornell.edu/dist/0/7265/files/2017/01/strgraymouldcontrol-25ra0uf.pdf

Ward NA, Hartman JR (2012) Strawberry Anthracnose. University of Kentucky-College of Agriculture, Cooperative Extension Service.

Williamson B, Tudzynski B, Tudzynski P, Van Kan JAL (2007) *Botrytis cinerea*: The cause of grey mould disease. [Online] Molecular Plant Pathology. Wiley/Blackwell.

25

Postharvest Diseases of Pineapple and Their Management

Pankaj Baiswar, Tasvina R. Borah and Akoijam R. Singh
Plant Pathology, Division of Crop Protection, ICAR Research Complex for NEH Region, Umiam, Meghalaya, India

CONTENTS

25.1 Introduction

Pineapple (*Ananas comosus*) is considered to be a major commodity crop in many countries of the world. Major postharvest diseases which affect pineapple are black rot (*Thielaviopsis paradoxa*), fruitlet core rot and yeasty fermentation. Since in pineapple fruitlets are fused to form a fruit, fruit is not sterile and contains bacteria and yeast. Postharvest losses in general are huge and drastically reduce the income of farmers and other stakeholders. Ningombam et al. (2019) have reported postharvest losses of 34.49% in the case of pineapple in Manipur, India. Postharvest losses are high in the supply chain of fruits and vegetables but exact assessment of the loss requires in-depth analysis, and several intricate factors act on the system level that need to be assessed (Tröger et al. 2020). Sustainable development involves reduction or minimization of postharvest losses. Many attempts have been made to prioritize the reduction of postharvest losses in order to increase income for farmers.

Postharvest losses are mainly due to infection by different phytopathogenic fungi (mostly members of Ascomycota and Deuteromycota) and bacteria. Infection is normally through wounds, or sometimes through the intact surface of the fruits. Although in several cases phytopathogenic fungi enter directly into the host, phytopathogenic bacteria always enter through wounds or natural openings. In pathogenic fungi, entry of the pathogen is through the intact fruit surface using lysing enzymes (enzymatic activities dissolve host cell wall and then inter- and intracellular growth occurs) which aid in penetration, or in some cases the infection is harboured from the cropping stage but is asymptomatic (quiescent) until the fruit matures and starts showing symptoms. The infection also spreads through flowers in many cases.

Components of the disease triangle, viz., host, pathogen and environment, play an important role in aggravating postharvest losses. Time of harvesting and physiological stage during harvest are some important host-related traits, and prevailing temperature, humidity, and other environmental factors during harvesting, storage and transportation play a role in disease development. Pathogen aggressiveness, physiological specialization, etc. are some pathogen-related factors affecting disease development.

25.2 Postharvest Diseases and Their Management

25.2.1 Black Rot of Pineapple (Butt Rot, Top Rot)

25.2.1.1 Symptoms

Black rot is a watery, soft rot which starts at the stem end. Black rot is a problem of fresh fruit wherever pineapple is grown. The whole fruit becomes black in advanced stages of infection due to the presence of a huge amount of black-coloured conidia, mycelium and chlamydospores of the fungus.

25.2.1.2 Causal Organism

The disease is caused by fungus *Thielaviopsis paradoxa* (=*Ceratocystis paradoxa*). This is a widespread soilborne pathogen infecting many crops worldwide. Some notable examples are pineapple rot (black rot, butt rot, top rot), sett rot of sugarcane, bud and trunk rot of many palms and black head disease on banana in tropical and subtropical areas (Girard and Rott, 2004). This pathogen causes black rot wherever pineapple is grown (Rohrbach, 1983). Association of this pathogen with black rot of pineapple fruit was first discovered by de Seynes (1886) in France.

The pathogen Ceratocystis is a member of the family *Ceratocystidaceae*, order Microascales (de Beer et al. 2013). Main identifying characters are ascomata with swollen bases and long necks which release ascospores in sticky masses (Upadhyay 1981; Wingfield et al. 1993; Seifert et al. 2013). Reported species of this genus have tubular, phialidic conidiophores; a few species also possess thick-walled, pigmented conidia. The anamorphic states of *Ceratocystis* species have recently been included in *Thielaviopsis* (Paulin-Mahady et al. 2002). DNA sequence analyses have revealed the presence of distinct phylogenetic lineages in *Ceratocystis* (Johnson et al. 2005; van Jacobs et al. 2007). Possibility of some of these lineages representing discrete genera has also been stated (Wingfield et al. 2013). Four described species in genus *Ceratocystis* are *C. paradoxa*, *C. radicicola* (Bliss 1941), *C. musarum* (Riedl 1962) and *Thielaviopsis euricoi* (Paulin-Mahady et al. 2002). *Ceratocystis paradoxa* and *C. radicicola* are well established as pathogens of monocots (Abdullah et al. 2009). Growth and sporulation potential of *T. paradoxa* on different media was observed by Majumdar and Mandal (2018) and maximum sporulation was reported on oatmeal agar medium at 30°C.

25.2.1.3 Disease Cycle and Epidemiology

Major factors which influence development of this disease are injuries during harvesting, storage, packaging and transportation, storage temperature during storage and transportation and also the amount of inoculum of the phytopathogenic fungus present on fruits during harvesting. Infection occurs within 8–12 hours of wounding. Under humid conditions, conidia are produced abundantly and these conidia spread through wind to healthy fruits and infect them through wounds or bruises. Environmental conditions which generally favour the disease and also inoculum buildup are high rainfall (resulting high humidity) at the time of harvesting or just a few days before harvesting (Rohrbach and Schmitt 1994). Pathogen growth is very rapid at 21–32°C. Normally, the infection spread is stopped at low temperatures during storage and transportation, but it resumes once the fruits are returned to normal temperature (Rohrbach and Phillips 1990). Susceptibility varies in different varieties but in general most of the varieties are susceptible.

Bioaccumulation of calcium in the infected fruit has also been reported as a result of induced calcium hydroxide formation in infected fruits (Oniah and Tawose 2018). Infected fruits after sometime do not show the symptoms or the symptoms are masked due to fermentation induced by yeasts.

25.2.1.4 Disease Management

The best way to manage this disease is by minimizing injury during harvesting, transportation and storage since injury becomes the entry point for the pathogen. Postharvest sanitation is also equally important since it reduces the inoculum on the fruit. Normally this is achieved by washing the fruits with chlorinated water.

Sanchez et al. (2007) have reported antagonistic potential of *Trichoderma longibrachiatum* against *T. paradoxa*. They reported that aleuriconidia of *T. paradoxa* were parasitized by *T. longibrachiatum* and it produced proteases, glucanases and chitinases which resulted in disintegration of the hyphae of the pathogen. Wijesinghe et al. (2010) have reported antagonistic isolate of *Trichoderma asperellum* against black rot pathogen (*T. paradoxa*). The possibility of using some yeast isolates as bio-control agents against *C. paradoxa* has also been reported (Oniah and Tawose 2018).

Fungicide treatment can also be given prior to packaging and transportation with approved fungicides. Thiabendazole and benomyl are commonly used fungicides for managing black rot. For this the fruits should be dipped in the fungicide solution within 6 hours of harvest. Sunburnt and damaged fruits with minor skin cracks should be discarded since these can act as inoculum sources.

25.2.2 Fruitlet Core Rot

25.2.2.1 Symptoms

Symptoms consist of areas that differ in colour from light to dark brown extending down the fruitlet core. The symptoms manifested by *Fusarium ananatum* include delayed maturity of the infected fruitlets which remain green. This is seen as a typical external symptom. Internally the affected fruitlets exhibit browning of the centre with the dry type of rot, sometimes spreading to the centre of the fruit and usually confined to the solitary diseased fruitlet.

Barral et al. (2019) elucidated the role of fruitlet anatomy and lignification in resistance against fruitlet core rot pathogens. They reported that lignin deposition in the hypodermis was responsible for resistance against penetration by *F. anantum*. In susceptible cultivars, cell walls are not lignified as much at the edges and the outside layer has fissures, and this allows for easy penetration by pathogenic fungi (Barral et al. 2019). X-ray, fluorescence and multiphoton microscopy conducted by Barral et al. (2019) revealed that sepals and bracts in susceptible cultivars were not perfectly fused with each other, and this allowed easy entry of the pathogen.

The fungus *Talaromyces funiculosum* causes the disease by its presence in pineapple fields on emaciated flowers, the plant heart, pineapple waste, insects and mites (Lim and Rohrbach 1980). The accrual of pineapple tarsonemid mite numbers, *Steneotarsonemus ananas*, in the plant heart in the period of growth marks the start of disease cycle of *T. funiculosum*. Mites feed on the emerging leaf trichomes in the heart of the plant, bracts and sepals of the flowers developing in the inflorescence. Mite population is at the peak after 6-7 weeks of the emergence of inflorescence. *T. funiculosum* colonize the mite-injured parts and infection ensues through emerging flowers at 1–2 weeks preceding regular anthesis. The fungus gradually develops in the internal flower parts. Symptoms are characterized by appearance of usually a grey water-soaked area and later blue-green sporulation in the central part of individual fruitlets. Leathery pocket and inter-fruitlet corking symptoms are closely associated with this disease due to *Talaromyces* infection (Rohrbach and Apt 1986). The severity of the symptoms depends on the time of infection, the pathogen or different pathogens present, the cultivars and the temperature and humidity prevailing at that time. Mean daily temperature ranging from 16 to 21°C favours mite population and infection by *T. funiculosum* (Petty 1978). Similar cool temperatures for about 10–15 weeks after flower induction further enhance infections (Joy and Sindhu 2012).

25.2.2.2 Causal Organisms

Fruitlet core rot is a disease complex of fungus-yeast-mite infestations. The causal pathogens are the fungi *Talaromyces funiculosum* (=*Penicillium funiculosum*) and *Fusarium subglutinas* (=*Fusarium moniliforme* var. *subglutinas*), the round yeast *Pichia guilliermondi* (=*Candida guilliermondi*) and association of the fruit mite *Steneotarsonemus* sp. and the red mite *Dolichotetranychus* sp. (Rohrbach and Apt 1986). In South Africa and few other areas, *F. ananatum* is reported to be the cause of fruitlet core rot (Jacobs et al. 2010). Association of *F. ananatum* with the disease was also reported from China (Gu et al. 2015). The disease is known to occur in several countries producing pineapples and causes losses of fresh as well as processed fruits. However, epidemic levels are sporadic in the main commercial pineapple-producing areas of the world. Low-acid cultivars which are being cultivated commercially are most susceptible. Infection can frequently lead to misshapen fruit.

Talaromyces funiculosum is the main causal organism of this disease (Mourichon 1997; Rohrbach and Johnson 2003). Pre-flower infections by *T. funiculosum* that initiate at the floret prior to the flower opening leads to fruitlet core rot. Numerous factors appear to implicate the pathogenesis of *T. funiculosum*, rendering the disease very intricate. Some aspects of the aetiology of fruitlet core rot are (i) contamination or infection at inflorescence developmental stage (before anthesis), (ii) association of mites, (iii) environmental conditions and (iv) physio-biochemical characteristics of the maturing fruits. Rohrbach et al. (1981) reported the association of pineapple fruit mite (*S. ananas*) with fruitlet core rot and it seems to enhance the pathogenesis of the fungi that cause the disease.

Dry type of rot is caused by *Fusarium* species, contrary to the moister *Talaromyces* and yeast rots. Very scanty published information exists regarding the role of the yeasts as pathogens of fruitlet core rot disease. Yeast infections symptoms are usually light brown. In Hawaii, epidemics of *Fusarium* fruitlet core rot were observed to be associated with excessive population of pineapple red mite, *D. floridanus*. However, the association between *D. floridanus* and *F. subglutinas* fruitlet core rot is not understood.

The symptom responses of the pineapple cultivars vary with infection by *Talaromyces* and *Fusarium*. Rohrbach et al. (1981) observed that main pineapple-producing areas appear to have typical pathogens related to the disease symptoms, perhaps due to the environmental conditions prevailing in the area. *Talaromyces* and *Fusarium* species are usually linked with fruitlet core rot in Hawaii (Rohrbach and Apt 1986). In South Africa, *Talaromyces* species are dominant (Keetch 1977), while in Brazil, *Fusarium* species are mostly dominant (Bolkan et al. 1979).

25.2.2.3 Disease Cycle and Epidemiology

Fusarium subglutinas is a soilborne fungus which can survive for up to 12 months in pineapple tissues and colonize the heart of the growing plants. A modelling approach has been used by Fournier et al. (2015) for accurate forecasting of the periods of high risk for fruitlet core rot by establishing a link between pluviothermic index and fruitlet core rot. Stepien et al. (2013) have reported that the fumonisin B1 and B3, moniliformin and beauvericin were produced *in vitro* by Fusarium species associated with pineapple.

Barral et al. (2020) reported about the ability of various pathogens to produce mycotoxins on pineapple medium and revealed that few species of Fusarium, viz. *F. anantum* produced fumonisin B1 and B2, *F. oxysporum* produced beauvericin and *F. proliferatum* produced the highest concentration of all the three reported mycotoxins. *Fusarium oxysporum* and *P. proliferatum* were able to induce mycotoxin production in artificially infected fruits but *F. anantum* could not produce mycotoxins in plant. Mycotoxin contamination by these three mycotoxins was also evident in natural fruitlet core rot. They also reported that in the La Reunion island of Indian Ocean, 21% of the isolated fungi from pineapple fruits belonged to *Talaromyces stollii* and 79% belonged to Fusarium (*F. anantum*, *F. proliferatum* and *F. oxysporum*). Among all species, *Fusarium anantum* was found to be the most dominant species.

25.2.2.4 Disease Management

The disease can be managed to a great degree by using organic insecticide sprays, specifically to control the mites that spread the disease. A deeper knowledge of the pathosystem will help find an effective means of control of fruitlet core rot.

25.2.3 Yeasty Fermentation

25.2.3.1 Symptoms

In the infection process of this disease, sugar is fermented and alcohol is produced along with release of carbon dioxide. Hence, the first symptoms observed are bubbling of gas and release of juice through cracks. It is a major issue in overripe fruits. Sometimes green fruits can also be affected but in the majority of the cases only ripe or overripe fruits are affected. It has also sometimes been linked with fruit sunburn (Lim 1985).

25.2.3.2 Causal Organism

The yeast species such as *Saccharomyces* spp., *Candida* spp. and *Hanseniaspora valbyensis* have been associated with yeasty fermentation. *Candida intermedia* var. *alcoholophila* has also been reported to cause yeasty fermentation (Rohrbach and Johnson 2003). Molecular identification studies were conducted by Cadez et al. (2006) based on RAPD-PCR, electrophoretic karyotyping and RFLP of the PCR-amplified ITS regions for precise identification of *Hanseniaspora* and *Kloeckera* species. They clearly demonstrated that among these approaches, PCR-RFLP analysis of ITS region along with restriction enzymes can be used for rapid identification of the species in these genera. Chanprasartsuk et al. (2013) reported that a large subunit of rDNA in combination with ITS sequencing analysis can be reliably used for identification of the yeasts associated with pineapple fruit.

25.2.3.3 Disease Management

The best strategies for managing this disease are (i) reduce sunburn,(ii.) fruits should be harvested before they are overripe and (iii) fruits should be covered with paper bags, especially in frost-prone areas.

25.2.4 Pink Disease

This disease was described from Hawaii in 1915 for the first time (Lyon 1915). This is a very typical disease that normally occurs at low levels and remains undetected until the fruits are processed for canning since no external symptoms are visible. This reduces its market value to a great extent. This disease is present in Hawaii, Mexico, the Philippines, South Africa and Taiwan (Marín-Cevada et al. 2006). Cha et al. (1997) also characterized the biochemical pathway, which leads to formation of pink colour because of this disease. Glucose is oxidised into 2,5-diketogluconic acid by a two-step conversion process. Glucose is first converted to gluconate by gluconate dehydrogenase. Two genes are responsible for production of gluconate dehydrogenase. The first is *gdhA* which is common in many bacteria, and the second is *gdhB* gene which is specifically produced by *P. citrea*. Finally, gluconate is further oxidized to 2,5-diketogluconate by 2-ketogluconate dehydrogenase (Pujol and Kado 1999). When this is heated it produces dimers which impart a red colour to the slices.

25.2.4.1 Symptoms

Typical symptoms of red or rusty red discolouration occur when fruit slices are heated during the canning process. In most cases fruits are asymptomatic until heated.

25.2.4.2 Causal Organism

This disease is caused by a bacterial pathogen, *Tatumella citrea* (=*Pantoea citrea*). Earlier many bacterial genera were thought to be associated with this disease, viz., *Pantoea agglomerans* (=*Erwinia herbicola*), *Acetobacter aceti* and *A. liquefaciensi* (Rohrbach and Pfeiffer 1976.). *A. liquefaciens* was reported by Gossele and Swings (1986). This bacterium causes pinkish brown to dark brown discolouration of fruits and in many cases it may not produce any external symptoms (Kontaxis and Hayward 1978). This bacterium produces a chemical 2,5-diketogluconic acid, which forms brown to black pigments upon reaction with amino acids (Buddenhagen and Dull 1967). This disease poses a special kind of problem since it can only be detected when fruits are heated; otherwise it is in general asymptomatic. Hence, quality control in the cannery is of utmost importance to minimize the losses (Rohrbach and Apt 1986).

The bacterium *Tatumella citrea* (=*Pantoea citrea*) is considered to be the causal agent of this disease (Cha et al. 1997; Brady et al. 2010). This bacterium belongs to Enterobacteriaceae, is gram negative, non-spore-forming and bacilliform in shape. *Tatumella morbirosei* and *T. ptyseos* are also associated with pink disease of pineapple.

25.2.4.3 Epidemiology

A warm and humid climate is considered to be favourable for development of this disease, and *T. ptyseos* is the major causal agent in Mexico. In most countries this disease is more prominent in rainy season (Marín-Cevada et al. 2010). An excellent review on pink disease has also been published by Marín-Cevada et al. (2016).

25.2.4.4 Disease Management

Some antagonistic bacteria have been evaluated against this disease and have been found to be promising, like *Bacillus gordonae* 2061R (Cha et al. 1994), but production of the biocontrol agent in large quantities is a practical problem, limiting its use. Marín-Cevada et al. (2012) reported that bacterial exudates of *Burkholderia gladioli* (isolate UAPS07070) inhibited the growth of *T. ptyseos*.

Possibilities of use of transgenic lines/varieties for managing this disease have also been expressed by Kado (2003), suggesting two targets: (i) genes for lowering the substrate which leads to formation of 2,5-diketogluconate and (ii) genes that can inhibit the growth of the pathogen in fruit tissue.

25.2.5 Marbling

25.2.5.1 Symptoms

The most common symptoms of this disease are yellow to red brown to very dark brown discolouration of the diseased fruit. External symptoms are mostly absent and symptoms can be seen only if fruit is cut open. These symptoms may not be very diagnostic, as different kinds of symptoms have been associated with this disease. Other symptoms include hardening of infected tissue, diseased tissues become granulated and brittle and speckled appearance in vascular tissue in the centre of the fruit (Rohrbach and Apt 1986). Several other types of symptoms are also produced and hence different names have been suggested, like fruitlet brown rot (Serrano 1928), fruitlet black rot (Barker 1926) and brown and grey rot.

25.2.5.2 Causal Organism

Acetic acid bacteria, viz. *Acetobacter peroxydans*, *Pantoea ananatis* and *Erwinia herbicola* var. *ananas*, are the casual agents of this disease. This disease has been described from all pineapple growing regions of the world (Rohrbach 1983). *Pantoea ananatis* has also been mentioned by Kado (2006).

25.2.5.3 Epidemiology

Infection levels are generally low, but under certain conditions when temperatures linger above 21–27°C through fruit development, this disease reaches epidemic proportions (Rohrbach and Apt 1986). Low fruit acid and Brix are also linked with severity of this disease, and economic losses occur due to marbling during canning operations.

25.2.5.4 Disease Management

At present there are no clear-cut management methods which work specifically against this disease, but differences in susceptibility of various varieties have been recorded and this can be utilized to some extent by using moderately resistant varieties. The best method is to discard the symptomatic fruits before processing, and only firm fruits should be processed.

25.3 Physiological Disorders

25.3.1 Internal Browning

Internal browning is not a disease but a physiological disorder of pineapple fruit since no organism has been shown to be associated with the symptoms. Low ascorbic acid levels and increased polyphenol oxidase activity seem to be responsible for this disorder (Paull and Rohrbach 1985). The problem becomes severe if the fruits are stored at low temperatures during transit or storage or if they are harvested in conditions where low temperature prevails (0–10°C) (Paull and Rohrbach 1985). This disorder actually limits the use of refrigeration for prolonging the shelf life of the pineapple fruit. Histochemical staining and scanning electron microscopy studies by Luengwilaia et al. (2016) have clearly demonstrated that postharvest internal browning initiates at phloem region, and the highest activity of enzymes polyphenol oxidase, hydrogen peroxide and other phenolic compounds were found adjacent to vascular bundles.

Another term used for this disorder is "black heart", since in the advanced stages of the disorder the entire fruit turns black. Initial symptoms start at the base of the fruitlet as a translucent zone, and as the disorder progresses this becomes black (Paull and Rohrbach 1982).

25.3.3.1 Management

Management basically depends on reducing the activity of polyphenol oxides, which can be done by waxing with paraffin polyethylene (Rohrbach and Apt 1986), since polyphenol oxidase activity is reduced due to increase in internal CO_2 concentration after waxing (Paull and Rohrbach 1985).

25.3.2 Chilling Injury

Pineapple fruit is susceptible to low temperatures, especially below 7°C. Hence, if fruits are stored below 7°C, chilling injury occurs. Symptoms include discolouration of the crown; the green portion of the fruit does not change colour to yellow upon complete ripening. Large sunken lesions are formed on the fruit surface because of collapse of fruitlets. The flesh inside becomes dark and off flavoured. Green immature fruits are more vulnerable to chilling injury and the level of damage increases with increasing time of storage at low temperature. It has been reported by Dolhaji et al. (2018) that suboptimal cold storage conditions (4°C) induced chilling injury as well as internal browning. Total soluble solids, amino acid content, phenolics and antioxidant capacity are reduced due to suboptimal storage conditions. Vulnerability to suboptimal storage conditions varies in different cultivars. Nukuntornprakit et al. (2015) have reported the correlation between reactive oxygen species metabolism and chilling injury in pineapple fruit cultivars. The only management option available for this kind of disorder is waxing of the fruit surface since this can minimize the damage (Dolhaji et al. 2018).

25.4 Minor Physiological Disorders

25.4.1 Woody Fruit

The exposed side of the fruit shows symptoms of discolouration that ranges from yellow to black. Normally growers use straw, weeds or shredded paper to protect the open side of the fruit for managing this disorder (Lim 1985).

25.4.2 Sunscald and Frost Injury

The major symptoms of frost injury are shell discolouration and cracking between the eyes. Localized sunburn can also be caused due to exposure of fruit to full sun after harvest. Sunburnt fruits are more prone to postharvest deterioration. Hence, fruits after harvest should always be kept under shade.

25.4.3 Fasciation/Multiple Crowns

Abnormal development of the inflorescence and crown, i.e., growing of multiple crowns extending from two to many, is known as fasciation or multiple crown disorder. The major factor responsible for this disorder is exposure to high temperatures that results in multiple fruits on the crown region. This is a major cause of concern where fresh fruits are harvested for sale. It can cause economic loss since market value is reduced due to the appearance.

25.4.4 Flesh Translucency

This physiological disorder is more prevalent in fruits that are harvested at a highly immature stage and also in the fruits possessing small crowns. Symptoms include shiny and translucent appearance of the internal flesh. Fruits with this disorder are more susceptible to mechanical damage. The only management option which can minimize the loss caused by this disorder is waxing.

25.5 Postharvest Disease Management Strategies

1. Proper sanitation can help in eliminating or reducing the inoculum load on the fruit surface. In this process, normal apparently healthy fruits are washed with peroxide, ozone or chlorine (gas, sodium hypochlorite), chlorine dioxide, acidified hydrogen peroxide, borax, sodium bicarbonate, UV treatment and also with some permitted fungicides in postharvest stage. This helps in eliminating phytopathogenic organisms on the fruit surface. Ozone is a very strong antisporulant. However, there are many critical factors which need to be addressed during treatment, like concentration, contact time or treatment time, pH and temperature. Sodium bicarbonate and calcium chloride are also permitted in organic postharvest management.
2. Sorting and grading of fruits helps in eliminating the inoculum or reducing the inoculum load.
3. Through cultural practices, contact between host and pathogen can be avoided.
4. Sufficient air circulation during storage and transport should be provided to avoid warming.
5. Strong protective packaging should be used to minimize bruise damage, as injury at this stage will result in postharvest decay.
6. Harvesting of fruits should be done at the proper stage and according to the purpose (export purpose has different maturity indices than local market maturity indices).
7. Storage and transport of fruits should be done at low temperature (9–10°C),but temperature should not be lower than noted to avoid chilling injury.
8. Permitted bio-control agents that are effective against Penicillium rots or decays and also against grey mould should be used, but the majority of the bio-control agents give inconsistent results. Efficacy is highly variable because of many factors that contribute to pathogen suppression.
9. Resistant varieties, if available, or varieties having tolerance against certain important postharvest diseases should be cultivated.

25.6 Conclusion

Pineapple is an important fruit crop having a tendency of high perishability which reduces the profit margin of the farmers. Managing postharvest diseases will be successful only if integrated approaches are used. Spreading awareness related to management aspects is a key issue, since awareness about avoiding mechanical injury during harvesting, transportation and storage and harvesting at proper maturity are the key factors which can make a huge difference. Therefore the reduction of postharvest losses caused by postharvest diseases and physiological disorders is a key issue to ensure the profitability of pineapple production.

REFERENCES

Abdullah SK, Asensio L, Monfort E, Gomez-Vidal S, Salinas J, Lopez Lorca LV, Jansson HB (2009) Incidence of the two date palm pathogens, *Thielaviopsis paradoxa* and *T. punctulata*, in soil from date palm plantations in Elx, Southeast Spain. J Plant Prot Res 49: 276–279.

Barker HD (1926) Fruitlet black rot disease of pineapple. Phytopathology 16: 359–363.

Barral B, Chillet M, Doizy A, Grassi M, Ragot L, Léchaudel M, Durand N, Rose LJ, Viljoen A, Schorr-Galindo S (2020) Diversity and toxigenicity of fungi that cause pineapple fruitlet core rot. Toxins 12: 339. doi:10.3390/toxins12050339.

Barral B, Chillet M, Léchaudel M, Lartaud M, Verdeil JL, Conejéro G, Schorr-Galindo S (2019) An imaging approach to identify mechanisms of resistance to pineapple fruitlet core rot. Front Pl Sci 10: 1065. doi:10.3389/fpls.2019.01065.

Bliss DE (1941) A new species of *Ceratostomella* on the date palm. Mycologia 33: 468–482.

Bolkan HA, Dianese JC, Cupertino FP (1979) Pineapple flowers as principalinfection sites for *Fusarium moniliforme* var. *subglutinans*. Plant Dis Report 63: 655–657.

Brady CL, Venter SN, Cleenwerk I, Vandemeulebroecke K, De Vos P, Coutinho TA (2010) Transfer of *Pantoea citrea*, *Pantoea punctata* and *Pantoea terrea* to the genus Tatumella Emend. as *Tatumella citrea* comb. nov., *Tatumella punctata* comb. nov. and *Tatumella terrea* comb. nov. and description of *Tatumella morbirosei* sp. nov. Int J Syst Evol Microbiol 60: 484–494.

Buddenhagen IW, Dull GG (1967) Pink disease of pineapple fruit caused by strains of acetic acid bacteria. Phytopathology 57: 806.

Cadez N, Raspor P, de Cock AWAM, Boekhout T, Smith MT (2006) Molecular identification and genetic diversity within species of the genera *Hanseniaspora* and *Kloeckera*. FEMS Yeast Res 1(4): 279–289. https://doi.org/10.1111/j.1567-1364.2002.tb00046.x

Cha JS, Ducusin AR, Macion EA, Lucas, LN, Hubbard CH, Kado CI (1994) Biological control of the pink disease of pineapple using bacterial epistasis. Mol Ecol 3: 609.

Cha JS, Pujol C, Ducusin AR, Macion EA, Hubbard CH, Kado, CI (1997) Studies on *Panotea citrea*, the causal agent of pink disease of pineapple. J Phytopathol 145: 313–319.

Cha JS, Pujol C, Kado CI (1997) Identification and characterization of a *Pantoea citrea* gene encoding glucose dehydrogenase that is essential for causing pink disease of pineapple. Appl Environ Microbiol 63: 71–76.

Chanprasartsuk O, Prakitchaiwattana C, Sanguandeekul R (2013) Comparison of methods for identification of yeasts isolated during spontaneous fermentation of freshly crushed pineapple juices. J Agr Sci Tech 15: 1479–1490.

de Beer ZW, Seifert KA, Wingfield MJ (2013) The ophiostomatoid fungi: Their dual position in the Sordariomycetes. In: Seifert KA, de Beer ZW, Wingfield MJ (Eds.). The Ophiostomatoid Fungi: Expanding Frontiers. CBS Biodiversity Ctr, Utrecht, the Netherlands, pp. 1–19.

de Seynes J (1886) Recherches pour servir à l'histoire naturelle des végétaux inférieurs III. Masson, Paris.

Dolhaji NH, Muhamad II, Ya'akub H, Aziz AA (2018) Evaluation of chilling injury and internal browning condition on quality attributes, phenolic content, and antioxidant capacity during sub-optimal cold storage of Malaysian cultivar pineapples. Malaysian J Fund Appl Sci 14 (4): 456–461.

Fournier P, Benneveau A, Hardy C, Chillet M, Lechaudel M (2015) A predictive model based on a pluviothermic index for leathery pocket and fruitlet core rot of pineapple cv. 'Queen'. Euro J Plant Pathol 142(3): 449–460. doi:10.1007/s10658-015-0625-8.

Girard JC, Rott P (2004) Pineapple disease. In: Rott P, Bailey RA, Comstock JC, Croft BJ, Saumtally AS (Eds.). A guide to Sugarcane Diseases. Librairie du Cirad, Montpellier, France, pp. 131–135.

Gossele F, Swings J (1986) Identification of *Acetobacter liquefaciens* as causal agent of pink-disease of pineapple fruit. J Phytopathol 116: 167–175.

Gu H, Zhan RL, Zhang LB, Gong DQ, Jia ZW (2015) First report of *Fusarium ananatum* causing fruitlet core rot in China. Plant Dis 99(11): 1653.

Jacobs A, van Wyk PS, Marasas WFO, Wingfield BD, Wingfield MJ, Coutinho TA (2010) *Fusarium ananatum* sp. nov. in the *Gibberella fujikuroi* species complex from pineapples with fruit rot in South Africa. Fungal Biol 114: 515–527.

Johnson JA, Harrington TC, Engelbrecht CJB (2005) Phylogeny and taxonomy of the North American clade of the *Ceratocystis fimbriata* complex. Mycologia 97: 1067–1092.

Joy PP, Sindhu G (2012) Diseases of pineapple (*Ananas comosus*): Pathogen, symptoms, infection, spread and management. Technical Bulletin. Pineapple Research Station (Kerala Agricultural University), Vazhakuam-686670, Muvattupuzha, Emakulam, Kerala, India.

Kado CI (2006) *Erwinia* and related genera. Prokaryotes 6: 443–450.

Kado CI (2003) Pink Disease of Pineapple. APSnet Features. Online. doi: 10.1094/APSnetFeature-2003-0303.

Keetch DP (1977) H.1 Black Spot (Fruitlet Core Rot) in Pineapples. Pine Series H: Diseases and Pests, Government Printer, Pretoria, Republic of South Africa, 3 pp.

Kontaxis DG, Hayward AC (1978) The pathogen and symptomatology of pink disease, *Acetobacter aceti*, *Gluconobacter oxydans*, on pineapple fruit in the Philippines. Plant Dis Report 62: 446–450.

Lim WH (1985) Diseases and Disorders of Pineapples in Peninsular Malaysia. MARDI Report No. 97, Malaysian Agricultural Research and Development Institute (MARDI), Kuala Lumpur, Malaysia, 53 pp.

Lim TK, Rohrbach KG (1980) Role of *Penicillium funiculosum* strains in the development of pineapple fruit diseases. Phytopathology 70: 663: 665.

Luengwilaia K, Becklesb DM, Siriphanicha J (2016) Postharvest internal browning of pineapple fruit originates at the phloem. J Plant Physiol 202: 121–133.

Lyon, HL (1915) A survey of the pineapple problems. Hawaii Plant Rec 13: 125–139.

Majumdar N, Mandal NC (2018) Growth and sporulation physiology of postharvest pathogen *Thielaviopsis paradoxa* (De Seynes.) Hohn. Int J Curr Microbiol Appl Sci 7(7): 537–544.

Marín-Cevada V, Caballero-Mellado J, Bustillos-Crista Les R, Muñoz-Rojas J, Mascarua-Esparza MA, Casta Nedalucio M, Lopez-Reyes L, Marti NL, Fuente S-Ramírez LE (2010) *Tatumella ptyseos*, an unrevealed causative agent of pink disease in pineapple. J Phytopathol 158(2): 93–99.

Marín-Cevada V, Fuentes-Ramirez LE (2016) Pink disease, a review of an asymptomatic bacterial disease in pineapple. Rev Bras Frutic 38(3). Jaboticabal, SP. http://dx.doi.org/10.1590/0100-29452016949.

Marín-Cevada V, Muñoz-Rojas J, Caballero-Mellado J, Mascarua Esparza MA, Casta Neda-Lucio M, Carreño-Lopez R, Estrada-De Los Santos P, Fuentes-Ramírez LE (2012) Antagonistic interactions among bacteria inhabiting pineapple. Appl Soil Ecol 61: 230–235.

Marín-Cevada V, Vargas HV, Juárez M, López VG, Zaga Da, G, Hernández S, Cruz A, Caballero-Mellado J, Lópezreye SL, Jiménez-Salga Do T, Carcañomontiel M, Fuentes-Ramírez LE (2006) First report of the presence of *Pantoea citrea*, causal agent of pink disease, in pineapple fields grown in Mexico. Plant Pathol 55(12): 294.

Mourichon X (1997) Pineapple fruitlet core rot (black spot) and leathery pocket: Review and prospects. doi: 10.17660/ActaHortic.1997.425.54.

Ningombam S, Noel AS, Singh J (2019) Post-harvest losses of pineapple at various stages of handling from the farm level up to the consumer in Manipur. Intl J Agri Sci 11(22): 9235–9237.

Nukuntornprakit O, Chanjirakul K, van Doorn WG, Sirphanich (2015) Chilling injury in pineapple fruit: Fatty acid composition and antioxidant metabolism. Postharvest Biol Tech. 99: 20–26.

Oniah T, Tawose FO (2018) Fungi associated with black rot disease of pineapple (*Ananas comosus* L.) fruits and the effects of the disease on nutritional value of the fruits. ISABB J Food Agri Sci 8(3): 18–24.

Paulin-Mahady AE, Harrington TC, McNew D (2002) Phylogenetic and taxonomic evaluation of *Chalara*, *Chalaropsis* and *Thielaviopsis* anamorphs associated with *Ceratocystis*. Mycologia 94: 62–72.

Paull RE, Rohrbach KG (1982) Juice characteristics and internal atmosphere of waxed 'Smooth Cayenne' pineapple fruit. J Am Soc Hortic Sci 107: 448–452.

Paull RE, Rohrbach KG (1985) Symptom development of chilling injury in pineapple fruit. J Am Soc Hortic Sci 110: 100–105.

Petty GJ (1978) H.16 Pineapple Pests: Pineapple Mites. Pineapple Series H: Diseases and Pests, Government Printer, Pretoria, Republic of South Africa, 4 pp.

Pujol CJ, Kado CI (1999) *GdhB*, a gene encoding a second quinoprotein glucose dehydrogenase in *Pantoea citrea*, is required for pink disease of pineapple. Microbiology 145: 1217–1226.

Riedl H (1962) *Ceratocystis musarum* sp. n., die heuptfruchtform der *Thielaviopsis*–Art von bananenstielen. Sydowia 15: 247–251.

Rohrbach KG (1983) Pineapple diseases and pests and their potential for spread. In: Singh KG (Ed.). Exotic Plant Quarantine Pests and Procedures for Introduction of Plant Materials. ASEAN Plant Quarantine Centre and Training Institute, Serdang, Selangor, Malaysia, pp. 145–171.

Rohrbach KG, Apt WJ (1986) Nematode and disease problems of pineapple. Plant Dis 70: 81–87.

Rohrbach KG, Namba R, Taniguchi G (1981) Endosulfan for control of pineapple inter-fruitlet corking, leathery pocket and fruitlet core rot. Phytopathology 71: 1006.

Rohrbach KG, Pfeiffer JB (1976) The interaction of four bacteria causing pink disease of pineapple with several pineapple cultivars. Phytopathology 66: 396–399.

Rohrbach KG, Phillips DJ (1990) Postharvest diseases of pineapple. In: Paull RE (Ed.) Symposium on Tropical Fruit in International Trade. International Society for Horticultural Science, Honolulu, Hawaii, pp. 503–508.

Rohrbach KG, Johnson, MW (2003) Pests, diseases and weeds. In: Bartholomew DP, Paull RE, Rohrbach KG. (Eds.) The Pineapple: Botany, Production and Uses. CABI, pp. 203–251.

Rohrbach KG, Schmitt DP (1994) Part IV. Pineapple. In: Plotz, RC, Zentmyer, GA, Nishijima, WT and Rohrbach, KG (Eds.). Compendium of Tropical Fruit Diseases. APS Press, St. Paul, MN, pp. 45–55.

Sanchez V, Rebolledo O, Picaso RM, Cardenas E, Cordova J, González O, Samuels GJ (2007) *In vitro* antagonism of *Thielaviopsis paradoxa* by *Trichoderma longibrachiatum*. Mycopathologia 163: 49–58

Seifert KA, de Beer ZW, Wingfield MJ (2013) The ophiostomatoid fungi: Expanding frontiers. CBS Biodiversity Ser, Utrecht, the Netherlands, 12. 337 p.

Serrano FB (1928) Bacterial fruitlet brown rot of pineapple. Philippine J Sci 36: 271–305.

Stepien L, Koczyk G, Waskiewicz A (2013) Diversity of *Fusarium* species and mycotoxins contaminating pineapple. J Appl Genet 54: 367–380.

Tröger K, Lelea MA, Hensel O, Kaufman B (2020) Re-framing post-harvest losses through a situated analysis of the pineapple value chain in Uganda. Geoforum 111: 48–61.

Upadhyay HP (1981) A monograph of *Ceratocystis* and *Ceratocystiopsis*. University of Georgia Press, Athens, GA, 176 pp.

van Wyk M, Al Adawi AO, Khan IA, Deadman ML, Al Jahwari AA, Wingfield BD, Ploetz R, Wingfield MJ (2007) *Ceratocystis manginecans* sp. nov., causal agent of a destructive mango wilt disease in Oman and Pakistan. Fungal Divers 27: 213–230.

Wijesinghe CJ, Wijeratnum SW, Smarasekara JKRR, Wjesundera RLC (2010) Biological control of *Thielaviopsis paradoxa* on pineapple by an isolate of *Trichoderma asperellum*. Biol Control 53(3): 285–290.

Wingfield MJ, Seifert KA, Webber JF (1993) *Ceratocystis* and *Ophiostoma*: Taxonomy, ecology and pathogenicity. APS Press, St. Paul, Minnesota, 304 pp.

Wingfield BD, van Wyk M, Roos H, Wingfield MJ (2013) *Ceratocystis:* Emerging evidence for discrete generic boundaries. In: Seifert KA, de Beer ZW, Wingfield MJ (Eds.). The Ophiostomatoid Fungi: Expanding Frontiers. CBS Biodiversity Sr, Utrecht, the Netherlands, pp. 57–64.

26

Postharvest Diseases of Minor Fruits and Their Management

S.D. Somwanshi,[1,4] Sunita J. Magar[2,4] and A.P. Suryawanshi[3,4]
[1]Krishi Vigyan Kendra, Badnapur, Maharashtra, India
[2]Department of Plant Pathology, College of Agriculture, Latur, Maharashtra, India
[3]Department of Plant Pathology, College of Agriculture, Latur, Maharashtra, India and
[4]Vasantrao Naik Marathwada Krishi Vidyapeeth, Parbhani, Maharashtra, India

CONTENTS

26.1 Introduction

Globally, India ranks second in production of fresh fruits and vegetables after China. India's diversified climate and soil potential ensures cultivation/production of a wide range of fruits and vegetables year-round. Apart from many other major fruit crops produced in India, sapota (*Manikara zapota*), custard apple (*Annona reticulata*) and aonla/amla (*Phyllanthus emblicae*) are commercially grown on a large scale in various states. The ever escalating Indian population has created food scarcity, nutritional deficiency and environmental insecurity. Unfortunately, average Indians cannot meet their basic daily requirement of fruits and vegetables, which has resulted in a very low Human Development Index (HDI).

Fruits are a very important and indispensable food commodity, having high commercial and nutritive value. Fruits play a vital role in the daily diet by supplying necessary growth factors such as vitamins and minerals essential to human health. Due to infections caused by pathogens, fruit shelflife is reduced and this in turn affects its economic value, which is a limiting factor. It is estimated that even in developed countries, pathogens decay about 20–25% of the harvested fruits during postharvest handling (Droby 2006; Zhu 2006). Because of high levels of sugars and nutrient elements in fruits and their low pH values, fruits are more prone to fungal decay, and infection of fruits may occur at any stage, i.e., during growing season, harvesting, handling, transit, storage after harvest and prevailing market conditions (Grewal 1954; Ratnam and Neema 1967; Thakur and Chenulu 1970; Manoharachary and Rama Rao 1989; Singh and Sharma 2007; Sudha et al. 2007).

Despite popularization and adoption of Good Agricultural Practices (GAP), the growers of sapota, custard apple and aonla frequently encounter crop health maintenance and management problems, especially due to diseases and insect pests. Diseases and pests in the field, transit and storage account for over 25–30% loss of farm produce.

Due to high moisture content, fruits are inherently liable to deteriorate, especially under tropical conditions, rendering them unfit for consumption as well as unmarketable. More than 250 parasitic diseases cause decay and blemishes in fresh fruits and vegetables during transit, storage and marketing. Considering the economic importance, export potential of fresh as well as finished produce, quantitative and qualitative losses caused due to various diseases causing pathogens, details of postharvest diseases and their management of sapota, anole and custard apple are given next.

26.2 Sapota (*Manikara zapota* (L.) P. Royen)

The key factors responsible for the development of postharvest diseases in sapota (Sapodilla) fruit are high moisture and nutrient content. Symptoms of fruit rot as an infection of *Phytophthora palmivora* (Butler) may appear on the fruits on lower branches of the tree, if it comes into contact with water coinciding with high temperature and relative humidity (RH) (Balasubramanian et al. 1988). Some other common diseases include blue mould rot (*Penicillium italicum*), Cladosporium rot (*Cladosporium oxysporum*) and sour rot (*Geotrichum candidum*) (Mickelbart 1996) and black mould rot (*Aspergillus niger* (Van Teigh)). Soft rot caused by following fungi are also reported as *Botryosphaeria dothidea* (Mougeot Fries Cesati & Denotari), *B. ribis* and *Fusarium moniliforme* Sheldon (Jain et al. 1981).

Different species of *Pestalotiopsis* and *Phomopsis* may also cause infection on fruits as fruit rot. Some bacteria species are found associated with fruit latex (Pathak and Bhat 1952). *Colletotrichum gloeosporioides* causing anthracnose can be a serious problem in fields having high RH (Kader 2009).

26.2.1 Pestalotiopsis Fruit Rot

26.2.1.1 Symptoms

Affected fruits show water-soaked lesions that later become brown within 2–3 days, covering the entire fruit within 3–4 days. Rotted fruits turn soft and small tufts of mycelium are visible. Later, fluffy growth of mycelium appears, with numerous black acervuli in rotted zones.

26.2.1.2 Causal Organism

Srivastava (1967) reported *Pestalotiopsis sapotae* P. Henn as the causal pathogen of soft rot disease from Uttar Pradesh, India. The causal agent of the disease has been reported by Gupta and Sehgal (1974) as *P. versicolor* (Speg.) Steyeart from Rajasthan, India and Dhingra et al. (1980) as *P. glandicola* (Cast.) Steyeart from Haryana, India. However, Jain et al. (1981) reported *Pestalotiopsis mangiferae* causing a leaf spot of sapota. The fungal colonies appear yellowish white, with branched and septate mycelium showing black, globose to subglobose acervuli like fruiting bodies. Conidia are fusiform, four-septate, and arise on short and simple conidiophores.

26.2.1.3 Disease Cycle and Epidemiology

No injury on fruits as the pathogen is primarily a wound parasite. Spore germination is found maximum at 30°C and no germination of spores will be observed below 15°C or above 40°C with RH above 96%.

26.2.1.4 Disease Management

The disease can be managed by removing and destroying diseased fruits from the orchard. Before and after harvest, fruits should be covered with polythene bags.

26.2.2 Phytophthora Fruit Rot

26.2.2.1 Symptoms

During the rainy season, initial symptoms can be seen at the calyx disc of the fruit. The infected part shows whitish cottony growth and covers the entire fruit surface within 3–4 days if the weather is humid. Fruits near the soil line with dense canopy coinciding with prevailing high RH in the field are severely affected, and more damage can be seen in fallen fruits. The lower portion of the fruit skin becomes soft, turns light brown to dark brown and emits a characteristic foul odour. The fruit is covered with whitish cottony fungal growth on the surface (Rao et al. 1962), and eventually such fruits drop off from the tree.

26.2.2.2 Causal Organism

The disease is caused by *Phytophthora palmivora* Butler. From Maharashtra, Rao (1968) reported this fungus as a transit and storage rot.

26.2.2.3 Disease Cycle and Epidemiology

Numerous sporangia and spores are produced as a part of pathogen infection, on the surface of diseased tissues, when the temperature is near to 25°C. Spores spread from the infected portion or soil with the help of rain splashes or wind. High soil moisture with cool and wet weather is favourable for disease development. High RH, temperature between 28 and 32°C, ill-drained soils and fruit injuries are the predisposing factors for disease initiation and development.

26.2.2.4 Disease Management

Maintain clean orchards. Follow cultural, mechanical and biological approaches for disease management. Prune and destroy old, dead and decayed twigs and fruits. Avoid crowding in the orchard by following the recommended plant spacing. Manage fertilizer regimes to avoid dense foliage. Mummified fruits should be collected from orchard and destroyed to check the primary inoculum. While picking from the trees, care should be taken not to mix infected fruits with healthy ones. Storing the fruits in cold storage at 10°C is found effective. Fruit coating with essential oils, viz., eucalyptus, lemon grass or thymine 0.6–0.8% (V/V) could help in minimizing postharvest fruit rot of sapota.

26.2.3 Botryodiplodia Fruit Rot (Brown Pedicel End Rot)

B. theobromae Pat is the causal pathogen of this disease. It is a serious pathogen and causes a good amount of loss of sapota fruits in storage (Srivastava 1967; Mandal and DasGupta 1982). On affected fruits, brownish discolouration can be seen, mostly at the stem end, later proceeding downwards in a wavy manner irregularly. In severe infections, a number of small pycnidia are produced over the entire fruit surface. Fruit rot is soft and watery, and maximum infection occurs at about 30°C. The same fungus can also cause dieback or flat limb, called fasciation. The disease can be controlled by avoiding injuries and bruises to the fruits and by storing the fruits at 10°C.

26.2.4 Black Mould Rot

The pathogen *Aspergillus niger* Van Tieghm is the cause of this disease. The disease was observed during the hot summer days on injured or over-ripe fruits (Mandal and DasGupta 1981). Infection advances more or less radially on the surface. Black conidial heads cover the fruit surface and fruit becomes soft and cracks. The disease can be controlled by avoiding injuries and bruises to the fruits and by storing the fruits at 10°C.

26.2.5 Green Mould Rot

Mickelbart (1996) reported *Penicillium italicum* as the causal pathogen of this disease. The fungus causes extensive decay and affected tissue becomes soft and water-soaked. The disease can be controlled by avoiding injuries and bruises to the fruits and by storing the fruits at 10°C.

26.3 Aonla (*Emblica officinalis*)

Aonla (*Emblica officinalis* Gaertn. Syn. *Phyllanthus emblicae* L.) (Syn. Indian gooseberry) is an important fruit used in many Ayurvedic medicines and in foods (Soni and Verma 2010). Aonla has a high potential to contribute to the food security of India. Aonla is grown abundantly in many tropical and subtropical countries like India, China, Indonesia, Myanmar and Sri Lanka. In India, intensive cultivation of Aonla is found in Pratapgarh, Azamgarh, Varanasi, and Jaunpur districts of Uttar Pradesh. It is also cultivated throughout India.

Aonla requires sandy loam to clay soils and grows well in arid and semi-arid regions. It possesses a number of health benefits. It is very high in ascorbic acid (vitamin C), which has antioxidant properties and helps to remove free radicals. It also contains polyphenols (ellagic acid and gallic acid) and hydrolysable tannins (emblicanin A, emblicanin B, punigluconin and pedunculagin) having antioxidant properties. Aonla fruit has anti-inflammatory, antipyretic, hepatoprotective and anti-tumor properties as well as a wide range of hypolipidemic, hypoglycaemic and analgesic properties. Commercially, in India Aonla varieties/cultivars grown are 'Banarasi', 'Francis', 'Chakaiya', 'Kanchan (NA-4)', 'NA6'and 'NA7'. Aonla fruits are prone to a number of postharvest fungal pathogens causing heavy losses if harvesting/handling is not properly done (Samuh and Kusum 1990; Mishra 1988; Omprakash 2003; Arya and Arya 2004); the most prominent are blue and black moulds causing 10–18% losses. Considering the enormous health benefits and economic importance of Indian gooseberry, information on its postharvest diseases and their management strategies is given in the next section.

26.3.1 Postharvest Diseases

26.3.1.1 Rust

Also called ring rust disease, it is serious disease in aonla causing considerable losses, especially in Rajasthan and more particularly, in the Udaipur district. It has recently been reported in the areas around Lucknow and Pratapgarh districts in Uttar Pradesh, causing major losses in aonla orchards.

26.3.1.1.1 Symptoms

A few black pustules appear on fruits as initial symptoms. Later, these pustules turn into ring-like structures. The pustules coalesce covering maximum fruit area. After rupturing of a papery covering of the infected fruit, the black mass of spores is exposed (Figure 26.1). It is observed that symptoms produced by pathogen on fruits and leaves resemble difference from each other. Teliospores are the infecting spores causing infection on the fruit and leaf surface. Initially, the size of pustules is small (1–1.5 mm) and with advancement of the pathogen they enlarge (3–4 mm). Pinkish brown pustules on leaves are found in groups or scattered.

26.3.1.1.2 Causal Organism

Ravenelia emblicae (Syn. *Phakopsora phyllanthi*) is the cause of this disease.

26.3.1.1.3 Favourable Conditions

Dense plant population and high RH favours disease development.

26.3.1.1.4 Disease Management

Grow rust-tolerant aonla cultivars like Banarasi and Chakaiya. Follow training and pruning to facilitate proper aeration in the orchard. Prune and destroy severely infected plant parts. Spraying of Chlorothalonil 75% WP @ 0.2% or Zineb 75% WP @ 0.2% or wet table sulphur 80% WP @ 0.4% was found effective to control the disease. Three sprays of any of these fungicides during July–September can manage the disease. *Trichoderma asperellum* and *Pseudomonas fluorescens* used alone or in combination are also effective against the disease.

26.3.1.2 Anthracnose

26.3.1.2.1 Symptoms

During the months of August–September, symptoms appear on leaflets and fruits. Minute circular, brown to greyish spots with yellow-coloured margins appear on leaflets in early infection. The affected portion shows a greyish centre, which is raised with dot-like fruiting bodies. Depressed/sunken lesions formed on fruits show dark colour in the centre and form acervuli arranged in rings. Irregular lesions appear on fruiting bodies with spore masses (Figure 26.2). Finally, the fruits rot and become mummified (Misra and Shivpuri 1983).

26.3.1.2.2 Causal Organism

The pathogen *Colletotrichum gloeosporioides* (Perfect stage: *Glomerella cingulata* Stoenm) Spauld causes this disease. Acervuli on host issues remain punciform, submerged, scattered and black in colour. Conidiophores are short, single celled, hyaline and cylindrical to phial shaped. Conidia are hyaline, agglutinate, cylindrical, single-celled and are 10.5–17.5 μm × 3.0–4.5 μm in size (Misra and Shivpuri 1983).

26.3.1.2.3 Disease Cycle and Epidemiology

Hot humid weather with intermittent rains and temperature ranging from 24 to 32°C coinciding with flowering favours anthracnose infections in the field. The pathogen survives on infected branches, leaves, etc. Field infection in

FIGURE 26.1 Rust disease symptoms on aonla fruits.

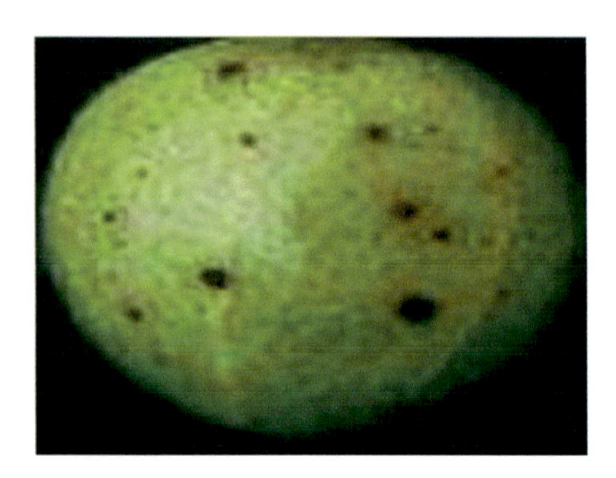

FIGURE 26.2 Anthracnose disease of aonla fruit.

developing fruits leads to quiescent infection/latent infection. Later, as ripening starts, lesions begin to develop under postharvest conditions, affecting fruit quality and causing economic loss.

26.3.1.2.4 Disease Management

Sanitation of the orchard, i.e., collection and destruction of affected fruits from the orchard, should be done. Spray Mancozeb 75% WP @ 0.25% or Carbendazim 50% WP @ 0.1% or copper oxychloride 50% WP @ 0.25% before harvesting of the fruits.

26.3.1.3 Soft Rot

26.3.1.3.1 Symptoms

Within 2–3 days of infection, smoky brown to black round lesions appear on fruits. After that, the affected portion shows olive brown discoloration with water-soaked areas and extends towards both ends of the fruits and forms an eye-shaped appearance (Figure 26.3). The infected portions of fruits show a dark brown colour and are crinkled with softening of underlying tissues and get reduced in size. Both young and mature fruits get infected, but mature fruits are more prone to the disease.

26.3.1.3.2 Causal Organism

The pathogen *Phomopsis phyllanthi* Punith is the cause of this disease. Pycnidia of fungus is abundant, immersed, becoming erumpent, black, conical shaped, measuring 200–500 μm, ostiolate with a prominent neck, pycnidial wall stromatic, composed of several layers of cells, thick walled and deeply pigmented on the outer side. Conidiogenous cell hyaline, simple, septate, rarely branched, phialidic, enteroblastic upto 20 μm long in chains.

26.3.1.3.3 Disease Cycle and Epidemiology

The pathogen also infects other hosts such as *Pyrus malus*, *Psidium guajava*, *Citrus sinensis*, *Musa Paradisiaca*, *C. medica*, *Lufa cylindrica*, *Abelmoschus esculentus* and *Citrullus vulgaris* by artificial inoculation. Hot and wet weather is more congenial for development of the disease. For fungal growth, the optimum temperature requirement is 29°C and grows well up to 32°C.

26.3.1.3.4 Disease Management

Treatment of fruits with Mancozeb 75% WP @ 0.25% or carbendazim 50% WP @ 0.1% or Difolatan (1500 ppm) just after the harvest is found effective to manage the disease (Lal et al. 1982). Avoid injury to fruits during harvesting. Smearing of fruits with mentha oil can check the fruit rot.

26.3.1.4 Sooty Mould

26.3.1.4.1 Symptoms

Black fungal growth with velvety covering appears on leaves, twigs and flowers, which is superficial and does not penetrate into leaves or affected plant parts.

26.3.1.4.2 Causal Organism

The disease is caused by *Capnodium* sp.

26.3.1.4.3 Disease Cycle and Epidemiology

Rain splashes spread conidia. During rain, sooty growth is washed away. High humidity and reduced ventilation within the orchard favour the disease. The honeydew secretion of insects sticking to the leaf surface will act as a medium for fungal growth.

26.3.1.4.4 DiseaseManagement

Spray Lambda Cyhalothrin @ 0.05% or wet table sulphur 80% WP @ 0.2% or starch @ 2%.

26.3.1.5 Blue Mould

26.3.1.5.1 Symptoms

On the fruit surface, mould causes brown patches and water-soaked areas. With the disease progression, three different types of colours develop in succession, i.e. bright yellow, purple brown and bluish green, respectively (Figure 26.4). Fruits emit a foul smell as yellowish liquid drops exudates on the fruit surface, and finally the entire fruit has a bluish-green beaded appearance.

26.3.1.5.2 Causal Organism

The pathogens *Penicillium citrinum* and *P. expansum* are the causes of this disease. Setty (1959) reported occurrence of blue mould disease caused by *Penicillium islandicum* Sopp.

FIGURE 26.3 Symptoms of the soft rot disease on aonla fruits.

FIGURE 26.4 Blue mould disease symptom on aonla fruit.

FIGURE 26.5 Black soft rot disease symptom on aonla fruit.

26.3.1.5.3 Disease Management

Careful handling of fruits and maintenance of strict sanitation in storage is required for controlling infection of postharvest diseases. Treat the harvested fruits with borax or sodium chloride (1%) to check the infection (Setty (1959). Treatment with carbendazim 50% WP @ 0.1% or thiophanate methyl 70% WP @ 0.1% should be given after the harvest. Smearing fruits with mentha oil can check the blue mould.

26.3.1.6 Black Soft Rot

26.3.1.6.1 Symptoms

Pre-injured fruits are more prone to pathogen infection. A number of minute, brown necrotic lesions with white mycelial growth appear on infected fruits (Figure 26.5). The lesions between fringes of mycelium and healthy tissue are surrounded with a pronounced halo of water-soaked, faded tissue. Spores with a black-coloured, powdery layer cover the rotted surface.

26.3.1.6.2 Causal Organism

The pathogen *Syncephalastrum racemosum* Cohn ex. J. Schrot is the cause of this disease.

26.3.1.6.3 Disease Management

Injury to fruits should be avoided during harvesting. Infected fruits should be discarded from the orchard. Mancozeb 75% WP @ 0.25% or carbendazim 50% WP @ 0.1%, should be sprayed before fruit harvesting.

26.3.1.7 Dry Fruit Rot

26.3.1.7.1 Symptoms

Disease symptoms appear as a light brown infected area from 1.0–2.0 cm on the fruit. Later, they appeared depressed and turn completely black. Occasionally, pycnidia are observed on older parts of the infected fruits (Jamaluddin et al. 1975).

26.3.1.7.2 Causal Organism

The pathogen *Phoma emblicae* Jamaliddin, Tand and Tand, is the cause of this disease. Mycelia are branched, septate, hyaline, hyphae changing to light brown with age. Pycnidia are abundant, large, pleurilocular, globose, sub-globose to irregular, ostiolate, carbonous, hard and stromatic in nature. Solitary pycnidia measure 547–900 × 525–7123.4 µm, while compound pycnidia range 1–1.5 mm. Diameter of the pore varies from 75 to 115 µm. Conidiophores are elongated, hyaline, 17.0–51.12 µm in height. Conidia are ovate to cylindrical, rod-like with pointed ends, hyaline, measuring 5.94–9.9 × 1.98–3.3 µm. The most suitable temperature for conidial growth as well as for infection is 27–30°C. The diameter of the fungal colony ranged from 5.3 to 6.2 cm on the eighth day under optimum conditions (Jamaluddin et al. 1975).

26.3.1.8 Fruit Rot (*Phoma putaminum* (Speg.))

Pandey et al. (1980) from Allahabad reported *Phoma putaminum* (Speg.) as the cause of fruit rot during the months of December–January in aonla. Minute pinkish-brown necrotic spots appear as an initial infection extending towards both ends of the fruit, forming an eye-shaped appearance. In severe cases of infestation, these minute lesions coalesce to form bigger pustules that are dark brown and show wrinkling. The internal tissues in the rotted fruits below the infection site become soft.

26.3.1.9 Fruit Rot (*Nigrospora sphaerica* (Sacc.))

Kamthan et al. (1981) recorded *Nigrospora sphaerica* (Sacc.) Mason as a cause of fruit rot during December from Kanpur, Uttar Pradesh. Due to this disease, loss was estimated 5–10% annually. The initial symptom of rot is a minute 2 mm diameter dot, which develops into a black ring spot upto 7 mm. Several such rings coalesce, resulting in complete rotting of the infected fruits (Kamthan et al. 1981).

26.3.1.10 Cladosporium Fruit Rot

Jamaluddin (1978) recorded fruit rot from Allahabad. Loss due to the disease is estimated to be 2–5%. *Cladosporium tenuissimum* Cooke and *C. cladosporioides* (Fries) de Vries are also causes of this disease. *C. herbarum* causing similar small spots of red colour on aonla fruits was reported by Tandon and Verma (1964).

26.3.1.10.1 *Cladosporium tenuissimum*

Infection of this pathogen is noticed during November–February. Infection starts as colourless area, slightly soft, and it subsequently progresses in a circular manner. Light brown mycelial growth of the fungus is also evident on the infected areas. It shows lesions of 1.0–1.5 cm in diameter within 7 days. For successful infection of this pathogen, injury to the fruit is essential (Jamaluddin 1978).

26.3.1.10.2 *Cladosporium cladosporioides*

Dark brown, necrotic lesions are observed as fruit rot infection. The disease is more severe only in matured and ripe fruits. Infection can also be seen in a few freshly harvested fruits. Lesions of 1week were 0.7–1.2 cm in diameter (Jamaluddin 1978).

26.3.1.11 Pestalotia Fruit Rot

Tandon and Srivastava (1964) recorded *Pestalotia cruenta* Syd. as a causal pathogen of fruit rot. They observed irregular brown-coloured spots on the fruits. Initially, brownish discoloured spot is observed on the fruit surface, which develops very slowly. Such spots become mummy brown and develop light brown discolouration of the skin around them. In severe cases, white fluffy aerial growth of the fungus is observed on the infected portion with dry dark brown discoloration of the internal parts of the diseased fruit (Tandon and Srivastava 1964).

Hyphae of the fungus are branched, septate, hyaline, 4–6 µm in thickness. Acervuli are black and gregarious. Conidia are fusiform, five-celled, 16.5–24 × 5.5–7.0 µm in size. Three intermediate cells coloured, versicolours and olivaceous in colour. Exterior cells are hyaline, septate, usually 2–3, rarely 4, 10–18 µm long, divergent, pedicels usually 4–6 µm long (Tandon and Srivastava 1964).

26.3.1.12 Alternaria Fruit Rot

During January–February 1981, from local fruit orchards of Allahabad, Pandey et al. (1984) reported Alternaria fruit rot disease. They revealed that *Alternaria alternata* was responsible for the fruit drop of aonla. Initially, a small brownish spherical necrotic spot appears on the fruit surface, and size of the spots gradually increases in circular fashion with advancement of the disease. These spots later show dark brown to black discolouration and the adjoining spots coalesce with soft and pulpy centre of the infected tissues. Pandey et al. (1984) proved the pathogenicity with artificial inoculation of *A. alternata*, using the pinprick method, and observed completely rotted fruits within 15–20 days of inoculation.

26.3.1.13 Cytospora Fruit Rot

Tandon and Verma (1964) reported *Cytospora* sp. causing rotting of aonla fruits. Small spots of ochre red colour are formed.

26.4 Custard Apple (*Annona reticulata*)

Annona squamosa L. is a dryland horticultural crop belonging to the family *Annonaceae* and most widely grown Annona species. It is called custard apple (Ahmed et al. 2008). As per Mukerji and Bhasin (1986), in India, fruit rot caused by *Fusarium*, *Phytophthora*, *Phoma*, *Pestalotia*, etc., *C. gloeosporioides* (*G. cingulata*) and *Diplodia* rot (*B. theobromae* Patouillard) are major postharvest diseases in custard apple. Under field conditions in custard apple, fruit rot (*C. gloeosporioides*) is most devastating, widespread and an economically important disease.

26.4.1 Fruit Rot and Anthracnose

Colletotrichum gloeosporioides, a causal pathogen of dry fruit rot, was found to be badly infected in the custard apple, causing huge losses of about 60–70%during the last 13–15 years. Nowadays, it has become a severe melody to this crop. In neglected gardens, the losses may go up to 100% (Gaikwad and Salunkhe 1999).

26.4.1.1 Symptoms

Initial infection starts at the blossom-end portion of the fruit and later covers the entire fruit surface. Diseased fruits are malformed in size and may remain hanging on the tree or fall down. Necrotic spots of 2–10 mm diameter appear on unripe fruits and later show dark brown to blackish spots, which coalesce to cover the entire fruit (Figure 26.6).

26.4.1.2 Causal Organism

Colletotrichum gloeosporioides (Teleomorph: *Glomerella cingulata*) is the causal agent of this disease. The pathogen is a hemi-biotroph that produces hyaline, septate branched mycelia and acervuli with black setae. Conidia produced are single celled, hyaline and thin walled with oil globules.

26.4.1.3 Disease Cycle and Epidemiology

Unharvested fruits of earlier seasons left on the trees provide inoculum for primary infection. Conidia will be spread through air as a secondary infection. Wet and windy conditions with temperature ranging from 28 to 32°C favour disease development.

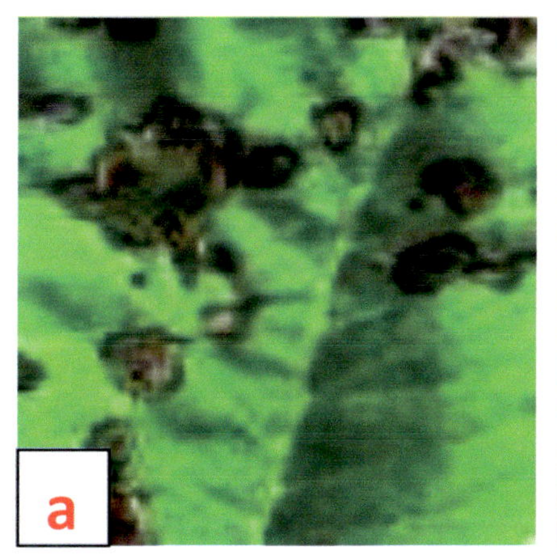

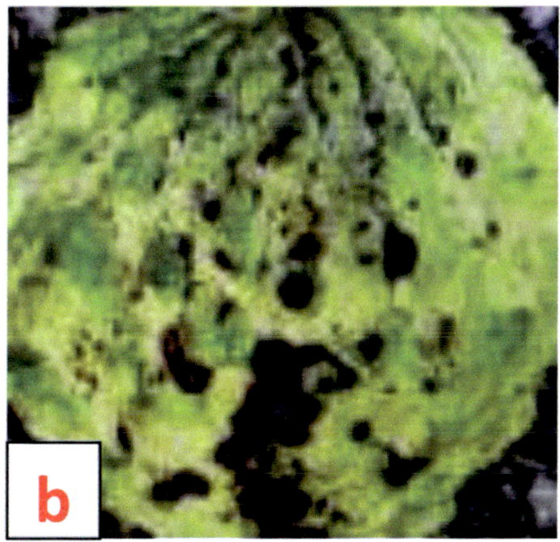

FIGURE 26.6 Symptoms of Anthracnose disease on leaf (a) and on fruit (b).

26.4.1.4 *Disease Management*

Mulching is the effective practice to reduce soil splash. Dead twigs and mummified fruits should be collected from the orchard and destroyed to reduce primary inoculums of the pathogen. Trees 50 cm above the ground should be pruned. Carbendazim 50% WP @ 0.1% or chlorothalonil 0.2% should be sprayed according to the requirement.

26.4.2 Stem-End Rot/*Diplodia* Rot

26.4.2.1 *Symptoms*

Purplish to black spots or blotches covered with white mycelia and black pycnidia are produced as primary infection on diseased fruit surface. Typical symptoms of Diplodia rot are dark internal discolouration with extensive corky rotting on fruit surface (Figure 26.7). As a part of secondary infection, penetrated flesh either softens or hardens and cracks as per the pathogen inoculum.

26.4.2.2 *Causal Organism*

The fungus *B. theobromae* is a causal agent of this disease. Pycnidial fungi produce hyaline thin-walled pycnidiospores which become brown, thick-walled, two-celled with longitudinal striations.

26.4.2.3 *Disease Cycle and Epidemiology*

The pathogen present on infected plant parts will serve as a source of primary inoculum. Temperatures ranging from 25.9 to 31.5°C and 80% RH favour the disease incidence.

26.4.2.4 *Disease Management*

Carbendazim 50% WP @ 0.1% or tebuconazole 25.9% EC @ 0.1% or trifloxystrobin 50% WG @ 0.1% or Zineb 75% WP @ 0.2% should be sprayed as per requirement.

FIGURE 26.7 Symptoms of stem-end rot on custard apple fruit.

26.4.3 Black Canker

26.4.3.1 *Symptoms*

Initially, small specks to large blotches of irregular shape appear on the fruits. These spots show an individual feathered edge. Beneath the spots, internal tissue damage is not more than 10 mm deep.

26.4.3.2 *Causal Organism*

The fungus *Phomopsis annonacearum* is a causal agent of this disease. Mycelium is septate, pale brown colour. Conidia are oblong and oval shaped.

26.4.3.3 *Disease Cycle and Epidemiology*

Chlamydospores and dormant mycelium in plant debris serve as a source of primary infection. Windborne conidia will be a source of secondary infection. For disease development, extreme wet weather and cool to moderate temperatures (15–20°C) are favourable conditions. Prolonged rainy periods and at least 6hours of continuous wetness, and crowded orchards increase the chances of infection.

26.4.3.4 *Disease Management*

Spraying of Tricyclazole 70% WG @ 0.1% or Pyraclostrobin 25% EC @ 0.1% should be done as per requirement.

26.4.4 Purple Blotch/Fruit Rot: *Phytophthora capsici*

26.4.4.1 *Symptoms*

Initially, small purple spots appear on small sized fruits and later cover the entire fruit surface. On large fruits, spots or blotches of 2–5 cm diameter develop, with a distinct margin and sometimes a halo around the spots. Under moist conditions, white fungal growth covers the fruits. Affected fruits drop off. Internal flesh becomes discoloured.

26.4.4.2 *Causal Organism*

The fungus *Phytophthora capsici* is a causal organism of this disease.

26.4.4.3 *Disease Cycle and Epidemiology*

The pathogen survives in a dormant state in the soil, as oospores or chlamydospores. Undersoil moisture oospores germinate and produce zoospores, which along with rain splash are spread to leaves, fruits, and branches on trees and initiate the infection. High rainfall, high soil moisture, low temperature and high RH favour disease development.

26.4.4.4 *Disease Management*

Thin out the dense canopy of the tree. Orchard sanitation should be a priority. It is required to remove excess soil water/

moisture. Spray the fungicides copper oxychloride 50% WP @ 0.3% or Mancozeb 75% WP @ 0.25% or Metalaxyl MZ 72% WP @ 0.2% or combi product fungicide Metalaxyl 4% + Mancozeb 64% Wg @ 0.25% should be done as per requirement. Spraying should be done at an initial fruit setting and later provide 2–3 sprays at an interval of 15 days.

REFERENCES

Ahmed ZU, Hassan MA, Begum ZNT, Khondker M, Kabir SMH, Ahmed M, Ahmed ATA, Rahman AKA, Haque EU (Eds.) (2008). Encyclopedia of Flora and Fauna of Bangladesh. Vol. 6. *Angiosperms: Dicotyledons (Acanthaceae – Asteraceae)*. Bangladesh Asiat. Soc., Dhaka, pp. 1–408.

Arya A, Arya C (2004) A new fruit rot pathogen of aonla. *J Mycol Plant Pathol* 34(1): 154–155.

Balasubramanian P, Ponnuswami V, Irulappan I (1988) A note on susceptibility of sapota varieties and hybrids to leaf spot disease (*Phaeophleospora indica* Chinnappa). *South Indian Hortic* 36(1–2): 72–73.

Dhingra R, Mehrotra RS, Aneja KR (1980) A new Pestalotiopsis fruit rot of sapodilla. *Indian Phytopathol* 33: 330.

Droby S (2006) Improving quality and safety of fresh fruits and vegetables after harvest by the use of bio-control agents and natural materials. *Acta Hortic* 709: 45–51.

Gaikwad AP, Salunkhe GN (1999) Custard apple – Integrated disease and pest management. A folder in Marathi.ADR, NARP, Ganeshkhind, Pune 7, pp.1–4.

Grewal JS (1954) Cultural and pathological studies of some fungi causing diseases of fruits. D Phil thesis. University of Allahabad, Allahabad.

Gupta IJ, Sehgal SP (1974) A new Pestalotiopsis rot of *Achras sapota* from India. *Indian Phytopathol* 27: 616.

Jain SK, Saxena AK, Saksena SB (1981) Two new fruit rot diseases of *Achras sapota*. *Indian Phytopathol* 34: 403.

Jamaluddin A (1978) Cladosporium rot of fruits of *Phyllanthis emblica*. *Proc Natl Acad Sci India* 48: 62.

Jamaluddin A, Tandon MP, Tandon RN (1975) A fruit rot of Aonla (*Phyllanthus emblicae* L.) caused by Phoma. *Proc Natl Acad Sci* India 45B: 75–77.

Kader AA (2009) Sapotes (sapodilla and mamey sapote): Recommendations for maintaining postharvest quality. Department of Plant Sciences, Univ California, Davis, CA. Available from: http://postharvest. ucdavis.edu/Produce/Produce Facts/Fruit/sapotes.shtml (accessed 15 January 2010).

Kamthan, KP, Mishra R, Shukla AK (1981) Nigrospora fruit rot of *Emblica officinalis* Gaertin, a new record. *Sci Cult* 47: 371–372.

Lal B, Arya A, Rai RN and Tiwari DK (1982) A new soft rot of aonla caused by *Phomopsis phyllanthi Punith* Punith and its chemical control. *Nat Acad Sci Lett* 5: 183–185.

Mandal NC, DasGupta MK (1981) Postharvest diseases of vegetables in West Bengal. *Ann Agric Res* 2: 73–85.

Mandal NC, Dasgupta MK (1982) Postharvest diseases of perishables in West Bengal. *Indian Phytopathol* 35: 645–649.

Manoharachary C, Rama Rao P (1989) *Survey and Pathophysiological Studies of Fruit Rot Diseases. Perspectives in Plant Pathology*. Today and Tomorrow's Printers and Publishers, New Delhi, India.

Mickelbart MV (1996) Sapodilla: A potential crop for subtropical climates. In: Janick J (Ed.). *Progress in New Crops*. ASHS Press, Alexandria, VA, pp. 439–446.

Mishra N (1988) Studies on fungi deteriorating stored fruits of Emblica officinalis G. *Intl J Trop Plant Dis* 6(1): 1995–1997.

Misra A, Shivpuri A (1983) Anthracnose, a new disease of aonla. *Indian Phytopathol* 36: 406–407.

Mukerji KG, Bhasin J (1986) *Plant Disease of India: A Source Book*. Tata McGraw-Hill, New Delhi, India, pp: 21–265.

Omprakash (2003) Fungal diseases of aonla and integrated management assuring the same use of produce. *Indian Phytopathol* 56(3): 320.

Pandey RS, Shukla DN, Kghatti DVS, Bhargava SN (1980) A new fruit rot of *Phyllanthus emblica*. *Indian Phytopathol* 33: 491.

Pandey RS, Bhargava SN, Shukla DN, Dwivedi DK (1984) Two new fruit diseases causedby Alternaria alternate. *Int J Trop Plant Dis* 2: 79–80.

Pathak S, Bhat JV (1952) Studies on the microorganisms associated with *Achras sapota*. *J Univ Bombay* 20(5): 14–18.

Rao VG (1968) Studies on market and storage diseases of fruits and vegetables in Maharashtra. *J Univ Poona Sci Technol* 34(Supplement): 21–50.

Rao VG, Desai MK, Kulkami NB (1962) A new Phytophthora of *Achras sapota* from India. *Plant Dis Reptr* 46: 381–382.

Ratnam CV, Neema KG (1967) Studies on market diseases of fruits and vegetables. *Andhra Agric J* 14: 60–65.

Setty KGH (1959) Blue mould of Amla (*Phyllanthus emblicae* L.). *Curr Sci* 28: 208.

Singh D, Sharma RR (2007) Postharvest diseases of fruit and vegetables and their management. In: Prasad D (Ed.). *Sustainable Pest Management*. Daya Publishing House, New Delhi, India.

Soni KK, Verma RK (2010) A new vascular wilt of aonla (*Emblica officinalis*) and its management. *J Mycol Plant Pathol* 40(2): 187–191.

Srivastava MD (1967) Outbreaks and new records—India. *FAO Plant Prot Bull* 15: 59–60.

Sudha R, Amutha R, Muthulaksmi S, Rani WB, Indira K, Mareeswari P (2007) Influence of pre- and postharvest chemical bibliography 244 treatments on physical characteristics of sapota (*Achras sapota* L.) Var. PKM 1. *Res J Agri Biol Sci*, 3(5): 450–452.

Samuh G, Kusum B (1990) New records of fungal species associated with fruit rot of *Phyllanthus emblica* L. *Indian J Mycol Plant Pathol* 20(2): 202–203.

Tandon RN, Srivastava MP (1964) Fruit rot of *Emblica officinalis* Gaertn. caused by *Pestalotia cruenta* Syd. in India. *Curr Sci* 33: 86–87.

Tandon RN, Verma A (1964) Some new storage diseases of fruits and vegetables. *Curr Sci* 33: 625–627.

Thakur DP, Chenulu VV (1970) Chemical control of soft rot of apple and mango fruits caused by *Rhizopus arrhizas*. *Indian Phytopathol* 23: 58–61.

Zhu SJ (2006). Non-chemical approaches to decay control in postharvest fruit. In: Noureddine B, Norio S (Eds.). *Advances in Postharvest Technologies for Horticultural Crops*. Research Signpost, Trivandrum, India, pp. 297–313.

27

Postharvest Diseases of Potato and Their Management

Rajesh Kumar Ranjan,[1] **Dinesh Singh**[2] **and Dinesh Rai**[1]
[1]*Department of Plant Pathology, RPCAU, Pusa, Samastipur, Bihar, India*
[2]*Division of Plant Pathology, ICAR-Indian Agricultural Research Institute, New Delhi, India*

CONTENTS

27.1 Introduction

Potato is the third most important food crop after rice and wheat in the world. Potato is cultivated throughout India with an area of 2.14 million ha and production of 51.31 MT (National Horticulture Board, Govt. of India, 2019) and after China, India is the second largest producer of potato in the world. However, productivity of potato in India is low due to biotic stress, including diseases caused by various groups of pathogens and poor management practices. Among these, the postharvest diseases of potato are some of the most important constraints in India.

It is not feasible to grow potato year-round; hence, it is essential to store the potato long-term for year-round delivery of fresh market and chip processing industry. Similarly, seed potato tubers are also required to be stored for more than 6 months before being used for planting (Olsen 2010). Potato tubers are infected by several postharvest pathogens including bacteria and fungi, and they can destroy potatoes during storage. Although healthy potatoes show resistance against postharvest diseases, damaged tubers are most susceptible. The disease-causing pathogens quickly infect injured tubers in the field as well as during storage. Postharvest fungal diseases of potato include late blight (*Phytophthora infestans*), dry rot (*Fusarium sambucinum* and *Fusarium* spp.), pink rot (*P. erythroseptica*), Pythium leak (*Pythium ultimum*), and silver scurf (*Helminthosporium solani*). Dry rot disease caused by species of *Fusarium* in the field and also in storage result in severe losses in all major potato growing places (Secor and Salas 2001). The pathogen infects the periderm through a wound and contaminants, particularly *Erwinia* species, occur on tuber surfaces and in the soil and causes spoilage in combination with *Fusarium*. Perombelon (1987) reported that *Erwinia carotovora* subsp. *carotovora* macerates parenchymatous tissue of tubers. Bacterial soft rot disease induces faster and more decays during storage, resulting in heavy losses (Harrison and Nielsen 1990) and is a threat to the potato industry. Losses up to 100% have been reported in storage both in developed and developing countries (Wale et al. 2008) resulting to insufficient planting material available for the following season. Potatoes are propagated throughout the world by use of potato tubers (Secor and Rivera-Varas 2004). Infected seed tubers in storage are the primary source of inoculums once the tubers are planted at warmer temperatures (Johnson and Cummings 2009). Studies have shown that planting infected potato seed tubers results in disease development in the field (Johnson 2010). The results indicated that the higher the moisture content the lower the pH value of the potato tubers cause more fungal spoilage of tubers. Togoshi et al. (1988) found 52 isolates of bacteria from 15 different soft rotted vegetable crops from Tsuruoka district of Japan, out of which 36 isolates belong to *E. carotovora* subsp. *carotovora* and 16 were *E. carotovora* subsp. *atroseptica*.

Besides postharvest losses, the safety of the potato for consumption is also greatly influenced by managing postharvest practices and storage conditions, because potato contains natural toxins, glycoalkaloids (GAs) (steroidal toxic secondary metabolites) which are found in all parts of the potato and it is synthesized as a form of defence against parasites and diseases. These metabolites have antimicrobial, insecticidal and fungicidal properties (Nema et al. 2008). These toxins, if consumed in large quantities, may cause harm to humans (Korpan et al. 2004) such as gastrointestinal disturbances and impaired nerve function (Mensinga et al. 2005; Milner et al. 2011), and high doses may cause coma and even death (Friedman 2006).

27.2 Major Postharvest Diseases

27.2.1 Late Blight

Late blight is the most destructive disease of potato crop and causes dry or wet rot disease in tubers. The disease was first observed in 1845 at Courtrai in Belgium. Late blight disease famously caused the worst ever Irish potato famine. The British introduced the potato in the subcontinent (1870) and its traces were observed first in Nilgiris Hill (Butler and Jones 1961). After that, many disease outbreaks have been reported from different parts of India, including Assam, Bengal and Bihar.

Potato late blight disease affects both foliage and tubers, resulting in high yield losses (Kirk et al. 2004). Potato tubers in the field are infected by sporangia and zoospores produced on the plant foliage, and then developing tubers may become blighted in a very short period. These pathogen propagules continue the infection process on the tubers during storage to cause significant postharvest losses in storage.

27.2.1.1 Symptoms

On the leaves there are pale green, irregular spots at tips or margins in the beginning. The spots enlarge rapidly and central tissue turns into necrotic and dark brown or black spots under moist weather conditions. A white cottony growth of fungus forms around the dead areas on the lower sides of the leaves. The lesions are not restricted by leaf veins; several lesions coalesce and entire leaves become blighted and killed within a few days. The lesions also appear on petioles and stems of the plant. The water-soaked areas of spots dry up and turn into brown lesions under dry weather conditions. On tubers, irregular, slightly depressed areas appear, with brown to purple lesions of variable size. A tan to reddish-brown, dry, granular rot is observed under the skin in the discoloured areas. It extends into the tuber less than half inch deep. The development of rotting in a tuber is based on the susceptibility of the cultivar, temperature and length of time after initial infection.

27.2.1.2 Causal Organism

Late blight disease of potato is caused by *Phytophthora infestans* (Mont.) de Bary and belongs to the oomycetes group. The pathogen is a heterothallic fungus having two mating types for sexual reproduction, referred to as A_1 and A_2. The fungus is identified by its characteristic lemon shaped, hyaline sporangia emerging through leaf stomata. Sporangia (21–38 × 12–23 µm) may germinate directly through germ tube or under cool moist conditions through zoospores, which are up to eight in number. Each biflagellate zoospore swims in a film of water for some time, sheds flagella, become encysted and initiates new infections. In nature, sexual reproduction in *P. infestans* is uncommon because of its heterothallic nature. In India, however, only A_1 mating type was observed. However, occurrence of A_2 mating types of *P. infestans* has been reported from hills (Singh et al. 1994). The antheridium is amphigynous and the oogonium is spherical. The germination of oospores has not been seen and is presumed to germinate by producing germ tube. Races viz. *P. infestans* 1.3.4.7.11, 3.11 and 4.7 were identified when tested on R0–R11 gene differentials of potato (Sokhi et al. 1993).

27.2.1.3 Disease Cycle and Epidemiology

The pathogenic fungus mainly overwinters through infected seed tubers of potato from one season to another. In the hilly areas, the temperatures in the country stores are quite favourable for survival in seed tubers up to the next season. In the plains, the pathogen overwinters in starch seed tubers, volunteer plants and crop residue of potato and tomato present in the field and gardens. Once primary infection occurs, secondary spread takes place on the plant by airborne or rain-splash sporangia. The mycelium of the fungus grows between the cells and produces haustoria, which feed off the interiors of the cell, causing cell lysis. The mycelial growth into healthy tissue continues, leaving dead, discoloured cells. The mycelium emerges through stomata on the dorsal layer of the leaf and produces sporangiophores. The sporangia then develop on these sporangiophores. A large number of sporangia are produced and infect leaves and stems of the same and/or neighbouring plants. The sporangia produced on the foliage are washed off by rain or irrigation water and enter into the soil, and they can germinate at between 6.7 and 12.7°C temperatures (Stevenson et al. 2008). The sporangia move in a water film down the stems and stolons of potato, or through cracks in the soil (Johnson 2010). When sporangia come into contact with the progeny tubers, the infection takes place either through lenticels or the eyes of potato. The whole tuber later becomes infected and reduces the quality of the harvested potato, which leads to a total crop loss (Yuen et al. 2008). The first infection often occurs soon after the plants' emergence when favourable moisture and temperature prevail. The growth of fungus is so fast under favourable environmental conditions that the pathogen can kill the whole plant within 1–2 weeks or 15 days.

High humidity at about 90% favours germination of sporangia, whereas high temperature promotes the development of mycelia. Eight to twelve biflagellate motile zoospores are released during wet and cool weather and can penetrate directly into tissue. They also infect the potato tubers near the soil surface. Foliage infection causes premature death of plants which decreases the yield, while infected tubers start rotting in the field as well as during storage. The most favourable temperature for successful sporangial germination either via zoospores or directly by germ tubes is 12 and 24°C, respectively. Higher temperature and low relative humidity favour direct germination of sporangia, while low temperature and high humidity facilitate indirect germination with formation of zoospores after infection; subsequent disease development is most rapid at 21°C. Production of sporangial is most rapid at 100% relative humidity and 21°C temperature. The sporangia are more sensitive to desiccation. After dispersal by wind or splashing rain these need free water to germinate. The fungus remains alive in host tissues at air temperature at 0–28°C. Zoosporangia develop at temperature at 9–22°C. Development of the fungus within leaf takes place at 16–18°C. Temperatures above 30°C are considered unfavourable for disease development. Continuous high temperature has been found to be more damaging as compared to intermittent intervals. Lesion formation on the leaves incurs infection at 17–25°C within 3–5 days.

27.2.1.4 Disease Management

27.2.1.4.1 Cultural Practices

Late blight disease affects potato in the field and during storage; hence management practices should be followed during the growing season of the crop and in storage. Proper management of the disease depends mainly on the reduction of infection in both foliar and tuber stages. Cultural practices have an important role in reducing the source of primary inoculum (Bhattacharya et al. 2002). Healthy and certified seed tubers should be used for sowing to decrease incidence and spread of the disease. Infected plant residues from the field should be removed and destroyed. Certified seed tubers should be allowed to sprout and be introduced directly into the field. Other cultural practices such as soil covers and hilling are based on the ability to directly suppress pathogen inoculum or filter it out before it reaches the tubers. This can be effective in managing the disease (Nyankanga et al. 2008).

27.2.1.4.2 Resistant Varieties

Host resistance was introduced from the wild species (*S. demissum, S. phureja, S. andreanum* and *S. edinense*) and other non-host resistance from related species after the Irish famine (Gevens and Seidl 2013). Some resistant varieties like Kufri Sherpa, Kufri Badshah, Kufri Jeevan, Kufri Mithun, Kufri Jawahar, Kufri Giriraj, Chipsona-1 and Chipsona-2 are recommended for cultivation.

27.2.1.4.3 Chemical Control

Late blight disease is managed mainly by using chemical fungicides. In chemical control, Bordeaux mixture was the first fungicide used for managing late blight disease (Fernández-Northcote et al. 2000). Later, dithiocarbamates and 1,2-bis-dithiocarbamate fungicides were introduced between 1940 and 1960 (Kaur and Mukerji 2004), followed by chlorothalonil and phthalimides (Captan and Folpet) as broad-spectrum fungicides (Whisson 2010). More specific fungicides like phenylamides and metalaxyl, which penetrate the plant tissue and eradicate established infections, were also introduced (Fernández-Northcote et al. 2000) and were highly effective to control the disease until the appearance of mefenoxam/metalaxyl resistant *P. infestans* genotypes (Stevenson et al. 2008). Other fungicides containing azoxystrobin, chlorothalonil, copper hydroxide, fenamidone, mancozeb, metiram, mandipropamid and triphenyltin hydroxide were also developed. To reduce dose and interval of fungicide application, the most economical strategy is to manage the disease in combination with the use of host resistance (Kirk et al. 2001). The efficacy of fungicides is highly dependent on the prevailing weather conditions during the disease development (Fernández-Northcote et al. 2000).

Phosphorus acid materials such as Phostrol, Rampart and ProPhyt showed effectiveness for postharvest disease manage of late blight and pink rot (Johnson 2007). Integrated cultural practices with chemical control is the most commonly used strategy to control late blight disease and also reduce direct contact of inoculum of pathogen with tubers (Stevenson et al. 2008).

27.2.1.4.4 Biological Control

Postharvest treatment of potato tuber with 0.8 mL of *Pseudomonas fluorescens* S22:T:04, P22:Y:05, S11:P:12 and *Enterobacter cloacae* S11:T:07 either alone or combination @ ~5×10^9 cfu per each 90 unwounded potatoes after pathogen inoculation of *P. infestans* (4×10^4 sporangia/mL) by spraying @ 1.6 mL reduced disease with the range of 35–86% in the first year and 35–91% in the second year (Slininger et al. 2007).

27.2.2 Pythium Leak Disease

Pythium leak disease of potato, also known as watery rot or shell rot, is caused by the soilborne pathogen of oomycetes *Pythium* spp. (Salas and Secor 2001). *Pythium* species including *Pythiumultimum* Trow, *P. debaryanum* Hesse and *P. aphanidermatum* (Edson) Fitzo (Salas and Secor 2001; Platt and Peters 2006) infect potato tubers. Among them, *P. ultimum* is the most common species causing Pythium leak in potato (Salas and Secor 2001). The fungus infects potato tubers in the field before harvest, at harvest or during storage, and causes severe yield losses (Salas and Secor 2001; Salas et al. 2003). *P. ultimum* is found in most potato growing areas and is endemic to most soils. It has a wide host range including peas, corn, carrots, onions, beans and cereals (Salas and Secor 2001; Paulitz and Adams 2003; Broders et al. 2007).

27.2.2.1 Symptoms

The diseased potato tubers appear discoloured, smoky grey to black and develop water-soaked lesions on surfaces of the tubers (Platt and Peters 2006). These lesions enlarge in size and become watery rots when the tuber is cut open under favourable conditions during storage. The rotten internal tissue becomes grey to brown in colour when it is exposed to air. A reddish-brown to black boundary line limits the rotten zone in tuber. The rotted tubers released semi-liquid contents. When rotted tubers are squeezed, an empty shell is left, thus the disease is called shell rot.

27.2.2.2 Causal Organism

Pythium leak disease is mainly caused by *Pythium ultimum* var. *ultimum* Trow. Although another *Pythium* species has been reported to cause the disease in potato tubers, i.e., *P. sylvaticum* Campbell & Hendrix (Peters et al. 2005), *P. ultimum* is a species complex which includes *P. ultimum*var. *Ultimum* and *P. ultimum* var. *sporangiiferum*. The main distinct feature is that *P. ultimum* var. *ultimum* produces porangia and zoospores only rarely. However, both species of *Pythium* make oospores having thick-walled structures produced by sexual recombination. Both varieties are homothallic (self-fertile), which means that a single strain can mate with itself. Additionally, oospores of *P. ultimum.* var. *ultimum* also make hyphal swellings and germinate like sporangia to form plant-infecting hyphae. Sporangia and zoospores are short lived, whereas oospores can persist for years within the soil and survive even winter freezes. Fungal mycelia and oospores present in the soil can infect potato seeds or roots.

27.2.2.3 Disease Cycle and Epidemiology

Pythium ultimum is not able to infect unwounded tubers because it is unable penetrate the periderm tissue (Taylor et al. 2004). The pathogen sometimes may enter the tuber through lenticels or the stem end, but infection predominantly originates by the fungus from wounds. Thus, the potato tubers are more prone to infection at the time of harvesting, transport and loading into storage. Sporangia or oospores germinate and enter the tubers through cut tuber seed pieces (Powelson and Rowe 2008) or wounds inflicted during harvesting. A germ tube grows and invades the inner tissues, resulting in tissue disintegration (Platt and Peters 2006). The development of potato leak disease is affected by potato cultivar, soil moisture level, and temperature when the tubers are still in the field. Seed tubers are generally planted when the soil moisture is near saturation and the soil temperature is more than 21°C, which readily gets infected and become a soft watery mass, resulting in delayed crop emergence as well as non-uniform establishment of crop (Powelson and Rowe 2008). Similarly, Lui and Kushalappa (2003) reported that wet conditions and high temperatures during harvesting favour tuber infection and increase disease severity. However, Pythium leak infection is started in the field but the symptoms become more severe during storage.

The pathogen is soil borne by nature. It penetrates tuber through wounds or abrasions caused during harvesting. Rotting of tubers can be increased at high temperatures of about 22°C. Tuber rot can spread quickly at 21°C during storage. However, disease infection may be reduced by minimum injuring during handling and harvesting of potato, removal of soil, and proper storage conditions.

27.2.2.4 Disease Management

Pythium leak disease can be managed by reducing any conditions such as field selection to avoid fields with a history of the disease or with poorly drained soils, which favours infection and disease development (Salas and Secor 2001). Crop rotation with non-host crops also reduces the disease infection (Powelson and Rowe 2008). Excessive irrigation should not be done at the end of the crop season to give sufficient time to kill the vine and a good skin set which is essential to reduce wounding during harvesting of tuber. Potato tubers should be harvested during cool dry conditions, with tuber pulp temperature <21°C and when the skin is properly set (Salas and Secor 2001). If the disease is noted in storage, the temperature should be maintained at about 12–15°C, air circulation should be increased, and dehumidifiers employed immediately to control the disease (Platt and Peters 2006).

Potato tubers should be kept warm about 10–13°C before cutting them to reduce bruising of tuber at the time handling after harvest, boost fast healing of cut surfaces, and also increase sprouting before sowing in the field (Secor and Johnson 2008). Storage temperature should not be kept above 13°C because it may cause excessive sprouting and has a negative effect on yield (Secor and Johnson 2008), and also favours the development of tuber spoilage (Powelson and Rowe 2008). Treatment of tuber with fungicide was found effective to control decay caused by *P. ultimum* (Platt and Peters 2006). A mixture of fludioxonil and mancozeb (Maxim TM MZ) reduced seed decay before sowing (Wharton et al. 2007b). Fungicides applied in furrows during sowing and sprayed on foliage at tuber initiation stage reduce the spread of the disease to the progeny tubers (Platt and Peters 2006). Mefenoxam and metalaxyl fungicides effectively reduce Pythium leak disease. Powelson and Rowe (2008) reported development of resistance against these fungicides over time in populations of *P. ultimum*, but the population of the pathogen is still largely sensitive to these fungicides (Taylor et al. 2002; Porter et al. 2009).

27.2.3 Pink Rot

Phytophthora erythroseptica var. *erythroseptica* was first described by Pethybridge (1913; 1914) as the causal agent of pink rot disease of potato in Ireland, and now the disease has been reported from all potato growing areas of the world. Pink rot disease is caused by *Phytophthora erythroseptica* (Lambert and Salas 2001). Other *Phytophthora* spp., including *P. crytogea* and *P. parasitica*, are associated with this disease (Grisham et al. 1983). Pink rot disease is so-called due to the pink coloration which develops on infected tissue once it has been cut and unveiled to air. Pink rot is found worldwide in potato growing areas (Lambert and Salas 2001). Pink rot disease affects roots and tubers in the field and tubers in storage (Powelson and Rowe 2008) and causes significant yield losses both pre-harvest and postharvest (Salas et al. 2003).

27.2.3.1 Symptoms

The disease appears first on the surface of roots and tubers, and a rubbery texture along with dark lenticels also appears in potato. Infected tubers show white colour inside and watery fluid exude on squeezing. The white tissue turns salmon pink colour and finally black. Infected tubers produce the smell of vinegar in response to the pathogen. Infected tubers release more oospore in soil and infect all underground parts.

27.2.3.2 Causal Organism

The disease is caused by *Phytophthora erythroseptica* Pethybr. Hyphae are uniform, fine, and <8 µm, rounded or angular swelling occurs in water. Sporangiophores are sympodial and branch immediately below the sporangium. Sporangium proliferates internally through empty sporangia. Sympodia form in water. Sporangia are non-papillate, vary from ellipsoid or ovoid to a distorted shape and are often constricted in the middle. They are non-caducous. Sporangia are 26–47 × 43–69 µm (average 27.2 × 44.2 µm) with a round or tapered base. The length-breadth ratio is greater than 1.6. *P. erythroseptica* is homothallic. Antheridia are amphigynous, elongated, ellipsoidal, or cylindrical and 13 × 14 µm in diameter. Oogonia with multiple antheridia (two to four) have been observed. Oogonia are smooth walled and 30–46 µm in diameter (average 34.9 µm). Oospores are aplerotic and 28–35 µm in diameter (average 30.5 µm) with a thick wall (2.5 µm). Chlamydospores have not been found. Minimum, optimum and maximum temperatures are 2.5, 27.5 and 34°C for growth of pathogen, respectively.

27.2.3.3 Disease Cycle and Epidemiology

P. erythroseptica survives in soil for several years as oospores (Lambert and Salas 2001; Wharton and Kirk 2007). The oospores germinate into sporangia, when soils have near saturation level, and then release zoospores. The zoospores move in the soil with water film. They infect tubers when tuber has temperature of >20°C (Wharton and Kirk 2007; Powelson and Rowe 2008). They infect the progeny tubers through lenticels and eyes, but mostly infection occurs through stolons

(Powelson and Rowe 2008). Benson et al. (2009) reported that infection and colonization of pathogen was increased at low pH in Russet Norkotah potato roots. Disease development is increased by excessive irrigation in wet and warm summer months. Field infection leads to tuber decay during storage (Wharton and Kirk 2007; Powelson and Rowe 2008). Tuber that remains in the field after harvesting may harbour oospores which later serve as the initial inoculum for the new crop.

27.2.3.4 Disease Management

For management of pink rot disease, good soil drainage is an essential strategy at the end of the growing season (Lambert and Salas 2001). Cultural practices that reduce the disease incidence include crop rotation, maximizing skin set, avoiding wounding of the tubers, and harvesting when tuber pulp temperature is less than 21°C. Disease may be managed better by integration of these techniques with host resistance. Unfortunately, there are no cultivars resistant to pink rot disease. Therefore, the disease can be managed only by applying metalaxyl/mefenoxam fungicides (Taylor et al. 2004), and more recently cyazofamid (Ranman) (Griffiths et al. 2008). However, metalaxyl-resistant strains of *P. erythroseptica* were first reported in 1993 in Maine (Lambert and Salas 1994) and North America (Taylor et al. 2002). Anenantiomer of metalaxyl, mefenoxam, released in 1997, was found effective against the oomycetes. Although mefenoxam-resistant isolates of *P. erythroseptica* have been reported in North America, there is still a good percentage of sensitive isolates (Venkataramana et al. 2010). Mefenoxam is still effective to control the disease by applying tuber treatment at sowing time (Taylor et al. 2007). Tuber treatment with fungicide at sowing time was found more effective to control the pink rot than foliar spray on foliage (Taylor et al. 2004; Taylor et al. 2006; Al-Mughrabi et al. 2007). Alternative chemicals like hydrogen peroxide (OxiDate^{TM}) were evaluated for pink rot disease management and inhibited growth of *P. erythroseptica in vitro* (Al-Mughrabi 2006). Phosphorous acid (phosphonate, phosphite) was used to control pink rot and late blight disease of potato (Zitter 2010) and was found effective to manage pink rot disease (Miller et al. 2006; Mayton et al. 2008).

Enterobacter cloacae strain S11:T:07 reduced lesion size more than the other antagonists (19% and 32% reduction versus the control) though *Pseudomonas fluorescens* bv. VS11:P:14, *Pseudomonas* sp. S22:T:04, and *Enterobacter* sp. S11:P:08 also significantly reduced disease incidence (Schisler et al. 2009).

27.2.4 Fusarium Dry Rot

Fusarium dry rot of potato incited by *Fusarium* species is one of the major postharvest diseases worldwide (Secor and Salas 2001; Wharton et al. 2005). Thirteen species have so far been implicated in causing dry rot in potato worldwide (Cullen et al. 2005). The disease affects tubers both in the field and in storage. The losses caused by dry rot were estimated at 6–25%, sometimes with losses as high as 60% during longer periods of storage (Secor and Salas 2001). Additionally, *Fusarium* species produce mycotoxins causing harm to humans and animals (Desjardins et al. 1993). Variations in dry rot incidence were found based on the presence of *Fusarium* species (Peters et al. 2008b). For example, Desjardins et al. (1992) reported that *F. sambucinum* has the ability to detoxify the phytoalexins produced by the potatoes, thus its virulence was increased.

27.2.4.1 Symptoms

Dry rot disease is initiated by inoculum present in infected seed tubers or from infested soils (Secor and Salas 2001). The initial symptom appears on the tuber surface as a shallow brown lesion, which later enlarges slowly and eventually becomes sunken and wrinkled. Necrotic areas shaded from light to dark chocolate brown colour or black characterize internal symptoms. This necrotic tissue is usually dry (hence the name dry rot) and may develop at an injury such as a cut or bruise on tuber. The pathogen enters the tuber, often rotting out the centre. Rotted cavities may be lined with mycelia and spores of various colours (Figure 27.1) from yellow to white to pink (Wharton et al. 2007b). In the field, *Fusarium* dry rot in seed tubers appears as germination gaps or severely stunted, chlorotic and necrotic stems as well as abnormal growth of roots and stolons of potato crop (Wharton et al. 2007a). Varying levels of aggressiveness among species and within isolates has been reported in *Fusarium* species (Daami-Remadi et al. 2006b), and this may have an implication on management, especially if the predominant species happens to be the most virulent.

27.2.4.2 Causal Organism

Dry rot is caused by *Fusarium* spp. (*F. coeruleum*, *F. eumartii*, *F. oxysporum* and *F. sulphureum*), and in response their infection small brown areas appear on the surface of potato tubers. Other species of Fusarium have been associated with dry rot, such as *F. avenaceum* (Fr.) Sacc., *F. culmorum* (Wm. G. Sm.) Sacc., *F. acuminatum* Ellis & Everh. *F. crookwellense* Burgess, Nelson & Toussoun, *F. equiseti* (Corda) Sacc., *F. graminearum* Schwabe, *F. sambucinum, F. scirpi* Lambotte & Fautrey, *F. semitectum* Berk. & Ravenel, *F. sporotrichioides* Sherb. and *F. tricintum* (Corda) Sacc. (Hide

FIGURE 27.1 White cottony fungal mycelium growth on rotted potato tuber.

et al. 1992; Stevenson et al. 2001; Sagar et al. 2011). On the surface of infected tubers, mycelial cushions of white fluff may develop. The pathogen has branched mycelium and septate hyphae and they present inter- or intracellularly in the host tissue. Conidiophore arises from the mycelium and produces sickle-shaped, one- to four-celled conidia of different sizes. Chlamydospore which is resting structure of pathogen, survive in soil, plant debris and develop pustules on tubers. These may be intercalary or terminal. Molecular identification through DNA sequencing targeting different genes and comparing with known species is more precise and ideal (Geiser et al. 2004). The combination of molecular and morphological characterization is the best option for correct identification of *Fusarium* spp. (Geiser et al. 2004).

27.2.4.3 Disease Cycle and Epidemiology

The fungus can survive in soil for 9–12 months. *Fusarium* spp. due to its good saprophytic ability. Due to poor storage facilities and prolong storage periods, blue, white, purple, black or pink spore masses of fungus can be seen on the tubers and it can survive well in infected tubers (Pinzon-Perea et al. 1999). The pathogen is unable infect intact periderm and lenticels of the tubers. However, it can enter through cuts and wounds on tubers during harvesting, grading, transport and storage. It has been reported that by increasing the interval between haulm destruction and the harvest enhances tuber skin strength, which is generally believed to reduce dry rots, but the contrary view is also reported (Carnegie et al. 2001). Development of dry rot is affected by several factors like tuber damage, degree of curing, tuber size and storage conditions. The pathogen enters the tubers through wounds; thus proper wound healing may decrease the chance of infection. The major source of spread of *Fusarium* spp. inoculum is infected potato tubers, soil infestation (Choiseul et al. 2001) from soil that adheres to farm implements, through wind and irrigation water, etc., in new areas or fields.

The potato tubers are cured at 21°C for wound healing, and adequate aeration develops wound periderm within 3–4 days, but at lower temperatures it takes longer. Development of disease is also affected by moisture and temperature. The pathogen can grow well at 15–28°C. However, disease development continues at low temperature in cold stores. A positive correlation with dry rot disease and storage period and relative humidity was recorded, whereas negative correlation was found with maximum temperature (Singh 1986). It has also been reported that large tubers are more susceptible to the disease than small tubers during storage. Lacy-Costello et al. (2001) reported that some volatile compounds are produced by dry rot affected tubers, and an early warning system has also been developed based on sensors to detect these volatile compounds.

27.2.4.4 Disease Management

Management of dry rot in storage still has limited options. For effective control of the disease, long-duration crop rotations can be used (Little and Bell 2009). However, dry rot incidence was increased when cereals and forage crops were included in crop rotation (Peters et al. 2008). It was also shown that some of the *Fusarium* spp.-infected cereals (*F. graminearum* and *F. sporotrichioides*) and forage crops (*F. avenaceum* and *F. oxysporum*) are pathogenic to potato crop (Peters et al. 2008).

There is no commercially grown potato cultivar available to show resistance against dry rot disease. However, Leach and Webb (1981) reported that the susceptibility level varied from one cultivar to another. They observed that clones showed high levels of resistance against *F. sambucinum* and *F. solani* in potato crop.

Losses caused by *Fusarium* spp. can be reduced by integrating physical and chemical treatments. The disease can be controlled primarily by reducing tuber bruising during harvesting, providing proper conditions for rapid wound healing (Secor and Salas 2001; Secor and Johnson 2008) and by applying fungicide thiabendazole (TBZ), during storage or before sowing the tuber. Studies have also shown varying responses of different isolates of *F. sambucinum* against TBZ, with some isolates being resistant and others being sensitive to TBZ. Another fungicide, fludioxonil, was found effective to reduce tuber decay disease as well as sprouts (Wharton et al. 2007b) and can help to produce healthy progeny tubers.

Cut tubers should be avoided for sowing or treated with dithiocarbamates to reduce *Fusarium* seed tuber decay. The other fungicides fludioxonil alone (Maxim™ Seed Potato, phenylpyrrole) registered in the United States for seed treatment or in combination with fludioxonil + mancozeb (Zitter 2010) are found effective to control of *Fusarium* dry rot. Chlorine dioxide (ClO_2) has been evaluated as a disease suppressant for stored potatoes. Benzothiadiazole (BTH), as Bion WG50 (100 mg *a.i.*/L) and acetylsalicylic acid (400 mg *a.i.*/L) treatments on harvested tubers decreased the severity of dry rot disease in field-grown tubers in some postharvest wound inoculated treatments (Bokshi et al. 2003).

Fusarium dry rot can be controlled by using various bioagents like *Trichoderma* spp. (Pinzon-Perea et al. 1999), *Pseudomonas fluorescens* (Schisler et al. 2000), *P. aeruginosa* (Gupta et al. 1999), *B. subtilis* (Kim-Byung Sup Choand Cho 1995), *Enterobacter* and *Pantoea* (Al-Mughrabi 2010). Ermakova and Shterushis (1994) developed commercial *P. fluorescens*-based bioproducts for disease control. Combination of bio-control genera *Enterobacter* and *Pseudomonas* and two chitinolytic enzymes from *T. harzianum* inhibited spore germination of *F. solani* (Lorito et al. 1993), which indicates a possibility that certain bacteria capable of binding to the fungal cell walls and expressing fungal genes coding cell wall degrading enzymes may act as powerful bio-control agents. Recep et al. (2009) reported that *Burkholderia cepacia* is antagonistic to *F. sambucinum*, *F. oxysporum* and *F. culmorum* causing dry rot of potatoes in storage.

27.2.5 Wart Disease

Potato wart disease caused by a fungus *Synchytrium endobioticum* (Schilberszky) Percival has been reported in Asia, Africa, Europe, Oceania, North America and South America (EPPO 2006). It has caused great damage to potato in Europe and serious disease of potato worldwide. The disease has quarantine issues, and once established is difficult to eradicate because resting sporangia of pathogen can survive inter-host periods for up to 20 years. In India, the disease is found in the Darjeeling hills. It has been managed by enforcement of strict quarantine legislation.

27.2.5.1 Symptoms

The disease appears as rough, warty, mostly spherical outgrowths or protuberances on buds and eyes of tubers, stolons or underground stems or at stem base. Underground galls are white to light pink at young stage, and later with age become brown or light black. Above-ground stems, leaves or flowers are also affected to form galls which are green to brown or black. The wart tissues are soft and spongy. The tubers are mostly replaced by warts which desiccate or decay at harvest.

27.2.5.2 Causal Organism

The disease is caused by *Synchytrium endobioticum* (Schilberszky) Percival, which is a member of Chytridiales order. The fungal pathogen is an obligate, holocarpic, endobiotic parasite. The fungus has several pathotypes. Baayen et al. (2006) have reported at least 43 pathotypes from Europe. The fungus does not have mycelium but has a thin-walled summer sporangium stage and a thick walled "winter" or resting sporangium stage which produce an extended vesicle called sorus from where zoospores are produced. The pear-shaped zoospores possess a posterior flagellum.

van den Boogert et al. (2005) sequenced ITS regions of the multi copy rDNA gene of *S. endobioticum* and used specific PCR primers and probes for identification. The 18S region of rDNA of fungus was used to develop a detection method for *S. endobioticum* (Abdullahi et al. 2005). van Gent-Pelzer et al. (2010) developed a quantitative detection method using TaqMan PCR technology to detect *S. endobioticum* from winter spores in soil.

27.2.5.3 Disease Cycle and Epidemiology

Warty growths of tubers disintegrate to release abundant resting sporangia in the soil. Resting sporangia of the pathogen are endogenously dormant and they can survive for 40–50 years up to 50 cm depth in soil. The resting sporangia germinate to release haploid uninucleate zoospores under wet soil conditions and 10–27°C temperature. The zoospores swim in soil, encysted. They infect with infection pegs which form within 1–2 h to epidermal cells of meristematic tissues of growing buds, stolon tips or leaf primordia. After successful infection, a uninucleate thallus develops within the infected cell which then enlarges to form a prosorus. A vesicle develops from prosorus, and contents of the prosorus pass on to the vesicle to form a sorus within an infected cell and it divides repeatedly to form several sporangia in which zoospores develop. Finally, the wall of the sorus breaks and releases sporangia and zoospores in soil. The zoospores infect new hosts and continue throughout the growing season of the crop. The fungus within the host stimulates hypertrophy and hyperplasia of neighbouring host cells without actively infecting them and result in increase in meristematic activity. The development of warts of variable sizes may depend upon the degree of stimulation. Under stress conditions, possibly under water stress, the haploid zoospores fuse in a pair to form a zygote. This zygote invades the host tissue and converts into a thick-walled "winter sporangia" which is echinulated with prominent exterior ridges, and they are released in soil by decaying warty growths and serve as primary inoculum.

The pathogen spreads through infected seed tubers, infested soil adhering tubers, machinery and other carriers. The resting sporangia survive passage through the digestive track of animals fed infected potatoes, and the contaminated manure can thus disperse the inoculum. Earthworms are another means of dispersing the inoculums, i.e. resting spores. They can also be disseminated by wind-blown soil or by flowing surface water.

Periodic flooding followed by drainage and aeration favours wart disease, because free water is needed to germinate sporangia and also for dispersal of zoospores. Temperature in the range of 14–24°C favours germination of resting sporangia to zoospores. Both summer sporangia and resting spores can germinate at 12–28°C. Disease development can be favoured by mean temperature below 18°C and about 70 cm annual precipitation.

27.2.5.4 Disease Management

27.2.5.4.1 Cultural Practices

Intercropping of potato with maize or crop rotation with bean and radish are found suitable to reduce viable resting spores in the soil (Singh and Shekhawat 2000). The pathogen population can also be reduced by amending infested soil with 4% and 8% crab shell (w/w) reduces (Hampson and Coombes 1995).

27.2.5.4.2 Host Resistance

Several resistant or immune varieties to wart have been developed throughout the world. In resistant varieties, the pathogen infects the plants but symptom development is suppressed, while in immune varieties, a hypersensitive reaction occurs upon infection with zoospores of the fungus which gets killed in the process. Kufri Jyoti, Kufri Bahar, Kufri Sherpa and Kufri Kanchan are wart immune varieties which are introduced into the wart-infested region of Darjeeling hills coupled with domestic quarantine, and this had a great impact in containing wart in the region (Singh 1998).

27.2.5.4.3 Integrated Disease Management

Wart disease can be successfully managed by field sanitation, following long crop rotation and growing resistant and immune varieties. Plant quarantine legislation should be enforced strictly in the countries of EPPO region, Canada, Maryland in the United States and India. However, periodic surveys are required to monitor viability of the pathogen in soil and efficiency of the quarantine measures (Arora and Sagar 2014).

27.2.6 Charcoal Rot

Charcoal rot disease is mostly prevalent in the Mediterranean region, warmer areas of India and Peru, Hawaii and the southern United States. High soil moisture along with temperature

FIGURE 27.2 Symptom of rotting development in potato tuber.

>28°C favours the disease. The disease occurs in tubers both in the field and during storage. It can cause severe losses under unusually warm, wet weather.

27.2.6.1 Symptoms

The fungus attacks the growing plants and tubers at the time of harvest and during storage. The infected plants show stem blight or shallow rot in the field similar to black leg and cause the foliage to wilt and turn yellow. Early symptoms appear as dark light grey, soft, water-soaked lesions on tubers around the eyes, lenticels and stolon ends (Figure 27.2). Later, the lesions become filled with black mycelium and sclerotia of the fungus. Secondary organisms may develop in such lesions, especially under wet conditions, which cause significant losses. The lesions may shrink and develop symptoms similar to dry rots under low moisture.

27.2.6.2 Causal Organism

The disease is incited by *Macrophomina phaseolina* (Tassi) Goidanich Syn. *M. Phaeoli* Maubl. and black, smooth, hard, 0.1–1.0 mm sized sclerotia of the fungus develop within roots, stems, tubers and leaves. *Botryodiplodia Solani-tuberosi* Thiram (Thirumalachar and O'Brien 1977) is the perfect stage of the fungus which may develop in stems of potato, jute, sun hemp and maize. Pycnidia may also develop on the leaves and stems depending upon the strain of the fungus. Conidia are hyaline, single and ellipsoid to obovoid in shape.

27.2.6.3 Disease Cycle and Epidemiology

The pathogenic fungus is a weak parasitic soil fungus and survives in the soil as sclerotia in plant debris, weeds and alternate host crops. *M. phaseolina* persists on dead plant tissues and survives the unfavourable periods by forming microsclerotia. Both soil and infected tubers served as sources of inoculum. Optimum temperature of about 30°C favours growth and infection of the fungus, and rot is slow at 20–25°C and stops at 10°C or below. Charcoal rot disease is predisposed by poor plant nutrition and wounds the plants. Fungal growth stops in tubers kept in cold rooms but resumes growth when it is removed from the cold storage. The pathogen can spread through the infected seed tubers and also through the infested soil carried along with the implements.

27.2.6.4 Disease Management

27.2.6.4.1 Cultural Practices

Early-maturing cultivars should be planted. Frequent irrigation should be given to the crop to keep the soil temperature down. Harvesting of tubers before the soil temperature exceeds 28°C can reduce the disease incidence in the eastern plains of India. Crop rotation with non-host and use of disease-free seeds and avoiding cuts and bruises during harvesting can reduce disease incidence. Resistance against charcoal rot has been located in certain clones of *Solanum chacoense* and could be utilized in developing resistant varieties. *Bacillus subtilis* used as a biocontrol agent through tuber treatment is found to reduce disease incidence (Thirumalachar and O'Brien 1977).

27.2.7 Silver Scurf

Silver scurf disease produces blemishes on tuber surfaces, causing them to look "dirty." Potatoes having silver scurf are safe to eat but consumers are less likely to purchase them due to the dirty appearance. Some tubers are initially infected at the field level, but the maximum damage occurs in storage. If the storage period of tubers is longer, the damage of the tuber is greater, although the pathogen does not reduce yield of the crop. Culling of tubers with unsightly surface infections and increased labour charges for inspection and sorting requirements for damaged potato lots can cause substantial economic losses. It has been reported that russet-skin cultivar potato is less susceptible to infection as compared to smooth-skin cultivars, particularly yellow and red, which show the disease symptoms.

27.2.7.1 Symptoms

Disease symptoms appear as circular or irregular, tan to silvery grey and occasionally show blackish lesions on the tuber periderm. These lesions generally contain a definite margin and vary in size from pinhead to patches which cover most of the tuber surface. As the disease progresses, individual lesions coalesce and form large patches. The silvery appearance of older lesions is most obvious when the tuber is wet. During storage, circular lesions produce a random "measles" appearance over the surface of the potato. The lesions usually remain superficial and do not damage the underlying tissues.

27.2.7.2 Causal Organism

The disease is caused by *Helminthosporium solani* Dur. & Mont. Mycelium is dark colour. Conidia are obclavate, straight, or curved, pseudosepta (2 to 8) and arranged in whorls on the sides of black, multiseptate conidiophores and attached to the conidiophore from its broad end.

27.2.7.3 Disease Cycle and Epidemiology

Diseased seed tubers are the major source of inoculum to cause disease, particularly in fields. The pathogen survives in the soil about 2 years. Silver scurf incidence increases by increasing the generation number of seed tuber. The pathogen spores are present on the surface of diseased tuber in the soil. The spores move in the soil by rain or irrigation water or can grow down to the roots or stolons, and they infect developing tubers through lenticels or directly through the periderm. The tubers are highly susceptible to infection after the periderm has started to mature. Disease severity and damage of tuber may be increased when the tubers are left for longer periods in the ground after vine death and skin set, although more problems often arise during storage, and a vital amount of infection and damage can already be present on smooth-skinned cultivars during harvesting. During storage infection, spores generally arise from infected tubers, but sometimes the disease also can originate from contaminated soil brought into storage. Other sources like contaminated wood, concrete and organic material may also be sources of spores.

Helminthosporium solani spores form on the surface of diseased tubers at relative humidity of more than 90–95% and temperature above >4°C. Spores are easily removed from the conidiophores. They spread through the air system of the storage unit to infect other tubers. Storage infection is a secondary infection. When free moisture is present on tuber surfaces due to fluctuation temperatures and high relative humidity, the spores germinate and cause infection the tuber. A large proportion of tubers in storage can become infected under adequate humidity and time. The environment of soil is often favourable for development of the disease in late summer and early fall. However, late-season infection in mature tubers, even after the vines are dead, can increase the silver scurf level. Secondary infection as small, randomly found lesions first becomes visible after 3–4 months in russet cultivars during storage.

27.2.7.4 Disease Management

- Crop rotation for at least 3 years of potato crops will greatly reduce the chance of survival of fungal spores in the soil and infect tubers.
- Always try to use silver scurf-free tuber seed, and it is required to test silver scurf disease before purchasing of seed.
- Storing different generations of seed in the same storage facility should be avoided.
- Treat the seed tubers with registered fungicides at the time of planting.
- Maintain vines healthy until frost or vine kill.
- It is preferable to grow russet skin varieties because they show resistance against the disease.
- Harvesting of potato should be done when skins are adequately set.
- Plant material, debris and straw should be removed from storage spaces and the spaces should be thoroughly cleaned and disinfected before storing tubers.
- Curing of tubers under high relative humidity of about 95%, 50–55°F temperature, and good ventilation reduce temperature immediately to the desired level for storage.
- Do not open the storage facility frequently if long-term storage of the produce is to be done.

27.2.8 Brown Rot

Bacterial wilt or brown rot caused by *Ralstonia solanacearum* (Smith) has worldwide distribution in tropical and subtropical parts of Asia, Africa and South and Central America (Shekhawat et al. 2000; Elphinstone 2005). In India, the disease is endemic on the west coast from Thiruvananthpuram in Kerala to Khera in Gujarat, Karnataka, Western Maharashtra, Madhya Pradesh, eastern plains of Assam, Orissa and west Bengal, Chhota Nagpur plateau and Andaman and Nicobar Islands. It is one of the most damaging pathogens to potato worldwide (Hayward 1991) and has been estimated to affect potato crop in 3.75 million acres in approximately 80 countries with global damage estimates exceeding $950 million per year (Floyd 2007). Losses up to 75% have been recorded under extreme conditions (Gadewar et al. 1991; Singh et al. 2010). It is likely to spread to new areas and affect potato cultivation due to increase of global temperature. Seed produced in bacterial wilt-infested areas cannot be used domestically or exported. Hence the disease spreads to new seed production areas and can provide a great setback to the seed industry.

27.2.8.1 Symptoms

Disease symptoms are slight wilting or drooping of foliage, especially at ends of branches, during the hot daytime period, which recover at sunset. Later, the leaves become yellow, the wilting becomes permanent and the plant collapses. Cross-section of the stem reveals browning of vascular bundles and bacterial slime oozes out from vascular region. The disease affects potato tubers, and brownish vascular discoloration appears which extends to eyes and other buds. Bacterial mass emerges through the affected eyes of the tubers (Figure 27.3) and soil adheres to the tuber at harvest. Freshly cut tubers show glistening droplets of bacterial ooze coming out from the vascular ring.

27.2.8.2 Causal Organism

Ralstonia solanacearum (Smith) Yabuuchi et al. (1996) causes brown rot and is a single-celled, small rod with rounded ends, 0.5–0.7 × 1.5–2.5 µm, Gram-negative, and flagella when present are polar. The bacterium has an oxidative metabolism,

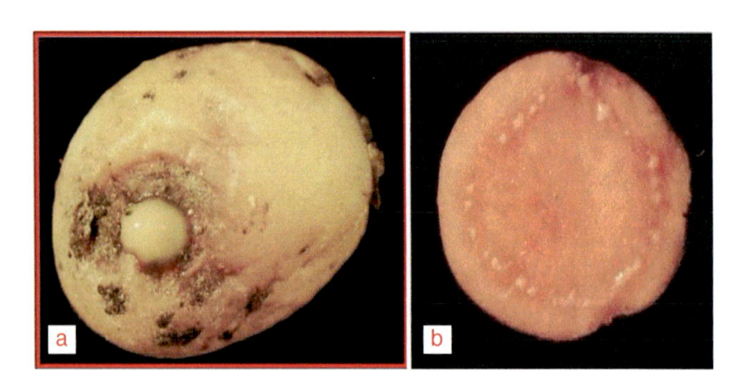

FIGURE 27.3 (a) Bacterial oozing from eye of potato. (b) Bacterial oozing from vascular bundle of cut potato.

strict aerobe and growth limited or slow growth when bacterial cells are not in direct contact with the air. The bacterium produces poly-β-hydroxybutyrate granules as cell energetic reserve. *R. solanacearum* strains from tropical areas require optimum temperatures of 35°C, while the strains occurring at higher altitudes in the tropics and in subtropical and temperate areas is lower temperature at 27°C. There is no growth at 40°C and also at 4°C. In general, *R. solanacearum* growth is limited in acid media but favoured in alkaline conditions and can grow in 1% NaCl liquid media, but in 2% NaCl there is little or no growth of bacteria. Two types of *R. solanacearum* colonies can typically be observed on TZC agar medium as fluidal or mucoid and a-fluidal or nonmucoid or butyrous type. The mucoid substance is due to accumulation of exopolysaccharide produced by bacteria which causes these mucoid colonies to show a typical irregularity of their surfaces with pinkish colour at centre and whitish at periphery. *R. solanacearum* colonies are non-fluorescent, although a diffusible brown pigment can be produced on some media. Out of five races of *R. solanacearum*, potato is affected mainly by race 1 biovar 1, 2 and 4 and race 3 biovar 2 and T2 (Denny 2006). However, there is no general correlation between races and biovars established. However, biovar 2 strains belong to race 3 and vice versa. Race 1 is a poorly defined group with a very wide host range and is endemic to the southern United States as well as Asia, Africa and South America. Race 3 is distributed worldwide and has primarily been associated with potato. *R. solanacearum* has four phylotypes, i.e., Phylotype I, Phylotype II, Phylotype III and Phylotype IV (Fegan and Prior 2005). *R. solanacearum* race 3 biovar 2 (R3b2) belongs to Phylotype II (sequevars 1 and 2). R3b2 is highly pathogenic on potato and tomato (primary hosts) and can also infect and eventually kill other solanaceous plants (e.g., eggplant, nightshade weeds) as well as geraniums.

Ji et al. (2007) used *egl* sequencing studies for determining the genetic diversity of *R. solanacearum* strains isolated from ornamental and vegetable crops from Florida. Horita et al. (2010) divided the Japanese strains of *R. solanacearum* isolated from potato into Phylotype I and Phylotype IV by *egl* gene sequence analysis. Prior and Fegan (2005) described the *mutS* gene sequence analysis for the classification of *R. solanacearum* into the four phylotypes. Sequence data of *mutS* and *egl* genes was used to determine the phylogenetic position of Phylotype II strains of *R. solanacearum* infecting diverse crops from Martinique (Wicker et al. 2007). Toukam et al. (2009) carried out the sequence analysis of *mutS* and *egl* genes from *R. solanacearum* strains from Cameroon to determine their phylogenetic relationships.

Rapid methods to detect pathogen in potato have been developed by Elphinstone et al. (2000), Lyons and Cruz (2001) and Weller et al. (2000). The bacterial pathogen can be detected in conventional PCR assay ranging from 10^4 cfu mL^{-1} (Khakvar et al. 2008) to 10^5 cfu mL^{-1} (Fegan et al. 1998). PCR-based methods for the detection of *R. solanacearum* are usually based on the amplification of ribosomal sequences (16S RNA or ITS region) (Weller et al. 2000; Pastrick et al. 2002). Grover et al. (2009) reported detection of as low as 1 cfu mL^{-1} of bacteria within 8 h including DNA isolation. The sensitivity of the technique is comparatively greater than the standard PCR amplification.

27.2.8.3 Disease Cycle and Epidemiology

Infected tubers and plant debris of potato in infested soil are a major source of inoculum. Additionally, symptomless plants may harbour the bacterium and transmit it to progeny tubers as latent infection which leads to severe disease outbreaks when the infected tubers are grown in disease free fields. The survivability of the pathogen is affected by high soil moisture, temperature, oxygen stress and soil type. The pathogen population declines gradually in soil due to absence of host plants as well as their debris. The pathogen infects roots of healthy plants through wounds caused by mechanical injury or nematodes such as *Meloidogyne incognita* infestation which affect potato roots and tubers and increase incidence of wilt disease. *R. solanacearum* is transmitted from one area to another through infected seed tubers, irrigation water and farm implements. Inoculum potential of *R. solanacearum* about 10^7 cfu/g soil favours infection but is also dependent on other predisposing factors. Soil temperatures below 15°C and >35°C do not favour the disease development.

27.2.8.4 Disease Management

27.2.8.4.1 Cultural Practices

Healthy seed tubers should be use for effective control of the disease. Hence, tubers must be tested for latent infection of *R. solanacearum*.

27.2.8.4.2 Host Resistance

In potato, breeding varieties resistant to bacterial wilt has not been successful because of the enormous variability found in the bacterial pathogen. Cultivars derived from *S. phureja* which show resistance under cool highland subtropics failed to disease under high temperature in the tropics. Therefore, variability in *R. solanacearum* related to temperature is needed to investigate and develop suitable breeding strategies.

27.2.8.4.3 Chemical Control

Bacterial wilt incidence can be declined by using bleaching powder @ 12 kg/h mixed with fertilizer or soil drenching after first earthing up (Shekhawat et al. 1988a, 1988b) along with the use of healthy seed (Gadewar et al. 1991). Application of copra and pea nut meal has also been reported to reduce wilt disease (Shekhawat et al. 1982).

27.2.8.4.4 Biological Control

Bio-control of bacterial wilt by using antagonistic bacteria such as *P. fluorescens*, *Bacillus* spp., *Pantoea agglomerans*, avirulent *P. solanacearum* and actinomycetes has been found to be effective in some countries (Mclaughlin and Sequeira 1988; Aspiras and de la Cruz 1986). In India, bio-control of bacterial wilt by use of antagonists such as *P. flourescens*, *Bacillus* spp., has been found to be effective at some locations (Shekhawat et al. 1993a). Similarly, *B. subtilis* (strain B5) recovered from rhizosphere soil of potato plants from bacterial wilt-infested fields of Bhowali, Uttarakhand controlled tuber-borne *R. solanacearum* under different agro climate conditions enhanced the crop yield (Sunaina et al. 2006).

27.2.8.4.5 Integrated Disease Management

The disease can be managed by using an integrated approach such as pathogen-free seed, reduction of field inoculum, growing crop under optimal environmental conditions and chemical and biological control (Lemaga et al. 2001). More such integrated approaches need to be identified to manage the disease and prevent its spread to healthy fields using the components suitable to subtropical conditions.

27.2.9 Soft Rot and Black Leg

Soft rot disease in potato caused by *Erwinia carotovora* subsp. *carotovora* are next in economic importance to bacterial wilt (van der Wolf and De Boer 2007). Soft rot disease represents a significant threat not only to reduce production but also postharvest losses during storage. Soft rot of potato is the result of infection caused by species of *Bacillus*, *Pseudomonas* and *Erwinia*. Among them, *E. carotovora* subsp. *carotovora* (now known as *Pectobacterium carotovorum* subsp. *carotovorum*) is, however, considered major soft rot-causing bacterium (Larka 2004). The postharvest losses range from 15 to 30% of the harvested crop caused by bacterial soft rot. Bacterial pathogen has wide host range and has the ability to survive both inside and on the hosts under different environmental conditions (Sledz et al. 2000).

27.2.9.1 Symptoms

27.2.9.1.1 Soft Rot

On tubers, the disease symptoms appear as small, soft, water-soaked spots around lenticels and enlarge under high humidity conditions. The pith of infected tuber is rotted beyond the boundary of external lesion, which turns into cream to tan-brown in colour. The tissues become soft and granular (Figure 27.4a, 4b). A brown to black pigment may develop around the lesion. Immature, large tubers damaged by bruises during harvesting, late blight-infected tubers and the crop grown under high nitrogen fertilizer, particularly ammonium chloride, are more predisposing factors for soft rot disease development. The soft rot-affected tubers become slimy and produce a foul smell, and brown liquid oozes out.

27.2.9.1.2 Blackleg

The disease symptoms appear on foliage of the plant at any stage. However, the disease is more common in dense canopy under wet and warm weather. Soft and black lesions appear at the base of stem and extend from the rotted mother seed tuber in soil to slightly above the ground level. The affected plants are stunted, exhibit yellow, chlorotic foliage and ultimately the plant wilts and dies without producing daughter tubers. Sometimes symptoms like leaf vein necrosis, brown to black lesions on petioles and succulent stems may appear. The lesions enlarge into stripes on stems and petioles. They cause soft rot as well toppling of stems and leaves. Hélias et al. (2000) reported that the infection of seed tubers by pectinolytic *Erwinia* species can lead to the development of various symptoms like non-emergence of plants, chlorosis, wilting, haulm desiccation and blackleg during vegetative growth of crops. Degefu et al. (2013) reported blackleg outbreaks in North Finland during the years 2008 and 2010, and they observed that *Dickeya* spp. was the major causal agent in blackleg complex.

27.2.9.2 Causal Organism

Pectobacterium atrosepticum (van Hall) Gardan et al. 2003 (syn. *Erwinia carotovora* subsp. *atroseptica*), *Pectobacterium carotovorum* subsp. *carotovorum* (Jones) Hauben et al. (syn. *Erwinia carotovora* subsp. *carotovora*), *Dickeya* spp. (including *D. dianthicola*, *D. dadantii*) (syn. *Erwinia chrysanthemi*), *Bacillus polymyxa*, *B. subtilis*, *B. mesentericus*, *B. megaterium* de Bary, *Pseudomonas marginalis* (Brown) Stevens, *P. viridiflava* (Burkholder) Dowson, *Clostridium* spp., *Micrococcus* spp. and *Flavobacterium* are also causal agents of soft rot disease in potato.

Pectobacterium atrosepticum are Gram-negative bacteria, rod shaped with peritrichous flagella and can grow under both aerobic and anaerobic conditions. They produce pectolytic enzymes and degrade pectin in middle lamella of host cell wall, breakdown tissues and cause soft rot and decay of tubers. The rotted tissue becomes slimy and foul smelling. The brown liquid oozes come out from the soft rot-affected tubers. Lyew et al. (2001) reported that bacterial pathogen produces certain volatile compounds like ammonia, trimethylamine and several volatile sulphides. These volatile compounds can be detected early in storage which could be used as a detection method of the disease at initial stage.

27.2.9.3 Production of Enzymes

Soft rot *Erwinias* are able to produce several cell wall-degrading enzymes including pectinases, cellulases, proteases and xylanases. These have different properties like acidic or basic

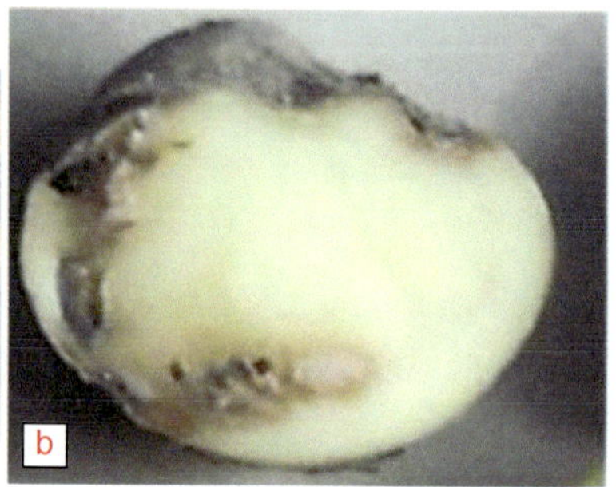

FIGURE 27.4 Symptoms of soft rot of potato tuber. (a) Rotted potato tuber. (b) Cross section of rotted potato tuber.

pH, high and low optimum pH, periplasmic or extracellular and exo- or endo-mode of action). The bacteria produce four main types of enzymes, and three have high optimum pH (~8), viz., pectatelyase (Pel), pectin lyase (Pnl) and pectin methyl esterase (Pme), and the fourth, polygalacturonase (Peh), has low optimum pH (~6). Perombelon (2002) reported that if more than one pathogen is present at the infection site, competition within rotting mother tubers, environmental conditions and temperature especially determines which pathogen will predominate to cause soft rot. The soft rot bacteria can also interact with other pathogens, particularly vascular pathogens like *R. solanacearum*, *Fusarium* spp., *Vertcillium* spp. and *R. solani* (Tsror et al. 1990; Perombelon 2002). One pathogen may favour the development of another by weakening of host resistance. Lehtimaki et al. (2003) studied the role of extracellular proteins for virulence of Gram-negative plant pathogenic bacteria. *E. carotovora* subsp. *carotovora* used type II secretion system to secrete pectolytic enzymes which are responsible for tissue maceration of soft-rot disease.

The pectinolytic *Dickeya* spp. are soft-rotting enterobacteriaceae family and Gram-negative bacteria causing severe diseases in several plant species. *Dickeya* spp. causes soft-rot symptoms due to the secretion of cell wall degradative enzymes that degrade plant cell walls (Reverchon and Nasser 2013).

27.2.9.4 Disease Cycle and Epidemiology

The bacteria can survive the winter on plant residues. The pathogenic bacteria are carried in lenticels, wounds and on surface of tubers. The bacteria can survive in places where rotten potatoes and vegetables are dumped. However, no survival has been reported in fields rotated with non-host crops. The bacteria may also be carried latently in tubers without any visible symptoms.

Soft rot-infected seed tubers are the primary source of inoculum, and the bacteria may enter vascular tissues of developing stems and cause blackleg disease under favourable conditions. The pathogen can reach daughter tubers through stolons from blackleg-infected plants and start tuber decay at the point of tuber attachment. These rotted tubers in the soil serve as a source of infection for healthy tubers. The bacteria spread to healthy tubers in stores, during seed cutting, handling, grading and planting (Weber 1990) and in the potato fields by irrigation water, insects, rain or bacterial aerosols. Similarly, Romberg et al. (2002) reported that the soft rotting bacteria are present on the surface of potato tubers, in soil and in surface irrigation water. Insects, especially maggots of *Hylemyia* spp., may also transmit the bacteria from one tuber to another. The pathogen may also spread through water during washing of the produce with contaminated water. Water film on surface of tuber causes proliferation of lenticels and creates anaerobic conditions, and injury on surface of tuber predisposes potatoes to soft rot. The threshold level for disease development is about 10^3 cells of *E. carotovora* spp. *atroseptica* per tuber (Perombelon 2000).

Tubers harvested in wet soil, poor ventilation in transit and storage promote the rot. In warm climates, the bacteria can pass easily from one crop to the next, especially in poorly drained soil. Predisposing factors like harvest injury, freezing injury and insect wounds are helpful to cause soft rot disease. Besides these factors, plenty of moisture at the wound site of tissue is essential to cause infection. After infection, high relative humidity is essential for disease progress. Bacteria smear the tubers from decayed potato during postharvest handling of produce (Farrar et al. 2009). At the packing site, potato tubers are first kept in a wash tank for cleaning, and surface bacteria infect potato tubers through lenticels by hydrostatic (exerted by water) pressure in the tank.

Walker (1998) reported that the pathogen is capable of causing more infection at high relative humidity with 26.7°C temperature. The optimum and maximum temperature is 29.4°C and about 37.7°C for the growth of pathogen, respectively. Maximum growth of pathogen was recorded at 30°C (Bhat et al. 2010). Raju et al. (2008) reported that soft rot ability of *E. carotovora* subsp. *carotovora* was enhanced by increasing the temperature of 20–30°C. They also reported that rotting was enhanced by increasing the relative humidity level. The maximum rotting was observed when radish discs were incubated at 35°C with 100% relative humidity.

27.2.9.5 Disease Management

27.2.9.5.1 Cultural Practices

Management of soft rot disease based on physical, chemical and biological methods has not been completely successful under field conditions (Czajkowski et al. 2012). Since no chemical control of the pathogen exists, use of certified pathogen-free seed tubers is a good option for disease management (Janse and Wenneker 2002). Disease-free seed tuber production under a seed certification scheme is widely used and has been partially successful. For production of bacteria-free progeny tubers, potato stem cuttings were initially used. After that, microplants were produced axenically, and presently, mini-tubers are produced in *in vitro* conditions. The mini-tubers are generally grown in a controlled pathogen-free environment using aeroponic and hydroponic cultures or in artificial soil systems (Farran and Mingo-Castel 2006). Improved storage management can reduce bacterial load on tubers and tuber rotting.

Well-drained soil reduces the risk of tubers which is surrounded by a water film and check to create an aerobiosis resulting reduces tuber decay in the field - Well drained soil reduces the risk of tuber decay in the field due to presence of water film layer surrounding the tuber which create an aerobiosis condition to the pathogen. Late harvesting of potato must be avoided because it cannot allow bacterial multiplication on leaves and in debris left on the ground following haulm flailing. This may result in control contamination of progeny tubers underground during wet weather conditions. Machines used in planting, spraying, haulm flailing, harvesting and grading in store should be washed and disinfected to reduce the risk of entry of soft rot bacteria in a disease-free crop (Perombelon 2002). Sanitation and removing of infected tubers from the field reduce the risk of bacteria spreading and smearing in a seed lot. Avoidance of wounding by using proper machinery with suitable adjustments should be used during postharvest handling (van Vuurde and de Vries 1994). Mature tubers having well-developed periderm will also reduce risk of injury. Disease-free true seeds may be used because they can be easily produced in large numbers (Chujoy and Cabello 2007).

Physical factors such as hot water, steam, dry hot air and UV and solar radiation (Ranganna et al. 1997; Bdliya and Haruna 2007) can be applied to manage soft rot disease in potato. Hot water treatment of potato tubers was first applied in 1983 to manage and control soft rot disease (Mackay and Shipton 1983). *P. carotovora* subsp. *carotovora* and *P. atroceptica* were detected in tuber peel after dipping naturally infected potato tubers in hot water at 55°C for 10 min. Similar results were obtained by Shirsat et al. (1991), who noted that tubers treated at 44.5°C for 30 min or at 56°C for 5 min significantly reduced the periderm and lenticel contamination of potatoes and consequently incidence of blackleg in the field. Afek and Orenstein (2002) reported that infection of tuber periderm from 26–59% to 1–3% was reduced by using steam treatment. Ranganna et al. (1997) tested the efficacy of UV radiation for managing *P. carotovora* subsp. *carotovora* in potato tubers. They reported that when tubers were inoculated by vacuum infiltration 6 h before radiation, the pathogenic bacteria were totally eliminated by a relatively low UV dose of 15 kJ per m^2.

27.2.9.5.2 Host Resistance

None of the commercial potato cultivars are naturally immune to blackleg and soft rot caused by *Dickeya* and *Pectobacterium* species, probably because of the narrow range of genetic diversity in parental breeding material used. However, some cultivars show partial resistance (Lyon 1989). Attempts have been made to breed for resistance against these bacteria in potato using wild *Solanum* spp., but with no success thus far (Birch et al. 2012).

Sexual hybrids between *S. tuberosum* and *S. phureja* are commonly used in breeding programmes and exhibit relatively more resistance, but they reduced tuber yields (Rousselle-Bourgeois and Priou 1995). When the wild *Solanum* species *S. commersonnii* (resistance to frost, nematodes and fungi) was crossed with *S. tuberosum*, the hybrids exhibited high resistance against both *Pectobacterium* spp. and *Dickeya* spp. (Laferriere et al. 1999). Similarly, the hybrids of commercial potato and *S. stenotomum* showed higher resistance to blackleg and soft rot bacteria (Fock et al. 2001). Somatic hybrids obtained by fusion of protoplasts of *S. tuberosum* and *S. brevidans* demonstrated a high level of resistance against blackleg and soft rot diseases due to a higher degree of esterification of cell-wall-binding pectin (McMillan et al. 1994).

27.2.9.5.3 Chemical Control

Once disease has been initiated, it is very difficult to control it because of fast multiplication and spread of bacteria and the inability of the chemicals to penetrate the inner tissues of the host. Streptomycin and its derivatives, inorganic and organic salts or combinations of these compounds are commonly used. Earlier, streptomycin was considered a promising control agent against blackleg and soft rot diseases in potato. Immersion of seed tubers in a mixture of streptomycin and oxytetracycline oxytetracycline hydrochloride or streptomycin, kasugamycin or virginiamycin (Bartz 1999) before planting reduced the incidence of blackleg in the field and tuber decay in storage. Organic compounds like hydroxyquinoline and 5-nitro-8-hydroxyquinoline are also found effective to control of soft rot in wounded potato tubers. Similarly, chlorine-based compounds such as bronopol (2-bromo-2-nitropropane-1, 3-diol) and the synthetic bactericide, 7-chloro-1-methyl-6-fluoro-1, 4-dihydro-4-oxo-3-quinolinic carboxylic acid are also showed promising result to control the disease (Bartz and Kelman 1986).

Chlorine dioxide in different concentrations (2–155 ppm) inhibits the growth of bacteria, depending upon the method of activating and diluting sodium chlorite solutions. Chlorine dioxide reacts quickly with the tuber and associated organic matter, thereby reducing its effectiveness. It may be applied at higher than currently registered rates, which may be necessary to achieve measurable disease suppression (Olsen et al. 2003).

Plant nutrition is a major component of natural disease resistance and affects the growth of plants as well as their interactions with pathogens and other microorganisms. Deficiency of essential elements affects the susceptibility of plants against diseases. Among these elements, calcium plays a vital role in the resistance of plants against soft rot bacteria (Berry et al. 1988). High calcium content in crops is usually positively correlated to increase resistance against potato blackleg (Berry et al. 1988). Adding $CaSO_4$ (gypsum) in Ca-low soils increased resistance to blackleg as well as soft rot of daughter potato tubers (Bain et al. 1996).

27.2.9.5.4 Biological Control

Although several attempts have been made to manage *Pectobacterium* spp. and *Dickeya* spp. on potato using bioagents, they are mostly restricted to *in vitro* conditions. Kastelein et al. (1999) reported that *Pseudomonas* spp. showed good results under *in vitro* where used to control of blackleg and soft rot diseases under field conditions. Potato tubers treated with fluorescent *Pseudomonas* spp. reduced populations of blackleg and soft rot bacteria on potato roots and inside progeny tubers (Kloepper 1983), and bacterial suspension applied on periderm tuber directly reduced soft rot disease. 2,4-diacetylphloroglucinol-producing *Pseudomonas fluorescens* strain F113 controlled *P. atroseptica in vitro* as well as on potato tubers (Cronin et al. 1997). Kastelein et al. (1999) applied *P. fluorescens* strains to protect wounds and cracks on tubers from *P. atroseptica* colonization. Application of *Pseudomonas* spp. either individually or in combinations of strains reduced the infection of potato tuber peel by 85% and 60–70%, respectively, which indicates the potentiality of bioagents to control soft rot-incited *P. atroseptica*. Other antagonistic bacteria such as *Lactobacillus plantarum*, *L. acidophilus*, *L. buchneri*, *Leuconostoc* spp. and *Weissellacibaria* isolated from fresh fruits and vegetables showed antagonistic activity against *P. carotovorum* subsp. *carotorumin vitro* due to production of hydrogen peroxide and acidification of the medium (Trias et al. 2008).

Gram-positive *Bacillus subtilis* BS107 was used as a biocontrol agent against soft rot- and blackleg-causing bacteria (Sharga and Lyon 1998). *B. licheniformis* P40 produced bacteriocin-like substance that was bactericidal to *P. carotovorum* subsp. *carotorum* and found effective to protect potato tubers against soft rot disease under storage conditions (Cladera-Olivera et al. 2006).

27.2.10 Common Scab

27.2.10.1 Symptoms

Symptoms of common scab appear as brown, raised, rough patches of skin on the tuber surface. The symptoms of shallow or surface scabs appear as superficial roughened areas, at times raised above but more often slightly below the surface of healthy skin. The lesions consist of corky tissue which arises from abnormal proliferation of the cells of the tuber periderm as a result of pathogen invasion. The size and shape of the lesions may vary greatly and show darker than the healthy skin. When lesion size increases, the periderm cracks due to formation of cork layer around lesions, and assume various shapes such as reticulate, shallow or deep pitted (Figure 27.5). The lesions coalesce to affect the large area on the tuber surface. The lesions on mature tubers may be mere abrasions; star shaped with corky depositions, concentric wrinkled layers of cork around a central black core; raised and rough corky pustules or 3- to 4-mm-deep pits surrounded by hard corky tissue. Tubers that protrude above ground level are not invaded by the pathogen. Tubers that grow faster than others develop higher infection. Quick-bulking varieties in general suffer more than slow-bulking varieties (Shekhawat et al. 1993). The tubers having corky and hard lesions are generally not affected by secondary organisms and they do not affect storage life of the potato. Russet scab of potato is a major problem in the north-western plains, especially in early crop (Arora and Sharma 2003). The disease appears on potato tuber as brown netted or irregular superficial abrasions. Lesions of russet scab can also appear on emerging sprouts, stolons and stem base besides potato tubers (Sharma et al. 2006).

27.2.10.2 Causal Organism

The disease is caused by *Streptomyces scabies* (Thaxter) Lambert and Loria. However, Hao et al. (2009) reported 13 different species of *Streptomyces* to cause common scab on potato, and among them *Streptomyces scabies* (Thaxter) Lambert and Loria, *S. acidiscabies* Bambert and Loria, *S. turgidscabis* Takeunchi, *S. collinus* Lindenbein, *S. griseus* (Krainsky) Waksman & Henria, *S. longisporoflavus*, *S. cinereus*, *S. violanceoruber*, *S. alborgriseolus*, *S. griseoflavus*, *S. catenulae* and others are most common. *Streptomyces* form vegetative substrate mycelium-like fungus that develops aerial filaments. However, the filaments are of smaller dimensions than the true fungi. *S. scabies* form grey, spiral spore chains on several media and produce brown pigment, whereas *S. acidiscabis* produce peach-coloured wavy chains of spores and brown pigment in the medium. The identification and taxonomy of *Streptomyces* spp. has been based on morphological and physiological characteristics combined with thaxtomin production and pathogenicity tests *in vitro* and *in vivo* (Wanner 2004). Ability to produce thaxtomin toxin is positively correlated with pathogenicity of bacteria. Boucheck-Mechiche et al. (2000) reported three groups of pathogenic *Streptomyces* and they differ in their ecological requirements and also for production of various symptoms on host under different soil temperatures.

Taxonomy of *Streptomyces* is complex; however, molecular techniques like 16S ribosomal (r)RNA gene sequence analysis and DNA-DNA hybridization have been used successfully for species-level identification (Song et al. 2004; Hao et al. 2009). The highly conserved genes responsible for pathogenicity among genetically diverse *Streptomyces* isolates are ideal molecular markers for determination of pathogenicity. For example, *nec1* gene is thought to be highly conserved in both structure and function among unrelated pathogenic *Streptomyces* spp., which is not present in non-pathogenic species (Park et al. 2003). Joshi et al. (2007) showed that *nec1* was necessary for pathogenicity, but others assume that it has a subsidiary role in pathogenicity because it is missing from some other pathogenic strains of *Streptomyces* (Wanner 2004).

27.2.10.3 Disease Cycle and Epidemiology

The *Streptomyces* spp. causing scab disease are both soil and tuber borne. The pathogenic bacteria can survive for several years in plant debris and infested soil. Tuber-borne inoculums play an important role in the spread of new species or their strains (Stevenson et al. 2001). Potato is physiologically most susceptible to *Streptomyces* spp. during initial tuber formation. They infect the newly formed tubers through immature lenticels and stomata. It has been observed that once the periderm is differentiated, the tubers are no longer susceptible to the pathogen. Soil conditions greatly influence the pathogen, and pH in the range of 5.2–8.0 or more, temperature 20–30°C and low soil moisture favour growth of the pathogen (Singh and Singh 1981).

Wang and Lazarovits (2005) reported the influence of tuber-borne pathogenic *Streptomyces* on potato common scab

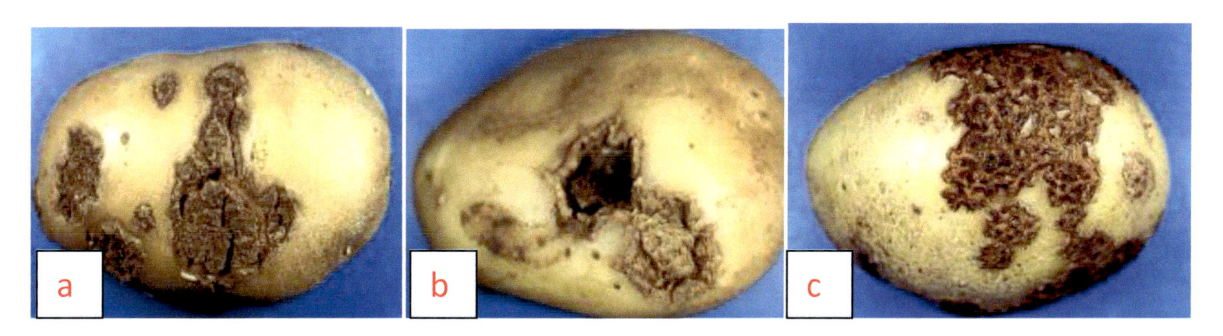

FIGURE 27.5 Different common scab symptoms on potato tuber. (a) Superficial and raised. (b) Deep pitted. (c) Brown netted.

incidence in daughter tubers under micro-plot conditions. They planted visually healthy tubers, surface-disinfested healthy tubers and tubers having 25% scab coverage in pasteurized soil. They observed that on the 30th day, population of pathogenic *Streptomyces* was below detectable levels while at day 93, measurable levels of bacteria were present on the belowground plant parts and in soil adjacent to scabby mother tubers at 10^4–10^5 cfu/g soil. However, scab incidence in the daughter tubers was 89% at the time of harvest. Scab incidence of 60% in progeny tubers was observed in visually healthy seed tubers, and substantial populations of pathogenic *Streptomyces* (10^4 cfu/g soil) were present in the zone near the mother tuber.

27.2.10.4 Disease Management

27.2.10.4.1 Cultural Practices

Healthy and certified seed tubers for sowing should be selected. Infected plant debris should be removed and destroyed from the field. The pathogen is aerobic in nature; hence maintaining high soil moisture for 10–20 days after tuber initiation can help to reduce common scab disease. The field should be irrigated as per requirements frequently at weekly intervals from tuberization until maturity (Singh and Singh 1981, Sharma and Singh 2005).

Soil solarization with transparent polyethylene mulching during the hot summer season was effective for control of russet scab. The disease severity by soil solarization reduces to almost one-third as compared to the unsolarized plots (Arora 2005; Arora et al. 2006). Rotational crops such as maize, cotton, grain sorghum, wheat, cabbage and onion have been observed to reduce incidence of common scab (Shekhawat et al. 1991). Green manuring also helps to reduce incidence of disease in the field. Larkin and Lynch (2018) reported that mustard blend, ryegrass, and other Brassica rotations also reduced common scab, silver scurf, and black scurf diseases at various sites, results were not consistent at all sites. At one site, mustard blend and barley/ryegrass rotations reduced black scurf (by 21–58%) and common scab (by 13–34%).

27.2.10.4.2 Host Resistance

Cultivars of potato differ in their susceptibility against common scab disease but as yet, no cultivar is totally resistant to the disease. Highly resistant cultivars may develop symptoms under highly conducive conditions (Wale et al. 2008).

27.2.10.4.3 Chemical Control

Seed tubers should be dipped in the suspension of 3% boric acid for 20–25 min. Acidic fertilizers such as ammonium sulphate, single super phosphate and potassium chlorides (Heald 1993) are found effective to reduce the disease. Synthetic auxin 2,4-dichlorophenoxyacetic acid (2,4-D) applied to the foliage of potato plants can reduce common scab. However, higher doses @ 200 mg/L of 2,4-D resulted in phytotoxic side effects and also reduced tuber yield and quality. Minimal significant threshold rates from 8.3–23.6 mg/L 2,4-D reduced disease incidence under pot conditions and from 10.8–41.0 mg/L minimized disease severity in both pot and field conditions (Thompson et al. 2014). 2,4-D was applied as early as 5 days after emergence plant provided better disease control (Thompson et al. 2013)

27.2.10.4.4 Biological Control

Antibiotic producing strain of bacteria was antagonistic against *S. scabiei* which were isolated from the soil of a potato-cultivating area (Han et al. 2005). The liquid culture of *Bacillus* sp. *sunhua* suppressed common scab disease under pot assay and decreased the infection rate from 75 to 35%. Hiltunen et al. (2009) used nonpathogenic *Streptomyces* strain (346) for controlling of *S. turgidiscabies* and *S. scabies*. The strain also reduced incidence of *S. turgidiscabies* in scab lesions on potato tubers in the field. *Pseudomonas* sp. LBUM 223s inhibited growth of *S. scabies*, by repression of thaxtomin biosynthesis genes (txtA and txtC) and protected potato against common scab disease (St-Onge et al. 2011).

27.3 Conclusion

Despite the best efforts to grow and harvest a healthy potato crop, several diseases can cause mild to severe postharvest losses of crop either in the field or in storage. The majority of the postharvest diseases of potato carry from the field only, hence management strategies should be laid down according to their occurrence. Although several control strategies against postharvest diseases of potato, like late blight (*Phytophthora infestans*), early blight (*Alternaria solani*), dry rot (*Fusarium sambucinum* and *Fusarium* spp.), brown rot (*Ralstonia solanacearum*), soft rot (*Erwinia carotovora* subsp. *carotovora*), pink rot (*Phytophthora erythroseptica*), Pythium leak (*Pythium ultimum*) and silver scurf (*Helminthosporium solani*) have been employed, effective control of these diseases has not yet been achieved. Until highly resistant cultivars become available, disease control measures will continue to rely primarily on avoidance of contamination in the production of healthy certified seed, because the pathogens are mainly soil and seed tuber borne.

For storage management of the produce, three basic tools – temperature, relative humidity and airflow – play important roles in managing postharvest disease storage. Postharvest treatment using physical, chemical and biological methods is available to manage storage diseases of potato. These methods should be employed in combination with good basic storage management. Potato seed should be derived from disease-free areas using strict hygienic practices. Knowledge of the pathogen sources and contamination pathways should be explored to apply hygienic measures, especially during harvest and postharvest. Control strategies can be supported also by tuber treatments, as discussed above. Integrated disease management strategy is the only way to succeed in reducing postharvest diseases of potato in the field as well as during storage.

REFERENCES

Abdullahi I, Koerbler M, Stachewicz H, Winter S (2005) The 18S rDNA sequence of *Synchytrium endobioticum* and its utility in microarrays for the simultaneous detection of fungal and viral pathogens of potato. Appl Microbiol Biotechnol 68: 368–375.

Afek U, Orenstein J (2002) Disinfecting potato tubers using steam treatments. Can J Plant Pathol 24: 36–39.

Al-Mughrabi K (2010) Biological control of Fusarium dry rot and other potato tuber diseases using *Pseudomonas fluorescens* and *Enterobacter cloacae*. Biol Control 53: 280–284.

Al-Mughrabi KI (2006) Sensitivity to hydrogen peroxide *in vitro* of North American isolates of *Phytophthora erythroseptica*, the cause of pink rot of potatoes. Plant Pathol J 5: 7–10.

Al-Mughrabi KI, Peters RD, Platt HW, Moreau G, Vikram A et al. (2007) In-furrow applications of metalaxyl and phosphite for control of pink rot (*Phytophthora erythroseptica*) of potato in New Brunswick, Canada. Plant Dis 91: 1305–1309.

Arora RK (2005) Efficacy of soil solarization in reducing tuber russet and improving yield of different cultivars. Potato J 32: 185–186.

Arora RK, Sharma J (2003) Management of russet scab of potato through soil solarization. J Indian Potato Assoc 30: 139–140.

Arora RK, Sagar V (2014) Major tuber borne diseases of potato. In: Singh D, Chowdappa BN, Sharma P (Eds.). Diseases of Vegetable Crops IPS, New Delhi, pp. 1–55.

Arora RK, Sharma J, Garg ID, Singh RK, Somani AK (2006) Boric Acid for Control of Tuber Borne Diseases. Technologies in Aid of Healthy Seed Production. Part II. Technical Bulletin No. 35. Central Potato Research Institute, Shimla, pp. 1–4.

Aspiras RB, de la Cruz AR (1986) Potential biological control of bacterial wilt in tomato and potato with *Bacillus polymixa* FU6 and *Pseudomonas solanacearum*. In: Persley GJ (Ed.). Bacterial Wilt Disease in Asia and the South Pacific. ACIAR Proceedings 13, Canberra, Australia, pp. 89–92.

Baayen RP, Cochius G, Hendriks H, Meffert JP, Bakker J, Bekker M, van den Boogert PHJF, Stachewicz H, van Leeuwen GCM (2006) History of potato wart disease in Europe – a proposal for harmonisation in defining pathotypes. Eur J Plant Pathol 116: 21–31.

Bain R, Millard P, Pe'rombelon M (1996) The resistance of potato plants to *Erwinia carotovora* subsp. *atroseptica* in relation to their calcium and magnesium content. Potato Res 39: 185–193.

Bartz J, Kelman A (1986) Reducing the potential for bacterial soft rot in potato tubers by chemical treatments and drying. Am J Potato Res 63: 481–493.

Bartz J (1999) Suppression of bacterial soft rot in potato tubers by application of kasugamycin. Am J Potato Res 76: 127–136.

Bdliya B, Haruna H (2007) Efficacy of solar heat in the control of bacterial soft rot of potato tubers caused by *Erwinia carotovora* ssp. *carotovora*. J Plant Protec Res 47: 11–18.

Benson JH, Geary Brad, Miller JS, Hopkins BG, Jolley VD, Stevens MR (2009) *Phytophthora erythroseptica* (pink rot) development in Russet Norkotah potato grown in buffered hydroponic solutions. II. pH Effects. Am J Potato Res 86(6): 472–475.

Berry S, Madumadu G, Uddin M (1988) Effect of calcium and nitrogen nutrition on bacterial canker disease of tomato. Plant Soil 112: 113–120.

Bhat KA, Masoodi SD, Bhat NA, Ahmad M, Zargar MY, Mir SA, Ashraf-Bhat M (2010) Studies on the effect of temperature on the development of soft rot of cabbage (*Brassica oleracea* var. *capitata*) caused by *Erwinia carotovora* subsp. *carotovora*. J Phytopathol 2(2): 64–67.

Bhattacharya RC, Vishwakarma N, Bhat SK, Kirti PB, Chopra VL (2002) Development of insect transgenic cabbage plants expressing a synthetic Cry 1 A (b) gene from *Bacillus thuringiensis*. Curr Sci 25: 146–150.

Birch PRJ, Bryan G, Fenton B, Gilroy EM, Hein I, Jones JT, Prashar A, Taylor MA, Torrance L, Toth IK (2012) Crops that feed the word 8: Potato: Are the trends of increased global production sustainable? Food Secur 4: 477–508.

Bokshi AI, Morris SC, Deverall BJB (2003) Effects of benzothiadiazole and acetylsalicylic acid on [beta]-1,3-glucanase activity and disease resistance in potato. Plant Pathol 52(1): 22–27.

Boucheck-Mechiche K, Pasco C, Andrivon D, Jouan B (2000) Difference in host range, pathogenicity to potato cultivars and response to soil temperature among *Streptomyces* species causing common and net scab in France. Plant Pathol 49: 3–10.

Broders KD, Lipps PE, Paul PA, Dorrance AE (2007) Characterization of *Pythium* spp. associated with corn and soybean seed and seedling disease in Ohio. Plant Dis 91: 727–735.

Butler EJ, Jones SG (1961) Plant Pathology. MacMillan, London and New York, 979 pp.

Carnegie SF, Cameron AM, Haddon P (2001) The effect of date of haulm destruction and harvest on the development of dry rot caused by *Fusarium solani*. Phytopathology 84: 1387–1393.

Choiseul JW, Allen L, Carnegie SF (2001) The role of stem inoculum in the transmission of *Fusarium sulphureum* to potato tubers. Potato Res 44: 165–172.

Chujoy E, Cabello R (2007) The canon of potato science: true potato seed (TPS). Potato Res 50: 323–325.

Cladera-Olivera F, Caron GR, Motta AS, Souto AA, Brandelli A (2006) Bacteriocin-like substance inhibits potato soft rot caused by *Erwinia carotovora*. Can J Microbiol 52: 533–539.

Costelloa B. P. J. de Lacy; Evansa , P.; Ewena, R. J.; Gunsonb, H. E.; Jonesa, P. R. H.; Ratcliffea*2, N. M. and Spencer-Phillipsb, P. T. N. Gas chromatography-mass spectrometry analyses of volatile organic compounds from potato tubers inoculated with Phytophthora infestans or Fusarium coeruleum. Plant Pathology (2001) 50: 489–496.

Cronin D, Moe¨nne-Loccoz Y, Fenton A, Dunne C, Dowling DN, O'Gara F (1997) Ecological interaction of a biocontrol *Pseudomonas fluorescens* strain producing 2,4-diacetylphloroglucinol with the soft rot potato pathogen *Erwinia carotovora* subsp. *atroseptica*. FEMS Microbiol Ecol 23: 95–106.

Cullen DW, Toth IK, Pitkin Y, Boonham N, Walsh K et al. (2005) Use of quantitative molecular diagnostic assays to investigate *Fusarium* dry rot in potato stocks and soil. JPhytopathol 95: 1462–1471.

Czajkowski R, Perombelon MCM, Van Veen JA, Van der Wolf JM (2012) Control of blackleg and tuber soft rot of potato caused by *Pectobacterium* and *Dickeya* species: A review. Plant Pathol 60: 999–1013.

Daami-Remadi M, Jabnoun-Khiareddine H, Ayed F, El-Mahjoub M (2006b) Effect of temperature on aggressivity of Tunisian *Fusarium* species causing potato (*Solanum tuberosum* L.) tuber dry rot. J Agronom 5: 350–355.

Degefu Y, Potrykus M, Golanowska M, Virtanen E, Lojkowska E (2013) A new clade of *Dickeya* spp. plays a major role in potato blackleg outbreaks in North Finland. Ann Appl Biol 162(2): 231–241.

Desjardins AE, Gardner HW, Weltring KM (1992) Detoxification of sesquiterpene phytoalexins by *Gibberella pulicaris* (*Fusarium sambucinum*) and its importance for virulence on potato tubers. J Indian Microbiol Biotech 9: 201–211.

Desjardins A, Hohn T, McCormick S (1993) Trichothecene biosynthesis in *Fusarium* species: Chemistry, genetics, and significance. Microbiol Mol Biol Rev 57: 595.

Elphinstone JG (2005) The current bacterial wilt situation: A global overview. In: Bacterial Wilt Disease and the *Ralstonia solanacearum* Species Complex. Allen C, Prior P, Hayward AC (Eds.). American Phytopathological Society, St. Paul, MN, pp. 9–28.

Elphinstone JG, Stead DE, Caffier D, Janie JD, Lopez MM, Santo MS, Stefani E, van Verenbergh J (2000) Standardization of methods for detection of *Ralstonia Solnacearum* in potatoes. EPPO Conference on Diagnostic Techniques for Plant Pests, Waddington, Netherlands, 1-4 Feb 2000. Bulletin–OEPC 30, pp. 391–395.

EPPO (2006) A2 list of pests recommended for regulation as quarantine pests (version 2006-09). Online. Quarantine Info. European and Mediterranean Plant Protection Organization (EPPO), Paris, France.

Ermakova NE, Shterushis MV (1994) New biological preparation of RITs against plant diseases. Zashchita Rastonii Moskva 12: 18.

Farran I, Mingo-Castel A (2006) Potato mini tuber production using aeroponics: Effect of plant density and harvesting intervals. Am J Potato Res 83: 47–53.

Farrar JJ, Nunez JJ, Davis RM (2009) Losses due to lenticel rot are an increasing concern for Karen County potato growers. Calif Agric 63: 127–130.

Fegan M, Prior P (2005) How complex is the *Ralstonia solanacearum* species complex? In: Allen C, Prior P, Hayward AC (Eds.). Bacterial Wilt: The Disease and the Ralstonia solanacearum Species Complex. APS Press, St. Paul, MN, pp. 449–461.

Fegan, M., Holoway, G., Hayward, A.C. and Timmis, J. (1998) Development of a diagnostic test based on the polymerase chain reaction (PCR) to identify strains of R. solanacearum exhibiting the bovar 2 genotype. In Bacterial Wilt Disease; Molecular and ecological aspects ed, P. Prior, C. Allen, and J. Elphinstone pp. 34–43 Berlin, Germany, Springer-Verlag.

Fernández-Northcote E, Navia O, Gandarillas A (2000) Basis of strategies for chemical control of potato late blight developed by PROINPA in Bolivia. J Phytopathol 35: 137–149.

Fock I, Collonnier C, Luisetti J et al. (2001) Use of *Solanum stenotomum* for introduction of resistance to bacterial wilt in somatic hybrids of potato. Plant Physiol Biochem 39: 899–908.

Friedman M (2006) Potato glycoalkaloids and metabolites: Roles in the plant and in the diet. J Agric Food Chem 54(23): 8655–8681.

Gadewar AV, Trivedi TP, Shekhawat GS (1991) Potato in Karnataka. Tech. Bull No. 17, Central Potato Research Institute, Shimla, 33 pp.

Geiser D, del Mar Jiménez-Gasco M, Kang S, Makalowska I, Veeraraghavan N et al. (2004) Fusarium-ID v. 1.0: A DNA sequence database for identifying *Fusarium*. Eur J Plant Pathol 110: 473–479.

Gevens AJ, Seidl AC (2013) First report of late blight caused by *Phytophthora infestans* clonal lineage US-24 on potato (*Solanum tuberosum*) in Wisconsin. Plant Dis 97: 152.

Griffiths H, Zitter T, Deahl K, Halseth D (2008) *Phytophthora erythroseptica*, isolate sensitivity to metalaxyl and disease control in potato in New York and Pennsylvania. J Phytopathol 98.

Grisham MP, Taber RA, Barnes LW (1983) Phytophthora rot of potatoes in Texas caused by *Phytophthora parasitica* and *Phytophthora cryptogea*. Plant Dis 67: 1258–1261.

Grover A, Azmi W, Gadewar AV, Pattanayak D, Naik PS, Shekhawat GS, Chakrabarti SK (2009) Genotypic diversity in a localized population of *Ralstonia solanacearum* as revealed by random amplified polymorphic DNA markers. J Appl Microbiol 101: 798–806.

Gupta CP, Sharma A, Dubey RC, Maheshwari DK (1999) *Pseudomonas aeruginosa* (GRC_1) as a strong antagonist of *Macrophomina phaseolina* and *Fusarium oxysporum*. Cytobios 99: 183–189.

Hampson MC, Coombes JW (1995) Reduction by potatoes wart disease with crushed crab shell: Suppression or eradication. Can J Plant Pathol 17: 69–74.

Han JS, Cheng JH, Yoon TM, Song J, Rajkarnikar A, Kim WG, Yoo ID, Yang YY, Suh JW (2005) Biological control agent of common scab disease by antagonistic strain *Bacillus* sp. sunhua. J Appl Microbiol 99(1): 213–221.

Hao JJ, Meng QX, Yin JF, Kirk WW (2009) Characterization of a new *Streptomyces* strain, DS3024, that causes potato common scab. Plant Dis 93: 1329–1334.

Hayward AC (1991) Biology and epidemiology of bacterial wilt caused by *Pseudomonas solanacearum*. Annu Rev Phytopathol 29: 65–89.

Heald FD (1993). Manual of Plant Diseases. McGraw-Hill, New York and London, 953 pp.

Hélias V, Andrivon D, Jouan B (2000) Internal colonization pathways of potato plants by *Erwinia carotovora* ssp Atroseptica. Plant Pathol 49: 33–42.

Hide GA, Read PJ, Hall SM (1992) Resistance to thiabendazole in *Fusarium* species isolated from potato tubers affected by dry rot. Plant Pathol 41: 745–748.

Hiltunen LH, Ojanpera T, Kortemaa H, Richter E, Lehtonen MJ, Valkonen JPT (2009) Interactions and bio-control of pathogenic *Streptomyces* strains co-occurring in potato scab lesions. J Appl Microbiol 106(1): 199–212.

Horita M, Suga MYA, Ooshiro A, Tsuchiya K (2010) Analysis of genetic and biological characters of Japanese potato strains of *Ralstonia solanacearum*. J Gen Plant Pathol 76: 196–207.

Janse JD, Wenneker M (2002) Possibilities of avoidance and control of bacterial plant diseases when using pathogen-tested (certified) or -treated planting material. Plant Pathol 51: 523–536.

Ji P, Momol T, Olson S, Meister C, Norman D, Jones J (2007) Evaluation of phosphorous acid-containing products for managing bacterial wilt of tomato. Phytopathology 97(7): S52.

Joshi M, Rong X, Moll S, Kers J, Franco C, Loria R (2007) *Streptomyces turgidiscabies* secretes a novel virulence protein, Nec1, which facilitates infection. Mol Plant Microbe Interact 20: 599–608.

Johnson SB (2007) Post-harvest application of phosphorous acid materials to reduce potato storage rots. Phytopathology 97: S178–S179.

Johnson D (2010) Transmission of *Phytophthora infestans* from infected potato seed tubers to emerged shoots. Plant Dis 94: 18–23.

Johnson D, Cummings T (2009) Latent infection of potato seed tubers by *Phytophthora infestans* during long-term cold storage. Plant Dis 93: 940–946.

Kastelein P, Schepel E, Mulder A, Turkensteen L, Van Vuurde J (1999) Preliminary selection of antagonists of *Erwinia carotovora* subsp. Atroseptica (Van Hall) Dye for application during green crop lifting of seed potato tubers. Potato Res 42: 161–171.

Kaur S, Mukerji K (2004) Potato diseases and their management. In: Mukerji K. (Ed.). Fruit and Vegetable Diseases. Springer, Netherlands, pp. 233–280.

Khakvar R, Sijam K, Wong MY, Radu S, Jones J, Thong KL (2008) Genomic diversity of *Ralstonia solanacearum* strains isolated from banana farms in west Malaysia. Plant Pathol J 7: 162–167.

Kim-Byung Sup Cho KC, Cho KY (1995) Antifungal effects of Plant pathogenic fungi and characteristics of antifungal substances produced by *Bacillus subtilis* SJ-2 isolated from sclerotia of *Rhizoctonia solani*. Korean J Plant Pathol 11: 165–172.

Kirk WW, Wharton P, Hammerschmidt R, Abu-El Samen F, Douches D (2004) Late blight. Michigan State University, Extension Bulletin, E-2945, June 2004.

Kirk WW, Felcher K, Douches D, Coombs J, Stein J, et al. (2001) Effect of host plant resistance and reduced rates and frequencies of fungicide application to control potato late blight. Plant Dis 85: 1113–1118.

Kloepper JW (1983) Effect of seed piece inoculation with plant growth promoting rhizobacteria on populations of *Erwinia carotovora* on potato roots and in daughter tubers. Phytopathology 73: 217–219.

Korpan YI, Nazarenko EA, Skryshevskaya IV, Martelet C, Jaffrezic-Renault N, El'skaya AV (2004) Potato glycoalkaloids: True safety or false sense of security? Trends Biotechnol 22(3): 147–151.

Laferriere LT, Helgeson JP, Allen C (1999) Fertile *Solanum tuberosum + S. commersonii* somatic hybrids as sources of resistance to bacterial wilt caused by *Ralstonia solanacearum*. TheorAppl Genet 98: 1272–1278.

Lambert DH, Salas B (1994) Metalaxyl insensitivity of *Phytophthora erythroseptica* isolates causing pink rot of potato in Maine. Plant Dis 78: 1010.

Lambert DH, Salas B (2001) Pink rot. In: Stevenson W, Loria R, Franc GD, Weingartner DP (Eds.). Compendium of Potato Diseases. APS Press, St. Paul, Minnesota, pp. 33–34.

Larka BS (2004) Integrated approach for the management of soft rot (*Pectobacterium carotovorum* subsp. *carotovorum*) of radish (*Raphanus sativus*) seed crop. Haryana J Agron 20: 128–129.

Larkin RP, Lynch RP (2018) Use and effects of different *Brassica* and other rotation crops on soilborne diseases and yield of potato. Horticulturae 4, 37; doi:10.3390/horticulturae4040037

Leach S, Webb R (1981) Resistance of selected potato cultivars and clones to Fusarium dry rot. J Phytopathol 71: 623–629.

Lehtimaki S, Rantakari A, RouttuJ, Tuikkala A, Li J, Virtaharju O, Palva ET, Romantschuk M, Saarilahti HT (2003) Characterization of the *hrp* pathogenicity cluster of *Erwinia carotovora* subsp. *carotovora*: high basal level expression in a mutant is associated with reduced virulence. Mol Gene Genom 270(3): 263–272.

Lemaga B, Siriri D, Ebanyat P (2001) Effect of soil amendments on bacterial wilt incidence and yield of potatoes in south western Uganda. Afr Crop Sci J 9: 267–278.

Little G, Bell S (2009) Controlling dry rot in seed potatoes. Retrieved February 23, 2011 from http://www.dardni.gov.uk/ruralni/dry_rot_leaflet_september_2009_chdb.pdf.

Lorito M, Di Pietro A, Hayes CK, Woo SL, Harman GE (1993) Antifungal, synergistic interaction between *Trichoderma harzianum* and *Enterbacter cloacae*. Phytopathology 83: 721–728.

Lui LH, Kushalappa AC (2003) Models to predict potato tuber infection by *Pythium ultimum* from duration of wetness and temperature, and leak-lesion expansion from storage duration and temperature. Postharvest Biol Technol 27: 313–322.

Lyew D, Gariepy Y, Raghavan GSV, Kushalappa AC (2001) Changes in volatile production during an infection of potatoes by *Erwina carotovora*. Food Res Intl 34: 807–813.

Lyon GD (1989) The biochemical basis of resistance of potatoes to soft rot *Erwinia* spp.—a review. Plant Pathol 38: 313–339.

Lyons N, Cruz L, Santos MS (2001) Rapid field detection of *Ralstonia solnacearum* in infected tomato and potato plants using the *Staphylococcus aureus* slide agglutination test. Bull OEPP 31: 91–93.

Mackay JM, Shipton PJ (1983) Heat treatment of seed tubers for control of potato blackleg (*Erwinia carotovora* subsp. *atroseptica*) and other diseases. Plant Pathol 32: 385–393.

Mayton H, Myers K, Fry WE (2008) Potato late blight in tubers - The role of foliar phosphonate applications in suppressing pre-harvest tuber infections. Crop Prot 27: 943–950.

McLaughlin RJ, Sequeira L (1988). Evaluation of an avirulent strain of *Pseudomonas solanacearum* for biological control of bacterial wilt of potato. Am Potato J 65: 255–267.

McMillan GP, Barrett AM, Perombelon MCM (1994) An isoelectric focusing study of the effect of methyl-esterified pectic substances on the production of extracellular pectin isoenzymes by soft rot *Erwinia* spp. J Appl Microbiol 77: 175–184.

Mekameh MA, Razavi M, Kasra S, Rasoul Z (2009) Investigation on genetic diversity of *Fusarium oxysporum* causing potato fusarium wilt by pathogenicity tests and RAPD markers. Iran J Plant Pathol 45: 9–24.

Mensinga TT, Sips AJAM, Rompelberg CJM et al. (2005) Potato glycoalkaloids and adverse effects in humans: An ascending dose study. Regul Toxicol Pharmacol 41(1): 66–72.

Miller JS, Olsen N, Woodell L, Porter LD, Clayson S (2006) Postharvest applications of zoxamide and phosphite for control of potato tuber rots caused by oomycetes at harvest. Am J Potato Res 83: 269–278.

Milner SE, Brunton NP, Jones PW, OBrien NM, Collins SG, Maguire AR (2011) Bioactivities of glycoalkaloids and their aglycones from *Solanum* species. J Agric Food Chem 59(8): 3454–3484.

Nema PK, Ramayya E, Duncan E, Niranjan K (2008) Potato glycoalkaloids: Formation and strategies for mitigation. J Sci Food Agric 88(11): 1869–1881.

Nyankanga RO, Wien HC, Olanya OM (2008) Effects of mulch and potato hilling on development of foliar blight (*Phytophthora infestans*) and the control of tuber blight infection. Potato Res 51: 101–111.

Olsen N (2010) Storage. In: Bohl WH, Johnson SB (Eds.). Commercial Potato Production in North America (Potato Association of America Handbook). Potato Association of America, p. 81.

Olsen NL, Kleinkopf GE, Woodell LK (2003) Efficacy of chlorine dioxide for disease control on stored potatoes. Am J Potato Res 80(6): 387–395.

Park DH, Yu YM, Kim JS, Cho JM, Hur JH, Lim CK (2003) Characterization of *Streptomycetes* causing potato common scab in Korea. Plant Dis 87: 1290–1296.

Pastrik, K. H., Elphinstone, J.G. and Pukall, R. (2002). Sequence analysis and detection of Ralstonia solanacearum by multiplex PCR amplification of 16S-23S ribosomal intergenic spacer region with internal positive control. Eur. J Plant Pathol. 108: 831–842.

Paulitz TC, Adams K (2003) Composition and distribution of *Pythium* communities in wheat fields in eastern Washington state. J Phytopathol 93: 867–873.

Perombelon MCM (1987) Pathogenesis by pectolytic Erwinias. In: Civerolo EL, Collmer A, Davis RE, Gillaspie AG (Eds.). Plant Pathogenic Bacteria. Martinus Nijhoff, Dordrecht, Netherlands, pp. 109–120.

Perombelon MCM (2000) Blackleg risk potential of seed potatoes determined by quantification of tuber contamination by the causal agent and *Erwinia carotovora* spp. *atroseptica*: A critical review. Bull OEPP 30: 413–420.

Perombelon MCM (2002) Potato diseases caused by soft rot *Erwinias*: An overview of pathogenesis. Plant Pathol 51: 1–12.

Peters RD, Platt HWB, Levesque CA (2005) First report of *Pythium sylvaticum* causing potato tuber rot. Am J Potato Res 82(2): 173–177.

Peters R, MacLeod C, Seifert K, Martin R, Hale L et al. (2008) Pathogenicity to potato tubers of *Fusarium* spp. isolated from potato, cereal and forage crops. Am J Potato Res 85: 367–374.

Pethybridge GH (1913) On the rotting of potato tubers by a new species of Phytophthora having a method of sexual reproduction hitherto undescribed. Sci Proc R Dublin Soc 13: 529–565.

Pethybridge GH (1914) Further observations on *Phytophthora erythroseptica* Pethybr. and on the disease produced by it in the potato plant. Sci Proc R Dublin Soc 14: 179–198.

Pinzon-Perea L, Salgado R, Martineg-Lopez G (1999) Antagonism between different isolates of *Trichoderma* spp. and *Fusarium oxysporum* f. sp. *dianthi* (Prill and Del.) Synd. & Hans. Fitopatol Colomb 23: 7–11.

Platt HW, Peters RD (2006) Fungal and oomycete diseases. In: Gopal J, Khurana SMP (Eds.). Handbook of Potato Production, Improvement, and Postharvest Management. Food Product Press, an imprint of The Haworth Press, Binghamton, NY, pp. 315–350.

Porter L, Hamm P, David N, Gieck S, Miller J et al. (2009) Metalaxyl-M-resistant *Pythium* species in potato production areas of the Pacific Northwest of the U.S.A. Am J Potato Res 86: 315–326.

Powelson ML, Rowe HC (2008) Managing diseases caused by seedborne and soilborne fungi and fungus-like pathogens. In: Johnson DA (Ed.). Potato Health Management. APS Press, St. Paul Minnesota, pp. 183–195.

Prior P, Fegan M (2005) Recent developments in the phylogeny and classification of *Ralstonia solanacearum*. Acta Hortic 695: 127–136.

Raju MRB, Pal V, Jalali I (2008). Inoculation method of *Pectobacterium carotovorum* subsp. *carotovorum* and factors influencing development of bacterial soft rot in radish. J Mycol Plant Pathol 38: 311–315.

Ranganna B, Kushalappa AC, Raghavan GSV (1997) Ultraviolet irradiance to control dry rot and soft rot of potato in storage. Can J Plant Pathol 19: 30–35.

Recep K, Fikrettin S, Erkol D, Cafer E (2009) Biological control of the potato dry rot caused by *Fusarium* species using PGPR strains. Biol Control 50: 194–98.

Reverchon S, Nasser W (2013) *Dickeya* ecology, environment sensing and regulation of virulence programme. Environ Microbiol Reports 5(5): 622–636.

Romberg MK, Davis RM, Nunez JJ, Farrar JJ (2002) Sources and prevention of *Erwinia* early drying of potato in Karen country. Calif Phytopathol 92: S70–S70.

Rousselle-Bourgeois F, Priou S (1995) Screening tuber-bearing *Solanum* spp. for resistance to soft rot caused by *Erwinia carotovora* ssp. *atroseptica* (van Hall) Dye. Potato Res 38: 111–118.

Sagar V, Sharma S, Jeevalatha A, Chakrabarti SK, Singh BP (2011) First report of *Fusarium sambucinum* causing dry rot of potato in India. New Dis Repor 24: 5.

Salas B, Secor GL (2001) In: Stevenson WR, Loria R, Franc GD, Weingartner DP (Eds.). Compendium of Potato Diseases. APS Press, St. Paul, MN, pp. 30–31.

Salas B, Secor GA, Taylor RJ, Gudmestad NC (2003) Assessment of resistance of tubers of potato cultivars to *Phytophthora erythroseptica* and *Pythium ultimum*. Plant Dis 87: 91.

Schisler D, Slininger P, Kleinkopf G, Bothast R, Ostrowski R (2000) Biological control of Fusarium dry rot of potato tubers under commercial storage conditions. Am J Potato Res 77: 29–40.

Schisler DA, Slininger PJ, Miller JS, Woodell LK, Clayson S, Olsen N (2009) Bacterial antagonists, zoospore inoculum retention time and potato cultivar influence pink rot disease development. Am J Potato Res 86(2): 102–111.

Secor G, Rivera-Varas V (2004) Emerging diseases of cultivated potato and their impact on Latin America. Latin Am J Potato (Suppl) 1: 1–8.

Secor GA, Johnson SB (2008) Seed tuber health before and during planting. In: Johnson DA (Ed.). Potato Health Management. APS Press, St. Paul, Minnesota, pp. 43–54.

Secor GA, Salas B (2001) Fusarium dry rot and Fusarium wilt. In: Stevenson WR, Loria R, Franc GD, Weingartner DP (Eds.). Compendium of Potato Diseases. APS Press, St. Paul, Minnesota, pp. 23–25.

Sharga BM, Lyon GD (1998) *Bacillus subtilis* BS 107 as an antagonist of potato blackleg and soft rot bacteria. Can J Microbiol 44: 777–783.

Sharma J, Singh RK (2005) Effect of soil texture and irrigation on development of "russet scab" on potato tubers. Indian Phytopatolh 58: 357.

Sharma J, Singh JP, Arora RK (2006) Effect of irrigation and nitrogen on russetting and superficial cracks of potato tubers. Indian Phytopathol 58: 358.

Shekhawat GS, Gadewar AV, Chakrabarti SK (1993a) Potato bacterial wilt in India. Tech Bull No 38. C.P.R.I., Shimla, 56 pp.

Shekhawat GS, Gadewar AV, Chakrabarti SK (2000) Potato bacterial wilt in India. Tech Bull No. 38 (Revised). C.P.R.I., Shimla, 45 pp.

Shekhawat GS, Gadewar AV, Bahal VK, Verma RK (1988b) Cultural practices for managing bacterial wilt of potato. Report of the Planning Conference on Bacterial Diseases of the Potato, 1987. International Potato Centre, Lima, Peru, pp. 65–84.

Shekhawat GS, Kishore V, Singh DS, Khanna RN, Singh R, Bahal VK (1982) Perpetuation of *Pseudomonas solanaclarum* in India. In: Nagaich BB et al. (Eds.). Potato in Developing Countries. Indian Potato Association C.P.R.I., Shimla, pp. 117–121.

Shekhawat GS, Kishore V, Suinaina V, Bahal VK, Gadewar AV, Verma RK, Chakrabarti SK (1988a) Latency in the management of *Pseudomonas solanaclarum*. Ann Sci Rept C.P.R.I., Shimla, pp. 117–121.

Shekhawat GS, Singh BP, Jeswani MD (1993b) Soil and tuber borne diseases of potato Tech. Bull No. 41. C.P.R.I., Shimla, 47 pp.

Shekhawat GS, Singh R, Gadewar AV (1991) Ecology and management of common scab and soft rot. Effect of cropping sequences on the incidence of common scab. Annual Scientific Report *1990-91*, Central Potato Research Institute, Shimla, pp. 82–83.

Shirsat S, Thomas P, Nair P (1991) Evaluation of treatments with hot water, chemicals and ventilated containers to reduce microbial spoilage in irradiated potatoes. Potato Res 34: 227–231.

Singh BP (1986) Studies on Fusarial wilt and Dry Rot of Potatoes (*S. tuberosum*). PhD thesis. Aligarh Muslim University, Aligarh, India.

Singh BP, Roy S, Bhattacharyya SK (1994) Occurrence of A2 mating type of Phytophthora infestans in India. Potato Res 37: 227–231.

Singh D, Sinha S, Yadav DK, Sharma JP, Srivastava DK, Lal HC, Mondal KK, Jaiswal RK (2010) Characterization of biovar/races of *Ralstonia solanacearum*, the incitant of bacterial wilt in solanaceous crops. Indian Phytopath 63(3): 261–265.

Singh N, Singh H (1981) Irrigation – a cultural practice to control common scab of potato. J Indian Potato Assoc 8: 35–36.

Singh PH (1998) Present status of wart diseases of potato in Darjeeling hills. J Indian Potato Assoc 25: 135–138.

Singh PH, Shekhawat GS (2000) Wart diseases of potato in Darjeeling hills. Tech Bull No. 19, 73 pp.

Sledz W, Jafra S, Waleron M, Lojkowska E (2000) Genetic diversity of *Erwinia carotovora* strains isolated from infected plants grown in Poland. EPPO Bull 30: 403–407.

Slininger PJ et al. (2007) Biological control of postharvest late blight of potatoes. Biocon Sci Technol 17. (6)1: 647–663.

Sokhi SS, Thind TS, Dhillon HS (1993) Late Blight of Potato and Tomato. Punjab Agricultural University, India, 19 pp.

Song J, Lee SC, Kang JW, Baek HJ, Suh JW (2004) Phylogenetic analysis of *Streptomyces* spp. isolated from potato scab lesions in Korea on the basis of 16S rRNA gene and 16S-23S rDNA internally transcribed spacer sequences. Int J Syst Evol Microbiol 54: 203–209.

Stevenson WR, Loria R, Franc GD, Weingartner DP (2001) Compendium of Potato Diseases. 2nd ed. American Phytopathology Society, St. Paul, MN.

Stevenson RW, Kirk WW, Atallah KZ (2008) Managing foliar diseases: Early blight, late blight, and white mold. In: Johnson DA (Ed.). Potato Health Management. APS Press, St. Paul, Minnesota, pp. 209–222.

St-Onge R, Gadkar VJ, Arseneault T, Goyer C, Martin F (2011) The ability of *Pseudomonas* sp. LBUM 223 to produce phenazine-1-carboxylic acid affects the growth of *Streptomyces scabies*, the expression of thaxtomin biosynthesis genes and the biological control potential against common scab of potato. FEMS Microbiol Ecol 75(1): 173–183.

Sunaina V, Pandy SK, Singh BP (2006) Bio-B5, an ecofriendly biopesticide-cum-biofertilizer for management of soil and tuber borne diseases of potato and other crops. Central Potato Research Institute Campus, Modipuram, Meerut, Extension Bull, p. 10.

Taylor RJ, Pasche JS, Gudmestad NC (2006) Biological significance of mefenoxam resistance in *Phytophthora erythroseptica* and its implications for the management of pink rot of potato. Plant Dis 90: 927–934.

Taylor RJ, Pasche JS, Gudmestad NC (2007) Susceptibility of eight potato cultivars to tuber infection by *Phytophthora erythroseptica* and *Pythium ultimum* and its relationship to mefenoxam-mediated control of pink rot and leak. Ann Appl Biol 152: 189–199.

Taylor RJ, Salas B, Gudmestad NC (2004) Differences in etiology affect mefenoxam efficacy and the control of pink rot and leak tuber diseases of potato. Plant Dis 88: 301.

Taylor RJ, Salas B, Secor GA, Rivera V, Gudmestad NC (2002) Sensitivity of North American isolates of *Phytophthora erythroseptica* and *Pythium ultimum* to mefenoxam (metalaxyl). Plant Dis 86: 797.

Thirumalachar MJ, O'Brien MJ (1977) Suppression of charcoal rot of potato with a bacterial antagonist. Plant Dis. Repor. 61: 543–546.

Thompson HK, Tegg RS, Corkrey R, Wilson CR (2014) Optimal rates of 2,4 dichlorophenoxyacetic acid foliar application for control of common scab in potato. Ann Appl Biol 165(2): 293–302.

Thompson HK, Tegg RS, Davies NW, Ross JJ, Wilson CR (2013) Determination of optimal timing of 2,4-dichlorophenoxyacetic acid foliar applications for common scab control in potato. Ann Appl Biol 163(2): 242–256.

Togoshi J, Takahashi S, Shibata Y (1988) Bacterial soft rot in red turnip *Brassica compertris* L. (rapifera group) cv. Atsumikabu, grown in burnt fields. Ann Phytopathol Soc Japan 54: 616–619.

Toukam MSG, Cellier G, Wicker E, Guibaud C, Kahane R, Allen C, Prior P (2009) Broad Diversity of *Ralstonia solanacearum* Strains in Cameroon. Plant Dis 93: 1123–1130.

Trias R, Baneras L, Montesinos E, Badosa E (2008) Lactic acid bacteria from fresh fruit and vegetables as bio-control agents of phytopathogenic bacteria and fungi. Int Microbiol 11: 231–6.

Tsror L, Nachimias A, Livescu L, Pe´rombelon MCM, Barak Z (1990) *Erwinia carotovora* subsp. *atroseptica* infection promotes *verticillium* wilt development in potato in Israel. Potato Res. 33: 3–11.

Van der Wolf JM, De Boer SH (2007) Bacterial pathogens of potato. In: Vreugdenhil D, editor. Potato Biology and Biotechnology, Advances and Perspectives. Oxford, UK: Elsevier; pp. 595–619.

van den Boogert PHJF, van Gent-Pelzer MPE, Bonants PJM, de Boer SH, Wander JGN, Levesque CA, et al. (2005) Development of PCR-based detection methods for the quarantine phytopathogen *Synchytrium endobioticum*, causal agent of potato wart disease. Eur J Plant Pathol 113: 47–57.

van Gent-Pelzer MPE, Krijger M, Bonants PJM (2010) Improved real-time PCR assay for detection of the quarantine potato pathogen, *Synchytrium endobioticum*, in zonal centrifuge extracts from soil and in plants. Eur J Plant Pathol 126: 129–133.

van Vuurde JWL, de Vries PM (1994) Population dynamics of *Erwinia carotovora* subsp. *atroseptica* on the surface of intact and wounded seed potatoes during storage. J Appl Bacteriol 76: 568–75.

Venkataramana C, Taylor R, Pasche J, Gudmestad N (2010) Prevalence of mefenoxam resistance among *Phytophthora erythroseptica* Pethybridge isolates in Minnesota and North Dakota. Am J Potato Res : 1–10.

Wale S, Platt HW, Cattlin N (2008) Diseases, pests and disorders diagostics. In: Diseases, Pests and Disorders of Potatoes. Academic Press, Boston, San Diego, p. 10.

Wale S, Platt B, Cattlin ND (2008a). Diseases, Pests and Disorders of Potatoes: A Color Handbook. Manson Publishing, London, pp. 26–27.

Walker JC (1998) Bacterial Soft Rots of Carrot. Discovery Publishing House, New Delhi, p. 78.

Wanner LA (2004). Field isolates of *Streptomyces* differ in pathogenicity and virulence on radish. Plant Dis 88: 785–796.

Wang A, Lazarovits G (2005) Role of seed tubers in the spread of plant pathogenic *Streptomyces* and initiating potato common scab disease. Am J Potato Res 82(3): 221–230.

Weber J (1990) *Erwinia*– a review of recent research. EAPR proceedings of the 11th Triennial Conf. European Assoc. Potato Res, Edinburg, U.K., 9–13 July 1990, pp. 112–121.

Weller SA, Elphinstone JG, Smith NC, Boonham N, Stead DE (2000) Detection of *Ralstonia solancearum* strains with a quantitative multiplex real time, fluorogenic PCR (Taqman) assay. Appl Environ Microbiol 66: 2853–2858.

Wharton P, Kirk W (2007) Pink rot. Michigan State University, Extension Bull E-2993, 2007.

Wharton P, Berry D, Kirk W (2005) Evaluation of seed piece fungicides for control of seed-transmitted Fusarium dry rot of potatoes. J Phytopathol 95: S110–S111.

Wharton P, Hammerschmidt R, Kirk W (2007a) Fusarium dry rot. Michigan State University. Extension Bull E-2995.

Wharton P, Kirk W, Berry D, Tumbalam P (2007b) Seed treatment application-timing options for control of *Fusarium* decay and sprout rot of cut seed pieces. Am J Potato Res 84: 237–244.

Whisson SC (2010) *Phytophthora*. In: Encyclopedia of Life Sciences. Wiley Online Library.

Wicker E, Grassart L, Coranson-Beaudu R, Mian D, Guilbaud C, Fegan M, Prior P (2007) *Ralstonia solanacearum* strains from Martinique (French West Indies) exhibiting a new pathogenic potential. Appl Environ Microbiol 73: 6790–6801.

Yuen J, Nielsen B, Ravnskov S, Kessel G, Evenhuis A, et al. (2008) The role of oospores in the epidemiology of potato late blight. Acta Hortic 834: 61–68.

28

Postharvest Diseases of Tomato and Their Management

V. Devappa,[1] C.G. Sangeetha,[1] M.R. Vinay,[1] Allada Snehalatharani,[2] N. Jhansirani[1] and P. Srinivas[1]
[1]Department of Plant Pathology, College of Horticulture, UHS Campus, GKVK, Bengaluru, India
[2]Horticultural Research Station, Kovvur, Andhra Pradesh, India

CONTENTS

28.1 Introduction

Tomato (*Solanum lycopersicum* L.) belongs to the family Solanaceae having a diploid chromosome number of 24. Western South America is considered the centre of diversity for tomato, and the cherry tomato *S. lycopersicum* var. *cerasiforme* is considered the probable ancestor of the cultivated tomato. It is the most commonly grown fresh market vegetable worldwide, covering up to an area of 7,89,000 ha with the production of 19.01 million tonnes in India (http://agricoop.gov.in/statistics/state-level). It has important vitamins like A and C. It is also rich in fibre, organic acids and antioxidants that protect against cancer. Tomatoes are susceptible to high postharvest losses ranging from 10 to 20%, so attention is needed to reduce this.

Postharvest diseases commonly occur during postharvest handling, transportation and storage. In general, crops held under any condition often pave way to diseases during the time of storage. Tomato, belonging to the *Solanaceous* family, has fruits which are rich in ascorbic acid. Tomatoes are prone to both fungal and bacterial infections before harvest as well as after harvest and a few reports have been published. These reports indicate the major postharvest diseases of tomato are Rhizopus rot (*Rhizopus stolonifer*), grey mould (*Botrytis cinerea*), anthracnose (*Colletotrichum coccoides*, *C. gloeosporoides*, *C. dematium*), bacterial soft rot (*Erwinia carotovora* pv. *carotovora*), early blight (*Alternaria solani*), southern blight (*Sclerotium rolfsii*), *Phoma* rot (*Phoma destructive*) and Fusarium rot (*Fusarium oxysporum* f. sp. *lycopersici* races 1–3).

28.2 Postharvest Diseases of Tomato

28.2.1 Grey Mould Rot

28.2.1.1 Symptoms

A localized abnormal structural change in a somatic part occurs. This abnormality is characterized by a tan to light brown watery area, which then converts to soft watery mass. Soon the skin is damaged; greyish mycelium and groups of spores develop within a short period. Later, halo, a small whitish ring, is formed around the point of entry. Such rings are usually single but turn into solid white spots whose centre contains dark-brown specks. These blotches get covered with dusty mould which causes buds, flowers and fruits to rot (Figure 28.1). *Botrytis*, a necrotrophic fungus, produces a grey-brown spore mass on infected tissue. Stem infections mainly occur when the humidity is high. The infection is seen wherever there are cracks and pruning wounds. The fungal spores can survive in the dormant form for up to 12 weeks in the pruning leaf scars and germinate when the plants are under stress. The stem lesions expand in concentric rings and later girdle the entire stem, leading to wilting just above the infection point. The petals of the flowers are very susceptible and will initiate infection to pedicels and to the immature fruit. Fruit lesions are whitish in colour and the skin usually ruptures near the middle portion of the decaying area, leading to the formation of an exposed tissue. Whitish rings appear on

FIGURE 28.1 Grey mould.

the infected fruit surface and the fruits are aborted, leading to the formation of ghost spots which reduce the market value of the crops. (Rodríguez et al. 2011).

28.2.1.2 Causal Organism

Botrytis cinerea Pers., which causes grey mould rot, is necrotrophic fungi which is mainly airborne and can attack over 200 different hosts around the globe. It produces huge amounts of asexual spores which when they come in contact with a host surface, germinate and produce an appressorium which in turn produces a penetration peg that enters the plant through the cuticle.

28.2.1.3 Disease Cycle and Epidemiology

The spores of Botrytis are mainly spread through air, waters plashes and wind. The pathogen over seasons as mycelia or sclerotia on the crop debris and organic matter. Other hosts present in and around the field may serve as pathogen reservoirs. This agent of grey mould requires high relative humidity for better spore production. The most favourable temperature required for infection is 18–24°C. When the favourable temperature exists, the infection can occur in 5 hours. Higher temperatures of 28°C and above suppress the growth and spore production.

28.2.1.4 Disease Management

28.2.1.4.1 Cultural Controls

Avoid the occurrence of blossom end rot or fruit cracking as these factors predispose the tomato to attack by the pathogen. Green tomatoes should not be stored below 15°C.

28.2.1.4.2 Chemical Control

Fungicide is recommended to manage early blight and target spot in the field (Sharma et al. 2009). Grey mould on flowers and fruit can be controlled by using antagonistic microbes. Destroying infected plant material, lowering the humidity level in storage plants and glass housing can be good measures.

28.2.1.4.3 Resistant Cultivars

Growing of resistant cultivars is the best option to avoid the occurrence of the disease (Bartz 1991).

28.2.2 Rhizopus Rot

Rhizopus stolonifera (Ehrenb.) Vuill. causes soft rot disease on many vegetables and fruits including tomato, and it rarely attacks growing plants. *R. stolonifera* grows well on crop debris, and the spores become ubiquitous and hence can be easily carried through wind. At room temperature it can rapidly spread from infected fruit to healthy fruits.

28.2.2.1 Symptoms

Initially water-soaked spots appear on fruits, and later exudation of liquid from these spots can be observed. Under wet surroundings, the lesions are covered with thin, cottony mycelia growth. The mycelium can cause infection through natural openings or mechanical wounds. The infected tomato is often covered by hairy mycelia with a cluster of black sporangia at their tips (Figure 28.2). In a short period, the infected fruit becomes colonized by other microbes (Agrios 2001). The fungus can disperse to nearby fruits, destroying the entire fruit. The entry of spores into the host usually requires wounds caused by insects or cracks or any other openings which can occur during harvest. According to Sommer (1982) the rotting progression is mainly related to temperature, with an optimum temperature of 27°C and high relative humidity. Thus the disease progress is greater during rainy season. No spore germination takes place at temperature of 4°C (Silvia et al. 2008).

28.2.2.2 Causal Organism

Rhizopus produces non-septate mycelia sporangiospores and it is a fast-growing and spreading type of fungi with white mycelia and black-coloured sporangia.

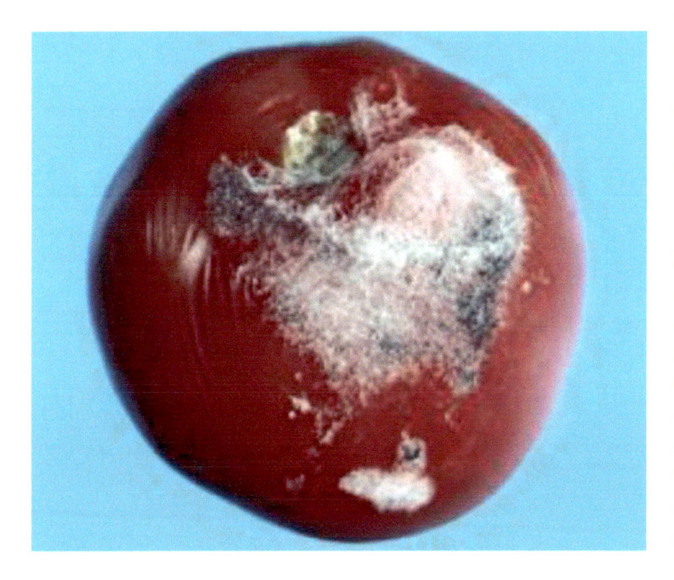

FIGURE 28.2 Rhizopus rot.

28.2.2.3 Disease Cycle and Epidemiology

The pathogen is a saprophyte which can survive for 30 years in mummified fruits. The fungus is a wound invader and can directly penetrate an uninfected fruit only when an infected fruit comes in contact with a healthy fruit. The most favourable temperature required for the infection to occur is 75–80°F. The pathogen mainly enters through cracks in the fruit made by fruit flies, and long-distance dispersal occurs by wind.

R. stolonifer is a polyphagous and saprophytic. The pathogen can grow under a wide range of temperatures, with 25°C being the optimum. However, a range of 5–25°C may favour the disease incidence to reach up to 97–100%. During cold storage the pathogen can also grow very aggressively. The pathogen can also survive for several months in fruit residues left in the storage containers.

28.2.2.4 Disease Management

28.2.2.4.1 Pre-harvest Treatments

To reduce the losses due to *R. stolonifer*, many options are available during crop production, including the selection of genetic material to be planted. Research efforts to determine if reducing field inoculum would be an effective means of controlling decay have been minimal. Management techniques to reduce loss throughout the growing season should reduce problems with postharvest pathogens that infect fruit during the growing season and survive in quiescent infections until the fruits mature. Inoculum reduction by removal of the fruit and crop debris from the infected field is another method. Field practices such as proper air circulation throughout the plant canopy, harvesting time, maturity index of tomato, disinfection, drying of containers, disinfection of harvesting tools, chlorination of water tanks and reduction of wounds and cracks on fruit are suggested to avoid or reduce the infection of the pathogen (Vigneault et al. 2000; Bartz et al. 2001).

28.2.2.4.2 Postharvest Treatments

Rhizopus infection occurs in the field and can be reduced by cleaning the fruits. In general, prevention is most important management method of *R. stolonifera* infection. To reduce the incidence, fruit injuries should be avoided. The storage containers and hydrocooling water should also be disinfected to avoid buildup of the inoculum. The harvested fruits should be transported to the storage facility as quickly as possible to avoid the occurrence of infection.

28.2.2.4.2.1 Chemical Treatments Treating the fruits with chemicals can ensure protection of the produce, but it is allowed in only a very few species. Of the different fungicidal treatments, HOCl (chlorine) is the most widely applied treatment. It is used as a general disinfectant (dip, spray or drench) but the feasibility of its commercial use has not been established. Furthermore, it is not capable of killing all spores located in injuries. In tomatoes, there is insufficient evidence on the effective results of chemical application in controlling of Rhizopus rot disease. However, the fungicide iprodione has been reported to reduce incidence of Rhizopus rot up to 59%. Yet in some countries, this fungicide has been banned as its registration was never obtained (Mallek et al. 1995). Dichloran is another fungicide used to reduce *Rhizopus* rot. Due to the consumer resistance to consume chemically treated horticultural commodities, control of postharvest diseases has taken new directions and products are being developed which are less harmful to human health and environment. Further, due to pathogen resistance to chemicals and occurrence of new races, a search for different methods to manage the plant diseases and capable of integrating or replacing the chemicals is needed (Mari and Guizzardi 1998). For the pathosystem *R. stolonifera* tomato, alternative methods such as biological control, application of essential oils and plant extracts, chitosan, UV-C and heat treatment are initiating or still under experimentation.

Washing the harvested tomatoes with at least 150 ppm free chlorine at pH 6.5–7.5 is recommended. Sanitizing the packing line equipment and packing containers with concentrated solutions of bleach (diluted to 0.5–1% NaOCl) for several minutes will reduce the pathogen infection.

28.2.2.4.2.2 Nitrous Oxide Qadir and Hashinaga (2001) reported the use of nitrous oxide for suppression of the disease. They also recommended low temperature storage and rapid pre-cooling for suppression of the pathogen, Low temperature as the first instance, low storage temperature has been recommended to reduce Rhizopus rot disease. Temperature seems to reduce the rapid development of *R. stolonifera* in tomatoes. However, tomatoes are chilling-sensitive and they cannot be stored bellow 12°C. There are different recommendations about the best storage temperature for this commodity according to cultivar and tomato stage of maturity. Rapid pre-cooling temperatures are also recommended to reduce infection by this fungus (Qadir and Hashinaga 2001).

28.2.2.4.2.3 Heat Treatment Heat treatment has been used in tomato to manage the disease. The fruits subjected to hot air at various temperatures decreased the disease severity. The best treatment was when tomatoes were exposed to 60°C for 6 min and 70°C for 1–15 min. The study reported correlation between the temperature and time (Aborisade and Ojo 2002).

28.2.2.4.2.4 UVC Radiation The use of ultraviolet (UV) radiation in controlling postharvest disease has advanced as an alternative technology to chemical fungicides. Low dose UV-C (100–280 nm) is reported to reduce postharvest disease and induce resistance in various horticultural commodities. Stevens et al. (1997) reported that the integration of UV-C treatments at 254 nm (dose 3.6 kJ/m^2) with other alternative methods such as yeast application (*Debaromyces hanseii*, 10^8 cell/mL) and calcium chloride (2%) decreased the percentage of infection caused by *R. stolonifer* more than four times in inoculated tomatoes (Silvia et al. 2008).

28.2.2.4.2.5 Biological Control To date, for Rhizopus rot disease on tomatoes there is no existing commercial bioagent. Nevertheless, Schena et al. (2000) reported effective control of Rhizopus rot in cherry tomatoes by various antagonists (isolates 182, 304, 495, LS15 and 320) obtained from different fruits and vegetables in Italy and Israel.

28.2.2.4.2.6 Plant Derivatives Most plants produce an array of antimicrobial metabolites in reaction to pathogen entry or stress. The essential oils produced are biologically active and endowed with antimicrobic properties. These antimicrobial compounds are present in different plant parts. Hexanal and benzaldehyde, formed by stone fruits, have antifungal activity and control *R. stolonifer* effectively.

28.2.2.4.2.7 Phytoalexins The phytoalexins are antimicrobial and antioxidative which are usually lipophilic and nonspecific in their antifungal activity (Smith 1996). Phytoalexins are not reported in healthy plant tissue and are usually produced only after pathogen attack. The production of phytoalexins is the induced defence response noticed in disease resistance in the crop system. Steroidal glycoalkaloids are reported in the crops belonging the family *Solanaceae*, which mainly include tomato and potato. Tomato plants contain α-tomatine which mainly protects tomato plants against attack by α-tomatine-sensitive fungi.

28.2.2.4.2.8 Chitosan Chitosan is a linear polysaccharide made by treating chitin shells of crustaceans, which is used for managing the postharvest rots of crops during storage. It delays the initiation of the infection and restricts the spread of the disease (Herna´ndez-Lauzardo et al. 2006). Coating the fruits with chitosan coating at 2% delays ripening and reduces decay incidence. Application of chitosan delayed Rhizopus development compared to the untreated tomato fruit (Bautista and Luna 2004).

28.2.2.4.2.9 Cultural Control To control *R. stolonifer*, certain prevention measures must also be followed. These practices include burying the crop debris from previous crops and damaged fruit from the present crop, cleaning and disinfecting fruit containers, removing fruit debris remaining in the containers, cleaning and disinfecting the packing equipment and removing plant debris from around packing lines.

28.2.2.4.2.10 Resistant Cultivars Cultivars which exhibit resistance to cracking, roughness or water uptake are less susceptible to *Rhizopus* rot.

28.2.3 Anthracnose

Anthracnose, which causes dark sunken lesions on leaves, stem, flowers and fruits, occurs on many crops in a destructive manner. Anthracnose causes significant losses in the tropical and subtropical areas of the world. The losses caused by *Colletotrichum gloeosporioides* in various crops are substantial throughout the world.

28.2.3.1 Symptoms

Anthracnose can cause lesions of different colours which are visible on ripe to overripe fruits. Fruit infection occurs during immature stage of the fruit, but the symptom development occurs as the fruits ripen. These lesions are small, sunken, circular spots which enlarge in size up to a half inch in diameter with zonate markings. Lesion surface appears

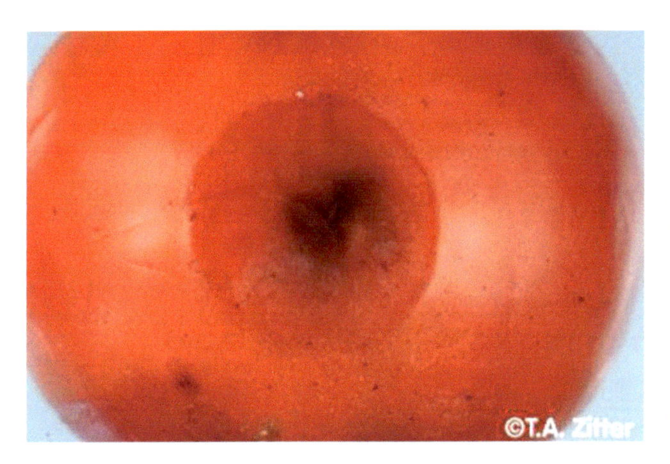

FIGURE 28.3 Anthracnose.

salmon-coloured due to the production of spores, and the centres of older spots become blackish (Figure 28.3). It is caused by *Colletotrichum* (*Gloeosporium*) or *Glomerella* fungi.

28.2.3.2 Causal Organism

The disease is caused by *Colletotrichum coccoides*, *C. gloeosporioides* and *C. dematium*. The colonies appear white with aerial mycelium with dark pigmentation. The colonies consist of numerous black sclerotia which are spherical. The conidia are straight, fusiform, attenuated at the ends, and the appressoria looks like clavate, tan and vary in shape.

28.2.3.3 Disease Cycle and Epidemiology

The pathogen mainly survives as sclerotial and hyphae in the crop debris. In late spring the lower leaves and fruit may become infected by germinating sclerotia and spores in the soil debris. The primary infections which occur on the lower leaves of the plants serve as important sources for secondary spread throughout the growing season. The older leaves with early blight infections and flea beetle injury also serve as sources of infection because the fungi can colonize and produce new spores in these wounded areas.

High temperature coupled with high relative humidity, wounds and cracks on the fruits are predisposing factors for the spread of this disease. Temperature of 30°C and relative humidity of 85–100% are most favourable for conidial germination.

28.2.3.4 Disease Management

28.2.3.4.1 Cultural Practices

Field sanitation is the most effective cultural practice to keep majority of the diseases at the economic threshold level. One of the most important methods is use of disease-free seeds which are certified. Since anthracnose is a seed-borne disease, healthy seeds should be used for sowing. Follow rouging of plants which are severely infected. Collect rotten fruits those that have dropped and bury them in a deeper layer of the soil. Remove all the crop residues after harvest and clean all farm tools.

28.2.3.4.2 Hot Water Seed Treatment

Since the pathogen is seed borne, hot water treatment can reduce the fungal infection by soaking the seeds in hot water at 50°C for 25 minutes.

28.2.3.4.3 Biopesticides and Physical Methods

There are many liquid copper fungicides which are recommended for the management of the disease. Seeds may also be treated well before planting. Neem oil, an organic spray, can also be used to prevent fungal attack. The most common copper compounds that can be used for the management of the disease are copper hydroxide and copper oxychloride.

28.2.4 Early Blight

28.2.4.1 Symptoms

Alternarias solani produces "bulls-eye" patterned spots on fruits. The disease is commonly observed in storage and usually injured fruits are affected. *Alternaria alternata* survives and reproduces on organic matter, and it can also survive on the dead and older leaves in the field. The infected area around the infection or the entire leaf may turn yellow or chloratic, as *A. alternata* is a weak pathogen which will colonize only the moribund tissue. The pathogen infects the immature fruit, and the symptoms become visible only after the fruit has ripened. As the disease advances, the symptoms are seen on the plant stem as well as on the fruits. Sometimes, though, the infections occur on the green fruit – only one or two epidermal cells become infected and the spots do not spread further even after the fruit ripens (Figure 28.4).

During warm, humid weather, the pathogen sporulates and forms a black, velvet-like layer on the surface of the sunken lesions. Spores are often formed on the shallow surface spots. Infected fruit frequently drop, and losses of 50% of the immature fruit may occur.

FIGURE 28.4 Early blight fruit rot.

28.2.4.2 Causal Organism

The mycelium is haploid and septate, and later becomes dark with pigmentation. The asexual spore is conidia which can be borne singly or in a chain of two on distinct conidiophores. The beaked conidia usually have 9–11 transverse septae.

28.2.4.3 Favourable Conditions

Free water is a prerequisite for the *Alternaria* spores to germinate; spores are not able to cause infection on dry leaves. The spores can germinate within 2 hours at a wide range of temperatures, but at a temperature range of 26.6–29.4°C it will only take a half-hour. Depending on temperature, another 3–12 hours are required for the fungus to penetrate the plant. After penetration, lesions may form within 2–3 days or the infection can remain dormant awaiting proper conditions. *Alternaria* sporulates best at about 26.6°C, when abundant moisture is present. Poorly nourished plants are more susceptible to infection.

28.2.4.4 Disease Cycle and Epidemiology

Alternaria solani is a necrotrophic fungus which reproduces asexually by producing conidia. The fungus overwinters in the plant debris or on wild relatives of the *Solanaceous* family. The overwintering multicellular conidia are splashed by water or by wings from the infected plants. The infection takes place by the conidia, which enters through small wounds or through direct penetration. The primary infections usually start on the older leaves which are present close to the ground level. The fungus gradually grows and forms a spot. From this spore, more conidia are produced which are released. The released conidia infect the plants and other plant tissues in the same growing season. This is especially important when fruits or tubers are infected, as they can spread the disease. In general, development of the pathogen can be aggravated by an increase in inoculum from alternative hosts such as weeds or other *Solanaceous* species. Mature plants show higher disease severity and prevalence.

28.2.4.5 Disease Management

28.2.4.5.1 Cultural Control

Cultural practices which encourage dense leaf canopies and growing of varieties which develop and retain a bigger canopy aid in preventing the disease by preventing the formation on dew on the fruits. Harvesting time is very crucial, as the more time the fruit remains in the field after ripening, the more chance of infection. Harvesting the tomatoes as soon as the fruit ripens is a good practice to minimize infection. The late harvest varieties are more vulnerable to severe losses.

The following cultural practices help to control the spread of Alternaria blight in tomato,

- Clear the plant debris from field to minimize the inoculum available for the next crop.
- Water the plants in the morning so that the plants remain wet for only a short time.
- Use a drip irrigation system to reduce leaf wetness, which is the most crucial condition for growth of the fungus.
- Use soil mulch so that spores in soil cannot splash on the leaves from the soil.
- Follow crop rotation with a non-solanaceous crop for a minimum of 3 years.
- Avoid wild population of Solanaceae.
- Closely monitor the field, especially in warm damp weather when it grows quickly to reduce loss of crop, and spray a chemical in a timely fashion.
- Plant available resistant cultivars.
- Increase the air circulation in the field. Wet conditions are desirable for the growth of *A. solani* and the disease spreads more rapidly when such conditions exist. This can be achieved by following proper spacing and by trimming the leaves.

28.2.4.5.2 Chemical Control

Since it is not possible to predict the weather parameters, use of chemicals for disease management is very necessary. Numerous fungicides are available in the market for managing the disease, *viz*., azoxystrobin, pyraclostrobin, chlorothalonil, copper products, hydrogen dioxide (Hydroperoxyl), mancozeb, potassium bicarbonate and ziram. The recommendations for each of the products should be read and followed carefully. Quinone outside inhibit or fungicides, e.g., azoxystrobin, are widely used due to their broad spectrum activity.

28.2.5 Phoma Rot

28.2.5.1 Symptoms

Phoma rot can be observed on all the parts above the ground. On leaves the symptoms appear as numerous small, dark brown to black sports which develop concentric rings as they enlarge. The first infection starts on the older leaves and can spread to all the other leaves. However, all the leaves are susceptible, and defoliation can result when the disease is more severe. The leaf spot looks very similar to that caused by early blight, except that the Phoma lesions contain numerous minute black fungal fruiting bodies (pycnidia). Dark brown lesions with concentric rings appear on the stems and both the green and ripe fruit can be infected. Fruit lesions usually develop at the calyx end as small sunken lesions which later develop into sunken, black, leathery lesions with numerous pycnidia in the centre.

28.2.5.2 Causal Organism

The disease is caused by *Phoma destructive* and the pathogen produces septate, branching and hyaline mycelium which becomes dark with age. It produces pycnidia, which is subcutaneous and later becomes erumpent and dark subglobose and produces hyaline pycnospores.

28.2.5.3 Disease Cycle and Epidemiology

The fungus mainly survives in the soil from one season to another and also on infected plant debris and closely related weeds. Injury to the plant due to pruning, insect feeding, mechanical damage or cracking often makes the plant susceptible to fungal infection. A temperature of 20°C and optimum humidity leads to the production of masses of fungal conidia which are released from the pycnidia. Such released conidia can easily disseminate by rain, overhead irrigation and on workers' clothing and equipment. Decreased soil nitrogen and phosphorus levels make the plants susceptible (Farr and Rossman 2016). Moderate temperature and high humidity are favourable conditions for the spread of this disease.

28.2.5.4 Disease Management

The recommended fungicide spray schedule with good sanitation practices will help in reducing the losses due to this disease. Preventing fruit injury at the time of harvest and picking dry fruits to minimize the spread of the disease in storage containers can also help reduce losses. Maintaining good soil fertility, following long crop rotations and removing all related weeds from the field can help in reducing the losses as well.

28.2.6 Southern Blight

28.2.6.1 Symptoms

Initially symptoms appear as a dark brown lesion on the stem near the soil surface. The lesion girdles the stem, causing leaf yellowing and wilting. White mats of fungal growth are produced on the stem and nearby in the soil on any organic debris. Wilting in infected plants may be more evident when soils begin to dry out. After a few days, mustard seed-sized, round, dark brown overwintering structures known as sclerotia appear on the white fungal growth. The abundant sclerotia that form on the outside of the stem tissue are a good diagnostic feature.

They are round, soft, and smaller in diameter, and lighter in colour than those caused by another fungus, *Sclerotinia sclerotiorum*, which also causes a disease in tomato. The sclerotia of the latter are found inside tomato stem tissue.

28.2.6.2 Causal Organism

The disease is caused by *Sclerotium rolfsii*. *S. rolfsii* is a necrotrophic, soilborne fungal plant pathogen that produces white mycelium on infected plants. Cells are hyaline with thin cell walls and sparse cross-walls. Main branch hyphae may have clamp connections on each side of the septum. Mature sclerotia are hard and have a distinct rind. Although most sclerotia are spherical, some are slightly flattened or coalesce with others to form an irregular sclerotium. *S. rolfsii* does not form asexual fruiting structures or spores.

28.2.6.3 Disease Cycle and Epidemiology

Sclerotia serve as the principle overwintering structures and primary inoculum for disease. Persisting near the soil surface, sclerotia may exist free in the soil or in association with plant debris. Those buried deep in the soil may survive for a year or less, whereas those at the surface remain viable and may germinate in response to alcohols and other volatiles released from decomposing plant material. Thus, deep ploughing serves as a cultural control tactic by burying sclerotia deep in the soil. High temperatures and moist conditions are associated with germination of sclerotia. High soil moisture, dense planting, and frequent irrigation promote infection. *S. rolfsii* can overwinter as mycelium in infected tissues or plant debris. It usually persists as sclerotia. Sclerotia are disseminated by cultural practices, infested transplant seedlings, water, wind, and possibly on seeds. In addition, a small percentage of sclerotia may survive passage through sheep and cattle, and thus could be spread through fertilizers.

The fungus is soilborne and also survives on machinery or water-moved infested soil, numerous weeds and crop hosts. Wet periods of high temperatures (85–95°F) favour the spread. The fungus affects many crops, including tomato, other Solanaceous crops such as potato, pepper, and eggplant, legumes and cucurbits. The pathogen persists on crop residues and as dormant sclerotia. The fungus infects plants either directly or through wounds caused by nematodes or insects. The fungus is spread into a field by infested soil or cultivating tools, infected transplants, running water, and as sclerotia mixed with seeds. High temperatures (above 30°C) and high soil moisture favour disease development, while low soil moisture favours survival of the sclerotia. The germination of sclerotia is most abundant at the soil surface and drops off with depth in the soil.

Optimum temperature required for penetration and infection is 27–30°C if moisture and relative humidity are present on the host surface. The long distance spread of the disease occurs through infested soil and infected planting material. Sclerotia usually survive below freezing temperatures, whereas mycelium does not survive, and the sclerotia will spread from infected to uninfected are as by soil, animals, wind and water.

28.2.6.4 Disease Management

28.2.6.4.1 Cultural Practices

Survival of the pathogen depends upon the method of crop rotation followed. Some cultural practices include growing of cereal crops and cotton, which are nonhosts for the pathogen. Application of green manure to the crop is important and allows few months' time for breakdown of green manuring before planting. Crop debris, including infected materials, should be buried in the soil to a depth of 1 foot. Follow the recommended spacing and avoid dense planting to minimize the relative humidity. The disease incidence levels have been reduced by application of ammonium nitrate before planting or side dressing at monthly intervals. Under field conditions, all the equipment used for cultural operations, etc. must be cleaned and free from pathogens. Select the fields that are free from the Sclerotium infection; since the sclerotia survive in the soil fora period of 3–4 years, avoid distribution of inoculum from infected areas to noninfected areas.

28.2.6.4.2 Soil Treatment

Manipulation of soil by adding organic amendments, application of biological agents, and also exposing the soil to sunlight (soil solarization) helps in management of the disease. If the crops are grown in greenhouses or commercial nurseries, treat the nursery beds with aerated steam, and areas must be brought to a temperature of 170°F for 30 minutes. Soil solarization reduces the sclerotia present in the soil and also reduces the population of other soilborne disease causing pathogens, plant parasitic nematodes and weeds.

28.2.6.4.3 Chemical Control

Application of azoxystrobin applied before and after planting for furrow treatments significantly reduces the southern blight of tomato. Soil fumigants or fungicides are commonly used for management of the disease. However, application of fungicides for large areas is not economical and practical under Indian conditions. Even after application of fumigants to the soil, the sclerotia survive in the soil and they have to be applied every year.

28.2.6.4.4 Biological Control

The use of biological agents has been shown to be very effective against soilborne pathogens under *in vitro* conditions, whereas they are less effective under field conditions for various reasons. Biological management of *S. rolfsii* has been achieved by using *Bacillus subtilis*, mycorrhizal fungus and a few species of *Trichoderma*. The quantity required for field application is very high and not economical under field conditions, whereas these management practices are well suited for commercial crops or economically important crops.

28.2.6.4.5 Resistant Cultivars

The most important disease management strategies are use of resistant varieties or cultivars. Unfortunately, the fungus has a wide host range and it is very difficult to expect high levels of resistance. Management of the pathogen is very difficult when the conditions are favourable for disease development, and inoculum quantity is more in the field. Avoid fields which are sick or infected in the initial stage only. Growing of non-host crops like millet for 2 years will minimize the pathogen population in the soil.

28.2.7 Bacterial Soft Rot and Hollow Stem

28.2.7.1 Symptoms

The symptoms are initially very small spots which are watery, and later soft decay of fruit can be observed. The infected area may be surrounded by yellow halo, and during wet conditions the infected area appears to be scorched. The lesions present on fruits are raised, brown in colour and sunken and later give rise to scab appearance. Finally, the epidermal layer of the fruit ruptures and leads to secondary infection by *Erwinia carotovora*. When the disease advances, it enlarges rapidly and the entire fruit becomes a soft, watery mass. Pathogen liquefies fruit tissue by breaking down the pectate "glue" that holds plant cells together, and affected fruits become slimy and collapse (Figure 28.5).

FIGURE 28.5 Soft rot.

28.2.7.2 Causal Organism

Bacterial spot of tomato caused by *Erwinia carotovora* subsp. *carotovora* is an important disease of tomatoes in humid conditions (Opara and Obani, 2009). It is a Gram-negative rod-shaped bacterium, non-spore forming and having peritrichous flagella. It is facultative anaerobic bacteria producing a number of extracellular cell wall degrading enzymes.

28.2.7.3 Disease Cycle and Epidemiology

The pathogen overwinters in the infected debris in the field condition, on contaminated tools and equipment used for intercultural operations, and also in certain insects. The bacteria primarily enters through wounds made during cultivation of the crops. When free moisture is present on host and in high humid conditions, uninjured tissues may become infected. Rains, ill drained soils and warm temperatures favour the disease severity under field conditions and humidity during transport and storage also increase the pathogen development. The tomato infected with Oomycets fungi (*Pythium* and *Phytopthora* spp.) creates favourable conditions for the development of soft rots.

28.2.7.4 Disease Management

28.2.7.4.1 Pre-Storage Heat Treatments

The use of pesticides in the management of pests and diseases has brought about some major problems. Hence, heat treatment is an alternative method which has been strongly recommended for disinfestations and disease management for fruit-borne pathogens (Opara and Amadioha 2006).

Fruits treated with mild heat stress recorded higher efficacy and the method was more effective and safer than other methods of management with beneficial effects during preservation of fruits in storage (Aborisade and Oguntemahin 2005). Gradual heating process of the fruits improved with disinfestations resulted in maintaining the quality of the fruits and also increased the colour development of the fruits (Williams et al. 1994).

28.2.7.4.2 Use of Botanicals

Botanicals or plant products are being used for the management of stored pests of agricultural crops around the globe (Golob and Webley 1980). This tradition has been neglected by the farmers with the entry of synthetic pesticides. However, the ecological problems and the increase in pesticidal resistance led to a search for new group of management practices which have lower mammalian toxicity and minimum persistence in the environment. Research has now focused on plant-based fungicides (Olajede et al. 1993). The most important compound extracted from this is a zadirachtin extracted from *A. indica*. Microbial activity of the different botanicals has shown the importance of natural chemicals which are nonphytotoxic in nature and easily biodegradable and these are alternatives to synthetic pesticides. Further, exploitation of plant-based pesticides is increasing in the field of plant pathology (Olajede et al. 1993; Amadioha 2004). Synthetic pesticides are costly in addition to the hazards involved, hence plant-based pesticides are readily available and cost effective in developing countries. Using these synthetic pesticides is beyond the reach of resource-poor farmers who produce over 98% of the food consumed. Development of pesticides of plant origin will be in expensive and readily available to poor farmers (Amadioha 2004).

28.3 Conclusion

Most fruit rot fungi and bacteria also cause foliar diseases, so management throughout the season is critical not only to produce a healthy plant and high yield, but also to provide high-quality fruit. The symptoms of postharvest diseases may not appear until after harvest; many of the disease-causing organisms that cause fruit rot infect tomatoes before they ripen. Once fruit is infected, little can be done during harvest or storage to reduce the damage. Management measures must be employed by vegetable growers, marketers and consumers at the time of harvesting, transportation, handling, storage and processing of tomato fruits.

REFERENCES

Aborisade AT, Oguntimehin, MO (2005) Efficiency of heat treatment on storage life extension ripening and decay of pineapple (*Ananas cosmosus* L.) fruit by *Trichodema harizianum*. In: Proc Nigerian Soc Plant Protec. Port Harcourt, Nigeria. p. 20.

Aborisade AT, Ojo FH (2002) Effect of postharvest hot air treatment of tomatoes (*Lycopersicon esculentum* Mill) on storage life and decay caused by *Rhizopus stolonifer*. *Z. Pflanzenkr. Pflanz* 109: 639–645.

Agrios GN (2001) *Fitopatología*. Second Edition. Limusa. Mexico, D.F. 809 pp.

Amadioha AC (2004) Control of black rot of potato caused by *Rhizoctoria bataticola* using some plant extracts. *Arch Phytopathol Plant Protect* 37: 111–117.

Bartz JA (1991) Postharvest diseases and disorders of tomato fruit. In: Jones JB, Jones JP, Stall RE, Zitter TA (Eds.). *Compendium of Tomato Diseases*. APS Press, St. Paul, MN.

Bartz JA, Eayre, CG, Mahovic MJ, Concelmo DE, Brecht JK, Sargent SA (2001) Chlorine concentration and the inoculum of tomato fruit in packinghouse dump tanks. *Plant Dis* 85: 885–889.

Bautista BS, Luna BL (2004) Evaluación del quitosano en el desarrollo de la pudricíónblanda del tomatedurante el almacenamiento. *Rev Iberoamer Tecnol Postcos* 1: 63–67.

Farr DF, Rossman AY (2016) Fungal databases. In: Systematic Mycology and Microbiology Laboratory. ARS USDA. Available via: http://nt.ars-grin.gov/fungaldatabases/. Accessed 10 June 2016.

Golob P, Webley SJ (1980) The use of plants and minerals as traditional protectants of stored products. *Tropical Products Institute Report*. Tropical Products Institute, Great Britain, 32 pp.

Hernández-Lauzardo AN, Bautista-Baños S, Velázquez-del Valle MG, Trejo-Espino JL (2006) Identification of Rhizopus stolonifer (Ehrenb.: Fr.) Vuill., causal agent of Rhizopus rot disease of fruits and vegetables. *Mex J Phytopathol*. 24: 65–69.

Mallek AY, Hemida SK, Bagy MK (1995) Studies on fungi associated with tomato fruits and effectiveness of some commercial fungicides against three pathogens. *Mycopathologia* 130: 109–116.

Mari M, Guizzardi M (1998) The postharvest phase: Emerging technologies for the control of fungal diseases. *Phytoparasitica* 26:59–66.

OISAT. Online Information Service for Non-Chemical Pest Management in the Tropics. Available at: www.oisat.org

Olajede F, Engelhradt G, Wallnofer PR, Adegoke GO (1993) Decrease of growth and toxic production in *Aspergillus parasiticus* caused by spices. *World J Microbiol Biotechnol* 2: 60–66.

Opara EU, Amadioha AC (2006) Effect of heat treatment of seed on disease severity, germination and yield of tomato in Umudike, *Nigeria*. *J Sustain Trop Agric Res* 17: 47–52.

Opara EU, Obani FT (2009) Performance of some plant extracts and pesticides in the control of bacterial spot diseases of *Solanum*. *Agric J* 5(2): 45–49.

Qadir A, Hashinaga F (2001) Inhibition of postharvest decay of fruits by nitrous oxide. *Postharvest Biol Technol* 22: 279–283.

Rodríguez A, Acosta A, Rodríguez C (2011) Fungicide resistance of *Botrytis cinerea* in tomato greenhouses in the Canary Islands and effectiveness of non-chemical treatments against gray mold. *World J Microbiol Biotechnol* 30(9): 397–406.

Schena L, Ippolito A, Nigro F, Pentimone I, Salerno M (2000) Efficacy of endophytic isolates of *Aureobasidium pullulans* in controlling storage rots of sweet cherries. In: Proceedings of the Fifth Congress of the European Foundation for Plant Pathology, Taormina-Giardini Naxos (Catania), Italy, pp. 527–530.

Sharma RR, Singh D, Singh R (2009) Biological control of postharvest diseases of fruits and vegetables by microbial antagonists. A review. *Bio Control* 150(3): 205–221.

Silvia B, Miguel G, Valle VD, Ana N, Lauzardoa H, Barka AT (2008) The *Rhizopus stolonifer* tomato interaction. *Plant Microbe Interact* 3(4): 269–289.

Smith CJ (1996) Accumulation of phytoalexins: Defence mechanism and stimulus response system. *New Phytol* 132: 1–45.

Sommer NF (1982) Postharvest handling practices and postharvest diseases of fruit. *Plant Dis* 66: 357–364.

Stevens C, Khan VA, Lu JY, Wilson CL, Pusey PL, Igwegbe CK, Kabwe K, Mafolo Y, Liu J (1997) Integration of ultraviolet (UV-C) light with yeast treatments for control of postharvest storage rots of fruits and vegetables. *Biol Control* 10: 98–103.

Vigneault C, Bartz JA, Sargent SA (2000) Postharvest decay risk associated with hydrocooling tomatoes. *Plant Dis* 84:1314–1318.

Williams MH, Brown MA, Vesk M, Brady C (1994) Effect of postharvest heat treatment on fruit quality, surface structure, and fungal disease in valence organs. *Aust J Exp Agric* 34: 1183–1190.

29

Postharvest Diseases of Capsicum *sp. and* Brinjal *and Their Management*

C. G. Sangeetha,[1] V. Devappa,[1] Allada Snehalatharani,[2] S. E. Navyashree[1] and Tirukovalur Sundeep[1]
[1]*Department of Plant Pathology, College of Horticulture, UHS Campus, Bengaluru, India*
[2]*Horticultural Research Station, Kovvur, Andhra Pradesh, India*

CONTENTS

29.1 Introduction

Capsicum has been known since the beginning of civilization in the Western Hemisphere. It has been a part of the human diet since about 7500 BC (MacNeish 1964). The genus *Capsicum* originated in the American tropics and is propagated throughout the world (Pickersgill 1997). Capsicum (chillies and other peppers) belongs to the family *Solanaceae* (tribe Solaneae, subtribe Capsicinae), which also includes other economically important crops such as tomato, potato and tobacco (Dias et al. 2013). They consist of annual or perennial herbs or shrubs and are native to South and Central America and the Galapagos (Walsh and Hoot 2001). They are predominantly diploid ($2n$ = 24, infrequently $2n$ = 26), except for a few (Moscone et al. 2003). The genus *Capsicum* can be grouped into different categories based on the ability of members to successfully interbreed. These include Annuum, made up of the species *C. annuum* (varieties *glabriusculum* and *annuum*), *C. frutescens*, *C. chinense*, *C. chacoense* and *C. galapagoensis*; the baccatum group, which consists of the species *C. baccatum* (varieties baccatum, pendulum and praetermissum) and finally *C. tovari*, and the pubescens group, which is also made up of the species *C. cardenasii*, *C. eximium* and *C. pubescens* (Pickersgill 1997). The five major domesticated species are *C. annuum*, *C. baccatum*, *C. chinense*, *C. frutescens* and *C. pubescens*, of which *C. annuum* is the most widely cultivated species worldwide followed by *C. frutescenes* (Bosland and Votava 2003).

The fruit of *Capsicum* is referred to by a variety of names, including "chilli", "chilli pepper" or "pepper", depending on the place. Capsicum species can also be divided to in several groups based on fruit/pod characteristics ranging in pungency, colour, shape, intended use, flavour, and size. Despite their vast trait differences most cultivars of peppers commercially cultivated in the world belongs to the species *C. annum* L. (Smith et al. 1987).

Capsicum annum L. is mainly cultivated in tropical and subtropical countries and is one of the most important constituents of the Indian cuisine, grown in almost all the states of India. Plants are bushy and grow up to 60–80 cm tall. They are semi-perennials but are usually grown as annuals in cultivation. Fruits are rich in vitamins (A, C and E), folic acid and potassium but low in sodium and calories. Chilli is used in many cuisines and is also found to have many medicinal properties. Green chillies are a rich source of vitamins, especially vitamin A, C, B_1 and B_2 and is also rich in vitamin P (rutin), which is of immense pharmaceutical importance. Pungency in chilli is due to the presence of capsaicin, a digestive stimulant and a cure for rheumatic troubles.

Major chilli producing states in India are Andhra Pradesh, Maharashtra, Karnataka, and Madhya Pradesh. Apart from being a large consumer and producer of chilli, India is also the largest exporter of the crop. Over 30% of the chilli produced in India is exported to countries of west Asia, east Asia, the United States, Sri Lanka and Bangladesh, most commonly in dried form. Chilli is a vegetable as well as a spice and is one of the most important cash crops of India. It is also used for industrial purposes for extraction of oleoresin.

29.2 Postharvest Diseases of Capsicum

29.2.1 Anthracnose

Anthracnose is considered to be the most economically important disease and a major constraint in the production of chilli and capsicum throughout the world. A loss of 20–80% has been accounted for from Vietnam (Don et al. 2007) and about 10% from Korea (Byung 2007). An annual loss of about 29.5%, amounting to US$491.67 million (Garg et al. 2014); calculated loss of 10–54% in yield (Lakshmesha et al. 2005; Ramachandran and Rathnamma 2006) and yield losses of up to 50% (Pakdeevaraporn et al. 2005) have been reported from various parts of India. Anthracnose is mainly a problem in mature fruits, causing severe losses due to both preharvest and postharvest fruit decay (Hadden and Black 1989; Bosland and Votava 2003). *Colletotrichum capsici* is the most important plant pathogen worldwide, causing the economically important disease anthracnose, which is also referred to as die-back or fruit rot, leaf spot and wilt.

29.2.1.1 Symptoms

Fruit lesion is the most economically important aspect of the disease, as even a small lesion on the fruit may reduce the market value, thereby affecting the profitable yield of the crop (Manandhar et al. 1995). The disease is reported to affect almost all aerial parts of the plant. It mainly causes fruit rot at both green and red stages, primarily attacking ripe fruits, hence it is also known by the name ripe fruit rot (Agrios 2005). The disease is seed-, soil-, water- and airborne. Fruit rot symptoms are more conspicuous and are referred to as ripe fruit rot. Fruit rot causes severe damage to mature fruits in the field as well as during transit and storage. The first sign of infection is the appearance of a small, black, circular spot which is generally sharply defined, or at times diffused. Typically, the symptoms first appear on mature fruits as small, circular, yellowish to pinkish, sunken water-soaked lesions that rapidly expand. These spots spread in the direction of the long axis.

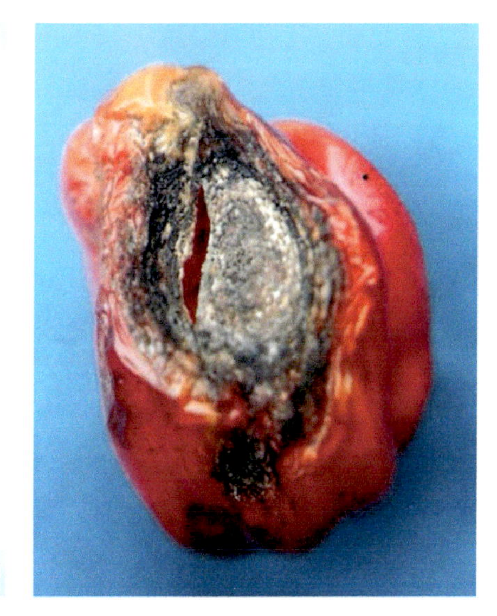

FIGURE 29.1 (a) Anthracnose symptoms on chilli. (b) Anthracnose symptoms on bell pepper.

As the fruit matures, the spots become brownish to black. Severely infected fruits look straw coloured and bear numerous dots like acervuli in concentric rings. The seeds produced in such fruits are discoloured and covered with mycelial mat (Figure 29.1a and 29.1b).

29.2.1.2 Causal Organism

Several species of *Colletotrichum* are associated with anthracnose disease, including *C. acutatum*, *C. coccodes*, *C. dematium* and *C. gloeosporioides* in Korea (Park and Kim 1992). *Colletotrichum capsici* is found to be prevalent in red chilli fruits, whereas *C. acutatum* and *C. gloeosporioides* cause infections in both young and mature chilli fruits (Hong and Hwang 1998; Kim et al. 1999, 2004; Park et al. 1990; Than et al. 2008). Among these species, *C. gloeosporioides* and *C. acutatum* are the most destructive and widely distributed (Sarath Babu et al. 2011; Voorrips et al. 2004). However, in India, primarily three important species, namely, *C. capsici*, *C. acutatum* and *C. gloeosporioides*, have been reported to be linked with the disease, with *C. capsici* Syd. Butler and Bisby causing major damage at the ripe fruit stage of the plant (Ranathunge et al. 2012; Saxena et al. 2014).

Colletotrichum capsici (Syd.) Butler and Bisby (Tel: *Glomerella cingulate* [Stoneman] Splaud and Schrenk) mycelium is septate, intercellular as well as intracellular, and aerial mycelium appears light to dark grey in colour. Acervuli are round and elongated in shape. Setae are scattered, brown, 1–5 septate, rigid and swollen at base and acute at apex. Conidiophores are short, hyaline to faintly brown, cylindrical, septate or aseptate. Conidia are falcate, fusiform with acute apices and narrow truncate base. They are one-celled, hyaline and uninucleate (Figure 29.2).

29.2.1.3 Disease Cycle and Epidemiology

The pathogen survives in the infected plant debris as well as in the infected seeds. The fungus can survive in plant debris in the soil for a minimum of 9 months and serve as source of primary infection, whereas secondary infection takes place through windborne conidia. Generally, infection occurs during warm, wet weather. Temperatures around 27°C and high humidity (a mean of 80%) are optimum for anthracnose disease development (Roberts et al. 2001).

29.2.1.4 Disease Management

29.2.1.4.1 Cultural Practices

The first and foremost step to manage the disease is to collect and destroy all the infected plant debris and solanaceous weed hosts from in and around the field. Healthy seedlings should be used. Affected fruits should be sorted out and stored separately. They should not be mixed with healthy fruits during drying and storage.

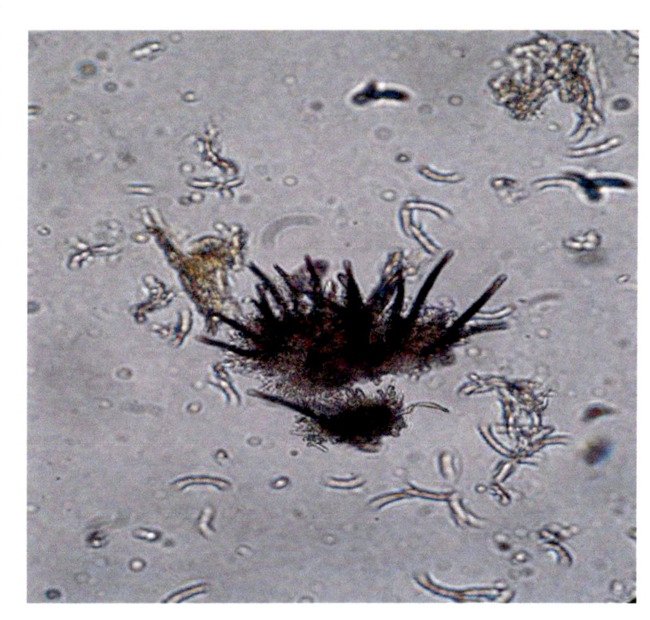

FIGURE 29.2 Acervulus of *Colletotrichum* sp.

FIGURE 29.3 (a) Disease symptom on leaf. (b) Disease symptom on fruit.

29.2.1.4.2 Chemical Control

Spraying the crop with carbendazim (0.1%) is effective, or thiophanate methyl (0.1%) or a combination of mancozeb (0.25%) and carbendazim (0.05%), repeated at 10- to 14-day intervals.

29.2.1.4.3 Biological Control

An effective eco-friendly approach is the combined application of plant extract of neem (*Azadirachta indica*), mahogany (*Swietenia mahagoni*) and garlic (*Allium sativum*). The combination of extracts from these plants showed significant impact on disease reduction as well as on yield of chilli (Rashid et al. 2015). Application of sweet flag crude extract twice when the majority of the plants are at the first bloom stage and at the mature bloom stage is also found effective (Charigkapakorn 2000). *Trichoderma* species have been reported to effectively manage *Colletotrichum* species in chilli with concomitant disease reduction (Boonratkwang et al. 2007). Other biological control agents that have been tested for efficacy against *C. acutatum* include *Bacillus subtilis* and *Candida oleophila* (Wharton and Dieguez-Uribeondob 2004).

29.2.2 Bacterial Leaf Spot

Bacterial spot is also one of the most devastating diseases on *Capsicum* sp. The disease was first reported from the United States in 1912 and now is prevalent worldwide where pepper is grown in warm, moist areas. In India, the disease occurs very commonly in all the regions where chillies are cultivated on a large scale. It occurs soon after transplanting when the weather conditions remain favourable for disease development, leading to complete crop loss.

29.2.2.1 Symptoms

The symptoms on leaves initially appear as water-soaked, circular or irregular spots which become necrotic with brown centres and thin chlorotic borders. These lesions are generally sunken on the upper surface of the leaf and slightly raised on the bottom. Under favourable environmental conditions, these spots coalesce and give a blighted appearance. Such leaves turn yellow and fall down prematurely (Figure 29.3a). Stem lesions are narrow, elongated and raised. On young fruit, the lesions are small, circular, green spots reaching a diameter of 2–3 mm. Fruit lesions initially begin as green spots, which enlarge and become brown in colour. These spots are raised with a cracked, roughened wart-like appearance. During periods of high humidity, fruit around the lesions may start rotting. Ripe fruits are rarely infected (Figure 29.3b).

29.2.2.2 Causal Organism

The bacterium responsible for this disease is *Xanthomonas campestris* pv. *vesicatoria* (ex Doidge) Vauterin. The bacterium is motile, strictly aerobic, Gram-negative, rod-shaped with round ends, measures 1.5–2.0 × 0.5–0.75 μ in size, are non-spore forming and possess a single polar flagellum. Bacteria occur singly, but occasionally in pairs and rarely in short chains.

29.2.2.3 Disease Cycle and Epidemiology

The bacterium persists in infected seed and plant debris. Moderate temperatures along with high precipitation and relative humidity favour the disease development. Infection of the plants can take place at a wide range of temperatures, i.e., 15–35°C. The optimum temperature lies between 22 and 34°C, and maximum disease develops in between July and September. The bacterium is disseminated within a field by wind-driven rain droplets, clipping of transplants and aerosols. Plants at the age of 40–50 days are more prone to attack by the bacteria in the main field. The bacteria also attack tomato and a few other species of *Capsicum*.

29.2.2.4 Disease Management

29.2.2.4.1 Cultural Practices

Since the disease requires high precipitation and high relative humidity, fields should be well drained and free of low-lying areas to minimize waterlogged conditions. The field should be rotated with non-solanaceous crops, and bell pepper/chillies should not be rotated with tomato. Use of disease-free seeds also reduces the immediate availability of primary inoculum and ensures disease-free transplants. Sprinkler or overhead irrigation of the field should be limited to keep the disease in check.

29.2.2.4.2 Chemical Control

Treating the seeds by dipping in streptomycin sulphate (100 ppm) for 30 minutes or Captan or thiram at 4g per kg of seeds at least 24 hours prior to sowing; hot water treatment of seeds at 50°C for a continuous period of 25 minutes eliminates the bacterial inoculum in the seeds. Spraying the crop with streptomycin sulphate (100 ppm) or Agrimycin 100 at 800–1000 ppm at fortnightly intervals during the monsoon period is effective in reducing the disease.

29.2.3 Phytophthora Leaf Blight and Fruit Rot

Phytophthora fruit rot is an important disease in those areas where fruiting coincides with the onset of monsoon rains. Due to favourable climatic conditions, the disease is so severe that it sometimes causes complete defoliation of the plants.

29.2.3.1 Symptoms

Symptoms can appear as water-soaked bleached spots on any portion of the leaf, resulting in premature leaf fall (Figure 29.4a). Small water-soaked spots also appear on the fruits, and the flesh below the skin becomes soft, and usually there is a distinct line of demarcation between the invaded tissue and healthy tissue. Whitish mould appears on the rotten fruits under humid conditions (Figure 29.4b). Completely rotten fruits may fall to the ground. Symptoms also appear on the collar region of adult plants as water-soaked areas with whitish growth of mycelium engirdling the collar region and the point of contact of the soil line. The rot often progresses downwards to the roots in the affected plants and there is sudden drooping of leaves, giving the appearance of sudden wilt.

29.2.3.2 Causal Organism

Two species of *Phytophthora*, namely *P. capsici* Leon and *P. nicotianae* (Breda de Hann) var. *nicotianae* Waterhouse have been associated with the disease. In *P. capsici* mycelium is hyaline, branched and non-septate, but a few septa are found in the case of old hyphae. The sporangia are hyaline, ovoid to pyriform or sometimes round to lemon shaped, non-pedicillate, with a predominant hemispherical papilla at the apex. The oospores are circular to spherical and germinate either by germ tubes or by stalked or sessile germ sporangia. In *P. nicotianae* var. *nicotianae*, the mycelium is hyaline branched and non-septate with branches at right angles. The sporangia are hyaline, non-pedicellate, globose to subglobose with a prominent beak-like papilla at the apex. The sporangia germinate by production of 15–20 zoospores. Oospores are formed abundantly on aerial and submerged mycelium in culture. The oospores are thick-walled and golden brown in colour.

29.2.3.3 Disease Cycle and Epidemiology

Both pathogens survive in the form of oospores in the soil as well as in infected seed. Presence of abundant rainfall, high relative humidity and warm weather are essential for initiation of this disease. Heavy rains help in sporangial formation and their germination by zoospores which are splashed with spattering rains. After infection, sporangia are produced on the fruit surface and are carried by wind to the adjoining fruits and foliage. Temperatures ranging from 22 to 25°C along with high humidity (>80%) favours disease development (Figure 29.5).

29.2.3.4 Disease Management

29.2.3.4.1 Cultural

Collecting and destroying the infected leaves and fruits regularly is the first step for disease management. Maintenance of proper drainage is very important.

FIGURE 29.4 (a) Disease symptom on leaf. (b) Disease symptom on fruit.

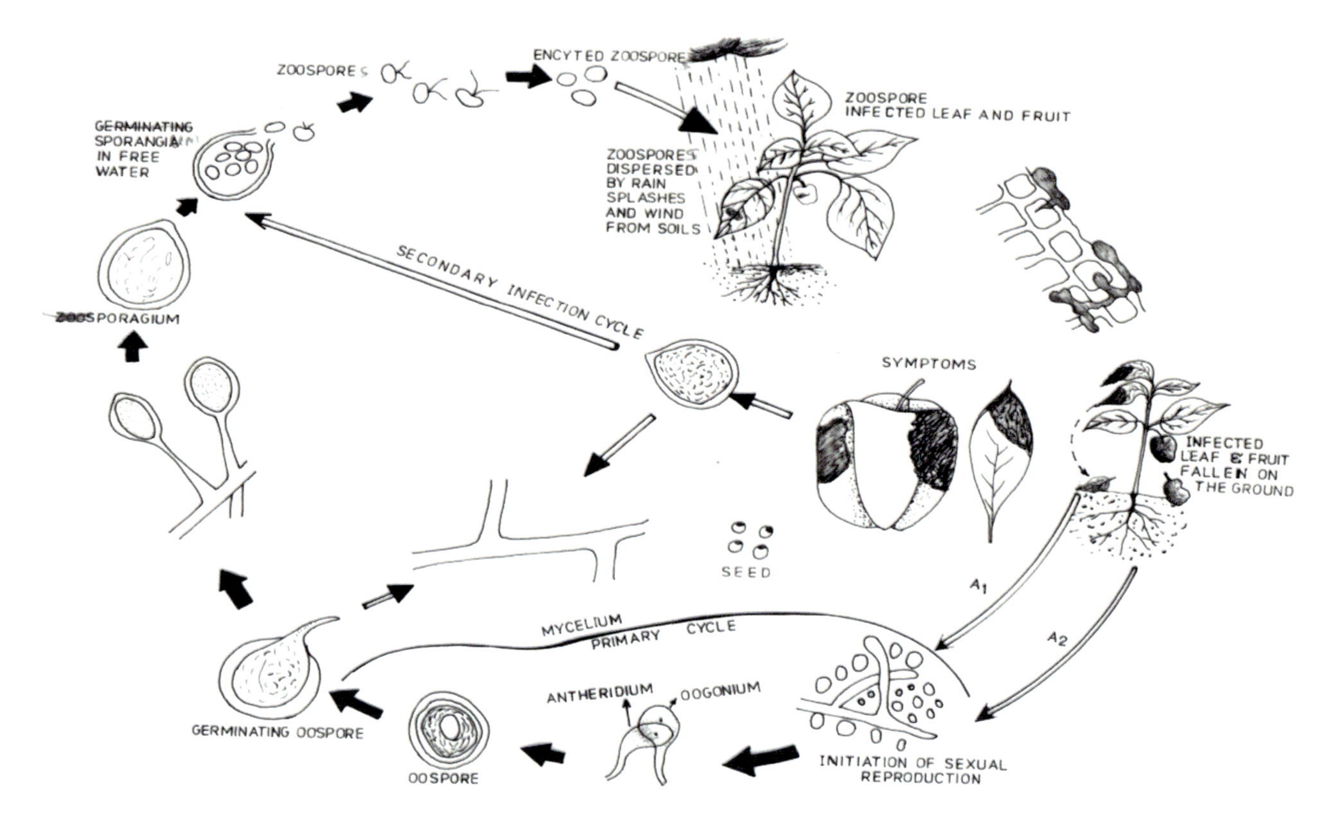

FIGURE 29.5 Disease cycle of *Phytophthora*.

29.2.3.4.2 Chemical Control

Spraying the crop with metalaxyl + mancozeb (0.25%) with the onset of monsoon rains followed by sprays of either mancozeb (0.25%) or copper oxychloride (0.3%) or Bordeaux mixture (4:4:50) at 7–10 day intervals.

29.2.4 Grey Mould

Grey mould, also called Botrytis fruit rot or ash mould, caused by *Botrytis cinerea* Pers. is an economically important disease and causes postharvest losses of upto 20–25% (Fallik et al. 1998). *B. cinerea* remains latent until environmental conditions are suitable and ripening causes physiochemical and biochemical changes in the fruit leading to the development of grey mould (Drobyand Lichter 2007), making the fruit unmarketable (Utkhede and Mathur 2003).

29.2.4.1 Symptoms

Initial symptoms of grey mould have been reported by Chang et al. (2001) as irregular water-soaked lesions which gradually enlarge, coalesce, turn brown and ultimately result in the death of infected plant parts. Grey to greyish brown, velvety mouldy growth consisting of several conidiophores with numerous conidia often appears on the lesions under moist conditions. On ripe fruits, brown firm areas spreading from calyx to other parts of the fruit are observed. On the affected fruits, a velvet-like mycelium and clusters of conidiophores grow; as a result, the infected fruit becomes mummified (Figure 29.6). The fungi can survive in the dead plant parts or the soil as resting bodies (sclerotia and mycelia) or in infected plant debris. The conidia are disseminated from conidiophores by wind, rain-splash, irrigation and by human hands to fruit (Barkai-Golan 2001).

29.2.4.2 Causal Organism

Botrytis cinerea Pers. produces white, dense, cottony mycelium, which after sporulation turns to greyish brown after 5days of incubation. The conidia are single celled with 7.0–14.0 × 5.25–10.5 µm size, hyaline to pale brown in colour, ovate, ellipsoidal, globose to subglobose and pyriform in shape.

FIGURE 29.6 Disease symptom on the fruit.

29.2.4.3 Disease Cycle and Epidemiology

The fungus produces sclerotiain addition to other types of spores. These structures survive in soil, dead plant material or on different host plants. The pathogen is easily dispersed by wind, splashing rain, tools and workers. The disease is favoured by cool and wet weather. The fungus requires a water film of several hours for spore germination, and a longer period of surface wetness for symptom development. Optimum relative humidity for spore production is about 90%. Temperatures of 17–23°Care ideal for disease development. The length of the surface wetness period needs to be longer at the lower temperatures for disease development. The fungus generally infects plants through wounds. Fruit can be infected through the stem scar, growth cracks, or other breaks in the skin. Excessive application of nitrogen makes plants susceptible to grey mould.

29.2.4.4 Disease Management

29.2.4.4.1 Cultural

Removal and burning of the decaying infected plant debris and yellowing plant tissues from the plant bed. Maintaining proper spacing to allow free flow of air is important in greenhouses. Maintaining a relative humidity of less than 80% and a greenhouse temperature higher than that of outdoors during the night is also important. Wounding the plants or the fruits should be avoided.

29.2.4.4.2 Chemical Control

Pre-harvest fungicide application is an effective way to reduce infections initiated in the fields (Barkai-Golan 2001) and to prevent the formation of latent infections in the young fruit. Postharvest fungicide application should be done as soon as possible after harvest to prevent mycelial growth in the host tissue. The antagonist yeast *Candida oleophila* and *Trichoderma* have been more effective in managing the disease (Mercier and Wilson 1995). The antagonistic yeasts combined with calcium chloride, salicylic acid, sodium bicarbonate, silicon, boron and glycine betaine can significantly improve their bio-control efficacy in control of postharvest diseases in various fruits. The best example is a combination of hot water treatment with *Candida guilliermondii* or *Pichia membranae faciens* which showed a better control efficacy against *B. cinerea* (Zong et al. 2010). Treatment with hot air at 38°C for 48–72 hours or hot water at 50–53°C for 2–3 minutes will reduce the disease.

29.2.5 Alternaria Rot

Alternaria fruit rot generally occurs in fruit having a predisposing injury. The *Alternaria* species responsible is a weak pathogen and rarely spreads in harvested fruit. The fungus has been reported on a wide range of fruits and vegetables but is especially common on pepper fruit which has been injured due to sunscald.

29.2.5.1 Symptoms

Symptoms of *Alternaria* rot on chilli begins as water-soaked, grey lesions on fruit which further become dark and become covered with spores; internal necrosis and mycelial growth occurs on the seeds, placenta and pericarp (Meiri and Rylski 1983;Wall and Biles 1993). Lesions of *Alternaria alternata* on chilli fruit are darker in colour and covered by mouldy growth of fungus with heavy sporulation (Shivakumara 2006). Typical fruit symptoms are circular or angular, depressed sunken lesions, with concentric rings of acervuli. Under severe disease pressure, lesions may coalesce. Conidial masses may also occur scattered or in concentric rings on the lesions (Figure 29.7) (Shivakumara 2006; Than et al. 2008; Akhtar et al. 2009).

FIGURE 29.7 Symptom of *Alternaria* rot on bell pepper fruit.

29.2.5.2 Causal Organism

The conidia of *Alternaria alternata* are pale brown to light brown, obclavate to obpyriform or ellipsoid short conical beak at the tip, or beakless with smooth surface to verruculose. The size varies from 20 to 63 × 9 to 18 μm in size. The conidia have several vertical and around eight transverse septa and are produced in an often branched, long chain more than five conidia. The conidiophore is pale brown to olive brown of 25–60 × 3–3.5 μm in length and is straight or flexuous. Individual conidiophores arise directly from substrate forming bushy heads consisting of four to eight large catenate conidia chains. Secondary conidiophores are generally short and one-celled.

29.2.5.3 Disease Cycle and Epidemiology

The pathogen exists naturally as a saprophyte and is one of the first colonizers of dying plants. Primary inoculum is airborne and originates on debris of a wide variety of plants. *Alternaria* fruit rot is favoured by warm temperatures and is greatly reduced by refrigeration.

29.2.5.4 Disease Management

29.2.5.4.1 Cultural Practices

Avoid conditions that predispose fruit to infection in the field and after harvest. Planting should be scheduled in such a way that harvesting does not occur during very hot or very cold weather. Select cultivars that provide good shading of developing fruit. Ensure that fertilization provides adequate calcium. Avoid rough handling during harvest. Harvested fruit should be properly cooled. Treatment with hot air at 38°C for 48–72 hours or hot water at 50–53°C for 2–3 minutes is recommended for management of the disease

29.3 Postharvest Diseases of Brinjal

Brinjal or eggplant is an important solanaceous crop of the subtropics and tropics. It is an agronomically important non-tuberous crop native to southern India and widely grown in America, Europe, and Asia (Sekara et al. 2007). It is botanically known as *Solanum melongena* ($2n$=24) and categorized into cultivated and wild type. The cultivated types are grouped into three main varieties based on fruit shape, i.e., round-shaped cultivar under *S. melongena* var. *esculentum*, the long slender type under *S. melongena*var. *serpentinum* and the dwarf brinjal under *S. melongena* var. *depressum*. The cultivated varieties can be distinguished from the wild relatives on the basis of phenotypic characters. It is one of the most popular and principal vegetable crops widely grown in the tropics and subtropics (Roychowdhury and Tah 2011), especially in Asia, Europe, Africa and America (Demir et al. 2010). The Indian subcontinent and China are its primary centres of diversity (Kashyap et al. 2003; Singh et al. 2014). Due to its versatility in use in Indian food, brinjal is described as the king of vegetables. Brinjal has an important nutritional value due to its composition, which includes minerals like potassium, calcium, sodium and iron (Mohamed et al. 2003; Raigon et al. 2008) as well as dietary fibre (Sanchez-Castillo et al. 1999). In India, it is one of the most common and versatile crops adapted to different agroclimatic regions and can be grown throughout the year from sea level to snowline for its immature, unripe fruits which are used in a variety of ways as cooked vegetable and in curries (Singh et al. 2014). Brinjal also possesses medicinal values for curing diabetes, hypertension, liver malfunction and obesity.

29.3.1 Phomopsis Blight

Phomopsis blight is one of the most important biotic factors limiting eggplant production in brinjal-producing areas of India, as it reduces yield and marketable value of the crop by 20–30% (Das 1998; Khan 1999). In India, the yield losses due to fruit rot ranged from 10 to 20%in Punjab and Delhi (Panwar et al. 1970). In an advanced stage of disease, seed quality is also adversely affected and infected seed becomes discoloured with poor germinability and reduced seed viability. Seed infection results in pre-emergence and post-emergence damping off of seedlings; approximately one-third of the plants are lost at each stage. The fungus is seedborne and can survive in crop debris in the absence of the host. It can cause seedling damping-off, leaf spot and stem canker on eggplant. The disease affects eggplant production severely and reduces fruit yield by 40–70% of the total harvest (Cristina 2002).

29.3.1.1 Symptoms

Phomopsis blight attacks the stems of eggplant, causing plants to wilt. The disease can also penetrate into fruit, creating a soft rot. Phomopsis blight can infect above-ground plant parts at all stages of development. The fungus can survive for more than a year in fields where a diseased crop is grown and is mainly favoured by warm, wet weather. The disease occurs on fruit and occasionally on leaves or stems. Infection of foliage and stem is less important than fruit infection. When seedlings are infected, the stems exhibit dark brown lesions that turn grey in the centre just above the soil line. Eventually these lesions girdle the stem and kill the plant. When the leaves are attacked, leaf spots first appear as small (less than 0.4 inches) grey to brown lesions with light centres. Lesions often become numerous and cover large areas of leaves. Severely infected leaves become torn, yellow and wither. On older plants, round or oval spots develop on the leaf and stem, which enlarge and become more irregular.

Fruit lesions are sunken, discoloured and soft with a surrounding margin of black fruit bodies. If conditions become dry, infected fruit become shrivelled, dry and form black mummies (Schwartz and Gent 2007). Fruits are infected while on the plant and exhibit spots that appear as pale, sunken areas which may cover the entire fruit if not treated. The small black pycnidia are present in abundance in the fruit spots (Pernezny and Palmateer 2009). Fungal fruiting bodies form concentric circles inside the fruit, reducing the grade of the eggplant. Symptoms first appear as pale, sunken, circular to oval areas on the fruit surface. These later turn brown and enlarge to up 2–3 inches in diameter; often two or more lesions merge to cover much of the fruit surface. Affected fruits became soft and watery at first; decay may penetrate rapidly throughout the fruit, causing a light brown discoloration of the flesh. Later black, pin-like structures, the pycnidia, develop in the centre of the old lesions. As the infection progresses the spots are markedly delimited by a thick and sharp outline enclosing a lighter black or straw-coloured area. When a diseased fruit is cut open, the lower surface of the skin is covered with minute, elevated, spherical, black stromatic masses of the fungus. The characteristic conidiomata appear as black pinhead-sized structures, which are often concentrically arranged on fruits. Infected fruits are soft and mushy or mummified and black. Affected fruits become wrinkled and deformed. Finally the diseased fruits are shrivelled and dried. The infected fruits either remain on the plant or drop off (Figure 29.8).

29.3.1.2 Causal Organism

The fungus responsible for this disease is *Phomopsis vexans* (Sacc. and Syd.) Harter. *Diaporthe vexans* Gratz is the perfect stage. The mycelium is hyaline and septate. The conidiophores (phialides) in the pycnidium are hyaline, simple or branched, sometimes septate and arise from the innermost layer of cells lining the pycnidial cavity. The pathogen is reported to produce two types of conidia, *viz.*, alpha and beta, in its pycnidium. Formation of conidia in pycnidia of *P. vexans* is temperature dependent. At 10–16°C temperature the pathogen produces beta conidia, and at 25–28°C, the alpha conidia. These two

FIGURE 29.8 Disease symptom of fruit rot on brinjal.

forms of conidia become inter-converted when subjected to specific temperature and alpha and beta are two forms of the same conidium. *P. vexans* produces only one type of conidia in its pycnidia which are hyaline, one celled, and subcylindrical during summer months and gradually change into beta form. In the beta form of conidia, the stylospores are filiform, curved, hyaline and septate. These spores normally do not germinate, but inoculation of the host with beta conidia causes interveinal necrosis. Perithecia are observed only in culture which is usually in clusters, beaked, carbonaceous, sinuate, and with irregular ostiole. Asci are clavate, sessile and contain eight ascospores. Ascospores are hyaline, narrowly ellipsoid to bluntly fusoid, one-septate and constricted at the septum.

29.3.1.3 Disease Cycle and Epidemiology

The pathogen is seedborne and also survives in plant debris. The fungus produces abundant fruiting structures along the killed tissue and galls called pycnidia. It spreads when spores are released from the pycnidia and dispersed by splashing rain, insects, and contaminated equipment (Howard and Gent 2007). In plant debris, fungus survives both as mycelium and pycnidia. Seed infections directly lead to diseased seedlings. High relative humidity coupled with high temperatures is favourable for disease development. Maximum disease development takes place at about 26°C under wet weather conditions. Temperature of 25°C, pH 4.0–9.0, favours growth and pycnidial formation. The pathogen is internally as well as externally seedborne and is transmitted to the seedlings. The inoculum on stem and fruits survives for 24 months, while inoculum on leaves survives for only 2 months in field (Figure 29.9).

29.3.1.4 Disease Management

29.3.1.4.1 Cultural Practices

Collection and destruction of the diseased plant debris, crop rotation and use of disease-free seed are recommended to reduce the initial inoculum. Crop rotation for 3years with non-host crop.

29.3.1.4.2 Chemical Control

Seed treatment with carbendazim (0.2%) or thiophanate methyl (0.2%). With the initiation of the disease, spray the crop with carbendazim (0.1%) or combination of mancozeb (0.25%) and carbendazim (0.05%) or copper oxychloride (0.3%) and repeat at 10-14-day intervals. Hot water treatment to seeds at 50°C for 30 minutes is also found effective.

29.3.2 Leaf Spot and Fruit Rot

The disease occurs worldwide wherever brinjal is cultivated.

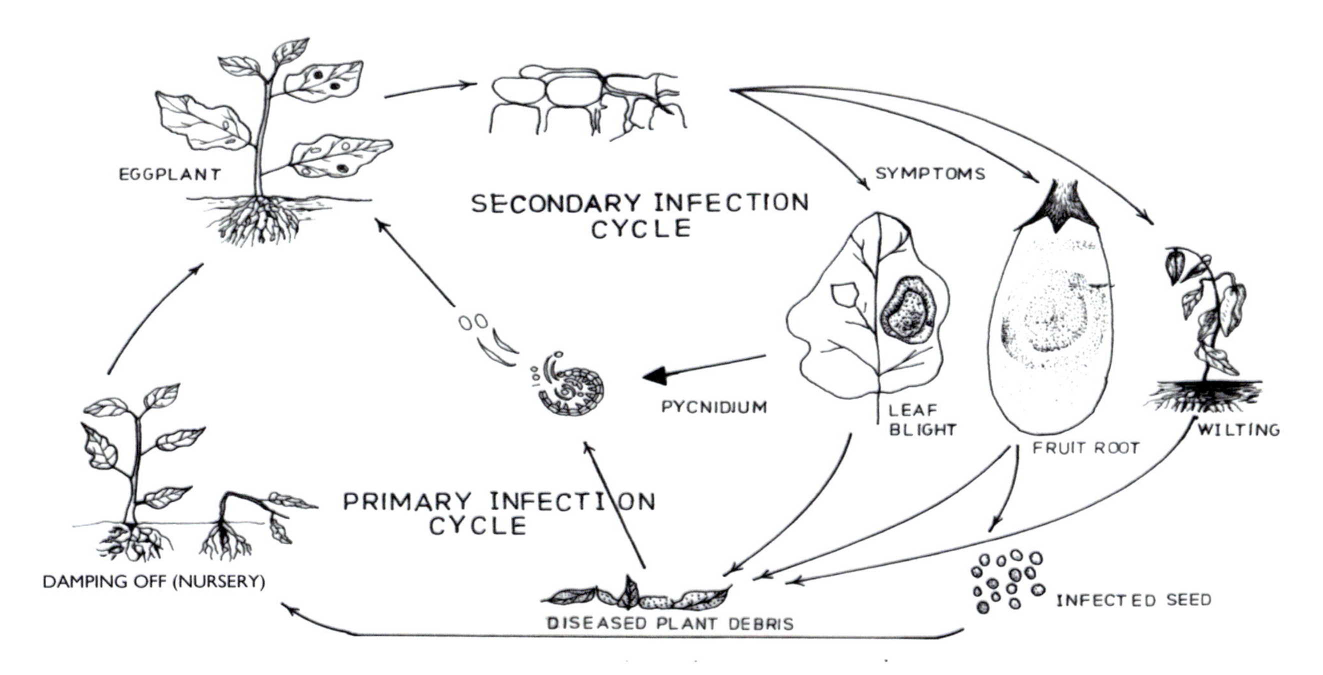

FIGURE 29.9 Disease cycle of *Phomopsis vexans*.

29.3.2.1 Symptoms

Initial symptoms appear as small, scattered, circular, pale brown, necrotic spots, mostly on the older leaves, and gradually spread to the younger leaves. The spots enlarge forming concentric rings and appear as target boards. The centre of the spot becomes necrotic and sometimes the central dried portion disintegrates and falls off, leaving a shot hole. As spots enlarge further, they become irregularly circular in shape and attain a diameter of upto 4.0–8.0 mm. Adjacent spots may coalesce and cover large areas of the leaves, and as a result the leaves turn yellow and drop off prematurely. Similar spots may develop on the fruits also. The spots on the fruits are fewer and are usually larger in size. They are sunken, scald-like, the skin becoming hard and leathery. Sometimes a single spot may enlarge and cover a major portion of the fruit or the entire fruit. Affected fruits turn yellow and drop off prematurely.

29.3.2.2 Causal Organism

Alternaria melongenae and *A. solani* are involved in causing the disease. The mycelium of the fungi consists of septate, branched, light brown hyphae, which become darker. The hyphae are intercellular in the beginning, becoming intracellular later on and penetrate into the cells of the invaded tissues. In *Alternaria melongenae*, conidia are produced in chains of 2–5. They are dark brown in colour, 28–84 × 7–17.5 μ in size with rather short beak of length 10.5–52.5 μ and have 7–14 cross septa and 0–5 vertical septa. Under moist conditions, they germinate by producing 5–10 germ tubes.

In *Alternaria solani*, the conidiophores are dark-coloured, 50–90μ in length and bear conidia singly in nature and in short chains of 2–3 in culture. The conidia are beaked, multiform, dark coloured, with 5–10 transverse septa and few longitudinal septa. They measure 120–296 × 12–20 μ in size, including the beak, and germinate by putting forth 5–10 germ tubes.

29.3.2.3 Disease Cycle and Epidemiology

The pathogens are mainly soilborne. The mycelium of the fungus remains dormant in dry, infected leaves and fruits for more than a year. The mycelium and conidia which remain in the soil in the diseased plant debris cause primary infection. Collateral hosts such as tomato and chillies play an important role in perpetuation of the disease. Under warm and moist conditions, the conidia germinate in about 35–45 minutes. Secondary spread of the disease is through conidia dispersed by wind, rain and by some insects mechanically. Moderate temperatures between 28 and 30°C and high atmospheric humidity favour the occurrence and spread of the disease.

29.3.2.4 Disease Management

29.3.2.4.1 Cultural Practices

Collection and destruction of the diseased plant debris fallen on the ground. Avoid growing alternate hosts in the vicinity of the field. Pusa Purple, which is moderately resistant, can be used.

29.3.2.4.2 Chemical Control

Seed treatment with Captan or Thiram at 4 g per kg of seeds. Spraying with copper oxychloride 750 g or dithiocarbamates – 600 g or organotin compounds 450 g in 300 L of water per acre.

29.3.3 Phythpthora Fruit Rot

Although the entire plant may be susceptible, fruit rot is the primary symptom caused by *P. capsici* in eggplant. It begins as a round, dark brown area on any part of the fruit at any stage of maturity.

29.3.3.1 Symptoms

The initial lesion is surrounded by a rapidly expanding light tan region. White to grey fungal-like growth may appear during wet, humid periods, starting on the oldest part of the fruit lesion. Phytophthora fruit rot in eggplant lacks the concentric patterns and dark fruiting structures present with Phomopsis rot, a fungus that also causes fruit rot on eggplant in Florida. Fruit rot in eggplant may also be caused by other *Phytophthora* spp. and *Phomopsis* spp.

29.3.4 Anthracnose Fruit Rot

Anthracnose fruit rot is present in all parts of the temperate and subtropical zones.

29.3.4.1 Symptoms

Fruit lesions vary in size. They may be upto 0.5 inches wide and may coalesce. The lesion tissue is sunken. A tan-coloured ooze of fungus spores appears on the lesions, especially during moist weather. There may be one to many spots per fruit. When many spots form, they can cover the fruit surface. Seriously affected fruit drop from pedicel to the ground. The eggplant fruit dries and becomes black. Frequently, the soft rot bacteria enter and cause a soft watery decay.

29.3.4.2 Causal Organism

Colletotrichum melongenae (Ellis and Halsted) Averna is known to produce circular, wooly and cottony colonies with pale brown to greyish white. The mycelium of growing culture is hyaline, septate and branched. The Conidiomata are acervular, composed of hyaline to dark brown septate hyphae. Conidia are hyaline, one celled, oval or oblong in shape.

29.3.4.3 Disease Cycle

Colletotrichum melongenae usually attacks only fruit on plants that are in a weakened condition or fruit that are overripe. Pathogen survives as acrevuli on crop debris and spreads through rain splashes.

29.3.4.4 Disease Management

Removal of crop debris after harvest will reduce the likelihood of the fungus surviving from one season to the next. Rotation of non-solanaceous (tomato, pepper, potato and eggplant) plants is advisable, as the fungus can survive for years on infected residues. Elimination of all reservoir hosts, destruction of debris, crop rotation.

29.4 Conclusion and Future Directions

Management of pre-harvest pathogens at the field level is of utmost importance in reducing postharvest diseases. A wide variety of fungal and bacterial pathogens cause postharvest disease in fruit and vegetables. Some of these infect produce before harvest and remain latent until the conditions become favourable for disease development. In the development of integrated disease management strategies, it is important to look into the production practices and postharvest handling system. Many pre-harvest factors influence the development of the postharvest diseases. Since many years, chemicals have played an important role in the postharvest disease strategy. However, now the trend is towards reduced chemical usage and use of safe economical and viable alternatives which can be exploited to the maximum. Hence there is a need for development of safer management practices to reduce postharvest losses.

REFERENCES

Akhtar J, Singh MK, Chaube HS (2008) Effect of nutrition on formation of acervuli, setae and sporulation of the isolates of Colletotrichum capsici. Pantnagar J Res 6(1): 110–113.

Barkai-Golan R (2001) Postharvest Diseases of Fruits and Vegetables: Development and Control. Elsevier Press, Amsterdam, the Netherlands.

Boonratkwang C, Chamswarng C, Intanoo W, Juntharasri V (2007) Effect of secondary metabolites from *Trichoderma harzianum* strain Pm9 on growth inhibition of *Colletotrichum gloeosporioides* and chilli anthracnose control. Proceedings of the 8th Natl Plant Protec Conf Naresuan Univ, Phisanulok, Thailand, pp. 323–336.

Bosland PW, Votava EJ (2003) Peppers: Vegetable and spice capsicums. CAB International, Oxford, Wallingford.

Byung SK (2007) Country report of Anthracnose research in Korea. In First International Symposium on Chili Anthracnose, Hoam Faculty House, Seoul National University, Seoul, Vol. 24.

Chang SW, Kim SK, Yi ES, Kim JW (2001) Occurrence of gray mold caused by *Botrytis cinerea* on *Cryptotaenia japonica* in Korea. Mycobiol 29(4): 227–229.

Charigkapakorn N (2000) Control of Chilli Anthracnose by different biofungicides, Thailand. Available at http://www.arc.avrdc.org/pdf_files/029-Charigkapakorn_18th.pdf

Cristina MB (2002) Pesticide management of major pests and diseases of eggplant. Veggies Today 2: 12–15.

Das BH (1998) Studies on Phomopsis Fruit Rot of Brinjal. MS thesis submitted to the Dept. of Plant Pathology, Bangladesh Agricultural University, Mymensingh, 106 pp.

Demir K, Bakir M, Sarıkamiş G, Acunalp S (2010) Genetic diversity of eggplant (*Solanum melongena*) germplasm from Turkey assessed by SSR and RAPD markers. Genet Molec Res 9(3): 1568–1576.

Dias GB, Gomes VM, Moraes TM, Zottich UP, Rabelo GR, Carvalho AO, Moulin M, Gonçalves LSA, Rodrigues R, Da Cunha M (2013) Characterization of Capsicum species using anatomical and molecular data. Genet Mol Res 12(4): 6488–6501.

Don LD, Van TT, Phuong VY, Kieu PTM (2007) "*Colletotrichum* spp. attacking on chilli pepper growing in Vietnam, Country Report," in Abstracts of the First International Symposium on Chilli Anthracnose, eds. DG Oh and KT Kim (Seoul: Seoul National University), p. 24.

Droby S, Lichter A (2007) "Postharvest botrytis infection: Etiology, development and management," in Botrytis: Biology, Pathology and Control, eds. Y Elad, B Williamson, P Tudzynsky, and N Delen (Springer, Dordrecht), pp. 349–367.

Fallik E, Archbold DD, Hamilton-Kemp TR, Clements AM, Collins RW, Barth MM (1998) (E)-2-hexenal can stimulate *Botrytis cinerea* growth *in vitro* and on strawberries *in vivo* during storage. J Am Soc Hort Sci 123(5): 875–881.

Garg R, Loganathan M, Saha S, Roy BK (2014) Chilli Anthracnose: A review of causal organism, resistance source and mapping of gene. In: Microbial Diversity and Biotechnology in Food Security. Springer, New Delhi, pp. 589–610.

Hadden JF, Black LL (1989) Anthracnose of pepper caused by *Colletotrichum* spp. Proceedings of Symposium on Integrated Management Practices. Tomato and Pepper Production in the Tropics, pp. 189–199.

Hong JK, Hwang BK (1998). Influence of inoculum density, wetness duration, plant age, inoculation method, and cultivar resistance on infection of pepper plants by *Colletotrichum coccodes*. Plant Dis 82(10): 1079–1083.

Howard FS, Gent DH (2007) Eggplant, pepper and tomato. Phomopsis fruit rot (Phomopsis blight). High Plains IPM Guide, a cooperative effort of the University of Wyoming, University of Nebraska, Colorado State University and Montana State University, pp.157–168.

Kashyap V, Kumar SV, Collonnier C, Fusari F, Haicour R, Rotino GL, Rajam MV (2003) Biotechnology of eggplant. Sci Hortic 97(1): 1–25.

Khan NU (1999) Studies on epidemiology, seed-borne nature and management of Phomopsis fruit rot of brinjal. MS thesis submitted to the Department of Plant Pathology, Bangladesh Agricultural University, Mymensingh, pp. 25–40.

Kim KD, Oh BJ, Yang JM (1999) Differential interactions of a *Colletotrichum gloeosporioides* isolate with green and red pepper fruits. Phytoparasitica 27(2): 97–106.

Kim KH, Yoon JB, Park HG, Park EW, Kim YH (2004) Structural modifications and programmed cell death of chili pepper fruit related to resistance responses to *Colletotrichum gloeosporioides* infection. Phytopathology 94(12): 1295–1304.

Lakshmesha KK, Lakshmidevi N, Mallikarjuna SA (2005) Changes in pectinase and cellulase activity of *Colletotrichum capsici* mutants and their effect on anthracnose disease on capsicum fruit. Arch Phytopathol Plant Protec 38(4): 267–279.

MacNeish RS (1964) The origins of New World civilization. Scientific American 211(5): 29–37.

Manandhar JB, Hartman GL, Wang TC (1995) Anthracnose development on pepper fruits inoculated with Colletotrichum gloeosporioides. Plant Dis 79(4): 380–383.

Meiri HA, Rylski I (1983) Internal mold caused in sweet pepper by *Alternaria alternata*. Fungal ingress. Phytopathology 73(1): 67–70.

Mercier J, Wilson CL (1995) Effect of wound moisture on the bio-control by *Candida oleophila* of gray mold rot (*Botrytis cinerea*) of apple. Postharv Biol Technol 6(1–2): 9–15.

Mohamed AE, Rashed MN, Mofty A (2003) Assessment of essential and toxic elements in some kinds of vegetables. Ecotoxicol Environ Safety 55(3): 251–260.

Moscone EA, Baranyi M, Ebert I, Greilhuber J, Ehrendorfer F, Hunziker AT (2003) Analysis of nuclear DNA content in capsicum (solanaceae) by flow cytometry and Feulgen densitometry. Ann Bot 92: 21–29.

Pakdeevaraporn P, Wasee S, Taylor PWJ, Mongkolporn O (2005) Inheritance of resistance to anthracnose caused by *Colletotrichum capsici* in *Capsicum*. Plant Breed 124(2): 206–208.

Panwar NS, Chand JN, Singh H, Paracer CS (1970) Phomopsis fruit rot of brinjal (*S. melongena* L) in the Punjab viability of the fungus and role of seeds in the disease development. J Res Ludhiana 7: 641–643.

Park KS, Kim CH (1992) Identification, distribution and etiological characteristics of anthracnose fungi of red pepper in Korea. Plant Pathol J 8(1): 61–69.

Park HK, Kim BS, Lee WS (1990) Inheritance of resistance to anthracnose (*Colletotrichum* spp.) in pepper (*Capsicum annuum* L) I genetic analysis of anthracnose resistance by diallel crosses. J Korean Soc Hortic Sci 31: 91–105.

Pernezny K, Palmateer A (2009) Florida Plant Disease Management Guide: Eggplant, Lewis W Jett, 2005. Eggplant Production. G6369.

Pickersgill B (1997) Genetic resources and breeding of *Capsicum* spp. Euphytica 96(1): 129–133.

Raigon MD, Prohens J, Munoz-Falcon JE, Nuez F (2008) Comparison of eggplant landraces and commercial varieties for fruit content of phenolics, minerals, dry matter and protein. J Food Comp Anal 21(5): 370–376.

Ramachandran N, Rathnamma K (2006) *Colletotrichum acutatum*- a new addition to the species of chilli anthracnose pathogen in India. Paper presented at the Annual meeting. Symp Indian Phytopathol Society, Central Planation Crops Research Institute, Kasaragod, 27–28 Nov 2006.

Ranathunge NP, Mongkolporn O, Ford R, Taylor PWJ (2012) *Colletotrichum truncatum* pathosystem on *Capsicum* spp: infection, colonization and defence mechanisms. Australas Plant Pathol 41(5): 463–473.

Rashid MM, Kabir MH, Hossain MM, Bhuiyan MR, Khan MAI (2015) Eco-friendly management of chilli anthracnose (*Colletotrichum capsici*). Int J Plant Pathol 6(1): 1–11.

Roberts PD, Penny KL, Kucharek TA (2001) Anthracnose caused by *Colletotrichum spp.* on pepper. Available at: http://edisifasufledu.Accessed on 12 May 2013.

Roychowdhury R, Tah J (2011) Differential response by different parts of *Solanum melongena* L. for heavy metal accumulation. Plant Sci Feed 1(6): 80–83.

Sanchez-Castillo CP, Englyst HN, Hudson GJ, Lara JJ, Solano ML, Munguía JL, James WP (1999) The non-starch polysaccharide content of Mexican foods. J Food CompAnal 12(4): 293–314.

Sarath Babu BS, Pandravada SR, Rao RP, Anitha K, Chakrabarty SK, Varaprasad KS (2011) Global sources of pepper genetic resources against arthropods, nematodes and pathogens. Crop Protec 30(4): 389–400.

Saxena A, Raghuwanshi R, Singh HB (2014) Molecular, phenotypic and pathogenic variability in *Colletotrichum* isolates of subtropical region in north-eastern India, causing fruit rot of chillies. J Appl Microbiol 117(5): 1422–1434.

Schwartz HF, Gent DH (2007) Eggplant, pepper, and tomato Phomopsis fruit rot (Phomopsis blight) high plains. IPM Guide, a cooperative effort of the University of Wyoming, University of Nebraska, Colorado State University and Montana State University, pp.157–168.

Sekara A, Cebula S, Kunicki E (2007) Cultivated eggplants–origin, breeding objectives and genetic resources. A review. Folia Hort 19: 97–114.

Shivakumara AP (2006) Management of postharvest diseases of chilli. MSc (Agri) thesis. University of Agricultural Sciences, Dharwad, India.

Singh MK, Yadav JR, Singh BM (2014) Genetic variability and Heritability in brinjal (*Solanum melongena*L). Hort Flora Res Spectrum 3(1): 103–105.

Smith PG, Villalon B, Villa PL (1987) Horticultural classification of peppers grown in the United States. Hort Sci USA 22: 11–13.

Than PP, Jeewon R, Hyde KD, Pongsupasamit S, Mongkolporn O, Taylor PWJ (2008) Characterization and pathogenicity of *Colletotrichum* species associated with anthracnose on chilli (*Capsicum* spp.) in Thailand. Plant Pathol 57: 562–572.

Utkhede R, Mathur S (2003) Fusarium fruit rot of greenhouse sweet peppers in Canada. Plant Dis 87(1): 100–100.

Voorrips RE, Finkers R, Sanjaya L, Groenwold R (2004) QTL mapping of anthracnose (*Colletotrichum*spp.) resistance in a cross between *Capsicum annuum* and *C chinense*. Theory Appl Genet 109(6): 1275–1282.

Wall MM, Biles CL (1993) Alternaria fruit rot of ripening chile peppers. Phytopathology 83(3): 324–328.

Walsh BM, Hoot SB (2001) Phylogenetic relationships of capsicum (solanaceae) using DNA sequences from two noncoding regions: the chloroplast atpB-rbcL spacer region and nuclear waxy introns. Int J Plant Sci 162(6): 1409–1418.

Wharton PS, Dieguez-Uribeondo J (2004) The biology of *Colletotrichum acutatum*. Anales Jardín Botanico Madrid 61(1): 3–22.

Zong Y, Liu J, Li B, Qin G, Tian S (2010) Effects of yeast antagonists in combination with hot water treatment on postharvest diseases of tomato fruit. Biol Control 54(3): 316–321.

30

Recent Advances in the Management of Postharvest Diseases of Onion and Garlic

Gurudatt M. Hegde, J. U. Vinay and Jahagirdar Shamarao
Department of Plant Pathology, University of Agricultural Sciences, Dharwad, Karnataka, India

CONTENTS

30.1 Introduction

Onion and garlic belong to the family *Alliaceae*, and *Allium* species are grown for food and ornamental purposes. *Allium cepa* L. (bulb onion) is the widely cultivated species of the genus *Allium*. It is grown from Scandinavia to the humid tropics, and the major production is seen in temperate and subtropical regions of world. According to FAO (2001), onions are ranked second after tomato in cultivation worldwide. The edible parts of onion and garlic are expanded leaf scales. In garlic, the storage tissues are specifically modified as bladeless leaves, and these are called "bulb scales". In onion, outer swollen sheaths of the bulb develop from bladed leaves, and the inner sheaths are bulb scales. Onions are generally propagated through seeds and planted directly into the field. In some instances, onion sets and transplants are also used for planting. Garlics are vegetative, propagated through cloves (Brewster 1994). The arrangement between bulb scales and leaves is most important in the development of bacterial diseases of onion because infections started on the leaves spread to the bulb.

Pathogens infect the onions and garlic in all the developmental stages of the crop. After germination of onion seeds, single cotyledon emerges as loop or knee form and both the ends in the soil. The end of haustoria pulls from the ground to form a flag or whip shaped structure. The primary true leaves emerge from a pore and form at the base of the cotyledon, while subsequent leaves emerge from pores in the centre of the plant at the base of the previously formed leaf (Pooler and Simon 1994). Onions are threatened by various fungi, bacteria, viruses and phytoplasmas in the field, but fungi and bacterial pathogens are predominant in storage (Schwartz and Mohan 1995). This chapter details the major postharvest diseases of onion and garlic and their management, with an emphasis on various types of of disease management research for effective crop protection systems that must be put into practice. Short descriptions of the causal organisms and their disease symptoms are detailed to help in detection of specific diseases.

30.2 Postharvest Losses

Onion is threatened by various diseases from pre-harvest to postharvest period. In onion production, about 35–40% loss was noticed due to various diseases (Gupta and Verma 2002). Various kinds of pathogens are causes of onion bulb rot. Fungi are the major causal agents and result in pre-harvest and postharvest losses in the onion (Currah and Proctor 1990). In garlic, loss due to postharvest diseases was 35–40%. The major postharvest diseases are white rot, neck rot, soft rot, blue mould, black mould, brown rot and among others. *Aspergillus niger* is a saprophytic fungi causing black mould disease of garlic, while *Penicillium* spp. and *Fusarium* spp. cause blue mould and white rot, respectively. Due to infection of fungi, garlic bulbs spoil and ultimately decrease the qualitative attributes and quantity of the produce (Borude and Fatima 2017).

30.3 Diseases of Onion and Garlic

The main reasons for the success of onion and garlic cultivation are storability of the produce for long periods of time, and their utility. If proper bulb treatment is provided before storage, onion and garlic can be stored eight to ten months. Annually there is 35–40% of postharvest loss in onion production due to storage diseases. Among incurred loss, fungal bulb rot disease results in 15–30% loss during storage. Various kinds of fungal and bacterial pathogens viz., *Aspergillus* spp., *Colletotrichum* spp., *Rhizopus* spp., *Penicillium* spp., *Pseudomonas* spp., *Lactobacillus* spp., *Erwinia* spp. and *Botrytis* spp. infect onion bulb in the storage condition (Table 30.1). Among postharvest pathogens, *Aspergillus niger* is the most predominant and aggressive pathogen in both the field and storage. Onion and garlic are prone to several field diseases also. The major field diseases are basal rot, white rot and downy mildew, which carry infection to storage resulting in damage of bulbs. The average losses due to reduction in weight, sprouting and rotting (decay) were 20–25, 4–5 and 10–12%, respectively (Pandey 1989). In India, presently 35–40% of postharvest loss is incurred in onion production during various postharvest operations which includes handling and storage. Fusarium basal rot imparts 30–40% of storage loss of onion. Storage diseases are the major reasons for deterioration of export quality of onion and garlic (Barnoczkine 1986; Gupta et al. 2009).

30.4 Fungal Diseases

30.4.1 Basal Rot

Basal rot disease is one of the major threats of onion cultivation and results in maximum yield loss in major growing areas of the world (Coskuntuna and Ozer 2008). Basal rot imparts up to 50% of yield loss in susceptible cultivars (Everts et al. 1985) and 90% loss due to infection at seedling stage (Davis and Reddy 1932). In India, basal rot disease was first reported by Mathur and Shukla (1963).

30.4.1.1 Symptoms

On garlic: The symptoms of basal rot develop gradually. The symptoms include yellowing and gradual dieback of the leaves. Prominent white fungal growth is noticed at the base of the bulb which results in both pre-harvest and postharvest rotting. Postharvest rotting may be seen in single to entire cloves of the garlic bulb.

On onion: The leaves become yellow or chlorotic and gradually dry up. Affected plant showed dieback symptoms, i.e., drying of the leaf tip downwards (Figure 30.1). Later, complete foliage drying is noticed. Infected bulbs show soft rotting symptoms, and further rotting of roots can be noticed. Prominent whitish mycelial growth can be seen on the scale of the bulb. The disease initiates in the field and carried to storage, resulting in rotting (Sumner 1995).

30.4.1.2 Causal Organism

The disease is caused by *Fusarium oxysporum* f. sp. *cepae*. The pathogen belongs to class Deuteromycete, and teleomorphic stage is unknown. The fungus has septate mycelium and produces three types of asexual spores, viz., microconidia,

TABLE 30.1

Postharvest Diseases of Onion and Garlic

Sl. No.	Disease	Pathogen
	Fungal Diseases	
1	Basal rot	*Fusarium oxysporum* f. sp. *cepae*
2	White rot	*Sclerotium cepivorum*
3	Downy mildew	*Peronospora destructor*
4	Botrytis rot or neck rot	*Botrytis* spp.
5	Penicillium decay or blue mould rot	*Penicillium* sp.
6	Purple blotch of onion	*Alternaria porri*
7	Black rot or mould	*Aspergillus niger*
8	Onion smudge	*Colletotrichum circinans*
9	Southern blight (Sclerotium bulb and stem rot)	*Sclerotium rolfsii* (teleomorph: *Athelia rolfsii*)
	Bacterial Diseases	
10	Bacterial brown rot	*Pseudomonas aeruginosa*
11	Bacterial leaf streak and bulb rot	*Pseudomonas viridiflava*
12	Bacterial soft rot	*Erwinia chrysanthemi*, *Pectobacterium carotovorum* subsp. *carotovorum*
13	Sour skin	*Burkholderia cepacia*

FIGURE 30.1 Symptoms of basal rot disease of onion bulbs caused by *Fusarium* sp.

macroconidia and chlamydospores. Microconidia are the predominantly produced single-celled (without septa) and oval to kidney shaped spores measuring 5–12 μm in length. Macroconidia is falcate shaped and have three to four septa. Generally chlamydospores are produced in older mycelium and thick cell wall which defends against the problem of cell degradation and parasitism of antagonists

30.4.1.3 Disease Cycle and Epidemiology

Basal rot pathogen is classified as weak or passive because the pathogen infects majorly wounded plants by other diseases or insect pests. Infection starts from the basal plate and spreads to bulbs. Some infected bulbs are symptomless. The pathogen spreads through infected seeds or movement of soil and debris or transfers from tools, infected onion sets and irrigation water. The pathogen produces chlamydospores as resting spores in soil and these survive for several years. Pathogen is mainly soilborne, and favourable temperature for disease development is 28–32°C. Optimum temperature for maximum disease spreading is 27°C and infection is restricted at lower temperatures (below 15°C). The disease is favoured by higher temperatures. Pathogen infects the plants through root hairs directly or through wounds indirectly. Direct infection of pathogen can be noticed at all stages of crop. Injury to the roots, basal plate or bulbs by onion maggots or insect pests results in rapid increase of disease incidence and severity.

30.4.1.4 Disease Management

In 1980 Katan found that soil solarization effectively decreased the basal rot incidence. Management includes the removal and burning of severely infected plants. Planting of disease-free seeds or bulbs helps to avoid the disease. Hot water treatment of the garlic cloves reduced the disease incidence up to 50.00%. Crop rotation must be practiced if soil is already sick. Four years continuous crop rotation with wheat and maize reduced the pathogen inoculum level in soil which resulted in 90.00% control of disease (Higashida et al. 1982). Harvested bulbs are perfectly cured to reduce moisture content to safer levels. Onion crop is more sensitive to low soil copper, hence application of copper-based fungicides to soil decrease the disease incidence. Basal rot disease is effectively managed by organic amendments, especially oil cakes. Soil application of mahua cake (10.00%) showed significant reduction in basal rot incidence (Malathi and Mohan 2011). Rajendran and Ranganathan (1996) showed that seed treatment with a combination of *T. viride* and *Pseudomonas fluorescens* was most effective in reducing the incidence of basal rot disease under field conditions. Coskuntuna and Ozer (2008) showed that seed treatment of bio-agent *Trichoderma harzianum* strain KUEN 1585 (Sim Derma) at 10 g/kg concentration decreased the basal rot incidence under field conditions. Prachi et al. (2013) reported that soil drenching of copper oxychloride (0.25%) showed fungistatic activity against *F. oxysporium*. Antifungal activity of carbendizim (bavistin) is increased when combined with copper nanoparticles against of basal rot pathogen. Malathi (2015) showed that the combined application of bacterial and fungal bio-control agents (*Pseudomonas fluorescens* Pf12 + *P. fluorescens* Pf27 + *T. harzianum* TH3) resulted in maximum (85.00%) disease reduction. Flori and Roberti (1993) found that *T. harzianum* and *T. viride* decreased the incidence of onion basal rot by 89.00% and 77.00%, respectively. Soil application of granosan 200 (benomyl 15% + mancozeb 60%) reduced the disease incidence by 77.00% (Naik and Burden 1981). Behrani et al. (2015) reported that carbendazim and antracol at 10000 ppm were more effective in managing the basal rot disease. Dugan et al. (2007) revealed the application of carbendazim 50% WP @ 1.40 g/L decreased the rotting losses in field and storage by 35.00 and 40.00%, respectively. Gupta and Gupta (2013) developed and validated integrated disease management module (NPK fertilization @ 100:50:50 + farmyard manure @ 10 t/ha + vermicompost @ 1 t/ha + *Pseudomonas fluorescens* @ 5 kg/ha + copper oxychloride @ 0.3%) which showed effective reduction in incidence of basal rot disease of onion.

30.4.2 White Rot

30.4.2.1 Symptoms

On garlic: Symptoms of white rot are confused with basal rot disease, but in white rot, the process of pathogen invasion and plant death is faster. Initial symptoms like whitish fluffy mycelial growth are noticed on the stem region and further extend to the base of the bulb. Later, sclerotial bodies are formed on the rotted tissue of bulb.

On onion: Foliage yellowing and dieback symptoms are more prominent in the initial stages. The leaves become yellow and dieback, and when the plants are pulled up, roots are found to be rotten and the base of the bulb covered with a white or grey fungal growth. Later, numerous small black spherical sclerotia are produced. The bulb of the onion completely rots (Figure 30.2).

30.4.2.2 Causal Organism

The causal organism of the disease is *Sclerotium cepivorum*, Berk. Pathogen produced black coloured sclerotia as dormant structure and these are survived in soil for 20 years. Sclerotia are inactive in the absence of susceptible host.

FIGURE 30.2 Symptoms of white rot disease of onion caused by *Sclerotium cepivorum*.

30.4.2.3 Disease Cycle and Epidemiology

Sclerotial bodies spread within and between fields through irrigation water, tillage equipment and plant materials. Sclerotial bodies' germination is stimulated by organic sulphur compounds released by roots of onion and garlic. Root exudates of *Alliums* species stimulate the germination of sclerotia in soil. Cooler weather is a prerequisite for sclerotial germination and mycelial growth. The range of soil moisture required for onion root growth and sclerotia germination is similar. Mycelium grows in the soil until it reaches to roots of Allium species and initiates the infection process. After finding the roots in soil, pathogen forms the aspersoria which helps in attachment and penetration to the host, further spread to other plants. Later sclerotial bodies are formed and persist in the soil. Sclerotia can survive in soil for 8–10 years in the absence of onion and garlic. Once plants get infected, it results in development of disease in the field or rotting in storage. The primary inoculum is sclerotial bodies from the previous season infesting plant debris in field. Secondary spread of sclerotial bodies is from field to field though irrigation or flood water. If bulbs are properly dried before storage the disease does not spread to other bulbs (McLean and Stewart 2004).

30.4.2.4 Disease Management

White rot is difficult to manage through a single approach; hence, integrated disease management is needed. Use of healthy planting material, avoidance of planting in sick soil and crop rotation are effective cultural practices to mitigate the disease. Removal and burning of infected plants. Field flooding, soil solarization and use of natural or synthetic sclerotia germination stimulants before planting may reduce the inoculum density in the soil. Application of iprodione fungicide (Rovral) at the time of planting may reduce disease incidence. Clove or bulb treatment before planting helps to reduce white rot incidence in the field. Hot water treatment, i.e., dipping of cloves or bulbs in hot water before planting, helps to kill the pathogen. Application of organic manures decreases the disease incidence in field. Seed dressing of carbendazim, benomyl or thiophanate-methyl (100–150 g/kg) helps to reduce white rot disease (McLean and Stewart 2004). *Trichoderma harzianum*, *T. viride*, *T. atroviride*, *T. pseudokoningii* and *T. koningii* were reported as the best bio-agents against white rot disease of onion and garlic under both *in vitro* and *in vivo* conditions (Paris and Cotes 2002; Metcalf et al. 2004; Clarkson et al. 2004; McLean et al. 2005). Application of silver nanoparticles (liquid formulation) at 7.00 ppm concentration reduced the incidence of white rot of onion (Jung at al. 2010).

30.4.3 Downy Mildew

30.4.3.1 Symptoms

On garlic: Whitish fluffy growth is noticed on the surface of leaves with prominent yellow discoloration. Death of younger plants and stunting of older plants are noticed. Drying and collapse of leaf tips and infected tissue can be observed. In storage, the necks of the bulbs become blackened, shrivelled and scales show a water-soaked appearance. Infected bulbs sprout prematurely.

On onion: Brownish-purple or velvet coloured fungal growth is noticed on the leaves. As the disease develops, lesions become pale yellow to brown necrotic areas which result in girdling and collapse of the leaf tissue. Infected seed stalks remain pale yellow and are invaded by other pathogens like *Stemphylium* or *Alternaria*. Field infections start in small areas and progress rapidly to the entire field. Infected bulbs will rot in storage. If a diseased bulb is planted, the foliage becomes pale green.

30.4.3.2 Causal Organism

The disease is caused by *Peronospora destructor* Berk. Sporangiophores are coenocytic, elongated and swollen at the base. Sporangia are pyriform to fusiform shaped and attached to the sterigmata by their pointed end. Sporangia germinate through one or two germ tubes. The mycelium is coenocytic, intercellular and bears filamentous haustoria. Oogonia are produced in the intercellular spaces.

30.4.3.3 Disease Cycle and Epidemiology

The pathogen has ability to survive in soil as oospores for many years. Moist conditions or free water are necessary for infection and spread of the pathogen. Sporangiospores are spread through rain with strong wind. The pathogen infects seed stalks in seed crops and grows on seeds in the form of mycelium but does not carry to the next season. The pathogen overwinters primarily as mycelium in diseased onion bulbs that remain in onion fields. The primary sources of survival are diseased planting bulbs and oospores in the diseased crop debris. Oospore in soil directly infects the roots of young onion plants and becomes systemic. If diseased bulbs are used for planting, the pathogen is grown along with the plant and when favourable environmental conditions exist, the overwintering pathogen mycelium in systemically infected plants produces Sporangiospores. These spores spread the disease to neighbour plants. Sporangia are produced at night time when humidity is high and temperature is 4–25°C (mean temperature must be 13°C). Sporangia mature in early morning and are spread in the daytime. Newly produced sporangia are viable for four days. Germination of sporangia occurs in the presence of free water at 7–16°C mean temperature. Continuous occurrence of heavy dews during the night and morning hours is sufficient for the infection process. Initially, the upper portions of the leaves are killed, the pathogen then infects next-lower part of the leaves in each successive cycle of spore formation. These cycles are repeated several times until the leaves are completely killed.

30.4.3.4 Disease Management

Use of healthy planting material, providing good air circulation, elimination of plant debris/cull piles and wide row spacing are the best cultural practices to decrease downy mildew incidence. Planting rows must be in the direction of prevailing winds and furrow irrigation rather than sprinkler irrigation should be practices; crop rotation for 3–4 years can reduce disease incidence. Three foliar sprays of metlaxyl +mancozeb at 0.2% concentration are effective to manage the disease. Spraying must be started 20 days after transplanting and repeated at every 10–15days. Onion bulbs (seed crop)must be exposed to sunlight for 12 days to kill the pathogen. Foliar application of zineb (0.2%), karathane (0.1%) and tridemorph (0.1%) also manages the disease.

30.4.4 Botrytis Rot or Neck Rot

30.4.4.1 Symptoms

On garlic and onion: Botrytis is the major disease in cool climate regions. Initially, water-soaked lesions develop on stems, and later grey-coloured fluffy mycelial growth is seen on lesions. Hence this disease is called neck rot. Many white specks are seen on the foliage. Disease spreads rapidly and whole crop gets killed.

30.4.4.2 Causal Organism

The disease is caused by Botrytis *acclada* Fresen and *B. allii* Munn. Genus *Botrytis* is characterized by grape bunch–like appearance of conidiophores. The conidiophores are dark coloured, elongated, erect, septate and dichotomously branched. Terminal cells swell to produce sporogenous ampullae. On each ampulla numerous conidia arise simultaneously on short denticles. Conidia are hyaline, tinted, aseptate and globose to ovoid in shape (Yohalem et al. 2003). Presently, eight *Botrytis* species are reported to cause onion diseases (Walker 1925; Chilvers and du Toit 2006). *B. acclada* and *B. allii* are the major speciesand cause neck rot, umbel blight, bulb rot and scape blight in onion (Vincelli and Lorbeer 1989).

30.4.4.3 Disease Cycle and Epidemiology

The pathogen infects onion and garlic plants in warm and wet weather conditions. In cooler growing seasons, the disease not appears in the field but develops on stored bulbs. The pathogen overwinters as resting structure sclerotial bodies. Sclerotia are produced on diseased onion bulbs left in cull piles or field and mother bulbs (stored for seed production). Sclerotia are produced on infected leaves, neck or on bulbs prior to or at the postharvest period. Presence of sclerotia in the soil is due to disintegration of infected leaves on which sclerotia were previously formed. Sclerotia on the surface of the soil in onion fields are able produce conidia that can infect the leaves of surrounding onion plants. Sclerotia present on leaf debris can produce conidia (asexual) and ascospores (sexual spores) that can infect plants. The sclerotia germinate and produce conidia frequently (four to five times) resulting in outbreak of disease. Sclerotia formed on the bulbs of volunteer onions also play a vital role in initiation of disease. If the inoculum of seed fields and culls piles is not influenced, sclerotia present in the soil and on volunteer plants are the primary source of inoculum for outbreaks of disease (Yohalem et al. 2003). Conidia spread from diseased seed fields and cull piles to other fields serves as a secondary mode of spread. Leaves of onion plants infected by conidia that developed from sclerotia in the soil serve as secondary sources of inoculums.

30.4.4.4 Disease Management

Harvest the fully matured crop, i.e., at least half of the leaves become brown. Harvested onion must be cured properly on the ground for 6–10 days. Confirm that onions are properly dried and necks are tight before topping. Botrytis neck rot and bacterial diseases spread if bulb tissue is green; hence proper matured bulb harvesting and curing is required. Avoid mechanical injury during topping and storage of bulbs. Store the bulbs in properly ventilated rooms at 32°F or slightly higher. Practice crop rotation for three years to reduce the inoculum. Remove and burn previous season debris in the field. Cold storage may help avoid postharvest disease. Adequate row spacing and use of disease-free bulbs are important. Bulb treatment with Captan or Thiram 0.25% and foliar spraying of mancozeb or mancozeb or chlorothalonil (at 5- to 7-day intervals) managed the disease. Falisolan (carbendazim 60% + bronopol 6%) is the most effective chemical against *Botrytis* rot in storage (Garcia et al. 1997). Proper drying of cull piles helps to reduce pathogen inoculums in the field. Bulb treatment and foliar application of *Trichoderma viride* inhibited the growth of *B. allii* (Hussein et al. 2014). Seedling dip and foliar spray treatments combination with the mixture of *T. harzianum* and *T. koningii* also provide protection against neck rot disease (El Neshawy et al. 1999).

30.4.5 Penicillium Decay (Blue Mould Rot)

30.4.5.1 Symptoms

On garlic: Decaying of the clove results in yellowing of plants with stunted growth, and finally wilting. The pathogen sporulate on the diseased cloves appears as a bluish-green mass.

On onion: Blue mould disease usually appears during the harvesting and storage period. The preliminary symptoms are water-soaked lesions on the outer surface of scales. Further, green to bluish green powdery mass develops on the lesions (Figure 30.3). Infected bulbs are tan or grey on the inner portion.

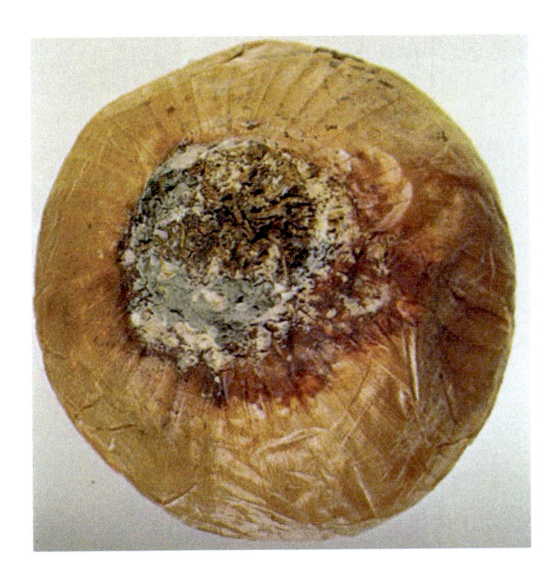

FIGURE 30.3 Symptoms of blue mould rot diseases of onion caused by *Penicillium* spp.

Diseased bulbs finally disintegrate into watery rot. Various species of *Penicillium* are reported to cause blue mould disease.

30.4.5.2 Causal Organism

The disease is caused by various *Penicillium* spp. Among them, *P. digitatum*, *P. oxalicum*, *P. luteum* and *P. expansum* are predominant. The pathogen produces an enormous number of conidia on broom-like conidiophores. The branch conidiophore further forms a structure, i.e., metulae. Phialides are formed on the metulae and chains of conidia are developed on the metulae in a basipetal manner. Conidia are oval to globose, single celled and non-motile.

30.4.5.3 Disease Cycle and Epidemiology

Pathogen survives as saprophytes on plant debris and dead plant tissues. Pathogen invades host through wounds and uncured neck tissue. Temperature of 21–25°C and higher relative humidity helps for multiplication and spread of pathogen. Conidia are present in the air for a long time and travel long distances by strong wind. In the presence of moist air, conidia germinate immediately. Development of symptoms is slow on bulbs. Planting diseased bulbs results in spreading of blue mould disease. In the field, infection occurs through the basal plate of the bulb. Invasion of pathogen in onion bulbs and garlic is most commonly through wounds and uncured neck tissue. After infection, the mycelium of pathogen grows through fleshy scales and gradually sporulates enormously on the surface of lesions or wounds. *Penicillium* spp. can survive in soil, plant or animal debris. Infection of bulbs also commonly occurs via damaged tissues due to freezing injury or sunscald (Sang et al. 2014).

30.4.5.4 Disease Management

Planting the cloves immediately after cracking and minimization of wounding on bulbs are the best ways to manage the disease. Proper drying of bulbs prior to storage helps to manage the disease. Minimum bruising and wounding of bulbs is recommended while harvesting. Storage must be done at low temperature (5°C) and relative humidity to reduce storage losses. Bulb treatment with fungicide is the cheapest way to manage the disease. Spraying of carbendazim (Bavistin) at 0.1% reduced the disease incidence by 93.20% (Sabale and Kalebere 2004).

30.4.6 Purple Blotch of Onion

30.4.6.1 Symptoms

Older leaves are more susceptible to the disease than younger leaves. Initially water-soaked lesions are formed and further transformed into prominent lesions with whitish centres and brown to purple margins with yellowing of the surrounding areas. Later, lesions become dark brown to black with prominent concentric rings, and sporulation occurs. Finally, lesions may girdle, resulting in collapse and death of leaves. Similar types of symptoms appear on seed stalks resulting in collapsing of stalk with shrivelled seeds. Infection of bulb occurs through the neck region. Initially, infected areas of bulb are bright yellow and later turn to characteristic red wine colour.

30.4.6.2 Causal Organism

The disease is caused by *Alternaria porri* (Ellis) Cif. The mycelium of pathogen is branched, septate and coloured. Conidiophores arise as single or in groups. Conidiophores are straight or flexuous or geniculate. Rands (1917) described that conidia of pathogen is obclavate in shape and borne as solitary on the tip of conidiophores but rarely formed in chains. The conidia were brown coloured and have both transverse as well as longitudinal septa. The numbers of transverse septa are in the range of 10–12. The beak on the conidia was simple or forked and sub-hyaline.

30.4.6.3 Disease Cycle and Epidemiology

The pathogen overwinters as dormant mycelium in plant debris and cull piles. Conidia are formed when prolonged humid nights and leaf wetness period are more than 12 hours. When dew dries, conidia become airborne and are disseminated. After infection, visible symptoms are noticed after 4 days. Disease development is greater during prolonged leaf wetness period.

30.4.6.4 Disease Management

Disease-free planting materials must be used. Seed treatment with Thiram at 4 g/kg seed provides initial protection against disease. The field must have a proper drainage system. Foliar sprays (three) with copper oxychloride 0.25% or chlorothalonil 0.2% or Zineb 0.2% or Mancozeb 0.2% are found effective against diseases. A fungicide spray schedule with broad spectrum fungicides applied prior to infection can provide excellent protection against disease. Minimizing the leaf wetness period by adapting drip of surface irrigation rather than sprinkler irrigation helps to avoid pathogen infection. The crop rotation with non-host crop for 4–6 years can reduce disease. Nidagundi (2013) reported that application of neem seed kernel extract (15%), *Trichoderma harzianum* (1.5%) and Difenconazole (0.2%), was effective against purple blotch of onion. Yadav et al. (2017) reported the foliar spray (three for bulb crop and four for seed crop) of tebuconazole 25 EC (Folicur) at 0.1% concentration at 15-day intervals is effective against disease with highest benefit:cost ratio. Seed treatment with *T. harzianum* provides good protection against disease with higher yield (Kumar and Palakshappa 2008; Chethana et al. 2012). Application of clove extract of *A. sativum* (10%) and leaf extract of *Aloe vera* (10%) are found to be effective against purple blotch disease (Mishra and Gupta 2012). Foliar application of bio-agents like *Pseudomonas fluorescens*, *Bacillus subtilis* and *T. viride* provides good control against disease. Seed treatment, seedling dip and foliar sprays (three) of *T. harzianum* reduced disease incidence and promoted plant growth (Yadav et al. 2013). Seed treatment with *P. fluorescens* (5 g/kg) followed by two foliar sprays of Difenconazole (0.1%) interspersed with spray of *P. fluorescens* (0.5%) at 15-day intervals provided good protection against disease (Savitha et al. 2014). Foliar application of neem oil (3%) immediately after onset of disease and 15 days later showed significant reduction of disease. Application of rhizome extract of *Acorus calamus* (10%) and leaves extract of *Mentha arvensis* significantly reduced disease incidence (Ramjegathesh et al. 2011).

30.4.7 Black Rot

30.4.7.1 Symptoms

Black mould usually develops on injured or necrotic tissue in the neck of the bulbs. The pathogen develops on injured parts of the bulb. Black discoloration at the neck region is prominent in diseased bulbs. A black spore mass is usually formed along with veins and between the outer papery scales of bulbs. Initially, a water-soaked appearance is noticed, and finally bulbs are dry and shrivelled. Some of the infected bulbs do not show external visible symptoms (Marziyeh Tolouee et al. 2010).

30.4.7.2 Causal Organism

The disease is caused by *Aspergillus niger* (Tiegh.)The pathogen belongs to class Eurotiomycetes and family Trichocomaceae. Pathogen forms filamentous hyphae and looks like small plants. Conodiophore arises from hyaline and septate hyphae. Conidiophore produces globose vesicles at the top on which metulae and phialides are covered. Metulae gives anchorage to phialides on conidiophores. On phialides, dark brown the conidia are formed.

30.4.7.3 Disease Cycle and Epidemiology

The pathogen is soil borne in nature and survived on plant debris in the field. The disease is spread through direct contact (bulb to bulb), wounds, mechanical means and air borne conidia. During high relative humidity, conidia germinate in 3–6 hours. If RH is below 75%, conidial germination is inhibited. The temperature range required for the growth of pathogen is from 28 to 34°C, and growth is inhibited if temperature is below 17 and above 47°C (Sumner 1995). Sporulation can be noticed in 24 hours after infection (Salvestrin and Letham 1994). Conidia of the pathogen are more frequent in the air and soil. The disease is severe when temperature is more than 30°C in the field and 24°C in storage. Presence of free moisture on the onion surface for more than 6 hours initiates infection process.

30.4.7.4 Disease Management

Usage of disease-free bulbs for planting is the cheapest method to reduce the occurrence of the black mould rot. Crop rotation must be practiced for 3–4 years with non-host crop. Soil solarization should do with 100-gauge LLDPE transparent plastic film for 15 days in the mid-summer season. Deep ploughing of soil in peak summer helps to kill the pathogens. Avoid irrigations to the crop 10–14 days prior to harvest. Moisture in soil at the time of harvest results in bulb infections. Avoid application of nitrogen to the crop 4–5 weeks before harvest since nitrogen stimulates growth of pathogen and makes onion tissues are susceptible to pathogen infections (Anonymous 2009). Seed treatment should be done with Thiram @ 4 g/kg seeds. The proper drainage should be maintained in the field. Three times foliar application of copper oxychloride (0.25%) or chlorothalonil (0.2%) or zineb (0.2%) or mancozeb (0.2%) provide protection against the disease. Soil drenching of copper oxychloride (0.25%) is also effective against disease. Bulb should be stored in cool and dry conditioned rooms (Tolouee et al. 2010). Postharvest dip with carbendazim (0.1%) was effective. Postharvest fumigation with sulphur dioxide for 4 hours or postharvest dip in acetic acid (0.4%) significantly reduced the black mould incidence (Srinivasan and Shanmugam 2006). Application of leaf extract of *Moringa oleifera* (75%) reduced the disease incidence (Arowora and Adetunji 2014). Othman et al. (2014) showed the antifungal effect of *Aspergillus terreus*-based silver nanoparticles (AgNPs) against black mould at 150 ppm concentration. AgNPs inhibited growth and aflatoxin production of *A. flavus*. Gajbhiye et al. (2009) showed that combination of silver nanoparticles and fungicide, fluconazole has a synergetic antifungal effectiveness against pathogen. Pre-harvest (before 20 days of harvest) foliar application of carbendazim at 0.1% (Bavistin) showed more than 90% reduction of disease in storage (Raju and Naik, 2006). Pre-harvest spraying of carbendazim + mancozeb (0.2%) or iprodione 50WP (0.2%) was effective against black mould rot (Ahir and Maharshi 2008). The top portion of the bulbs should cut 1/2 to 3/4 inch from the bulb to allow proper neck drying. Practice good field and storage sanitation to minimize occurrence of the disease. Remove diseased and damaged bulbs during harvest or grading and packing to reduce contamination in storage. Minimize mechanical damage to the bulb during harvesting. Proper curing must be done before storage to reduce postharvest losses. Onions are considered as cured when the neck of the bulb is tight and dry. Ideal temperature and relative humidity for long storage are 0–2°C and 65–70%, respectively.

30.4.8 Onion Smudge

30.4.8.1 Symptoms

Disease infects mature bulbs and continues in storage. Dark green coloured fruiting bodies of the pathogen are developed on the bulbs, and when they mature, turns to black colour with concentric rings on the neck and outer bulb scales (Figure 30.4). In cases of higher humidity, the disease spreads to the inner scales of bulbs with yellow lesions. Finally, bulbs are shrivelled and prematurely sprouted. In alternate warm and wet conditions, the pathogen causes damping-off and leaf spot symptoms.

30.4.8.2 Causal Organism

The disease is caused by *Colletotrichum circinans* Berk. Conidia are acrogenously borne and individually bud off. Conidia are fusiform shaped, hyaline to slightly ochraceous in colour and a prominent vacuole is present in the centre of the conidium, but under some conditions the conidium may contain a large number of vacuoles. Chlamydospores are thick walled, dark brown in colour, egg-shaped and normally terminal but occasionally intercalary in position. The fruiting bodies, acervulli, are formed on the stromata, which forms beneath the cuticle of the host. Conidiophores are short, hyaline and form in palisade layers and rupture the cuticle of the host for emergence.

30.4.8.3 Disease Cycle and Epidemiology

The pathogen overwinters in the soil and on infected bulbs. Warm and moist conditions favour conidial production. Wind and rain splashes are secondary spreading agents of the disease. Conidia usually infect mature bulbs and symptoms are expressed when favourable conditions like free moisture and 20–26°C of temperature prevail.

30.4.8.4 Disease Management

Proper curing of the bulbs should be done immediately after harvest. Bulbs should be stored in properly ventilated rooms.

FIGURE 30.4 Symptoms of onion smudge caused by *Colletotrichum circinans*.

Yellow and red skinned onion varieties should be used in regions of higher disease pressure. Usage of disease-free planting materials and crop rotation with non-host crops are good cultural approaches to reduce disease severity. Harvesting of crop should be done in dry weather conditions, and immediate curing helps to reduce disease incidence.

30.4.9 Southern Blight or Sclerotium Bulb and Stem Rot

30.4.9.1 Symptoms

The pathogen infects the outer scales of bulbs and leads to development of whitish lesions. Bulb and neck tissues become soft in diseased bulbs and further watery rot symptom is noticeable. Finally, white mycelial growth develops on the surface of the bulbs and pale brown coloured sclerotia are formed on the diseased bulbs or debris or in soil.

30.4.9.2 Causal Organism

The disease is caused by *Sclerotium rolfsii*. The sexual stage of pathogen is *Athelia rolfsii*. The pathogen cause root diseases over 500 plant species. The pathogen produces resting structure, i.e., sclerotia, at the end of its life cycle.

30.4.9.3 Disease Cycle and Epidemiology

The causal organism has wide host range, i.e., about 500 plant species. The pathogen survives as sclerotia in the soil and also in plant debris. Spread of pathogen is mainly through the movement of soil and irrigation water. Diseases become severe in 25–30°C temperature and in moist soils. Pathogen multiplication rapidly decreases at temperatures below 15°C.

30.4.9.4 Disease Management

Deep ploughing should be done in peak summer months to reduce soilborne inoculums. Field sanitation like removal and burning of crop residues should be carried out. Following soil fumigation or soil solarization approaches helps to reduce disease on succeeding crops. Crop rotation with non-host crop helps to decrease the soilborne inoculum. Postharvest application of fungicides can reduce storage losses. Bulbs should be stored at 10°C or lower to avoid storage losses due to disease (Eckert and Ogawa, 1988).

30.5 Bacterial Diseases

Bacterial diseases have major economic importance and result in huge losses (qualitative and quantitative) of produce in field, transport and storage. Rotting due to bacteria can occur in both transport and storage. Bacterial infected bulbs are non-marketable. Bacterial pathogens cause rots, necrosis and blights of onion, garlic and other *Allium* species as primary or secondary pathogens. It is followed by secondary infection of fungal pathogens, viz., *Aspergillus*, *Alternaria*, *Fusarium* and

Botrytis sp. Bacterial diseases usually occur in wet periods with warm temperature.

30.5.1 Bacterial Brown Rot

30.5.1.1 Symptoms

Bacterial brown rot is a destructive storage disease of onion. Rotting initiates at the neck of the bulb and foul smell is noticed in the neck of the bulb when squeezed. The pathogen mainly enters through wounds and causes infection.

30.5.1.2 Causal Organisms

The disease is caused by *Pseudomonas aeruginosa* (Schroeter) Migula. The pathogen is a Gram negative, rod-shaped and polar flagellated organism. It is an opportunistic pathogen that infects wounds, surgical parts, burns and sites of catheterization in humans. The pathogen has ability to grow in temperature range of 4–43°C, and average growth is noticed at 37°C. Virulence of the pathogen is mainly due to production and secretion of various lytic enzymes that can degrade polysaccharides of plant cell wall.

30.5.1.3 Disease Management

Proper curing and drying of the bulbs should be done immediately after harvest to minimize the occurrence of disease. Diseased bulbs must be removed prior to storage. If rainfall occurs during the maturity period of bulbs, foliar spraying of antibiotic streptocycline (0.02%) should be done.

30.5.2 Bacterial Leaf Streak and Bulb Rot

30.5.2.1 Symptoms

The disease is initiated with development of oval shaped water-soaked lesions on leaves and tip burn symptoms on leaves. Leaf streaking is a characteristic symptom of the disease with varying lengths of streaks. The green leaf streaks gradually become black. As the disease progresses, whole leaves dry and collapse. Leaf distortion and twisting symptoms are also noticed in the severe form of disease. Bulb infection is seen as characteristic dark spots on the outer scales, while reddish brown discoloration is seen on the inner scales. Later, a ring-like pattern of rot symptom develops due to restriction by the scales.

30.5.2.2 Causal Organism

The disease is caused by *Pseudomonas viridiflava* (Burkholder) Dowson. The pathogen causes leaf blight on broad range of hosts, viz., cauliflower, cabbage, eggplant, grapevine, melon, pumpkin, poppy, carrot, tomato, lettuce, pea and turnip. The pathogen is also associated with pith necrosis of pepper and tomato (seedborne).

30.5.2.3 Disease Cycle

The pathogen is associated with weeds and survives over seasons. Humid and rainy conditions are favourable for pathogen growth and multiplication. The secondary spread of the pathogen is through raindrops or splashes. Prolonged rain periods resulted in epidemic of the disease. Excessive nitrogen fertilization favours disease development. Frost damage predisposes onion plants to the disease.

30.5.2.4 Disease Management

Foliar applications of antibiotic streptocycline along with copper fungicides restrict the spread of the disease. Excessive nitrogen fertilization should be avoided. Harvest the bulbs at proper maturity stage without wounding or damaging. Proper curing and drying of bulbs should be done with forced hot air (McDonald et al. 2004).

30.5.3 Bacterial Soft Rot

Bacterial soft rot usually affects onion bulbs in transit and storage periods worldwide. Many of the bacterial species of various genera have the ability to produce enzymes to macerate parenchymatous tissues of plants, but only a few bacterial species cause rotting of plant tissues.

30.5.3.1 Symptoms

Bacterial soft rot symptoms are only noticed on mature bulbs. Initially, infected scales become water-soaked and pale yellow to light brown (in the case of *Dickeya chrysanthemi*) or greyish to white (in the case of *Pectobacterium carotovorum* subsp. *carotovorum*). As rotting invades, fleshy scales of the bulbs become soft and sticky along with breaking of inner portion of bulb. When diseased bulbs are squeezed, foul-smelling thick watery liquid comes from the neck of the bulb (Rahman et al. 2017).

30.5.3.2 Causal Organisms

Dickeya chrysanthemi (syn. *Erwinia chrysanthemi*), *Pectobacterium carotovorum* subsp. *carotovorum* (syn. *E. carotovora* subsp. *carotovora*) (Jones) Bergey et al.

30.5.3.3 Disease Cycle and Epidemiology

The infection occurs in the field prior to harvest if heavy rains prevail and leaves are dry. The pathogen survives in the soil and crop residues. The pathogen spreads through rain splashes, irrigation water and some insects. The pathogen enters bulbs through wounds which are caused by mechanical injuries and sunscald. In addition, onion maggots are able to carry pathogen while feeding. The disease is favoured by a warm and humid climate with temperature range of 20–30°C. If temperature is above 3°C in transport and storage, soft rot disease develops gradually (Perombelon and Kelman 1980).

30.5.3.4 Disease Management

Avoid sprinkler irrigation in the field. Manage onion maggots to avoid spread of the disease. Application of copper-based fungicides manages the disease. The top portion of the onion should be properly matured before harvesting. Avoid damaging of bulbs while harvesting. Before storage, onion bulbs should be properly dried. Provide proper temperature and relative humidity with good ventilation in storage to prevent bacterial decay due to soft rot.

30.5.4 Sour Skin of Onion

30.5.4.1 Symptoms

In the field, the disease symptom appears as light brown colouration of one or two leaves. Watery rot develops at the base of the leaves and spreads to the neck regions. This allows the leaves to be easily pulled off from the diseased bulb. The outermost scales and inner scalesof the bulb are not infected. This distinguishes sour skin disease from slippery skin, in which the inner scales are first infected. Slimy pale yellow to light brown decay is noticed. Infected bulbs smell like vinegar due to secondary invaders like yeasts.

30.5.4.2 Causal Organism

The disease is caused by *Burkholderia cepacia* (Burkholder) Palleroni & Holmes (syn. *Pseudomonas cepacia*). The pathogen is obligately aerobic, Gram-negative, rod-shaped (1.6–3.2 × 0.8–1.0 µm), polar flagellated (tuft) and inhabits soil and water. The pathogen usually produces nonfluorescent yellowish to greenish pigments, and the colour of pigments may vary among strains. The pathogen is negative for denitrification and arginine dihydrolase activities. It is positive for oxidase activity and gelatin liquefaction.

30.5.4.3 Disease Cycle and Epidemiology

The pathogen is spread through heavy rains, overhead irrigation and flooding. Infection occurs through wounds. Infection starts when water drops on the upright leaves and flows into leaf axils and carries the pathogen with it. The optimum temperature for growth of pathogen is 30–35°C. No growth is noticed at 4°C, and most of the strains grow at 41°C. Sour skin disease is favoured by rainstorms and warm weather (temperatures above 30°C) (Hussain et al. 1977).

30.5.4.4 Disease Management

Following furrow or drip irrigation rather than sprinkler irrigation. Avoid wounding or damaging of bulbs while harvesting and handling. Crops should be harvested at proper maturity stage and properly dried before storage. Storage of onions at 0°C with proper ventilation helps to prevent development of sour skin disease.

30.6 Factors Favourable to Storage Diseases

- A long period of wet conditions during the period of crop maturity.
- Injury or wounding due to other pre-harvest insect pests and diseases.
- Improper curing of bulbs and cloves.
- Temperature of 27°C with a relative humidity of 80% while curing.
- Leaving mature crop in the field for long period of time.
- Damage due to improper handling while harvesting or grading.
- Higher temperature with higher humidity during the storage period.

30.7 General Control for Storage Diseases of Onion and Garlic

- Maintenance of good crop hygiene like periodic crop rotations and proper crop residue management minimizes the occurrence of disease. Plant at proper spacing (For onions: 2–4 rows/bed, 9–18″ between rows, and 6–9 plants/foot or plants 3–4″ apart in the row. For garlic: 2-row beds 30″ apart on centre with 6″ spacing in and between rows; 3–4 rows per bed, with 6–8″ between and within rows; single rows spaced 24–30″ with 6″ spacing in the row).
- Avoid nitrogen fertilization in the late season of crop.
- Manage other insect pests and diseases in the field to prevent wounding of bulbs.
- Onion and garlic are properly cured in hot and dry conditions.
- Turn the onion and garlic frequently during field curing.
- Avoid the windrowing in wet weather conditions.
- Proper handling of bulbs while harvesting and grading to minimize physical damage to bulbs.
- After harvest bulbs should dry continually.
- Before storing of bulbs, inspect and remove diseased bulbs.
- Maintain below 20°C temperature and less than 80%relative humidity during storage.
- Storage rooms should have good airflow and ventilation facilities.
- Forced air curing should be practiced with temperature of 27°C with less than 80% relative humidity.
- Fungicidal treatment (if recommended only) of bulbs should be done. Selection of fungicides should be careful.

30.8 Conclusion

Agricultural production depends not only on cultivation area and genotype but also on various biotic stresses, including diseases and pests that threaten the crop during the cropping and postharvest periods. Onion and garlic suffer from various diseases, most of which cause economic loss through damage both in field and storage. It is difficult to manage many diseases through a solo approach; hence an integrated disease management approach is a prerequisite for successful disease management. Screening of genotypes and pesticides should consider the racial pattern of pathogen in geographical region because virulence of pathogens varies with the race. Management strategies should be harmless to humans since these are storage diseases. Control the diseases and insect pests by avoiding bruising and other mechanical injury in the field, during harvest and transport. Examine the garlic and onion before storage and discard all diseased bulbs. Proper spacing, recommended dose of fertilizer application and storing bulbs under ideal conditions will help to minimize postharvest storage losses. Despite use of recent synthetic fungicides and bactericides, there exists a tremendous scope to use new molecules of biopesticides and bio-rationales for eco-friendly and effective tackling of the postharvest diseases

REFERENCES

Ahir RR, Maharshi RP (2008) Effect of pre-harvest application of fungicides and bio-control agents on black mold (*Aspergillus niger*) of onion in storage. Indian Phytopathol 61(1): 130–131.

Anonymous (2009) Onion Diseases in New Mexico. Circular 538. Revised by Stephanie Walker, Natalie Goldberg, Christopher Cramer, College of Agricultural, Consumer and Environmental Sciences, NM State University.

Arowora KA, Adetunji CO (2014) Antifungal effects of crude extracts of *Moring oleifera* on *Aspergillusniger* v. Tieghem associated with postharvest rot of onion bulb. SMU Med J 1(2): 214–223.

Barnoczkine S (1986) Possibilities to control *Fusarium* on onions. Zoldsegtermesztesi kutato Intezet Bulletinje 19: 33.

Behrani GQ et al. (2015) Pathogenicity and chemical control of basal rot of onion caused by *Fusarium oxysporum* f. sp. *cepae*. Pakistan J Agric Eng Vet Sci 31: 60–70.

Brewster JL (1994) Onion and Other Vegetables: Alliums. CAB International, Willingford, UK. p. 236.

Borude D, Fatima S (2017) Studies on market diseases of garlic (*Allium sativum* L.). Epitome Int. J Multidisc Res 3(5) ISSN: 2395–6968.

Chethana BS, Ganeshan G, Rao AS, Bellishree K (2012) *In vitro* evaluation of plant extracts, bioagents and fungicides against *Alternaria porr*i (Ellis) Cif., causing purple blotch disease of onion. Pest Manag Horti Ecosyst 18(2): 194–198.

Chilvers MI, du Toit LJ (2006) Detection and identification of *Botrytis* species associated with neck rot, scape blight, and umbel blight of onion. Plant Health Progr 7(1): 1127-01-DG.

Clarkson JP, Mead A, Payne T, Whipps JM, Warwick HRI (2004) Effect of environmental factors of *Sclerotium cepivorum* isolate on sclerotial degradation and biological control of white rot by *Trichoderma*. Plant Pathol 53: 353–362.

Coskuntuna A, Ozer N (2008) Biological control of onion basal rot disease using *Trichoderma harzianum* and induction of antifungal compounds in onion set following seed treatment. Crop Prot 27: 330–336.

Currah L, Proctor FJ (1990) Onion in Tropical Region. Bulletinno.35, Natural Resources Institute, Chatham, Maritime, Kent, UK, 79 p.

Davis GN, Reddy CS (1932) A seeding blight state of onion bulb rot. Phytopathology 22: 8.

Dugan FM, Hellier BC, Lupien SL (2007) Pathogenic fungi in garlic seed cloves from the United States and China, and efficacy of fungicides against pathogens in garlic germplasm in Washington State. Am Plant Pathol 52: 426.

Eckert JW, Ogawa JM (1988) The chemical control of postharvest diseases: Deciduous fruits, berries, vegetables and root tuber crops. Annu Rev Phytopathol 26: 433–439.

El Neshawy S, Osman N, Okasha KH (1999) Biological control of neck rot and black mould of onion. Egypt J Agric Res 77: 125–137.

Everts KL, Schwartz HF, Epsky ND, Capinera JL (1985) Effects of maggots and wounding on occurrence of *Fusarium* basal rot of onion in Colorado. Plant Dis 69(10): 878–882.

FAO (2001) http://faostat3.fao.org

Flori P, Roberti R (1993) Treatment of onion bulbs with antagonistic fungi for the control of *Fusarium oxysporum* f. sp. *cepae*. Dif Piante 16: 5–12.

Gajbhiye M, Kesharwani J, Ingle A, Gade A, Rai M (2009) Fungus-mediated synthesis of silver nanoparticles and their activity against pathogenic fungi in combination with fluconazole. Nanomedicine 5: 382–386.

Garcia JL, Lopez MT, Perera ET, Berkmunn H, Gonzalez LA (1997) Investigations on chemical control during storage of onion bulbs. Agrotec Cuba 27(1): 8–10.

Gupta RC, Gupta RP (2013) Effect of integrated disease management packages on disease incidence and bulb yield of onion (Allium cepa L.). SAARC J Agric 11: 49–59.

Gupta RC, Pandey, Sujay, Srivastava KJ (2009) Integrated crop management for the production of export quality onion bulb (*Allium cepa* L.). Proc Int Conf on Horti-Horti for Livelihood Security & Economic Growth. Nov 09–12, 2009, Bangalore, pp. 687690.

Gupta RP, Verma LR (2002) Problem of diseases during storage in onion and garlic and their strategic management. In: Singh DP (Ed.). Implication of Plant Diseases on Produce Quality. Kalyani Publishers, Ludhiana, pp. 55–62.

Higashida S, Oshaki I, Narita Y (1982) Effects of crop rotation on onion yields and its microbial factors. J Bull Hokkaido Agric Exper Station 48: 1–9.

Hussain FN, Abd-Elrazik FA, Darweish A, Rushdi MH (1977) Survey of storage diseases of onion and their incidence. Egypt J Phytopathol 9: 55.

Hussein SAA, Abd-El-Razik AA, Abd-El-Rahman TM, Eraky AMI (2014) Influence of certain carbon and nitrogen sources on antagonistic potentiality of *Trichoderma harzianum* and *Bacillus subtilis* against *Botrytis allii* the incitant of onion neck rot. J Phytopathol Pest Manag 1(2): 9–16.

Jung J-H, Kim S-W, Min J-S, Kim Y-J, Lamsal K, Kim KS, Lee YS (2010) The effect of nano-silver liquid against the white rot of the green onion caused by *Sclerotium cepivorum*. Mycobiology 38(1): 39–45.

Kumar PT, Palakshappa MG (2008) Management of purple blotch of onion through bioagents. Karnataka J Agric Sci 21(2): 306–308.

Malathi S, Mohan S (2011) Evaluation of bio-control agents and organic amendments against onion basal rot caused by *Fusarium oxysporum* f. sp. *cepae*. Madras Agric J 98(12): 382–385.

Malathi S (2015) Biological control of onion basal rot caused by *Fusarium oxysporum* f. sp. *cepae.* Asian J Bio Sci 10(1): 21–26.

Mathur BL, Shukla HC (1963) Basal rot of onion. Phytopathology 32(9): 420.

McDonald MR, Jaime MA, Hovius MHY (2004) Management of diseases of onions and garlic. In: Naqvi SAMH (Ed.). Diseases of Fruits and Vegetables. Kluwer Academic Publishers, The Netherlands, pp. 149–200.

McLean KL, Stewart A (2004) Application strategies for control of onion white rot by fungal antagonists. N Z J Crop Hortic Sci 28: 115.

McLean KL, Swaminathan J, Frampton CM, Hunt JS et al. (2005) Effect of formulation on the rhizosphere competence and bio-control ability of *Trichoderma atroviride* C52. Plant Pathol 54: 212–218.

Metcalf DA, Dennis JJC, Wilson CR (2004) Effect of inoculums density of *Sclerotium cepivorum* on the ability of *Trichoderma koningii* to suppress white rot of onion. Plant Dis 88: 287–291.

Mishra RK, Gupta RP (2012) *In vitro* evaluation of plant extracts, bio-agents and fungicides against purple blotch and *Stemphylium* blight of onion. J Medicinal Plants Res 6(48): 5840–5843.

Naik DM, Burden OJ (1981) Chemical control of basal rot of onion in Zambia. Trop Pest Manage 27: 4554–60.

Nidagundi PP (2013) Studies on purple blotch of onion caused By *Alternaria porri* (Ellis). MSc (Horticulture) thesis, University of Horticultural Sciences, Bagalkot, Karnataka.

Othman ARM, Aziza AE, Mahmoud MA, Eifan SA, El Shikh MS, Majrashi M (2014) Application of silver nanoparticles as antifungal and antiaflatoxin B1 produced by *Aspergillus flavus.* Digest J Nanomat Biostruc 9(1): 151–157.

Pandey UB (1989) Problems in postharvest handling of onion and current status of research work done by AADF in the field of postharvest technology. AADF News Lett 9(3&4): 12–15.

Paris MA, Cotes A (2002) Evaluation of microbial isolates for control of *Sclerotium cepivorum* in onion. Bulletin-OILB/SROP 25: 311–314.

Perombelon MCM, Kelman A (1980) Ecology of soft rot Erwinia. Annu Rev Phytopathol 18: 361–387.

Pooler MR, Simon PW (1994) True seed production in garlic. Sex Pl Repr 7: 282–286.

Prachi K, Sonal B, Swapnil G, Aniket G, Amedea B et al. (2013) *In vitro* antifungal efficacy of copper nanoparticles against selected crop pathogenic fungi. Materials Lett 115: 13–17.

Rahman MM, Khan M, Mian IH, Akanda, Alam MZ (2017) Characterization of onion soft rot bacteria in Bangladesh. Bangladesh J Sci Ind Res 52(3): 209–220.

Rajendran K, Ranganathan K (1996) Biological control of onion basal rot (*Fusarium oxysporum* f. sp. *cepae*) by combined application of fungal and bacterial antagonist. J Biol Cont 10: 97–102.

Raju K, Naik MK (2006) Effect of pre harvest spray of fungicides and botanicals on storage diseases of onion. Indian Phytopathol 59(2): 133–141.

Ramjegathesh R, Ebenezar EG, Muthusamy M (2011) Management of onion leaf blight by *Alternaria alternata* (FR.) Keissler by botanicals and bio-control agents. Plant Pathol J 10: 192–196.

Rands RD (1917) The production of spores by *Alternaria solani* in pure culture. Phytopathology 7: 316–317.

Sabale A, Kalebere S (2004) Storage behaviour of onion (*Allium cepa* L.) varieties under the influence of pre-harvest and postharvest treatment of maleic hydrazide and carbendazim. Acta Bot Hungarica 46(34): 395400.

Salvestrin J, Letham D (1994) The control of *Aspergillus niger* in Australia. Acta Hortic 358: 289–293.

Sang MK, Han GD, Oh JY (2014) *Penicillium brasilianum* as a novel pathogen of onion (*Allium cepa* L.) and other fungi predominant on market onion in Korea. Crop Prot 65: 138–142.

Savitha AS, Ajithkumar K, Ramesh G (2014) Integrated disease management of purple blotch [*Alternaria porri* (Ellis) Cif] of onion. Pest Manag Hortic Ecosyst 20(1): 97–99.

Schwartz HF, Mohan SK (1995) *Compendium of Onion and Garlic Diseases.* The American Phytopathological Society. APS Press, Minnesota, p. 54.

Srinivasan R, Shanmugam V (2006) Postharvest management of black mould rot of onion. Indian Phytopath 59(3): 333–339.

Sumner DR (1995) Fusarium basal plate rot. In: Schwartz HF, Mohan SK (Eds.). Compendium of Onion and Garlic Diseases. APS Press, St. Paul, Minnesota, pp. 10–11.

Sumner DR (1995) Diseases of bulbs caused by fungi-black mold. In: Schwartz HF, Mohan SK (Eds.). Compendium of Onion and Garlic Diseases. APS Press, St. Paul. Minnesota, pp. 26–27.

Tolouee M, Alinezhad S, Saberi R, Eslamifar A et al. (2010) Effect of *Matricaria chamomilla* L. flower essential oil on the growth and ultrastructure of *Aspergillus niger* van Tieghem. Int J Food Microbiol 139: 127–133.

Vincelli PC, Lorbeer JW (1989) Blight-alert: A weather-based predictive system for timing fungicide applications on onion before infection periods of *Botrytis squamosa.* Phytopathology 79: 493–498.

Walker JC (1925) Two undescribed species of Botrytis associated with the neck rot disease of onion bulbs. Phytopathology 15: 708–713.

Yadav R, Singh A, Jain S, Dhatt A (2017) Management of purple blotch complex of onion in Indian Punjab. Int J Appl Sci Biotechnol 5(4): 454–465.

Yadav PM, Rakholiya KB, Pawar DM (2013) Evaluation of bioagents for management of the onion purple blotch and bulb yield loss assessment under field conditions. Bioscan 8(4): 1295–1298.

Yohalem DS, Nielsen K, Nicolaisen M (2003) Taxonomic and nomenclatural clarification of the onion neck rotting *Botrytis* species. Mycotax 85: 175–182.

31

Postharvest Diseases of Cucurbitaceous Fruits and Their Management

Dinesh Singh and Mehjabeen Afaque
Division of Plant Pathology, ICAR-Indian Agricultural Research Institute, New Delhi, India

CONTENTS

31.1 Introduction

Cucurbit vegetable crops belong to the *Cucurbitaceae* family and are grown extensively in tropical and subtropical regions of the world, including throughout India. In this family, about 118 genera and 825 species are recorded. India has rich biodiversity of these crops. It is believed that India is a primary and secondary centre of origin of many gourds and melon crops. Under the *Cucurbitaceae* family, cucumber, gourds, melons and squashes are the main crops of cucurbits. These vegetables are well known for their nutritional and medicinal values and as potential sources of crop diversity. They are used in different forms, such as salad (cucumber, gherkins, long melon), pickles (gherkins), sweet (ash gourd and pointed gourd), and desserts. Mostly cucurbits are annuals, directly sown and propagated through seed. However, the vegetative propagation process is used for pointed gourd, Ivy gourd and chow-chow, the perennials. Majority of cucurbits species possess the bitter principle called cucurbitacin at one stage of development or another. Bitterness is more prominent in crops like cucumber, bottle gourd and long melon. Some are acknowledged for their unique medicinal properties, e.g., bitter gourd. The world's most widely cultivated cucurbit is watermelon, followed by melons, cucumber, pumpkin and squashes. Since cucurbits are grown broadly throughout India and other tropical and subtropical regions globally, a variety of problems are faced by farmers daily. The severity of certain pest is such that it reduces the yield to more than 50%, resulting in huge losses to the grower and the economy.

31.2 Postharvest Losses

A significant postharvest loss of quality and quantity of horticultural produce has been noted during storage and transportation if proper postharvest management practices are not applied in a timely manner. It has been recorded that fruit rots due to fungal causes is a key factor for deterioration of melons. In Australia, Morris and Wade (1983) reported that *Fusarium* spp., *Geotrichum* spp., *Rhizopus* spp., *Cladosporium* spp., *Alternaria* spp. and *Pseudomonas* spp. are responsible for decay in melon. It has also been estimated that the average disease loss after transport to distant markets on the east coast is 30–50% and has even reached 80% in extreme cases. Heavy postharvest losses may occur due to cultivar susceptibility to disease, improper postharvest handling of the produce, lack of proper facilities for packaging and storage temperature and prolonged transport time (Mayberry and Hartz 1992).

31.3 Postharvest Diseases

Cucurbitaceous crops suffer from a wide range of pathogens that constitute >200 diseases, which affect production as well as productivity of the crops (Zitter 1996). The pathogens such as fungi, bacteria and viruses infect crops during fruit set until harvest (Babadoost 2004; Hausbeck and Lamour 2004; Koike et al. 2007; Kazi et al. 2019). However, some diseases may cause infection to cucurbits during storage and transit stages. The major rotting of pumpkin fruits is due to infection of *Phytophthora capsici* (Phytophthora rot), *Fusarium* spp. (Fusarium rot), *Didymella bryoniae* (black rot), *Sclerotinia sclerotiorum* (*Sclerotinia* rot) and *Xanthomonas campestris* pv. *cucurbitae* (bacterial spot) (Babadoost 2004). Several fungi including *Alternaria tenuis*, *A. alternata*, *Botrytis cinerea*, *Choanephora cucurbitarum*, *D. bryoniae*, *Fusarium oxysporum*, *Sphaerotheca fuliginea*, *Phytophthora capsici*, *Rhizopus nigricans*, *Penicillium oxalicum* and *Geotrichum candidum* infect cucumber plants and affect the quantity of yield and quality (Blancardn et al. 2005; Farrag et al. 2007; Mohapatra et al. 2014; Sani et al. 2015; Ziedan and Saad 2016). Ziedan et al. (2018) obtained six fungal genera –*A. tenusinium*, *Fusarium* spp. (*F. fujikuroi*, *F. verticiolides*, *F. solani*, *F. geraminearium* and *F. incarnatum*), *G. candidium* and *P. alli*–to infect produce of cucurbits. Among them, species of *Fusarium* and *Galactomyces* were the ones mainly associated with fruits of cucumber in open field cultivation from El-Gharbeia; 50 and 25%, respectively. Various pathogenic fungi were isolated from cucumber and they showed isolation frequency differently; for example, *A. alternata* (52.0% isolation frequency), *F. equiseti* (40.0%), *Cladosporium tenuissimum* (27.0%), *B. cinerea* (6.0%), *F. solani* (6.0%), *Corynespora cassiicola* (3.0%), *Aspergillus* spp. (2.0%), *Curvularia* sp. (1.0%) and *Bipolaris* sp. (1.0%). Among them, *F. equiseti* was the most destructive, whereas *C. cassiicola*, *B. cinerea* and *A. alternata* were found the least effective to cause disease on the fruits of cucumber (Al-Sadi et al. 2011). Bitter gourds are susceptible to many fungi and bacteria to cause postharvest diseases, including *A. alternata*, *R. solani*, *Pythium* sp., *R. stolonifer*, *B. theobromae*, *Fusarium* sp., *G. candidum* and *Erwinia carotovora* (Naureen et al. 2009; Sharma and Sharma 2018). In melon fruits, *Alternaria* rot, *Fusarium* rot, blue mould rot (*Cladosporium* sp.) and pink mould rot are major postharvest diseases (Snowdon 1992). The major postharvest diseases of cucurbitaceous crops are described below.

31.3.1 Black Rot or Gummy Stem Blight

Black rot disease of cucurbits was first recorded in 1891 from Europe and the United States. The disease causes significant during pre-harvest and postharvest rotting in winter squashes as well as pumpkin, cucumber and cantaloupe. *Didymella bryoniae* infects at the vegetative stage of pumpkins and other crops of cucurbits, and is called gummy stem blight disease. *Stagonosporopsis cucurbitacearum* causes 100% field loss in cantaloupe under favourable environment for infection (Nuangmek et al. 2018),whereas in watermelon, gummy stem blight and black rot may cause significant losses in production in the field as well as after harvest of the produce (Maynard and Hopkins 1999). Keinath (2017) reported that commercial cultivars of cucurbits did not show resistance against gummy stem blight disease.

31.3.1.1 Symptoms

31.3.1.1.1 Black Spot

The disease symptoms appear before harvest of the produce in the field and it will continue to develop disease during storage as well as in transit. Lesions develop as dark-brown, small and water-soaked on fruit (Zitter 1996). Later lesions expand and they become sunken and black. On infected areas of fruits, fungal fruiting bodies as pycnidia develop. The disease is called black rot due to the presence of blackened fungal tissue on the rind. Injury to the rind during harvesting or handling after harvest may accelerate development of black rot on pumpkin. The infected fruits can rot within 2–3 days under favourable conditions. Keinath (2011) also reported that black mycelia of *S. cucurbitacearum* developed inside melon and pumpkin fruit.

31.3.1.1.2 Gummy Stem Blight

The spots have a ringed appearance. Spots appear as water-soaked reddish-brown on the mid vein of leaves and petioles. The infection often starts at margins of the leaf. The leaves become cracked after drying and may tear off, giving a ragged appearance. Stem infections consist of oblong, water-soaked brown lesions. The main stem lesions enlarge and slowly girdle the stem. Red or brown gummy fluid oozes out from the wounded or canker portion, and whole plant shows wilting. The fungus also produces tiny black pimple-like fruiting bodies on the stem or nodes.

31.3.1.2 Causal Organism

Stagonosporopsis cucurbitacearum (syn. *Didymella bryoniae* (Auersw.) Rehm (anamorph *Phoma cucurbitacearum* (Fr.:Fr.) Sacc.) is a causal agent of gummy stem blight and black rot disease of cucurbits (Sitterly and Keinath 1996; Keinath et al. 1995; Moumni et al. 2019). *D. bryoniae* produces dark pseudothecia, which are globose, immersed, becoming erumpent, with size of 125–213 μm dia. The pseudothecia open by apical papillate ostiole and produce numerous bitunicate asci. The asci are cylindrical to subclavate, short stipitate or sessile. Each ascus has eight ascospores with size of 60–90 × 10–15 μm. The ascospores are hyaline, ellipsoid, having mostly rounded ends, monoseptate. They are slightly constricted at the septum and size of ascospore is 14–18 × 4–7 μm. The lower cell of the ascospore is thinner than the upper cell. *Phoma cucurbitacearum* produce dark pycnidia on the surface of infected tissues and they are dark, with size of 120–180 μm diameter, solitary or agregarious, immersed, becoming erumpent. The fungus produces conidia, which are hyaline, 1-septate, cylindrical with rounded ends. However, a small percent of conidia are unicellular with size of 6–13 × 3–4 μm.

31.3.1.3 Disease Cycle and Epidemiology

The fungal pathogen survives on diseased crop residues and seeds between two crop growing seasons (Lee et al. 1984; Koike 1997; Keinath 2002, 2008). The pathogen can also survive in the soil for 2 years in absence of the host. Keinath (2002) reported that in most cases *D. bryoniae* survived for less than 9 months in diseased host debris. However, it can survive in crowns of muskmelon plants infected by gummy stem blight (Keinath 2008). Pathogen survived on seed of cucumber and pumpkin, particularly the seed coat, perisperm, and cotyledon seed (Lee et al. 1984). Van Steekelenburg (1983) reported that airborne ascospores are considered to be another source of inoculum. The fungal conidia are disseminated through irrigation water, splashing rain as well as winds from the disease source to healthy crops. Various reports with valuable information on the epidemiology of gummy stem blight cucurbits and cucumber fruit rot are available (Van Steekelenburg 1986). Striped cucumber beetles (*Diabroticaundecim punctatahowardii* Barber and *Acalymma vittatum* Fabricius) are believed to transmit *D. bryoniae* in a non-persistent manner. The beetles cause injury to healthy plants, which provide opportunities for infection.

Temperature 20–24°C and free moisture favour disease development. However, the pathogen is often devastating in warm and humid weather. For disease development, moisture is of greater importance than temperature. Peak ascospore dispersion occurs after rain and during dew periods at night. For infection of the pathogen, free moisture on leaves for at least 1 hour is necessary, and further continuous leaf wetness is required for lesion development. Fruit is penetrated either through wounds or through flower scars during pollination. Fruit rot begins to develop about 3 days after infection. The optimum conditions for the disease development are>90% humidity, leaf wetness and 16–24°C temperature (Park et al. 2006; Seebold 2011).

31.3.1.4 Disease Management

The management of black rot disease should commence by effectively managing gummy stem blight in the field (Van Steekelenburg 1986).

- Proper diagnosis of the disease is required because charcoal rots as well as fungal wilt show similar symptoms.
- Certified disease-free seed should be purchased from authentic sources.
- All plant debris should be destroyed from within and around the farms.
- Two-year crop rotation with non-host crops should be followed for management of the disease.
- Chilton, Gulfcoast, AUrora, and AC-70-154 varieties of cantaloupe showed resistance to gummy stem blight disease (Norton and Cosper 1989).
- Irrigation through overhead sprinklers should be avoided.
- For field sanitation, infected fruits and other plant debris should be removed and destroyed from the field after the season is complete.
- Seeds should be treated with Thiram or Captan @ 3.0 g/kg of seed.
- Spray Carbandazim (0.2%) as soon as the disease is noticed. If the disease is not controlled, spray Mancozeb (0.25%) or Propiconazole (0.1%).
- Zhou and Everts (2008) reported that the disease can be managed by combined application of green manure along with fungicide such as Chlorothalonil (Bravo Ultrex 82.5 WDG at 3.0 kg/ha), biopesticide (*B. subtilis*, Serenade 10WP at 4.5 kg/ha) and Pyraclostrobin plus boscalid (Pristine 38 WG at 1.0 kg/ha). To minimize use of synthetic fungicide, biopesticide may be substituted.
- Safe harvesting of fruits is required to reduce the chance of injury to the fruits.
- Antagonistic activity was recorded by *Bacillus subtilis* against the pumpkin pathogen *Didymella bryoniae* (Fürnkranz et al. 2012; Glassner et al. 2015).
- Curing of pumpkins and winter squashes fruits at temperature of 20–25°C or higher temperatures for 7–15 days should be done to harden the rind.
- The fruits should be stored at 11–16°C with optimum temperature of 13°C and 55–75% relative humidity (RH) with optimum 60% RH (Zitter 1996).

31.3.2 Phytophthora Fruit Rot

Phytophthora capsici Leonian was first reported on cucurbits in 1937 in Colorado and California as Phytophthora blight (Tompkins and Tucker 1937). Since then, the disease has been noted worldwide under temperate, subtropical, and tropical environments where cucurbits are grown. It is now one of the most devastating diseases of cucurbit crops and can cause total crop loss. Foliar blight as well as fruit rot can cause total loss of the produce. The pathogen has wide host range infecting 49 plant species (Erwin and Ribeiro 1996). The major hosts are red and green peppers (*Capsicum annuum*), watermelon (*Citrullus lanatus*), cantaloupe and honeydew melon (*Cucumis melo*), cucumber (*C. sativus*), blue Hubbard squash (*Cucurbita maxima*), *C. pepo* varieties, pumpkin (*C. moschata*), tomato (*Solanum lycopersicum*), black pepper (*Piper nigrum*), eggplant (*S. melongena*), etc. Six additional hosts are reported: sugar beet, Swiss chard, lima bean, turnip, spinach, and one weed species (*Abutilon theophrasti*) (Tian and Babadoost 2004).

31.3.2.1 Symptoms

The pathogen infects seedlings, leaves, vines and crown and produces numerous sporangia. In cucurbit crops, root and crown rot appears suddenly and shows as a permanent wilt due to root infection following irrigation or rain. Roots and stems at the base of soil appear water-soaked, dark brown in colour and soft. The plants or vines collapse rapidly. Small water-soaked areas appear as symptoms which expand quickly. The disease area is covered by fungal sporangia under high humidity and sporangia-covered lesions show yeast-like with grey to white appearance. The development of rotting is very fast until the fruit is completely rotted.

31.3.2.2 Causal Organism

Phytophthora capsici Leonian is a causal agent of blight disease. The pathogen is coenocytic hyphae and has sexual oospores, asexual sporangia and zoospores. *P. capsici* fungus is isolated from pumpkins and it grows from 10 to 36°C temperature and with optimum temperatures of 24–33°C (Islam et al. 2005). Colonies of fungus are cottony, petaloid, rosaceous, and stellate. Sporangia are papillate in different shapes and have a long pedicel (35–138 µm). Sporangia are subspherical, ovoid, obovoid, ellipsoid, fusiform, and pyriform and 32.8–65.8 × 17.4–38.7 µm. Zoospores are produced inside sporangia and are biflagellate. Oospore diameter ranges from 22 to 34 µm and amphigynous antheridia. *P. capsici* is heterothallic, and both A1 and A2 types may exist in the same field (Islam et al. 2005). Oospores are resistant to desiccation, cold temperatures, and other extreme environmental conditions. They can survive in the soil in the absence of a host plant for 3 years or longer (Pavón et al. 2008). Oospores germinate to produce sporangia and zoospores. Zoospores are released from sporangia in water. They are dispersed by surface water or irrigation. The zoospores swim for some hours and then infect susceptible plant tissues.

31.3.2.3 Life Cycle and Epidemiology

The disease is polycyclic under favourable conditions. The pathogen reinfects the crops many times throughout the growing season. *P. capsici* has two mating types, A1 and A2, which are not distinguishable on the basis of morphology but are distinct genetically. In the presence of both mating types in one field, they may mate to produce oospores. Oospores can work as resting spores and can survive in the soil for many years and can provide the primary inoculum for disease initiation under favourable conditions. Oospores are spread from field to field in infested soil adhering to machinery or humans. The asexual stage of *P. capsici* is responsible for initiating infection and mainly depends on water for infecting plants. Disease will nearly always begin in low spots of fields or in areas that do not drain readily. Infected soils that are saturated for several hours and temperatures moderately warm favour *P. capsici* to form sporangia having zoospores. Zoospores are released from sporangia into the saturated soil and are attracted to living plant parts in the soil and swim towards them. Zoospores can germinate after finding a suitable host plant. They infect any plant part either in the soil (roots, crowns) or via splashing water (leaves, fruit). They are disseminated primarily through water splashing, irrigation or rain, or water running through fields.

The disease develops rapidly under conducive environmental conditions. Soil moisture conditions are very important for development of the disease. When soil moisture is at field capacity, the sporangia of *P. capsici* form. While under saturated soil, they release zoospores. The disease is generally associated with heavy rainfall, excessive irrigation, or poorly drained soil. The disease incidence increases under frequent irrigation.

31.3.2.4 Disease Management

31.3.2.4.1 Cultural Practices

There are several management practices which help to reduce disease incidence, described here.

- Growing susceptible crops in *Phytophthora*-infested soil should be avoided.
- Crop rotation of at least 3 years with non-host crops such as corn, small grains, brassicas, alliums, or other non-hosts.
- Susceptible crops should be separated from the main crop, which restricts the movement of water from one planting to another.
- Susceptible crops should not be grown at low level areas of land.
- It is better to always work in clean fields before working in infested fields to restrict entry of *Phytophthora* into uninfected fields. Soil should always be washed from equipment after working in *Phytophthora*-infested soil.
- The following methods should be followed to maintain proper drainage and to avoid prolonged soil saturation.
    - Irrigation systems should be checked regularly to find and fix leaks.
    - Do not allow water stagnation in the field; ensure there is proper drainage.
    - Avoid soil compaction.
    - Water should be able to flow out of the field by creating breaks in raised beds and clearing away soil at the ends of rows to prevent damming.
    - Mulching of crops should be done with straw, or cover crop stubble should be left, to reduce splash dispersal of inoculum.
    - Follow good sanitation of the field and discard culled fruit outside of the field.
    - If symptoms are localized in a small area, disking the area may be useful.

31.3.2.4.2 Resistant Varieties

It is good practice to grow resistant varieties whenever possible. Pumpkins with hard, gourd-like rinds are less susceptible to *Phytophthora* fruit blight. Varieties such as Lil Ironsides, Apprentice, Iron Man, Rockafellow and Cannon Ball are moderately resistant to the disease.

31.3.2.4.3 Chemical Control

- Fungicide application alone cannot manage Phytophthora blight. The disease can be successfully managed only by managing the drainage system of the field as well as utilizing other cultural practices.
- To manage Phytophthora blight, fungicide programs must be preventive, which requires multiple applications throughout the crop season.

- Mefenoxam or Metalaxyl fungicide can be used as seed treatment to protect seedlings of pumpkin for 5 weeks after sowing (Babadoost and Islam 2003).
- Treating the seed with Chitosan @ 500 ppm showed disease resistance against damping off disease in cucumber caused by *P. capsici* (Zohara et al. 2019).
- Several fungicides have been reported to apply more than 10 applications to control the disease in pumpkin fields (Babadoost 2008; Holmes 2007).
- Due to development of resistance by pathogens against Mefenoxam/Mancozeb (e.g., Ridomil, Apron XL), broad spectrum fungicides like Copper compounds (Kocide, Basic Copper 53), phosphorous acid fungicides (e.g. Prophyt, K-phite, Agri-fos) can be used as drench treatments at plant and also foliar application.
- Newer products such as fluopicolide, cyzofamid, mandipropamid and ametoctradin/dimethomorph are registered for *Phytophthora* in cucurbits to control the disease. The fungicides should be applied before appearance of disease symptoms, when the risk of infection is high depending on weather conditions and local soil (McAvoy 2014).

31.3.2.4.4 Bio-control Measures

- Biofungicides such *as Trichoderma* spp. (e.g. Root Shield Plus), *Bacillus* spp. (e.g. Serenade, Double Nickel, Taegro) or *Streptomyces* spp. (Actinovate, Mycostop) may be used through drip irrigation or foliage spray. However, their efficacy is generally lower than conventional fungicides.
- Application of *Pseudomonas fluorescens* CV6 and V11 as a soil treatment reduced disease incidence 85.71 and 69.39% caused by *P. drechsleri* in cucumber under glasshouse conditions, respectively (Maleki et al. 2011). Seed treatment of cucumber with PGPR strains as *P. stutzeri* (PPB1), *Bacillus subtilis* (PPB2, 5, 8, 9 and 11), *Stenotrophomonas maltophilia* (PPB3), and *B. amyloliquefaciens* (PPB4) @ 10^8 CFU/seed, significantly suppressed Phytophthora crown rot under greenhouse. Among them, *B. subtilis* PPB11 reduced the disease maximum (Islam et al. 2016). Another strain of *B. subtilis* ME488-controlled pathogens of cucumber might be due to the release of antibiotics (Chung et al. 2008). In an experiment, two *Streptomyces* species, i.e., *S. rimosus* C 201 and *S. monomycini* C 801 isolates had the highest antagonistic activity against *P. drechsleri* (Sadeghi et al. 2017). Plant growth promoting rhizobacteria may indirectly enhance seed germination and vigour index due to reduction of disease incidence caused by seed mycoflora (Begum et al. 2003).

31.3.3 Anthracnose

In 1867, the disease was first reported from Italy on gourd crop. *Colletotrichum orbiculare* (Berk. & Mont.) Arx is the causal agent of anthracnose disease of cucurbits in warm and rainy summers. The pathogen infects melon, watermelon, squash, pumpkins and cucumbers. Gopinath et al. (2006) reported that pathogenic specialization is based on the ability of pathogen to infect cucurbit genera, species, and cultivars. It occurs in most humid regions of the world where the crop is grown. In India, occurrence of anthracnose disease has been reported from Punjab, Haryana, Assam, Karnataka and other states wherever cucurbits are grown in rainy weather. Muskmelon, bottle gourd and cucumber are most susceptible, whereas bitter gourd is least affected and pumpkin is rarely affected. In temperate regions heavy losses occur on fruits of watermelon. Ullasa and Amin (1986) reported yield losses of marketable fruits up to 99.5% due to this disease.

31.3.3.1 Symptoms

The disease symptoms appear on the seedling stage as cotyledons wilt. When the fungus is seed-borne, stem lesions near the soil line are visible. Small pale yellow, water-soaked spots appear near veins on mature leaves and enlarge rapidly, turn tan to dark brown spots. The spots coalesce and form blighting symptoms, distortion, and finally the entire leaves die. The spots dry and often crack at dead centres of old lesions. They tear and give a ragged appearance to the foliage. On petioles and stems, the lesions are elongated and slightly sunken. Young fruits may turn black and die due to infection of their pedicels, while on older fruit, circular, noticeably sunken, and dark-green to black lesions develop and may show salmon coloured exudates under moist weather conditions (Gupta and Thind 2006).

31.3.3.2 Causal Organism

The disease is caused by fungus *Colletotrichum orbiculare* (Berk. and Mont.) Arx (Syn. *C. lagenarium* (Pass.) Ellis and Halst (Teleomorph: *Glomerella lagenarium* Stevens). The teleomorph stage is rarely found in nature. Tsay et al. (2010) reported a new species, *Glomerella*, i.e., *G. magna*, causing anthracnose on cucurbitaceous crops like cucumber, calabash gourd (*Lagenaria siceraria*) and *Luffa* (*Luffa cylindrica*, *L. aegyptiaca*) in Taiwan. The mycelium is septate, hyaline when young and dark when old. Stromata (acervuli) are brown to black and variable in size, bearing black setae. Conidiophores are hyaline on the host surface. Setae are brown, thick-walled, 2–3 septate, and 90–120 µm long. Conidia are produced one at a time at the tip of the conidiophores and accumulate in a slimy, pinkish mass. Individually the conidia are hyaline, oblong to ovate oblong, 1-celled, and measure 13–19 µm × 4–5 µm. They germinate through germ tube on contact with a hard surface. Seven physiological races (1–7) of the pathogen have been reported and race 1 and 2 are common and virulent (Zitter 1996). Cucumber and melons are primarily infected by races 1 and 3, whereas watermelon is mainly infected by race 3.

31.3.3.3 Disease Cycle and Epidemiology

C. orbiculare is mainly soil-borne but it becomes seed-borne when the fruits are infected and mycelium reach the seed. The pathogen survives in diseased crop debris, on volunteer plants, or in weeds of the cucurbitaceous family between two cropping seasons and serve as the primary source of inoculum. The disease spreads by rain splash, irrigation water, workers, animals, beetles and other insects and machinery moving through the crop in the field under moist conditions which lead to secondary infection. Conidia are dispersed to the healthy plants and cause infection under favourable weather conditions.

The spores of the pathogen are germinated best at 22–27°C and relative humidity is 100% for at least 24 h. Humid rainy weather is essential for infection. If moisture is present, spores germinate and penetration occurs effectively up to 72 h after conidial deposition. The fruiting bodies of the pathogen are visible on the lesions in next few days. Development of disease can occur at 20–30°C temperature with optimum temperature of 25°C and 100% relative humidity for at least 18 h. The pathogen can be passed on seed harvested from diseased fruits. They can be spread by cucumber beetles after feeding. Disease development can be brought about by frequent rains, high humidity and warm temperatures.

31.3.3.4 Disease Management

- The primary inoculums of the pathogen are reduced by deep ploughing of crop residue without delay after harvest.
- Crop rotation of 1–2 year periods with non-cucurbitaceous crops should be included.
- Cucurbitaceous weeds and volunteer crops should be removed from the field and the surrounding area.
- Disease-free seed must be used to reduce the primary inoculum of the pathogen.
- Resistance to races 1 and 3 can be found in Africa 8, Charleston Gray and Congo while African Citron W-695, Charleston Gray, Fairfax, Candy Red, Verona, Gray Ball and Super Sweetare resistant to race 2. Chauhan and Bhatia (2013) reported that 3 germplasm lines (GH-3, GH-9 and Ghiya-I GH-10 GH-25) of bottle gourd showed resistance against anthracnose disease caused by *C. lagenarium*.
- Treatment of seed with fungicides like Thiram is recommended.
- Sprays of Carbendazim followed by Captafol can best control the disease.
- Combined spray of Copper oxychloride and wet table sulphur was found most effective to reduce disease incidence and increase yield in bitter gourd with maximum cost-benefit ratio (1:2.82) (Fugro and Mandokhot 2000). The fungicide Propiconazole is effective to inhibit mycelial growth of the fungus even at low concentration (Gopinath et al. 2006).
- Foliar application of Carbendazim, Copper oxychloride, Mancozeb and Iprobenfos on bottle gourd cv. PSPL (highly susceptible) to control the disease. The best disease control (89.6%) was found by spraying of Carbendazim followed by Copper oxychloride (79.7%) (Chauhan and Bhatia 2012).
- *Bacillus mycoides* BmJ and *B. mojavensis* 203-7 controlled *G. cingulata* var. *orbiculare* by inducing systemic acquired resistance in the plant in the greenhouse. They also delayed disease onset and reduced total and live spore production of lesion area (Neher et al. 2009).
- Infected crop debris after harvest should be collected and destroyed.
- Avoid wounding of the fruits during harvesting.
- Fruits should be immersed in clean, fresh water containing chlorine (120 ppm).
- Eradicant or protective fungicides may be included in the spraying schedule to control the disease in the field.
- Neem leaf extract (50%) reduced disease severity (11.70%) when sprayed at the first appearance of symptoms of bottle gourd anthracnose (Mukund 2006).

31.3.4 Fruit Rot (Cottony Leak)

Certain species of *Pythium* such as *Pythium aphanidermatum*, *P. debaryanum* and *P. butleri* cause fruit rot of many crops of cucurbits. *Pythium* fruit rot (cottony leak or watery rot) often occurs during moist weather or poorly drained areas of the fields (Deadman et al. 2002). *Phytophthora* spp., *Fusarium* spp. and *Rhizoctonia* spp. are also responsible for this cottony leak disease of cucurbits. The pathogen infects many cucurbit fruits such as sponge gourd, snake gourd, parwal pumpkin, cucumber, kheera, bottle gourd, bitter gourd, etc. (Chand et al. 2014).

31.3.4.1 Symptoms

Disease symptoms of small, water-soaked lesions on immature or mature fruit often start near or in contact with the soil. The whole fruit is spoiled within 72 h. The epidermis is ruptured, and the fruit collapses. A white, cottony growth of fungus may be apparent on the lesion surface of the fruit under high moisture. In watermelons, the symptoms start at the blossom end of the fruit. The cottony leak is a mycelial growth of the pathogen which may be confused with that produced by species of *Phytophthora*. However, *Phytophthora* spp. felt-like mycelium on rotted fruits instead of cottony mycelium.

31.3.4.2 Causal Organism

The disease is caused by *Pythium* spp. including *Pythium aphanidermatum*, *P. debaryanum* and *P. butleri*. Besides *Pythium* spp., other fungi are also allied to cause cottony leak disease such as species of *Phytophthora*, *Fusarium* and *Rhizoctonia* in *Cucumis sativus* L. to give the name cottony leak. In 1928, Mitra and Subramaniam (1928) also reported a fruit rot of various cultivated cucurbits in India, caused by *P. aphanidermatum* (Eds.) Fitz. Infection was observed on cucumber, *Cucumis melo* L. var. *momordica*, white-flowered gourd (*Lagenaria leucantha* Rusby), *Luffa acutangula*, *L. aegyptiaca*, balsam-pear (*Momordica charantia* L.), serpent or snakegourd (*Trichosanthes anguina*), and *T. dioica*. The symptoms were in close agreement with that described by Drechsler for cucumber (Drechsler 1923).

31.3.4.3 Disease Cycle and Epidemiology

Pythium species survive for a longer period on various organic substrates in the soil, on many crops and weeds. The pathogen survives in the soil through thick-walled oospores as resting spores. The fungus produces sporangia and zoospores under free moisture available in the soil. The fungus infects the fruit by any means of vegetative mycelium, sporangia, zoospores, or oospores. Fruit exudates attract the zoospores and the zoospores swim towards the fruit. Although the pathogen is capable of infecting the fruit by direct penetration, wounds enhance the infection. Highly succulent plants are more susceptible to infection, especially excessive nitrogen due to deprived growing conditions. The pathogens grow quickly through diseased tissues. The pathogens are primarily disseminated by irrigation water as well as contaminated equipment within and among fields. Fruits on or near the soil level are readily attacked, and the disease has also been observed to spread in storage.

Cultures grown on corn meal, oatmeal, and potato-dextrose agar developed a copious amount of white cottony aerial mycelium, filling the space between the slope and the wall of the tube after 4 days at room temperature. Cultures on steamed corn meal produced a profuse aerial mycelium, filling the basal portion of the Erlenmeyer flasks after 7 days. Hyphae are cylindrical, smooth, and non-septate when young, measuring 1.7–6.2 μm in diameter, average 3.4 μm, and becoming septate with age. Sporangia were present on the aerial growth in all cultures, being most numerous on potato-dextrose agar, moderately abundant on corn meal and oatmeal, and rare on steamed corn meal Intramatrical sporangia were often produced, though never abundantly. Sporangia were spherical when acrogenous, measuring 14.5 μm × 28.3 μm in diameter, average 20.4 μm. Occasional intercalary bodies were formed which varied in size from 15.7 μm in width and 19.3 μm × in length to 23.3 μm × 27.3 μm, average 19.6 μm × 23.9 μm. Tufts of mycelium from potato-dextrose agar were placed in pea broth and allowed to grow at room temperature. Sporangia were produced abundantly in both solutions. Germination always occurs by 1–5 germ tubes production. To study the formation of oogonia, antheridia, and oospores, the fungus was grown in clear media. Agar squares containing the organism were cut from 2-day-old potato-dextrose agar plates and transferred to Petri dishes containing cornmeal decoction agar and plain water agar. Oogonia, antheridia, and oospores were produced freely in these media in 3–4 days. Oogonia were largely intramatrical, acrogenous, spherical with a smooth thin wall, and measured 17.3–24.2 μm in diameter, average 20.2 μm. Antheridia were generally androgynous, usually one to an oogonium, though occasionally two were present, in which case one was always of diclinous origin. Typical antheridia arose immediately below the oogonium on the oogonial stalk. They were somewhat clavate, short, upcurved, the tip in narrow contact with the oogonial wall. At maturity the contents of the oogonium differentiated into a peripheral layer, the periplasm, the remaining portion becoming the oosphere. A penetration tube was produced by the antheridium, permitting its contents to pass into the oosphere. After fertilization the oosphere further differentiated to form the oospore. When mature the oospore was spherical, smooth, and thick-walled, not filling the oogonial cavity. A reserve globule was present. The oospores measured 15.7 μm to 22.9 μm in diameter, average 17.4 μm, including the oospore wall, which was approximately 1.4 μm thick. Germination occurred by the production of normal hyphae arising from the tips of the enlarged threads. Of the species reported from cucurbits only *Pythium debaryanum* and *P. ultimum* produce spherical sporangia. Both have acrogenous spherical sporangia; the intercalary sporangia produced sparingly in *P. ultimum* are moderately abundant in *P. debaryanum*. Germination of the sporangia by germ tubes is common to both; however, zoospores are frequently produced by *P. debaryanum* but not by *P. ultimum*. The oogonia of *P. debaryanum* are smooth-walled and somewhat larger in diameter than those of *P. ultimum*. The intercalary oogonia frequently found in *P. debaryanum* are absent or extremely rare in *P. ultimum*. Several antheridia are produced in *P. debaryanum* which may be either diclinous or androgynous, but if the latter, they do not originate immediately below the oogonium as in *P. ultimum*. Mature oospores of both species lie free in the oogonial cavity. A reserve globule is present in oospores of each species, though somewhat smaller in size in *P. ultimum*. A small refrainment body is frequently associated with the globule. The oospore wall is approximately 1.0 μm thick in *P. debaryanum* and 1.5 μm in *P. ultimum*.

Pythium spp. requires warm weather conditions for growth. High soil moisture, high temperature and juvenile tissues of the host are important factors determining development of the disease. On the fruits, infections always occur when there is some injury to the skin, by soil particles, excessive wetness around fruits or insect bites.

31.3.4.4 Disease Management

- Grow high quality seed under optimum temperature, moisture and nutritional conditions.
- Plastic mulch reduces disease in semi-arid production environments.
- Overhead sprinkler irrigation or in high rainfall areas should be avoided to grow the crop due to accumulation of water on the plastic which promotes fruit decay.
- Excessive watering and low, poorly drained areas of fields should be avoided.
- Staking and/or mulching may be done to prevent fruit contact with the soil.
- Soil disinfection is not advisable.
- Hygiene transportation and storage is the only methods to evade this fruit rot.

31.3.5 Belly Rot

Belly rot incited by fungus *Rhizoctonia solani* is soil-borne, affecting mainly cucumber crops and is a major problem for cucumbers grown for pickling. It rarely infects other cucurbits. The belly rot disease incidence is significantly elevated during the rainy season. Fruit initially becomes infected in the field where they contact the soil.

31.3.5.1 Symptoms

The rot symptoms develop in fruits that come into contact with the soil. Yellowish brown, superficial discoloration occurs on young infected fruit, and later sunken, irregular spots on the underside or belly develop. On mature fruits, large water-soaked decayed areas may develop.

31.3.5.2 Causal Organism

Belly rot is caused by *Rhizoctonia solani* Kühn (teleomorph = *Thanatephorus cucumeris* (Frank) Donk). Fungal colonies on malt extract agar medium are light grey to brown in colour with profuse mycelial growth. A septum is always present in the branch of hyphae near the originating point with a slight constriction at the branch. No conidiophores or conidia are produced by the *R. solani*.

31.3.5.3 Disease Cycle and Epidemiology

R. solani is a soil-borne fungus and survives in the soil and infested crop debris as mycelia and sclerotia. Warm temperatures, high humidity, and excessive moisture favour infection and development of the disease. The disease symptoms can become visible within 24 h of infection and within 72 h entire fruits spoil under favourable conditions. Temperature is more important than moisture for disease development; however, high humidity and excessive moisture favour the infection. The pathogen is disseminated by moving soil, infected plant parts by equipment and also by irrigation water within fields.

31.3.5.4 Disease Management

- Filed should be ploughed deeply prior to planting.
- Crop rotation of 3 years or longer should be followed between cucurbit crops.
- Black plastic mulch should be used between the fruit and the soil to provide a physical barrier.
- Irrigation should be given as per requirement and avoid excessively wet soils.
- Crop debris should be removed from the field promptly.
- Sprays of the fungicides Azoxystrobin (Quadris), Chlorothalonil (Bravo), and Thiophanate-methyl (Topsin-M) pre-harvest provide protection against belly rot infection. Storing the fruit at 10°C will retard disease development during transit and storage.

31.3.6 Choanephora Fruit Rot

Choanephora flower and fruit rot is incited by *Choanephora cucurbitarum*. The pathogen infects many vegetable crops including beans, cantaloupe, eggplant, okra, peas, pumpkin and squash.

31.3.6.1 Symptoms

Symptoms appear on flowers and blossom ends of the fruit as a soft wet rot. Whisker-like fungal growth is observed on the blossoms and fruit. The fruit becomes soft and watery and perish rapidly. On infected tissues, a furry fungal mycelium and bulky black spore masses form. The pathogen can be distinguished by the appearance of numerous small black-headed pins sticking out of a pincushion, which indicate appearance for the disease.

31.3.6.2 Causal Organism

Choanephora cucurbitarum (Berk, et Rav.) Thaxt. is a causal agent of the disease. On PDA, the fungus grows very fast. It produces white colonies which later turn yellow or pale brown with abundant sporangia. Sporangiophores are longitudinally finely striate, brown, unbranched, non-septate, with clavate vesicles formed at their apices. Appendiculate bipolar spores are 5–13 μm × 1–10 mm, erect, solitary. Sporangia are monosporous brown to dark brown, longitudinally coarsely striate, ellipsoid to ovoid (Pornsuriya et al. 2017; Žerjav and Schroers 2019). On potato dextrose agar, they germinate to produce single-spore, fast-spreading colonies with profuse white mycelium and reverse side bright yellow.

31.3.6.3 Disease Cycle and Epidemiology

The fungal pathogen survives on dead plant tissue in the soil or as dormant spore structures. Spores of the fungus release in the spring and they are spread by wind and insects to squash blossoms. The pathogen infects wilted blossoms and later spreads to the attached fruit. For disease development, high relative humidity as well as wet conditions favour. About 28–34°C temperature and 75–90% relative humidity favour the growth of the fungus (Das et al. 2016). When conditions are favourable, spores are released and dispersed to blossoms by wind, splashing water, bees, cucumber beetles or other insects. The symptom development as well as establishment of disease can be triggered by moisture status in the macro and micro level.

31.3.6.4 Disease Management

- Disease incidence can be reduced by application of fungicides.
- Avoid overhead sprinkler irrigation or reduce overhead watering time to allow for leaf drying and also avoid wetness of canopy.
- Planting should be done on raised beds with plastic mulch which helps to reduce fruit contact with moist soil and infested plant debris.

- Diseased plant parts should be removed from the planting.
- All crop residues should be cleaned up and destroyed after over the crops.
- It is helpful if fruit and flowers do not touch the ground.

31.3.7 Fusarium Rot

In 1932, Fusarium crown and fruit rot disease in cucurbits was first described in detail in South Africa (Martyn, 1996). It is a pre-harvest and postharvest disease of pumpkins and other cucurbit crops (Bruton and Duthie 1996). Many species of *Fusarium* such as *F. graminearum*, *F. acuminatum*, *F. culmorum*, *F. moniliforme*, *F. semitectum*, *F. equiseti*, *F. scirpi* and *F. solani* are reported to cause cucurbit fruit rot disease (Bruton and Duthie 1996). Elmer et al. (2007) reported that *F. solani* f. sp. *cucurbitae* Mart. was the main causal agent of fruit rot of pumpkins in the United States. In Brazilian melon, 10 to 30% of postharvest losses due to corky dry rot caused by *F. semitectum* Berk. & Ravenel has been reported (Terao et al. 2008).

31.3.7.1 Symptoms

The disease symptoms are tiny, pitted, corky spots to large, sunken areas covered with a white or grey mould appearing on the pumpkin fruits. Most infections take place in the areas where the fruits are in contact with the soil. Wyenandt et al. (2010) observed three kinds of fruit rot symptoms: Type 1: Slow-expanding rot just below the rind surface of the infected fruit (*F. oxysporum* and *F. acuminatum*). Type 2: Slightly sunken, irregular rot of the rind surface (*F. graminearum*), and Type 3: Spherical sunken lesions on the fruit surface bearing white to tan sporodochia (*F. solani*). The pathogen also causes a crown rot (Elmer et al. 2007).

31.3.7.2 Causal Organism

Fusarium solani f. sp. *cucurbitae* (Teleomorph: *Nectria haematococca* Berk. & Broome (syn. *Hypomyces solani* (Rke. & Berth.) W.C. Snyder & H.N. Hans) causes rot in cucurbits. The fungus produces white to cream mycelium in heterothallic nature. The pathogen forms all three asexual spore types like microconidia, macroconidia, and chlamydospores. Sexual stage is found when suitable mating types are present. Microconidia are single-celled but some are two-celled, hyaline, cylindrical wedge shaped, or allantoid, 9–16 × 2–4 µm and form from elongated lateral phialides. Macroconidia are abundant in cream sporodochia. The macroconidia are often 3 to 4 septate, elongated (40–100 × 5–7 µm). They are slightly curved, relatively wide, and thick-walled with a short, circular, and sometimes hooked apical cell. They are also referred to as "sausage shaped" because they are rounded at each end. Conidiophores are branched or sometimes unbranched monophialides. Chlamydospores are produced plentiful in most of the cultures within 15–21 days. They are usually globose to elliptical, smooth to rough walled, 10–11 × 8–9 µm, either intercalary or terminal. Teleomorph *N. haematococca* produces reddish orange or white perithecia, and each perithecia contain eight 2-celled ascospores in each. The pathogen has two races (Elmer et al. 2007). Race 1 occurs worldwide and causes disease on fruit, stem, and root rot. Race 2 is found the United States and causes fruit rot.

31.3.7.3 Disease Cycle and Epidemiology

F. solani can overwinter in the soil as well as on/in seed. It is more ample in higher rainfall areas or in irrigated soils. Usually, soil-borne spores come into contact with the flower or fruit during rain splash or irrigation or when the fruit touches the soil. Infested seed can also be the cause of Fusarium rots of winter squash. Infected pumpkins may be a key source of primary inoculum. About 30–50% of the macroconidia are converted to chlamydospores in 1–8 weeks, which are produced from sporodochia or conidia on the stem near the ground surface either deposited in the soil by moisture or by mechanical means. Elmer (1996) reported that these chlamydospores did not survive for >1year. The pathogen may directly penetrate fruit under moist conditions. However, wounds facilitate entry of pathogen into the fruit.

The disease severity increases with temperature increases from 15–25°C, while the lesions are smaller when the temperature is higher than 25°C. The optimum temperature below 30°C for the development of corky dry rot favours the disease development in melon fruits caused by *F. semitectum* (Terao et al. 2006).

31.3.7.4 Disease Management

- Three years or longer crop rotations with non-cucurbit crops should be followed to reduce the corky rot of pumpkin.
- Avoid wounding during postharvest handling and packing. Store the produce in dry conditions.
- Remove infected fruit from the field.
- Provide a physical barrier (mulch) between the fruit and soil to reduce fruit infection.
- Pre-harvest fungicides do not seem to affect the spread to a high degree, but postharvest applications are helpful.
- Treating melon fruits with hot water (at 45°C for 10, 15, 20 and 25 min) reduced germination of spores and reduced fruit decay (Sui et al. 2014).
- Pre-cooling of melon fruits before export to lower their temperature to 10–15°C is considered to be an efficient control of corky dry rot at postharvest.
- Sweet melon fruits treated with hot water dip (53–55°C, 5–15 min) followed by irradiation (0.75–1.5 kGy) reduced decay caused by *Fusarium* sp. and *R. stolonifer*. In another case, 'Galia' melon fruits treated with hot water dip at 52°C for 5 min or at 55°C for 2 min with irradiation at 0.5 kGy showed 0–10% decay during storage (Barkai-Golan et al. 1994).

- Fungicide treatment postharvest prevents the spread of the disease to the rest of the harvest.
- Risse et al (1985) reported that Imazalil, waxing and film wrapping reduced weight loss and decay of Florida cucumber. Coated melon fruit with wax containing 2000 ppm Imazalil protects fruit from decay.

31.3.8 Sclerotinia Rot

Sclerotinia rot of cucurbits incited by *Sclerotinia sclerotiorum* (Lib.) de Bary exists throughout the world (Latin and Rane 1999; Babadoost et al. 2004). The pathogen infects about 300 different plant species including horticultural crops.

31.3.8.1 Symptoms

The symptoms of the disease appear on the cucurbits including pumpkin fruit and stem of fruit. The fungal pathogen causes rotting in pumpkin fruits in the field or during storage after harvest. The fruit areas contacting soil or on the stem end may be infected, or where water stands in the depressed area. The fungus infects dead tendrils, and withered flowers still attach to developing fruit. White, cottony mycelium growth of pathogen develops around the water-soaked infection site. Moreover, hard black sclerotia are produced among the mouldy growth.

31.3.8.2 Causal Organism

The disease is incited by *Sclerotinia sclerotiorum* (Lib.) de Bary and it grows a white colony on PDA medium. The overwintering structures of *S. sclerotiorum* produce apothecia in the soil. They are tan, mushroom-like structures that host the ascospores which are utilized for dispersion and infection. The overwintering structure of *S. sclerotiorum* is a hard, black structure called sclerotia and can be found inside or outside the affected plant (Mueller et al. 2015). Another characteristic symptom of *S. sclerotiorum* is mycelium, a white and cotton-like structure that will later form clumps and turn into the sclerotia (Heffer and Johnson 2007). The apothecia tend to mimic the *Cyathus striatus*, bird's nest fungus, or other fruiting fungal structures and the mycelial growth that is a natural part of decomposition may be mistaken for the mycelium of *S. sclerotiorum* (Giesler 2016). Black round or elongated, up to 1 cm across sclerotia develops on the colony surface mainly near the edges of petri dishes. They are broadly reniform in vertical section, with a smooth or shallowly pitted surface. The fungus produces apothecia which contains asci and ascospores.

31.3.8.3 Disease Cycle and Epidemiology

The pathogen overwinters in soil in the form of sclerotia for many years and in plant debris as mycelium. The sclerotia of the pathogen germinate to form either mycelium or apothecium depending on the availability of nutrients. Mycelium of the fungus infects the host plants which can occur at or beneath the soil line. Fungus penetrates the host cuticle by mechanical pressure. Carpogenic germination produces an apothecium at the optimum temperature of 10–20°C and it releases ascospores which are later dispersed by wind. Ascospores germinate within 3–6 h of release under favourable conditions including adequate moisture. Generally, cool and moist conditions favour disease development. Ascospores infect non-living host tissues, germinate, and deluge non-living plant parts with mycelium. The fungus mycelium then attacks healthy plant tissues. After plant or plant part dies, the sclerotia are formed either on or in the plant. The sclerotia return back to the soil for a longer period, which is a "resting" period for weeks or years before they become active under suitable environmental conditions. Mycelium again from sclerotia can also cause local infection. The pathogen is disseminated by contaminated soil (on farm equipment, shoes, infected seedlings) and fertilizers with manure from animals fed infected plant debris, irrigation water from one place to another. It has been reported that pH temperatures and moisture of the soil had little effect on their survival directly; however, a combination of high temperatures and high moisture appears to spur the degradation of sclerotia near the soil surface. Seed is also an infective source, either from contaminating sclerotia or internal mycelium. Once the pathogen enters a field, the chance for future outbreaks will remain high.

31.3.8.4 Disease Management

- Crop rotation with non-host crops (e.g., cereal grains) should follow, which will limit the damage to subsequent crops of vegetable.
- Deep ploughing of the field should be done immediately after harvest, which may help to reduce incidence of the disease.
- Resistant varieties of pumpkin against Sclerotinia rot are not reported so far.
- Good drainage facility in the field should be maintained cultivation of cucurbits.
- Overhead sprinkler irrigation should be done during the daytime, when the leaves will dry before dew forms.
- Fungicides may be used effectively on young plants which could be threatened during cool, wet summers.
- Postharvest treatment chemicals (heated dicloran or thiabendazole) by dipping method.
- The fruits should be stored in controlled atmospheres of oxygen and carbon dioxide at 2°C to reduce wastage.

31.3.9 Scab (Gummosis)

Scab disease has been reported as an important foliar and fruit disease in some parts of the world. It affects generally cucurbit crops and losses caused by this disease is sometimes exceed above 50%. Though the resistant cultivars of cucumber against the disease are available, this makes scab less important for production of crop.

31.3.9.1 Symptoms

The disease appears as water-soaked spots on leaves, petioles, stems and fruits of the plant and later spots turn grey to white. The "shot-hole" appearance is given on the leaf due to drop-out of the centres of the spots. The spots on fruit are generally confused with anthracnose disease; ooze a gummy substance, measuring about 3–4 mm in diameter. The spots are invaded by secondary rotting bacteria that cause the spots to smell.

31.3.9.2 Causal Organism

Fungus *Cladosporium cucumerinum* Ell. & Arth. is a causal agent of scab disease in cucurbit crops. The pathogen grows on the medium as olivaceous brown to greenish and black at maturity. The mycelium is septate, hyaline at young stage and becomes greenish to black on maturity. The conidia are formed in long, acropetal chains on dark conidiophore, are olive-brown in colour and are branched. Two types of conidia are formed as lower conidia, which are larger in size with one or two septa and called ramoconidia. The second type is upper conidia, which are smaller, one celled, ellipsoidal to cylindrical to fusiform with size 4–9 × 3–5 μm. Another *Cladosporium* species, *C. tenuissimum* Cooke, has been reported by Batta (2004), which is associated with small (3.0 mm dia), circular swelling on cucumber fruit grown in Israel.

31.3.9.3 Disease Cycle and Epidemiology

The fungi are seed-borne; however, they also survive on un-decomposed plant debris in the soil where they can grow as a saprophyte. The pathogen is disseminated by air in moist conditions. Cool, wet weather, including rain, dew and fog favours the disease development. The growth and sporulation of the fungus on agar medium is about 2–35°C, with optimum temperature about 21°C. The pathogen needs 100% RH or free moisture for infection.

31.3.9.4 Disease Management

31.3.9.4.1 Cultural Practices

Practicing 2–3 years crop rotation may reduce the incidence of disease with non-cucurbit crops like corn. Field should be ploughed under plant debris after harvest. Sprinkler irrigation should be avoided to reduce moisture on the plant surface.

31.3.9.4.2 Host Resistance

The use of scab-resistant cultivars such as Dasher II, Raider, Encore, Sprint, Poinsett 76, Turbo, Regal, Flurry, Calypso, Quest, Gemini, Marketmore, Pioneer, SMR-58, and SMR-18 may control the disease.

31.3.10 Alternaria Leaf Spot

Alternaria species infect the foliage of cucurbits and cause leaf spots and blight and on the fruits of musk melon they usually cause dry rot. Leaf spot disease is one of the limiting factors that affects and lowers the production of cucurbitaceous vegetables.

31.3.10.1 Symptoms

Disease symptoms emerge as small tan spots on the upper surface of muskmelon; the leaves sometimes develop into larger roughly circular areas and often coalesce to involve most of the leaf, showing blight symptoms. On the leaves of cucumber and watermelon, the spots become dark coloured with age and these lesions might have concentric rings. Complete defoliation or sometimes partial defoliation may occur, which exposes the fruits to sunscald injury. On over-ripe fruits that remain in the field, sunken spots about 3.0 cm in diameter are found and the spots are often covered with olive-green mass of conidia of the fungus. Small, brown, globular or oval spots enlarge up to 6.0 diameter with definite margins and often with alternating light and dark zones emerging on melon fruits during transit and storage. Due to sporulation of the fungi, the colour of the spot changes to black. The internal decay is dry and tough, but if it progresses deeply it may become moist and spongy.

31.3.10.2 Causal Organism

The disease is incited by fungus *Alternaria alternata* (Fr.) Keissler, *A. cucumerina* (Ellis and Ev) Elliott and *A. tenuissima* (Nees ex.Fr.) in cucurbits. The conidia are obclavate, muriform and dark (15–25 × 30–75 μm).

31.3.10.3 Disease Cycle and Epidemiology

The pathogen survives on plant debris in the soil. The seeds are also primary sources of inoculums for new infection to the host. Rain, fog and heavy dew by moist and wet conditions favour the disease. The disease is therefore more common in higher rainfall areas.

31.3.10.4 Disease Management

- Dithane Z78 and Perzate can be used for control of the disease. Chemical fungicides were evaluated under laboratory and also tested under field conditions against the pathogen. The disease was successfully managed by spraying of Indofil M-45 (0.2%) in the field conditions.
- *Bacillus subtilis* (B1 strain) was used for bio-control of many pathogens of fruit rot of muskmelon including *A. alternata*, *F. semitectum*, *R. stolonifer* and *Trichothecium roseum* (Young et al. 2006). Another strain of *B. subtilis* EXWB1 suppressed postharvest Alternaria rot (*A. alternata*) in melon (Wang et al. 2010).

31.3.11 Rhizopus Rot

Rhizopus stolonifer is one of the most important postharvest diseases of fruits and vegetables including cucurbits. Fruits of cucurbits are badly affected by Rhizopus soft rot and it causes huge losses in a very short period under favourable environmental conditions for infection (Agrios 2005).

31.3.11.1 Symptoms

The symptoms of soft rot appear on fruit as large water-soaked areas which turn soft and sunken. On the fruit surface, grey whiskery moulds with dusty black spores grow within the seed cavity. There is steady moisture loss by tissue until the fruit shrivels into a mummy (Agrios 2005).

31.3.11.2 Causal Organism

The disease is incited by *Rhizopus stolonifer* (Ehrenb.:Fr.) Vuill. (Synonym: *R. nigricans* Ehrenb.). Mycelium as aerial, arching stolons, often several cm long, arising from the points of contact with the substrate, where tufts of repeatedly branched rhizoids form. Sporangiophores arise from stolons and are generally unbranched, globose to oval or angular and smooth. Sporangia are terminal, large, globose, and have many spores. Zygospores are formed from the rhizoids or stolons.

31.3.11.3 Disease Cycle and Epidemiology

Rhizopus species are generally saprophytic. They penetrate into fruits through injuries or bruises caused during handling of produce (Lunn 1977). The pathogen survives on plant debris. It grows quickly and sporulates readily. Spores of the pathogen are disseminated by wind and water as well as by insects. The pathogen invades dying blossoms; after that it gains entry into the fruit during moist weather. Mature melons are especially susceptible due to their high sugar content. Temperature (22–25°C) and high relative humidity favour the growth of *R. stolonifer* during storage. Optimum growth occurs at 25°C and 35°C for *R. stolonifer* and *R. oryzae*, respectively. Growth is very slow at or below 10°C. However, *R. stolonifer* do not usually develop spores at temperatures below 5°C. However, it has also been recorded that a small percentage of the spores may germinate at temperatures as low as 2°C but their germ tubes cannot sustain growth. Hence, these fungi are not active on the host tissue and cannot develop disease (Barkai-Golan 2001).

31.3.11.4 Disease Management

- During harvesting and handling, injuring fruit should be avoided. Proper temperature control should be maintained during storage and transport.
- Rice chaff treatment equally showed significant inhibition compared to the control (Kazi et al. 2019).
- Imazalil fungicide with wax on 'Galia' melon cultivar prevented development of *Fusarium* spp. and *A. alternata*. However, this treatment showed a high level of residue (4–5 ppm) and prevailed after the storage period (Aharoni et al. 1993).

31.3.12 Blue Mould Rot

Several species of *Penicillium* are among the most important postharvest fungi of fruit. They are usually the most harmful of all postharvest diseases affecting major fruits and vegetables.

31.3.12.1 Symptoms

Penicillium first appears on any part of the fruit in the form of soft rot, watery and slightly discoloured spots of different sizes. Bluish-green or olive-green colours of *Penicillium* species are usually surrounded by white mycelium on water-soaked tissue. Some infections occur in the field, but the blue or green moulds generally infect after harvest diseases. They often account for 90% of decay in transit and storage of the produce.

31.3.12.2 Causal Organism

Penicillium species is the causal agent of blue mould rot. The pathogen that induces postharvest fruit decay in *C. africanus* and *C. myriocarpus* fruits was identified as *P. simplicissimum* (Oudem.) Thom. The disease was first reported in Korea on the melon (*Cucumis melo*) incited by *P. oxalicum* (Wwon et al. 2002) and in muskmelon in Thailand (Chaninunand Chitphithak 2018). On MEA and CYA media, colony was white colour. Conidia were ellipsoid and size 2.6–7.4 × 2.6–5.8 μm. Stipes were 86–320 × 2.8–4.3 μm in size. Metulae were 12.4–31.6 × 2.6–4.2 μm in size. Phialides were ampulliform to cylindroids, and size 8.2–15.4 × 3.6–4.6 μm (Kwon et al. 2002).

31.3.12.3 Disease Cycle and Epidemiology

The fungus is acidophilic. It proliferates in the juice of *C. africanus*, having acidic pH. Moist and warm air is conducive for the fungal development. Under dry conditions, the fruit may shrink and become mummified (Amari and Bompeix 2005). Infection is usually through wounds or directly through skin which is weakened by chilling injury or extended storage.

31.3.12.4 Disease Management

- Avoid injury on the fruits and store at 10°C.
- Handling of the produce should be done with care to minimize rind damage.

31.3.13 Bacterial Leaf Spot

The disease was first described on Hubbard squash in New York in 1926 (Burgess and Liddell 1983). Subsequently, the disease was reported from cucurbit growing areas in Asia, Australia, Europe and North America (Özdemir and Zitter 2006). Bacterial leaf spot disease is a very serious disease in some pumpkin growing areas. The pathogen *Xanthomonas cucurbitae* also infects other cucurbits like summer and winter squashes, gourds and cucumbers (Babadoost et al. 2004). The disease was reported on squash, pumpkin (Lamichhane et al. 2010) and watermelon (Pruvost et al. 2009; Dutta et al. 2013). Fruit rot caused by the bacterial pathogen exceeds 50% in some commercial fields in Illinois.

31.3.13.1 Symptoms

The symptoms start on leaf from the time of spreading vines until harvest. Angular, small spots (2–4 mm), yellow water-soaked or greasy areas appear on the underside of leaves, and as indefinite yellow areas appear on the upper side of the leaves. Later, within 5 days, the spots develop into round to angular with thin, brown, translucent centres as well as wide yellow halo. Young fruits, stems and petioles are also sometimes affected. On fruits, small spots, water-soaked areas (1–3 mm) sunken, circular, each with a beige centre and dark brown halo may produce light brown ooze which can extend into the seed cavity and cause seed infection. Bacteria penetrate of into the flesh, which leads to fruit rot in the field and storage.

31.3.13.2 Causal Organism

Xanthomonas cucurbitae (ex Bryan 1926) Vauterin et al. 1995 (Syn. *X. campestris* (Pammel) Dawson pv. *cucurbitae* (Bryan Dye.) is a causal agent of bacterial leaf spot disease of cucurbits. The colonies are mustard-yellow when cultured on beef agar. Yellow-pigmented *Xanthomonas* like bacterial colonies were isolated on KC semi-selective medium by Pruvost et al. (2009). They used advanced techniques such as AFLP and MLSA for identification of strains. The bacterium is rod shaped, motile with a single polar flagellum (monotrichous), gram-negative, O positive and F negative in the oxidative and fermentative test, and oxidase negative. The bacterium hydrolyzed esculin and starch. It grows on YDC agar at 33°C. *X. campestris* pv. *cucurbitae* is distinguished with *Pseudomonas syringae* pv. *lachrymans* by production of fluorescent pigment on King's B agar culture medium. Also, *X. campestris* pv. *cucurbitae* is recognized from *P. syringae* pv. *lachrymans* with its bright yellow colour on yeast extract-dextrose- $CaCO_3$ and on a specific medium for *X. campestris*. The pathogenic bacteria was identified by using primers RST2 (5′AGGCCCTGGAAGGTGCCCTGGA3′) and RST3 (5′ATCGCACT GCG TCC GCGCGCGA3′) in PCR assay (Leite et al. 1994), amplified at 1500-bp band (Babadoost and Ravanlou 2012).

31.3.13.3 Disease Cycle and Epidemiology

The disease is a seed-borne and the pathogen survives in infected crop residue. The infection of seed occurs internally in the seed coat. The bacterial pathogen infects pumpkin leaves and fruits at any stage of the crop. The cotyledons get infected after germination of the seed. The disease is transmitted through seeds and rain splash and the movement of people as well as machinery. Symptom expression usually occurs during periods of high temperature generally following rain or overhead irrigation (Saddler 2000).

High temperature and moist plat tissue favour leaf and fruit infection. For growth of the bacterium, the optimum temperature range is from 24 to 30°C and maximum 35°C.

31.3.13.4 Disease Management

- Disease-free seed or treated seeds should be used to reduce pathogen inoculums.
- One–two years crop rotation non-cucurbit crops may be used to reduce the incidence of disease. Plant debris should be destroyed by burning or ploughing into the soil.
- Seeds of pumpkin (*C. pepo* cv. New Rocket) with *X. cucurbitae* treated with in aqueous solutions of chemicals Copper plus Mancozeb (0.36 g + 0.27 g/100 mL of water), 1% peroxyacetic acid, and 1% sodium hypochlorite for 15 min was found effective in eliminating the pathogen on seed (Özdemir and Zitter 2006).
- To minimize seed-borne pathogen populations, production of seed should be done in arid regions under furrow irrigation.
- Avoid overhead irrigation for minimizing moisture on the plants to reduce disease incidence.
- Furrow irrigation should be used as much as possible.
- Application of copper fungicide from the commencement of fruit set until harvest may reduce disease incidence.
- Employ proper spraying of fungicide to cover the foliage and fruit thoroughly.

31.3.14 Bacterial Soft Rot

Bacterial soft rots cause huge crop loss in fruits and vegetables including cucurbits worldwide. Bacterial soft rot caused by *Bacillus carotovorus* was first reported on carrot in by Jones of Vermont in 1901. Later van Hall (1902) reported that *Bacillus atrosepticus* causes soft rot on potato in Holland. Appel (1902) reported that *Bacillus aroideae* was causal agent of soft rot on potato in Germany. Soft rot bacteria produces pectolytic enzymes that bind plant cells together to degrade plant structure to eventually macerate the tissues. The bacteria usually infect cucurbits such as cucumbers, melons, squash and pumpkins. These diseases can occur on crops in the field as well as post-harvested crops in storage. Soft rot can be found in many vegetables but is common in cucurbits. It is generally a fruit disease and occurs due to injury to fruits and poor, unhygienic transport and storage conditions. Rot can occur over a broad temperature range, particularly when oxygen is inadequate.

31.3.14.1 Symptoms

The soft rot symptoms appear initially as water-soaked spots on fruits which later enlarge and become soft and sunken. Interior tissues underneath the spots become mushy and tarnished and change colour from cream to black at anywhere on the fruits. There is leakage of water from affected tissue. Soft rots are known for maceration of tissue and a strong, unpleasant odour.

31.3.14.2 Causal Organism

Soft rot disease of cucurbits is caused by several bacteria including most commonly *Erwinia carotovora* subsp. *carotovora* (Syn. *Pectobacterium carotovorum* subsp. *carotovorum*). Cells of *E. carotovora* subsp. *carotovora* are rod shaped, gram-negative, facultative anaerobe with peritrichous flagella usually 0.6–1.8 × 1.7–5.1 µm in size. On potato agar colonies appear fine, blue and rounded with arose margins. Bacteria form blue colonies in big cup-like deepening on Logan's media, visible within 2–3 days. *E. carotovora* subsp. *carotovora* produces complex pectolytic and proteolytic enzymes, which are responsible for maceration tissues. The bacteria dilute gelatine and reduce (some strains can peptonize) litmus milk. They produce H_2S and NH_3, but they do not produce indole and not hydrolyze starch. Bacteria show catalase test positive and oxidase test negative. Optimum temperature of 24–28°C and maximum 37°C is required for growth of bacteria. *Bacillus polymyxa* (Pramzowski) Mace causes disease as large brown spots, occasionally hard and sometimes soft in melons in Israel (Volcani 1962). *Erwinia ananas* (now *Pantoea ananatis* (Serrano) Mergaert et al. 1993) causes smooth, firms, brown spots on honeydew melons and produces yellow pigments (Wells et al. 1987).

31.3.14.3 Disease Cycle and Epidemiology

Bacterial pathogen survives on contaminated vegetation residues and stumps, irrigation water, rhizosphere of vegetable and some weed plants, and insects. They usually penetrate into the plant body through wounds caused by equipment, insects and severe weather such as hail, freezing injuries and bruises during harvesting or through natural openings. The bacteria multiply initially in the intercellular spaces. They can never penetrate the intact epidermis nor, ordinarily, the natural openings. High relative humidity (about 90%) is essential for successful disease development. Once bacteria enter into the intercellular spaces, warm temperature and high moisture favour initiation of multiplication. The bacteria are unable to penetrate the living cells, but they produce pectolytic enzymes. These enzymes diffuse in advance of the bacteria and dissolve the middle lamella which, composed of pectates, cements the cells together. Separation of cells occurs due to destruction of this cementing substance. The plant cells die and disintegrate because they are poisoned by the pectolytic enzymes and other metabolic products of the bacteria. These cells provide nutrients to ever-growing hordes of bacteria. The bacteria can be disseminated by insects, contaminated equipment or by infested plant debris, soil, or contaminated water from one place to another. Such type of soft rot becomes more destructive during moist weather. It can be more severe when the plants lack adequate calcium.

31.3.14.4 Disease Management

31.3.14.4.1 Cultural Practices

Diseased plants must be removed and discarded from the field, but do not bury or compost this material. Wet conditions should be avoided to manage the disease. Cucurbits should be planted in proper drained soils. Corn, snap beans and beets are vegetables that are not considered susceptible to soft rot which can be used in rotation. The plants are watered adequately as per requirement and uniform. Proper spacing of plants should be maintained to avoid over crowding in the field. Harvesting should be done only during dry conditions. Infected fruits should be separated before transportation for distant market and storage. Fruits should be stored in cool, dry and well-ventilated place as per recommendation to reduce bacterial multiplication. Soft rot can be prevented by avoiding injury to the skins of the fruits at the time of handling.

31.3.14.4.2 Chemical Control

Sanitization of garden tools before and after use by treating them for at least 30 sec with 10% bleach or preferably 70% alcohol is important. Rubbing alcohol and many spray disinfectants usually contain about 70% alcohol.

31.3.14.5 Integrated Fruit Rot Disease Management

Management of fruit rots of cucurbits is challenging for several reasons.

- Lack of resistant cultivars against the fruit rots.
- The site of infection is mainly fruit contacting soil, which is very difficult to cover with fungicide at this site.
- Due to thick and dense canopy, it is typically difficult to reach fungicide because fruit is covered.
- Most of the fruit-rot pathogens are soil-borne and survive in the soil for long periods.
- The pathogens causing fruit rot also incite foliar infection in the field.
- The pathogens have wide host-ranges.

There is no single method that will provide successful management of all or any of the fruit rot diseases in cucurbits. Efficient management strategies are required to integrate all effective methods to fruit rots. The following recommendations may reduce the fruit rots of cucurbitaceous crops:

- Restrict entry of the pathogen to the field.
- Remove plant debris from the field.
- Follow crop rotations of 3 years or more with non-host crops.
- Use disease-free seeds from sowing.
- Manage foliar diseases to control fruit rot.
- Apply fungicide for disease control prior to occurrence of the disease as preventive measure.
- The fungicides may be used in different modes of action at alternate times.

REFERENCES

Agrios GN (2005) Plant Pathology. Academic Press, London.

Aharoni Y, Copel A, Fallik E (1993) Hinokitiol (β-thujaplicin), for postharvest decay control on 'Galia' melons. N Z J Crop Hortic Sci 21(2): 165–169.

Al-Sadi AM, Al-Said FA, Al-Kaabi SM, Al-Quraini SM, Al-Mazroui SS, Al Mahmooli IH, Deadman ML (2011) Occurrence, characterization and management of fruit rot of cucumbers under greenhouse conditions in Oman. Phytopathol Medit 50: 421–429.

Amari A, Bompeix G (2005) Diversity and population dynamics of *Penicillium* spp. on apple pre-and postharvest environment: Consequences for decay development. Plant Pathol 54: 74–81.

Babadoost M (2008) Fungicide evaluation for control of Phytophthora blight and downy mildew of processing pumpkin. Plant Disease Management. Reports (online). Report 2:V164. DOI: 10.1094/PDMR02. American Phyto Pathological Society, St. Paul, MN.

Babadoost M, Islam SZ (2003) Fungicide seed treatment effects on seedling damping-off of pumpkin caused by *Phytophthora capsici*. Plant Dis 87: 63–68.

Babadoost M, Ravanlou A (2012) Outbreak of bacterial spot (*Xanthomonas cucurbitae*) in pumpkin fields in Illinois. Plant Dis 96: 1222.

Babadoost M (2004) Phytophthora blight: A serious threat to cucurbit industries. APS net feature, Apr.-May, http://www.apsnet.org/ online/feature/cucurbit. American Phytopathological Society, St. Paul, MN, 3.

Babadoost M, Weinzierl RA, Masiunas JB (2004) Identifying and managing cucurbit pests. In: Diseases, insects, weeds. University of Illinois, College of ACES Extension, C1392, Urbana-Champaign, IL.

Barkai-Golan R, Podova R, Ross I, Lapidot M, Copel A, Davidson H (1994) Influence of hot water dip and Gamma-irradiation on postharvest fungal decay of Galia melons. Postharvest News Inform 5: 1136

Barkai-Golan R (2001) Postharvest Diseases of Fruits and Vegetables: Development and Control. Elsevier, Amsterdam, 418 pp. ISBN 0-444-50584-9.

Begum M, Rai VR, Lokesh S (2003) Effect of plant growth promoting rhizobacteria on seed borne fungal pathogens in okra. Indian Phytopath 56: 156–158.

Blancard D, Lecoq H, Pitrat M (2005) A Color Atlas of Cucurbit Diseases. Observation, Identification and Control. Manson Publishing, London, UK, 304 pp.

Bruton BD, Duthie JA (1996). Fusarium rot. In: Compendium of Cucurbit Diseases. Zitter TA, Hopkins DL, Thomas CE, eds. American Phytopathological Society, St. Paul, MN, pp. 50–51.

Bryan MK (1926) Bacterial leaf spot on Hubbard squash. Science 63: 165.

Burgess LW, Liddell CM (1983) Laboratory Manual for Fusarium Research. Department of Plant Pathology and Agric. Entomol., University of Sidney, NSW, Australia.

Chaninun P, Chitphithak I (2018) Blue mold caused by *Penicillium oxalicum* on muskmelon (*Cucumis melo*) in Thailand. Australas Plant Dis Notes 13: 46.

Chauhan RS, Bhatia JN (2012) Evaluation of disease control potentiality of certain fungicides in controlling anthracnose disease of bottle gourd. Plant Dis Res 27(2): 237–238.

Chauhan RS, Bhatia JN (2013) Screening of bottle gourd genotypes against anthracnose disease under natural as well as artificial epiphytotic conditions. Plant Dis Res 28(1): 92–93.

Chung S, Kong H, Buyer JS, Lakshman DK, Lydon J, Kim SD (2008) Isolation and partial characterization of *Bacillus subtilis* ME488 for suppression of soilborne pathogens of cucumber and pepper. Appl Microbiol Biotechnol 80: 115–123.

Das S, Dutta S, Mondal B (2016) Emergence of *Choanephora cucurbitarum*, the causal agent of twig blight disease of chilli, as a major pathogen under change in climatic condition in West Bengal. In: Climate Challenges: Status and Management of Plant Diseases, Sri Konda Lakshman Telangana, Horticultural University, Rajendranagar, Hyderabad.

Deadman ML, Al-Saadi AM, Al-Mahmuli I, Al-Maqbali YM, Al-Subhi R, Al-Kiyoomi K, Al-Hasani H, Thacker JRM (2002) Management of *Pythium aphanidermatum* in greenhouse cucumber production in the Sultanate of Oman. In: BCPC Conference: Pests and Diseases Brighton, UK, pp. 171–176.

Drechsler C (1923) A new blossom-end decay of watermelons caused by an undescribed species of *Pythium*. (Abstract) Phytopathology 13: 57.

Dutta B, Gitaitis RD, Lewis KJ, Langston DB (2013) A new report of *Xanthomonas cucurbitae* causing bacterial leaf spot of watermelon in Georgia, USA. Plant Dis 97: 556.

Elmer WH (1996) Fusarium fruit rot of pumpkin in Connecticut. Plant Dis 80: 131–135.

Elmer WH, Covert SF, O'Donnell K (2007) Investigation of an outbreak of Fusarium foot and fruit rot of pumpkin within the United States. Plant Dis 91: 1142–1146.

Erwin DC, Ribeiro OK (1996) Phytophthora Diseases Worldwide. American Phytopathological Society, St. Paul, MN.

Farrag ESH, Ziedan EH, Mahmoud SYM (2007) Systemic acquired resistance induced in cucumber plants against powdery mildew disease by pre-inoculation with tobacco necrosis virus. J Plant Pathol 6(1): 44–50.

Fugro PA, Mandokhot AM (2000) Fungicidal management of anthracnose of bitter gourd and its comparative economics. Pest Manag Ecol Zool 8(2): 197–199.

Fürnkranz M, Lukesch B, Muller H (2012) Microbial diversity inside pumpkins: Microhabitat-specific communities display a high antagonistic potential against phytopathogens. Microbiol Ecol 63: 418–428.

Giesler L (2016). Sclerotinia stem rot (White mold) in Soybean: What to Look For. Crop Watch. University of Nebraska–Lincoln. Available at: https://cropwatch.unl.edu/2016.

Glassner H, Zchicori-Fein E, Compant S, Sessitch A, Katzir N, Portnoy V, Yaron S (2015) Characterization of endophytic bacteria from cucurbit fruits with potential benefits to agriculture in melons (*Cucumis melo* L.) FEMS Microbiol Ecol 91(7): fiv074. doi:10.1093/femsec/fiv074.

Gopinath K, Radhakrishan NV, Jayaraj J (2006) Effect of propiconazole and difenconazole on the control of anthracnose of chilli fruits caused by *Colletotrichum capsici*. Crop Prot 25(9): 1024–1031.

Gupta SK, Thind TS (2006) Disease Problems in Vegetable Production. Scientific Publishers, Jodhpur, India, 576 pp.

Heffer LV, Johnson KB (2007) White mold. Plant Heal Instr., doi: 10.1094/PHI-I-2007-0809-01.

Holmes GJ (2007) Chemical control of Phytophthora blight caused by *Phytophthora capsici*: A summary of field experiments in the U.S. 24 Program of the First International *Phytophthora capsici* Conference, Islamorada, Florida.

Hausbeck M, Lamour K (2004) *Phytophthora capsici* on vegetable crops: Research progress and management challenges. Plant Dis. 88: 1292–1302.

Islam S, Akanda AM, Prova A, Islam MT, Hossain MM (2016) Isolation and identification of plant growth promoting rhizobacteria from cucumber rhizosphere and their effect on plant growth promotion and disease suppression. Front Microbiol 6: 1360.

Islam SZ, Babadoost M, Lambert KN, Ndeme A, Fouly HM (2005) Characterization of *Phytophthora capsici* isolates from processing pumpkin in Illinois. Plant Dis 89: 191–197.

Kazi N, Chimbekujwo IB, Anjili SM (2019) Identification and environmentally friendly control of post-harvest diseases of pumpkin. African J Plant Sci 13(9): 239–245.

Keinath AP (2002) Survival of *Didymella bryoniae* in buried watermelon vines in South Carolina. Plant Dis 86: 32–38.

Keinath AP (2008) Survival of *Didymellabryoniae* in infested muskmelon crowns in South Carolina. Plant Dis 92: 1223–1228.

Keinath AP (2011) From native plants in central Europe to cultivated crops worldwide: The emergence of *Didymella bryoniae* as a cucurbit pathogen. Hort Sci 46: 532–535.

Keinath AP (2017) Gummy stem blight. In: Compendium of Cucurbit Diseases and Pests. 2nd Ed. Keinath AP, Wintermantel WM, Zitter TA, eds. American Phytopathological Society, St, Paul, MN, p. 59.

Keinath AP, Farnham MW, Zitter TA (1995) Morphological, pathological, and genetic differentiation of *Didymella bryoniae* and *Phoma* spp. isolated from cucurbits. Phytopathology 85: 364–369.

Koike ST (1997) First report of gummy stem blight, caused by *Didymella bryoniae*, on watermelon transplants in California. Plant Dis 81(11): 1331–1331.

Koike ST, Gladders P, Paulus AO (2007) Vegetable Diseases. In: A Color Handbook. Academic Press, Boston, MA, 32 pp.

Kwon OE, Rho M, Song HY, Lee SW, Chung MY, Lee JH, Kim YH, Lee HS, Kim YK (2002) Phenylpyropene A and B, new inhibitors of acyl-CoA: cholesterol acyltransferase produced by Penicillium griseofulvum F1959. J Antibiotics 55: 1004–1008.

Lamichhane JR, Varvaro L, Balestra GM (2010) Bacterial leaf spot caused by *Xanthomonas cucurbitae* reported on pumpkin in Nepal. New Dis Rep 22.

Latin R, Rane, K (1999) Identification and Management of Pumpkin Diseases. Purdue Univ. (BP-17), West Lafayette, IN.

Lee DH, Mathur SB, Neergaard P (1984) Detection and location of seed-borne inoculum of *Didymella bryoniae* and its transmission in seedlings of cucumber and pumpkin. Phytopathology 109: 301–308.

Leite RP, Minsavage GV, Bonas U, Stall RE (1994) Detection and identification of phytopathogenic *Xanthomonas* strains by amplification of DNA sequences related to the hrp genes of *Xanthomonas campestris* pv. *vesicatoria*. Appl Environ Microbiol 60: 1068–1077.

Lunn JA (1977) *Rhizopus stolonifer*. CMI Description. Pathogens Fungi Bacteria 524: 1–2.

Maleki M, Lachin M, Somayyeh M (2011) Screening of rhizobacteria for biological control of cucumber root and crown rot caused by *Phytophthora drechsleri*. J Plant Pathol 27(1): 78–84.

Mergaert J, Verdonck L, Kersters K (1993) Transfer of Erwinia ananas (synonym Erwinia uredovora) and Erwinia stewartii to the genus Pantoea emend. as Pantoea ananas (Serrano, 1928) comb. nov. and Pantoea stewartii (Smith, 1898) comb. nov. respectively, and description of Pantoea stewartii. Int J Sys Bacteriol 43: 162– 173.

Mayberry KS, Hartz TK (1992) Extension of muskmelon storage life through the use of hot water treatment and polyethylene wraps. Hort Sci 27: 324–326.

Maynard DN, Hopkins DL (1999). Watermelon fruit disorders. Hortic Technol 9: 155–161.

Mitra M, Subramaniam LS (1928) Fruit rot disease of cultivated Cucurbitaceae caused by Pythium aphanidermatum (Edson). Fitz Mem Dep Agric India (Bot ser) 15: 79–84.

Mohapatra S, Das S, Chand MK, Tayung K (2014) Bio-control potentials of three essential oils against some postharvest pathogens. J Agric Tech 10(3): 571–582.

Morris SC, Wade NL (1983) Control of postharvest disease in cantaloupes by treatment with guazatine and benomyl. Plant Dis.67: 792–794.

Moumni M, Mancini V, Allagui MB, Murolo S, Romanazzi G (2019) Black rot of squash (*Cucurbita moschata*) caused by *Stagonosporopsis cucurbitacearum* reported in Italy. Phytopathol Medit 58(2): 379–383.

Mueller D, Bradley C, Chilvers M, Esker P, Malvick D, Peltier A, Sisson A, Wise K (2015) White Mold. Available via: https://www.ncsrp.com/pdf_doc/WhiteMold_CPN1005_2015.pdf.

Mukund B (2006) Epidemiology and management of anthracnose of bottle gourd caused by *Colletotrichum lagenarium* (Pass.) Ellis and Halsted. Thesis, Department of Plant Pathology, CCS Haryana Agrcultural University, Hisar, India, 50 pp.

Naureen F, Humaira B, Viqar S, Jehan A, Syed EH (2009) Prevalence of post-harvest rot of vegetables and fruits in Karachi, Pakistan. Pakistan J Bot 41(6): 3185–3190.

Neher OT, Johnston MR, Zidack NK Jacobsen BJ (2009) Evaluation of *Bacillus mycoides* isolate BmJ and *B. mojavensis* isolate 203-7 for the control of anthracnose of cucurbits caused by *Glomerella cingulata* var. *orbiculare*. Biol Contr 48(2): 140–146.

Norton JD, Cosper RD (1989) AC-70-154, a gummy stem blight-resistant muskmelon breeding line. HortSci 24: 709–711.

Nuangmek W, Aiduang W, Suwannarach N, Kumla J, Lumyong S (2018) First report of gummy stem blight caused by *Stagonosporopsis cucurbitacearum* on cantaloupe in Thailand. Can J Plant Pathol 40: 306–311.

Özdemir A, Zitter TA (2006) Bacterial leaf spot (Xanthomonas campestris pv. cucurbitae) as a factor in cucurbit production and evaluation of seed treatments for control in naturally infested seeds. In: Holmes GJ, ed. Universal Press, Raleigh, NC, pp. 498–506.

Pavón CF, Babadoost M, Lambert KN (2008). Quantification of Phytophthora capsici oospores in soil by sieving-centrifugation and real-time polymerase chain reaction. Plant Dis 92: 143–119.

Park SM, Jung HJ, Kim HS, Yu TS (2006) Isolation and optimal culture conditions of *Brevibacillus* sp. KMU-391 against black root pathogens caused by *Didymella bryoniae*. Korean J Microbiol 42: 135–141.

Pornsuriya C, Chairin T, Thaochan N (2017) Choanephora rot caused by *Choanephora cucurbitarum* on *Brassica chinensis* in Thailand. Australas Plant Dis 12–13.

Pruvost O, Robene-Soustrade I, Ah-You N, Jouen E, Boyer C, Wuster G, Hostachy B, Napoles C, Dogley W (2009) First report of *Xanthomonas cucurbitae* causing bacterial leaf spot of watermelon in the Seychelles. Plant Dis 93(6): 671.

Risse LA, Miller WR, Chun D (1985) Effect of film wrapping, waxing and Imazalil on weight loss and decay of Florida cucumbers. Proc Florida State Hortic Soc 98: 189–191.

Saddler GS (2000) *Xanthomonas cucurbitae*. Descriptions of fungi and bacteria. IMI Descriptions of Fungi and Bacteria 146: 1454.

Sadeghi A, Koobaz P, Azimi H et al. (2017) Plant growth promotion and suppression of *Phytophthora drechsleri* damping-off in cucumber by cellulase-producing *Streptomyces*. Bio Control 62: 805–819.

Sani MA, Usman N, Kabir F, Kutama AS (2015) The effect of three natural preservatives on the growth of some predominant fungi associated with the spoilage of fruits (mango, pineapple and cucumber). J Agri Sci 4(12): 923–928.

Seebold KW (2011) Gummy stem blight and black rot of cucurbits. Cooperative Extension Service, University of Kentucky, College of Agriculture, PPFS-VG-08.

Sharma DK, Maya Sharma (2018) Enumerations on seed-borne and post-harvest diseases of bitter gourd (*Momordica charantia* L.) and their management. Asian J Pharm Pharmaco 14 (6): 744–751.

Snowdon AL (1992) A Color Atlas of Postharvest Diseases and Disorders of Fruit and Vegetables. CRC Press, Boca Raton, FL, pp. 1–2.

Sitterly WR, Keinath AP (1996) Gummy stem blight. In: Compendium of Cucurbit Diseases. Zitter TA, Hopkins, DL and Thomas CE, eds. APS Press, St. Paul, MN, pp. 27–28.

Sui Y, Droby S, Zhang D, Wang W, Liu Y (2014) Reduction of Fusarium rot and maintenance of fruit quality in melon using eco-friendly hot water treatment. Environ Sci Pollut Res Int 21(24): 13956–13963.

Terao D, Oliveira SMA, Viana FMP, Rossetti AG, Souza CCM (2006) Integraçao de fungicidas à refrigeração no controle de podridãoemfrutos de meloeiro. Fitopatol Brasi 31(1): 89–93.

Terao D, Oliveira SMA, Viana FMP, Saraiva ACM (2008) Estratégias de controle de podridõesempós-colheita de melão: umarevisão. Embrapa Agroind Tropical, Fortaleza.

Tompkins CM, Tucker CM (1937) Phytophthora root rot of honeydew melon. J Agric Res 54: 933–944.

Tsay JG, Chen RS, Wang WL. Weng BC (2010) First report of anthracnose on cucurbitaceous crops caused by *Glomerella magna* in Taiwan. Plant Dis 94(6): 787.

Ullasa BA, Amin KS (1986) Epidemiology of bottle gourd anthracnose, estimation of yield loss and fungicidal control. Trop Pest Mgt 32(4): 277–282.

Van Steekelenburg NAM (1983) Epidemiological aspects of *Didymella bryoniae*, the cause of stem and fruit rot of cucumber. Neth J Plant Pathol 89: 75–86.

Van Steekelenburg NAM (1986) Factors influencing internal fruit rot of cucumber caused by *Didymella bryoniae.* Neth J Plant Pathol 92: 81–91.

Vauterin L, Hoste B, Kersters K, Swings J (1995) Reclassification of *Xanthomonas*. Int J Sys Evol Microbiol 45: 472–489.

Volcani Z (1962) Occurrence of two bacterial diseases on new hosts in Israel. Plant Dis Rep 46 (12): 893.

Wang Y, Xu Z, Zhu P, Liu Y, Zhang Z, Mastuda Y, Toyoda H, Xu L (2010) Postharvest biological control of melon pathogens using *Bacillus Subtilis* EXWB1. J Plant Pathol 92(3): 645–652.

Wells JM, Sheng WS, Ceponis MJ, Chen TA (1987) Isolation and characterization of strains of *Erwinia ananas* from honeydew melons. Phytopathology 77: 511–514.

Wwon JH, Kang SW, Kim JS, Park CS (2002) Blue mold on melon (*Cucumis melo*) caused by *Penicillium oxalicum*. Res Plant Dis 8(4): 220–224.

Wyenandt CA, Riedel RM, Rhodes LH, Bennett MA, Nameth SGP (2010) Survey of *Fusarium* spp. associated with fruit rot of pumpkin (*Cucurbita pepo*) in Ohio. Plant Health Progr 11(1): 9.

Young MJ, Theriault SS, Li M, Court DA (2006) The carboxyl-terminal extension on fungal mitochondrial DNA polymerases: Identification of a critical region of the enzyme from *Saccharomyces cerevisiae*. Yeast 23(2): 101–116.

Zerjav M, Schroers HJ (2019) First report of Cucurbita fruit rot caused by *Choanephora cucurbitarum*in Slovenia. Plant Dis 103(4): 760.

Zhou XG, Everts KL (2008) First report of *Alternaria alternata* f. sp. *cucurbitae* causing Alternaria leaf spot of melon in the Mid-Atlantic region of the United States. Plant Dis 92: 652.

Ziedan EH, Abd El-Hafez Kattab AE-N, Sahab A (2018) New fungi causing postharvest spoilage of cucumber fruits and their molecular characterization in Egypt. J Plant Protec Res 58 (4): 362–371.

Ziedan EH, Saad MM (2016) Efficacy of nanoparticles on seed borne fungi and their pathological potential of cucumber. Int J Pharm Tech Res 9(10): 16–24.

Zitter TA (1996) Black rot. In: Compendium of Cucurbit Diseases. Zitter TA, Hopkins DL,Thomas CE, eds. American Phytopathological Society, St. Paul, MN, p. 48.

Zohara F, Surovy MZ, Khatun A, Prince MFRK, Akanda MAM, Rahman M (2019) Chitosan biostimulant controls infection of cucumber by *Phytophthora capsici* through suppression of asexual reproduction of the pathogen. Acta Agrobot 72(1): 1763.

32

Postharvest Diseases of Cole Crops and Their Management

Raghavendra Mesta, Manjunath Hubballi and T. S. Archana
Department of Plant Pathology, University of Horticultural Sciences, Bagalkot, Karnataka, India

CONTENTS

32.1 Introduction

The vegetables belonging to the mustard family are broadly grouped as cole crops. The major crops included in the group are cauliflower, cabbage, brussels sprouts, broccoli, knol-khol (kohlrabi), kale, and Chinese cabbage. They are widely grown throughout the world, including tropical, subtropical, and temperate regions (Agrawal et al. 2010). Among various cole crops, cauliflower and cabbage are regarded as major vegetables, the rest, especially in India, are considered minor vegetables. The crops contain an appreciable amount of minerals including phosphorus, potash, calcium, sodium, and iron and they also have a good amount of vitamin A and C. There are many constraints in cole crop production. Plant pathogens, especially fungi and bacteria are one of the major constraints leading to severe crop loss. In addition, there are many postharvest diseases which reduce the quality as well as quantity after harvest at the time of transit and storage. This chapter briefly discusses the important postharvest diseases and their management practices.

Postharvest disease may be initiated during any stage after harvest until the end usage, inflicting loss in quality and also reduction in the market value of the product. Production of mycotoxins is a serious health issue associated with these diseases in addition to having a bearing on economic considerations. According to Coates and Johnson (1997), the extent of damage in postharvest diseases largely depends on the type of commodity, resistance potential of cultivar to particular diseases, and the temperature, relative humidity and composition of atmosphere.

32.2 Postharvest Diseases

The main causal organisms for postharvest diseases are bacteria and fungi. In case of cole crops, pathogen infections start before harvest, and sometimes infection develops more rapidly after harvest. Bacterial soft rot, Alternaria rot, Phytophthora rot, watery soft rot and grey mould are the important diseases inflicting losses from harvest to consumption of the cole crops. The scope of the chapter is to describe these diseases.

32.2.1 Bacterial Soft Rot

The most common disease appearing in storage in many vegetables, including cole crops, is soft rot incited by *Erwinias* sp. It affects both leading cole crops such as cabbage and cauliflower, and leafy cole crops like broccoli, kale and mustard. The disease can be seen under field conditions as well as under storage. Sometimes the disease may not show any external symptoms in field conditions but may worsen under storage, especially if the infection is seen in the later part of the crop, and worsens under storage conditions when the cool chain is not maintained properly and temperature is allowed to rise (Ren 1999; Ren et al. 2001).

32.2.1.1 Causal Organism

The disease is incited by species of *Pectobacterium* (Erwinia) and one bacterial pathogen belonging to genus *Pseudomonas*, viz. *Pseudomonas marginalis* pv. *marginalis* (Coates and Johnson 1997; Tsuda et al. 2016; Oskiera et al. 2017; Li et al. 2020; Bhat et al. 2012; Bhat et al. 2012).

32.2.1.2 Symptoms

Infected plants exhibit water-soaked lesions on young leaves, stems and underground parts which enlarge gradually in both diameter and depth. The infected tissue will change to brown colour and becomes soft and mushy. The slimy bacterial mass will ooze from the cracks in the tissue, and within 72 hours after infection the infected parts collapse as rotting starts. A foul odour emits from the decaying tissues that have been invaded by secondary pathogens (Cariddi and Bubici 2016).

32.2.1.3 Disease Cycle and Epidemiology

The pathogen, being saprophyte, survives in soil, especially in the root zone, as well as on the debris in the field, for a considerable amount of time. The infection starts with entry of pathogen through wounds caused at the time of planting, cultural operations, harvesting grading, packing freeze injuries, insects, hailstorms, sunscalds, etc. If there are no mechanical injuries to plants, the pathogen may also enter through natural openings such as hydathodes and stomata. In general, the uninjured parts become infected when there is 100% humidity and availability of free moisture on plant parts. Free moisture on the surface enables bacteria to penetrate, and after penetration the bacteria produce various enzymes that degrade the host tissues leading to a mass of unorganized free cells surrounded by bacteria and their enzymes. The host cells will lose water and are finally invaded by bacteria.

The generation time of bacteria is 20–60 min under ideal conditions. The temperature of 18–35°C with humid conditions favours the development of disease to a greater extent. The bacteria can tolerate wide range of temperature from 2 to 41°C, depending on the species involved.

The bacterial pathogens are spread by direct contact, tools, farm implements and machinery, water, rains and clothes. Maggot flies and other insects are also known to spread the disease. *Erwinia* spp. and *Pseudomonas* spp. also act as secondary pathogens, following other diseases such as black rot or blackleg. Cultivation practices, harvesting methods, handling during harvest and transition, freezing or insect injuries are often points of initial infection (Bhat et al. 2012; Oskiera et al. 2017).

32.2.1.4 Integrated Disease Management

To minimize soft rot losses, care has to be taken in the field as well as after harvest. It is important to manage insects during field conditions as well as during storage, and to avoid injuries from harvest to consumption. It is not advisable to pack the produce when it is wet. Further, efficient packing material that greatly suppresses postharvest soft rot is preferable during postharvest and until the product reaches consumer.

Treatment with hot water is a very simple, easy way to manage infections in the postharvest stage. This is the most practiced method in many countries. Galati et al. (2005) gave another option as chlorination by use of calcium or sodium hypochlorite to manage the disease. However, chlorination is proven to be only a preventive but not a curative measure. Napitupulu and Lubis (1987) reported reduction in soft rot disease in cabbage after treating with silica gel and lime postharvest.

The use of antibiotics for management of soft rot under field conditions is the last option for disease management. Several antibiotics have proved to be potent in arresting bacterial growth. The important antibiotics against *Pectobacterium carotovorum* sub sp. *carotovorum* are ciprofloxacin, oxytetracycline, tetracycline, amoxicillin, striplin, azithromycin and penicillin. Chemicals other than antibiotics that produced a moderate to low inhibition of bacterium include mercuric chloride, bleaching powder, sodium hypochlorite, and copper-oxychloride (Bhat et al. 2012).

32.2.2 Grey Mould

Grey mould of cole crops is an important postharvest disease incited by the cosmopolitan fungus *Botrytis cinerea*. The disease is not host specific and the pathogen can attack many hosts of several other vegetables. It is an important postharvest disease that reduces the quality as well as market demand of cole crops such as cabbage, cauliflower and broccoli. Under cool, wet conditions, it is an impediment from production to storage of cole crops. Grey mould of cabbages is seen after 2–3 months of storage. The disease is usually not visible at the time of harvest but develops rapidly under storage conditions. This disease is the most common cause of losses under storage conditions (Coley-Smith et al. 1980).

32.2.2.1 Symptoms

The characteristic symptoms of the disease vary according to part of the plant infected and also prevailing growth conditions. In moist, humid conditions, a thin web of mycelium initially develops on the plant surface that causes the damping-off in younger seedlings, and in mature plants, the fungus is reported to infect through wounds present in all aerial parts. Initially, elliptical water-soaked lesions appear and later these spots enlarge and cause girdling of plants and ultimately kill them.

On the other hand, if the humidity is too high the fungus can grow extensively on plant surfaces, leaving grey-coloured mycelia. The conidiophores are produced on the surface and contain numerous spores. Any slight disturbance of infected plants leads to release of spores.

32.2.2.2 Causal Organism

The disease is incited by *Botrytis cinerea*. This fungus produces grey-coloured septate mycelium with branched conidiophores that have round and ovoid conidia. The conidia may be coloured or colourless. The conidia and its bearing structures conidiophores look very similar to grape bunches. The fungus mostly produces the sclerotia, and few species were also reported to produce *Botryotinia* (Jarvis 1977; Holz et al. 2007).

32.2.2.3 Disease Cycle and Epidemiology

The life cycle of *Botrytis* produces macro conidia under the somatic stage and sclerotia under the sexual cycle in different environmental conditions (Beever and Weeds 2007). The sclerotia, being hard structures, act as important means of survival, and the fungus also survives as dormant mycelium in debris. In temperate regions, in the early spring sclerotia germinate and produce conidiophores and conidia which act as source primary inoculum (Williamson et al. 2007). The mycelium, if present in debris, also acts as primary inoculum.

The pathogen also produces micro conidia from phialides, which act mainly as spermatia. The spermatization of sclerotia leads to production of apothecia and asci. Each ascus contains eight ascopores, which are binucleate. The details on exactly how plasmogamy and apothecia form have not been discussed. Formation of apothecia is rare phenomenon in many of the crops infected by *Botrytis*. Hence, the role of the sexual cycle is only based on genetic variation through molecular analysis (Beever and Weeds 2007). Sexual reproduction of *B. cinerea* was first reported by Groves and Drayton (1939) and later studied *in vitro* (Faretra and Antonacci 1987; Faretra et al. 1988).

*B. cinerea*is a necrotrophic fungus; it kills the host plant and then colonizes dead tissues (Amselem et al. 2011). The fungus can tolerate temperatures as low as 0°C; the most favoured temperature is 20°C and the maximum temperature that can be tolerated by this fungus is 30°C. Hence, the disease easily develops well even in cold storage conditions. Poor air currents in storage also add to development of disease. Warm to hot dry weather tends to reduce or stop the growth and spread of the disease (Prins et al. 2000). The rate of spoilage in postharvest conditions depends on temperature, relative humidity, air speed and composition of oxygen, carbon dioxide and ethylene, and also cleanliness (Kader 2005).

32.2.2.4 Disease Management

The disease, being long-established, was traditionally managed with synthetic fungicidal molecules during crop growth as preventive measures. The ever-increasing public awareness on the hazardous effects of these molecules not only to human beings but also toot her beneficial microbes and non-target organisms, coupled with increasing reports of resistance development in pathogen against chemical molecules, have diverted research attention and brought alternatives in disease management. These alternatives are categorized as bioagents, natural compounds, compounds generally recognized as safe (GRAS), and physical methods alone or the combination of all four groups (Mari et al. 2009; Romanazzi et al. 2012).

Bacteria and yeast are the major bioagents used against this pathogen. The main mode of action of these two microbes is competition for space and nutrition, parasitism, antibiosis and induction defence in plants and also volatiles production to some extent (Jamalizadeh et al. 2011). Many of the microbes are registered for commercial usage against *B. cinerea*. The commonly used biofungicides are *Candida oleophila* (Nexy, BioNext, Paris, France), *Cryptococcus albidus* (YieldPlus, Canada), *Candida sake* (Candifruit, Spain), *Bacillus subtilis* (Serenade, Bayer, Germany), *B. amyloliquefaciens* (Biogard CBC, Italy) and *Pseudomonas syringae* (BioSave, the USA).

Natural compounds, otherwise called GRAS compounds, are substances that are eco-friendly and have no adverse effects on the health of human beings. Generally, these compounds are used for their antimicrobial properties or elicitation of defence in host plants. Essential oils and natural compounds form the major natural compounds being used worldwide to reduce postharvest losses. In addition to eco-friendliness, they also prolong the keeping quality of fresh commodities (Feliziani and Romanazzi 2013). Various inorganic salts also exhibit a range of antimicrobial properties against postharvest diseases. Bicarbonates have been proposed as a safe and effective alternative means to control postharvest rot of vegetables. Several sanitizers classified as GRAS have been applied to lengthen the postharvest storage of various products, including acetic acid, electrolyzed oxidizing water and ethanol.

Induction of resistance is another alternative way of managing the disease. The natural biopolymer of chitosan and synthetic elicitor benzothiadiazole have been shown to activate systemic acquired resistance (Feliziani and Romanazzi 2013; Romanazzi et al. 2013).

Heat treatment, hypobaric and hyperbaric treatments, UV-C light and ozone are some other measures. These can have dual effect on fruit by activating defence as well as being directly antimicrobial against pathogens.

The last resort in any disease management is fungicide. In grey mould, since fungus attacks heads, fungicides are seldom recommended. The major fungicides recommended are composed of azoxystrobin at the rate of 200 g/ha, boscalid at the rate of 267 g/ha + pyraclostrobin at the rate of 67 g/ha and tebuconazole at the rate of 250 g/ha (Surviliene et al. 2010).

32.2.3 Alternaria Rot

Another common problem in cole crops is Alternaria rot, including *Alternaria raphani*, *A brassicicola* and *A. brassicae*. Turnip, cauliflower, cabbage, broccoli, cabbage, cauliflower, Chinese cabbage, kohlrabi, kale and Brussels sprouts are infected by two species, *A. brassicicola* and *A. brassicae*. *A. raphani* is frequently noticed in radish; however, it can also be noticed on crops belonging to Brassica family.

32.2.3.1 Symptoms

The characteristic symptoms of the disease include appearance of dark brown-coloured circular leaf spots which rapidly enlarge, and as the disease progresses, the centre may fall off leaving a shot-hole appearance. In later stages, all lesions coalesce to form necrotic areas and eventually leaf drop and death of plants. Similar lesions can be found on all aerial parts of the plants, including seeds. The dark-coloured fuzzy mass occurring inside leaf spots is the result of spore production (Sabry et al. 2015).

32.2.3.2 Causal Organism

Three species of *Alternaria* are reported to infect cole crops: *Alternaria brassicicola*, *A. brassicae*, and *A. raphani*. *A. raphani* is primarily a pathogen of radish, however, it is also known to infect other cole crops.

32.2.3.3 Disease Cycle and Epidemiology

The species of Alternaria can survive saprophytically in the infected debris and act as the primary inoculum. Apart from this, resting spores such as microsclerotia and chlamydospores have also been reported to produce disease. Weeds and seeds are also sources of infection. The disease favours warm temperature of 28–31°C and relative humidity of >90% or more. The secondary spread is through water, wind, equipment and also through workers. The production of spores is generally seen at night, and release of spores occurs during the day. Cloudy weather for a long time is favourable for enormous sporulation. It takes 7–8 days under most favourable conditions to produce spores (Kohl et al. 2010). The seed borne nature of the fungus is considered to be the major cause of introduction of pathogens in new areas.

32.2.3.4 Disease Management

Since the pathogen survives in debris and also on weed host, the primary source of disease can be reduced by following strict sanitary measures. The debris present in the field should be managed to avoid infection (Humpherson-Jones and Maude 1982). Crop rotation with non-cruciferous vegetables is advised for at least 2 years. Irrigation management also plays major role in disease, as excess irrigation leads to weeds in crop plants and in turn they act as sources of infection.

Because the pathogen is seed borne in nature, the use of disease-free seeds or seeds which are reported to be resistant or tolerant can be used to manage the disease. In addition to this, various treatments like hot water and fungicidal treatment can also be adopted for management. Treatment of seeds with Captan, Thiram and Iprodione is reported to reduce disease incidence.

Aftercare of produce during storage and transit also plays a role in managing the disease. Any bruise and injuries will lead to entry of the opportunistic pathogen and initiates the disease. Hence, it is advised to clean and disinfect the storage facilities (Kohl et al. 2010).

32.2.4 Sclerotinia Stem Rot and Watery Soft Rot

This disease is caused by *Sclerotinia sclerotiorum* and affects several crucifer crops. It has been reported particularly on cauliflower and on the common field cabbage. *S. sclerotiorum* is one of the most devastative diseases of vegetables in relatively cold areas of India (Zewain et al. 2004). It has restricted the production of cabbage and cauliflower seed production in northern India (Sharma et al. 2005).

32.2.4.1 Symptoms

Appearance of water-soaked lesions on stems marks the beginning of infection. As the infection progresses, it spreads downward to root, and roots start decaying, and infection moves upward and leaves start to show wilting symptoms and eventually the plant collapses. Occurrence of white cottony and black sclerotia on infected tissues is a characteristic feature of this disease. If dry weather prevails, brown cankers can also be seen on stems without progressing further.

32.2.4.2 Causal Organism

The disease is caused by *Sclerotinia sclerotiorum*. It is a polyphagous pathogen infecting a range of crops. The pathogen is reported to attack many plants belongs to 64 families (Purdy 1953).

32.2.4.3 Disease Cycle and Epidemiology

The pathogen survives in a small black-coloured sclerotia produced in decomposing plant portions for many years. Sclerotia on congenial conditions germinate and produce ascospores which infect through stomatal openings, cuticle or epidermal cells. The fungus invades all plant parts, and moist conditions favour infection. Sclerotia can be distributed with seeds, farm implements, animals and with irrigation water. Ascospores can be carried with winds (Steadman 1975; Hims 1979). Development of disease is generally favoured by abundant soil moisture and temperature ranging from 10 to 25°C.

32.2.4.4 Disease Management

Good sanitation is the most important aspect in managing the disease. Crop rotations with non-host plants can reduce the inoculum level in field and flooding the field with sclerotia can be an option for destroying sclerotia in soil. Krishnamoorthy et al. (2017) reported the efficacy of trifloxystrobin + tebuconazole and carbendazim against *Sclerotinia sclerotiorum.*

32.2.5 Phytophthora Root Rot

This is a destructive root rot disease that occurs in many cole crops; however, severity is greater in cabbage and cauliflower. Considerable losses as a result of this disease have been observed in all varieties. The disease has also been reported in Brussels sprouts.

32.2.5.1 Symptoms

The disease is characterized by discoloration of the older leaves to a reddish colour followed by immediate wilting of infected leaves, which fall face down to the ground, leaving the head or curd exposed. The curd is not noticeably discoloured but becomes tough and rubbery and is not suitable for marketing. Plants of all ages are susceptible to the disease. The lower end of the taproot, along with the underground part of the stem, is badly rotted, and infected plants may be pulled from the soil very easily. The cortex of the taproot and lateral roots is softened and water-soaked and usually sloughs off and remains in the soil when the plant is pulled. In the case of Brussels sprouts and cauliflower, infection of the root is followed by leaves exhibiting red to purple colour from the margins inward, followed by wilting and death. Plants of all ages succumb to the disease. Rotting of the taproot is seen with necrosis. Adventitious roots may form above the rotted portion (Abd Allaand El Shoraky 2017).

32.2.5.2 Causal Organism

Tompkins et al. (1936) in the United States first described as *Phytophthora megasperma* as the causal agent of root rot of *Brasssica* sp. Presently it has been observed in many countries, where it is regarded as a significant disease.

32.2.5.3 Disease Cycle and Epidemiology

Fungus mainly survives in soil for a long period of time. The pathogen is not seed borne in nature. The primary source of inoculum is either chlamydospores or oospores, and the secondary source is zoospore movement through water.

32.2.5.4 Disease Management

Since the pathogen is soil borne, there are few options for managing the disease. Improvement of the drainage facility so as to avoid prolonged saturation of beds is the best way to manage the disease (Wilcox and Mircetich 1985). Drought stress also makes plants susceptible to Phytophthora, and an even supply of moisture without major fluctuations helps to suppress disease development. In addition to this, soil amendments and application of proper nutrients are recommended to boost resistance in plants.

32.3 Conclusion

The world population is estimated to reach 9.1 billion by 2050, and around 70% more food is required to feed the ever-increasing population around the world. The population increase is expected to be greater in developing countries than in developed ones. Therefore the problem of food security is more pronounced in countries like India. It is a herculean task to meet the food demand of the country, as the sectors of agriculture and horticulture are already facing resource problems. The per capita arable land in India has gone from 0.34 ha to 0.12 ha owing to urbanization, industrialization and other reasons. At this juncture, realizing horizontal expansion of arable land is next to impossible, so safeguarding what is produced is more important than producing more. It is reported that 6–13% of vegetables are lost at the postharvest stage. The diseases are among the major factors contributing to postharvest losses. Hence, effective management of diseases at the postharvest stage can go a long way in sustaining food production in conjunction with the ever increasing population.

REFERENCES

Abd Alla MA, El Shoraky FS (2017) Impact of biological agents and plant essential oils on growth, quality and productivity of cabbage and cauliflower plants correlated to some diseases control. J Sustain Agric 43: 27–38.

Agrawal N, Mehta N, Sharma HG, Dixit A, Dubey P (2010) Cultivation of cole crops under protected environment. Karnataka J Agri Sci 16: 2.

Amselem J, Cuomo CA, Van kan JAL (2011) Genomic analysis of the necrotrophic fungal pathogens *Sclerotinia sclerotiorum* and *Botrytis cinerea*. PLOS Genet 7(1): 27.

Beever RE, Weeds PL (2007) Taxonomy and genetic variation of *Botrytis* and *Botryotinia*. In: Botrytis: Biology, Pathology and Control. Springer, Dordrecht.

Bhat KA, Bhat NA, Masoodi SD, Mir SA, Zargar MY, Sheikh PA (2010) Studies on status and host range of soft rot disease of cabbage (*Brassica oleracea* var. *capitata*) Kashmir Valley. J Phytol 2: 55–59

Bhat KA, Bhat NA, Mohiddin FA, Mir SA, Mir MR (2012) Management of postharvest Pectobacterium soft rot of cabbage (*Brassica oleracea* var. *capitata* L.) by biocides and packing material. Afr J Agric Res 7: 4066-4074.

Cariddi C, Bubici G (2016) First report of bacterial pith soft rot caused by *Pectobacterium carotovorum* subsp. *odoriferum* on cauliflower in Italy. J Plant Pathol 3: 563–568.

Coates L, Johnson G (1997) Postharvest diseases of fruit and vegetables. Plant Pathogens Plant Dis 4: 533–548.

Coley-Smith JR, Verhoeff K, Jarvis WR (1980) The Biology of Botrytis. Academic Press, London.

Faretra F, Antonacci E (1987) Production of apothecia of *Botryotinia fuckeliana* (de Bary) Whetz. Under controlled environmental conditions. Phytopathol Medit 26: 29–35.

Faretra F, Antonacci E, Pollastro S (1988) Sexual behavior and mating system of *Botryotinia fuckeliana*, teleomorph of *Botrytis cinerea*. J Gen Microbiol 134: 2543–2550.

Feliziani E, Romanazzi G (2013) Pre-harvest application of synthetic fungicides and alternative treatments to control postharvest decay of fruit. Stewart Postharvest Rev 3: 1–9.

Galati A, McKay A, Soon CT (2005). Minimising postharvest losses of carrots. Farm note No.75/95. Department of Agriculture and Food. Government of Western Australia, South Perth.

Groves JW, Drayton FL (1939) The perfect stage of *Botrytis cinerea*. Mycologia 31: 485–489.

Hims MJ (1979). Damping-off of *Brassica napus* ('mustard and cress') by *Sclerotinia sclerotiorum*. Plant Pathol 28(4): 201–202.

Holz G, Coertze S, Williamson B (2007) The ecology of Botrytis on plant surfaces. In: *Botrytis*: Biology, Pathology and Control. Springer, Dordrecht.

Humpherson-Jones FM, Maude RB (1982) Control of dark leaf spot (*Alternaria brassicicola*) of *Brassicaoleracea* seed production crops with foliar sprays of iprodione. Annals Appl Biol 100(1): 99–104.

Jamalizadeh M, Etebarian HR, Aminian H, Alizadeh A (2011) A review of mechanisms of action of biological control organisms against postharvest fruit spoilage. EPPO Bull 41(1): 65–71.

Jarvis WR (1977) *Botryotinia* and *Botrytis* species: Taxonomy, Physiology, and Pathogenicity. Hignell Printing Limited, Winnipeg, Canada.

Kader AA (2005) Increasing food availability by reducing postharvest losses of fresh produce. Acta Hortic 682: 2169–2175.

Kohl J, Van Tongeren CA, Groenenboomde Haas BH, Van Hoof RA, Driessen R, Van Der Heijden L (2010) Epidemiology of dark leaf spot caused by *Alternaria brassicicola* and *A. brassicae* in organic seed production of cauliflower. Plant Pathol 59: 358–367.

Krishnamoorthy KK, Sankaralingam A, Nakkeeran S (2017) Compatibility between fungicides and Bacillus amyloliquefaciens isolate B15 used in the management of *Sclerotinia sclerotiorum* causing head rot of cabbage. Intl J Curr Sc. 5: 239–243

Li X, Fu L, Chen C, Sun W, Tian Y, Xie H (2020) Characteristics and rapid diagnosis of *Pectobacterium carotovorum* ssp. associated with bacterial soft rot of vegetables in China. Plant Dis 104(4): 1158–1166.

Mari M, Neri F, Bertolini P (2009) Management of important diseases in Mediterranean high value crops. Stewart Postharvest Rev 5: 1–10.

Napitupulu B, Lubis MH (1987) Experiments to test the effectiveness of several chemicals (lime, alum and silica gel) to control soft rot development during transport and storage period. Bull Penal Hortic 15: 332–336.

Oskiera M, Kałuzna M, Kowalska B, Smolinska U (2017) *Pectobacterium carotovorum* subsp. *odoriferum* on cabbage and Chinese cabbage: Identification, characterization and taxonomic relatedness of bacterial soft rot causal agents. J Plant Pathol 99: 149–160.

Prins TW, Tudzynski P, von Tiedemann A, Tudzynski B, Ten Have A, Hansen ME, Tenberge K, van Kan JA (2000) Infection strategies of *Botrytis cinerea* and related necrotrophic pathogens. In: Fungal Pathology. Springer, Dordrecht.

Purdy LH, Bardin R (1953) Mode of infection of tomato plants by the ascospores of *Sclerotinia sclerotiorum*. Plant Dis Rep 37: 361–362.

Ren J (1999) Resistance to bacterial soft rot (*Erwinia carotovora* subspecies *carotovora*) in *Brassica rapa* vegetables. Cornell University, 177 pp.

Ren J, Petzoldt R, Dickson MH (2001) Screening and identification of resistance to bacterial soft rot in *Brassica rapa*. Euphytica 118: 271–280.

Romanazzi G, Lichter A, Mlikota Gabler F, Smilanick JL (2013) Recent advances on the use of natural and safe alternatives to conventional methods to control postharvest gray mold of table grapes. Postharvest Biol Technol 61: 141–147.

Romanazzi G, Lichter A, Gabler FM, Smilanick JL (2012). Recent advances on the use of natural and safe alternatives to conventional methods to control postharvest gray mold of table grapes. Postharvest Biol Technol 63(1): 141–147.

Sabry S, Ali AZ, Abdel-Kader DA, Abou-Zaid MI (2015) Control of cabbage Alternaria leaf spot disease caused by *Alternaria brassicicola* Zagazig. J Plant Pathol 42: 55–60.

Sharma P, Zewain QK, Bahadur P, Sain SK (2005) Effect of soil solarization on sclerotial viability of *Sclerotinia sclerotiorum* infecting cauliflower (*Brassica oleracea* var. *botrytis* subvar *cauliflora*). Indian J Agric Sci 75: 90–94.

Steadman JR. (1975) Nature and epidemiological significance of infection of bean seed by *Whetzelinia sclerotiorum*. Phytopathology 65(11): 1323–1324.

Surviliene E, Valiuskaite A, Duchovskiene L, Kavaliauskaite D (2010) Influence of fungicide treatment on grey mould of cabbage. Veg Crops Res Bull 73: 133–142.

Tompkins CM, Tucker CM, Gardner MW (1936) Phytophthora root rot of cauliflower. J Agric Res 53: 685–692.

Tsuda K, Tsuji G, Higashiyama M, Ogiyama H, Umemura K, Mitomi M, Kubo Y, Kosaka Y (2016) Biological control of bacterial soft rot in Chinese cabbage by *Lactobacillus plantarum* strain BY under field conditions. Biol Control 100: 63–69.

Wilcox WF, Mircetich SM (1985) Effects of flooding duration on the development of Phytophthora root and crown rots of cherry. Phytopathology 75: 1451–1455.

Williamson B, Tudzynski B, Tudzynski P, van Kan JAL (2007). *Botrytis cinerea*: The cause of grey mould disease. Mol Plant Pathol 8: 561–580.

Zewain QK, Bahadur P, Sharma P (2004) Effect of fungicides and neem extract on mycelial growth and myceliogenic germination of *Sclerotinia sclerotiorum*. Indian Phytopath 57: 101–103.

33

Postharvest Diseases of Leguminous Vegetable Crops and Their Management

A. N. Tripathi,[1] D. Singh,[2] K. K. Pandey[1] and J. Singh[1]
[1]ICAR-Indian Institute of Vegetable Research, Varanasi, Uttar Pradesh, India
[2]ICAR-Indian Agricultural Research Institute, New Delhi, India

CONTENTS

33.1 Introduction

India is the second largest producer of vegetables in the world after China, and shares about 16% of global vegetable production (Tripathi et al. 2020). Processed vegetables have been exported at a compounded annual growth rate in volume of 16% and in value of 25% (Chikkasubbanna 2006). Broad bean (*Vicia faba*), cowpea (*Vigna unguiculata*), cluster bean (*Cyamopsis tetragonoloba*), French bean (*Phaseolus vulgaris*), Indian field bean (*Dolichos lablab*), Indian sword bean (*Canavalia gladiate*), vegetable pea (*Pisum sativum*) and winged bean (*Psophocarpus tetragonolobus*) are major leguminous vegetable crops cultivated under different climatic conditions around the world. They have a significant role in enhancing farm income, sustainable global food as well as nutritional security. These vegetables provide as high as 14% protein, which is more in dry seeds (Choudhury 2006). In addition to protein, they are rich in vitamins, especially vitamin A and vitamin C, minerals (P, Ca, Fe) and dietary fibre. Leguminous vegetables suffer with several pre- and postharvest diseases, viz. anthracnose, *Ascochyta* blight, *Sclerotinia* blight, bacterial soft rot, bacterial spot, bacterial blight, canker, etc. Postharvest losses in vegetables are reported up to 30–40% owing to poor postharvest practices (Ahsan 2006). However, the losses in leguminous vegetables due to diseases are slightly lower about 16–18% than weeds and animal pest (34%).

33.2 Impact of Postharvest Diseases of Vegetables

Fresh vegetables are highly perishable, and they have relatively short shelf lives. Fresh vegetables are living, respiring tissues that start senescing immediately after harvest. They are mostly comprised of water, with most having 90–95% moisture content. Losses due to postharvest disease may occur during handling of produce from harvest to consumption. Postharvest diseases cause qualitative and quantitative losses of leguminous vegetables and make them unfit for human consumption due to potential health risks. Primary and secondary agricultural practices are also important and costs such as harvesting, packaging and transport must be taken into account when estimating the value of the produce lost as a result of postharvest wastage. Because of the perishable nature of vegetables, special skills are required for postharvest handling. A large number of postharvest diseases are caused by fungi-derived carcinogenic mycotoxins and mutagenic secondary metabolites (Klich 2007). *Aspergillus flavus* is a saprophytic soil inhabitant fungus which infects and contaminates pre-harvest and postharvest vegetable crops and produces carcinogenic secondary metabolite aflatoxin in tropical, subtropical and temperate geographic regions of the world. It also causes animal and human diseases (causing aflatoxicosis and/or liver cancer) due to consumption of contaminated food and feed and through invasive growth (causing aspergillosis), which is often fatal to humans who are immunocompromised (Hedayati et al. 2007). A holistic approach is needed for regulating aflatoxins under the trade/export market with biosecurity including bio-warfare, biodiversity and biosafety for liberalized trade under the World Trade Organization (WTO) (Tripathi et al. 2013).

33.3 Issues and Challenges of Postharvest Losses in Vegetable Crops

Losses in vegetable beans are the result of (i) poor knowledge about the right harvesting index; thus, a large proportion of the harvested beans are usually over-mature, (ii) poor handling practices, such as the use of plastic sacks for bulk packaging and transportation which results in mechanical damage that serves as entry points for disease-causing organisms leading to rotting of the pods, (iii) poor transport practices such as the use of trucks that have no cover, thus exposing the produce to direct sunlight and high temperature, (iv) the absence of low temperature storage facilities and transport systems, and (v) rough handling practices during distribution in retail markets (Table 33.1). Application of good postharvest management practices which are supported by good technologies and also improving postharvest systems will maintain the quality of beans and reduce quantitative losses.

33.4 Causes of Postharvest Losses

Postharvest losses are caused by diseases (biotic) and abiotic factors.

33.5 Postharvest Parasitic Diseases

Fungi, bacteria, insects and biological factors are major causes of postharvest losses in leguminous vegetables. Microbes infect horticultural produce and spread rapidly due to lack of natural defence mechanisms in the tissues of the produce. Management of postharvest spoilage is becoming a very difficult task because the pesticides/chemicals available are rapidly declining with consumer concern for food safety.

33.6 Types of Postharvest Diseases

In general, postharvest diseases of leguminous vegetables are incited by fungi and bacteria. Postharvest diseases are often classified on the basis of the infection as "quiescent" or "latent", where the pathogen infects before harvest in the field. Examples of postharvest diseases arising from quiescent infections include anthracnose of various vegetables caused by *Colletotrichum* spp. and grey mould rot caused by *Botrytis cinerea*. Some pathogens infect leguminous vegetables after harvest during storage and transport, which is called postharvest infection e.g. nesting disease of pea caused by *Pythium* species or *Rhizopus* species.

33.6.1 Postharvest Fungal Diseases

Common postharvest diseases resulting from wound infections initiated during and after harvest includes blue and green mould (caused by *Penicillium* spp.) and transit rot (caused by *Rhizopus stolonifer*). Important fungal genera of anamorphic postharvest pathogens include *Penicillium*, *Aspergillus*, *Geotrichum*, *Botrytis*, *Fusarium*, *Alternaria*, *Colletotrichum*, *Phomopsis*, *Rhizoctonia*, *Sclerotium* and *Sclerotinia*. The most important pathosystem of postharvest leguminous vegetables are grey mould (*Botrytis* spp.), white mould and watery soft rot (*Sclerotinia* spp.), cottony leak (*Pythium* spp.), powdery mildew discoloured pod of cluster bean (Fig. 33.10) and Sclerotium rot (*Sclerotium rolfsii*) (Table 33.2).

33.6.1.1 Anthracnose

Anthracnose (*Colletotrichum lindemuthianum*, *C. orbiculare*) symptoms appear on immature pods. Sunken cankers with lighter or grey central areas of about 5–7 mm size are seen. The spots on leaves and pods are enlarged and produce tiny black acervuli in the centres which in humid conditions ooze viscous droplets consisting of a mass of pinkish spores.

33.6.1.2 White Mould or White Rot

White mould (*Sclerotinia sclerotiorum*) appears in warm and moist weather (>95% relative humidity) and favours fungal growth which develops as a white cottony mat of mycelium on pods (Fig. 33.1–33.8) Within the superficial mycelium, initially white but later hard dark black sclerotia are formed. Infected pods show brown discoloration and soft rot.

TABLE 33.1

Issues/Problems Related to Postharvest Handling, Storage and Marketing of Vegetables

Operation	Issue/Problem	Strategy
Harvesting index	There is no established maturity index for some commodities. There is no specific maturity index for local as well as export markets. Adoption of established indices is low and price and distance to market influence the adoption rate.	• Research and development with emphasis on quality, safety and sustainability, development of farmer-friendly harvest indices are required. • Extension activities must be conducted
Harvesting methods and time of harvesting	Rough handling and untimely harvesting are major problems. Appropriate and/or poorly designed harvesting tools, equipment and harvest containers are issues in vegetable crops.	• Creation of awareness of appropriate methods and time of harvesting is needed. • Research and development should be focused on design and efficiency of harvesting tools and equipment.
Field sorting, grading and packing	Inadequate field sorting, grading and packing protocols for commodities that lend well to field packing are issues.	• Sorting, grading and packing protocols for certain commodities should be established. • Farmers and stakeholders must be educated.
Pre-cooling	The problems in pre-cooling are due to lack of pre-cooling facilities and their higher cost. There is limited knowledge of precooling technology on a commercial scale and information on cost–benefit ration of pre-cooling technology.	• A good policy environment should be created for promotion of investment. • Alliances/commodity-based clusters should be formed to overcome problems with limited facilities. • For commercial scale using pre-cooling, focus must be on research and development.
Transportation	There is poor infrastructure on roads and bridges. Appropriate transport systems and refrigerated transport are not available. Poor temperature management and poor loading and unloading practices are common during transportation.	• Investment from the private sector and policy support from government should be encouraged. • There is a need to create awareness of proper transport system management.
Storage	Storage facilities at the farm level and refrigerated storage at the markets and ports are not readily available. Temperature is not maintained properly in storage, including sanitation of the storage room and facilities. There is a limited knowledge of temperature requirement during storage. Ethylene sensitivities of different commodities for mixed loading area major concern.	• Favourable policy environment for investment should be created. • Focus is needed on research and development for determining the cost–benefit ratio of storage systems. • Awareness must be created of correct operations and management of storage facilities. • Research and development focusing on temperature, relative humidity, ethylene sensitivities of different commodities under storage are needed.
Grading	National standards and poor enforcement of standards are not available. There is lack of skill and awareness.	• National standards need to be developed. • Capacity building of staff is required.
Procurement of centres, packinghouses, grading facilities	Lack of collection centres, packinghouses and grading facilities.	• Government support is needed for clustering.
Packaging and labelling	Packing technology/suitable packaging (for transportation, storage, and consumers) are not sufficiently available.	• The existing technologies should be developed/ adapted.
Minimally processed; fresh-cut; ready-to eat/ cook	Proper knowledge about appropriate technologies for anti-browning and protection from microbial contamination is lacking.	• Capacity building and information dissemination, especially to small processors, are required.
Secondary processing	Fewer suitable varieties for processing are available.	• Suitable varieties for processing purposes must be developed. • Information to small processors should be disseminated.
Marketing	Limited market information is available and there are no marketing strategies.	• National/ regional information networking system should be established.

TABLE 33.2

Postharvest Diseases/Pathosystems of Leguminous Vegetable Crops

Pathogen	Disease	Symptom
Sclerotinia sclerotiorum	Watery soft rot or white stem rot	Disease symptom initially appears in the form of water-soaked lesions on pods and stems. Later, infected tissues become whitish and covered with white mycelia mats and black-coloured sclerotia.
Colletotrichm lindemuthianum	Anthracnose	Disease symptoms appear in the form of brown to black sunken spots and lesions on leaves, stems and pods. The centre of anthracnose lesions on pods is covered with numerous black dot-like acervuli.
Ascochyta pisi	*Ascochyta* blight	Black spot symptoms on pods result in the production of round tan-coloured sunken spots bearing dark margins with pycnidia on pods.
Macrophomina phaseolina (Rhizoctonia solani)	Charcoal rot or ashy stem blight	Disease symptoms appear in the form of dark brown to black charcoal-coloured lesions covered with black dot-like fruiting bodies (resting microsclerotia and pycnidia) on pods.
Sclerotiorum rolfsii	*Sclerotiorum* rot	Whitish growth with mustard-like sclerotia on pods.
Pythium spp.	Cottony leak	White mycelia growth on pods.
Pseudomonas syringae pv. *Phaseolicola* (halo blight) and *P. syringae* pv. *Syringae* (brown spot blight)	Bacterial blight	Disease symptoms appear as translucent, water-soaked lesions along with a chlorotic halo on pods.
Erwinia, *Pseudomonas*, *Bacillus*, *Lactobacillus*, and *Xanthomonas*	Bacterial soft rot	Disease symptoms appear as water-soaked lesions finally rotten pods.

33.6.1.3 *Ascochyta Blight*

Ascochyta blight (*Ascochyta pisi*) black spot symptoms on pods result in the production of round tan-coloured sunken spots bearing dark margins (Fig. 33.9). Pycnidia develop in the centres of such spots on pods.

33.6.2 Postharvest Bacterial Diseases

Phytopathogenic bacteria cause postharvest diseases of economically important leguminous vegetables. They are unable penetrate directly into plant tissue; however, they enter through wounds or natural plant openings (Table 33.2). Wounds can be caused by insects and tools during operations like pruning and picking of the produce. The bacteria only become more active and cause infection when factors are conducive. Factors conducive to infection are high humidity, crowding, poor air circulation, plant stress caused by overwatering, under watering, or irregular watering, poor soil health and deficient or excess nutrients. The bacteria multiply quickly when free moisture and moderate temperatures are available. The major causal agents of bacterial soft rots are various species of *Erwinia*, *Pseudomonas*, *Bacillus*, *Lactobacillus* and *Xanthomonas*. *Psuedomonas syringae* pv. *syringae*, *P. syringae* pv. *pisi* and *P. syringae* pv. *phaseolicola* cause diseases in leguminous vegetables. The symptoms appear as water-soaked spots on pods which become sunken and dark-brown in colour with distinctive reddish-brown margins.

FIGURE 33.1 *Sclerotinia* white rot of cowpea.

FIGURE 33.2 Anthracnose spot on pod of cowpea.

FIGURE 33.3-33.4 *Sclerotinia* white rot on French bean.

FIGURE 33.5-33.7 *Sclerotinia* white rot on dolichos bean/Indian field bean.

FIGURE 33.8 *Sclerotinia* white pod rot on peas.

FIGURE 33.9 Ascochytaspot on pod of pea.

FIGURE 33.10 Powdery mildew on pod of cluster bean.

33.7 Mechanical Injury

Leguminous vegetables are highly susceptible to mechanical injury due to their soft texture and high level of moisture content. Improper handling, unsuitable packaging material and poor packing during transportation are the main cause of bruising, cutting, breaking, impact wounding, and other types of injury in the produce.

33.8 Physiological Deterioration

Tissues of vegetable produce are still alive after harvest and they continue their physiological activity during postharvest handling. Many physiological disorders occur due to mineral deficiency, low or high temperature injury, or undesirable environmental conditions like high humidity.

33.9 Postharvest Disease Management

Postharvest losses in leguminous vegetables are found due to fungal and bacterial infection worldwide. New challenges are faced under trade liberalization and globalization, and serious efforts are needed to reduce these losses in vegetables.

33.9.1 Chemical Control

Chemical fungicides are commonly used for management of postharvest disease in vegetables. For post-harvest pathogens which infect produce before harvest the fungicides should be applied at field level during the crop season, and/or strategically applied as systemic fungicides. At the postharvest level, the fungicides are often applied to reduce infections already established in the surface tissues of produce or they may protect infections occurring during storage and handling. Fungicides used during postharvest are actually fungistatic in nature rather than fungicidal under normal usage. Sodium hypochlorite as a disinfectant is used to kill spores of pathogens present on the surface of the produce. The fungicides are applied on the produce as dips, sprays, fumigants, treated wraps and box liners or in waxes and coatings. Dip and spray methods are very common in postharvest treatments. The fungicides generally applied as dip or spray method are benzimidazoles (e.g. benomyl and thiabendazole) against anthracnose, and triazoles (e.g. prochloraz and imazalil) and fumigants, such as sulphur dioxide, for the control of grey mould used for postharvest disease control (De Waard et al. 1993; Ampatzidis et al. 2017). Dipping in hot water (at 50°C for 5–10 min, depending on the size of produce in combination with the fungicide) is also used for effective control of disease.

33.9.2 Biological Control through Microbes

International markets reject produce containing unauthorized pesticides, with pesticide residues exceeding permissible limits, and with inadequate labelling and packaging. Hence, biological control of postharvest diseases has great potential because postharvest environmental conditions like temperature and humidity which can be strictly controlled to suit the needs of the bio control agent. Much information has been provided in relation to postharvest bio control and the problems faced for the development of commercial products (Droby et al. 2000, 2001). Biological control is used through microbes such as fungi, bacteria, actinomycetes and viruses (bacteriophages) to control postharvest disease of vegetables (Mohamed and Benali 2010; Pandey et al. 2016; Loganathan et al. 2016; Chaurasia et al. 2018; Tripathi et al. 2020). The degree of disease control or disease suppression achieved with these bio agents can be comparable to that achieved with chemicals. As per estimates, market of Indian bioagents is equivalent to 2.89% of the overall pesticide market in India with worth of rupees 690 crores. It is expected to show an annual growth rate of about 2.3% in the coming years (Thakore 2006; Cheng et al. 2010). In India, so far only 16 types of bio pesticides have been registered under the Insecticide Act of 1968. Among agriculturally important microbes, *Trichoderma viride*, *T. harzianum*, *Pseudomonas fluorescens* and *Bacillus subtilis* are the most potential bio agents which as act as producers of biologically active metabolites like antibiotics and bacteriocin, elicitors and inducers of systemic resistance in plants. Bio control-mediated pathogen inhibition is found to be more effective when the antagonist is applied prior to infection taking place. Antagonists which act against postharvest pathogens of vegetables by competitive inhibition at wound sites include the yeasts *Pichia* and *Debaryomces* species. Chitosan, for example, is not only an elicitor of host defence responses but also has direct fungicidal action against a range of postharvest pathogens. *Trichoderma* have potent antifungal activity against *Botrytis cinerea*, *S. sclerotiorum*, *Corticum rolfsi* and other important biotic stresses. Microbial pesticide active ingredients of *Streptomyces griseoviridis* K61 against bacterial soft rot, grey mould, *Phytophthora*; *Gliocladium catenulatum* against grey mould; *Candida oleophila* strain against postharvest diseases; *Coniothyrium minitans* against *Sclerotinia sclerotiorum*, *Sclerotinia minor*; *Trichoderma aspellerum* (formerly *T. harzianum*) against *Pythium*, *Phytophthera*, *Botrytis*, *Rhizoctonia*; *Trichoderma atroviridae* against *B.*

cinerea and *B. subtilis* against *Botrytis* spp. are the most commonly used bio-control agents for postharvest diseases. Characteristics of an "ideal antagonist" for the postharvest environment are listed here.

- It should be genetically stable.
- It should be effective at low concentrations of bio agent.
- It should not be fastidious for its requirement of nutrients.
- It should have the ability to survive adverse environmental conditions (including low and high temperature and controlled atmosphere storage).
- It should have a wide range, to control postharvest pathogens on a variety of produce.
- It can be produced on an inexpensive medium for mass multiplication.
- The formulation should have a long shelf life.
- The product should be easy to dispense.
- The bio agent does not produce metabolites which are harmful to human health and beneficial microbes.
- The bio agent should show resistance to pesticides.
- It should be compatible with commercial processing procedures.
- It should not grow at 37°C and should not infect humans.
- It should be non-pathogenic to the host commodity.

Antagonistic yeast forms a biofilm to stick pathogen and parasitize on the hyphae of pathogen. Bar-Shimon et al. (2004) reported that bio-control efficacy of yeast correlates with the production of lytic enzymes and their ability to tolerate high concentration of salts. Further, molecular approaches were used to examine the role of glucanases in bio-control activity of the yeast *C. oleophila* and bio-control activity was enhanced by overexpression of antimicrobial peptides. By early 2000, three postharvest biological products, Aspire™ (the USA and Israel), Bio-Save™ (the USA) and Yield Plus™ (South Africa) were available in the market. However, Aspire was initially involved to combine the product with a low concentration of postharvest fungicide (Droby et al. 1998) or salt solutions (1–2%) of calcium chloride or sodium bicarbonate and also with other additives which are commonly used in the food industry (Droby et al. 2003b). These products were also combined with physical treatments like hot air, curing, hot-water brushing and combinations of the above with pressure infiltration of calcium for improvement of efficacy (Droby et al. 2003a). To increase bio-efficacy, the antagonists can also be combined with a sugar analogue (2-deoxy-D-glucose).

An effort has been made to develop two new products based on yeast antagonist *C. saitoana* and a derivative of either chitosan (Biocoat) or lysozyme (Biocure). These products had been evaluated worldwide. They showed strong eradicative activity. The two commercial products based on the use of heat-tolerant strain of *Metschnikowia fructicola* also contain other additives such as sodium bicarbonate. The additives are found highly effective to increase bio-control efficacy to levels equivalent to those found with available postharvest fungicides. The product is marketed under the name ProYeast-ST and ProYeast-ORG in Israel by the company Agro Green and found effective against rots incited by *Botrytis*, *Penicillium*, *Rhizopus* and *Aspergillus*.

33.9.3 Botanicals/Plant Essential Oils

Botanical pesticides cause no adverse effects on non-target biota with biodegradability. It should be noted that most of the crops sprayed with botanical pesticides are quite safe for consumption after a short period after spraying. A large number of defensive of rich chemicals such as terpenoids, alkaloids, phenols, tannins, coumarins, flavonoids, etc. present in plants which cause physiological effects on pathogens. These compounds have already been identified in the extracts/exudates of many plants. They have antimicrobial activities and are used for postharvest disease control.

The use of natural botanical products would be a supplement or an alternative to synthetic fungicide to reduce the indiscriminate use of pesticides. So that it is called "reduced-risk" pesticides which are favoured by the Environmental Protection Agency in the United States. Examples include 1,8-Cineole, the major constituent of oils from rosemary (*Rosmarinus officinale*) and eucalyptus (*Eucalyptus globus*),eugenol from clove oil (*Syzygium aromaticum*),thymol from garden thyme (*Thymus vulgaris*) and menthol from various species of mint (*Mentha* species). The majority of research is progressing in this regard to develop plant oil–based pesticides. Therefore, essential oil-based formulations have great scope in the future to use as green pesticides as plant protectants in the integrated pest and disease management of value-added agriculture and horticulture crops. The exact mechanism of antifungal and antibacterial action of essential oils is still not fully clarified. However, it is speculated to involve a broad range fungitoxic spectrum for its possible exploitation as botanical fungitoxicant/fungicide.

Many exhaustive studies have been carried out on the utility of neem oil against various fungal pathogens. Its efficacy has been evaluated against *F. moniliforme*, *Aspergillus niger*, *Drechslera rostrata* and *M. phaseolina*. During an antifungal study, neem oil was found to be on par with the fungicide hymexazole in the control of the soil pathogens *F. oxysporum*, *F. ciceri*, *R. solani*, *S. rolfsii* and *S. sclerotiorum*. Researchers have reported *in vitro* inhibition of 16 aromatic compounds against five major seedborne fungal pathogens (*D. sorokiniana*, *P. sojae*, *F. solani*, *C. graminicola* and *M. phaseolina*) in the concentration range of 100–8000 ppm and minimum inhibitory concentration (MIC) value for all the test fungi was 270–1704 ppm. Essential oils were used against five seed borne fungal pathogens, *viz.*, *D. sorokiniana*, *P. sojae*, *F. solani*, *C. graminicola* and *M. phaseolina* and reported MIC value for all the test fungi was 439.8–2200 ppm. Essential oils under commerce used as bio pesticides have many problems, such as non-tariff barrier, scarcity of natural resources, need of quality control and difficulties of registration. Some plant products have been commercialized. SPIC Science Foundation has developed a fungistatic product "Wanis" which has a single monoterpene as an ingredient and it is reportedly very effective in controlling more than 30 different types of phytopathogenic fungi. It is nontoxic to human beings and livestock. Recently, an antifungal agent by the name "TALENT", containing carvone as the active ingredient, derived from essential oil of Carumcarvii, was commercialized. Mycotech Corporation product Cinnamite™, based on cinnamon oil, has been developed as a fungicide/miticide for glasshouse and horticultural crops. World leading essential oil-based pesticide producing EcoSMART technologies developed EcoPCORR under the name Bioganic™ as insecticide and miticide for nursery crops, horticultural crops and value-added crops under glasshouse conditions. The EcoSMART formulation based on rosemary oil, viz. EcoTrol™ (insecticide/miticide), Sporan™ (fungicide) and eugenol oil formulation Mataran™ (weedicides) were classify as generally recognized as safe (GRAS) by the U.S. Food and Drug Administration, and these are exempt from EPA registration. Thus plant essential oils are safe to the user and the environment and have a good potential as crop protectants and integrated pest management under organic farming and value-added agricultural and horticultural crops (Tripathi et al. 2014). These chemicals of biological origin are safe to use, and in a few cases can even be produced by farmers and rural communities.

33.9.4 Hygiene Practices

Maintenance of hygiene in all stages of postharvest handling is critical to minimize the source of primary inoculum for postharvest diseases. For post-harvest diseases which arise from pre-harvest infections, the practices which make the crop environment less favourable to the pathogens will help reduce the amount of pathogen inoculum. Field sanitation and removal of infected plant parts is also an important way to minimize inoculum build-up. Chlorinated water is commonly used for washing of vegetables. Sterilized packing and grading equipment, particularly brushes and rollers, are used.

33.9.5 Management of Abiotic Factors

Postharvest disease losses are affected by several factors, *viz*., commodity type, cultivar susceptibility to postharvest disease, maturity and ripeness stage, harvesting, treatments used for disease management, produce handling methods, storage environment (temperature, relative humidity, atmosphere composition, etc.) and postharvest hygiene.

33.9.6 Temperature Management Practices

Management of temperature is the most important factor to extend shelf life of fresh vegetables after harvest. It begins with rapid removal of the field heat by using any of the following cooling methods: hydro-cooling, in-package ice, top icing, evaporative cooling, room cooling, forced air cooling, serpentine forced air cooling, vacuum cooling and hydro-vacuum cooling. Beans are sensitive to chilling injury and russeting if kept at <40°F temperature and symptoms of chill damage may appear within 3 days. If they are kept at 32°F, they will russet and become pitted and lose moisture rapidly. Optimum temperature for storage of beans is 45–50°F. Beans stored for too long or at too high a temperature are more prone to various soft rot diseases. Beans should be washed before refrigerating to help to retain moisture content. Snap beans require air circulation during storage. Containers should be stacked in such a way as to allow maximum air circulation.

33.9.7 Control of Relative Humidity

The relative humidity during storage should be maintained about 85–95% for most fruits and 95–98% for vegetables. Relative humidity may be controlled by addition of moisture to air by humidifiers, regulation of air movement in relation to produce, maintaining coil temperature to 1°C difference to air temperature, wetting the floor in the storage room and addition of crushed ice.

33.10 Postharvest Handling Operations of Leguminous Vegetable Crops

For beans, the appropriate time to harvest is when the morning dew is off and the plants are thoroughly dry. The pods are best harvested manually using a cutting tool (such as scissors or shears). The pod must be held by hand and cut off from the plant. Harvested pods must be directly placed in a field collection container for transporting to a collection point. Plastic crates are one example of a good field collection container. Postharvest handling operations prepare harvested produce for marketing. Workers who handle the beans must observe good hygiene. The sorting of culling out produce which is unmarketable should be done at the field level to reduce the cost of hauling and also to minimize disease infection. Where possible, snap beans should be sorted at the farm in order to remove low quality (diseased) or damaged (insect- or rat-damaged beans). At the packing area, snap beans may be sorted to remove beans with harvest-related defects such as cuts, severe abrasions, etc. before grading. Sorting and grading can be done at a collection centre or in a packinghouse having sorting tables and weighing scales. The area in which these operations are performed must be clean. Workers must observe good hygiene and must be able to sit comfortably in order to engage in sorting operations. Plastic crates are the best bulk packaging containers for snap beans because of their smooth inside finish, good ventilation that prevents the build-up of heat produced during respiration of the bean pods and because of the ease with which they can be cleaned. Due to their uniform shape, plastic crates can be stacked securely inside transport vehicles. For best results, beans must be loosely packed in plastic crates in order to allow air flow through the beans inside the package.

General practices to prevent postharvest losses of leguminous vegetables are as follows.

- Protruding nails or staples should be removed and any rough edges on field containers should be smoothed.
- Harvest workers should not have long, sharp fingernails.

- Care should be taken during dumping products from one container to another.
- Padding should be used on all impact areas when possible.
- Sand and all debris out of all containers should be cleaned.
- Containers should not be overfilled, which causes severe damage during stacking.
- Produce should be harvested during the day because many products are more turgid in the early morning and cause bruise more easily.
- Start with clean water each day.
- Washing of the produce should be done by using about 100 ppm of free chlorine. This can be done with chlorine gas or with either liquid or granular hypochlorite. pH should be adjusted to the range of 6.0–7.0. Chlorine gas has little effect on pH, but hypochlorite raise pH, so some adjustment is usually necessary.
- The respiration rate of produce should be kept slow by maintaining optimal gaseous environment.
- Water loss of produce should be slow by maintaining optimal relative humidity.
- Resistant varieties should be selected.
- Transport vehicles should always be cleaned and sanitized before loading.

33.11 Conclusion

Leguminous vegetables are highly perishable in nature and are susceptible to many types of fungal and bacterial pathogens that cause postharvest disease during storage and transport. Some of these infect produce before harvest and others after harvest. For postharvest disease management, various strategies such as postharvest handling systems, sanitation and integration of botanicals/plant essential oil, microbial bio agents and safe chemicals need to be integrated and develop integrated postharvest diseases management techniques under WTO regime. Among them, it is expected that the knowledge of bio control will lead to new, innovative approaches to minimize postharvest decay of the produce and it presents the best hope for the future of postharvest disease management of vegetable produce.

REFERENCES

Ahsan H (2006) India (1). In: Rolle RS (Ed.). *Postharvest Management of Fruit and Vegetables in the Asia-Pacific region*. Asian Productivity Organization, Tokyo, pp. 131–142.

Ampatzidis Y, DeBellis L, Luvisi A (2017) Pathology: Robotic applications and management of plants and pant diseases. Sustainability 9(6): 1010. https://doi.org/10.3390/su9061010.

Bar-Shimon M, Yehuda H, Cohen L, Weiss B, Kobeshnikov A, Daus A, Goldway M, Wisniewski M, Droby S (2004) Characterization of extracellular lytic enzymes produced by the yeast bio-control agent Candida oleophila. *Curr Genet* 45: 140–148.

Chaurasia A, Meena BR, Tripathi AN, Pande KK, Rai AB, Singh B (2018) Actinomycetes: An unexplored microorganisms for plant growth promotion and bio-control in vegetable crops. *World J Microbiol Biotechnol* 34(9): 132.

Cheng XL, Liu CJ, Yao JW (2010) The current status, development trend and strategy of the bio-pesticide industry in China. *Hubei Agric Sci* 49: 2287–2290.

Chikkasubbanna V (2006) India (2). In: Rolle RS (Ed.). *Postharvest Management of Fruit and Vegetables in the Asia-Pacific region*. Asian Productivity Organization, Tokyo, pp. 143–151.

Choudhury ML (2006) Recent development in reducing postharvest losses in the Asia-Pacific region. In: Rolle RS (Ed.). *Postharvest Management of Fruit and Vegetables in the Asia-Pacific region*. Asian Productivity Organization, Tokyo, pp. 15–22.

De Waard, Georgopoulos MA, Hollomon SG, Ishii DW, Leroux P (1993) Chemical control of plant diseases: Problems and prospects. *Annu Rev Phytopathol* 31: 403–421.

Droby S, Cohen L, Daus A, Weiss B, Horev E, Chalutz E, Katz H, Keren-Tzour M, Shachnai A (1998) Commercial testing of Aspire: A bio-control preparation for the control of postharvest decay of citrus. *Biol Control* 12: 97–101.

Droby S, Cohen L, Wiess B, Daus A, Wisniewski M (2001) Microbial control of postharvest diseases of fruits and vegetables – current status and future outlook. *Acta Hortic* 553: 371–376.

Droby S, Wilson C, Wisniewski M, El Ghaouth A (2000) Biologically based technology for the control of postharvest diseases of fruits and vegetables. In: Wilson C, Droby S (Eds.). *Microbial Food Contamination*. CRC Press, Boca Raton, FL, pp. 187–206.

Droby S, Wisniewski M, El Ghaouth A, Wilson CL (2003a) Biological control of postharvest diseases of fruits and vegetables: Current advances and future challenges. *Acta Hortic* 628: 703–713.

Droby S, Wisniewski M, El-Ghaouth A, Wilson C (2003b) Influence of food additives on the control of postharvest rots of apple and peach and efficacy of the yeast-based biocontrol product Aspire™. *Postharvest Biol Technol* 27: 127–135.

Hedayati MT, Pasqualotto AC, Warn PA, Bowyer P, Denning DW (2007) *Aspergillus flavus*: Human pathogen, allergen and mycotoxin producer. *Microbiology* 153: 1677–1692.

Klich MA (2007) *Aspergillus flavus*: The major producer of aflatoxin. *Mol Plant Pathol* 8: 713–722.

Loganathan M, Rai AB, Pandey KK, Nagendran K, Tripathi AN, Singh B (2016) PGPR *Bacillus subtilis* for multifaceted benefits in vegetables. *Indian Hortic* 61 (1): 36–37.

Mohamed B, Benali S (2010) The talc formulation of *Streptomyces* antagonist against *Mycosphaerella* foot rot in pea (*Pisumsativum L.*) seedlings. *Arch Phytopathol Plant Prot* 43: 438–445.

Pandey KK, Nagendran K, Tripathi AN, Manjunath M, Rai AB, Singh B (2016) Integrated disease management in vegetable crops. *Indian Hortic* 61(1): 66–68.

Thakore Y (2006) The biopesticide market for global agricultural use. *Industrial Biotechnol Fall* 2006: 194–208.

Tripathi AN, Sharma P, Agarwal PC, Dev Usha, Hazarika BN, Tripathi SK, Singh US, Khetarpal RK, Satpathy S (2013) Aflatoxins: Threat for agricultural trade and food safety. In: Prasad D, Ray DP (Eds). *Biotechnological Approaches in Crop Protection*. Biotech Books, New Delhi, India, pp. 490–499.

Tripathi AN, Gotyal BS, Sharma PK, Tripathi RK, Dev Usha, Biswas C, Agarwal PC (2014) Essential Oils: as a Green Biopesticide for Organic Farming. In: Biswas SK, Pal S (Eds.). *Organic Farming and Management of Biotic Stresses*. Biotech Books, New Delhi, India, pp. 548–554.

Tripathi AN, Meena BR, Pandey KK, Singh J (2020) Microbial bioagents in agriculture: Current status and prospects. *New Frontiers in Stress Management for Durable Agriculture*. 1st ed. Rakshit A, Singh HB, Kumar Singh A, Singh US, Fraceto L (Eds.). Springer Nature, Singapore, pp. 490–499; 361368.

34

Postharvest Diseases of Tropical Tuber Crops and Their Management

S. S. Veena, C. Visalakshi Chandra, M. L. Jeeva and T. Makeshkumar
ICAR-Central Tuber Crops Research Institute, Sreekariyam, Thiruvananthapuram, Kerala, India

CONTENTS

34.1 Introduction

Roots and tuber crops are plants produced for their starchy roots, tubers, rhizomes, corms and stems which generally develop in the soil. They are rich in carbohydrates and are used as staple foods for human beings and animal feed as well as industrial raw materials for the production of different products such as starch and fermented beverages, including beer and alcohol. Root and tuber crops can be consumed fresh or processed into various food products. The roots and tuber crops are grown for their starch-rich roots, rhizomes, stems, corms, etc., and in the growing countries, the tropical root and tuber crops perform a pivotal role in assuring food security to all. The energy and nutrition requirements of around two billion people round the world are taken care of by roots and tuber crops. In rural and marginal areas, tuber crops aid as a valuable source of income; they are climate resilient and serve as food security crops, staple food crops and cash crops; provide raw material for the industry; and serve as feed for livestock (Reddy 2015). The major portion of the root and tuber produced across the globe is being used as food (45%); the balance serves as animal feed as well as industrial products like starch, distilled spirits, biofilm and many other minor products. Tuber crops are imparted with resilience to global warming and climate change, are capable of performing well even in less fertile soils and unfavourable weather conditions, provide immense raw materials for diverse industrial applications, have high nutritional qualities, and most have exceptionally high carbon sequestration potential.

Cassava/tapioca (*Manihot esculenta*), sweet potato (*Ipomoea batatas*), yams (*Dioscorea* spp.), yams, taro (*Colocasia esculenta*), elephant foot yam (*Amorphophallus paeonifolius*) and tania (*Xanthosoma sagittifolium*) are the major tropical root and tuber crops. Three crops, cassava, sweet potato and yam, are the most important food crops globally and back a significant part of the world's food supply (they contribute about 6% of the world's dietary calories) and are equally important as sources of animal feed and raw materials for many industries.

Apart from field diseases caused by viruses and fungi, postharvest losses due to microbes lead to heavy financial losses to farmers. Unlike the pathogens that occur in the field, the majority of postharvest pathogens will not penetrate directly through the tuber surface, and need a wound for penetration. Various wounds/damage caused during harvest, transportation, packaging operations and storage processes pave way for postharvest pathogens (Barkai-Goland 2001). Several species of pathogens belonging to the genera *Alternaria*, *Aspergillus*, *Botrytis*, *Fusarium*, *Geotrichum*, *Gloeosporium*, *Mucor*, *Monilinia*, *Penicillium* and *Rhizopus* cause the most important postharvest diseases in tuber crops (Barkai-Goland 2001).

Reduction of extent of damage by postharvest losses increases food availability, decreases the demand for area for production and saves natural resources. Many treatment methods have been used for minimizing and managing postharvest diseases. Several physical, chemical, biological control and use of botanicals are being utilized for this purpose. The management strategies need refinement, since no management strategy is foolproof and may vary depending on various factors, viz., place, climate, genetic makeup of the crop, maturity, transportation, handling, packaging, etc. An integrated approach encompassing several techniques has proven beneficial for managing postharvest diseases.

Lessening mechanical injuries and following proper sanitation procedures during harvest has a pivotal role in mitigating postharvest losses. The use of synthetic fungicides continues as the prime form of management of postharvest diseases. However, due to increasing awareness about the environmental and toxicological hazards, the practice of applying postharvest fungicides is being curtailed. In addition, postharvest fungicide treatment has been banned in some countries, and uninterrupted use of fungicides has led to the elaboration of resistance in pathogens (Adaskaveg and Förster 2010). The challenging requirements in sustainable agriculture, integrated crop management and organic production necessitate development of alternate methods of control to combat postharvest decay.

Evolving safe and efficient replacements to synthetic chemical fungicides for the mitigation of postharvest losses has been the emphasis of abundant research for several decades. Commercial viability and consumer acceptance remain a challenge while selecting alternatives to chemical fungicides. Tapping the potential of heat energy, which is non-polluting, for the physical management of postharvest decay and insect disinfestations may be studied in detail.

The present application of bio-control agents as an alternative to synthetic fungicides to check postharvest pathogens has various problems that make implementation difficult and hamper their use as a practical control strategy. Elaborate studies on bioagents, the factors that favour their establishment, multiplication procedures, etc. will ensure the successful adoption of postharvest bio-control and integration of bio-control agents into joint pre-harvest and postharvest management systems. Two basic approaches may be helpful while using microbial antagonists against postharvest pathogens: utilization of the microbes that exist on the produce itself, and introduction of organisms which can act against the postharvest pathogens (Sharma et al. 2009). Microorganisms as a part of biological control are being engaged currently to suppress several diseases, especially those caused by fungi, and to some extent bacterial diseases on many tuber crops (Lebot 2008). Application of microbial bioagents against postharvest diseases of tuber crops is an excellent, sensible and handy method.

34.2 Cassava

The crop cassava (*Manihot esculenta*) is believed to have originated in South America, probably Brazil. It is popularly known by the name tapioca and has its place in the family *Euphorbiaceae*. It is believed that from Brazil, the crop was carried to the Pacific Islands, Africa, Asia, etc. The starch content in the starchy tuberous roots of cassava varies from 25 to 35%. In developing countries, cassava alone provide food for more than 500 million persons from small-scale and subsistence farming. Many value-added forms of products, viz., flour, chips, pellets and starch, are produced from cassava and thus the crop adds to the economy of exporting countries during international trade (Misganaw and Bayou 2020). The storage potential of cassava is very short, and the fresh root cannot be stored even for few days. The rapid postharvest deterioration of cassava hampers storage and limits its use as a vegetable. The postharvest deterioration not only causes direct physical loss of the crop but also causes a reduction in root quality, which ultimately leads to price reductions and thereby contributes to economic loss.

34.2.1 Postharvest Physiological Deterioration

The global expansion and industrial use of cassava is hampered due to the shortened shelf life of tubers immediately after harvest, called postharvest physiological deterioration (PPD) (Zainuddin et al. 2018). PPD is the major cause for loss of acceptability of cassava tubers, and the yield loss can extend from 30–40% to 100% within 5 days if the tubers are not consumed or processed (Naziri et al. 2014). PPD, or vascular streaking, is an endogenous physiological disorder where cassava root deteriorates accompanied by foul odour and taste within 2–3 days of harvest, making it unfit for consumption, processing and marketing (Figure 34.1).

FIGURE 34.1 PPD symptom development seen in field-grown cassava tuber.

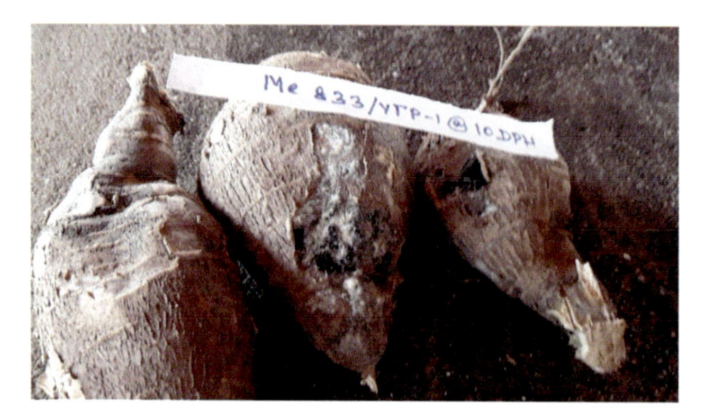

FIGURE 34.2 Microbial deterioration in cassava tubers.

34.2.1.1 Symptoms

The deterioration appears as brownish or dark bluish radial veins in xylem vessels originating from the site of injury and later spreads to storage parenchyma within 24–72 h after harvest, altering the structure of stored starch granules (Sánchez at al. 2006). The third tissue layer of the secondary growth (the edible portion), composed of secondary vessels and storage parenchyma, is where the PPD symptoms appear as a dark strip reported to be due to the formation of tylose occlusions in the xylem vessels (Djabou et al. 2017).

The deterioration of cassava tubers takes place in two stages, primary deterioration, or vascular streaking, followed by secondary deterioration, or microbial deterioration (Figure 34.2). Primary deterioration is marked by the appearance of dark bluish or brownish radial veins or streaks near the xylem vessels appearing within 2–3 days after harvest. This vascular streaking is due to physiological and biochemical changes, and no microbial activity is involved. The involvement of pathogenic microorganisms causes secondary deterioration, or microbial deterioration, and is generally observed in tubers with moderate to severe damage at the time of harvest and those already affected by primary deterioration, leading to softening of the tissues (Sánchez et al. 2013; García et al. 2013). The later phase of PPD involves formation of callus along with cell death.

Cassava tubers lack wound healing properties like other tuber crops once detached from the plant. The signals produced after detachment are insufficient to produce a wound-healing response and therefore the roots deteriorate rapidly (Salcedo and Siritunga 2011). The factors affecting PPD onset and development are mechanical damage during harvest, tuber shape and length, presence of neck, peel texture and adherence. PPD symptom development is greatly influenced by both genotype and environment. The PPD response is negatively correlated with starch and β-carotene content of roots and the high dry matter (Chavez et al. 2006).

34.2.1.2 Current Understanding of Postharvest Physiological Deterioration

Various researchers of PPD have suggested many biochemical changes in response to wounding/damage during harvest, accumulation of fluorescent compounds and secondary metabolites, alteration of starch:sugar ratio, cell respiration acceleration and enzymatic regulation, water loss leading to increase in dry matter and oxidative burst occur in cassava roots. PPD induction is also proposed as a result of cross-talk between calcium signalling, reactive oxygen species (ROS,) and programmed cell death (PCD) (Djabou et al. 2017). It is reported that occurrence of colourless occlusions, synthesis of hydroxycoumarins, accumulation of steroids and diterpenoids, decrease in phenolics, carotenoids, phospholipids and glyceroglycolipids with increase in flavonoids, anthocyanins, sterol-containing lipids leads to discolouration and deterioration of the roots during PPD symptom progression.

34.2.1.3 Management of Postharvest Physiological Deterioration

Certain pre-harvest and postharvest strategies have proven to be more effective in controlling PPD symptom development. Retaining the roots below ground for an extended period and pruning before harvest have been practiced to control PPD. Modified storage techniques, processing after harvest and chemical treatment are practiced, in which several storage techniques involving exclusion of oxygen has proven to be successful since PPD development is chiefly attributed to increased activity of ROS. Storage of tubers in polythene bags, boxes, freezing, wax coating, low temperature treatment, hot water treatment, use of chemicals, spraying of antioxidants, melatonin and conversion to stable intermediate products or flour are some of the strategies suggested by various workers to avoid PPD.

Development of PPD-tolerant genotypes through breeding and transgenic approaches has gained success in recent years. There are reports for wide genetic variation and heritability observed for PPD. Several genotypes with delayed PPD have been identified and developed. A few cassava genotypes with enhanced shelf life up to 20 days after harvest have been observed (Visalakshi et al. 2017). The transgenic approaches involve altered regulation of gene coding for enzymes responsible for PPD initiation and development which includes ROS scavengers, cytosolic glutathione peroxidase, melatonin biosynthesis and scopoletin biosynthesis (Liu et al. 2017). Extensive genetic and genomic studies and inclusive management strategies will augment our understanding of the complex phenomenon of PPD, which in turn will benefit farmers, traders and processors handling cassava roots.

34.2.2 Cassava Root Rot

The roots which are affected by cassava root rot disease show dry, wet or soft rots and cause massive postharvest losses and either infect the produce on-farm or develop during storage. The pathogens responsible for the root rot have been identified as *Botrydiplodia theobromae*, *Rhizopus stolonifer*, *Aspergillus niger*, *Penicilium oxalicum*, *Fusarium solani*, *F. oxysporium*, *Diplodia manihotis*, *Cylindrium clandestrium* and *Macrophomina phaseolina* (Shukla et al. 2012).

34.2.2.1 Disease Management

Application of chemicals, use of resistant varieties and curing of the tubers are the control measures suggested to minimize root rot disease. Even though many chemicals are recommended for management, farmers often fail to follow the recommendations. This is mainly due to the harmful effect fungicides pose to farmers and the environment, and the cost of chemicals remains unaffordable to many. Plant extracts are known to suppress many pathogens. They are chosen over many methods of disease management due to their easy availability, modest or no toxicity to humans, eco-friendly nature and ease in preparation and application procedures (Awurum and Enyiukwu 2013). It was found that the water and ethanol extracts of the plants *Tagetes erecta*, *Piper guineense*, *Casia alata* and *Ocimum graticimum* were able to arrest the mycelial growth of the pathogens responsible for postharvest rot of cassava under *in vitro* and *in vivo* conditions. Despite the fact that many of the botanicals effectively inhibit the growth of the pathogens, little effort has been made to locate the antifungal principle in these botanicals. Concrete efforts are needed for the appropriate utilization of these plant extracts, which are readily accessible and cost effective for the management of deterioration of cassava root rot imparted by pathogens, especially in developing countries (Amadioha and Chidi 2019). Different parts of the plants like leaf, seed, root, etc., have excellent antifungal properties. Amadioha and Markson (2007) evaluated the potential of leaf and seed extracts of *Piper nigrum* and *Aframomum melegueta* for mycelial growth suppression of postharvest rot pathogens of cassava. Among the botanicals, seed extracts of *P. nigrum* and *A. melegueta* could significantly inhibit the mycelial growth of *Botryodiplodia acerina* under *in vitro* conditions. Ethanol, water and petroleum ether extracts of the botanicals were used for the study, and ethanol extracts recorded the greatest mycelial growth inhibition, followed by water and petroleum ether extracts. The efficiency of the extracts to check rot development in uninjured cassava tubers was also tested, and the results clearly indicated the efficiency of the botanicals, particularly when they were applied previous to inoculation with *B. acerina*.

34.3 Sweet Potato

Sweet potato (*Ipomoea batatas*) ranks as the 11th most key food crop in terms of production, with yearly global production of over 106 million tonnes; it is extensively cultivated around the world and consumed as a staple and a supplementary food. The delicious and rich nature of the crop as well as its nourishing and therapeutic benefits makes the crop very popular worldwide (Mohanraj and Sivasankar 2014). The crop can be raised under less favourable ecological surroundings with reasonable production. However, like any other vegetable crop, sweet potato storage roots are also prone to several forms of postharvest loss during harvesting, shipping from farmers' field to market and in storage. Tuber rot/decay poses a grave risk to sweet potato production worldwide, causing substantial economic loss to growers. Approximately 32.5% of the annual yield of sweet potato is lost to postharvest biodeterioration, particularly in the humid tropics due to lack of suitable storage and processing facilities (Agu et al. 2015). Certain pre-harvest factors exist in the field like condition of the soil, contamination by microorganisms and invasion by insect pests, which may augment postharvest deterioration. Similarly, faulty postharvest handling practices, which are very common in tropical developing countries, can result in quantitative as well as qualitative losses (Rees et al. 2001). *Ceratocystis fimbriata*, *Fusarium* spp., *Botryodiplodia theobromae* and *Rhizopus oryzae* are the major microbes associated with storage rot in sweet potato. In addition to these, *Sclerotium rolfsii*, *Macrophomina phaseolina*, *Curvularia lunata*, *Plenodomus destruens* and *Rhizoctonia solani* also cause storage rot in sweet potato (Ray and Ravi 2005; Ray and Nedunchezhoyan 2012). The geographical location and environmental conditions prevailing during the period play a great role in deciding the comparative importance of the major pathogens.

34.3.1 Pre-harvest Infection

Many postharvest sweet potato diseases are due to the damage/infection occurring prior to harvest. This may be due to the use of infected plant material to raise the crop or a moderately low level microbial infection at the harvest still supplying an excellent source of inoculum. During handling, the inoculum gets mixed in and then spreads to healthy roots by contact.

34.3.1.1 Black Rot

Black rot disease of sweet potato is caused by the fungus *Ceratocystis fimbriata* and is a significant storage and field disease causing notable loss to farmers. The key sources of infection of the disease, black rot are seed sweet potatoes tuberous roots (TRs) with pathogens. Contaminated seed TRs or pathogen-invaded tubers without symptoms get mixed in with healthy tubers. Under congenial conditions, pathogens infect tubers mostly through wounds or bud eyes. The damage due to the disease can go as high as 20–50% and can direct sweet potato tubers to rot and cause great economic loss to farmers (Yang et al. 2013).

34.3.1.1.1 Symptoms and Spread

Tiny, slightly depressed spots appear on the tubers. The colour of the spots changes from brown to greenish black on advancement of the disease. The spots coalesce and spread to a major portion of the tuber and make the cooked roots taste bitter. Under favourable conditions, perithecia are seen attached to the infected portion, and they appear as tiny black structures with long necks. The pathogen mainly enters the tubers through wounds caused during harvest or transport. Even a trace of infection in the tubers can lead to rapid spread in storage and can cause great loss to farmers. Many of the postharvest pathogens attack the sweet potato plants in field, and the plants raised from black rot–affected tubers tend to produce diseased plants. *C. fimbriata* survives in the soil during off season and transmits from crop to crop. Moderate (warm) temperatures and moist soil favour disease development. The infected tuber and soil serve as the inoculum of the pathogen. Even though the pathogen attacks the crop in the field, the incidence and severity of the attack is greater in sweet potatoes under storage. This may be due to closeness of the tubers and the congenial conditions persisting in storage. The pathogen *C. fimbriata* spreads to roots in the seed tubers and sprouts, and because of the low infection it may go unnoticed and survive in field on crop residues. The ambient temperature in storage for the rapid spread of the pathogen is 24–26.5°C. Mite or other insect infestation on tubers makes the tubers more vulnerable for pathogen attack and aggravates the severity of the incidence.

34.3.2 Harvest Infection

Microorganisms in soil cause harvest infection and the organisms make their way to sweet potato roots through new wounds on the surface. The slightest scratch on the outer skin can act as an entry point for the pathogens and cause infection.

34.3.2.1 Java Black Rot

Java black rot occurs in many countries such as India, Bangladesh, the United States, Nigeria, the Philippines and Ghana (Ray and Ravi 2005). The fungus *Botryodiplodia theobromae* is responsible for this most destructive disease of sweet potato tubers in storage.

34.3.2.1.1 Symptoms and Spread

The affected tissue initially shows yellowish brown discolouration and the colour later changes to black. The rot increases from the proximal end of the tubers; it can also start from another wound site. Dark patches appear on the surface of the affected tubers after 6–8 weeks of storage. Many pycnidia grow on the dark patches; the internal tissue initially turns yellow in colour and slowly the colour changes to black. Because of the internal rotting, the tubers become shrivelled, mummified and breakable (Ray and Nedunchezhiyan 2012). Injuries to the tuber act as the most important predisposing factor for the disease. *B. theobromae* growth is favoured by relative humidity (RH) of 85–90% and temperature of 25–35°C.

34.3.2.2 Fusarium Root Rot

Fusarium root rot is also known as dry rot and end rot. As the name indicates, the soilborne fungi, *Fusarium* spp., are responsible for the disease. It is a major disease of sweet potato under storage in many countries (Wang et al. 2014). *Fusarium solani*, *F. pallidoroseum F. oxysporum* and *F. solani* are the prevalent species of *Fusarium* found linked with sweet potato dry rot in Brazil, the United States., China, Israel and India. However, *F. pallidoroseum* is reported from India. The type of decay varies with the pathogen. Wounding is a prerequisite for the infection. Injury caused during harvest or handling usually invites the pathogen.

34.3.2.2.1 Symptoms and Spread

Sunken lesions form on the outer part of the tubers which progress with time and reach the periderm of the tuberous root. Empty cavities can be seen under the lesions on surface with advancement of the rot, and white mycelial growth of the fungus is often visible inside the cavities. The symptoms of the diseases on the surface of tubers looks similar to the symptoms exhibited by Fusarium surface rot caused by *F. oxysporum* and makes it difficult to discern between the symptoms of surface rot (*F. oxysporum*) and root rot (*F. solani*). However, internally it can be seen that the damage caused by *F. solani* encompasses through the vascular tissue into the centre. In surface rot due to *F. solani*, the lesion will not be deep, remains superficially on the tuber and exhibits circular lesions which are pale brown in colour with white mould growth. Many factors favour the growth of the pathogens and help in development of disease. These factors include RH of above 80%; the pathogen thrives in a wide range of temperatures (13–35°C) and the inoculum density of the pathogen in soil. Clark et al. (2013) found that physical damage caused on the tubers and the inoculum density of the causal organism in soil combined dictate the disease incidence. Even under congenial conditions, penetration of the pathogen into the tuber and disease progression takes longer, ranging from weeks to months.

34.3.3 Postharvest Infection

Postharvest infection frequently occurs immediately after harvest or during storage. Even minor injury to the tuber while cleaning or in transport provides a site for infection.

34.3.3.1 Soft Rot

Regardless of the region, whether temperate or tropical, soft rot is observed in all sweet potato growing countries. Different species of *Rhizopus*, viz., *R. nigricans*, *R. oryzae* and *R. stolonifer*, cause soft rot in storage.

34.3.3.1.1 Symptoms and Spread

Soft rot is very common in all vegetables under storage. They are fast colonizers and capable of colonizing the entire tuber within 48 h if they are not stored properly and can cause heavy loss by destroying a considerable portion of the produce. The spread from tuber to adjacent tubers is very fast and in storage piles it forms a "nest of decay" (Ray and Balagopalan 1997).

As the name indicates, soft, spongy wet rot is caused by the fungi during transit or storage. The rot spreads quickly, and due to the nature of rot, affected roots typically decay completely. The tuber surface shows a luxuriant growth of grey incoherent mould. The nature of the symptoms varies with the environmental conditions. Tubers will not rot under dry weather. Shrinking of the tuber occurs; however, tuber tissues retain firmness, whereas in humid conditions, tubers not only shrivel but also become tender and soggy; the surface of the tuber ruptures and growth of white mycelial growth with globular head occurs. The factors which trigger development of the disease include temperature below 35°C; injury to the tuber and high RH of 75–85%. Factors such as infected storage tubers and unhygienic storage boxes, baskets or tools can aggravate the damage.

34.3.3.2 Charcoal Rot

The causal organism of charcoal rot is the fungus *Macrophomina phaseolina*.

34.3.3.2.1 Symptoms and Spread

At the beginning, the damage is limited to the portion adjacent to the surface of the tuber. The rot-affected tuber retains firmness, shows reddish brown to brown colouration and exhibits the symptoms of wet rot. On advancement of the disease, the pathogen travels from the periphery to the central portion of the tuber which in turn causes severe rotting. Due to the pattern of rotting, two dissimilar zones can be noted in affected tissue. The disease advancing end shows initial reddish brown to brown colouration with wet rotting, whereas the area where rotting is initiated will have black colour, as indicated by the name of the disease. Charcoal rot generally destroys the entire tuber and causes drying of the tuber which becomes very hard to fell and mummified. The pathogen responsible for the rot, *M. phaseolina*, is soilborne in nature and the propagules can persists in plant rubbles or in the soil. The pathogen enters through wounds caused during harvesting or handling. The rise in temperature favours the pathogen and thus the disease is more prevalent in storage houses that are overly hot.

34.3.3.3 Sclerotium Rot

The disease is caused by the fungus *Sclerotium rolfsii*, and the incidence has been recorded from many countries.

34.3.3.3.1 Symptoms and Spread

The mode of rotting shows variability. Spots on stored tubers initially occur in circular shape and start rotting. The tissue under the water-soaked lesions on the tuber surface turns firm, hard and stringy.

34.3.3.4 Bacterial Soft Rot

Bacterial soft rot is seen in both tropical and temperate regions. Soft rot in potato is caused by the bacterium *Erwinia carotovora* subsp. *carotovora* which is a very common pathogen responsible for storage rot in many vegetables. It was found that the bacterium that causes rot in sweet potato is not *E. carotovora*; the causal organism is identified as *E. chrysanthemi*, and the disease is also known as "Erwinia soft rot" (Ray and Ravi 2005).

34.3.3.4.1 Symptoms and Spread

Apart from *Erwinia*, the fungus *Rhizopus* also causes soft rot in sweet potato, and the rotting symptoms expressed by both pathogens are alike except for the presence of mycelial growth in Rhizopus rot. *Erwinia* causes black stripes in the vascular tissue of the tubers while rotting. Eventually the tubers become soft and show the symptoms of wet rot.

34.3.3.5 Spongy Rot

As the name indicates, the tubers become spongy on infection with the pathogen *Cochliobolus lunatus*. The disease is found in India, and the sweet potato tubers start swelling accompanied by spongy formation (Ray and Ravi 2005). Apart from this, the interior portion turns black.

34.3.3.6 Rhizoctonia Rot

Tubers on infection with the fungus *Rhizoctonia solani* shrivel after showing pale brown spots on the tuber surface. Once the disease advances, the growth of the fungus covers the surface of the tubers and appears brownish in colour. Rhizoctonia rot occurs in India (Ray and Ravi 2005).

34.3.3.7 Gliomastix Rot

High humidity of above 85% and moderate temperature (27°C) is preferred by the pathogen *Gliomastix novae–zelandiae*. Slight sunken corky lesions appear on the surface the tubers. Under congenial conditions, abundant growth of the fungus along with numerous spores can be seen on the surface and give a black colour to the infected portion.

34.3.3.8 Pythium Rot

The disease is prevalent during cool weather with high humidity. *Pythium ultimum* causes the disease and creates huge losses for sweet potato growers in countries such as the United States and Australia. Uniform rotting of the tubers occurs and banding in the tuber with dark sunken tissue is very common. The tubers upon infection become dark brownish grey in colour, dry and easily break off.

34.3.3.9 Yeast Infection

Due to yeast infection, brown or black lesions appear on the skin of the tuber. The infection not only causes colour change on the surface of the tuber but also affects the texture. The tubers dry, become mummified and very hard to the feel. In addition, the infection imparts a foul odour and anon-desirable flavour, thus making them unsuitable for consumption. Many yeasts have been found to be associated with sweet potato, including *Sporobolomyces marcillae*, *Saccharomycopsis fibuligera*, *Rhodotorula mucilaginosa*, *R. minuta*, *Pichia guilliermondii* and *P. anomala*. Oladoye et al. (2014) reported that many extracellular enzymes are produced by yeasts *R. minuta* and *P. anomala*. These enzymes are capable of disrupting the cell walls and spoiling the tuber. The enzymes include glucanase, xylanase, cellulose, amylase, polygalacturonase and ferulic acid esterases.

34.3.4 Disease Management

Most of the microorganisms which cause postharvest loss in sweet potato enter the tuber at the time of harvest. Injuries caused at the time of harvesting, by the insects during their infestation or the use of unhealthy planting material (used as propagating materials) initiate the infection. The following management practices are found effective in mitigating postharvest diseases in sweet potato.

34.3.4.1 Cultural Methods

34.3.4.1.1 Careful Handling

As per the findings of Rees et al. (2001), improper handling and transport can result in severe damage to the tubers which cause serious abrasions on the surface of the tuber. The proportion of the damage depends on the handling and varies from 20 to 86% of tubers. Thus handling at the time of harvest necessitates utmost care. All possible damage to the sweet potato tubers should be avoided during transport and a standard quality storage place must be assured for proper storage. These steps minimize the chances of coming in contact with the pathogens and disrupt the conditions required by the microbes for successful establishment.

In clonally propagated crops, use of disease-free planting material plays a major role in warding off the pathogen in field and storage. Sweet potato plants grown from cuttings may be preferred, since the pathogens associated with the tubers will affect only the portions below the soil, whereas the tubers formed from the plants which rose from the infected tubers directly provide continuity for the pathogen and aggravate the disease incidence.

34.3.4.1.2 Curing

In addition to other benefits which improve the quality of the tuber, the process of curing toughens the outer skin of the tuber and prevents advancement of injury caused during harvesting and handling. This in turn reduces the vulnerability of the tubers to pathogens and subsequent rot during storage (Sowley and Oduro 2002). Curing can significantly increase the shelf life compared to uncured tubers, and drastically reduced fungal infection to less than 10% (Ray and Ravi 2005). Pruning reduces skinning injury and root rotting during storage. The soil usually adheres to the tubers and brings pathogens along with them. Proper washing of the tubers reduces the inoculum of the pathogens. Ensuring adequate ventilation is important to remove CO_2 and replenish O_2 while curing the tubers. Many techniques of curing are being practiced, and one of the simple techniques is by covering the freshly harvested tubers with a polythene sheet. The polythene sheet must be arranged so that it is above 6–8 inches from the tubers spread open in a well-ventilated place. During the night, the polythene cover has to be removed. It generates a temperature range of 29–33°C and RH of 80–95%, the ideal conditions for proper curing, and this may be continued for 4–7 days for the desired effect.

34.3.4.1.3 Hot Water Treatment

Treating the product in hot water of suitable temperature for a specified time is a widely practiced technique to reduce postharvest loss. In sweet potato tubers, dipping them at 70–90°C for 2–10 sec can delay the incidence of initial rot (Ray and Ravi 2005).

34.3.4.2 Resistant Varieties

The genetic make-up of plants dictates their susceptibility to pathogens. Tubers of different cultivars significantly vary in storability and reaction to pathogens. Some cultivars show resistance to a few diseases such as *Fusarium* root rot, Sclerotium rot, Java black rot, Erwinia root rot and Rhizopus soft rot. Postharvest rots are usually complex rots and involve many pathogens. This necessitates developing varieties with wide-ranging resistance to major pathogens causing postharvest diseases. Biotechnological tools may be explored to combine resistant genes from various sources (Ray and Ravi 2005).

34.3.4.3 Chemical Control

As food or feed, sweet potato tubers are directly consumed (raw) in many countries. This hampers the application of chemicals to reduce microbial spoilage in tubers. Even then, fungicides continue as the key method to surpass the pathogen and reduce postharvest diseases of sweet potato. Sweet potato TRs have to be treated before postharvest preservation. It was found that dipping of tubers that were pre-inoculated with pathogens in thiabendazole fungicides could significantly check the disease development. The effect of thiabendazole and benomyl treatment was further improvised by increasing the temperature of the fungicidal solution to 44–54°C. Similarly, dipping of the tubers in the fungicide, ferbam suspension at 44–54°C, could significantly arrest black rot (Ray and Ravi 2005).

34.3.4.4 Environmentally Friendly Antifungal Agents

The application of chitosan (2 mg/mL) partially deacetylated product of chitin resulted in antifungal and eliciting properties against *C. fimbriata*. Under *in vitro* conditions, mycelial growth as well as spore germination of the pathogen was arrested by chitosan treatment. Disruption caused in the barrier properties of the cell wall and plasma membrane of *C. fimbriata* leads to cell necrosis. The effect of chitosan was validated with *C. fimbriata*-infected sweet potato tubers and a similar result was obtained for postharvest development of *C. fimbriata*. The ability of chitosan to induce defence enzymes has been suggested as the possible mode of action (Xing et al. 2018).

34.3.4.5 Biological Control

Victor (2019) treated sweet potato tubers with the bioagents *Trichoderma* species, *Trichoderma viride*, *T. harzianum*, *T. hamatum* and *T. pseudokoningii* prior to storage. It was seen that the treatments significantly inhibited storage rot in sweet potato tubers even after 4 months of storage. Among the species of *Trichoderma*, *T. harzianum* expressed maximum inhibition in terms of reduction of mycelial growth of the rot pathogens. The inhibition of rot incidence shown by the isolates ranged from 54.6 to 77.3% *in vitro* and 47.2–68.8% in *in vivo* conditions.

Volatile organic compounds (VOCs) from *Streptomyces lavendulae* (SPS-33) possess inhibition potential against *C. fimbriata*. It inhibited the growth and multiplication of *C. fimbriata* and significantly decreased the rate of increase of lesion size on tubers and loss of water in sweet potato tubers infected with *C. fimbriata*. It causes elevated activities of enzymes such as superoxide, peroxidase and catalase. The antifungal activity is attributed to the compound, pyridine, 3-methyl-1-butanol, 2-methyl-1-butanol and phenylethyl alcohol, which have convincing antifungal properties on *C. fimbriata* (Li et al. 2020).

34.3.4.6 Use of Botanicals

A secondary metabolite obtained from *Perilla frutescens* (Perillaldehyde) has antifungal properties and is capable of protecting tubers from the pathogens that attack during storage. Perillaldehyde suppresses the pathogens by inhibiting their growth and multiplication. Perillaldehyde vapour significantly reduced *C. fimbriata* both *in vitro* and *in vivo*. The potential of Perillaldehyde to mitigate the infection of *C. fimbriata* is associated with the elevated host defence response. Perillaldehyde vapour application imparts tolerance in sweet potato tubers to the pathogen *C. fimbriata* by the induction of number of defence-related enzymes such as ascorbate peroxidase, peroxidase, phenylalanine ammonia lyase, superoxide dismutase, catalase and polyphenol oxidase. It is evident that the mode of pathogen suppression by Perillaldehyde is by both direct interactions on the fungus and the defensive responses induced in the tuber tissue (Tian et al. 2019).

34.4 Yam

The genus *Dioscorea* forms an important staple food in the tropics and belongs to the family *Dioscoraceae* (Ogaraku and Usman 2008). Yam tubers are considered a treasure of calcium, phosphorus, carbohydrates, iron and vitamins, including riboflavin, thiamine and vitamin C. In addition, yams possess certain medicinal properties for managing hyperchosterolaemia and diabetes mellitus (Okigbo and Ogbonnaya 2006). The major portion of the yam tubers is utilized in the form of tuber. Therefore farmers prefer the option of storing a good portion of yams after harvest. Farmers adopt many technologies to store the yams, including traditional indigenous strategies. Storage in simple piles or trenches, delayed harvesting and storage in specially designed buildings are some of the techniques used. The factors responsible for postharvest loss in yam are pathogens, insects and rodents, respiration of the dormant tuber, nematodes, sprouting and loss of water by evaporation. Postharvest loss caused by microbes remains the most important issue related to storage of yams for farmers as well as traders. Yams are subject to several diseases (Figure 34.3). Rotting of tubers ranges from 10 to 20% in the initial 3 months of storage and jumps to 30–60% once the storage period exceeds 6 months (Ofor et al. 2010). The data collected from farmers' fields in Ghana shows that the percentage of loss due to diseases can go up to 70% during storage. The pathogens cause drastic reduction in quality and quantity of tuber, and among the pathogens, fungi accounts for the maximum loss more than any other single factor (Aidoo 2007).

Storage deterioration in yam tubers is caused by many fungal genera. They include fungi *Aspergillus niger*, *A. flavus*, *A. tamari*, *Lasiodiplodia theobromae*, *Fusarium moniliforme* var. *subglutinans*, *F. oxysporum*, *F. solani*, *Penicillium chrysogenum*, *P. sclerotigenum*, *P. oxalicum*, *Candida albicans*, *Rhizopus stolonifer*, *R. nodosus* and *Rhizoctonia* sp. (Nyadanu et al. 2014; Anwadike 2018). Among the pathogens, *P. oxalicum* and *A. niger* induced the most destructive loss. *Aspergillus fumigatus*, *A. niger* and *Penicillium chrysogenum* caused postharvest rot in white and water yam, and *P. chrysogenum* was the most destructive pathogen, causing severe loss to yam tubers. Other fungal species were not reported as destructive as *P. chrysogenum* (Okpogba et al. 2019). In the beginning, the infected tubers may not display any signs of external symptoms (Okigbo and Ogbonnaya 2006).

Internal microbial rot of yam tubers during storage is an important disease of yam in the tropics and has a great deal of influence on yam production. Due to internal rot, the internal content extensively degrades rapidly while the tuber looks healthy and attractive from the outside. As time passes, rotting spreads and invades the whole tuber. The rotting is due to the production of metabolites and extracellular enzymes by the pathogens responsible for the rot. These chemicals act on the cell wall polymers and cause maceration of parenchymatous tissues (Ogaraku and Usman 2008). Postharvest rot in yam usually starts when the crop is in the field and progresses during storage. Depending on the pathogen, rot symptoms vary in colour, and the infected portion becomes hard and dried out. Involvement of *Scutellonema bradys*, *Penicillium*, *Fusarium* and nematodes have been implicated with this symptom (Nyadanu et al. 2014; Nweke 2015).

34.4.1 Dry Rot

The affected tuber turns stiff and dry, and the symptoms of dry rot change with the pathogen responsible for the damage. *Penicillium cyclopium* and *P. oxalicum* infection on yam cause browning of the tissue combined with drying and stiffening of the tuber. In addition, green colour indicating fungal growth can be seen on the affected portion of the tuber. The microbes *Aspergillus tamari* and *A. niger* regularly attack the tubers and cause browning of the tissue with yellowish margins. Similarly, infection by *Botryodiplodia theobromae* and *Rosellinia bunodes* results in browning of the affected tissue. Slowly the colour of the affected portion changes to grey and then to black, and causes black rot. The affected tuber tissues remain dry, and the tuber becomes crumbly and breaks into small, dry particles (Amusa and Baiyewu 2003). In Nigeria, it was found that many species of *Fusarium* were also apparently associated with dry rot in yam tubers (Morse et al. 2000). The pathogens give a pinkish colour to the infected tissues with a yellowish border. *F. moniliforme*, *F. oxyporum* and *F. solani* are the species of Fusarium associated with dry rot in yam (Amusa and Baiyewu 2003).

Scutellonema bradys also causes dry rot disease in yam under storage. Mature tubers rot more in storage, and dry rot disease may occur along with general decaying of the tubers.

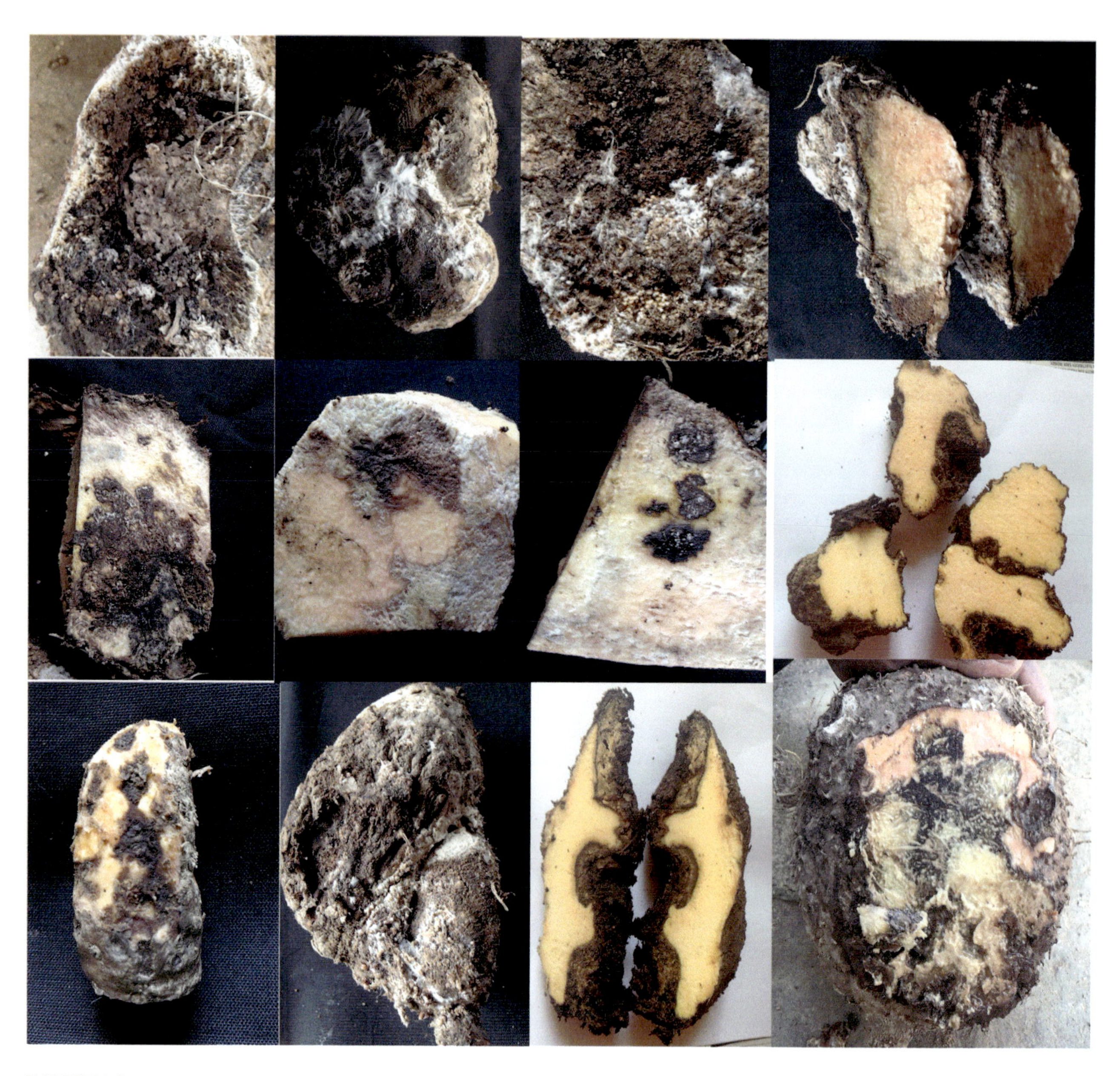

FIGURE 34.3 Elephant foot yam corms showing different types of rot.

S. bradys damages the outer portion of the tuber, extending to a depth of 1–2 cm, and forms cream- to light yellow-coloured lesions beneath the outer surface. Initially, the tubers will not show any external symptoms and the tuber looks healthy. Even after advancement of the infestation, the damage usually will not extend deep and is confined within 2 cm from the outer skin. Deep invasion by the nematode has been rarely observed. Infected tissues turn to light brown to dark brown and finally to black in later stages of the rot. Breaking of skin occurs and portions of the skin blow off, exposing the discoloured and rotten tissue beneath the skin (Amusa and Baiyewu 2003).

34.4.2 Soft Rot

Many pathogens, viz., *Rhizoctonia solani*, *Armillariella mellea*, *Mucor circinelloides*, *Rhizopus* spp. and *Sclerotium rolsii* have been associated with soft rot in yams (Amusa and Baiyewu 2003). As the name indicates, the affected tissues become soft and brown with circular or irregular margins which clearly demarcate the infected from the healthy portion. The sight of ramification on the affected portion by the causal organism is a regular feature. Like many other types of rot, no visible external signs will be seen on infected tubers. The internal tissue completely rots and tissues disintegrate and convert into a watery mass which can come out with even a little force. The internal white fluid oozes out when pressed gently and is a characteristic feature of wet rot, and can be used as a marker to differentiate wet rot from other rots (Morse et al. 2000).

34.4.3 Disease Management

Management strategies of storage diseases in yam have been comprehensively studied and many methods are practiced for reducing damage driven by various pathogens under storage.

These management strategies comprise of use of healthy planting material, following crop rotation with non-hosts, destruction of contaminated crop and use of synthetic chemicals, bioagents, botanicals, etc. (Amusa et al. 2003).

34.4.3.1 Physical and Cultural Methods

Wood ash possesses modest or no mammalian toxicity, and application of wood ash prior to storage is being advised for managing postharvest rot in yams (Amusa and Baiyewu 2003). The growers of yam in southwestern Nigeria process about 35% of the harvested produce into different products such as chips or cubes to store them for more than 6 months. This conversion helps reduce loss due to postharvest pathogens (Amusa and Baiyewu 2003). In India, it is a regular practice by the farmers to apply cow dung paste on tubers previous to planting. It is assumed that the application of cow dung slurry promotes sprouting and growth and guards the tubers from rotting.

34.4.3.2 Resistant Varieties

The use of resistant varieties is considered the most money saving, stable and eco-friendly way to protect tubers from postharvest pathogen attack. The differential response of yam genotypes to microbial rot suggests that resistance to postharvest microbial rot disease is under genetic control and should therefore be liable to genetic improvement. This provides an opportunity for selection of yam genotypes with higher resistance to internal microbial rot (Nyadanu et al. 2014).

34.4.3.3 Botanicals

Many botanicals show inhibitory action towards the growth and multiplication of organisms causing rot in yam. Leaf extracts of *Ficus exasperate*, *F. saussureana*, *F. sur* and *F. thonningii* (Oyelana et al. 2011); *Cinnamomum verum*, *Syzygium aromaticum* and *Curcuma longa* were able to check the growth of *Aspergillus niger*, *A. flavus*, *Fusarium solani* and *F. oxysporum* (Omaka et al. 2020). *Jatropha curcas*, *Moringa oleifera* (both seed and leaf), *Costus afar*, *Aloe vera* and *Tagetes erecta* were screened against *Aspergillus*, *Fusarium* and *Botryodiplodia* (Opara et al. 2015). Both ethanol and aqueous extracts of the plants *Syzygium aromaticum* and *Xylopia aethiopica* could arrest the growth of *Penicillium chrysogenum*, *Aspergillus fumigatus* and *A. niger*, fungi causing rotting in the yams. Maximum inhibition of mycelial growth was given by ethanol extracts of *S. aromaticum* fruit on all tested organisms. The inhibitory effect of botanicals varied with the pathogen. High mycelial growth inhibition of *A. fumigatus* was expressed by aqueous extract of *X. aethiopica* leaves, whereas high inhibition of *P. chrysogenum* and *A. niger* was produced by ethanol and aqueous extracts of *X. aethiopica* (Okpogba et al. 2019).

Leaf extracts of *Aframomum melegueta* and *Ocimum gratissimum* with ethanol, cold-water and hot water extraction generally subdued mycelial growth and germination of spores of the pathogens responsible for soft rot in yam tuber, viz., *F. oxysporum*, *Aspergillus flavus*, *A. niger*, *B. theobromae*, *R. stolonifer* and *P. chrysogenum*. Antifungal constituents in various extracts were responsible for the inhibition of growth and germination of spores *in vitro* and reduced development of rot in storage. The tubers without any injury, which were soaked in ethanol extracts for 6 h, were protected from the pathogens, hence the extracts derived from *A. melegueta* and *O. gratissimum* unveil a chance of using them as protectant pesticides to reduce rotting in storage. Application of these botanicals is economically feasible, has little or no environmental risk and the plants are easily accessible to farmers (Okigbo and Ogbonnaya 2006). Similarly, root and leaf extracts of false yam (*Icacina oliviformis*) treated tubers were protected from *A. niger*, *A. flavus* and *Penicillium sclerotigenum*, rot-causing organisms in white yam. Mycelial growth of all the three fungi were most inhibited by leaf extracts under *in vitro* condition. The possibilities of utilizing leaves and roots of false yam as surface protectants of yam in lieu of mancozeb may be explored (Sowley et al. 2019).

34.4.3.4 Chemical Control

The *in vitro* studies with the chemicals terbinafine HCl, ketoconazole, sodium propionate, fluconazole, and griseofulvin adopting zone of inhibition, minimum inhibitory concentration(MIC), minimum fungicidal concentration (MFC), fractional inhibitory concentration (FIC) and fractional fungicidal concentration (FFC) showed their potential to inhibit the fungi causing postharvest rot in yam, viz., *Penicillium citrinum*, *A. niger*, *A. flavus* and *R. stolonifer* (Otegwu et al. 2018). Different kind of rots in yam tubers can be significantly reduced by treating with Captan and orthiophenylphenate, maleic hydrazidine, bleach (sodium hypochlorite), lime and gin, borax and naphthalene acetic acid (Okigbo 2004; Okigbo and Nmeka 2005).

34.4.3.5 Biological Control

The use of bio-control agents may be an economically feasible technique for limiting the attack due to pathogens causing postharvest rots. *Bacillus subtilis* is the most exploited bacteria for managing postharvest diseases. *B. subtilis* were isolated from fresh cow dung and its antagonistic properties were evaluated against a few organisms. The mycelial growth of *B. theobromae* and *F. oxysporum* was by the bacterium *B. subtilis* under *in vitro* conditions. *B. subtilis* effectively arrested the invasion of *B. theobromae* and *F. oxysporum* in wound cavities of yam tubers by inhibiting >80% of the growth of the organisms under *in vivo* studies (Swain et al. 2008). Similarly, another isolate of *B. subtilis* strongly inhibited the mycelia growth of *A. niger*, *P. oxalicum* and *B. theobromae* up to 60% (Okigbo 2005). Liquid culture filtrates of bacterial isolates of palm wine origin showed antagonism against *P. oxalicum* and *A. niger* (Assiri et al. 2015). Spore suspension of *B. subtilis* was applied on yam tubers with a knapsack sprayer prior to storage. The application of bioagent drastically reduced pathogens in a traditional yam barn throughout the 5-month storage period.

Conidial suspension of *T. viride* was used to treat white yam (*Dioscorea rotundata* Poir.) tubers before storage under controlled environmental conditions. The treatment with *T. Viride* significantly reduced the population of *P. oxalicum*, *A. niger* and *B. theobromae*, the surface mycoflora which attack the tuber (Okigbo and Ikediugwu 2008).

Eradication or management of *Scutellonema bradys* from yam tuber is possible with hot water treatment. Difficulty in maintaining constant temperature is the main difficulty preventing its widespread use. It is practicable for limited operations for making the tubers free from nematode. *S. bradys* can be excluded without damaging tubers by keeping water temperature of 50–55°C for less than an hour (Amusa and Baiyewu 2003). The organic leaf extracts of *Alstonia boonei*, *Chromolaena odorata*, *Crotolaria retusa* and *Xylopia aethiopica* can effectively be used for prevention of bio-deterioration due to nematode infestation of yam tubers in storage (Aghale et al. 2016).

34.5 Elephant Foot Yam (*Amorphophallus paeoniifolius*)

The genus *Amorphophallus* is native to tropical Africa and Asia (Nedunchezhiyan and Byju 2005). Elephant foot yam (EFY) has a special place amid root and tuber crops due its elevated productivity, acceptance as vegetable and valuable medicinal properties. The major share of production of EFY comes from two Asian countries–India and China. In addition to the field diseases, EFY is highly susceptible to various postharvest rots. The major share of loss of corms attributed to the wounds form during harvesting time, transporting to storage yards as well as markets (Ray 2015). The incidence of postharvest losses in EFY is very high at all stages from harvesting to consumption.

34.5.1 Postharvest Handling of the Corms

During harvesting, special precautions are needed to avoid injury to the corms. The adhering soil and roots are removed and the damaged tubers and those showing symptoms of rotting/decay or any sort of infection are separated from the lot before storage.

34.5.2 Curing and Grading of the Corms

Even after taking all precautions to avoid injuries to the corms during postharvest handling, wounding always ensues with the separation of the corms from the pseudo stem. Curing of the corms can lessen losses during storage and provide a better product for marketing. Curing helps the skin of the corms to toughen and to heal the injuries made during harvesting and transportation (Ray and Ravi 2005). Curing of EFYs is practiced in tropical countries by keeping the corms in open, shaded places at a temperature of 28 ± 2°C for a week or two or until the moist soil adhering to the corms desiccates and drops off (Misra et al. 2007). Once the curing is complete, the corms are graded according to their size and shape. After grading, the corms are transported to the market packed in bamboo baskets or in gunny bags, with minimal physical injury. Care should be taken to maintain ventilation for the packed corms. The corms can also be kept in layers sandwiched with paddy straw/dried palm/banana leaves to avoid bruising during transportation (Misra et al. 2007).

34.5.3 Storage of Corms

The corms of EFY can be stored safely for 3–4 months. The corms should be stored in a well-ventilated, cool and rain-protected place. Research studies carried at RC, ICAR-Central Tuber Crops Research Institute (CTCRI), Bhubaneswar, India indicated that the corms could be effectively stored by scattering them in a layer with the apical portion upwards and then covering with coarse dry sand, followed by periodic elimination of damaged tubers (Ray 2015).

34.5.4 Technologies for Extending Shelf Life of Tubers

34.5.4.1 In India

To extend shelf life of corms, farmers in Kerala and Andhra Pradesh treat the corms in cow dung slurry and allow the tubers to dry under shade. The tribal farmers of Tripura convert the corms into cakes to extend shelf life.

34.5.4.2 In China

The indigenous people of China practice several storage methods for extending shelf life of EFY tubers. The practice includes storing the corms (*A. nanus*) in cellars near their houses, piling up corms (*A. yuloencis* and *A. krausei*) near the fireside, cutting corms into pieces and drying them in the sun before storage, making dry konjac cakes and eliminating the corm skins and processing the clean core corm into purified konjac cakes dried outside during the winter season (Long 1998).

34.5.5 Postharvest Diseases

The tubers are prone to several postharvest diseases due to their high moisture content and starch. In addition to damaging seed corms, these pathogens also inhibit sprouting and render the plants more prone to field diseases. Mechanical injury to the corms during harvesting and transportation makes them vulnerable to various rotting fungi and bacteria. Pre-harvest infection in the field, infected soil adhering to the corms, nematode damage and high moisture content of the corms stimulates physiological and microbiological spoilage (Misra et al. 2007). Postharvest rot can directly cause 25–30% crop loss, which amounts to Rs. 3,50,000 to 4,00,000/ha and additional loss by increasing the susceptibility of the plants to pathogens in field.

34.5.5.1 Corm Rot/Storage Rot

Storage rot of the EFY tubers is a serious disease causing severe economic losses to the farmers of India and elsewhere.

Misra et al. (2007) identified a few microbes which caused rotting of EFY tubers under storage, including *Sclerotium rolfsii*, *Botryodiplodia theobromae*, *Phytophthora colocasiae*, *Rhizopus* species and *Fusarium*. One bacterial pathogen, *Erwinia carotovora*, was the only bacterium which caused serious damage to EFY corms stored under poor aeration and warmer temperatures. Later, ICAR-CTCRI (2018) identified 14 organisms, viz., *Sclerotium rolfsii*, *Lasiodiplodia theobromae*, *Rhizopus oryzae*, *Cunninghamella elegans*, *Rhizoctonia solani*, *Ceratobasidium* sp., *Fusarium* spp.(3), *Colletotrichum gloeosporioides*, *Aspergillus* (3) and *Penicillium citrinum* that cause corm rot in EFY and *Sclerotium rolfsii*, *Lasiodiplodia theobromae* and *Fusarium* spp. cause maximum damage to the produce.

34.5.5.2 Symptoms and Spread

The corms affected with fungi show discoloration, softening of the tissue and rotting. External symptoms may or may not be there. The symptom varies with the organism(s) involved with the rotting (Figure 34.4). In some cases, a white powdery appearance will be seen on the outer surface of the corm. The surface of the tuber often becomes completely rotten. Inside tissue shows brown/black colour spots, and adjoining spots coalesce and form bigger, irregular patches. In some cases, the inner portions of the corms show putrefaction of the tissue with brown to black discolouration. Chocolate brown-to-brown colour oozing from the lesions was also observed. The affected portion may turn to powdery mass of tissue. Discolouration of the corms and softening and rotting of tissue are the most common symptoms of corm rot.

Erwinia rot, caused by the bacterium *Erwinia carotovora* subsp. *carotovora*, has been reported to create spoilage to EFY corms when they are stored at higher temperatures (<40°C) with reduced ventilation. Tubers infected by *E. carotovora* turn watery and produce a foul odour, and 80–100% tuber loss occurs.

34.5.6 Disease Management

34.5.6.1 Proper Storage

Use of healthy planting material plays a key role in the general health of plants in the field, and the pathogens in seed tubers can be transmitted to the next crop. This necessitates proper storage of seed tubers to ensure healthy and disease-free planting material for the next crop. Storage of corms in a well-ventilated place and frequent monitoring and culling out of the unhealthy corms has been found effective in checking tuber rot. The infected tissues need to be removed with a sharp knife in such a way that no infected portion is left on the tubers. While removing the infected portion, even the healthy tissues adjourning the infected portions should be removed.

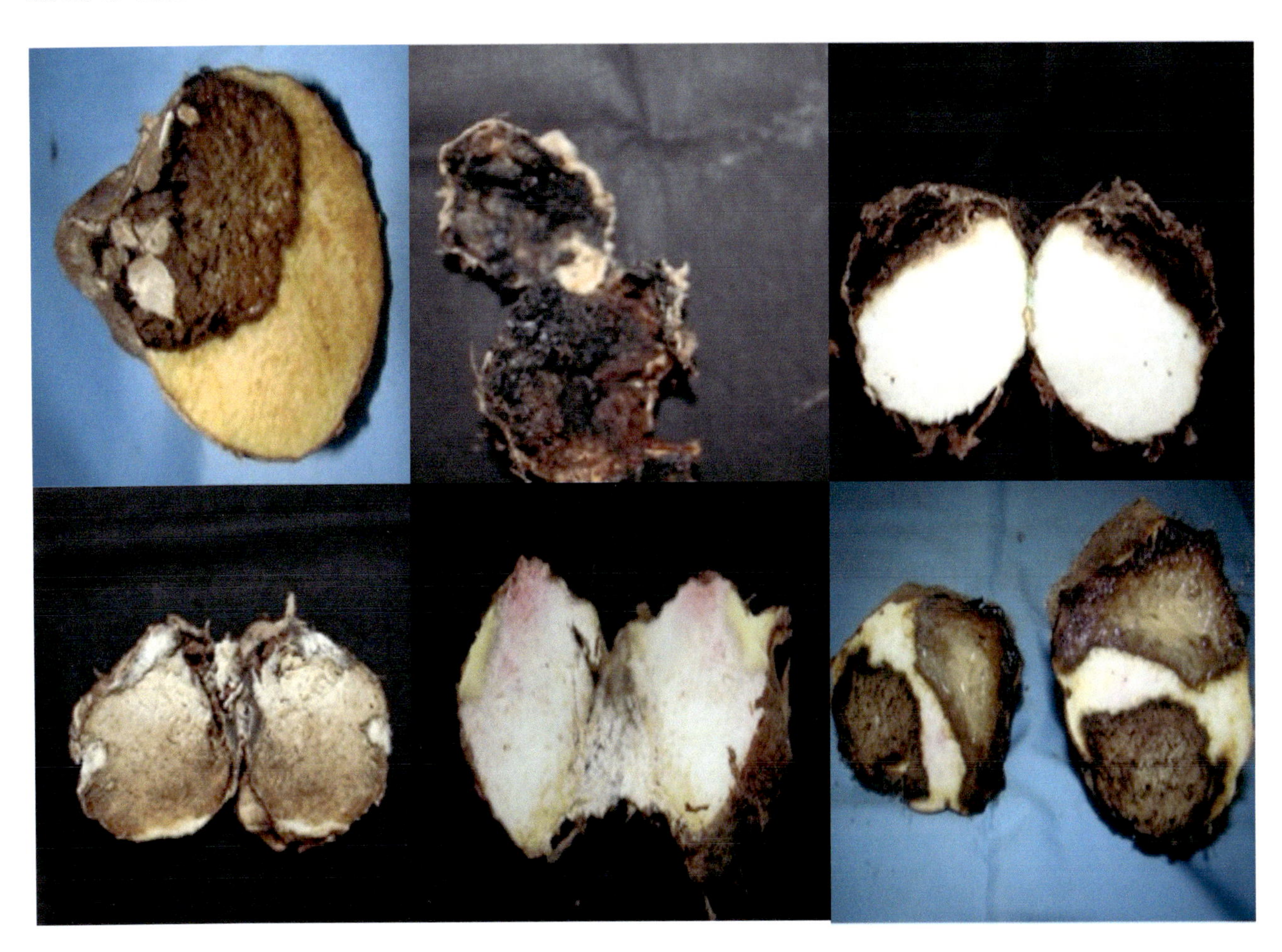

FIGURE 34.4 Taro cormels with different types of postharvest infection.

34.5.6.2 Chemical Control

Murthy et al. (2008) treated the plants with carbendazim (0.1%) + streptomycin (150 ppm) at 30 days before harvest, and treatment of the corms after the harvest with 0.3% copper oxychloride and 100 ppm streptomycin was found effective in checking storage rot. The treatment of cut corm pieces before planting with mancozeb (0.2%) + monocrotophos (0.1%) was found effective against storage rot (Naskar et al. 2008). Corm treatment with mancozeb or combination fungicide containing mancozeb and carbendazim (0.2%) before storage gave minimal corm rotting (<5%) (ICAR-CTCRI 2018).

34.5.6.3 Biological Control

The treatment of cut tuber pieces before planting with cow dung slurry mixed with *Trichoderma* @ 5 gKg^{-1}) was found effective against storage rot (Naskar et al. 2008). Treating the corms with cow dung slurry or cow dung slurry amended with *T. asperellum* @ 5 gKg^{-1} or ICAR-CTCRI–developed biopesticide Nanma @ 7 mL^{-1} resulted in very low incidence of storage rot (ICAR-CTCRI 2018).

34.6 Taro

Taro (*Colocasia esculenta*) is of tropical nature and is extensively cultivated throughout Asia, Africa and the Pacific. The crop is considered one of the oldest crops grown by human beings for consumption. Taro comes under the family *Araceae*, and many plants in this family are grown for their leaves, petioles and corms. It is grown in tropics, subtropics and some hotter areas of temperate regions for a main or subsistence crop. In comparison with other tuber crops, taro can be cultivated even in fields with more water, like paddy fields. Upland it grows with support of rainfall or the help of irrigation depending on the nature of the field. The field diseases taro leaf blight, caused by *Phytophthora colocasiae*, and mosaic disease greatly affect the productivity. Certain facts about taro make the produce highly vulnerable for postharvest diseases. The major portion of the taro produced is consumed as fresh, and the high moisture content of the corms lessens the storability of the produce at ambient temperatures. Hence, loss encountered due to diseases under storage is substantial.

34.6.1 Physiological Problems

Bulky corms of taro are usually very sensitive to chilling injury. When the corms are exposed to 4°C for 10 days, the internal portion of the corm shows browning due to chilling injury. This causes noteworthy weight loss if the corms are stored at low RH. If stored at high RH, sprouting occurs and creates a problem if the top is still intact. Maintaining the corms at a temperature of 7–13°C and RH of 85–90% ensures longer storage life (Holcroft 2018).

34.6.2 Postharvest Rot

The high moisture content in the corms coupled with poor handling and storage creates congenial conditions for the entry and establishment of pathogens, which is a major contributor for postharvest loss in taro (Figure 34.4). Many pathogens are involved in causing corm rots, and the pathogens alone or in combinations create havoc. Combined infection is more common and causes severe loss to the farmers. *Erwinia*, *Fusarium*, *Athelia rolfsii*, *Ceratocystis*, *Aspergillus flavus*, *Rhizopus oryzae*, *A. niger*, *Geotrichum candidum* and *Botryodiplodia* are involved with the rot in storage (Paull and Chen 2015). Among the pathogens, *G. candidum* and *R. oryzae* are very frequent, followed by *A. niger*, whereas *A. flavus* is the least frequent one. High virulence and quick crumbling of the corms were recorded with *A. flavus* and *R. oryzae* in a span of 20 days after inoculation (Khatoon et al. 2015). *Verticilium lateritium* and *Rhizopus stolonifer* were also responsible for postharvest rot in *C. esculenta* var. *antiquorum* and *C. esculenta* var. *esculenta* and *R. stolonifer* induced the maximum rot (Garba et al. 2016). Apart from these, *Phytophthora colocasiae* and *Erwinia crysanthemi* were also reported to cause postharvest rots. Most of the organisms make their entry with field infection through wounds. Although the pathogens do a great deal of harm to the corms, only in the case of *Lasiodiplodia* infection we may not be able to diagnose the disease until the corm is cut open.

34.6.2.1 Pythium and Phytophthora Rot

Widespread and fast damage is created by *Pythium* and *Phytophthora*; the corms often are completely destroyed within a period of 10 days. *Pythium* produces white, crumbly rot. *Phytophthora* infection produces a firm big patch of rot. The colour of the rotten area appears in dark grey to blue colour with imprecise borders.

34.6.2.2 Lasiodiplodia Rot

With invasion of *L. theobromae* into the taro corms, at the beginning the invaded area turns to whitish cream in colour and in time the colour changes to bluish black. Although *Lasiodiplodia* shows quick invasion, generally the fungus follows the infection by *Pythium* and *Phytophthora*. For this reason, isolation of *Lasiodiplodia* is positively reported from the rotten areas created by *P. colocasiae* and *Pythium splendens*. However, the damage by other pathogens is not a prerequisite for the entry of *Lasiodiplodia*. The injury made to corms during harvesting and handling can facilitate entry and cause complete rotting of the tubers within 2 weeks. *L. theobromae* seldom causes drying and converts the tissues into powdery form after causing a spongy rot. The rotting is demarcated from the healthy portion with an undefined margin. The affected portion has a strong unpleasant odour and the interior portion becomes completely black with lot of spore masses (Jackson 2017).

34.6.2.3 Athelia Rot

In addition to causing field disease, the soilborne fungus *Athelia rolfsii* attacks the corms in storage. The attack is distinguishable from others because of the pinkish corm rot it develops. Injuries made during detachment of suckers during harvest allow the entry of pathogens in field and storage. The disease incidence is greater when harvesting is delayed and the corms over-mature.

34.6.2.4 Erwinia Rot

The infection pattern of *Erwinia* is the same as with all tuber crops. It causes severe damage to the corms. The corms rot rapidly and more aggressively when stored in plastic bags and the humidity is very high. Erwinia, usually cause soft rot and the rotten corms become very soft and emit foul smell.

34.6.3 Disease Management

34.6.3.1 Harvesting and Handling

The maturity of taro corm is indicated by the change of colour of leaves to yellow and withering of the plants. To minimize postharvest loss in taro, harvesting can be delayed for few weeks in dry weather conditions (Holcroft 2018). Leaving the "tops" and suckers attached while harvesting the taro corms can prolong storage life. Removal of foliage, shortening the petioles and placing the produce in pits lined and covered with banana leaves can also lengthen storage life. The harvested corms can stay fresh and retain good quality and flavour for a month. In transport, use of plastic-lined cardboard boxes prevents damage to the corms. Washing the corms and dipping for 2 min in 1% household bleach (sodium hypochlorite) can further reduce deterioration during storage and transport (Jackson 2017).

34.6.3.2 Chemical Control

High humidity and lack of aeration flare up rotting of corms, hence care should be taken to dry the corms sufficiently before putting them in bags made of plastic. Dipping the corms for 2 min in 1% sodium hypochlorite (bleach) prior to storage or taking to market can save the produce from rotting caused by *Pythium* and *Phytophthora*. Soaking of corms in benomyl can mitigate storage rot due to *Lasiodiplodia*. Benomyl is ineffective in reducing the damage by *Pythium* and *Phytophthora*, whereas sodium hypochlorite can reduce all common decay inducing organisms.

34.7 Tannia

Tannia (*Xanthosoma sagittifolium*) is also known as new cocoyam and is an important sustenance crop of subtropical and tropical emerging nations. The origin of this crop is listed as the American tropics. Both the leaves and subterranean tubers are eaten. On attaining maturity, the older leaves become yellowish in colour. Staggered harvesting is not a rare practice in Tannia, since the corms on maturity do not get damaged even if the plants are retained for a longer time. Hence harvesting is done as per the demand. The mechanical damage occurring on corms at the time of harvest attracts microbial attack later and causes great concern to the producers by causing deterioration during the storage period. Wounding causes the cormels to develop serious rots if stored. Much attention is needed to avoid any abrasions on corms while harvesting.

The different types of rotting that occur during storage seriously affect the quality of the tuber. Retention of the quality of the corms is very important for the purposes of planting and consumption. Corms have short span of storage life and start rotting very early, and as high as 90% losses have been recorded. Storage rots in cocoyam are caused by the pathogens causing postharvest damage, viz., *Fusarium solani*, *F. oxysporum*, *Mucor circinelloides*, *Aspergillus niger*, *A. tamari*, *A. flavus*, *Botryodiplodia theobromae*, *Penicillium citrinum* and *Rhizopus stolonifer* (Agu et al. 2014).

34.7.1 Spongy Black Rot

Rots incited by *Botryodiplodia theobromae* are pale and soft, but as the disease progresses the colour changes and the tissue turns dark-brown. Infected tissues later became spongy-black and produce a putrefying odour.

34.7.2 Fusarium Rot

Cocoyam rots caused by the two species of *Fusarium* are generally soft, but in the early stages the cream–dirty white rot caused by *F. moniliforme* is easily distinguishable from the bluish-grey soft rot caused by *F. solani*.

34.7.3 Sclerotium Rot

Brown rotting of cocoyam corms and cormels is caused by *Sclerotium rolfsii*. Damage to the tissue of corms is trailed by oxidation of polyphenols catalysed by polyphenol oxidases and forms a complex. The complex in turn cause the production of pigments, which causes brown discolouration (Owusu-Darko et al. 2014).

34.7.4 Disease Management

By providing sufficient ventilation, the storage span can be lengthened and the product can be stored for 6 months without any significant reduction in quality. Treatment such as sodium hypochlorite and fungicide soaking of cormels within a day after harvest prevents the development of decay (Owusu-Darko et al. 2014). It was reported that application of fungicide, viz., mancozeb, benomyl, and iprodione @ 200 ppm prevents storage rot. Rot development in cocoyam cormels can be effectively checked by treating the corms with leaf powders made from pepper fruit (*Dennettia tripetala*) and candle bush (*Cassia alata*).

34.8 Conclusion and Future Prospects

Storage of tuber crops provides an opportunity to market the product, ensures a year-round supply and helps the growers fetch good prices. Most of the tuber crops are clonally propagated and thus storage of tubers is an essential factor. By mitigating the losses encountered by postharvest diseases, the food accessibility to the growing human population can be increased, the extent of area under cultivation can be reduced and exploitation of natural resources can be restricted. Understanding different postharvest diseases, precise identification of the pathogens responsible for various diseases, preharvest and postharvest environmental conditions that favour the disease along with proper handling and curing help to alleviate the huge losses encountered by postharvest disease.

In recent decades, high priority is being given to research on developing eco-friendly and economically feasible alternatives to management strategies using chemicals to reduce postharvest losses. The growing concern about mycotoxins and foodborne pathogens increases the necessity to find viable alternatives. Correct information is essential in formulating effective and eco-friendly management strategies and thus ensure tubers that are safe to eat at affordable prices.

Development of varieties resistant to postharvest diseases through appropriate breeding programs, transgenic plants and selection of varieties suitable for storage and transport can improve the shelf life of roots.

Postharvest loss has not received as much attention as the problem demands. Mitigating postharvest loss can not only increase the quantity of saleable product but also ensure its quality. Use of stored tuber/corm for planting is a common procedure being followed for cultivation. Many pathogens that attack tuber crops in the field are capable of causing huge loss in storage also. Disease-free quality planting material ensures better crop stand and lower spread of pathogen from season to season. Considering all these points, utmost care should be given during handling, and mitigation of postharvest loss warrants urgent attention.

REFERENCES

Adaskaveg JE, Förster H (2010) New developments in postharvest fungicide registrations for edible horticultural crops and use strategies in the United States. In: Prusky D, Gullino ML (Eds.). Postharvest Pathology Series: Plant Pathology in the 21st Century, Vol 2. Springer, the Netherlands, pp. 107–111.

Aghale DN, Umeh OJ, Onyenobi FI (2016) Postharvest studies on Yam tubers *Dioscorea rotundata* Poir. pre-treated with bio-pesticides against soil pests and pathogen. J Plant Sci Res 3: 1–4.

Agu KC, Awah NS, Sampson PG, Ikele MO, Mbachu AE, Ojiagu KD, Okeke CB, Okoro NCN, Okafor OI (2014) Identification and pathogenicity of rot-causing fungal pathogens associated with *Xanthosoma sagittifolium* spoilage in South Eastern Nigeria. Int J Agric Innov Res 2(6): 1155–1159.

Agu KC, Nweke GU, Awah NS, Okeke BC, Mgbemena ICC, Okigbo RN, Ngenegbo UC (2015) Fungi associated with the post-harvest loss of sweet potato. Int J Res Stud Biosci 3(9): 32–37.

Aidoo KA (2007) Identification of yam tuber rots fungi from storage systems at the Kumasi Central market. A dissertation submitted to Faculty of Agriculture, Kwame Nkrumah University of Science and Technology, 205 pp.

Amadioha A, Markson A (2007) Postharvest control of cassava tuber rot by *Botryodiplodia acerina* using extracts of plant origin. Arch Phytopathol Plant Protect 40: 359–366.

Amadioha AC, Chidi KP (2019) Efficacy of extracts of some plants against post harvest fungal deterioration of cassava root (*Manihot esculenta* Crantz) in Nigeria. Microbiol Res J Int 27(1): 1–9.

Amusa NA, Baiyewu AR (2003) Storage and market diseases of yam tubers in South-Western Nigeria. Ogun State. J Agric Res 11: 211–255.

Anwadike BC (2018) Fungal rot of yam (*Dioscorea alata* Lin.) sold at Nsukka markets in Nigeria. Annu Res Rev Biol 28(2): 1–9.

Assiri PK, Koutoua S, Diallo HA, Aké S (2015) Antifungal activity of bacteria isolated from palm wine against postharvest rot fungi on yam (*Dioscorea* spp.). Int J Adv Res 3(1): 231–240.

Awurum AN, Enyiukwu DN (2013) Evaluation of the seed-dressing potentials of phytochemicals from *Carica papaya* and *Piper guineense* on the germination of cowpea (*Vigna unguiculata* L. Walp) seeds and incidence of the seed-borne fungi. Cont J Agric Sci 7(1): 29–35.

Barkai-Goland R (2001) Postharvest Diseases of Fruit and Vegetables. Development and Control. Elsevier, Amsterdam.

Chavez L, Sánchez T, Jaramillo G, Bedoya JM, Echeverry J, Bolaños E, Ceballos H, Iglesias C (2006) Variation of quality traits in cassava roots evaluated in landraces and improved clones. Euphytica 143(1–2): 125–133.

Chavez AL, Sanchez T, Ceballos H, Rodriguez-Amaya DB, Nestel P, Thome J, Ishitani M (2007) Retention of carotenoids in cassava roots submitted to different processing methods. J Sci Food Agric 87: 388–393.

Clark CA, Ferrin DM, Smith TP, Holmes GJ (2013) Compendium of Sweet Potato Diseases, Pests, and Disorders. APS Press, St. Paul, MN.

Djabou ASM, Carvalho LJCB, Li QX, Niemenak N, Chen S (2017) Cassava postharvest physiological deterioration: A complex phenomenon involving calcium signaling, reactive oxygen species and programmed cell death. Acta Physiol Plant 39(4): 91.

Garba PS, Alexander PG, Ada, AC (2016) Pathogenicity of fungi associated with postharvest deterioration of two cocoyam varieties (*C. esculenta* var. *antiquorum* and *C. esculenta* var. *esculenta*) Schott in some parts of Jos. Int J Phytopathol 5(1): 29–34.

García JA, Sánchez T, Ceballos H, Alonso L (2013) Non-destructive sampling procedure for biochemical or gene expression studies on post-harvest physiological deterioration of cassava roots. Postharvest Biol Technol 86: 529–535.

Holcroft D (2018) Curing and storage of tropical roots, tubers and corms to reduce postharvest Losses. In: PEF White Paper No. 18-02, Oregon.

ICAR-CTCRI Annual Report (2018) ICAR-Central Tuber Crops Research Institute, Sreekariyam, Thiruvananthapuram.

Khatoon A, Mohapatra A, Satapathy KB (2015) Fungi associated with storage rots of *Colocasia esculenta* L. tubers in Bhubaneswar city, Odisha. Br Microbiol Res J 12(3): 1–5.

Lebot V (2008) Tropical Root and Tuber Crops Cassava, Sweet Potato, Yams and Aroids. CAB International, Cambridge, MA.

Li X, Beibei L, Shurui C, Yu Z, Mingjie X, Chunmei Z, Bo Y, Ke X, Sheng Q (2020) Identification of Rhizospheric Actinomycete *Streptomyces lavendulae* SPS-33 and the inhibitory effect of its volatile organic compounds against *Ceratocystis fimbriata* in postharvest sweet potato (*Ipomoea batatas*). Microorganisms 8: 319.

Liu S, Zainuddin IM, Vanderschuren H, Doughty J, Beeching JR (2017) RNAi inhibition of feruloyl CoA 6′-hydroxylase reduces scopoletin biosynthesis and postharvest physiological deterioration in cassava (*Manihot esculenta* Crantz) storage roots. Plant Mol Biol 94(1–2): 185–195.

Long CL (1998) Ethnobotany of Amorphophallus in China. Acta Bot Yunnanica Suppl 10: 89–92.

Misganaw CD, Bayou WD (2020) Tuber yield and yield component performance of Cassava (*Manihot esculenta*) varieties in Fafen district, Ethiopia. Int J Agron 2020: 1–6.

Misra RS, Nedunchezhiyan M, Acharya M, Ranasingh N (2007) Post harvest management of Amorphophallus tubers. In: Padmaja G, Premkumar T, Edison S, Nambisan B (Eds.). Root and Tuber Crops: Post Harvest Management and Value Addition. Proceedings of the National seminar on Achievements and Opportunities in Post Harvest Management and Value Addition in Root and Tuber Crops (NSRTC2). CTCRI, Thiruvananthapuram, Kerala, pp. 150–154.

Mohanraj R, Sivasankar S (2014) Sweet potato (*Ipomoea batatas* [L.] Lam)–A valuable medicinal food: A review. J Med Food 17: 733–741.

Morse S, Acholo M, McNamara N, Oliver R (2000) Control of storage insects as a means of limiting yam tuber fungal rots. J Stored Prod Res 36(1): 37–45.

Murthy N, Bhagavan BVK, Ramanandam G, Babu Ratan P, Madhava Rao D, Reddy RVSK, Suryanarayana Reddy P (2008) Amorphophallus (var. Gajendra) in Andhra Pradesh. In: Palaniswami MS, Anil SR, Sajeev MS, Unnikrishnan M, Singh PP, Choudhary BC. National Seminar on Amorphophallus: Innovative technologies. Abstract Book: Status Papers and Extended Summary. Central Tuber Crops Research Institute, Thiruvananthapuram, pp. 15–18.

Naskar SK, Nedunchezhoyan M, Sivakumar PS (2008) Status of Elephant Foot Yam (Amorphophallus spp) cultivation in Orissa. In: Palaniswami MS, Anil SR, Sajeev MS, Unnikrishnan M, Singh PP, Choudhary BC (Eds.). National Seminar on Amorphophallus: Innovative technologies. Abstract Book: Status Papers and Extended Summary, Central Tuber Crops Research Institute, Thiruvananthapuram, pp. 62–67.

Naziri D, Quaye W, Siwoku B, Wanlapatit S, Viet T, Bennett B (2014) The diversity of postharvest losses in cassava value chains in selected developing countries. J Agric Rural Dev Trop Subtrop 115(2): 111–123.

Nedunchezhiyan M, Byju G (2005) Productivity potential and economics of elephant foot yam based cropping system. J Root Crops 31(1): 34–39.

Nweke FU (2015) Some fungal pathogens of Yam (*Dioscorea* spp.) in storage and the effects of their infection on the nutrient composition. J Biol Agric Healthcare 5(12): 153–157.

Nyadanu D, Dapaah H, Agyekum AD (2014) Resistance to postharvest microbial rot in yam: Integration of genotype and storage methods. Afr Crop Sci J 22(2): 89–95.

Ofor MO, Oparaeke AM, Ibeawuchi II (2010) Indigenous knowledge systems for storage of yams in Nigeria: Problems and prospects. Researcher 2(1): 51–56.

Ogaraku AO, Usman HO (2008) Storage rot of some yams (*Dioscorea* spp.) in Keffi and environs Nasarawa State, Nigeria. Prod Agric Technol 4: 22–27.

Okigbo RN, Ogbonnaya UO (2006) Antifungal effects of two tropical plant leaf extracts (*Ocimum gratissimum* and *Aframomum melegueta*) on postharvest yam (*Dioscorea* spp.) rot. Afr J Biotechnol 5(9): 727–731.

Okigbo RN (2004) A review of biological control methods for postharvest yams (*Dioscorea* spp.) in storage in South Eastern Nigeria. KMITL Sci Technol J 4(1): 207–215.

Okigbo RN (2005) Biological control of postharvest fungal rot of yam (*Dioscorea* spp.) with *Bacillus subtilis*. Mycopathologia 156: 81–85.

Okigbo RN, Ikediugwu FEO (2008) Studies on biological control of postharvest rot in yams (*Dioscorea* spp.) using *Trichoderma viride*. J Phytopathol 148: 351–355.

Okigbo RN, Nmeka IA (2005) Control of yam tuber rot with leaf extracts of *Xylopia aethiopica* and *Zingiber officinale*. Afr J Biotechnol 4(8): 804–807.

Okpogba TC, Sobowale AA, Gbadamosi IT (2019) Control of some *Penicillium* and *Aspergillus* rots of *Discorea alata* Poir and *Discorea rotundata* L. using extracts of *Xylopia aethiopica* (Dunal.) Linn. and *Syzygium aromaticum* (Linn.) Merr. Afr J Plant Sci 13(5): 113–124.

Oladoye CO, Connerton IF, Kayode RMO, Omojasola PF (2014) Investigation on the activities of yeasts in the postharvest spoilage of sweet potato (*Ipomea batatas L.*). Ethiopian J Environ Stud Manage 7(5): 499–507.

Omaka O, Okafor CC, Ifediba FA, Nwankwegu AC (2020) Control of yam rot with plant extracts of *Curcuma longa*, *Syzygium aromaticum* and *Cinnamomum verum*. Int J Agric Biol Environ 1(1): 11–23.

Opara EU, Nwokocha, NJ (2015) Antimicrobial activities of some local medicinal plants against post harvest yam rot pathogens in humid south eastern Nigeria. J Microbiol Res Rev 3: 1–9.

Otegwu TC, Adeshina GO, Ehinmidu JO (2018) Antifungal activity of selected chemical agents against phytopathogenic fungi spores. Eur J Med Plants 25(2): 1–7.

Owusu-Darko PG, Paterson A, Omenyo EL (2014) Cocoyam (corms and cormels)–An underexploited food and feed resource. J Agric Chem Environ 3(1): 22–29.

Oyelana OA, Durugbo EU, Olukanni OD, Ayodele EA, Aikulola ZO, Adewole AI (2011) Antimicrobial activity of *Ficus* leaf extracts on some fungal and bacterial pathogens of *Dioscorea rotundata* from Southwest Nigeria. J Biol Sci 11: 359–366.

Paull RE, Chen CC (2015) Taro: Postharvest quality-maintenance guidelines. Veg Root Crops 5: 1–3.

Ray RC, Nedunchezhiyan M (2012) Postharvest fungal rots of sweet potato in tropics and control measures. Fruit Veg Cereal Sci Biotechnol 6(1): 134–138.

Ray RC (2015) Post harvest handling, processing and value addition of elephant foot yam–An overview. Int J Innov Hort 4(1): 1–10.

Ray RC, Ravi V (2005) Post harvest spoilage of sweet potatoes in tropics and control measures. Crit Rev Food Sci Nutr 45: 623–644.

Ray RC, Balagopalan C (1997) Post Harvest Spoilage of Sweet Potato. Tech. Bull. Ser. 23, Central Tuber Crops Res. Inst, Trivandrum, India.

Reddy PP (2015) Plant Protection in Tropical Root and Tuber Crops. Springer, New Delhi.

Rees D, Kapinga R, Mtunda K et al. (2001) Effect of damage on market value and shelf life of sweet potato in urban markets of Tanzania. Trop Sci 41: 142.

Salcedo A, Siritunga D (2011) Insights into the physiological, biochemical and molecular basis of postharvest deterioration in cassava (*Manihot esculenta*) roots. Am J Exp Agric 1(4): 414–431.

Sánchez T, Chávez L, Ceballos H, Rodriguez-Amaya DB, Nestel P, Ishitani M (2006) Reduction or delay of post-harvest physiological deterioration in cassava roots with higher carotenoid content. J Sci Food Agric 86(4): 634–639.

Sánchez T, Dufour D, Moreno JL, Pizarro M, Aragón IJ, Domínguez M, Ceballos H (2013) Changes in extended of cassava roots during storage in ambient conditions. Postharvest Biol Technol 86: 520–528.

Sharma RR, Singh D, Singh R (2009) Biological control of postharvest diseases of fruits and vegetables by microbial antagonists: A review. Biol Cont 50: 205–221.

Shukla AM, Yadav RS, Shashi SK, Dikshit A (2012) Use of plant metabolites as an effective source for the management of postharvest fungal pest: A review. Int J Curr Disc Innov 1(1): 33–45.

Sowley ENK, Oduro KA (2002) Effectiveness of curing in controlling fungal-induced storage rot in sweet potato in Ghana. Trop Sci 42: 6.

Sowley ENK, Kankam F, Nsarko RM (2019) Evaluation of efficacy of false yam (*Icacina oliviformis*) as surface protectant against rot pathogens of white yam (*Dioscora rotundata* Poir). Ghana J Agric Sci 54(1): 1–9.

Swain MR, Ray RC, Nautiyal CS (2008) Bio-control efficacy of *Bacillus subtilis* strains isolated from cow dung against postharvest yam (*Dioscorea rotundata* L.) pathogens. Curr Microbiol 57: 407–411.

Tian J, Pan C, Zhang M, Gan YY, Pan SY, Liu M, Li YX, Zeng XB (2019) Induced cell death in Ceratocystis fimbriata by pro-apoptotic activity of a natural organic compound, perillaldehyde, through Ca2+ overload and accumulation of reactive oxygen species. Plant Pathol 68: 344–357.

Victor DV (2019) Bioefficacy of *Trichoderma* species against important fungal pathogens causing post-harvest rot in sweet potato *(Ipomoea batatas (L.)* Lam). J Bangladesh Agri Univ 17(4): 446–453.

Visalakshi CC, Sheela MN, Mukherjee A, Raju S, Hegde V (2017) Postharvest physiological deterioration of cassava – An inherent miscue. ICAR-CTCRI Newsletter 34(4): 12–14.

Wang RY, Gao B, Li XH, Ma J, Chen SL (2014) First report of *Fusarium solani* causing Fusarium root rot and stem canker on storage roots of sweet potato in China. Plant Dis 98(1): 160.

Xing K, Teng JL, Yuan FL, Zhang J, Zhang Y, Shen XQ, Li XY, Miao XM, Feng ZZ, Peng X, Li ZY, Qin S (2018) Antifungal and eliciting properties of chitosan against *Ceratocystis fimbriata* in sweet potato. Food Chem 268: 188–195.

Yang D, Sun H, Zhao Y, Xu Z, Zhang C, Xie Y (2013) Biological characteristics of *Ceratocystis fimbriata* and selection of fungicides in laboratory. Southwest China J Agric Sci 26: 2336–2339.

Zainuddin IM, Fathoni A, Sudarmonowati E, Beeching JR, Gruissem W, Vanderschuren H (2018) Cassava post-harvest physiological deterioration: From triggers to symptoms. Postharvest Biol Technol 142: 115–123.

35

Postharvest Handling and Disease Management of Cut Flowers

Mast Ram Dhiman, Raj Kumar and Sandeep Kumar
[1]*ICAR-Indian Agricultural Research Institute, Regional Station, Katrain, Kullu, Himachal Pradesh, India*

CONTENTS

35.1 Introduction

Flowers have constantly remained a vital part of the social fabric of human life due to their essence and aroma, being important in social, cultural and religious functions of all societies since time immemorial. Today, cut flowers play a significant role in the floriculture business. In addition to home decoration, cut flowers are also used for cultural ceremonies and are used a symbol of societal expression. Cut flowers are generally comprised of multiple organs, with stems typically bearing leaves are generally highly perishable floral structures. There is profound variation in inflorescence structure and individual blooms. Flower development is often initiated by plant sensitivity to varying temperature or photoperiod, followed by bud growth until flower opening, pollination and finally senescence. Respiration as well as ethylene production during cut flower senescence vary among genotypes from typically climacteric(e.g.some carnation cultivars) to non-climacteric (e.g. certain rose cultivars). The major cut flowers having commercial value are rose, carnation, chrysanthemum, lilium, gerbera, orchid, tuberose, gladiolus, anthurium, etc. The traditional or loose flowers are marigold, jasmine, chrysanthemum, desi rose, China aster and crossandra.

Preservation of turgidity and increment in vase life are essential requirements for postharvest management of flowers. The highly perishable nature of flowers and foliage plants exposes them to high postharvest losses. Due to their firmness and softness, the flowers are more vulnerable to various biotic and abiotic conditions during and after harvest. Even after harvesting, cut flowers are metabolically active and hold on to all life processes at the cost of stored food material in the form of carbohydrates, proteins and fats, extending their longevity for a few more days. Approximately 10–30% of the total crop yield of cut flowers produced in the world is lost due to postharvest diseases (Dole and Wilkins 2004). These losses caused by postharvest diseases may reduce the quality or cause reduction in the shelf life of cut flowers. The postharvest qualities of cut flowers are judged by their final flower size and shape, number and size of florets/spikes, stem strength and length, changes in fresh weight, petal colour, turgidity and freshness of flower/appearance and foliage characteristics. The quality and longevity of cut flowers depends upon the conditions of cultivation, harvesting time, transportation facilities and postharvest handling, which are dependent on the different stresses administered upon them, *viz*., decline in water uptake, water content of flowers and water potential, transpiration, hydraulic conductivity, and fresh weight.

The flower business is one of the fastest growing and vibrant global enterprises today. In this industry, postharvest management of ornamentals is an attractive and worthwhile subject of study. Longevity of cut flowers affected by several pre-harvest and postharvest factors, including genotype, environmental (light, temperature, photoperiod, relative humidity, air composition and pressure) and management (soil condition, nutrition, fertilization, irrigation, plant protection, etc.) factors. Recent advances made in postharvest physiology and senescence of cut flowers have enumerated that many proven methods of lengthening the life of cut flowers are still not known outside the research community. One of the objectives of this chapter is to increase awareness of these methods among flower lovers.

35.2 Factors Affecting Postharvest Life of Cut Flowers

35.2.1 Pre-harvest Factors

35.2.1.1 Light

Light intensity has a direct effect on photosynthesis, which ultimately affects the carbohydrate content of flowers. Flowers having high carbohydrate, i.e. mobile sugars, will have longer vase life (Pandya and Saxena 2003). Treatments of cut flowers with external sucrose increase the vase life. Carnations and chrysanthemums grown under high light intensity exhibited longer vase life as compared to those grown under low light intensity. Similarly, if carnation, gerbera, or rose are grown under cool temperatures, low light intensity and short day conditions, they will exhibit low vase life (Davarynejad et al. 2008). Low light intensity will result in elongation of stems and delay stem hardening. Bent necks in roses and stem bending in gerbera and carnation are due to less hardening of stems. Light intensity also affects the colour of petals, as excessive shading in roses causes bluing of petals. Excessive light intensity also affects the quality of flowers, as it will produce red colouration of tissues, leafs pots and leaf drops. Proper spacing of plants and appropriate light intensity should be kept in mind for production of good quality of flowers.

35.2.1.2 Temperature

The differences in day and night crop growing temperature have a significant influence on quality through their effects on production and utilization of biochemical compounds and energy. Higher temperature during crop cultivation can decrease the shelf life and quality of flowers and can result in faster consumption of stored carbohydrates in the plant parts and faster water loss from the plants. The respiration rate of flowers under increased temperature can be so high that net photosynthesis in flowers harvested on bright days may actually be lower than those cut on cloudy days. In bulbous plants, the quality of flowers is good when night temperature is about 10°C. In carnations, vase life is shorter when cultivated under higher temperature, i.e. >25°C. Most of the physiological disorders are the result of fluctuations in day and night temperatures. Temperatures along with light largely influence the pigmentation in flowers. Various cultivars of roses grown at lower temperature of <15°C develop a greenish tint due to partial conversion of chloroplasts to chromoplasts.

35.2.1.3 Season

The season of crop cultivation has a large effect on growth of plants and longevity of cut flowers. Seasonal fluctuation results in considerable variations in the postharvest life of cut flowers. For example, when chrysanthemum is grown in the spring and summer months, flowers have a longer vase life than those which mature in mid-winter, which may be due to greater accumulation of photosynthates during the summer and spring period.

35.2.1.4 Nutrition

In addition to carbon, oxygen and hydrogen supplied by air and water, plants require nitrogen, phosphorus, potassium, calcium, magnesium, sulphur, iron, manganese, boron, copper, zinc, molybdenum, chlorine, etc. An optimum fertilization program is required for growing good quality flowers. Excessive application of nitrogenous fertilizers can reduce the shelf life of cut flowers and increase the risk of disease incidence. Soil salinity and chloride content of soil also reduce vase life. Foliar spray of various biofertilizers and organic growth regulators enhances the vegetative growth as well as postharvest life of gerbera cut flowers (Palagani and Singh 2017).

35.2.1.5 Watering

Like other horticultural crops, water is essential for flower crops. Generally, less than 5% of water entering the plant system is utilized by plants in different physiological processes, the rest being lost through transpiration. The water content of the floral crops varies significantly with the species, growth phase, types of tissue or organ, time of the day, season and environmental conditions influencing absorption and transpiration of water. Stress conditions due to either high or low water content reduce the quality and vase life of flowers. A stress condition at any stage of growth may affect leaf and flower senescence by influencing photosynthesis, stomatal closure, protein synthesis and cell division. Salt content of water also affects the quality and hastens the senescence rate.

35.2.1.6 Humidity

High humidity in the air results in the development of fungal and bacterial diseases such as grey mould which will cause loss during storage and transportation. As a result of this, infected flowers lose water more rapidly and produce more ethylene (Fanourakis et al. 2015). In protected conditions, adequate ventilation is required for maintaining the proper humidity level and to reduce disease incidence (Byung-Chun et al. 2016). Cut flower roses grown at high relative humidity (85%) often have a very short vase life. Sensitive cultivars, when cultivated under elevated relative humidity, show precocious postharvest senescence symptoms which are normally related to water stress, including premature flower and leaf wilting as well as pedicel bending. An incomplete capacity to reduce water loss due to stomatal malfunctioning may be the main reason for the vase life reduction in plants grown under long-term high relative humidity.

35.2.1.7 Diseases and Insect Pests

Control of insect pests and diseases in the nursery is essential for production of good quality flowers with longer vase life of cut flowers. Discoloration of flowers and leaves are the main cause of damage due to insect pests and diseases which deteriorate the quality of plants (Gupta and Dubey 2019). Damaged tissues lose water more rapidly, resulting in wilting and faster ethylene production. Ethylene causes senescence, leaf and petal fall in flowers. *Botrytis*, *Alternaria*, *Puccinia* and *Diplocarcon* are the major species which cause great loss to flowers, which should be controlled.

35.2.1.8 Sanitation

Proper sanitation is necessary to eliminate decaying debris from the field and greenhouse. The field should be weed free and free from decaying material, as this will result in the production of ethylene and accelerate senescence.

35.2.2 Harvest Factors

35.2.2.1 Stage of Harvest

The optimum stage of harvesting is of considerable importance for long vase life of flowers. Flowers harvested at a suitable stage of development can result in longer shelf life. Early harvesting stage can result in short shelf life. The stage of harvesting will depend upon the plant variety, species, growing season, marketing place and consumer preference. In carnation, rose and chrysanthemum, harvesting should be done at early stages during the summer as compared to the winter season. Bent neck is the major problem in roses and gerberas when they are harvested in the early stage; this is due to less maturity and hardening of vascular tissues. Generally, flowers harvested too early do not develop properly in vase water;

TABLE 35.1

Optimum Stage of Maturity for Harvest of Fresh Cut Flowers

Crop	State of Harvesting
Alstroemeria	4–5 florets open
Anthurium	Spadix fully developed
Cattleya orchid	3–4 days after opening
Chrysanthemum (standard)	Outer petals fully elongated
Chrysanthemum (spray)	Fully opened flowers
Cymbidium orchid	Fully opened florets
Dendrobium orchid	Fully opened florets
Dahlia	Fully opened flowers
Carnation (standard)	Paintbrush stage
Carnation (spray)	Fully opened flowers
Freesia	First buds start opening
Gerbera	Outer petals start unfurling
Gladiolus	2–4 buds showing colour
Gypsophila	Fully opened flowers
Iris	Buds develop colour
Lilium	Buds showing colour
Narcissus	Gooseneck stage
Ornithogalum	Buds develop colour
Phalaenopsis orchid	2–3 florets open
Rose	First 2 petals start unfolding
Marigold	Fully opened flowers
Tulip	Half coloured buds

these are low in quality and short lived. Harvesting of flowers at the bud stage is generally preferred as the buds continue to open over time and hence have extended vase life. In addition, buds are less sensitive to ethylene, easy to handle during storage and transportation and are less prone to damage by diseases and pests. The optimum stage of maturity for harvesting of fresh cut flowers is summarized in Table 35.1.

35.2.2.2 Time of Harvest

Time of harvest is also a critical factor for harvesting of flowers. Morning is the best time for harvesting of flowers, as cells are more turgid. However, flowers remain wet with dew during the morning hours and are more prone to postharvest diseases. Evening harvesting has the advantage of greater carbohydrate accumulation in the stems during the day time (Ahmad et al. 2014). It is better to harvest flowers when the temperature is mild, as high temperature leads to rapid respiration, greater consumption of carbohydrates and excessive water loss. If the cut stems are immediately placed in the solution of preservatives, then the time of harvesting will not matter. Morning harvesting is suggested in flowers which lose water more quickly after harvesting, e.g. chrysanthemum, gerbera and rose. Harvesting of flowers at high temperatures and strong light intensity conditions must be avoided as it results in faster utilization of accumulated stored carbohydrates from the cut stems.

35.2.2.3 Mode of Harvest

Preferably, sharp tools are used for cutting the flowers because dull cutting tools result in crushed stem that reduces water uptake. The cut angle on the stem should be slanting and smooth in woody plants. Crushing of stems at the cut end should be avoided as it will result in the exudation of sap along with sugars, and encourage the growth of microorganisms, which may cause plugging of stems (Hannweg 2008). Flowers are usually cut close to the soil to provide the longest possible stem.

35.2.3 Postharvest Factors

The postharvest life of cut flowers is influenced by a number of factors that induce the senescence of various organs. Therefore, optimized growth environment, production season, and postharvest handling (temperature, humidity, light, ethylene inhibitors, preservatives, water quality, etc.), are all important to maintain the quality and improve the longevity of cut flowers after harvesting.

35.2.3.1 Temperature

Ambient temperature is the most essential factor affecting postharvest flower quality. High temperature after harvesting increases the rate of respiration, which ultimately increases the rate of development and senescence (Celikel and Reid 2005). Low temperature can reduce the respiration rate and also reduces the use of stored food material in the plant tissues. Lower temperature also reduces water loss, ethylene production and development of microorganisms. After harvesting, the cut flowers should be shifted to the cold store immediately. Flowers produced in the temperate zone should be stored at slightly higher temperature than the freezing point, and flowers of tropical regions at temperature between 8 and 10°C. Storing tropical flowers at low temperature may cause discoloration of petals, and sometimes flower buds fail to open.

35.2.3.2 Relative Humidity

Relative humidity is also an important postharvest factor which plays a significant role in the maintenance of tissue hydration, one of the main factors preserving the postharvest ornamental quality of cut flowers. Cut flowers contain a great amount of water and when subjected to low humidity conditions, they lose the water more rapidly and there is a reduction in the initial fresh weight (Doi et al. 2000). A 10–15% reduction in fresh weight resulted in wilting of flowers. At high relative humidity flowers will lose water slowly. High humidity and high temperature increase the risk of fungal and bacterial diseases and it is advised that along with ambient humidity and low temperature maintains the moderate air circulation to limit pathogen development.

35.2.3.3 Light

Light does not have any significant effect on longevity of flowers treated with preservatives. Under low light conditions for long-distance transportation, yellowing of leaves in alstroemeria, chrysanthemum, dahlia, and gladiolus has been reported. High light intensity is necessary for opening of flowers at the retailer or consumer.

35.2.3.4 Ethylene

Ethylene is the major gaseous compound that has harmful effect on the longevity and quality of cut flowers. Ethylene content is high during autumn and winter in the atmosphere due to a low rate of photochemical degradation. Transportation of flowers along with fruit crops should be avoided as they produce more ethylene. Some of the flower commodities produce more ethylene as compared to others with the passage of time, e.g. carnation, lilium, etc. (Wagstaff et al. 2005).

Very Sensitive Flowers	Alstroemeria, Carnation, Freesia, Iris, Lilium, Daffodil, Orchid, etc.
Relatively Insensitive Flowers	Anthurium, Gerbera, Tulip, etc.

35.2.3.5 Pre-Marketing Treatments

The quality of cut flowers after harvest can be improved by treating them with floral preservatives, particularly when they cannot be kept at low temperatures during storage and transportation. Floral preservative solutions containing both surfactant and antimicrobial ingredients are generally used to extend the longevity of cut flowers. These treatments can be given at each step of the marketing chain, especially from farmers to wholesalers, retailers and final consumers. These preservatives maintain the colour of petals and leaves, increase vase life and promote flower opening. Chemical preservatives like 8-HQC, 8-HQS, alkyl ethoxylate, Tween-20, Tween-80, Agral-LN, Triton X-100 and ethylene inhibitors like aminooxyacetic acid (AOA) have been used to increase shelf life, delay flower senescence and improve the keeping quality. In order to prevent microbial growth and proliferation in vase solutions of cut flowers, various biocidal compounds have been used. Some of the most commonly used biocides are sodium hypochlorite, silver nitrate, physan-20, hydroxyquinoline citrate, silver thiosulfate, aluminium sulphate and chlorine dioxide. Despite their benefits, the response is dosage- and duration of exposure-dependent because in many ornamental flower species, higher concentration or long exposure results in toxicity and thereby reduces shelf life by dehydration of tissues, browning, leaf chlorosis and premature flower abscission.

- **Composition of preservatives:** Most preservatives contain carbohydrates in the form of sucrose, germicides, ethylene inhibitors, growth regulators and few mineral compounds.
- **Carbohydrates:** Carbohydrates are the most important source of energy for performing biochemical reactions and physiological processes in cut stems. Sugars mainly maintain the mitochondrial structures and functions and maintain the water balance by regulating transpiration and increasing water uptake (Adugna et al. 2012). Sucrose is commonly used but in some formulations glucose and fructose can also be used. Concentration of sugar in the solution may vary according to the cultivar and species to be treated.
- **Germicides:** Sugars are present in the preservative solutions which can promote the growth of microorganisms and result in the plugging of water vessels in the cut stems. The main microorganisms are bacteria, yeast and mould in the solutions and they block the xylem. Most commonly the salts of 8-hydroxyquinoline are used as germicides in the preservative solutions (Sudaria et al. 2017). Silver salt, mainly silver nitrate, is used as bactericide but the disadvantage of this is that it reacts with the chlorine present in tap water and get oxidized in the presence of light. The different germicides used as floral preservatives are given in Table 35.2.

TABLE 35.2

Germicides Used as Flower Preservatives

Name of the Compound	Concentration(s)
8-hydroxyquinoline citrate	200–600 ppm
8-hydroxyquinoline sulphate	200–600 ppm
Silver thiosulphate	0.2–4.0 mM
Silver nitrate	10–200 ppm
Thiabendazole	5.0–300 ppm
Aluminium sulphate	200–300 ppm
Quaternary ammonium salts	5.0–300 ppm
Slow-release chlorine compounds	50–400 ppm of CI

- **Ethylene inhibitor**: Silver thiosulphate (STS) is used as the inhibitor of ethylene and it will also act as antimicrobial inside the plant tissues (Hassan and Ali 2014).
- **Growth regulators**: Synthetic growth hormones are used in the solutions; they may be applied alone or in mixture with other chemicals. Growth regulators may initiate, accelerate or inhibit various biochemical and physiological processes in the plants. Cytokinins are the most commonly used growth regulators for extending the vase life of plants. They have a major role in inhibition of ethylene production. Cytokinins can be used in dip or spray treatment to inhibit the yellowing of leaves in gladiolus. Auxins are also used in preservatives in some crops, but in carnation they accelerate ethylene production and senescence (Mutui et al. 2006). Gibberellic acids in solution can accelerate flower opening in carnation and gladiolus. ABA treatment in roses delays the wilting of cut stems by reducing the water loss through stomatas. Growth retardants such as daminozide (B-9) and chlormequat (CCC) can delay senescence in carnation, roses, etc., and CCC plus 8-HQS and sucrose extend the vase life of tulips, carnations, gerbera, etc. Ethylene inhibitors like AVG, MVG and AOA can be used to delay senescence and extend the vase life of flowers (Table 35.3).

35.2.3.6 Postharvest Treatment Techniques

35.2.3.6.1 Conditioning

Conditioning is the process of restoring the turgor pressure of cut flowers after cutting from the main plants. This is done by treatment of the cut stem with demineralized water along with germicide and citric acid to maintain the pH as 4.5–5.0.

TABLE 35.3

Growth Regulators Used as Floral Preservatives

Growth Regulators	Concentration Range (ppm)
BA	10–100
Kinetin	10–100
IAA	5–100
NAA	1–50
GA	10–350
ABA	1–10
B-9	10–500
CCC	10–50
AVG	10–100
MVG	10–100
AOA	50–500

Tween-20, a wetting agent, is also used in solution at concentration of 0.01–0.1%. Harvested stems must be placed in slightly warm water at room temperature or cold storage conditions for 10–12 h. In flowers where wilting occurs more frequently (gerbera, chrysanthemum, etc.), treatment with hot water (80–90°C) for a few seconds, and after that treatment with cold water is recommended for restoring the turgor pressure of cut stems.

35.2.3.6.2 Impregnation

Treatment with silver nitrate @ 1000 ppm for 8–10 minutes can protect the blockage of stem vessels by microorganisms. After the treatment, stems should not be cut again from the below, as this chemical only acts at the site of application. This is beneficial in flowers such as aster, gerbera, carnation, chrysanthemum and orchid.

35.2.3.6.3 Pulsing

The pulsing is a short time treatment given to freshly harvested flowers for extending the vase life and to maintain postharvest quality. Sucrose is the main constituent of pulsing solutions and its concentration vary from 2 to 20%, depending upon the crop. Some cut flowers are also fumigated with 1-MCP or pulsed with silver thiosulphate to reduce the adverse effects of ethylene. Flowers can also be pulsed with STS for short duration at warm temperatures (e.g.10 min at 21°C) or for longer durations at cool temperatures (e.g. 20 h at 2°C). Short pulse treatments (10 sec) in a solution of silver nitrate are helpful for some crops.

35.3 Storage of Cut Flowers

Storage is a vital practice during postharvest handling of ornamental crops. The storage of cut flowers can help in the regulation of flowers and other planting material supply to market demand, and enables the accumulation of large quantities of flowers and planting materials for distant markets. The cut flowers can be stored by various methods.

35.3.1 Cold Storage

The storage of ornamentals under cold conditions may regulate the supply of flowers and other valuable planting materials to meet the market demand and maintain quality for commercialization. Treating cut flowers with low temperatures during cold storage or in shipment delays the entire metabolism in the plant tissue and slows the respiration rate. Flowers harvested at the proper stage of maturity must be pre-treated with a suitable floral preservative solution, and pre-cooling must be done for rapid packaging. A stable and uniform temperature during storage and shipment is essential. Generally, temperate cut flowers (rose, carnation) are stored at a temperature of 0–1°C, whereas subtropical flowers (gladiolus, jasmines, proteas, gloriosa, etc.) at 4–7°C and tropical cut flowers (anthurium, cattleya, vanda, euphorbia) are stored at 7–15°C. Depending upon the requirements of flower crops, they can be stored dry or wet in the cold storage.

35.3.2 Dry Storage

Dry storage has various advantages over wet storage but this method is labour intensive. This type of storage will also depend upon the plant to be stored, as dry storage is not suitable for dahlia, gerbera, freesia, gypsophila, etc. In this, the flowers are susceptible for grey mould disease, and before storage the flowers must be treated with fungicides. After that, cut stems should be pulsed with the pulsing solution and placed in the cold store for removal of heat. Packaging should be done in CFB boxes and must be air-tight (i.e. low O_2 and high CO_2). Some moisture-absorbing material can also be placed between the boxes as moisture will deteriorate the quality of flowers. In some flowers, (gladiolus) geotropic bending is the problem, so they should be packed in a vertical position. The recommended dry storage temperature and durations for important cut flower crops are presented in Table 35.4.

35.3.3 Wet Storage

Storage of flowers in containers containing the water or preservative solution is the most commonly practiced method. The disadvantage of this method is that it will require more space in the cold stores. Flowers are generally kept at 3–4°C temperature, slightly higher than the dry storage method. In this, flowers are stored for shorter periods, as developmental

TABLE 35.4

Optimal Recommended Dry Storage Temperature and Duration for Important Flower Crops

Crop Name	Storage Temperature (°C)	Storage Duration (Weeks)
Anthurium	10–13	04
Carnation	0–1	16–24
Chrysanthemum	1.0	03
Gladiolus	4.0	04
Lilium	1.0	06
Narcissus	1.0	02
Rose	1–3	02
Tulip	0–1	08
Bird of paradise	8–10	04

TABLE 35.5

Optimum Recommended Wet Storage Temperature and Duration for Some Important Flower Crops

Crop Name	Storage Temperature (°C)	Storage Duration (Days)
Antirrhinum	0–2	7–14
Orchid (Cymbidium)	−0.5 to 4	14
Carnation	4	21–28
Lilium	0–1	28
Narcissus/Daffodil	0–1	10
Paphiopedilum	−0.5 to 4	14–21
Statice	2–4	21–28
Gerbera	4	21–28
Tulip	−0.5 to 0	14–21
Rose	4	28

processes are faster as compared to low temperature storage (0°C). In this, lower leaves must be removed to avoid contact with water that will result in the development of fungal diseases. The water used in the containers should be disinfected at regular intervals. Some bacterial fungal spores are present on the stem and foliage of cut stems and they can easily multiply in the water and cause plugging of vessels. Sodium hypochlorite (0.005%) is commonly used for disinfecting the water, and sometimes treatment with UV lights is also done. The recommended wet storage temperature and durations for important flower crops is shown in Table 35.5.

35.3.4 Controlled Atmosphere Storage

Controlled atmosphere (CA) storage of cut flowers is based on the three basic factors, temperature, O_2 and CO_2, which significantly affect the quality and marketability of cut flowers. CA storage extends the storage life of the commodity by suppressing the synthesis and action of ethylene and reduction in respiration rate. Storage of cut daffodils and roses in nitrogen enrichment (100% N_2 and 99% N_2 + 0.5–1% O_2), have promising results. Cold chambers are filled with higher concentrations of CO_2 and lower of O_2. The main objective of this is to lower the respiration rate, which reduces the rate of consumption of stored food material in the cut stems (Macnish et al. 2009). The chambers should be airtight and provisions for cooling should be made. This method is more expensive as compared to other normal refrigerating methods. A higher concentration of CO_2 during CA storage controls disease infections in cut carnation, daffodil and anthurium (Table 35.6).

35.3.5 Modified Atmospheric Storage

Modified atmospheric (MA) storage is a less well-defined form of controlled atmospheric storage. The cut flowers are packed in sealed bags which results in the reduction of oxygen and elevation in CO_2 concentration due to respiration of flower tissues. The elevation in the level of CO_2 concentration decreases the ethylene biosynthesis and prolongs flower longevity. It is a cost-effective method of storage as it does not require very specific storage conditions.

35.3.6 Low Pressure Storage/Hypobaric Storage

Cut flowers are stored under low pressure, low temperature and cool, moist air conditions. The CO_2 and ethylene are produced much faster through stomata and intercellular spaces under low pressure compared to normal pressure. The beneficial outcome of low pressure storage (LPS) is ascribed to the decline of ethylene production at low oxygen concentration and other volatile compounds. Earlier studies have revealed that LPS at 40–60 mm Hg is beneficial for extending longevity of many cut flowers. Cut flowers of rose held at 180–210 mm Hg pressure can be stored up to 3 weeks with the retention of 63% of the original vase life.

35.4 Transportation/Shipping of Cut Flowers

In earlier times, flowers were cultivated near the market, which required only a few hours to reach the market by road. Currently, flowers are transported for long distances via cargo

TABLE 35.6

Controlled Atmospheric Storage Requirements of Some Flower Crops

Crop Name	CO_2 Concentration (%)	O_2 Concentration (%)	Temperature (°C)
Lilium	10–20	21	1.0
Freesia	5	1–3	1.5
Carnation	5	1–3	0–1
Rose	5–10	1–3	0
Tulip	5	21	1.0

planes, ships and trucks. Most floriculture commodities are transported to Europe by cargo planes and then exported to different countries either by planes or ships. Different modes of transportation of flowers are discussed here.

35.4.1 Shipment by Road

For short-distance markets (15–20 h), the cut flowers are transported in the insulated trucks without refrigeration. Before loading in the trucks pre-cooling is done and the flowers are pre-cooled and packed in airtight boxes. For distance markets (>20 h), flowers are transported in refrigerated trucks and are packed in ventilated boxes for proper air circulation. Placement of the boxes in the trucks is also important as this will affect the air circulation inside the truck.

35.4.2 Shipment by Air

This is the quickest way to transport flowers overlong distances. In this method, there is no control of the temperature inside the flight, as pre-cooling should be done prior to shipment (Reid 2009). There is high ethylene concentration at airports, so anti-ethylene (STS) treatment should be provided to the cut flowers.

35.4.3 Shipment by Sea

This mode of transportation has is come into practice during the recent years. It takes a longer time to reach the final destination. Low cost is the major advantage of this mode of transportation. Proper air conditioning facilities should be required during the duration of transportation (Reid 2001). In this method, flowers can also be transported under wet as well as dry storage conditions, as space is not the limiting factor.

35.5 Handling of Flowers at Retailer Site

Packs received at the retailer end should be unpacked immediately and checked for physical damage. When flowers are transported at optimal temperature, they should be simply put into water or preservative solution, but when transported at low temperature, first they should be kept at 5–10°C temperature for 10–12 h. The lower-most leaves should be removed, as they will easily decay in the vase solution/water. Outside damaged petals must be removed carefully without damaging the whole flower. The entire flower stems removed from the packs are kept in the containers with water, and the water should be replaced daily to avoid any microorganism growth. If cold room facility is available, then flowers should be kept in cold rooms along with water containers to increase their longevity.

35.6 Postharvest Diseases of Cut Flowers

Postharvest diseases of ornamental flower crops are caused by a comparatively small number of fungi and ascomycetes and by a very few species of bacteria and Basidiomycetes. The bacteria generally belong to the genera *Xanthomonas*, *Pectobacterium*, *Dickeya* and *Pseudomonas*. Microorganisms build up and infect after crop harvest have weak pathogenic ability and are saprophytic in nature, even if they contaminate via natural openings and wounds found on the mature plant epidermis under favourable climate. Postharvest diseases of cut flowers develop during storage or transportation and/or at the consumer level. Floral products may or may not show any disease symptoms that were started during production periods. Conidia of *Botrytis cinerea* develop on flower petals inside water droplets and enter the epidermis with or without infection structures. Afterwards, the infection remains invisible until harvest and disease symptoms develop slowly during storage of flowers at low temperature and a high RH (Darras et al. 2006a). Symptoms are noticed either as grey mould or as hyper sensitive flecking on the host plant. Depending on postharvest conditions, tissue maturity, and host resistance, the dormant period may last up to **6** weeks.

35.6.1 Fungal Diseases

35.6.1.1 Grey Mould

The causal organism of this disease is *Botrytis cinerea* and is usually called grey mould or Botrytis blight. It is considered the most serious disease of many cut flower crops. The fungus is pathogenic to the majority of cultivated flowering potted plants and cut flowers. The symptoms of Botrytis on gerbera and freesia flowers appear as small necrotic, dark-brown lesions or spots. These pathogens also infect rose flowers and develop necrotic spots or blister-like patches on petal surface. Damping off, crown rot, stem cankers, and stem, leaf and flower blight-like symptoms appear on Lisianthus flowers. Grey mould may also cause postharvest deterioration of flowers (Wegulo and Vilchez 2007).

Grey mould is also a major threat for lilium growers. To combat this problem, resistant cultivars with prolonged post-harvest life must be grown for commercial purposes, which results in fewest losses from the disease and less dependence on fungicide applications (Daughtrey and Bridgen 2013). Different field experiments were conducted from 2008 to 2011, and *Lilium* spp. resistance against *B. elliptica* infection was evaluated. The most susceptible cultivars were Connecticut King, Chianti, Côte d'Azur, and Gironde, and among the *Lilium tigrinum*, the cvs. Pink, Red Alert, Royal Fantasy, Menorca, Sweet Kiss, Vermeer, and White Heaven are highly susceptible to Botrytis blight.

Infection of Botrytis on gerbera, freesia, and rose plants occurs during cultivation under the glasshouse, but after passing through a dormant period the disease symptoms become visible during storage or transportation (Darras et al. 2004; Harkema et al. 2013). Fluctuation in temperature and RH after harvesting results in faster development of disease (Darras et al. 2006a; Harkema et al. 2013). However, clear symptoms of Botrytis injury were seen during prolonged transport of cut roses cvs. Red Naomi, Aqua and Avalanche at low temperatures and high relative humidities (Harkema et al. 2013). These results revealed significant positive effect of dry cut rose transport on product quality with distinct emphasis on Botrytis development (Harkema et al. 2013).

35.6.1.2 Alternaria Rot

Diseases caused by *Alternaria* are relatively common on ornamental plants and may infect landscape as well as cut flowers crops. The most frequent species is *Alternaria alternata*, which infests vinca, dahlia, gerbera, hibiscus and geranium (Thomma 2003). Spores are dark brown to black in colour and appear in black masses on leaves and as spots on petals. This disease is spread through water splashes or air movement. Disease symptoms lead to the appearance of necrotic lesions, and is caused by unfavourable climatic conditions and diurnal cycles. *Alternaria dianthi* and *Alternaria dianthicola* infect both pinks and carnation and are characterized by grey-brown leaf and petal spots through purple margins. Generally, *Alternaria* species are more widespread during wet and warm conditions. Conidia of *Alternaria* germinate within 35–45 min when the temperature ranges between 28 and 30°C. A longer duration of wetness (>6 h) is essential for symptoms to develop.

35.6.2 Bacterial Diseases

Bacteria are generally found in the rotting tissues either in field or during storage. Bacterial rot in plants is generally categorized by soft and watery tissue; sometimes cellular debris also normally oozes watery liquid from the infested cracks. In different soft rots, the bacteria involved are not the plant pathogens but they reside somewhat saprophytically as secondary parasites. This is primarily the case for harvested ornamentals. Normally, the natural resistance responses of harvested commodities are low and sharply decline during handling or storage. At that period, bacteria may develop and macerate mature and susceptible plant tissues. On the other hand, a few bacteria can attack the plant and infect living tissues and lead to development of soft rot in the field or in storage. The bacterial genera that commonly cause postharvest rots of flower crops are *Dickeya dadantii* (syn. *E. chrysanthemi*), *Pectobacterium carotovorum* spp. *carotovorum* (syn. *E. carotovora* pv. *carotovora*), *Xanthomonas campestris* and *Pseudomonas fluorescens*.

35.6.2.1 Dickeya and Pectobacterium Species

Dickeya and *Pectobacterium* are closely related bacterial species of straight, rod-shaped, 0.5–1.0 by 1.0–3.0 µm, and are motile, having several to numerous peritrichous flagella. Bulbous crops like hyacinthus, dahlia, iris, muscari, freesia and zantedeschia are more susceptible to soft rot diseases that can damage plant tissue and affect flower production.

35.6.2.2 Xanthomonas Species

The cells of *Xanthomonas* bacterial species are straight rods, 0.4–1.0 by 1.2–3.0 µm in size, and are motile by way of a polar flagellum. In the 1980s, the *Xanthomonas campestris* pv. *dieffenbachiae* damaged the whole anthurium crop in the French Antilles (Anais et al. 2000). The symptoms of *Xanthomonas* on Anthurium cv. 'Kansako Red' appear as angular, pale brown, necrotic spots, 1–3 mm in diameter with dissimilar chlorotic halos. Scales of bacterial exudate occur on the under-surface of older leaves, whereas on younger leaves, the scales were less apparent and the lesions were more wide and dark brown to black.

35.6.2.3 Pseudomonas Species

The *Pseudomonas* is straight to curved; rod shaped, 0.5–1.0 by 1.5–4 µm. They are motile through one or more polar flagella. *Pseudomonas fluorescens* is the most pathogenic species to infect ornamental plants. It develops yellow-green, diffusible, fluorescent pigments on a low-iron medium. Temperature and relative humidity play crucial roles in the development of Pseudomonas disease.

35.6.2.4 Stem Blockage of Cut Flower

Bacteria may also cause stem blockage which interrupts water uptake and wilting of ornamental flowers. Large numbers of bacteria have been detected using light microscopy on the cut surface of dahlia cut stems placed in water for a 2–4 days. Ultrastructural study of cut roses also revealed that the population of bacteria at the cut surface was responsible for vascular occlusions. In most flower crops, bacterial population in vase solutions is normally low at the grower's storage level but increases progressively during handling at the auction centres and at retailers. Microorganisms present within the xylem of flower stems leads to a decrease in hydraulic conductivity of the stems, primarily in the basal stem section. The development of occlusions in rose stems was correlated with the increasing numbers of microorganisms in stems. Besides the bacteria, yeasts and filamentous fungi can also lead to vascular blockage. In carnation, increased numbers of bacteria and yeasts in the holding solutions resulted in a significant decrease in flower longevity. *Pseudomonas* bacteria were the most devastating among those microorganisms that significantly decreased the vase life of carnations. Pompodakis et al. (2004) found that solution turbidity of cut cv. 'Baccara' roses attributed to bacterial growth in the vase was negatively correlated with the vase life ($P < 0.01$; $r^2 = -0.48$) and the vase solution treatment ($P = 0.05$; $r^2 = -0.65$). Reduction in postharvest vase life of rose cv. 'Sonia' was observed when holding solutions were inoculated with 106 cfu/mL *Bacillus subtilis*, *Enterobacter agglomerans*, *P. fluorescens* or *P. putida*. Similarly, the shelf life of carnation cv. 'White Sim' was also decreased when vase water was inoculated with 5–108 cfu/mL suspensions, a mixed bacterial population of *Pseudomonas* spp., *Acinetobacter calcoaceticus* and *Alcaligenes* spp. Bacterial counts in the vase solution of cut gerbera cvs. 'Liesbeth' and 'Mickey' that showed a curvature of more than 90° were found with 106 and 108 cfu/mL. When cut flowers of rose cv. 'Sonia' were keep in vases for 1–4 days, the bacterial count in the basal 5.0 cm of the stems was linearly correlated with the number of bacteria in the vase solution. In the vase solution, in which rose cut stems were placed, the prime bacterium observed was *Pseudomonas*, while *Enterobacter* was a minor associated genus. Bacterial extracellular polysaccharides and globular proteins canal so result in vascular blockage of cut flower stems. Due to the presence of pectic enzymes, microorganisms might also block the xylem cells and form numerous loose vessel fragments.

35.7 Postharvest Disease Management of Flower Crops

In the majority of the cases, control of postharvest diseases of flower crops must be practical before harvesting either grown under polyouse or open field conditions (Darras et al. 2006a, 2007). This is important because most of the pathogens complete their life cycles in the soil. Disease inoculum loads are mostly present inside the polyhouses and can infect host plant tissues under favourable environmental conditions. Pre-harvest infections may become latent and develop postharvest infections in a compatible host–pathogen interaction (Darras et al. 2006b). For that reason, sources of contamination must be eliminated and a precautionary preventive management practice through fungicidal treatments has to be applied. However, precaution must be taken as common postharvest pathogens including *Botrytis cinerea*, *Penicillium* spp. and *Alternaria* spp. may increase fungicide resistance. Careful handling of the final product is essential as it minimizes mechanical damage and notably reduces successive wastage caused by postharvest fungal or bacterial infection. The following management practices have been suggested for the control of postharvest diseases of ornamental crops.

35.7.1 Use of Chemicals (Fungicides/Bactericides)

There are different classes of fungicides or chemicals used to control pre-harvest and postharvest pathogens. The commonly used chemicals for the management of *Botrytis cinerea* are chlorobenzenes, benzimidazoles and dicarboximides. It was observed that when cut rose flowers previously sprayed with pyrimethanil were wrapped in packing paper strips or cellophane bags, the Botrytis blight infection was delayed. In field conditions, various fungicides including carbendazim, hexaconazole, mancozeb, propineb, chlorothalonil and some biotic agents such as *P. fluorescens*, neem oil and garlic clove extracts were tested against *Alternaria alternata* infesting chrysanthemum crops (Kumar et al. 2011). It was found that hexaconazole applied @ 0.1% was very effective against the disease and showed the lowest disease index of 4.49 as compared to the control plants, followed by those treated with chlorothalonil @ 0.2% and mancozeb @ 0.2%. In addition, chrysanthemum plants sprayed with hexaconazole also recorded the highest yield (76.25 q/ha) (Kumar et al. 2011). Nagrale et al. (2012) reported that protective sprays of gerbera plants grown under polyhouse with recommended fungicides against *A. alternata* were more effective as compared to curative applications. They also found that preventive sprays of gerbera plants with Bordeaux mixture (0.6%), tricyclazole (0.1%) and iprodione (0.1%) + carbendazim significantly reduced *A. alternata* disease up to 95.85%, 96.59% and 95.88%, respectively, as compared to the untreated plants.

35.7.2 Management of Environmental Conditions, Inhibition of Ethylene and Irradiation

35.7.2.1 Storage and Transport

Differences in damage caused by *Botrytis* on cut roses of cvs. 'Red Naomi', 'Aqua' and 'Avalanche' were studied by Harkema et al. (2013) involving wet and dry transportation at low temperatures and high relative humidity. They observed that *Botrytis* develop during dry or wet transport but the rate of infection was greatly lower in dry transport conditions. These results concluded that dry transportation of rose cut flowers had a positive significant effect on final product quality by means of eliminating *Botrytis* development. Packaging types of cut flowers during storage or transportation may also affect postharvest disease development. For example, package designs have a prominent effect on the ratio of flowers of cut rose cv. 'Sweet Promise' with spotting caused by *Botrytis cinerea*. Large ventilation holes in the packaging boxes facilitate effective air circulation in packed flowers that helps in reduction of relative humidity and avoidance of condensed water dispersion on the buds, and therefore eliminates the chance of conidial germination and subsequent infection.

35.7.2.2 1-Methylcyclopropane

1-Methylcyclopropane (1-MCP) is a gaseous ethylene inhibitor commonly used to stop the detrimental effects of ethylene on climacteric fruits, vegetables, and ornamental flowers. It was recently was found that 1-MCP can be used to control postharvest Botrytis blight of some cut flowers and ornamental plants (Table 35.7). Seglie et al. (2009, 2012) reported that 1-MCP (3.62 µL/L) significantly reduced the *Botrytis cinerea* infections of *Dianthus caryophyllus* cv. 'Idra di Muraglia' and *Cyclamen persicum* cv. 'Hallows Bianco Puro Compatto' petals. Application of lower concentrations of 0.38 µL/L 1-MCP delays *B. cinerea* development on *Rosa hybrida* cv. 'Ritz' by up to 3 days as compared to the untreated flowers. It was inferred that 1-MCP can limit B. cinerea development but its effects were dependent on plant species and concentration used (Seglie et al. 2009). However, this result was contradicted in a study conducted by Favero et al. (2015) on *Rosa hybrida* cv. 'Avant Garde' flowers which reported that 1-MCP did not protect cut roses that were artificially inoculated with B. cinerea.

TABLE 35.7

1-MCP for Management of *Botrytis Cinerea* in Ornamentals

Crop Name	Treatment	1-MCP Concentration (µL/L)	Reference(s)
Carnation	Postharvest	0.38–3.62	Seglie et al. (2009, 2012)
Cyclamen	Postharvest	0.38–3.62	Seglie et al. (2009, 2012)
Rose	Postharvest	0.38–5.34	Favero et al. (2015); Seglie et al. (2009)

TABLE 35.8

Irradiation Treatment for Management of *Botrytis Cinerea* in Ornamentals

Ornamental Crop	Treatment	Irradiation Doses	Reference(s)
Freesia	UV-C postharvest	0.5–10 kJ/m^2	Darras et al. (2010)
Geranium	UV-C postharvest	0.5–10 kJ/m^2	Darras et al. (2015)
Gerbera	UV-C postharvest	0.5–10 kJ/m^2	Darras et al. (2012a, 2012b)
Rose	Gamma postharvest	0.2–4 kGy	Chu et al. (2015)

35.7.2.3 Irradiation Treatments

Gamma irradiation is an alternative non-chemical treatment that has been used to control pest and postharvest diseases in more than 40 countries (Hallman 2011) (Table 35.8). Ionizing irradiation has been effectively used for many years to disinfect plant and flower surfaces by killing the pests without causing any damage to the plant tissues. Cia et al. (2007) found that gamma irradiation can improve the postharvest vase life of cut flowers, though its effects are dependent on fungal growth, treatment doses, moisture level, composition of treated products and storage conditions. Even though doses as low as 0.2 kGy were adequate to protect the plants against the most harmful insects, it is not possible to completely control postharvest fungal diseases of cut flowers. UV-C irradiation is one more type of physical energy used to manage postharvest spoilage. Exposure of products at low doses of UV-C can reduce storage rots in many fruits, vegetables, and cut flowers (Darras et al. 2010, 2012a) (Table 35.8). Reduction in diseases is the direct germicidal effect of UV-C on the pathogen or defence response induction in the exposed host tissue (Darras et al. 2012a, 2015). Previous research showed that exposure of gerbera and freesia flowers to low doses of UV-C irradiation (0.5–2.5 kJ/m^2) may decrease postharvest infection of *B. cinerea* up to 70 and 75% (Darras et al. 2010, 2012a, b). This reduction was attributed to direct fungicidal activity, while reductions of 55 and 24% were observed as a result of an induced defence response. Such types of defence responses often were associated with increased polyphenoloxidase (PPO) and/or phenylalanine ammonia lyase (PAL) activities (Darras et al. 2012a).

35.7.2.4 Organic and Inorganic Compounds

There are sufficient literature studies published on the use of alternative organic and inorganic compounds to control postharvest diseases of flower crops (Table 35.9). Sodium hypochlorite (NaOCl), a chlorine-containing compound, has been used to control bacterial and fungal diseases in flower vase solutions due to its strong oxidizing activities. A postharvest quick dip application of 200 µL/L NaOCl for 10 sec at 20°C provides significant control against *Botrytis cinerea* infecting rose flowers of cvs. 'Akito' and 'Gold Strike' (Macnish et al. 2010). Woltering et al. (2015) found that irrespective of the type of packaging material used and shipment conditions, rose cut flowers that received a pre-shipment treatment with 100–150 mg/L NaOCl significantly lower the *Botrytis* disease compared to the untreated roses. Pre-harvest treatments of calcium sulphate ($CaSO_4$) can be used to manage postharvest *Botrytis cinerea* infection on cut rose flowers (Capdeville et al. 2005). Treatment of 10–20 mM $CaSO_4$ for 24 h before harvesting resulted in 86% reduction of disease severity. Pulsing and spraying treatment of GA_3 @ 20 mg/L for 24 h has suppressed *Botrytis cinerea* in detached petals of rose cvs. 'Mercedes' and 'Sonata'.

35.7.2.5 Biological Agents

Over the past 100 years, research has repeatedly confirmed that phylogenetically diverse microorganisms can work as natural antagonists against different plant pathogens (McSpadden Gardener and Fravel 2002). Some of the microbial taxa that are effectively used and presently marketed as EPA-registered biopesticides bacteria belong to the genera *Bacillus*, *Agrobacterium*, *Pseudomonas* and *Streptomyces* and fungi of genera *Ampelomyces*, *Candida*, *Coniothyrium* and *Trichoderma* (McSpadden Gardener and Fravel 2002). With regard to ornamental crops, *Pseudomonas* spp. 677 successfully reduced the conidial germination and retarded germ tube elongation of *B. cinerea* (Beasley et al. 2001). It was also reported that *Pseudomonas* spp. 677 was the most antagonistic effect against *B. cinerea* by delaying waxflower abscission up to 3 days.

35.7.2.6 Elicitors of Defence Responses

35.7.2.6.1 Jasmonic Acid and Methyl Jasmonate

Postharvest efficiency of methyl jasmonate (MeJa) has been studied on potted and cut flower plants (Table 35.10). Preharvest and postharvest applications of MeJA on cut *Geraldton*

TABLE 35.9

Organic and Inorganic Compounds Used to Control *Botrytis Cinerea* of Ornamentals

Crop Name	Treatment	Concentration Used	Reference(s)
Geraldton waxflower	S-Carvone postharvest	0.64 and 5.08 mM	Hu et al. (2009)
Rose	$CaSO_4$ postharvest	10–20 mM	Capdeville et al. (2005)
Rose	NaOCl postharvest	100–200 mg/L	Macnish et al. (2010); Woltering et al. (2015)

TABLE 35.10

Elicitors of Protective Responses Tested against the Postharvest Pathogens of Ornamental Crops

Name of Pathogen	Host Plant	Elicitors Tested	Reference(s)
Alternaria spp. *B. cinerea*	*Geraldton waxflower*	SA, MeJA, ASM	Beasly et al. (2001); Dinh et al. (2007); Eyre et al. (2006); Zainuri et al. (2001)
B. cinerea	*Freesia*	ASM, MeJA	Darras et al. (2005, 2006a, 2006b, 2007)
	Peony	MeJA	Gast (2001)

waxflowers conferred variable protection against post harvest infections by *B. cinerea* (Dinh et al. 2007). Jasmonic acid (JA) and MeJA give systemic protection to different rose cultivars, namely 'Mercedes', 'Lambada', 'Frisco', 'Europa', 'Eskimo' and 'Sacha', against *B. cinerea*. Postharvest pulse treatment with MeJA significantly reduced lesion size of *B. cinerea* on detached rose petals. Darras et al. (2007) found that a postharvest treatment with MeJA as pulse, spray, or vapour @ 200 µM, 600 µM or 1 µL/L, respectively, significantly decreased *Botrytis cinerea* petal specking on cut *Freesia hybrida* inflorescences of cv. 'Cote d'Azur'. Applications of vapour MeJA to fresh cut peonyflowers resulted in the least disease severity and improved the postharvest life of cut flowers compared to controls (Gast 2001).

35.7.2.6.2 *Salicylic Acid and Acibenzolar-S-methyl*

Salicylic acid (SA) and acibenzolar-S-methyl (ASM) have been used as pre-harvest treatments on ornamental crops to control postharvest diseases caused by different pathogens. Beasley et al. (2001) reported that pre-harvest sprays of SA at 2000 µg/mL on *Geraldton waxflower* cv. 'CWA Pink' plants reduced the flower disease incidence caused by *Alternaria* spp. and *Epicoccum* spp. Pre-harvest and postharvest sprays of ASM on freesia cvs. 'Cote d'Azur' and 'Dukaat' control postharvest *Botrytis cinerea* infection (Darras et al. 2006b, 2007). Postharvest disease reduction with ASM on treated freesia flowers ranged from 30 to 45% as compared to the controls (Darras et al. 2006b).

35.7.2.7 Antimicrobial Compounds

The longevity of different cut flowers can be improved during storage by the use of antimicrobial and other compounds in the vase solution (Damunupola and Joyce 2008; Knee 2000). Different types of biocides have been used in ornamentals to circumvent multiplication of microorganisms in the vase solutions. However, their antimicrobial action may be confounded by their other physicochemical effects (Damunupola and Joyce 2008; Knee 2000) but their response on cut stems may vary among cut flower or foliage type, the involvement of the specific microorganism, and the use of other vase solution ingredients.

35.7.2.7.1 *Hydroxyquinoline*

The success of hydroxyquinoline (HQ) used as a biocide in cut flower treatment solutions has been known for decades (Damunupola and Joyce 2008). Hydroxyquinoline sulphate (HQS) and Hydroxyquinoline citrate (HQC) are the two most frequently used HQ compounds in the management of flower crops. Furthermore, 8-HQS promotes stomatal closure, whereas both HQS and HQC increase flower longevity by acidifying the vase solution. The acidic nature of quinolone esters reduces the solution pH and inhibits stem plugging.

35.7.2.7.2 *Silver Thiosulfate, Silver Nitrate and Silver Nanoparticles*

Silver is usually applied in the form of nitrate salt in vase solutions. It may act as an antimicrobial agent or an inhibitor of aquaporins in plants (Niemietz and Tyerman 2002), and as an inhibitor of ethylene-binding during ethylene synthesis and action (Serek et al. 2006). Pretreatment of cut stems with silver nitrate ($AgNO_3$) @ 170 mg/L for 30 min effectively decreased bacterial growth in the basal stem segments of rose cv. 'Sonia'. The effect of $AgNO_3$ in combination with other vase solutions on the longevity and bud opening of cut *Dendrobium* cv. 'Pompadour' was studied and it was found that $AgNO_3$ plus glucose was not as effective in increasing bud opening and vase life as HQS plus glucose solutions. Silver thiosulfate (STS) is generally used in the floral industry as dip, spray, or pulse treatment of cut flowers (Slater et al. 2001; Van Doorn and Cruz 2000). The ability of STS to delay the ethylene action was utilized to extend the vase life of various cut flowers (Premawardena et al. 2000; Yapa et al. 2000), but STS may not be very effective in reducing the bacterial count in holding solutions. Pure colloidal silver nanoparticles (SNP) are new-generation broad-spectrum antimicrobial agents. Pulse treatment of cut rose cv. 'Movie Star' with 10 mg/LSNP + 5% sucrose for 24 h and after that holding the samples in 0.5 mg/L SNP + 2% sucrose significantly delayed the vascular blockage caused by bacterial contamination and stomatal conductance. This resulted in the improvement of water balance and significantly improved the vase life of cut roses up to 11.8 days as compared to controls (Lü et al. 2010). Kim et al. (2004) investigated the effects of SNP as a dip or spray pretreatment on the subsequent vase life and physiology of cut Oriental hybrid *Lilium* cv. 'Siberia' and Asiatic *Lilium* hybrid cv. 'Dream Land'. They found that the vase life of *Lilium* cv. "Siberia" florets was increased with a treatment of 0.1% SNP + natural chitosan. When cut flowers of gerbera cvs. 'Double Dutch' and 'Red Explosion' were placed in a solution containing 6 mg/LSNP, their longevity was extended up to 7.8 days as compared to the untreated controls (Oraee et al. 2011). At the same time, the bacterial populations in the vase solutions also decreased drastically.

35.7.2.7.3 *Chlorine Compounds*

Different chlorine (Cl) compounds having antimicrobial activities are commercially available in the markets. The sodium salt of dichloroisocyanuric acid (Na-DICA) is a slow-release chlorine compound broadly used for sterilization of water in swimming pools (Joyce et al. 2000; Xie et al. 2007). Both Na-DICA and domestic bleach or sodium hypochlorite (NaOCl) are mostly used in experimentation and postharvest handling of cut flowers (Akoumianaki–Ioannidou et al. 2010; Knee 2000). Addition of 2.0 or 10.0 µL/L ClO_2 to clean the deionized water extended the shelf life of alstroemeria cv. 'Senna', antirrhinum cv. 'Potomac Pink', dianthus cv. 'Pasha', gerbera cv. 'Monarch', gypsophila cvs. 'Crystal' and 'Perfecta', lilium cv. 'Vermeer', matthiola cv. 'Ruby Red', and rose cv. 'Charlotte' flowers by 0.9–13.4 days (7–77%) compared to the untreated controls, respectively. The positive effects of ClO_2 treatment were related with the decrease of aerobic bacteria in vase water and on cut surfaces of flower stems. Knee (2000) reported that Na-DICA used alone at 200 mg/L was most effective in extending the longevity of cut alstroemeria, dianthus, and rose flowers up to 2.6, 2.9 and 1.5 days, respectively, as compared to the controls.

35.7.2.8 Acidifiers

An acidic solution helps in rehydration of cut flowers and provides unfavourable conditions for bacteria to grow. Increase in water flow through rose stem segments was observed with decreasing pH from 6.0 to 3.0. Decreasing solution pH from 8.0 to 6.0 improves flower-water relations, maintenance of fresh weight, and vase life of cut rose cv. 'Baccara' (Pompodakisetal. 2004). Akoumianaki–Ioannidou et al. (2010) found that application of citric acid @ 50 µg/L + 2 % sucrose had significant positive effects on vase life and flower opening scores of *Nerium oleander* inflorescences independent of the storage temperatures tested (i.e. 2, 5 or 20°C). Likewise, 50 µg/L citric acid +2% sucrose protected the flowers from low-temperature injuries observed under a temperature of 2°C.

REFERENCES

Adugna B, Belew D, Kassa N (2012) Effect of pulsing solution on postharvest performance of carnation (*Dianthus caryophyllus* L.) cultivars. Trends Hortic Res 2: 8–13.

Ahmad I, Dole JM, Blazich FA (2014) Effects of daily harvest time on postharvest longevity, water relations, and carbohydrate status of selected specialty cut flowers. Hortic Sci 49 (3): 297–305.

Akoumianaki–Ioannidou A, Darras AI, Diamantaki A (2010) Postharvest vase solutions and storage effects on cut *Nerium oleander* inflorescences. J Hortic Sci Biotechnol 85(1): 1–6.

Anais G, Darrasse A, Prior P (2000) Breeding anthuriums (*Anthurium andraeanum* L.) for resistance to bacterial blight caused by *Xanthomonas campestris* pv *dieffenbachiae*. Acta Hortic 508: 135–140.

Beasley DR, Joyce DC, Coates LM, Wearing AH (2001) Saprophytic microorganisms with potential for biological control of *Botrytis cinerea* on *Geraldton waxflower* flowers. Aust J Exp Agric 41: 697–703.

Byung-Chun I, Ji YS, Jin HL (2016) Pre-harvest environmental conditions affect the vase life of winter-cut roses grown under different commercial greenhouses. Hortic Environ Biotechnol 57(1): 27–37.

Capdeville G, Maffia LA, Finger FL, Batista UG (2005) Gray mold severity and vase life of rose buds after pulsing with citric acid, salicylic acid, calcium sulfate, sucrose and silver thiosulfate. Fitopatol Bras 28: 380–385.

Celikel FG, Reid MS (2005) Temperature and post-harvest performance of rose (*Rosa hybrida* L. 'First Red') and Gypsophila (*Gypsophila paniculata* L. 'Bristol Fairy') flowers. Acta Hortic 682: 1789–1794.

Chu EH, Shin EJ, Park HJ, Jeong RD (2015) Effect of gamma irradiation and its convergent treatment for control of postharvest *Botrytis cinerea* of cut roses. Radiat Phys Chem 115: 22–29.

Cia P, Pascholati SF, Benato AE, Camili EC, Santos CA (2007) Effects of gamma and UV-C irradiation on the postharvest control of papaya anthracnose. Postharvest Biol Technol 43: 366–373.

Damunupola JW, Joyce DC (2008) When is a vase solution biocide not, or not only, antimicrobial? J Jpn Soc Hort Sci 77: 211–228.

Darras AI, Bali I, Argyropoulou E (2015) Disease resistance and growth responses in *Pelargonium* x *hortorum* plants to brief pulses of UV-C irradiation. Sci Hortic 181: 95–101.

Darras AI, Demopoulos V, Kazana E, Tiniakou CA (2012b) Effects of UV-C irradiation on *Botrytis cinerea* floret specking and quality of cut gerbera flowers. Acta Hortic 937: 493–498.

Darras AI, Demopoulos V, Tiniakou CA (2012a) UV-C irradiation induces defence responses and improves vase life of cut gerbera flowers. Postharvest Biol Technol 64: 168–174.

Darras AI, Joyce DC, Terry LA (2004) A survey of possible associations between pre-harvest environment conditions and postharvest rejections of cut freesia flowers. Aust J Exp Agric 44: 103–108.

Darras AI, Joyce DC, Terry LA (2006b) Acibenzolar-S-methyl and methyl jasmonate of glasshouse-grown freesias suppress post-harvest petal specking caused by *Botrytis cinerea*. J Hortic Sci Biotechnol 81(6): 1043–1051.

Darras AI, Joyce DC, Terry LA (2010) Post-harvest UV-C irradiation on cut *Freesia hybrida* L. inflorescences suppresses petal specking caused by *Botrytis cinerea*. Postharvest Biol Technol 56(3): 186–188.

Darras AI, Joyce DC, Terry LA, Pompodakis NE, Dimitriadis CI (2007) Efficacy of postharvest treatments with acibenzolar-S-methyl and methyl jasmonate against *Botrytis cinerea* infecting cut *Freesia hybrida* L. flowers. Austra Plant Pathol 36: 332–340.

Darras AI, Joyce DC, Terry LA, Vloutoglou I (2006a) Postharvest infections of *Freesia hybrid*a L. flowers by *Botrytis cinerea*. Austr Plant Pathol 35: 55–63.

Darras AI, Terry LA, Joyce DC (2005) Methyl jasmonate vapour treatment suppresses specking caused by *Botrytis cinerea* on cut *Freesia hybrida* L. flowers. Postharvest Biol Technol 38: 175–182.

Daughtrey ML, Bridgen MP (2013) Evaluating resistance to *Botrytis elliptica* in field grown lilies. Acta Hortic 1002: 313–318.

Davarynejad E, Tehranifar A, Ghayoor Z, Davarynejad GH (2008) Effect of different pre-harvest conditions on the postharvest keeping quality of cut gerbera. Acta Hortic 804: 205–208.

Dinh SQ, Joyce DC, Irving DE, Wearing AH (2007) Field applications of three different classes of known host plant defence elicitors did not suppress infection of *Geraldton waxflower* by *Botrytis cinerea*. Austra Plant Pathol 36: 142–148.

Doi M, Hu Y, Imanishi H (2000) Water relations of cut roses as influenced by vapor pressure deficits and temperatures. J Jpn Soc Hortic Sci 69: 584–589.

Dole JM, Wilkins HF (2004) Floriculture. Principles and Species. Pearson Prentice Hall, Upper Saddle River, NJ.

Eyre JX, Faragher J, Joyce DC, Franz PR (2006) Effects of postharvest jasmonate treatments against *Botrytis cinerea* on Geraldton waxflower (*Chamelaucium uncinatum*). Austr Plant Pathol 46: 717–723.

Fanourakis D, Velez-Ramirez AI, In BC, Barendse H, Meeteren UV, Woltering EJ (2015) A survey of pre-harvest conditions affecting the regulation of water loss during vase life. Acta Hortic 1064: 195–204.

Favero BT, Benato EA, Dia GM, Cia P (2015) Gibberellic acid, ozone and 1-methylcyclopropene on the gray mold control in 'Avant Garde' Rose. Acta Hortic 1060: 177–182.

Gast K (2001) Methyl jasmonate and long-term storage of fresh cut peony flowers. Acta Hortic 543: 327–330.

Gupta J, Dubey RK (2019) Factors affecting postharvest life of flower crops. Intl J Curr Microbio Appl Sci 7(1): 548–557.

Hallman GJ (2011) Phytosanitary applications of irradiation. Compr Rev Food Sci Food Saf. 10: 143–151.

Hannweg KF (2008) Harvest and post-harvest treatment of gerbera cut flowers ensures optimum vase-life under hot conditions for farmers with limited resources. Acta Hortic 768: 437–443.

Harkema H, Mensink MGJ, Somhorst DPM, Pedreschi RP, Westra EH (2013) Reduction of *Botrytis cinerea* incidence in cut roses (*Rosa hybrida* L.) during long term transport in dry conditions. Postharvest Biol Technol 76: 135–138.

Hassan FAS, Ali E (2014) Physiological response of gladiolus flowers to anti-ethylene treatments and their relation to senescence. Intl J Advan Res 2(10): 188–199.

Hu J, Dinh SQ, Joyce DC (2009) S-carvone effects on *Botrytis cinerea* and harvested waxflower (*Chamelaucium uncinatum*). N Z J Crop Hortic Sci 37: 79–83.

Joyce DC, Meara SA, Hetherington SE, Jones PN (2000) Effects of cold storage on cut Grevillea 'Sylvia' inflorescences. Postharvest Biol Technol 18: 49–56.

Kim JH, Lee AK, Suh JK (2004) Effect of certain pre-treatment substances on vase life and physiological character of *Lilium* spp. Acta Hortic 673: 307–314.

Knee M (2000) Selection of biocides for use in floral preservatives. Postharvest Biol Technol 18: 227–234.

Kumar GSA, Kamanna BC, Benagi VI (2011) Management of chrysanthemum leaf blight caused by *Alternaria alternata* (FR.) Keisler under field condition. Plant Arch 11: 553–555.

Lü P, He H, Li H, Cao J, Xu HI (2010) Effects of nano-silver treatment on vase life of cut rose cv. Movie Star flowers. J Food Agric Environ 8: 1118–1122.

Macnish AJ, Morris KL, de Theije A, Mensink MGJ, Boerrigter HAM, Reid MS, Jiang CZ, Woltering EJ (2010) Sodium hypochlorite: A promising agent for reducing *Botrytis cinerea* infection on rose flowers. Postharvest Biol Technol 58: 262–267.

Macnish A, Reid M, Joyce D (2009) Ornamentals and cut flowers. In: Yahia E (Ed.). Modified and Controlled Atmospheres for the Storage, Transportation, and Packaging of Horticultural Commodities. CRC Press, Boca Raton, FL, pp. 491–506.

McSpadden Gardener BB, Fravel DR (2002) Biological control of plant pathogens: Research, commercialization, and application in the USA. Online. Plant Health Progress. doi:10.1094/PHP-2002-0510-01-RV.

Mutui TM, Emongor VE, Hutchinson MJ (2006) The effects of gibberellin $_{4+7}$ on the vase life and flower quality of Alstroemeria cut flowers. Plant Growth Reg 48 (3): 207–214.

Nagrale DT, Gaikwad AP, Goswami S, Sharma L (2012) Fungicidal management of *Alternaria alternata* (Fr.) Keissler causing blight of gerbera (*Gerbera jamesonii* H. Bolus ex J.D. Hook). J Appl Nat Sci 4(2): 220–227.

Niemietz CM, Tyerman SD (2002) New potent inhibitors of aquaporins: Silver and gold compounds inhibit aquaporins of plant and human origin. FEBS Lett 531: 443–447.

Oraee T, Zadeh AA, Kiani M, Oraee A (2011) The role of preservative compounds on number of bacteria on the end of stems and vase solutions of cut gerbera. JOP 1: 161–165.

Palagani N, Singh A (2017) Post-harvest quality and physiology of gerbera flowers as influenced by bio-fertilizers, chemicals and organic growth regulators. Res Crops 18(1): 116–122.

Pandya HA, Saxena OP (2003) Post-harvest light intensity and temperature on carbohydrate levels and vase life of cut flowers. Acta Hortic 624: 427–432.

Pompodakis NE, Joyce DC, Terry LA, Lydakis DE (2004) Effects of vase solution pH and abscisic acid on the longevity of cut 'Baccara' roses. J Hortic Sci Biotechnol 79: 828–832.

Premawardena PS, Peiris CN, Peiris SE (2000) Effects of selected post-harvest treatments on vase life of cut-flower Gladiolus (*Gladiolus grandiflorus*). Trop Agric Res 12: 325–333.

Reid MS (2001) Advances in shipping and handling of ornamentals. Acta Hortic 543: 277–284.

Reid MS (2009) Handling of cut flowers for export. Proflora Bull 1–26.

Seglie L, Spadano D, Devecchi M, Larcher F, Gullino ML (2009) Use of 1-methylcyclopropene for the control of *Botrytis cinerea* on cut flowers. Phytopathol Mediterr 48: 253–261.

Seglie L, Spadaro D, Trotta F, Devecchi M, Lodovica Gullino M, Scariot V (2012) Use of 1-methylcylopropene in cyclodextrin-based nanosponges to control grey mould caused by *Botrytis cinerea* on *Dianthus caryophyllus* cut flowers. Postharvest Biol Technol 64: 55–57.

Serek M, Woltering EJ, Sisler EC, Frello S, Sriskandarajah S (2006) Controlling ethylene responses in flowers at the receptor level. Biotechnol Adv 24: 368–381.

Slater AT, Blakemore MC, Faragher JD, Franz PR, Henderson B, Green K (2001) Leptospermum as an export cut flower crop. Rural Industries Research and Development Corporation (RIRDC), Barton.

Sudaria MA, Uthairatanakij A, Nguyen HT (2017) Post-harvest quality effects of different vase life solutions on cut rose (*Rosa hybrida* l.). Intl J Agricul Forest Life Sci 1(1): 2–20.

Thomma BPHJ (2003) *Alternaria* spp.: From general saprophyte to specific parasite. Mol Plant Pathol 4(4): 225–236.

Van Doorn WG, Cruz P (2000) Evidence for a wounding-induced xylem occlusion in stems of cut chrysanthemum flowers. Postharvest Biol Technol 19: 73–83.

Wagstaff C, Chanasut U, Harren FJM, Laarhoven LJ, Thomas B, Rogers HJ, Stead AD (2005) Ethylene and flower longevity in Alstroemeria: relationship between tepal senescence, abscission and ethylene biosynthesis. J Exper Bot 56(413): 1007–1016.

Wegulo SN, Vilchez M (2007) Evaluation of Lisianthus cultivars for resistance to *Botrytis cinerea*. Plant Dis 91: 997–1001.

Woltering EJ, Boerrigter HAM, Mensink MGJ, Harkema H, Macnish AJ, Reid MS, Jiang C-Z (2015) Validation of the effects of a single one second hypochlorite floral dip on *Botrytis cinerea* incidence following long-term shipment of cut roses. Acta Hortic 1064: 211–219.

Xie L, Joyce DC, Irving DE, Eyre JX (2007) Chlorine demand in cut flower vase solutions. Postharvest Biol Technol 47: 267–270.

Yapa SS, Peiris BCN, Peiris SE (2000) Potential low cost treatments for extending the vase-life of Anthurium (*Anthurium andraeanum* Lind.) flowers. Trop Agric Res 12: 334–343.

Zainuri A, Joyce DC, Wearing AH, Coates L, Terry LA (2001) Effects of phosphonate and salicylic acid treatments on anthracnose disease development and ripening of 'Kensington Pride' mango fruit. Aust J Exp Agric 41: 805–813.

Index

Note: Locators in *italics* represent figures and **bold** indicate tables in the text.

N

O

P

Q

R

S